Vegetation mapping

Handbook of vegetation science

FOUNDED BY R. TÜXEN

H. LIETH, EDITOR IN CHIEF

Volume 10

Vegetation mapping

Edited by

A. W. KÜCHLER and I. S. ZONNEVELD

Kluwer Academic Publishers

Dordrecht / Boston / London

Library of Congress Cataloging in Publication Data

Vegetation mapping.

(Handbook of vegetation science ; v. 10)
Includes index.
1. Vegetation mapping. 2. Phytogeography--Maps.
I. Küchler, A. W. (August Wilhelm), 1907-   .
II. Zonneveld, Isaak Samuel. III. Series.
QK911.H3 pt. 10 [QK63]   581 s [581.9]   86-7269
ISBN 90-6193-191-6

ISBN 90-6193-191-6

Kluwer Academic Publishers incorporates the publishing programmes of
Dr. W. Junk Publishers, MTP Press, Martinus Nijhoff Publishers, and D. Reidel
Publishing Company.

Distributors

for the United States and Canada: Kluwer Academic Publishers, 101 Philip Drive,
Norwell, MA 02061, USA
for all other countries: Kluwer Academic Publishers Group, P.O. Box 322, 3300
AH Dordrecht, The Netherlands

Copyright

PRINTED IN THE NETHERLANDS

Contents

Series Editor's Preface

The volume on vegetation mapping treats one of the major objectives of vegetation science. In several other volumes in the handbook series some aspects of mapping were covered. This volume, however, has this topic as the prime objective. This made it necessary that auxiliar methods for selecting mapping units had to be covered in this volume, although these topics were extensively treated already in other volumes or are planned to be treated in coming volumes of the series.

The volume was started by Dr. Küchler when Dr. Tüxen was still editor of the series. The response to his first manuscript was so much delayed that the new developments in automated mapping methods, remote sensing and new goals for vegetation mapping made it necessary to write a complete new manuscript. I was very fortunate that Dr. Küchler was willing to write and edit a new manuscript, jointly with Dr. Zonneveld of ITC* in Enschede, after a new

table of contents was generated. To get the cooperation of Dr. Zonneveld was another advantage for this volume. ITC in Enschede is one of the best vegetation mapping centers today, employing the latest developments in remote sensing, mapping routines, and field survey equipment.

I thank both editors for their fine work. Though it was sometimes difficult for them to select individual topics for inclusion, the final decision was always made in the interest of the book as a whole. I hope, therefore, that the volume provides a general source book for everyone interested in vegetation mapping, for mapping at large and small scales for special topics in agriculture, forestry, or ecology. The respective volumes for the application of vegetation science in this handbook are greatly enhanced by the appearance of this volume on vegetation mapping.

The two editors deserve thanks and respect for their work on behalf of our profession. I am sure the volume will be widely used in the future.

Osnabrück, April 1988   *H. LIETH*

---

* ITC is the acronym for the present "Institute for Aerospace Survey and earth sciences" (formerly International Training Centre for Aerial Survey) Delft/Enschede Netherlands. P.O. B.6.7500 AA.

1. Preface

A.W. KÜCHLER

The intimate intercourse between two or more fields of knowledge often bears interesting and valuable fruit. Vegetation maps are such fruit, resulting from the union of botany and geography. The work of botanists can be comprehensive only if it includes a consideration of plants in space, i.e. in different types of landscapes. At this point, the work of geographers becomes important through their development of maps as tools to determine and to analyze distributions in space. Our highly developed knowledge of vegetation is matched by the refinement of cartographic techniques, and maps can now be made that will show the extent and geographical distribution of vegetation anywhere on the surface of our planet with a remarkable degree of accuracy.

Vegetation maps are now being prepared on all continents, and it is interesting to see how a field of knowledge can become prominent almost overnight just because a certain phase in the long history of science has been reached. Less than a century ago, vegetation maps were largely unkown, but the stage was set, and by today, the attained heights of achievement are spectacular. Indeed, much energy is now being spent on vegetation maps, and great is the need for more and better maps.

The basis of the successful advances in the field of vegetation mapping is fourfold:

1. vegetation maps present an inventory of existing plant communities, their location, extent and geographical distribution in the landscape at the time of mapping;

2. vegetation maps are scientific tools for analyzing the environment and the relationships between vegetation and the site on which it occurs. This helps to explain the distribution of plant communities on the basis of the physical and chemical features of the landscape. On the other hand, plant communities allow conclusions on the nature of the environment;

3. vegetation maps are valuable standards of reference for observing and measuring changes in the vegetation, their direction and their speed, i.e. the rate of change. This is important because the character of vegetation is dynamic and is increasingly affected by man;

4. vegetation maps can serve as a scientific basis for planning future land-use, especially with regard to forestry, range management, and agriculture in all its forms and variations. Such ecologically based planning permits an optimal land-use, managing for highest yields on a sustained basis without damaging the environment.

These four observations suffice to explain the development of the field of vegetation mapping. But in fact, the circles are expanding steadily. Vegetation consists of plants. It seems logical, therefore, that botanists should be the prime producers and users of vegetation maps. But botanists are by no means the only ones interested in vegetation maps, and whether or not their actual number is increasing, their proportion among those who use vegetation maps is

A.W. Küchler & I.S. Zonneveld (eds.), Vegetation mapping. ISBN 90-6193-191-6.

declining. On the other hand, there are growing numbers of geographers, zoologists, geologists, plant and animal ecologists, pedologists, and an ever-increasing multitude of people concerned with land-use planning, forestry, agriculture and conservation, with climatology and communications, investments and fiscal problems, education and military matters who are discovering the practical value of vegetation maps as applied to their respective fields.

Vegetation is so closely tied to its environment that an appreciation of its character can reveal the qualities of the sites on which it occurs. 'A qualified analyst looks through a vegetation map even as a physician looks through an X-ray screen into the human body' (Tüxen, 1958).

The usefulness of vegetation maps is no longer questioned. It is, however, quite a different matter to produce a vegetation map. This requires a considerable expertise because vegetation has so many features and the maps are applied to so many different purposes that proper selection and adaptation become imperative. These latter must be based on a thorough understanding of the nature of vegetation, its environment, and also the nature of the purposes to which a vegetation map is to be applied, if the usefulness of the vegetation map is to be assured.

In this book, some basic considerations must necessarily precede the more technical aspects of vegetation maps and their preparation. Once these have been discussed, the reader will more fully appreciate the presentation of various mapping methods and techniques. When the basic features of vegetation mapping have thus been clarified, the section on what might be termed Applied Vegetation Mapping can reveal the wide variety of uses which vegetation maps are serving today.

The senior author published a volume on Vegetation Mapping in 1967. In the meantime, much progress has been made. Some material of the original work remains fully applicable today and is here reproduced in an up-dated form. The publisher's understanding concurrence is greatefully acknowledged. However, much of the work needed either a thorough reworking or indeed, an entirely new approach. Many sections of this book are therefore revised and rewritten or entirely new, especially all the chapters by the junior author and his collaborators. The present volume is therefore up-to-date and should be useful to all persons interested in vegetation maps for whatever purposes.

An effort has been made to help the reader find relevant material in the world literature. The bibliography has therefore been enriched over and beyond the actual citations in this book in order to provide a rich source and guide for vegetation mappers who wish to penetrate more deeply into the thought of their colleagues.

A number of questions has been answered in this volume but many more problems await solution. This book will have served its purpose if bringing together the numerous facets of vegetation mapping will stimulate scientists to expand their research and make new vegetation maps more profitable to a wider variety of users.

2. Historical Sketch

A.W. KÜCHLER

Our modern technology continues its rapid development. It makes such inventions as the telephone and the steam engine look old, and indeed their modern versions are vastly different from their prototypes of not too many generations ago. Vegetation maps, however, began as early as the 15th century. No man can claim to have invented them.

The evolution of vegetation maps is characterized by a very slow tempo, matched by an equally slow acceleration during most of its long history, a regionally spotty development and, finally, a much increased acceleration after the first quarter of the 20th century. The highly sophisticated vegetation maps of our days evolved from beginnings that are far removed in time and character.

When maps came into more general use during the 15th century, vegetation (sensu lato) was frequently shown on them. Foncin (1961) reports on maps I, V and VI in the Cosmographia of Ptolemy, published in Bologna, Italy, in 1447, on which named forests are indicated by showing little groups of trees in perspective. In the second edition (Rome, 1478), this practice is extended, and the Ardennes Forest is shown on the map of France. These are serious efforts to reveal the geographical location and extent of known vegetation types. However, simultaneously with these maps and for long afterward as well, many maps were published on which vegetation in one form or another was shown largely for decorative purposes or in order to fill in the empty spaces for which information was lacking.

In the 16th century, forests were recognized as items of significance in military matters, as hunting reserves, timber resources, or obstacles to communications. Forests were therefore shown on maps with growing frequency and accuracy. Thus Mraz and Samek (1963) report on a map by Claudianus (1518) and another by Criginger (1569) which they consider forerunners of modern vegetation maps. At the very end of the century (1599), three maps were published in widely separated areas: Nürnberg, Germany (Foncin, 1961); Trokavec, Bohemia (Mraz and Samek, 1963); and Muscovy, Russia (Foncin, 1961). The Nürnberg map shows the distribution of forests and even some cultivated fields. On the Bohemian map, Podolsky uses tree symbols for his forests and distinguishes four tree genera: oak, hornbeam, birch and willow. The Russian map separates forests from grasslands. The maps in the famous atlas by Blaeu (1640–1654) also show little tree symbols, but they appear unrelated to the landscape. In the course of the 17th century, cultural vegetation is shown, occasionally at first, but with increasing frequency. Its symbolization is established remarkably soon: stippling for meadows, lines (furrows) for ploughed fields, vines for vineyards, etc. Foncin (1961) describes the map of some Mediterranean coastal regions in France by Perelle (1680) who shows vineyards, orange groves, olive groves, pines, myrtle and straberry trees *(Arbutus unedo).*

A.W. Küchler & I.S. Zonneveld (eds.), Vegetation mapping. ISBN 90-6193-191-6.

During these early periods of development, the purpose of maps was primarily geographical. They therefore showed above all the location of places, administrative districts, rivers, moutains, etc. Information on vegetation is incidental. As cartography developed, especially in the form of detailed topographic maps with a high degree of accuracy and reliability, the symbolization of vegetation evolved, too. The first real success in a modern sense came with the famous topographic map of France at 1 : 80,000 by Cassini in the late 18th century. On these sheets, broadleaf deciduous forests were consistently distinguished from needleleaf evergreen forests. In addition, some sheets showed remarkable detail. Thus a sheet of the lower Rhone region included forests of *Quercus ilex, Quercus pubescentis, Populus alba, Pinus pinea,* scrub of *Quercus coccifera* and *Rosmarinus officinalis,* swamps of *Scirpus* and *Phragmites,* stands of *Salsola* and, finally, vineyards. On the topographic sheets of many European countries the actual vegetation is now shown to the extent of indicating needleleaf evergreen forests, broadleaf deciduous forests, shrub formations (heath), meadows and pastures, marshes, swamps and bogs, vineyards, orchards and similar items. In the United States, this advanced state of mapping has unfortunately not yet been attained, and it is considered a step forward if forests (sensu lato) are shown at all, even if only by a flat green color.

However, topographic maps are not vegetation maps, even though many of them show some kinds of vegetation. In spite of the excellence of many modern topographic maps and of efforts to include vegetation on them as an integral part of their content, topographic maps are vegetation maps only in a limited sense, if, indeed, vegetation is shown on them at all. In many instances it is not. It is therefore important to distinguish between maps such as these and vegetation maps sensu stricto, i.e., maps the primary purpose of which is to show the geographical distribution of various vegetation types in a given area.

Early vegetation map

It seems that the first genuine vegetation map is the one by Sendtner (1854) showing the geographical distribution of 'Spagnetum with *Pinus pumilio*'. Even though this map is quite limited in scope, it is nevertheless interesting because it builds on Cassini's approach to vegetation: it is a map showing the extent of a defined vegetation type and thus pioneering in a new direction. Soon afterwards, Martius (1858) published a map of the 'Florenreiche' of Brazil. This is the first successful attempt to map the vegetation of an entire country. The name implies an approach to floristic areas as used by his famous predecessors Schouw (1823) and Humboldt (Bromme, 1851). But the five floral realms of Brazil as transcribed by Hueck (1957) reveal that Martius showed in fact the physiognomic formations he had already described in 1824.

In the meantime, Harper (1857) published his map of the Prairies above Tibley Creek, and with it started the United States on a development that remained unmatched for decades. Curtis' (1860) vegetation map of North Carolina was followed by maps of New Jersey (Cook and Smock, 1878) and of Colorado (Hayden, 1878). Then came two maps of parts of Wisconsin (Sweet, 1880; Wooster, 1882), leading to the important vegetation map of the whole state of Wisconsin (Chamberlin, 1883), which was unsurpassed for detail and completeness. Only one year later, the development became almost explosive when, under the editorship of Sargent (1884), the tenth census of the United States published vegetation maps of North America and of more than a dozen states from Maine to Florida and westward to Texas and Minnesota. Nothing like it had ever been attempted before, and the vegetation of a larger area at similar scales elsewhere was not mapped until decades later. Then McGee (1889) added his map of northeastern Iowa and Leiberg (1899) his map of Montana to the already impressive score and by the turn of the century, the area of vegetation mapped in the United States corresponded to nearly all of Europe west of the Baltic and the Adriatic Seas. From then on, a steady stream of vegetation maps issued forth from various sources, and while its volume fluctuated, it never ceased. This period of American vegetation mapping culminated in the well known vegetation map of the United States by Shantz and Zon (1923), the manuscript of which has already been completed in 1912. This map was a truly remarkable achievement, considering the amount of information available at that time. For more than a generation, it remained the only significant vegetation map of the United States.

While vegetation mapping made rapid strides in the United States, a similar though more modest development took place in Argentina. The first vegetation map of the country was published by P.G. Lorentz (1876), accompanied by a more detailed vegetation map of the northwestern part of the country. Then Brackebusch (1893) presented another map of the Argentinian Northwest which Hueck considered an extraordinary achievement even for today. The census of 1895 included a vegetation map of the whole country, not unlike Sargent's earlier efforts in the United States, and Kurtz presented his vegetation map of the province of Córdoba in 1905. The early introduction of vegetation maps into Argentinian schools books was unequaled anywhere. In the United States, for instance, vegetation maps in school books are rare even today.

The spectacular development of vegetation mapping in the United States during the last decades of the 19th century was not matched by a similar evolution in Europe. Thus the bibliography of German vegetation maps by Tüxen and Hentschel (1954) does not contain a single entry for the nineteenth century or earlier. No bibliography can hope to be complete, and Sendtner's map seems to have escaped the authors' notice. Even so, this bibliography reveals what is characteristic for Europe, not just Germany.

Nevertheless, there were notable achievements, and prominent among these is the world map by Grisebach (1872). This was an advance from Humboldt and Schouw for the map shows such physiognomic formations as the Hylaea (the tropical rain forest of the Amazon Basin), prairies, pampas, etc. But the old floristic practices lingered on in items like the 'arctic flora', the 'Chinese-Japanese Region', or the 'Californian Region'. They do not reveal the character of the vegetation. Presumably, these items are due to a lack of adequate information. Basically, however, Grisebach's map is the first attempt to prepare a genuine vegetation map of the world.

Grisebach's map revealed a scientific recognition of formations and their relations to the environment, especially the climate. Ecological relations had long been known to exist, especially since Linné (1707–1778) and a few scientists had shown concern for them. However, at the end of the 19th century, the time was ripe for an ecological approach, and Schimper (1898) presented a new vegetation map of the world, the first one that may be considered modern, as is indicated by its legend:

Distribution of the most important types of formations of the earth, at an equatorial scale of 1:120,000,000
1. Tropical rainforests
2. Subtropical rainforests
3. Monsoon forests
4. Temperate rainforests
5. Summergreen broadleaf forests
6. Needleleaf forests
7. Sclerophyll woods
8. Savanna forests
9. Thorn forests and scrub
10. Savannas
11. Steppes and transitions between steppes and deserts (semi-deserts)
12. Heaths
13. Dry deserts
14. Tundra
15. Cold deserts

Haltingly at first, but with slowly increasing momentum, vegetation maps gained recognition as bases for research and as instruments for the formulation of problems and for finding solutions to them. Toward the end of the 19th century, Flahault (1894) presented a vegetation map of the area around Perpignan in southern France at the medium scale of 1:200,000 and discussed the possibilities of preparing similar maps for all of France. Schröter (1895) published his large scale vegetation map of the St. Antöniertal, which provided an extraordinary stimulus to the development of vegetation mapping in Switzerland. The century closed with the publication of a vegetation map of Russia by Tanfilev (Sochava, 1958).

The 20th century

The new century did not at first bring about any radical change in vegetation mapping. However, the handwriting was on the wall. Phytocenology and ecology as sciences were making great strides, and the formulation of principles for a systematic analysis and classification of vegetation advanced rapidly. Inevitably, the new ideas were expressed cartographically. Thus Drude (1907), on parts of three topographic sheets, mapped the vegetation of the eastern Ore Moutains in Germany at 1:25,000 and showed physiognomic-ecological formations, subdivided floristically into units of similar-species combinations.

As the world recovered from the catastrophe of World War I, vegetation maps began to appear in more rapid succession. By then, the situation had changed profoundly and the days of 19th century mapping were gone forever.

The new situation found its expression in at least three ways. First, the difference between the United States and Europe had changed. The United States no longer led the way but shared the honors of being a major producer of vegetation maps with Europe and the Soviet Union. Second, the European phytocenologists placed heavy emphasis on mapping at large scales. Third, the development of Schools of Thought in phytocenology found its expression also on

vegetation maps (Küchler, 1953a). The contending Schools of Thought were at times engaged in spirited debates which were most fruitful, resulting in an ever-greater refinement of methods and an increasing solidity of the scientific foundation of the more recent vegetation maps. The Europeans took over the leadership, based on a strictly scientific approach, reasoning that a vegetation map can be fully exploited only if it is based on the scientific method. Vegetation maps that are prepared for some utilitarian goal are not only limited in applicability, but rarely permit a later diversification of their uses.

European mapping institutes

The extraordinary success of European vegetation mappers rests on very clear and precise definitions and classifications of phytocenoses. This observation applies to all European Schools of Thought. It was recognized gradually that each of these had its own merits and should be promoted. Swiss phytocenologists were particularly active. Many governments, too, realized the value of vegetation maps and established special agencies devoted to vegetation mapping.

Thus the Centre National de la Recherche Scientifique in France sponsored two institutes for vegetation mapping above all in France but also overseas, and with their own journal, the Bulletin du Service de la Carte Phytogéographique. These institutes are the Service de la Carte de la Végétation de la France au 1:200,000 at Toulouse, founded by Henri Gaussen, and the Service de la Carte des Groupements Végétaux de la France au 1:20,000 at Montpellier, founded by Louis Emberger. The Toulouse institute is devoted above all to mapping all of France at the uniform scale of 1:200,000 according to the method developed by Gaussen (cf. Chapter 31). Well over 80% of all sheets have already been published, and the few remaining sheets should become available within the next few years.

On the other hand, the Montpellier institute maps the vegetation of specified critical areas at 1:20,000 or larger; no effort is made to map the entire nation. In the past, these maps were based on the classification by Braun-Blanquet, but important modifications in the approach now give these maps a more ecological character, and the institute has been renamed Centre d'Etudes Phytosociologiques et Ecologiques Louis Emberger. More recently, Ozenda has developed the botanical institute at the University of Grenoble into an important and highly productive center for vegetation mapping.

In Switzerland, Rübel founded the Geobotanical Research Institute at Zürich which soon became the center for phytocenological research and mapping. Schmid presented a new mapping method and produced a vegetation map of Switzerland at 1:200,000 (cf. Chapter 32). In the meantime, Germany established the Bundesstelle für Vegetationskartierung (Federal Institute for Vegetation Mapping), founded by Tüxen. This institute was moved to Bonn when Tüxen retired. Belgium has its Centre de Cartographie Phytosociologique where the whole country is being mapped at 1:20,000. Similarly, substantial numbers of vegetation maps are being published in Sweden, Norway, Italy, Austria, Poland, Czechoslovakia, Yugoslavia, Romania and Bulgaria. Both Scotland and England are currently being mapped at large scale. In addition, there are numerous individuals who prepare valuable vegetation maps, e.g. Pedrotti in Italy, Falinski in Poland, and many others.

In the Soviet Union, the Komarov Botanical Institute of the Academy of Sciences of the USSR at Leningrad has two establishments for mapping vegetation, one for large scale maps and one for small scale vegetation maps. The well known examples of vegetation maps (e.g. Kleopov and Lavrenko, 1942; Lavrenko and Rodin, 1956; Lavrenko and Sochava, 1956) issues from these institutes. Sochava (1958) reviewed the development of Soviet vegetation mapping. Upon inspecting the exhibit of vegetation maps at the Toulouse Colloquium, Fosberg (1961) had this to say: 'the most impressive exhibit was that of Professor Sochava, of small scale vegetation maps of the USSR and of various of its component parts. The walls of a large room were covered and a stack of sheets for which there was no room on the walls were available for examination on the table. The careful detail of this work must be seen to be believed. This was the product of the small scale mapping institute, headed by Professor Sochava, manned by ten scientists plus the necessary technicians and draftsmen. Another institute, not represented at the meeting, handles large scale mapping.'

Fosberg (1961) studied European vegetation mapping and concluded: "The most significant generalization from the observations made during this visit to Europe is that most of the maps seen deal with the basic scientific aspects of ecology. Although the ultimate objectives for which such large sums of money are being spent are certainly practical, it is an accepted assumption that the most economical way to serve a wide variety of practical purposes is by the production of the best possible scientific maps. From these, it is assumed, that many different sorts of practical correlations and special maps may be made very readily. The major effort, however, is put into the basic work. In this way, not only are the foreseen practical ends served, but there are likely to be frequent extra dividends of unanticipated practical consequences. Also, these maps can serve as bases for practical maps of subjects that, though not of present importance or urgency, may become important in the future. They also serve as reference points for measuring environmental change and for detecting unfavorable tendencies or deterioration in the environmental complex. Hence they can serve the entire growing field of conservation and research on natural resources. It seems clear enough that here is an area of basic scientific research in which we in the United States are lagging seriously, if not dangerously, behind the European countries on both sides of the 'Iron Curtain'."

The usefulness of scientific vegetation maps turned out to be so great that after World

War II, considerable sums of Marshall Plan (ERP) funds were devoted to mapping vegetation. This seemed surprising to many Americans but the European logic proved to be sound. The funds were well spent, serving the local economy and its reconstruction (cf. Buchwald, 1954; Krause, 1954b).

Developments in the United States

In the meantime, vegetation mapping in the United States took a different road which, as it turned out, was beset with serious obstacles to progress. Whereas the phytocenological ideas and methods of any European country were promptly tested by its neighbors and modified or adapted to new situations, leading to an ever greater refinement and sophistication, the American phytocenologists drifted under the influence of Clements. The basic difference between Clements and his European counterparts was, for vegetation mappers, that his widely accepted classification of vegetation lacked a precisely defined terminology for describing individual phytocenoses. The development toward scientific vegetation maps was thereby seriously retarded, and emphasis was placed on immediate applicability and utility. Not all American phytocenologists succumbed to the spell of Clements: such eminent scientists as Shantz and Shreve stand out among the exceptions. But they were too few in number and too often preoccupied with other problems. Furthermore, Europe was remote, and the urgently needed exchange of ideas was distinctly underdeveloped.

Nevertheless, the rapid growth of ecological thought brought the realization that vegetation maps can serve a useful function. As a result, the production of vegetation maps continued to grow. Notably, the U.S. Forest Service became interested in mapping its vast holdings. Fortunately, a good deal of freedom was permitted to the various regions, leading to a number of different systems. The Forest (and Range) Experiment Stations became important mapping cen-

ters and the vegetation of large areas under their jurisdiction was mapped. The station at Portland, Oregon, mapped the states of Washington and Oregon on eight sheets, four for each state, at 1:250,000 and with a common legend. All non-forest vegetation was ignored, and the forest units included only trees, never the entire phytocenoses. Some of the commercially more valuable stands were presented in different age classes. In addition, burned areas were shown but without indicating the type of forest that had been burned. Considering how close the vegetation of the Northwest is to climax conditions, this information would have enriched the maps appreciably.

The station at St. Paul, Minnesota, mapped the forests of Minnesota and Michigan. The maps did not cover the entire states and their vegetation units were somewhat oversimplified. The same can be said of the South where the stations at New Orleans, Louisiana, and Asheville, North Carolina, mapped all southern states at 1:1,000,000, using a uniform approach. But the scale was not fully exploited.

The California Forest and Range Experiment Station at Berkeley was the only one where large scale mapping was carried on under the leadership of A.E. Wieslander. It was also the only one where the success of the program led to its continuation down to the present. The large scale, not unlike those used in Europe, permitted a much more accurate and detailed grasp of the character of the vegetation and its relation to soil, fire and other features, and the vegetation units were not limited to treees. Wieslander's system therefore offered major advantages: the large scale permitted detail, and the detailed description of the vegetation permitted the recognition of ecological relations. It was the only system then sponsored by the Forest Service that could conceivably be developed into a scientific method for analyzing and mapping vegetation.

The value of the earlier vegetation maps of the Forest Service may be debatable. But these maps are expressions of the trial-and-error approach, and it is to the credit of the Forest Service that it continued to map and to raise its standards. Much mapping is now done on planimetric maps at 1 : 31,680 (2 inches = 1 mile) and, considering the vast areas involved, the extent of mapping already accomplished is impressive.

The Forest Service is by no means the only organisation that publishes vegetation maps. Both federal and state agencies became aware of their usefulness and, as the years passed, the number of vegetation maps grew enormously. The Bureau of Land Management, the Soil Conservation Service, fish and wildlife agencies, river basin authorities, planning agencies, state geological surveys, the National Park Service, and others realized the need for vegetation maps of the area under their jurisdiction.

The evolution of vegetation mapping in the United States is reflected in the evolution of vegetation mapping in the state of Kansas where Küchler (1969a) brought together all published vegetation maps of the state and presented them at a uniform scale. This facilitated comparisons and permitted a more effective tracing of the progress made. Unfortunately, this work was published before Küchler presented his new vegetation map of Kansas (1974) which, therefore, is not included in the earlier historical-comparative study.

At the same time, vegetation mapping aroused more interest among the scientist at universities and colleges, resulting, for instance, in the vegetation maps of Michigan (Veatch, 1953), Florida (Davis, 1967), Minnesota (Marschner, 1974), Nebraska (Kaul, 1975), Maryland (Brush et al., 1976), and many others. In the meantime, Holdridge and his collaborators mapped the vegetation of the Central American republics (Holdridge, 1959; Tosi, 1969). These people focused their attention on basic research and gradually became more intimately acquainted with European techniques. As a result, they learned to appreciate the value of vegetation maps as research tools. New vegetation maps were prepared in all parts of the Union, eventually leading to the new map of the whole country by Küchler (1964). There is today no

longer any reason for producing vegetation maps in the United States that are in any way inferior to the European products because the needed information and technology are available.

The present and the future

The United States, Europe and the Soviet Union are by no means the only areas where vegetation maps are being published in considerable numbers. All other continents are now sharing in the progress (cf. Küchler, 1960b). Argentina has long been prominent for mapping its vegetation (Küchler, 1981a) and Miyawaki (1979) is leading a spectacular development in Japan. Both India and Brazil (Küchler, 1982) are currently being mapped at the scale of 1 : 1,000,000.

Much of the evolution of vegetation mapping in Africa is due to the efforts of the former colonial powers, notably Great Britain, France, Germany, Belgium and Portugal. Scientists from these countries explored and mapped much of the African vegetation (Küchler, 1960b), and the development has continued. Thus Langdale-Brown (1959a, 1959b, 1960a, 1960b, 1964) and Wilson (1962a, 1962b; Küchler, 1965ff.) produced detailed vegetation maps of Uganda while Trapnell et al. (1969) presented the vegetation of Kenya. More recently, UNESCO (White, 1981) published a map of the vegetation of the whole continent in three sheets at the relatively large scale of 1 : 5,000,000.

In Australia, the Commonwealth Scientific and Industrial Research Organization (CSIRO) continues its important mapping work while Beard mapped the entire state of Western Australia at 1 : 1,000,000 (Beard, 1974–1981). Meanwhile Hou (1979) presented his new vegetation map of the People's Republic of China at 1 : 4,000,000 which is the same scale as that of the Soviet Union by Lavrenko and Sochava (1956), and nearly the same as that of the United States by Küchler (1964). In addition, the

United Nations Educational, Scientific and Cultural Organization (UNESCO) as well as the United Nations Food and Agriculture Organization (FAO) have been and continue to be a powerful stimulus to vegetation mapping, especially in the underdeveloped regions of the world.

The above examples illustrate the efforts to produce vegetation maps of large areas, sometimes whole countries. These maps are necessarily of small scale. In addition, there are innumerable vegetation maps of larger scale focusing on the vegetation of small areas almost anywhere on earth. Such maps are usually prepared by individual scientists, often at universities, rather than by government agencies. Numerous such examples can be cited from Canada, Mexico, France, Italy, and Japan. Clearly, the volume of published vegetation maps is growing steadily and rapidly.

Many mappers in many countries necessarily have different ideas and, as a result, develop different techniques. Their circumstances and their goals vary, and their maps reflect this heterogeneity (Küchler, 1960a). The evolution continues. At national and international botanical and geographical congresses, the vegetation mappers exhibit their latest works and discuss their problems. There have even been international conferences devoted exclusively to vegetation mapping, notably the 1959 Symposium at Stolzenau, Germany (Tüxen, 1963a), the 1960 Colloquium at Toulouse, France (Gaussen, 1961a) and the 1980 meeting at Grenoble, France (Ozenda, 1981). In 1984, Beard and Küchler organized an exhibit of vegetation maps at Perth, Western Australia, focusing on all regions with a mediterranean climate. Such exchanges of ideas are very fruitful indeed.

The period after World War II saw important advances in the field of vegetation mapping. One was the greatly improved technology of surveying the vegetation from the air. This includes above all aerial photography in black-and-white, in natural color, and in false color-infrared. Under the leadership of Robert Eliot Stauffer of the Kodak Research Laboratories,

the progress in emulsion research resulted in such a refined film quality that the recording even of details of vegetation is now practical.

In addition, the use of radar made significant strides, and the new maps of Brazil at 1:1,000,000, i.e. the Radambrasil project (Romariz, 1981), bear witness to the possibilities of this method.

Finally, surveying from space not only made its debut but rapidly developed into an important form of vegetation mapping. Satellite imagery covers large territories, thus greatly reducing the cost per unit area. From the point of view of the vegetation mapper, maps based on satellite imagery appear coarse and applicable at best to highly generalized, small scale vegetation maps. But again, the steady progress in a relatively short time promises a bright future for this kind of vegetation mapping.

There is a large literature available discussing the relations between remote sensing techniques and vegetation. Merchant (1983) has given a brief but useful summery of this development. An important contribution was made by Goetz, Rock and Rowan (1983) which reveals that remote sensing is now being developed even to the level of recognizing individual plant species, e.g. distinguishing one oak species from another. They describe the latest in instrumentation: the airborn imaging spectrometer, which permits an even more sophisticated and precise interpretation of remotely sensed data. The airborn imaging spectrometer is already in use but in addition, it is constantly undergoing improvements in its capabilities. It promises an ever-growing refinement in the exploitation of remote sensing for the analysis of vegetation. Vegetation mapping is thus fast becoming one of the main beneficiaries of this development. For more details see the later chapters in this book.

Another noteworthy development during the recent decades is the growing effort to use the mapped vegetation as an expression of ecological conditions. The vegetation maps tend thereby to develop towards maps of ecosystems. While ecosystems are much too complex to permit the portrayal of all their features on a map, the vegetation can be used successfully to reveal given relationships. The vegetation mappers of Grenoble have presented excellent examples of this approach. One need only peruse the various numbers of the Documents de Cartographie Ecologique to realize the variety of possibilities this method holds for the future.

Klinka (1977), on his vegetation map of central Vancouver Island published what so far may well be the most comprehensive presentation of environmental features as indicated by the vegetation (cf. Chapter 30). It is important to note here that the numerous efforts to map ecological features of the environment, to map habitat types or even ecosystems, nearly always converge on the vegetation, and all these maps are primarily vegetation maps. This is not astonishing when considering that vegetation is the integrated expression of the ecosystem.

If vegetation mapping had its childhood diseases, it has not only outgrown them, but indeed, it has developed into a healthy maturity. Slowly but perceptibly, the evolving ideas are converging toward the most rational and the most practical methods. This by no means weakens the originality of individual vegetation mappers, but it does replace the more dubious techniques of the trial-and-error period with the latest and most sophisticated methods. While phytocenological research is a never-ending process, the cartographic possibilities are beginning to be realized, and the utility of vegetation maps grows with the advances in our understanding of the vegetation and the progress of technology.

The International Bibliography of Vegetation Maps (Küchler, 1965ff) bears witness to this development. Its first volume covers Nort America (1965) and the second volume covers Europe (1966). There followed volume 3 containing the Soviet Union, Asia and Australia with Oceania (1968), and volume 4 covering Africa, South America and world maps (1970). A new volume on South America (Küchler, 1980) brought that region up-to-date, and the collection of material for further volumes continues.

3. The Nature of Vegetation

A.W. KÜCHLER

Webster's New World Dictionary defines vegetation as 'plant life in general'. An older edition (Webster's Collegiate Dictionary) called it 'the sum of vegetable life'. Such definitions are not useful in any scientific discussion of vegetation mapping. Indeed, the mapper must know exactly what the term means because only then can he use and apply it correctly, meaningfully, and to best advantage.

Vegetation consists of plants which can be identified in two ways: morphologically and taxonomically. Morphologically means that the identification is based on the general appearance of the plant, and the types of plants based on such a morphological description are referred to as growth forms or life forms. The two terms are synonymous. Life form is the older term but, more recently, growth form has increased in popularity because it is, in a sense, more logical. It refers to such items as trees, shrubs, dwarf shrubs, forbs, graminoids, etc.

The second manner of identifying plants is taxonomic. By international agreement, since 1950, a systematic (taxonomic) unit without reference to its rank is termed a taxon (plural taxa). The rank is determined according to the binomial system that goes back to the 18th century Swedish scientist Carl von Linné (Linnaeus). The major ranks of the hierarchy are classes – orders – families – genera – species and their various divisions and subdivisions. The vegetation mapper nearly always uses genera and/or species or the latter's divisions. The system is accepted throughout the world and needs

no further comment. In contrast, growth forms of individual plants and the resulting physiognomic and structural features of plant communities lack such a uniform approach and need elaboration.

Plants can thus be identified precisely.

Tolerance and competition

Species have given qualities which affect, even control, their geographical distribution. Of these qualities, two stand out above all others: tolerance and competition. Others may also assume significance, among them symbiosis, parasitism, protection, need, avoidance etc. Tolerance is here defined as the ability of a species to tolerate the conditions of its environment. Environments are complex, and a species must be able to tolerate every individual feature of the environment. Actually, a species can tolerate a range of conditions with definite maxima and minima beyond which it can not go. If only one feature of the environment rises above the maximum or falls below the minimum, the plant will die, as for instance, when temperature drops too low or the soil becomes too acid. A species may have a wide range of tolerance for one environmental feature and a narrow one for another, and to further complicate the matter, the ranges of tolerance of a given species change with the age of the plant, with its ontogenetic phases, and even with the seasons. Thus it is well established that plants will die during the

A.W. Küchler & I.S. Zonneveld (eds.), Vegetation mapping. ISBN 90-6193-191-6.

summer if exposed to a given low temperature, when this same species can tolerate much lower temperatures during the winter. Walter (1951, p. 57) quotes Ulmer's well known experiment with an alpine pine species (*Pinus cembra*). He found that this species will die in July if the temperature falls to $-7\,°C$ but that it can tolerate temperatures down to $-47\,°C$ in January.

The other feature which controls the geographical distribution of taxa is competition. The ability to compete is also a complex matter, and a given species may be able to compete easily with one species while it faces the greatest difficulties when competing with another. Successful competition means that a taxon can complete its entire life cycle while under the influence of competition. Plants do best where the physical and chemical conditions of the landscape are most favorable to their development. Some aggressive species can crowd out other species and force them to survive on sites with distinctly less favorable conditions, although still within their ranges of tolerance.

Ellenberg (1956) performed the classical experiments in which he showed the effects of competition on the productivity of given taxa and demonstrated that a taxon behaves according to the kinds of plants with which it must compete, and according to the character of the site. He distinguishes the physiological behavior of species, when plants are grown in isolation, from the ecological behavior of species, when growing under the effects of competition. The physiological behavior is relatively fixed but the ecological behavior is not. The ecological behavior of plants always deviates from the physiological behavior, usually resulting in a narrower range of tolerance. Just how it deviates depends on the competing taxa and the site qualities.

An example of symbiosis is the group of lichens that consist of a combination of green algae and fungi. Parasitism is illustrated by parasitic fungi on cereals such as rusts and smuts, while protection is afforded to shade-tolerant plants by a forest canopy.

Phytocenoses

The biotic and abiotic features of a site, quite especially tolerance and competition combine to exert a controlling and selective influence on the geographical distribution of taxa. This selectivity results in a limited number of species on a given site. Such a combination of species is called a plant community or phytocenose. Therefore, *a phytocenose may be defined as an aggregation of taxa which are capable of successfully competing with one another within the confines of a particular combination of environmental features they can tolerate.*

The geographical distribution of phytocenoses is, of course, controlled by the species of which they are composed and of the qualities of the site. The distribution of plant communities in the landscape is therefore by no means accidental; there is always a distinct pattern of distribution. For instance, a given phytocenose may occur along the banks of a river. Some distance from the river but still within the flood plain, there will be a different phytocenose. On the valley sides, the phytocenoses will be different again, and even differ among themselves according to the exposure of the slope toward north or south. Finally, on the uplands, there will be still another phytocenose which differs from all others. Each phytocenose has its place in this distribution pattern and can occur only in that particular place. These communities combine to form the vegetation of an area. *Vegetation may therefore be defined as the mosaic of plant communities in the landscape.*

The close relation of a plant community to the site on which it occurs led the British botanist Tansley (1935) to consider them together as a unit. He called such a unit an ecosystem. This term is now widely accepted in all English speaking countries. The Russian forester Sukachev (1947, 1954, 1960) followed a similar trend of thought and introduced the term biogeocenose which is synonymous with ecosystem. Sukachev's term is long and cumbersome but it offers an important advantage: it is more explicit. Bio- from the Greek term bios = life, i.e. both plant and animal life, is combined with geo-, i.e. ge, derived from the Greek word for earth, land. The o in geo is strictly for euphony. Biogeo- are then attached to cenose, derived from coenosis which, in turn stems from the Greek word koinos = common. A biogeocenose is therefore life, i.e. organisms, the site on which the organisms occur, treated or considered together as a unit.

Rowe (1961) presented the biogeocenose and its divisions graphically (Fig. 1). Accordingly, the biogeocenose is first divided into its organic and inorganic components: the biocenose or community of plants and animals, and the geocenose or the physical site on which the biocenose occurs.

Biocenose and geocenose can be further divided.* Thus, the biocenose consists of three divisions: the phytocenose or plant community, the zoocenose or animal community and the microbiocenose or community of plant and ani-

* Sometimes the term biotope is used for the same concept, but this term is more commonly used with a different meaning.

mal microbes. The geocenose can be divided into the climatope if the place is characterized primarily by its climate, and the edaphotope if the site is characterized more by its soil.

The phytocenose is of special interest here because the phytocenoses are the components of the vegetation. They are the units which are presented on vegetation maps. Fundamental in this scheme is the revelation that vegetation is, in fact, only a part of a greater and more complex entity, the ecosystem, within the framework of which it is inseparably linked with its biotope. It is an integral part of the ecosystem.

Parenthetically, as vegetation consists of phytocenoses, the correct term for the study of vegetation is phytocenology. Similarly, a student of vegetation is termed a phytocenologist. The term phytocenology is etymologically and philosophically more correct than 'phytosociology' and should therefore be preferred. Gams has emphasized this for many years and is supported by Schmid, Ellenberg, a number of French scientists, some Americans and Russians and many others. In the United States,

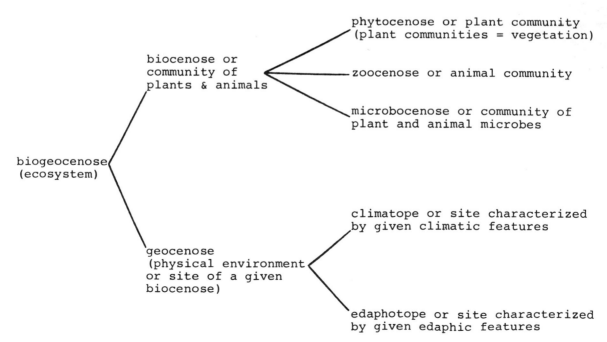

Fig. 1. The position of vegetation within the ecosystem (after Rowe, 1961, elaborated).

phytocenology is generally recognized as a valid term but it is rarely employed. There are even a number of followers of Braun-Blanquet, the founder of phytosociology, who agree that phytocenology should be given preference over phytosociology, but they usually persist in their habit. The term phytocenology should be promoted, especially at universities and colleges.

In order to avoid confusion, some phytocenologists prefer to distinguish any given, concrete stand from the more abstract community type. In many instances, there is no need for such a distinction. For instance, a given plant is called *Panicum virgatum. The plant is correctly named, i.e., this is indeed the name of this particular specimen. However, the name also applies to all other plants of the same species, regardless of their variations. If mappers want to apply the specimen versus species distinction to plant communities, then the individual specimen of a plant community is referred to as a stand, whereas the phytocenose corresponds to the species. It thus becomes the abstraction of many similar stands. If the same method is applied to the places on which the stands or phytocenoses occur, then the term site or habitat corresponds to stand while the term biotope corresponds to the phytocenose. These terms are precisely defined and their use should be encouraged.*

Structure and floristic composition

The vegetation mapper maps phytocenoses, their location, extent and distribution. It is they he must view critically in the field to permit meaningful distinctions between them, even when they are quite similar. Here it is important to observe that:
1. all phytocenoses consist of growth forms;
2. all phytocenoses consist of taxa.
This is a significant observation because it applies to all phytocenoses on earth. There is no exception to the rule.

Growth forms are so important in the study of vegetation that many scientists have focused their attention on them. Beginning with Humboldt (1807), growth forms became an object of study but progress languished for more than a hundred years. Du Rietz (1931) was the first phytocenologist of the 20th century to prepare a detailed classification of growth forms.

Raunkiaer (1934) presented a novel approach, his reasoning going somewhat as follows: all climates show a seasonal rhythm, but some seasons are more favorable to plant growth than others. It follows that one of the seasons must be the least favorable. In the hot and wet climates of what Humboldt called the 'equinoxial regions' the character of the least favorable season does not differ noticeably from any other season but everywhere else the least favorable season is such as to result more or less in a period of dormancy in the plants. The protection afforded the meristematic tissues during this critical season of drought or frost makes the location of the perennating buds the basic criterion for classifying growth forms. Accordingly, Rauniaer divided all plants into the following major classes:

1. Phanerophyta   buds more than 25–30 cm above the ground;
2. Chamaephyta   buds above ground but less than 25–30 cm;
3. Hemicryptophyta buds at the surface of the ground;
4. Geophyta   buds below the surface of the ground;
5. Therophyta   buds in the seed: annuals.

Raunkiaer's method remained unique as all other growth form classifications focused on the actual shapes of plants, their morphology. But his approach at once captured the imagination of phytocenologists and his terms have been adopted everywhere.

Other classifications followed and it is indeed difficult today to be a competent phytocenologist without a thorough acquaintance with a least some of the more recent systems such as those by Ginzberger and Stadlmann (1939), Dansereau (1961), Mueller-Dombois and Ellenberg (1974), Box (1981). The classification of growth forms by Mueller-Dombois and Ellenberg appears to be the most comprehensive and the most practical one.

Now that a precise description of all growth forms is available, it is possible to use it in the analysis of phytocenoses. It soon becomes clear that the growth forms are distributed within a phytocenose in a definite order. The grouping of certain growth forms permits a detailed analysis of a plant community and its division into individual sections, usually in the form of horizontal layers or strata.

Such a division is referred to as a synusia, a term introduced by Gams (1918), who defined it as 'a group of plants of one or several related life forms, growing under similar environmental conditions'. It may consists of few or many species but its over-all floristic composition is quite uniform. Synusias often correspond to the tree layers, the shrub layers, a ground layer of forbs or of mosses, etc. The stratification of phytocenoses is very common, even in grasslands, and the recognition and distinction of the individual synusias is of fundamental importance in analyzing a phytocenose.

Synusias are not limited to a vertical arrangement. They may also occur side by side and even interpenetrate one another without surrendering their identity. For instance, in the northwest European heath, a synusia of mosses will often develop under the protection of heather (the synusia above it). But where the dwarf shrubs of the heather plants stand far enough apart to permit more sunlight and wind to reach the ground, a synusia of lichens takes the place of the moss synusia. In sections of the tundra, a graminaceous synusia will frequently develop in addition to the dwarf shrub synusia as is well known in Alaska and other regions of high latitude.

The over-all appearance of the vegetation is called its physiognomy. The physiognomy is used to describe the broad features of the vegetation such as broad-leaved deciduous forest or grassland, etc. The analysis of individual phytocenoses and a study of its synusias leads to a description of the individual strata, usually characterized by height and density or coverage of the respective growth forms. The result of such an investigation is termed the structure of a plant community. *The structure is therefore defined as the spatial distribution pattern of growth forms in a phytocenose.*

A phytocenose, as observed above, consists of taxa. Usually, such taxa are unevenly distributed insofar as certain taxa may be common while others are less conspicuous. One or a few taxa are usually so prominent that they dominate the character of the phytocenose. They are therefore referred to as dominants. Plant communities on small scale maps are often named after a dominant genus, e.g. oak forest. This means only that oaks dominate the forest. It does not mean that there are no other taxa present. Species, too, can be dominants, e.g. a red oak-white oak forest, or a Douglas fir-ponderosa pine forest. All such names emphasize the taxa of the most prominent synusia while ignoring all others. Most plant communities consist of so many taxa that it is not practical to list them all in the map legend. The use of dominants to characterize a phytocenose is therefore common on vegetation maps.

Just as a phytocenose can be described in detail according to its structure, so it can be described by its component taxa. Such a description is termed *the floristic composition of the phytocenose.* When speaking of the floristic composition, all species are included, although it is not unusual for phytocenologists to characterize the floristic composition by listing the more conspicuous species but omitting the rare or incidental ones.

It is important to observe that:
1. all phytocenoses have a structure;
2. all phytocenoses have a floristic composition.

This is important because it applies to all plant communities on earth. There is no exception to the rule. By using the combined information on structure and floristic composition, any plant community can be described, identified and defined precisely. This method has the advantage that it permits direct comparisons of any plant community with any other one, and so reveal

18

even the finest differences between them. The combined use of structure and floristic composition is scientifically the most acceptable and precise method of dealing with vegetation and should be adopted by all vegetation mappers. It implies the most comprehensive analysis and description of phytocenoses; it is clear, definite and unambiguous.

In some parts of the world, especially in northwestern Europe, vegetation mappers work often with maps of a very large scale, showing great detail. They then limit their work to the floristic composition on the theory that the taxonomic name of plants implies their growth form. This is true but useful only to those who acquainted with the plants so named. Most phytocenologists are not acquainted with plant names in any detail anywhere except in their own region, and even there such knowledge is often incomplete. Using both structure and floristic compositions to identify phytocenoses permits a reader at least to visualize the vegetation, which is valuable.

In many tropical regions, the vegetation is mapped strictly on a physiognomic-structural basis. The explanation is that the mapper does not know the plants and/or has no opportunity to have them identified. Indeed, many taxa remain unidentified. Furthermore, some vegetation types, such as most tropical rain forests and the Brazilian Campo Cerrado are so rich in species that their number alone makes a detailed floristic description difficult. However, whether a mapper works in the tropics or in higher latitudes, he should always strive to present his vegetational units as comprehensively as possible with regard to both structure and floristic composition. The better he succeeds in this, the more valuable will be his vegetation map.

Dynamism

An important feature of the vegetation is its degree of stability. In a sense, no plant community can be entirely stable because it consists of living organisms which sooner or later must die and will then be replaced by other organisms, though not necessarily of the same species. However, a very high degree of stability can be attained in types of vegetation that are more or less free from human influence. This is truest of the so-called climax in the natural vegetation (see below). Stability means a lack of change or at least a minimum of change. However, even in the natural vegetation, abrupt natural changes can temporarily destroy the stability. When, for instance, a lightning starts an unchecked fire which sweeps through a mature forest and leaves nothing but ashes in its wake, it is most likely that the forest will then be replaced by so-called pioneer communities which in turn are replaced by a longer or shorter series of other communities until at last a community develops that is once again in harmony with the biotope and thus is likely to prevail over a longer period and remain in a steady state. Such a development, termed succession, usually follows any destruction of the vegetation. Where melting glaciers retreat, or after landslides, the freshly exposed area will experience a similar succession. A relatively stable community reproduces itself and thus assures its continuity.

Primarily as a result of human influences, much of the world's vegetation is today quite unstable. This means that it is in the process of changing, or else it is deliberately kept by man in a state which remains stable only as long as man keeps it so. As soon as man's influence diminishes or ends, succession will set in at once. The world is therefore filled with types of vegetation most of which are unstable. These unstable communities may be entirely artificial, depending on man for their continued existence or range from this extreme to the other when through a number of seral communities, climax is once again established.

In the field, the mapper finds himself surrounded by many different types of vegetation. All of these are termed actual. Actual vegetation is that which exists at the time of observation, be it natural or not, and regardless of the character, condition and stability of its component communities. But because there is such a variety of communities, it seems best to classify them (Fig. 3–2) and bring some order to what, at first sight, seems chaotic (Küchler, 1969b).

Fig. 2 reveals that vegetation can first be divided into two major types: natural and cultural. The natural vegetation is not appreciably affected by man, i.e. the presence of man does not lead to a change in the character of the vegetation. The natural vegetation may still be observed in many parts of the world, even near relatively densely populated areas as in California. But much of it has been destroyed, and this destruction continues unabatedly today.

Some phytocenologists distinguish between the natural vegetation and the original vegetation, which is also considered natural. It is, however, not often clear what the original vegetation is. For instance, the original vegetation of parts of North America is considered the vegetation that existed at the time of the arrival of the Europeans. Surely this does not mean that it was natural. Indians grew maize, beans and squash, and wherever this occurred, the vegetation must have been altered substantially.

The originally natural vegetation can only be compared with the natural vegetation of today. In most parts of the world, the latter no longer exists, for various reasons. Usually, man has become very active, destroying natural plant communities, changing them or replacing them with others. Hence we can speak only of the potential natural vegetation. This, however, is most important, and in order to obtain it, two assumptions are necessary: (1) that man be removed from the scene, and (2) that the resulting succession of plant communities be telescoped into a single moment in order to exclude the effects of climatic changes. This then, is the potential natural vegetation of today. In it, man's past activities may remain a factor. It is

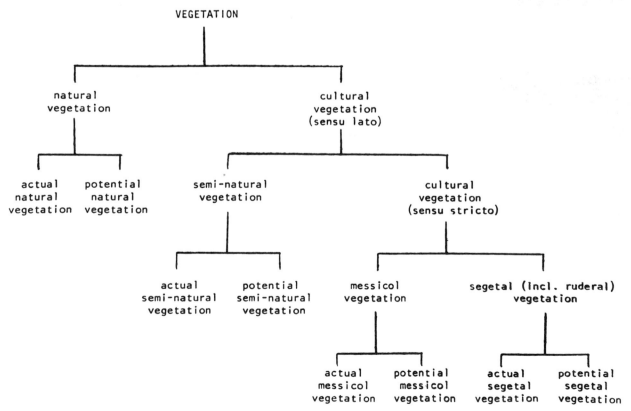

Fig. 2. The divisions and subdivisions of vegetation.

essentially the same as the climax vegetation, provided the term 'climax' is not used in the original Clementsian sense, but rather in the modern and more realistic way.

The original natural vegetation can differ appreciably from the potential natural vegetation of today. This is especially clear in areas where irreversible changes have taken place. For example, in the lower valley of the Weser River in northwestern Germany, the flood plain consisted during Roman times and much later of sands and gravel occupied by a natural vegetation of oaks and willows. In subsequent centuries, the forests around the headwaters of the Weser River were cleared and the resulting heavy erosion led to the deposition of a clay layer several meters thick on the flood plain of the lower valley. The natural vegetation of this clay, however, is not oak or willow but ash and elm (Tüxen, 1957). Irreversible changes may also take place where forests of spruce or pine replace broadleaf deciduous forests for more than three successive generations. They can also be induced by continuous heavy grazing, the artificial lowering of the water table, and in many other ways. Some parts of the world have been populated for a long time, at least since the last interglacial period. Climate and topography have changed significantly since then and so, of course, has the vegetation. The original vegetation and the potential natural vegetation of today have therefore little or nothing in common.

The vegetation is natural wherever it is in balance with the abiotic and biotic forces of its biotope. The biotic forces include man as long as he does not change the character of the vegetation. Just exactly when the character of the vegetation changes due to man's activities can be determined only arbitrarily.

When the natural vegetation is altered or destroyed, the resulting plant communities are unstable. They immediately enter upon the road of succession unless prevented by man. Presumably, the succession would eventually lead to a return of the natural vegetation. This has been interpreted as the natural vegetation still being there, but only potentially, not actually. The natural vegetation, be it actual or potential, is in harmony with its environment and therefore a precise indicator of prevailing conditions. Many mappers have produced maps of the potential natural vegetation because of their great usefulness.

The United States Forest Service distinguishes between the potential natural vegetation and the climax. Accordingly, potential natural vegetation allows for the present existing vegetation to be projected into the future, including accounting for the effects of past disturbances, the presence of naturalized exotic species, past species extinctions and the existing climate. The concept of climax deals with the theoretical past conditions of the vegetation (Schlatterer, 1983) without, however, indicating how long ago such a climax existed. It corresponds therefore to the above described original vegetation.

The natural vegetation stands in contrast to the cultural vegetation. These anthropogenic types are the result of man's interference in the balanced natural vegetation. Disturbing this balance brings change, and this may be indirect or direct. Accordingly, the cultural vegetation sensu lato, i.e. the cultural vegetation in the wider sense of the term, can also be divided into two sections: the seminatural vegetation and the cultural vegetation sensu stricto, i.e. the cultural vegetation in the strict sense of the term.

The seminatural vegetation is natural in so far as man did not plant it. It grew by itself. But man influenced it noticeably. This he may have done by grazing, possibly overgrazing the natural phytocenoses, or else he may have practiced selective logging in a natural forest, thereby changing the floristic composition. The seminatural vegetation is therefore the direct consequence of man's management, and it will change as the management changes. In this manner, man may change the seminatural vegetation deliberately or quite inadvertently. The actual seminatural vegetation is therefore one of a series of possible, i.e. potential, seminatural phytocenoses.

The cultural vegetation sensu stricto is man-made. Man deliberately removes the entire vegetation and replaces it by whatever seems most useful to him at a given time. The actual cultural vegetation therefore includes such types as wheat fields, vineyards, banana plantations, rice paddies, orchards, etc. This is called the messicol vegetation (from messis = crops harvested or to be harvested, and colens = cultivating, as in agriculture). It is entirely artificial and requires man's presence to survive.

The actual messicol vegetation is that which exists at the time of observation. It is interesting from an economic point of view but it is ecologically significant as well. The natural order of the landscape as expressed by the mosaic of plant communities is not removed by man's interference. Even modern technology does not permit man to raise any crop anywhere. For instance, he can not grow coffee in Kansas or coconuts in Minnesota. He must always select appropriate sites for his crops, and these sites are the biotopes of certain natural plant communities. Tüxen (1957) illustrates this with the following example from the areas around the southern North Sea, which were glaciated during the earlier part of the Pleistocene, leaving large areas covered with sandy moraines, others with heavy clay. On the sand, the potential natural vegetation consists primarily of forests of birch (*Betula pendula*) and oak (*Quercus robur*). Today's communities feature pine plantations, relatively few cultivated fields, mostly with rye and potatoes and their respective weed communities: the Teesdalio-Arnoseretum and the *Panicum crus galli-Spergula arvensis* association. Hay meadows of the Molinietalia type and pastures of the Lolieto-Cynosuretum are fairly common. Occasionally, there are fragments of *Calluna* heath (the Calluneto-Genistetum). The country roads are lined with birches, leading through heath and forests where the most common native tree is again this birch.

In contrast to the sandy sites, the potential natural vegetation on the river terraces or on ground moraines with heavy clay soil consists primarily of a forest of beach (*Fagus silvatica*) and oak (*Quercus petraea*). As a result, the present communities are quite different, too. The pine plantations have disappeared as, indeed, have most forests. There are a few spruce plantations, but the few remaining forests of vigorous oaks and beeches remain characteristic. Pastures and hay meadows are less common, too, but by no means absent. On the other hand, turnip fields are frequent with their specific weed communities of *Alchemilla arvensis, Matricaria chamomilla* and *Chrysanthemum segetum*. The country roads are lined with apple trees, linden and maple.

The important point here is that every given plant community of the potential natural vegetation is replaced by a definite variety of cultural communities, the so-called substitute or replacement communities. The various substitute communities of a given natural phytocenose may lie side by side in any order, as dictated by economic considerations. As any one of these may contact any other one of them, they are referred to as contact communities. Contact communities are therefore replacement communities of a particular phytocenose of the potential natural vegetation. The contact communities of one natural phytocenose will not lie next to those of another natural phytocenose unless there is a break in the environmental features, such as a change in the substrate, topography, microclimate, water conditions, and others. The contact communities of one natural phytocenose will not intermingle with those of another.

The actual vegetation consists of some pioneer communities, some seral phases, a host of more or less permanent substitute communities, and a very few types of natural vegetation. Even these are probably not entirely free from human influences. The actual vegetation which the mapper observes is therefore the result of the integrated effects of natural and man-made conditions which control the order of all phytocenoses in space and time, and give direction to their evolution. Hence the actual vegetation consists of a mosaic of phytocenoses which make up the natural vegetation of an area, or else of various sets of replacement communities, with each set corresponding to a particular phytocenose of the potential natural vegetation.

The potential messicol vegetation is becoming increasingly significant as the growing population of the world exerts an ever increasing pressure on the land and the highest productivity on a sustained yield basis becomes imperative. It is therefore important to recognize crop plants that may be introduced into a region from afar in order to raise the level of productivity. This is now being done on a growing scale, especially in developing countries. All crop plants that may be cultivated profitably in a given area other than those actually present compose the potential messicol vegetation.

Very closely related to the messicol vegetation is the segetal vegetation (from seges originally = grain field). It involves phytocenoses growing under or among cultivated and other crop plants, i.e., the weed communities. Like the messicol vegetation, the weeds depend entirely on man and on the crops he plants. Different crops require different forms of management and cultivation, and the segetal communities are closely adapted to this.

The ruderal vegetation (from rudus = crushed stone, rubbish, hence debris, ruins, etc.) also occurs on sites which are strongly influenced by man. It is characteristic of 'unused' sites like building lots, road sides, railroad yards, and waste places of various kinds. Many ruderal species occur also as weeds in cultivated fields where they are constituents of the segetal vegetation.

Robertson (1982) illustrates the above with an example from Scotland: 'A freely drained brown forest soil on the valley slopes of the Southern Uplands can carry – in its least disturbed state – a dry Atlantic heather-moor. A sufficient increase in grazing pressure can alter the vegetation to a sward of acid bent-fescue grassland and, where slope is not a limiting factor, arable fields or permanent pastures may be established. These three replacement communities are, however, all linked with brown forest

soils and are phases of management pressure on the vegetation. If this pressure is withdrawn, a reversal of the process can take place with the eventual reestablishment of a heather-moor'. In this case, the bent fescue grassland represents a semi-natural vegetation type whereas the crops and weed communities of the arable fields and the artificially established permanant pastures are messicol vegetation. The once actual natural vegetation of dry Atlantic heather-moor has become the potential natural vegetation. It becomes once again the actual natural vegetation when the land is abandoned and the process is reversed.

These terms help establish order among the innumerable types of phytocenoses. Vegetation mappers have often used them in the organization of the map content.

The nature of vegetation is now clear. Mappers can present it on their maps clearly and unambiguously, especially if their maps are based on structure, floristic composition, and an appropriate terminology.

Phytocenoses must be carefully analyzed if they are to be classified and meaningfully arranged in the legend of a vegetation map. So many life forms and so many taxa occur together in so many different combinations that, at first sight, the possible combinations approach infinity and chaos. However, by selecting certain criteria, it is possible to establish units of vegetation which are manageable and meaningful. Obviously, the number of criteria at the disposal of the mapper is large. However, he must remember that basically all phytocenoses have structure and floristic composition. These are the basic features and should always be employed in the greatest possible detail in order to characterize a phytocenose. Thus Aichinger (1954) observed quite correctly that 'in any analysis of the vegetation of an area, the mapper should first describe the physiognomy and/or structure of the plant communities in as much detail as is practical. The units so established can then be refined by observing their floristic composition'. Often the structural unit is relatively large and composed of one or more floristic units which are divisions and subdivisions of such a structural unit.

It is, of course, possible to analyze a plant community only by its structure or only floristically. Küchler (1956b) presented vegetation maps of southeastern Mount Desert Island, Maine, using the two methods separately, but they can of course be combined as on the vegetation maps of California and Kansas (Küchler, 1974, 1977).

4. Composition and Structure of Vegetation

I.S. ZONNEVELD

Introduction to morphology

In order to have data according to which vegetation can be classified in a classification system and be delineated on photographs and in the field, vegetation has to be analysed. The basis for that analysis is vegetation morphology.

Vegetation morphology deals with the composition, the three-dimensional form, shape and structure of the 'green mantle'. Without morphological definitions it is not possible to describe the vegetation or to make a classification. Form can be described in the 3-dimensional space. Time can be considered as the 4th dimension. The 'structure in time' is usually referred to as 'process' and as 'succession' or as phenological aspect. In the scheme below the items under consideration in spatial morphology are tabulated (Fig. 1). The various categories will be discussed below.

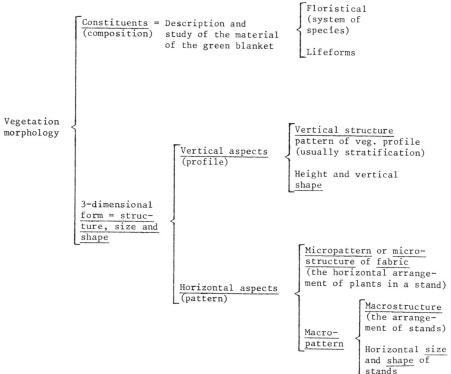

Fig. 1. Main aspects of structure.

A.W. Küchler & I.S. Zonneveld (eds.), Vegetation mapping. ISBN 90-6193-191-6.

Constituents (composition)

General

The 'green mantle' consists purely of plant individuals. These individuals can be classified (idio-systematics) in various ways.

The *Floristic system of species* is well-known. In it, the whole plant kingdom is ordered according to morphometric characteristics. The guiding principles of this system are derived from the phylogenetics. The units of the system are arranged in a hierarchical sequence: variety, species, genus, family, order, class.

For most parts of the world the not-too-rare plants can be described and placed into this system. It is one of the most logical natural and comprehensive systems that exist in science, although it is by no means perfect. The vegetation classification based on floristics is discussed in Chapter 6. In the next chapter the floristic analyses will be treated in detail.

Lifeforms

It is apparent that many plant individuals show morphological adaptations to the environment. For example, adaptations to drought or marshy conditions are widespread and well-known even to non-ecologists (e.g. leaf shape and size, succulency, thorny shrubs, spongy air tissues in *Scirpus* species, etc.).

Some adaptations are only phenotypical. More important are the genotypical adaptations. In the latter case the species may show the same form under all conditions. Often the adaptation is potential, but still specific to a single species of the floristic system. In this case it only occurs if 'necessary': if the environment stimulates it (e.g. airenchym of *Lycopus europeus,* various forms of amphybious plants). Danish scientists in particular have contributed to the study of these adaptations and their use in vegetation science (Warming, Raunkiaer, Iversen).

The concept 'lifeform' [1] is based on the existence of morphological adaptations that are easily observable. A lifeform is a group of plants having certain morphological features in common, which are usually assumed to be a hereditary adjustment to environment (Dansereau and Cain).

The characteristics are purely morphometrical and the guiding principles are totally different from that of the floristic system. In lifeform systems the (mesological) ecology of the environment guides the selection of characteristics and the arrangement in groups and also the hierarchy. So Raunkiaer uses as guiding principle the adaptation, as expressed in morphology, to the unfavourable season (too cold, too hot, too dry for life processes).

Iversen uses the adaptation to the factor water (too wet, too dry, etc.). The difference between a floristic system and a lifeform system can also be expressed as follows: in a floristic system plants are classified according to their ability which determines their job. So far especially climate and hydrology are used as such. The occurrence and abundance of lifeforms in a flora of a region or in the floristic species composition of a vegetation unit can be expressed statistically or in a simple graph and be used to indicate the environmental factors.

Lifeforms play an important role in physiognomic-structural classification systems. The analysis will be discussed further in this chapter. The lifeform composition of a certain vegetation unit can be expressed in spectra, e.g. a histogram with the presence (or frequence) of each lifeform as parameter. Other graphs can also be used in representing life form composition (c.f. Zonneveld, 1959, 1960; Biesbosch).

The formulas as proposed by Küchler are treated in chapter 3 as an expression means of physiognomic-structural analysis.

---

[1] The author prefers the term lifeform above the also used form growthform, because the latter is also used in phenotypical sense.

Three-dimensional form
(Structure, size and shape)

The vertical aspect

The vertical aspect of vegetation is studies in the 'Vegetation Profile'. The study of the vertical structure is mainly a study of vertical pattern. The most obvious aspect of this vertical 'pattern' is stratification. In a forest (natural, cultivated or planted), one can often distinguish between various layers or strata, but arable land also shows at least two vegetation strata, the crop itself and a weed stratum. These layers may be different in: (1) size, that is high (thick) or low (thin); (2) shape (straight, undulating, irregular, sharply defined of diffuse); (3) coverage or density (the layer may be dense and intercept much light and/or precipitation, or may be open and allow easy penetration of sun radiation and water to the lower strata; coverage is usually expressed as a percentage); (4) floristic composition; (5) composition of life-forms.

Usually all the five types of differences occur at the same time. For single strata that can be characterised by a certain life form composition the term 'synusium' is used (Gams). So the vegetation may be built up of various synusiae. In practice the words synusium and stratum (layer) are used almost as synonyms, because nearly always various layers can be distinguished with the help of life forms.

The various strata are not always horizontal: very distinct synusiae may occur on the stems of trees (epiphytic bryophyte and lichen communities). Lianes may spread vertical curtains through the forest.

Sometimes it is necessary to classify various layers individually, as separate communities, e.g. in savanna areas where not only seasonal periodicity, but also irregular changes occur over the years due to rainfall differences (annuals react differently from perennials).

The system of symbols of Dansereau (viz. Appendix A.4) is a useful tool to describe schematically the vertical structure with the help of the composition of the constituents.

Methods showing the plant in a more realistic (less schematical) way are also generally in use. Their usefulness depends on the artistic capacity of the authors. The vertical aspects of structure are very important for the physiognomic classification, and therefore also for airphoto-interpretation, as this is based on physiognomic classification. The vertical dimension can be estimated and measured on the photograph with help of stereoscopy.

The horizontal aspect

General remarks and micro pattern. The horizontal aspect is often neglected in general vegetation typologies but is most important in photo-interpretation.

The finest horizontal aspect is 'granulation'. This may be 'coarse' or 'fine'. It varies from the consideration of single plants in their mutual situation (micro pattern) to the consideration of the various stands (phytocoenoses) in their spatial relation (macro pattern). Study of the micro pattern is a subject favoured by those botanists with a primary interest in the constituents of the vegetation. The result of this study is of much interest in connection with one of the main problems of vegetation science, the *homogeneity* of the basic unit! Moreover, it is also the place where *autecology* and *synecology* have concrete contact.

In practical vegetation survey there usually is a danger of submerging into too much detail, if micro structure problems are treated. However, micro-structural data may be of interest for obtaining the diagnostic characteristics required in classification and mapping, and may even have physiognomic importance. The latter aspect provides the link with photo-interpretation: photo 'texture' and shade and colour are often a function of (micro) vegetation structure. Depending on scale between micro pattern and macro pattern (see below) a meso pattern can be distinguished.

28

Macro pattern. In macro pattern we can distinguish three main types. All of them exist of an arrangement of stands.

— *Mosaic patterns.* The most typical pattern shows a regular or more irregular chessboardlike arrangement of two or more vegetation units. The most simple form of a mosaic can be a *dot pattern.* Here usually only two elements are present one as matrix, one as dot. This can be a *random* or a systematic pattern. If not random, we can speak of a *grid pattern.* Also lines can form a grid (see Fig. 2).

— *Zonation.* This is an arrangement of units according to a distinct gradient. It may be a slope of a mountain with a climate gradient, or an area bordering a lake or river, etc.

— *Alternation.* This is related to zonation, but the units (at least two) alternate, so that there is not one distinct continuous gradient.

Between the three main types all kinds of combinations and transitions may occur, and many subdivisions according to size and form of the pattern elements can be made. In the next picture some examples are given (Fig. 2).

The macro horizontal aspect is of great importance to the landscape guided classification, and hence also to mapping. The concept of macro pattern is linked with *scale,* which is typically a horizontal size phenomenon. This is clear as soon as we consider the main types of pattern: zonation, alternation, mosaic, dots and grids. If the basic units of such a pattern are big enough in relation to scale they will appear on the map, and show clearly either the zonation or the mosaic nature. This is usually the case on large-scale maps. In map units of small-scale maps, however, the concept 'complex' is well-known. This concept is used in cases where distinctly different vegetation stands – (coenon, phytocoenose, etc.; in soil maps pedon or soil units, etc.) – can be distinguished in the field or even on the photo, but where the area is too small to map them. They are then taken together in one 'complex' unit.

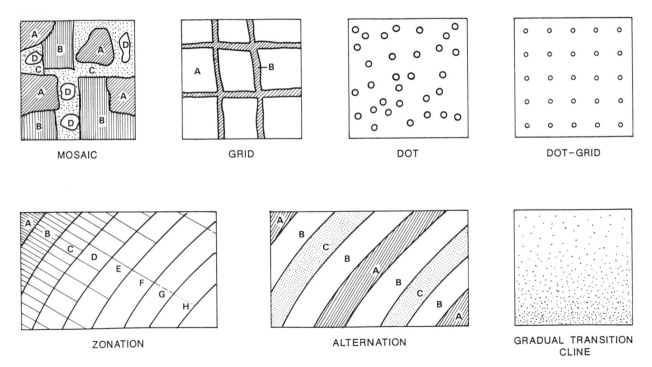

Fig. 2. Some examples of pattern types.

In soil survey, the word 'association' is often used for such a complex, especially if the components show a typical pattern that is repeated in a similar way in various places. It may be a zonal or a mosaic-like pattern. Typical zonal associations are also called catenas. This could also be used in vegetation.* However, there is confusion about the catena concept. For some it is just a zonation related to relief. For others it is only a zonation on the same parent rock material caused by relief and climate.

In vegetation survey we call a complex a mosaic, if it has a mosaic-like pattern. If it has a zonal aspect, it will usually be called on the map a (zonation) complex or catena (see above). In the case of a mapping unit, the term mosaic has also been used for a unit with zonal aspects. This should, however, be avoided.

The influence of scale is also clearly illustrated by the following example. Suppose there is a zonation of mosaic-like patterns. If the latter are too small to map the composing units, the total map may show a zonal picture. So, e.g. in a mountainous area in a very detailed map, we should see many mosaics. In a less detailed map, more zonations (according to physiographic and climate gradients) might appear. On a very small scale, the zones are too small to map separately and by consequence alternation of zonal complexes may dominate the total map picture on the same area. Zonation (given a certain scale) shows the existence of a clear environmental gradient, and the direction thereof. The mosaic- and alternation patterns show that there is a more complex set of factors, or complexity in one factor, e.g. stratification etc. in rock.

* The word 'association' cannot be used for vegetation patterns, because this term is already generally in use in the syntaxonomic classification for basic (not complex!) units. There it is comparable with 'mineral association' in petrography. The 'associated' things are in this case the composing items: plants (resp. minerals in petrography) and not the comprehensive units of the 'green mantle' (resp. of the 'earth's crust' in soil science). The different use of the word association may be one of the sources of misunderstanding between soil scientiests and ecologists!

Coarse-grained dot patterns are often associated with rather unfavourable environment situations like drought or waterlogging, pioneer vegetations (*Scrirpus* sp.), termitaria, salinity, etc. Straight lines and accurate geometric figures usually show the influence of the factor man. Also in other cases, the shape, form and size of pattern may show the main active factors, be it hydrology, landform, geology or soils. For the probable relation of coarse-grained and fine-grained structures to stability it is referred to the relation theory of Van Leeuwen (1966, 1973).

Structure analysis

The important features for classification systems according to physiognomy and structure are the stratification in layers and the arrangement and density within these layers. For the physiognomy, besides the above-mentioned pure structural features, the life (growth) form composition is most important be it in great detail or just (only) the dominating lifeform in each synusium.

Systematic analysis of structure and physiognomy of plant communities has been attempted by various authors. Thus Dansereau (1958, 1961) presented an interesting system of carefully drawn symbols, each of which representing a particular lifeform. On the other hand, African authors, see Van Wyngaarden (1986) and also Van Gils and Van Wyngaarden, (1984) analysed just layers of (synusiae) characterised by some dominant lifeforms (gras, shrub, trees) in order to be able to indicate some main practically important structural aspects. See also Chapter 29 App. C. Küchler (1949) developed a system based on letter combinations for lifeform analysis which eventually (1966) led him to a standardised printed form, the Phytocenological Record with the help of which plant communities can be analyzed accurately and rapidly. Such a standardized method offers the great advantage of making all phytocenoses directly comparable, thus revealing even fine differences. The printed forms greatly expedite the field work. However, if such forms are not available, the letter symbols can be combined into 'formulas' which are then recorded in a notebook. Their later transcriptions into the Phytocenological Records presents no problem.

Application from the very largest map scales to the very smallest (Küchler 1950, 1951b, 1953b, 1955a) has been fruitfull in clarifying trends and needs.

APPENDIX A

Lifeform systems

1. Raunkiaer
2. Iversen-Zonneveld
3. Iversen
4. Symbols Danserau

1. System of Raunkiaer

Raunkiaer used the adaptation to the *unfavourable season as the main* guiding principle. His system has since been amended by various authors to make it suitable as a general system that can be used for other purposes besides that of characterisation of the unfavourable season. Below we give a version of the Raunkiaer system and also of a revised scheme based on Raunkiaer by Ellenberg 1958.

The Life Form System of Raunkiaer (1903)

P: *Phanerophytes* (P): trees and shrubs with buds higher than 25 cm above the surface. Thins (buds) with which they survive the bad season are exposed to the climatic conditions.
The following sub-division can be made:
   a. mega phanerophytes (Pg): buds more than 25 m above ground
   b. macro phanerophytes (Pm): buds from 10–25 m above ground
   c. micro phanerophytes (Pp): buds from 2–10 m above ground
   d. nano phanerophytes (Pn): buds from 0.25–2 m above ground
   e. climbing phanerophytes (Ps): buds higher than 0.25 m above ground

CH: *Chamaephytes:* herbaceous or woody low plants with buds close to the ground but above surface. Abundant in drier, colder climates, in which they find protection in winter under a snow cover.

H: *Hemicryptophytes (H):* these plants winter with buds down to the soil at soil surface.
Sub-division:
Hc. caespitose plants: buds are hidden in winter under a clump of dead leaves.
Hr. rosette plants
Hs. scapose plants: plants with solid, scape-like roots.
Hg. climbing plants

G: *Geophytes:* plants winter in form of tubers or bulbs.

Th: *Therophytes:* annual plants, producing seed.
Later on other people (e.g. Braun-Blanquet 1951) have added some life forms to the above mentioned 5 groups:

E: *Epiphytes:* plants living upon another plant.

SS: *Stem-succulents:* plants with water tissue in their stem (can be phanerophytes and chamaephytes).

HH: *Hydrophytes:* plants living in water and which overwinter under water.

HL: *Helophytes:* swamp plants.

Raunkiaer used his system to characterize the flora, not the vegetation. He introduced the 'biological spectrum', which can also be used in vegetation science. This spectrum consists of numbers indicating the % of the flora of a certain region that belongs to that life form (Viz. Table 1).

Table 1.

Area	Climate	P	Ch	H	G	Th
Seychelles	tropical rain	*61*	6	12	5	16
Virgin isl.	climate	*61*	12	9	4	14
Alps (Switzerl.)	arctic, alpine	0	25	*68*	4	4
Spitzbergen	(polar)	1	22	*60*	15	2
Eastern Canada	climate	0	26	*57*	15	2
Paris Basin	moderate	8	7	*52*	25	9
Centr. Switzerl.	climate	10	5	*50*	15	20
Connecticut		15	2	*49*	22	12
Death Valley	arid	26	7	18	7	*42*
Cyrenaica	climate	9	14	19	8	*50*
Libyan Desert		12	21	20	5	*42*

The table 1 shows that tropical moist regions have a predominace of Phanerophytes, dry regions of Therophytes, and temperate regions of Hemicryptophytes, whereas the arctic or alpine regions show a dominance of Hemicryptophytes and Chamaephytes.

(The places on earth which have the same biological spectra of Raunkiaer may be called a certain type of *biochores*. Lines bounding the biochores on a map are isobiochores).

This method of Raunkiaer has proved to be a useful mean to a synecological characterisation of the vegetation zones and vegetation units, especially in combination with other classification methods. For detailed vegetation description and classification, even if the floristic data are lacking, it is also very useful.

It should be stressed, however, that the life forms in the sense of Raunkiaer, are mainly based upon climatological conditions, and that there can be many other factors influencing the life forms, e.g. soil factors, hydrological factors, human influence, etc.

The method has the disadvantage of being based mainly on regions with a pronounced cold or arid season: for many tropical areas it is less suited. Braun-Blanquet (1944) pointed out that for many alpine species their success in life depends not so much on their adaptation to the rigourous season as on their adjustment to the very short, cool summer.

Life Form System of Ellenberg Based on Raunkiaer, with Amendments (Ellenberg 1956)

This differs from the former because some aspects of the vegetation during the more favourable season are incorporated.

This classification creates the possibility of a broad ecological characterisation of plant communities, particularly in those parts of the world where the flora is little known. Each plant community consists of a definite group of life forms; each habitat favours certain groups of life forms and may conclude others. However, this system is also not fully comprehensive. Climate adaptations are still a dominant factor in a plant's selection of characteristics.

A. Rooting Plants (Radikante)

I. *Macro Phanerophytes (M or MP):* trees or tree-like plants with buds more than 2 m above ground.
1. Eu-macro-phanerophytes: real trees (M)
   a. evergreen trees
      a.a. evergreen rainforest trees: ombro macro phan. (oM)
      a.b. evergreen soft leaves: daphno macro phan. (dM)
      a.c. evergreen hard leaves: sklero macro phan. (sM)
      a.d. evergreen needle leaved trees: belonido macro phan. (bM)
   b. summer green trees
      b.a. summer broad leaved trees: there macro phan. (tM)
      b.b. summer needle leaved trees: therobelonido macro phan. (tbM)
   c. rainy green trees
      rainy broad leaved trees: cheimo macro phan. (cM)
2. Treegrasses: macrophanerophyta graminidea
      e.g. bamboo species (M gram)

3. Rosette trees: macrophanerophyta scaposa
   e.g. many palms (M scap)
4. Herb-stam trees: macrophanerophyta herbacea
   e.g. banana trees (M herb)
5. High-stem succulents: macrophanerophyta sukkulenta (M sukk)
   plants with water tissue in the stem
6. Lianas: macrophanerophytes scandentia (M scand)

II. *Nano Phanerophytes:* shrubs, buds between 0.25 and 2 m above ground same sub-division as I plus (NP or N)
7. Leafles shrubs (not succulents):
   e.g. Sarothamnus scoparius

III. *Chamaephytes:* buds lower than 0.25 m, but not on the ground.
1. Dwarf shrubs: Chamaephyta frutescentia (Ch. frut)
   a. evergreen dwarf shrubs (dCh. frut)
      a.a. evergreen soft leaved dwarf shrubs: daphno cham. frut. (dCh frut)
      a.b. evergreen hardwood dwarf shrubs: sklero cham. frut. (sCh frut)
      a.c. leafless dwarf shrubs (eCh frut)
   b. summer-green dwarf shrubs
      summer broad leaved dwarf shrubs: thero cham. frut. (tCh frut)
2. Semi shrubs: Chamaephyta suffrutescentia (Ch suffr.)
   same sub-division as 1.
3. Trailing shrubs: Chamaephyta velantia (Ch vel.)
   same sub-division as 1 except a.c.
4. Cushion dwarf shrubs: Chamaephyta frutescentia pulvinata
   (Ch frut pulv)
5. Cushion herbs: Chamaephyta pulvinata (Ch pulv)
6. Creeping herbs: Chamaephyta reptantia (Ch rept)
7. Low perennial climbers: Chamaephyta scandentia (Ch scand)
8. Low succulents: Chamaephyta sukkulenta (Ch sukk)

a. leaf succulents
b. stem succulents
9. Hard grasses: Chamaephyta graminidea (Ch gram)

IV. *Hemicryptophytes:* buds close to the earth surface and generally protected by a covering layer of dead leaves (H)
1. Tussock plants: hemicrypt. caespitosa (H caesp)
   grasses and grass-like plants forming tussocks
2. Creeping Hemicryptophytes: Hemicryptophyta reptantia (H rept)
3. Rosette plants: Hemicryptophyta rosulata (H ros)
4. Semi rosette plants: Hemicryptophyta hemirosulata (H hem)
5. Scapose plants: Hemicryptophyta scaposa (H scap)
6. Climbing plants: Hemicryptophyta scandentia (H scand)
7. Water-Hemicryptophytes: Hydro-Hemicryptophyta (Hyd H)

V. *Geophytes:* (or cryptophytes) (G)
1. Root-bud geophytes: Geophyta radicigemma (G. rad)
2. Rhizome-geophytes: Geophyta rhizomatosa (g rhiz)
3. Bulb geophytes: Geophyta bulbosa (G bulb)
4. Water-geophytes: Hydro-geophyta (Hyd G)
In 2 and 4: Subdivisions can be made into: green in spring, summer or rainy season.

VI. *Therophytes:* annuals (T)
1. Tussock therophytes: Therophyta caespitosa (T caesp)
2. Creeping therophytes: Therophyta reptantia (T rept)
3. Rosette therophytes: Therophyta rosulata (T ros)
4. Semi-rosette therophytes: hemirosulata (T hem)
5. Scaposa therophytes: Therophyta scaposa (T scap)

6. Climbing therophytes: Therophyta scadentia (T scand)
7. Water therophytes: Hydro-Therophyta (Hyd T)

All groups 1–6 can be subdivided according to the season to which they germinate and are green.

a. Winter annuals (Therop. hibernalis) (germinate in autumn, green in winter)
b. Summer annuals (Therop. aestivalis) (germinate in spring-autumn, but die in winter)
c. Rain annuals (Therop. pluvialis) (germinate and live only in rainy season)
d. In summerdress hibernating annuals (Therop. epigeios) (germinate in spring and summer)

B. Attached Plants (Adnate)

I. Vascular Epiphytes: Kormo-Epiphyta.
II. Thallo-Epiphytes: Mosses, lichens, fungi, algae living on other plants.
III. Thallo-Chamaephytes: mosses, lichens.
IV. Thallo-Hemicryptophytes: mosses, lichens, algae.
V. Thallo-Geophytes: fungi.
VI. Thallo-Therophytes: short living fungi and mosses.

C. Mobile Plants (Errante)

I. Higher floating plants: Kormo-hydrophyta natantia.
II. Lower floating plants: Thallo-hydrophyta natantia.
III. Hydroplankton: Hydro-planktophyta.
IV. Cryoplankton: Cryo-planktophyta.
V. Edaphophytes: Microscopic soil flora.

2. System of Iversen, 1936 and Zonneveld 1960

Iversen 1936 introduced his 'Hydrotypen' system with the factor water as guiding principle. Too much water reduces the O_2 content of the soil. O_2 for respiration has to be carried to the roots by *airenchym*.

Drought causes wilting (loss of turgor). Wilting causes damage of veins in stem and leaf by folding or breaking (wind etc.). If these veins were to remain intact, the plant could in most cases easily recover after rewetting. Plants with a construction that avoids this damage are more resistent against drought. Such plants are 'scleromorphic': they have abundant sclerencym (= woody tissue) around the veins, or elsewhere in stem and leaves.

Plants with water reservoirs (succulency) in their tissues, and with roots, and plants that have special structures such as wax or hairs on their leaf surfaces, cylindrial leaf shape, tufted growth etc. which avoid evaporation, (*xeromorphic* plants), are able to withstand greater drought than other species which are not adapted to it. Thus in this system of Iversen, (see below) these items play a deciding role; sclerenchym, xeromorphy, aerenchym, succulency, and root length. Most of these are easy to recognize. For the determination of sclerenchym a pseudoquantative measure was developed. The plant is kept for some time in steam (or boiling water). After that, three classes of stiffness are easily recognized. Some plants collapse immediately (ascleromorphic plants e.g. good vegetable), others remain absolutely stiff (scleromorph) while in between an intermediate group occurs (mesoscleromorph).

This method is also applicable to a special life form system of 'Sclerotypes', used to 'measure' the reaction on mechanical action of water or other agencies (see Zonneveld 1960). Here only the three groups of the scleromorphic classes are used: Schleromorphic plants, mesoscleromorphic plants, ascleromorphic plants.

3. Life Form System of Iversen (1936), Hydrotypes

A. Terriphytes (land plants)

Seasonal xerophytes: with no strongly developed root system but with special xermorph (succulency) or waterconservating structures which enable them to survive dry periods.

Euxerophytes:* (very strongly developed root system).

Hemixerophytes: Plants without special structures or strong root system but with strong scleromorphy resisting damage by wilting.

Mesophytes: Plants without special structures or roots but with some scleromorphy resisting slight damage by wilting.

Hygrophytes: Plants without any xeromorphic structures or root development, neither with scleromorphy, thus badly damaged by wilting.

B. *Telmatophytes:* Plants with assimilating parts adapted to air but also with aerenchym available to provide (and store) air for root respiration.

C. *Amphiphytes:*** Plants with adaptations to both the atmosphere and the hydrosphere.

D. *Limnophytes:*** Plants that are only adapted to the hydrosphere and may or may not be rooting in the bottom of the river, lake, etc.

Some plants classed under C and D which are not attached to the bottom of the lake, pond etc. may be 'Mobile Plants' (Errante) in the system of Ellenberg (Raunkiaer). Others may be Hydrophytes in the original system of Raunkiaer, or grouped under various sub-divisions in Ellenberg's system.

* We regret the choice of Iversen of the Name 'euxerophytes' for rather untypical drought-resistant plants. The season-xerophytes really are able to overcome drought. The so called *eu*xerophytes need a water supply and obtain this by their extra-ordinarily long roots. But without ground water they cannot live! However, it would cause confusion to change the names now.

** For plants floating at the water surface but rooting in the soil, there is a reason to call them amphiphytes, although Iversen includes those with the Limnophytes.

4. Life form Symbols (Dansereau 1957)

Because the variety of morphological adaptation is so great it is useful to have for the sampling stage of the survey a system of description of some basic aspects of plant morphology that may occur in different combinations. An example of such a system of symbols of basic forms and aspects has been introduced by Dansereau for terrestrial plants. Such a system can be useful in describing the structure of the vegetation even if the floristical analysis gives difficulties.

1. *Main life form*
   T   trees
   F   shrubs
   H   herbs
   M   bryoids
   E   epiphytes
   L   lianas

2. *Size*
   t   tall   (T: minimum 25 m)
                 (F: 2–8 m)
                 (H: minimum 8 m)
   m   medium(T: 10–25 m)
                 (F, H: 0.5–2 m)
                 (M: minimum 10 cm)
   l   low   (T: 8–10 m)
                 (F, H: maximum 50 cm)
                 (M: maximum 10 cm)

3. *Coverage*
   b   barren or very sparce
   i   discontinuous
   p   in tufts or groups
   c   continuous

4. *Function*
   d   deciduous
   s   semideciduous
   e   evergreen
   j   evergreen succulent or evergreen leafless

5. *Leaf shape and size*
   n   needle or spine
   g   graminoid
   a   medium or small

h   broad
v   compound
q   thalloid

6. *Leaf texture*
f   filmy
z   membranous

x   sclerophyl
k   succulent or fungoid

It would not be difficult to add more symbols for specific purposes adapted to the kind of vegetation. See also Küchler in the joining chapter.

4A. A Physiognomic and Structural Analysis of Vegetation

A.W. KÜCHLER

This chapter presents a system of analyzing the physiognomy and structure of vegetation in detail as it is used by the Vegetation mapper.

It began as a method based on combinations of letter symbols and evolved eventually into a system which is expressed on the 'Phytocenological Record' (Küchler, 1966). This is a printed form with the help of which plant communities can be analyzed accurately and rapidly. Such a standardized method offers the great advantage of making all phytocenoses directly comparable, thus revealing even fine differences. The printed forms greatly expedite the field work. However, if such forms are not available, the letter symbols can be combined into 'formulas' which are then recorded in a notebook. Their later transcriptions into the Phytocenological Records presents no problem. The system's application from the very largest map scales to the very smallest has been fruitful in clarifying trends and needs. The following paragraphs present a revised version of this system.

The idea of using symbols is not new and was employed with particular success by Wladimir Köppen (1931). His classification of climates has been applied all over the world and, although the climate of any place is something very elusive, he has nevertheless succeeded in expressing its major aspects simply and clearly. As Köppen was much influenced by the distribution of vegetation when he formulated his system, it was not a very long step from his climatic classification to a classification of vegetation along similar lines. The Köppen classification describes the climate of any region with satisfactory detail. It does not state what weather may be expected there on any given day, even though the climate of a place is ultimately the sum total of individual weather conditions. Likewise, a physiognomic method can describe the physiognomy and structure of the vegetation of any region with satisfactory detail but it does not give the species of which this vegetation is composed.

For the vegetation mapper, a flexible physiognomic system of analyzing vegetation has five major advantages: (1) it can be used on maps of any scale; (2) it can be used on maps of any country or region; (3) it can be expressed in a clear and unequivocal terminology; (4) it can be employed readily because it does not require taxonomic knowledge; and (5) it forms an excellent basis for studies in comparative phytocenology. The latter point is one of great significance and, regrettably, ignored all too often. In the following paragraphs, such a flexible system to describe the physiognomy and structure of vegetation is outlined.

This physiognomic system resembles Köppen's classification of climates insofar as it, too, relies on letters to describe types of vegetation. Some groups of letters consist of capital letters whereas others consist of lower case letters and numbers. The capitals are used to describe the basic character of a phytocenose and some special growth forms; the small letters and numbers are added as appropriate; they serve to

A.W. Küchler & I.S. Zonneveld (eds.), Vegetation mapping. ISBN 90-6193-191-6.

indicate structural details as the case may require. The capitals together with their associated small letters permit the mapper to present the vegetation with remarkable and often unexpected refinement (Küchler, 1950, 1956b) because the number of letter combinations is very large. This leads to the great flexibility which is the basic prerequisite and, indeed, one of the major virtues of this system.

Each type is, of course, physiognomic in character. The use of bamboos seems to introduce a floristic element, but bamboos are used strictly as a growth form, and no distinction is made between the numerous genera and species of bamboos.

Basic woody vegetation categories

The entire plant kingdom is divided into two major sections: woody plants and herbaceous plants. Woody vegetation is seemingly more varied in its appearance than non-woody or herbaceous vegetation. In accordance with physiognomy as the guiding principle, the woody vegetation is shown on the basis of leaf characteristics, i.e., whether it is evergreen or deciduous, broadleaf, needleleaf, or without leaves. This at once establishes five categories, each one with its particular capital letter.

B: *broadleaf evergreen*. The plants have broad leaves in contrast to needles (see below) and are not bare or without green leaves at any season. Broadleaf evergreen plants include the *Mora excelsa* of the tropical rainforests in Guyana, the mangrove (*Rizophora mangle*) of tropical tidal flats, the carob tree (*Ceratonia siliqua*) of the Mediterranean region, the Australian *Eucalyptus* tree, etc. Forests composed largely of broadleaf evergreen trees are common in the wet tropics, as in northern Borneo and along the Amazon River, in parts of Australia, in regions with a Mediterranean climate such as California, and elsewhere.

D: *broadleaf deciduous*. The plants have broad leaves as in the case of 'B', but defoliate periodically so that they carry no green leaves during a part of the year. The time during which the trees are bare many vary greatly in length. Representatives are the tulip tree (*Liriodendron tulipifera*) in Tennessee, the paper birch (*Betula papyrifera*) in Quebec, the kapok tree (*Bombax malabaricum*) in Indonesia, the boabab (*Adansonia grandidieri*) in the Sudan, and many others. Forests consisting mostly of broadleaf deciduous trees are well known in Kentucky, France, India, and elsewhere.

A further comment is needed concerning deciduousness which is a complex feature. How long must a tree be without green leaves to be deciduous? If a tree is bare only 2 weeks out of 52, it is deciduous? Extreme cases go even further: a tall *Ceiba pentandra* was observed in Jamaica which had lost its leaves on some branches. The other branches were bare at quite a different time.

Vegetation mappers from temperate middle latitudes are reminded that tropical vegetation is usually more complex in behavior than the vegetation in cooler climates. This implies that some terms, coined in the middle latitudes, may not always fit tropical conditions well, as for example, the term deciduous.

The contrast between 'evergreen' and 'deciduous' seems clear. The former means that plants bear green leaves throughout the year, and the latter means that they do not. For instance, in the beech-maple forests of Ohio, all trees will lose their leaves at about the same time and remain bare until, at the appropriate time in the spring, they all produce their new foliage. All members of a species act in unison: all have their leaves or none have leaves. This may be called the 'classical' form of deciduousness.

In the tropics, it is entirely normal that plants do not exhibit a common rhythm as do plants in the higher latitudes. For instance, in northern Thailand, three specimens of *Dipterocarpus tu-*

berculatus were observed simultaneously. The trees stood side by side and one had last year's foliage, one was bare and one was covered with fully developed new foliage. In the same area, *Dipterocarpus obtusifolius* was evergreen throughout the year although is it usually deciduous. On the other hand, the teak tree (*Tectona grandis*) is deciduous in northern Thailand but evergreen near Bangkok, implying a different behavior in different regions.

Another feature of tropical deciduousness is the difficulty of relating it to seasons. In the higher latitudes, the onset of the cold season with its shorter days induces plants to drop their leaves. In the low latitudes, such an adverse period is usually represented by the dry season. The length of day varies very little, especially in equatorial regions. The drought, on the other hand, may be forbidding, and it is easy to see that survival seems best assured by remaining leafless until the return of the rain ends the danger of desiccation. Yet, numerous plants will unfold their young leaves at the height of the dry season when heat and aridity are most intense.

Some plants will deviate substantially from the normal, i.e. the seasonal rhythm. Thus, Korriba (1958) found the periodicity of leaf fall in equatorial regions can vary from 4 or 8 months to 14, 15 and more months.

Unless the mapper spends the whole year in the tropical region to be mapped, he may be unable to judge whether the phytocenoses are deciduous or not, even though he is working during the dry season. This certainly differs from conditions in the middle latitudes where, during the fall, winter and early spring one look will tell the observer the whole story.

In an effort to clarify the situation, at least for the record, Küchler (1967a) proposed that the term 'deciduous' be reserved for its 'classical' form as described above. He further suggests that the term 'tropophyllous' be used for the irregular, indefinite and unreliable forms of deciduousness encountered in so many parts of the tropics. Such a distinction should often prove useful.

The term 'tropophyllous' is not new and is usually employed as a synonym for 'deciduous'. It is derived from the Greek word 'tropos' (turn), and phyllon (leaf). As proposed, tropophyllous refers to plants that may, and usually do, lose their leaves during some period of each year without specifying the exact time or duration of this period. Tropophyllous plants may shed their leaves on an individual basis rather than as members of a species. The term 'tropophyllous' can be applied to individual plants, to taxa, and to phytocenoses composed wholly or in part of such plants. Tropophyllous vegetation is common in the tropics, but does not occur in the higher latitudes.

Plant communities consisting of both evergreen and tropophyllous plants are found in all but the driest parts of the tropics, but are best developed in regions with a short dry season. Such phytocenoses are best referred to as being semi-evergreen, not semi-deciduous.

Where the degree of deciduousness of a given forest varies with individual synusias, the mapper can describe each synusia separately and so attain a high degree of accuracy. For instance, the type of 'seasonal evergreen forest' of the American tropics (Beard, 1944) is entirely evergreen except the uppermost synusia of emergent trees which are up to 25% deciduous.

E: *needleleaf evergreen.* The term 'needleleaf' is here understood to apply to the typicaly needle-shaped leaves of such trees as the pitch pine (*Pinus rigida*), the hemlock (*Tsuga canadensis*), the true cedars like *Cedrus libani*, etc. It also applies to leaves that are more scale-like in appearance, such as the leaves of the arbor vitae (*Thuja occidentalis*). It includes all plants with needle-like leaves, even though these are not conifers, e.g. the chamise (*Adenostoma fasciculatum*) of the California chaparral and some Australian acacias like *Acacia verticillata,* and *A. asparagoides.* The most magnificent and varied forests of needleleaf evergreen trees occur in the western coastal region of the United States.

N: *needleleaf deciduous.* The term 'needleleaf' is here used as in the case of 'E', and the term 'deciduous' has the same meaning as in 'D'. Some of the best-known examples of plants which shed their needles seasonally are the larch (e.g. *Larix laricina*) of northeastern North America and the bald cypress (*Taxodium distichum*) of the southeastern United States. Extensive forests of needleleaf deciduous trees (*Larix*) occur especially in eastern Siberia.

O: *leaves absent or nearly so.* Plants without leaves are termed 'aphyllous'; they have their chlorophyll in their stems, branches, and twigs, which are frequently succulent. The *Casuarina* formations of Australia and the *Euphorbia* forest of Ethiopia are examples. Aphyllous plants are most common in arid and semi-arid regions, and include such genera as *Ephedra, Tamarix*, and many others. Technically, some of these plants do have leaves, but they are either very shortlived or else they are extremely small, often reduced to scales or thorns; in all these cases, the leaves play a negligible role in photosynthesis as compared with the twigs and branches.

The five major categories are therefore those of B, D, E, N and O. As the vegetation mapper is usually pressed for space on his map, frequently occurring combinations may conveniently have their own symbol. To the above symbols, therefore, two more are added, namely:

M: *mixed.* Unless a phytocenose consists of a pure stand, all plant communities are mixed. However, the term is here employed in a much more restricted way and limited exclusively to a mixture of needleleaf evergreen ('E') and broadleaf deciduous ('D') plants. The combination of 'E' and 'D' is very common but in order to use the 'M', it is necessary that each of the two components occupies at least 25% of the area. Good examples occur in Michigan, Georgia, Manchuria, etc. If either 'D' or 'E' does not cover at least 25% of the plant community's area, the two letters are recorded separately.

S: *semi-deciduous.* This term applies to combinations of broadleaf evergreen ('B') and broadleaf deciduous ('D') plants in which each of these occupies at least 25% of the area. This type is particularly important in tropical and subtropical countries because the 'B'-forests of the humid tropics merge gradually with the 'D'-forests of the drier regions, but note the above comments on tropophyll.

Basic herbaceous vegetation categories

The second basic group of categories is applied to the non-woody or herbaceous vegetation. It happens frequently that herbaceous plants are decidedly seasonal in character (Küchler, 1954a). This is, of course, more true of some types than of others, but in any event, herbaceous vegetation is always shown on a map as it appears in the landscape at the time of its fullest development. As in the case of woody plants, the herbaceous categories are divided according to their appearance, their physiognomy. There are three categories.

G: *graminoids.* This term includes all herbaceous grasses. To these are added all plants which are grass-like in appearance even though they are not grasses in a taxonomical sense, such as sedges, reeds, cattails, and others. The bamboos are also grasses but they are here excluded because they are woody. Plants of the 'G' category are illustrated by the Indian grass (*Sorghastrum nutans*) of the North American prairie, sedges like *Mariscus jamaicensis* of the Florida Everglades, species of *Imperata* of the tropical savannas, and many others. Examples of vegetation types composed primarily of graminoid plants are the Argentine pampa, the Russian steppes, the North American prairie, the African savannas, etc.

H: *forbs.* The term 'forb' is applied to the numerous broadleaf herbaceous plants, in contrast to the narrow leaf graminoids. They are usually of the flowering type, but 'H' includes also all non-epiphytic ferns except tree ferns. 'H' is a common synusia in many broadleaf deciduous forests, especially when these are not very dense and their soil is more or less permanently moist. It is also common in many arid and semi-arid regions where these forbs are referred to as 'ephemerals' because their life cycle is very brief.

L: *lichens and mosses (bryoids).* This category includes all mosses and lichens which grow on the ground, whether soil or rock outcrops. Epiphytic lichens and mosses are here excluded. Lichens (e.g. *Cladonia* spp.) may cover large areas, especially in higher latitudes, as in the forest tundra and the tundra of North America and Eurasia. Mosses can be equally important, for instance the *Sphagnum* of the Irish blanket bogs that cover the landscape for miles. In moist climates of the middle and higher latitutes, mosses and lichens may form important elements of the lowest forest synusia.

Algae are here included if aquatic vegetation is to be mapped. Fungi and bacteria are not included in this system. They are physiognomically of no significance in the landscape and therefore need not be shown on a vegetation map.

Special growth form categories

The ten capital letters stand for broad categories, but the vegetation of the earth consists, obviously, of more than ten types. In order to refine the vegetational descriptions, a number of special categories is added, each of which has some very distinctive features. The features usually lend the physiognomy of the vegetation a new character and so lead to new physiognomic types. The special growth forms are not mentioned on a map unless they appreciably affect the general appearance of the vegetation. Just when this is the case is left to the judgment of the mapper.

C: *climbers (lianas).* The term 'climber' or 'liana' is here used for all woody plants that climb trees and shrubs. Herbaceous climbers belong to 'H'. Lianas are most common in tropical rainforests (e.g. *Paullinia cupana*) but are frequent in many North American flood plain forests, too, like the fox grape (*Vitis vulpina*).

K: *stem succulents.* These striking lifeforms of great variety are concentrated in the more arid or semi-arid regions of the world. Where they grow in appreciable numbers the general physiognomy of the vegetation is profoundly affected. The barrel cacti (*Echinocactus grandis*) in some Mexican semi-arid regions are well-known examples.

T: *tuft plants.* These plants have in common that they consist of a trunk (often unbranched) which carries at its apex a tuft of leaves, as do most palms. A forest with many palms certainly presents a different picture from one without them. This is even more true where palms dominate the open tree synusia of savannas, as for instance the *Mauritia* savannas of northern South America or the *Areca* palm groves at Angkor in Cambodia. Climbing palms are listed with lianas. Tree ferns (e.g. *Alsophila camerunensis*) are exceedingly graceful tuft plants with slender stems and very feathery leaves, and where they occur in large numbers, every observer becomes acutely aware of them. But tuft plants can also be quite stocky, as when short trunks bear a rosette of simple leaves, e.g. *Espeletia hartwegiana* and *Senecio keniodendron* on the high volcanoes of equatorial regions. Some plants have tufts of long grass-like leaves, e.g. the Australian *Kingia australis.*

V: *bamboos.* Taxonomically speaking, bamboos are grasses; but because they are woody and because of their peculiar growth forms they are given a separate letter. Bamboo groves usually consist of individual clumps of bamboo with tall graceful trunks and feathery twigs, e.g. *Gigantochloa maxima* in Java. Elsewhere, as in Chile, bamboos may form impenetrable thickets; thus the *Chusquea quila* represents a shrub synusia in temperate rainforests. The 'V' also applies to woody canes.

X: *epiphytes.* Epiphytes are plants which grow upon other plants, as for instance mistletoe on apple trees in New England or Spanish moss on live oaks in Florida. Strictly speaking, epiphytes are not growth forms. They range from mosses and lichens to ferns and many flowering plants. Obviously, they include a great variety of growth forms. For the purposes of mapping vegetation, they are grouped together. While they are therefore not at all uniform in appearance, they do introduce a new physiognomic element into the growth form of their host plants, and it is this changed look of the latter that is here significant.

These five additional types, the special growth form categories, help to further describe the ten basic types, and their presence or absence in a given plant community is often highly significant. They should therefore be shown wherever they attain any degree of prominence in the physiognomy of the vegetation.

The introduction of special growth forms is subjective and perhaps not always entirely logical. However, in analyzing and mapping phytocenoses it is often necessary to proceed quite pragmatically. The special growth forms listed above change the physiognomy of the vegetation appreciably. Yet without their own symbols, they would be submerged in other categories and the reader would remain unaware of their presence. Palms, for instance, would have

to be included with other broadleaf evergreen trees such as live oaks; stem succulents like barrel cacti would merge with Mormon tea (*Ephedra*), both being aphyllous. To classify bamboos would be baffling as they, too, may be evergreen. Should they perhaps be in one physiognomic class with mango trees and coconut palms? Obviously, the special growth forms have their place, and it is necessary to record them if important variations in the physiognomy of plant communities are not to be ignored.

Parenthetically, one of the finest collections of illustrations of tropical life forms may be found in The Tropics (Aubert de la Rüe et al., 1957); it is difficult to match this superb book.

Leaf characteristics

The growth forms, identified by capital letters, may be further described by adding small (lower case) letters to reveal some particular leaf features. This is important because the physiognomy of the vegetation may be profoundly affected by the character of the leaves. Five categories are here introduced but some of these are employed more frequently than others.

k: *succulent.* Fleshy leaves are always striking in appearance, as for instance some species of *Mesembrianthemum.* Due to their water-storing capacity, they usually occur on evergreen plants.

h: *hard; w: soft.* The degree of hardness of leaves is not usually significant to vegetation mappers; this is especially true of soft leaves. As a result, these letters are usually omitted. Hard leaves, however, may be important. They are termed 'sclerophyll' or 'leather-like' and are common in regions with a Mediterranean climate and in tropical regions. On medium and small scale maps, the letter 'w' may be considered implied and need not be used in describing broadleaf vegetation. Sclerophyll broad leaves should be indicated ('Bh'). On large scale maps with 'Bh', the letter 'w' may be used where applicable in order to make the contrast with hard leaves clearer.

1: *large; s: small.* Raunkiaer (1934) has presented a scale of six leaf size classes with quantitative values. While this scale is occasionally quoted in the literature (Cain and Castro, 1959), it is not usually applied on vegetation maps. Nevertheless, the more extreme size of leaves noticeably affect the physiognomy of phytocenoses and they should be recorded. For instance, in the Mojave Desert, there are areas covered with almost pure stands of creosote bush (*Larrea divaricata*). The shrubs are up to 2 m tall but their evergreen leaves are tiny, producing a remarkable physiognomic effect.

In general, leaves are considered to be of medium size if no reference is made in the record. Only the more extreme sizes are indicated. The question then arises as to the particular limits at which these extreme sizes begin. The sizes here proposed are as follows: large leaves ('1') cover at least 400 cm^2 and small leaves ('s') cover not more than 4 cm^2. Obviously, the size of most leaves remains unmentioned. However, large and small leaves as defined above should be recorded. Thus, the symbol for the creosote bush communities is 'Bs'.

When different species of one life form, e.g. 'B' (broadleaf evergreen) have leaves of different sizes, the large ('1') or small ('s') leaves are recorded only if their respective species together cover more than 25% of the area.

Raunkiaer has six leaf size classes; they may be used whenever details on leaf sizes are needed. However, it is usually adequate to limit the record to the extreme sizes. The limits here proposed fall well within the range of Raunkiaer's scale: both leptophyll and nanophyll size classes are less than 4 cm^2, and the upper limit of macrophyll and the whole megaphyll classes

is well above 400 cm^2. The class of small leaves ('s' < 4 cm^2) includes therefore all or Raunkiaer's leptophyll and nanophyll classes and the lower margin of the microphyll class. The large leaves ('1' > 400 cm^2) include most of the macrophyll and all of the megaphyll class. The unrecorded medium-sized leaves therefore include the upper part of the microphyll, all of the mesophyll, and the lower part of the macrophyll classes. It is as if, in the system here proposed, Raunkiaer's mesophyll class had been extended beyond its margins into each of the adjoining size classes.

It is tedious and often difficult to determine the exact size of a leaf. However, Cain and Castro (1959) have discovered a simple solution to the problem. They have shown that it is possible to obtain results of reasonable accuracy by taking two-thirds of the rectangle formed by the length and the width of the leaf. If. therefore, a leaf measures 40 cm in length and 28 cm in width, the area of the rectangle of length times width is $40 \times 28 = 1120$ cm^2. The area of the leaf is then approximately two-thirds of this area, or 746 cm^2. As this exceeds 400 cm^2, the minimium size of large leaves, the measured leaf is classified as large ('1').

It is often difficult to appreciate the ecological significance of leaf sizes. However, when the latter become extreme, it is usually reasonable to conclude that an extreme ecological condition may also prevail. The interpretation of extreme leaf sizes is therefore often easier and more correct that the interpretation of less extreme leaf sizes. This is the reason for proposing here to ignore all leaf sizes except the more extreme ones.

Compound (pinnate, palmate, etc.) leaves are considered in their entirety when their sizes are to be determined. Some authors prefer to express the size of a compound leaf by referring only to an individual leaflet, and sometimes this is convenient. For instance, no problem arises in clearcut cases, as with the common locust tree (*Robinia pseudoacacia*). However, it is often difficult to establish a leaflet. The matter becomes quite problematical when leaves are pinnately divided, as for instance in the case of the compass plant (*Silphium laciniatum*) of the Kansas prairie. Indeed, some plants may have simple and compound leaves simultaneously, such as the Rocky Mountain maple (*Acer glabrum) of the western United States, or the Japanese woodbine (Parthenocissus tricuspidata*). Even the same compound leaf may have distinct leaflets in its lower parts and be only pinnately divided in the upper parts, as for example in the case of the poison hemlock (*Conium maculatum*). There are numerous such cases, especially in the tropics. The problem is solved simply by always considering the surface of the whole leaf and not just one of its parts.

Structural categories

The physiognomy and structure of the vegetation are dependent not only on deciduousness, the presence of forbs or palms, etc. It is essential to describe two other characteristics because of their profound effect. They are height and coverage. Both can be measured accurately. Usually, however, it will suffice to estimate rather than measure them. Applied to the individual synusias, they enable the mapper to describe the structure of the vegetation in remarkable detail, especially on maps of large scale. On small-scale maps the mapper can show only the upper-most synusia and describe it in general terms.

As in the case of Köppen's classification of climates, here, too, it has been impossible to avoid the introduction of arbitrary delimitations. Whenever an arbitary choice is made, it is open to criticism, but this is inevitable. It is not possible to satisfy everybody, but the numerical values here proposed should be acceptable to most vegetation mappers.

The height of the vegetation is measured from the ground upward to the 'surface' or average height of the upper limits of a synusia. The mapper can ignore the fact that the height of some plants differs from the average; such a detail need not be important. In the tropics, however, many rainforests have a 'surface', albeit a most irregular one, which is pierced by widely spaced trees, the emergents, which tower above the rest, for instance, the Brazilnut trees (*Bertholletia excelsa*). They form a very open synusia of their own and should not be ignored. The mapper must acquaint himself with the characteristics of the vegetation before mapping it. This will assist him materially in all cases where he must use his own judgement and decide arbitrarily whether or not to show a given feature and how to show it.

The height classes, here proposed, are as follows (Table 1):

Table 1. Height Classes of vegetation.

Class		Height
8	=	> 35 m
7	=	20– 35 m
6	=	10– 20 m
5	=	5– 10 m
4	=	2– 5 m
3	=	0.5– 2 m
2	=	0.1–0.5 m
1	=	< 0.1 m

The height of woody vegetation does not usually present any problems. It can be measured or estimated without difficulty.

The reader may have noted that no specific distinction has been made between trees and shrubs. There are many examples where such a distinction is difficult or impossible to make. For instance, if the mapper is confronted with a dense stand of Douglas fir (*Pseudotsuga menziesii*) on a large recently burned-over area, and the Young Douglas firs are barely 1 m tall, does he record trees? Or perhaps shrubs? At what height does a plant change from a shrub into a tree? Some authors, e.g. Ellenberg (1956), consider a height of 2 m as an acceptable dividing line but many shrubs grow much taller than 2 m. On the other hand, Dansereau (1957), used 8 m as a dividing line between trees and shrubs. As many trees are less than 8 m tall, his proposal does not seem to offer an improvement over

Ellenberg. In some regions, especially in semi-arid areas (e.g. western Texas or much of the Gran Chaco in Paraguay), many square miles are covered with a vegetation that consists essentially of a transition between trees and shrubs. The palo verde (*Cercidium* spp.) in Arizona, the mesquite (*Prosopis* spp.) in Texas, and many others occur as shrubs, trees, and in every intermediate form. It is therefore often impractical to distinguish between them. Indeed, it seems that an upper limit for shrubs is undesirable.

A simple solution to the problem is offered by omitting the distinction between trees and shrubs altogether. Instead, a group of height classes has been introduced which is applicable to all growth forms. By establishing eight such classes, a description of the vegetation can be given in adequate detail and the vexing tree-shrub problem is eliminated.

In addition, the height classes are not expressed by letters as are all other categories. Numbers are used instead. This clarifies the formulas (see below) appreciably. A number stands out boldly among letters and makes it easier to read the formula. Dansereau (1961) has already pioneered in this direction.

An absence of symbols for height categories in woody vegetation implies that trees are the dominant growth form. This requires that height classes must be given for shrub synusias regardless of whether they occur alone, above lower synusias, or under trees. In the latter case, i.e. where both trees and shrubs are mentioned in the formula, both must be given their respective height class symbols.

The height of herbaceous vegetation is subject to strong seasonal fluctuations as many herbaceous plants die down to the ground during the period of dormancy, no matter how tall they may grow. Kansas sunflowers, for instance, will reach a height of 3 m during the growing season, but every winter they vanish. For the purpose of mapping, the herbaceous vegetation is recorded as of the time of its maximum development which usually begins with the flowering phase, (cf. Chapter 22).

The final group of small letters refers to the spacing of plants in the landscape, thus describing the density of the vegetation. These letters express the density by indicating the percentage of the ground covered by the respective growth forms, assuming that they are projected vertically to the ground. Thus the density is in reality the degree of coverage of the individual growth forms. There are six coverage classes, as follows:

c: *continuous.* This implies a continuous growth, and the plants often touch one another. The 'c' is usually omitted and the vegetation is assumed to be continuous unless it is qualified by any one of the other letters in this group. Where vegetation types of varying coverage occur together, the 'c' is used to advantage in order to show the contrast with more open stands. The coverage characterized by 'c' is 76–100%.

i: *interrupted.* The plants are standing rather close together and, from a distance, may give the appearance of continuous growth. However, they are usually spaced so widely apart that they do not touch. The mapper must not confuse the coverage with the distances between trees. For instance, spruces need not be far apart before the ground between them is flooded with sunlight, whereas many broadleaf trees may be much more widely spaced before the canopy is broken at all. Among herbaceous plants, the letter 'i' may be especially useful in connection with bunch grasses. The coverage characterized by 'i' is 51–75%.

p: *parklike or in patches.* When the 'p' is used in connection with woody vegetation, it signifies that the trees and shrubs grow singly or in small groves, as in so-called parklands and in savannas. In the case of herbaceous vegetation, the 'p' may signify disconnected patches. The coverage characterized by 'p' is 26–50%.

r: *rare.* This is applied to growth forms which are more widely scattered than in 'p'. It may, for instance, refer to a life form associated with another one of greater continuity. Thus, occasional, widely spaced medium-tall palms may rise out of a continuous cover of grass: 'GTr'. A more detailed description of this vegetation might be: 'G3cT6r'. The coverage characterized by 'r' is 6–25%.

b: *barren*. The vegetation is much reduced and the landscape appears barren. The coverage has now shrunk to 1–5%.

a: *almost absent*. Growth forms now cover less than 1% of the area, or else there is no plant life at all and the vegetation is absent. This is common on shifting sand dunes, in very rocky desert landscapes, etc. The letter 'p' attached to an 'a' ('ap') at the end of a formula implies that small patches of plant communities occur within a desert landscape.

The numerical values of the coverage classes are as follows (Table 2):

Table 2. Coverage classes of vegetation.

Class		Coverage
c	=	> 75%
i	=	50–75%
p	=	25–50%
r	=	5–25%
b	=	1–5%
a	=	< 1%

Both height and coverage classes vary in size, and in both cases the lower classes show a finer detail than the upper classes. The height classes increase steadily in size whereas the coverage classes increase only in the lower categories and the three higher classes are of uniform size. It

Table 3. Categories of growth form and structure.

Growth form categories				
Basic growth forms			*Special growth forms*	
– Woody plants:			Climbers (lianas)	C
Broadleaf evergreen	B		Stem succulents	K
Broadleaf deciduous	D		Tuft plants	T
Needleleaf evergreen	E		Bamboos	V
Needleleaf deciduous	N		Epiphytes	X
Aphyllous	O			
Semideciduous (B+D)	S		*Leaf characteristics*	
Mixed (D+E)	M		hard (sclerophyll)	h
– Herbaceous plants:			soft	w
Graminoids	G		succulent	k
Forbs	H		large (> 400 cm^2)	l
Lichens, mosses	L		small (< 4 cm^2)	s

Structural categories	
Height (stratification)	*Coverage*
8 > 35 m	c = continuous (> 75%)
7 = 20–35 m	i = interrupted (50–75%)
6 = 10–20 m	p = park-like, in patches (25–50%)
5 = 5–10 m	r = rare (6–25%)
4 = 2–5 m	b = barely present, sporadic (1–5%)
3 = 0.5–2 m	a = almost absent, extremely scarce (< 1%)
2 = 0.1–0.5 m	
1 < 0.1 m	

makes relatively little difference whether a forest canopy is 26 or 28 m high. But it makes a great difference whether a synusia is 8 cm tall or 160 cm. Similar reasoning applies to coverage.

The letter symbols of this system are conviently grouped together in Table 3.

Formulas

If a phytocenologist wishes to analyze vegetation, and has no forms (see below) on which to record his observations, he will find it convenient to combine and group the various letters and number symbols in formulas so as to produce the most meaningful description of the phytocenoses in the shortest form. For instance:

D = a forest of broadleaf deciduous trees;

D7 = a forest of broadleaf deciduous trees which are 20–35 m tall on the average;

D75 = a forest of broadleaf deciduous trees in two continuous layers; in one layer the trees are 20–35 m tall and in the other layer the trees are only 5–10 m tall.

There is no need to record a growth form more than once just because it occurs in more than one stratum. The growth form symbol is repeated when the individual synusias are described separately, perhaps written one below the other in the same order in which they occur in the field, as, for instance:

B8CX
B6p
T5r
B3i T3p

In all other cases one-line formulas are composed for a detailed description of the entire phytocenose.

There are eight height classes, each characterized by a one-digit number. No height class symbol ever consists of more than one digit. If therefore the formula reads 'D75', it means that there is a broadleaf deciduous forest consisting of two continuous strata, one layer in height class 7 and another in class 5. The reader may recall that the 'c' symbol for 'continuous' is usually omitted and used mostly to emphasize contrasts (see example immediately following).

'D73p' or 'D7c3p'. In this case, we find a patchy layer of broadleaf deciduous shrubs under a continuous canopy of broadleaf deciduous trees. The first of the two given versions is legitimate and there is no need for a 'c' symbol to follow the '7'. However, it is safer and, especially for beginners, more reassuring to use the second version and place a 'c' symbol after the '7'. The contrast in the degree of coverage of the two strata is then established beyond any doubt.

Symbols for lianas and epiphytes are inserted in the formula after the tallest height classes of woody growth forms in which they occur; they are given without further qualifications: 'B8CX B6p3i T5r3p'.

The most conspicuous feature of the vegetation is always placed at the beginning. For insurance: 'DG' is a 'forest of broadleaf deciduous trees with a layer of grass underneath the trees'. The trees are the most conspicuous feature and therefore are mentioned first. If, on the other hand, the vegetation consists of a grassland with thinly scattered trees, then the order is reversed because the grass is more prominent than the trees. The formula now reads 'GDp' or perhaps 'G3 D5r'.

The small letters are used to qualify the capital letters. Each capital letter has its own small letters and therefore must be followed by them immediately. If there is more than one capital letter in the formula, each will have its small letters independently of the other. For instance: 'D6H' implies that a broadleaf deciduous forest is 10–20 m tall, but the formula says nothing about the height of the forbs that grow in the forest. If the formula reads 'DH2', the implication is that, in a broadleaf deciduous forest (height not given), there grows a ground cover of forbs, 10–50 cm tall. If the height of both the trees and the forbs is wanted, then both must be given separately: 'D6H2', 'D7H2p', or whatever else the case may be.

The symbols are used in a prescribed sequence. First, the growth form is shown by its capital letter. If leaf size is to be recorded, it follows immediately after the growth form symbol, thus: 'H1' or 'Bs'. The height class is given next and the coverage is shown last. The sequence is always the same: growth form – leaf size (if any) – height – coverage. This sequence applies to each recorded growth form individually. Thus 'D6iE6b' shows a unistratal forest of broadleaf deciduous trees covering 50–75% of the stand's area. There is an admixture of needleleaf evergreen trees of similar height and covering only 1–5% of the area. Height and coverage are therefore given separately for both 'D' and 'E', which are members of the same synusia.

The number of letters used in a formula depends on the available information and on the scale of the map. Maps of very small scale should have short formulas. A little world map in an atlas will have vegetation types many of which can be described with formulas of minimum size, perhaps a single capital letter. The other extreme is a map of a small area, done at a very large scale. On such a map, the formulas should contain the features of the physiognomy and structure of every phytocenose just as completely as possible. This should certainly be so on the field map and in the field notes, or at least in one of these. It will give the mapper a complete record that can later be manipulated to suit the scale and the purpose of the map. Formulas can be very long, but the mapper should not hesitate to write long formulas because completeness of information is of crucial significance in several stages of map preparation, even though some of this information will not be used on the map. When mapping at large scales, it is always advantageous to establish separate formulas for each synusia of each phytocenose. This will contribute materially to the clarity of the field maps and notes. Two examples may illustrate such descriptions.

'S6i E8r B4p'. This formula describes the physiognomy and structure of the following type of vegetation in northern California. The major synusia consists of a mixture of broadleaf evergreen and broadleaf deciduous trees, 10–20 m tall on the average and covering from 50–75% of the ground ('S6i'). Towering above it are needleleaf evergreen trees more than 35 m tall; they cover only 5–25% of area ('E8r'). Then, below the major layer, there is a patchy layer of broadleaf evergreen shrubs of 2–5 m in height and covering 25–50% of the ground ('B4p').

Another complex example is 'D7c4r H2p L1r'. This formula describes the following vegetation in western New York. There are two layers of deciduous broadleaf plants, one continuous layer of trees 20–35 m tall ('D7c') and one layer of shrubs 2–5 m high and covering only 5–25% of the area ('D4r'). Under the shrub layer is a synusia of forbs averaging 10–50 cm in height, and covering about one-fourth to one-half of the ground ('H2p'). Finally, there is a very open synusia of mosses and/or lichens that are less than 10 cm tall and that cover 5–25% of the area ('L1r').

The generalization of vegetation types on small-scale vegetation maps results in the omission of details in the formulas. For instance, the first of the examples given above may be reduced by omitting the undergrowth. Before the reduction, the formula read 'S6i E8r B4p'; after the reduction it reads 'S6i E8r'. Further reduction leads to omission of the height classes: 'SiEr'. Ultimately even the coverage is dropped and the formula is reduced to its minimum: 'SE'.

The second example may be adjusted to the desired smaller scale in successive steps as follows: 'D7c4rH2pL1r' – 'D7c4rH2p' – 'D7c4r' – 'D7' – 'D'.

Generalizations can be carried out in more ways than one. If a vegetation mapper must generalize his data, he should do so in a manner that the purpose of his map is served best. For instance, if the vegetation in parts of the Rocky Mountains is to be mapped for foresters, forest economists and people of similar interest, then the generalizations will retain the various types of trees, their height and density. The lower synusias may be omitted and all herbaceous vegetation is ignored. However, if the vegetation of the same area is to be mapped for sheep herders, the tree synusia looses much of its significance while the ground cover of graminoids and forbs becomes all-important. Tree growth is then shown on the map simply as 'forest' and all available space is devoted to the details of the herbaceous vegetation.

In this and similar ways, the vegetation mapper selects from his data only those that serve a given purpose. While in the field, he collects information on all features of the vegetation. This is done as completely as possible and in the greatest possible detail. Once his field work is completed, he is in a position to prepare vegetation maps for a variety of purposes. All his maps will be based on the same original data, but these are used selectively and, indeed, can be so used in a great variety of combinations. The author experimented along such lines in Maine (Küchler, 1956b), and Dansereau (1961) produced interesting results in the region around Montreal, Canada.

It may be that the purpose of the vegetation map is to portray the general features of the vegetation of a region. This is especially true when the map scale is small. The mapper is the free to select what, in his opinion, is most characteristic of the prevailing vegetation types. Generalization always implies selection and selection is all too often a matter of judgment. The mapper will want to present the vegetation in the greatest possible detail that the scale permits without cluttering the map. This consideration and the amount and quality of the available information form the framework within which the mapper must manipulate his generalization. If he proceeds on the basis of well-reasoned arguments and if his map content is well organized, his results should be satistactory.

Recording data

The basis of all phytocenological work is, of course, the intensive observation and detailed analysis of plant communities in the field. Only this will permit the accumulation of information that can then be used in a great variety of theoretical and practical considerations. A simple form, the Phytocenological Record, has been devised to facilitate the standardized recording of physiognomy and structure of the vegetation in the field (see Chapter 5, Fig. 3).

The Phytocenological Record has proved satisfactory in the middle latitudes, for natural vegetation as well as in cultural types. It has also been employed successfully in tropical and arctic environments, with vegetation types ranging from a variety of tropical rain forests to scrubby low deciduous forests, tundra, and grasslands. It is therefore adapted to a wide variety of vegetation types. It takes very little time to record the desired information on this form.

The phytocenologist who wishes to analyse the vegetation does well to acquaint himself thoroughly with the individual phytocenoses by walking around in them. He then observes the stratification of a given plant community and proceeds to write down his findings for each synusia separately, one at a time. If he is in a forest that is 16 m tall, he observes everything in height class 6. If the forest is deciduous, he will go to the square on the Phytocenological Record where the vertical column of D meets the horizontal column of height class 6. In this square he records the coverage of D. If the canopy is closed, or nearly so, he writes a small 'c' into the square. If a few patches of deciduous shrubs cover a rather small portion of the area, and if these shrubs are approximately 0.5–2 m tall, they are recorded in the horizontal column of height class 3 where it meets the same vertical column of D. As the patches of shrubs cover so little ground, their coverage is more likely to be 'r' and, if the leaves of these shrubs happen to be small (less than 4 cm^2), coverage and leaf size are shown together: 'rs'.

In this way, any growth form in any height class is noted in the appropriate space.

Epiphytes are recorded as follows: they are indicated at every height level in which they occur. They are shown according to their growth form. Therefore, if the epiphytes consist mostly of mosses and/or lichens, then an L is shown in the proper space. It is therefore possible that an L occurs several times in the same column but at different levels. If ferns or flowering plants, e.g. mistletoe, occur, then an H is used. If both H and L occur together, then both should be shown. If a type of epiphyte is rare, then the proper letter is used, followed by an exponential minus: L^-. Conversely, an exponential $+$ is used where the epiphytes occur in large masses: H^+. Just when such quantitative signs should be used is left to the judgement of the mapper.

In contrast to epiphytes which are recorded in every level in which they occur, lianas or climbers are shown only in the highest level they attain, and where they are apt to spread out.

The lower part of the Phytocenological Record is used for notes on anything that seems significant. Comments on reproduction of forest trees, evidence of fire, logging, grazing or some other form of disturbance or exploitation are entered here as well as observations on the soil, rockiness, profile, etc. Additional comments on growth forms are also useful, such as stilt roots, buttresses, etc.

The method of describing the structure of phytocenoses on the Phytocenological Record is adequate for all practical purposes. But there may be occasions when the mapper wishes to show greater detail. If he, for instance, wishes to add height classes, then he should retain the existing ones but refine them. Thus height class 8 shows forests more than 35 m tall. In the great needleleaf forests of western North America, such a refinement may be useful. The mapper then does not add a height class 9 but rather increases the height classes by the introduction of primes. For instance, he may wish to show the height of the forests as $8' = 35$–50 m and $8'' =$ more than 50 m tall. In this manner, the mapper can enrich his descriptions of the vegetation. But he should use this method of refinement only when really necessary and when it does not lead to possible confusion. A comment in his notes will certainly be helpful.

5. Floristic Analysis of Vegetation

A.W. KÜCHLER * and I.S. ZONNEVELD **

Various methods are applied for the floristic analysis of vegetation. An extensive survey exists in Knapp (ed., 1984), Volume 4 of this handbook. In this volume we will treat, therefore, only five methods that are often used by vegetation mappers:

1. the quadrat (relevé) method;
2. Küchler's phytocenological Record;
3. the line transect method;
4. the step-point method;
5. the Curtis transect.

1. The quadrat (relevé) method

1.1. *Introduction*

The simplest floristic analysis is a list of names of plants occurring in a sample quadrat of a certain size (usually between 1 and 1,000 m^2) in a homogeneous representative part of the vegetation to be described. After the names, usually some estimate of cover and density aggregation, etc., of each species present can be given. Such a list is called a relevé. This list is made preferably on a form with preprinted headings, also allowing for the notation of structural and environmental data (see chapter 17. Fig. 1a. I.T.C. relevé data sheet). In detailed surveys of relatively small areas, it may be useful to preprint all existing plant names already on the form. This facilitates the field work and the statistical processing of the data later. The species may even be already grouped according to certain

guidelines. The selection of the sample sites is a special subject. Usually other data (structure, environmental data) are also indicated on the sheet. The collected data are used in the classification. If the total floristic content is used, such classification requires statistical treatment. This can be a manual table method as developed in the Zürich-Montpellier school (Braun-Blanquet), as demonstrated in chapter 6B and here under Küchler's Phytocenological Record, but may be elaborated by several computer routines as shown by Feoli et al. (1984) and Gauch (1985).

A special volume on this important development in vegetation science has been composed for the Handbook of Vegetation Science (Whittaker 1973).

1.2. *The generally used criteria*

The commonly used criteria (measuring standards) are:
— *cover* (by plant taxa or strata) in percentage estimated or measured;
— *abundance* (plant taxa) which may be expressed either as *frequency* (relative number/area), or as *density* (absolute number/area), or just as *occurrence* (yes or no); and

---

* Mainly page 57-62.
** Mainly page 51-56.

A.W. Küchler & I.S. Zonneveld (eds.), Vegetation mapping. ISBN 90-6193-191-6.

— *mass* (in kg/area of plant taxa or strata or vegetation as a whole, phyto-mass).*

The first two are commonly used in general classification systems, the latter sometimes, but mainly in applied studies for land evaluation (timber production, carrying capacity of grazing lands, etc.).

1.3. *Abundance, occurrence, density, frequency*

Most commonly, *abundance* is used in its most simple form: plain *occurrence* = the taxon *is or is not* represented on the sample plot. Statistical treatment over a group of relevés then gives the 'presence' = the percentage of relevés in which the species occurs. Abundance can be counted or estimated as *'density'* in a (preferably logarithmic) scale with boundaries expressed in absolute numbers per square metre, or simply qualitative in such terms as: rare, few, common, frequent, abundant, many.

One can also use a sequence of estimates (ranking) calling the species (or taxon) with the highest abundance 1, the second 2, and so on to the lowest (or group of lowest). This ranking can also be done in combination with cover or frequency or mass (De Vries, 1933).

Density is related to the distance between plants. So estimates or measurements of distance can be used in abundance assessment. For scales see Appendix A.

Frequency is a reasonable intermediate method between estimates and density measurement. Within the sample plot (relevé area) minor plots are selected, in sufficiently large numbers to give statistically significant results. In each of these plots the *occurrence* of taxa is noted. Afterwards the number of minor plots in which the taxon occurs is the frequency of that taxon. Frequency can best be expressed as a percentage

of the total number of minor plots. Thus frequency is comparable with presence. Presence, however, is determined for a series of samples of minimum area size. Frequency is determined within one sample or at least of minor sample plots each far below the minimum area size. It is clear that the absolute figure or frequency also depends on the size of the minor plots. The larger these are, the more chance there is of taxa being observed within it. A useful method is to throw a ring randomly in a plot (preferably using a random figures coordinate table) and to note the species (taxa) within the ring. This can be repeated e.g. 20 times. Reasonable results were obtained on arable land by the author using a ring of a size of 1/2000 of the sample area (Banning c.s. 1974). With twenty observations 20/2000, about 1% of the area was sampled. Afterwards the area between rings was scrutinized to see whether plants occurred which were not caught in the ring; they were noted in the group of lowest frequency. One can also estimate cover per taxon in the ring area or do a mass determination. This can be done in order to carry out a relative classification, but also for absolute mass determination. For most vegetation types the mentioned sampled area (1%) is a bit too low. In 'coarse grained' vegetations minor samples should be much larger than in 'fine grained' ones. The total area of the sample plot varies in the same way.

In fine grass swards a sample plot may be a few m^2, the minor samples may then equal 1 dm^2 or less. The minor samples should at least exceed tree crown or shrub size, if shrubs and trees are to be included. For details and mathematical aspects see Cain and Castro (1959) and Brown (1954), Mueller Dombois and Ellenberg 1974 and 1979.

For very detailed work *point contact sampling* is executed with pins which, by means of special devices, are placed on the vegetation. This is only practical in herbaceous vegetation. The pin usually touches only one individual plant. The pins are randomly arranged, and the number of individuals of each taxon touched by a pin is noted. The number of pins touching

---

* A plant taxon is any kind of classification unit, usually the scientific concent 'species' is used here, but also varieties or higher units (genus) may be used, depending the state of taxonomical knowledge in the study area (see chapter 3).

one individual plant of a taxon is a measure of the frequency of that taxon. (With a dense pin pattern cover can also be estimated.) For details see Can and Castro (1959) and Brown (1954).

The *point-centered quadrate method* is a plot-less density estimate that, within (large) quadrates, also may be used to measure cover (see page 54). The line intercept method (see below) may also be used to provide abundance and frequency figures. Examples of abundance scales can be found in Appendix A.

1.4 *Cover* (area)

The cover per species is hardly ever measured in detail for a whole sample plot except for special, very detailed, work. The most common practice is *estimation*. An experienced surveyor can easily estimate in tenths of the total area. For classification purpose this is more than enough. If one cannot see enough of the sample plot at one glance, accuracy can be increased by dividing it into equal parts and taking an average of the findings in each part. One can use external cover (= the cover within the projection circumference of a plant), or internal cover (= the real projection, e.g. shade by a vertical light source). For classification use the first method is preferable. For examples of cover scales see Appendix A. Two (semi-) quantitative cover estimate methods are described below: *line intercept* and *point-centred quarter*. See also *point contact sampling* above.

Note. In the literature the term 'dominance' is sometimes used as a synonym for cover, but it is also (more logically) used to indicate that a dominant species has the highest scores in either cover, abundance or frequency. In between is the definition that a dominant species is one plant species which occupies a larger area of the sample plot than all the others together. If a small group of species together cover more than half the area, then they may be called co-dominants. In the above mentioned use of dominant, meaning the highest score (no matter how high) the second highest or group of second highest scores can be called co-dominants. Hence the

words 'dominant' or 'dominance' should never be used without explanation.

The line intercept is a widely used simple, fairly accurate and quick method to measure cover, particularly useful in rather open vegetations like "steppes", open grasslands, but also in shrublands like chapparal, garigue, heath, etc., especially to measure the shrub cover.

A measuring tape (a steel one of 50 m is very practical) is laid on the ground and preferably tightly stretched (using pins at both ends). The tape intercepts individual plants, clusters of plants and bare areas.

One simply reads, beginning at one end, the number of cm or dm interception of each plant individual or cluster, and bare areas. The individual figures are noted in a table. The total length of interception per species and for the bare areas, is then calculated. These figures will then be calculated as percentages of the total length of the tape (if this is 100 m every metre is equivalent to 1 %). This percentage equals the cover percentage of the species in the whole sample plot, provided the latter is sufficiently

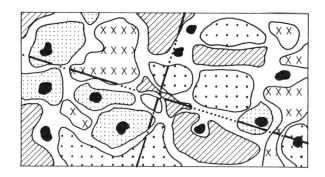

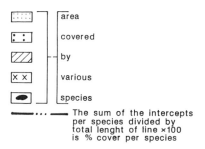

The sum of the intercepts per species divided by total lenght of line ×100 is % cover per species

Fig. 1. The line intercept method (see text).

homogeneous. In order to increase the accuracy, one should make at least two measurements along both diagonals of the plot or if one has not delineated a precise sample plot simply in a cross. The cross reduces bias; at least one of the lines will cross possible horizontal strata if zonations occur. On can combine this measurement with height estimates or measurements of the individual plants. If the interceptions are noted separately for each plant individual and/or per distance unit, a sort of relative abundance or frequency can also be calculated (see also transect method page 59). In vertically stratified vegetations the measurements can be done separately for the various layers. The method is not recommended for strata with relatively small herbs and grasses.

Point-centered quarter method. A semi-quantitive method focusing on density, and cover in relative open vegetation is the point-centered quarter method (see Mueller-Dombois and Ellenberg, 1979). Within the sample area (quadrat or mapping unit) a cross is randomly thrown a number of times (20 or 30). At each spot the distance (d') to the central point of the nearest plant individual from the centre of the cross is measured in each quarter of the cross (see fig. 2). The average distance between the plant individual centres can be calculated with this data.

$$md = \frac{\sum d^1}{4\,A} \times \sqrt{2} \qquad (I)$$

(md = average distance between plant centres in meters

d' = distance between central point of plants and cross centre in meters

A = number of times the cross is thrown)

The density of the plant population can now be calculated as follows in number of plants per hectare.

$$N = \frac{10,000}{md^2} \qquad (II)$$

hence

$$N = \frac{10,000}{\left(\dfrac{\sum d^1}{4\,A} \times \sqrt{2}\right)^2} = \frac{80,000\,A}{(\sum d^1)^2} \qquad (III)$$

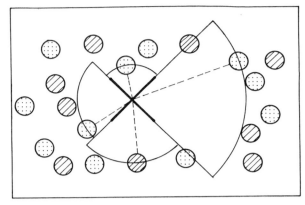

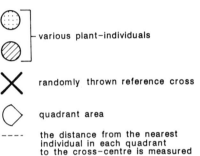

various plant–individuals

$\times$ randomly thrown reference cross

$\diamondsuit$ quadrant area

---- the distance from the nearest individual in each quadrant to the cross–centre is measured

Fig. 2. Point-centered quarter method.

(N = number of individuals/ha).

One can also calculate the number of individuals per species/ha or as a % of total N. Also cover (total or per species) can be calculated by taking diameter of each individual of which the distance was measured and apply $CC = \frac{1}{4}\pi D^2$ (if more or less circular), then:

total $CC = N \times \frac{1}{4}\pi D^2 \qquad (IV)$

(D = average diameter per plant individual in m^2, CC = crown cover, or basal cover etc.)

For each species one can calculate total cover/ha or cover % etc.*)

* The method is an approximation because the formula is only strictly valid if the centre of the cross is such that the $\sqrt{2}$ function is justified. Mueller Dombois and Ellenberg (1979), dot not apply the $\sqrt{2}$ factor, but also then the approximation character remains. In practice, when applied for medium open "steppe" vegetation types, the cross will be thrown never purely random but tends to fall in the centre on open places justifing the $\sqrt{2}$ factor. The method can also be used without quadrats ("plotless"). See also the related Curtis transect (1.5 p. 61)

1.5 *Mass (weight, volume, bulk)*

Unlike abundance, cover and frequency, mass is less often used in classification. There are two reasons for this:
a) mass is a rather unstable factor, because it changes more rapidly during the seasons then abundance and frequency, or even cover;
b) estimation requires more skill and direct measurement is very time-consuming and moreover, often destroys the object of study.

Nevertheless it is important in more detailed scientific and applied ecosystem studies to know the mass. If mass data is necessary for land evaluation such as carrying capacity in rangeland, energy availibility, and timber volume, then it is advisable to do special sampling for it in well selected spots *after classification* and preliminary mapping are ready. Usually then the map-legend will not be expressed in species but, if not in the total mass, in groups (palatable/non palatable and some special interests groups such as legumes etc. see Chapter 37).

1.6. *Combined estimations of cover and abundance*

For pure classifications, frequency and cover data are mainly used in a relative sense, and need not be very accurate. In modern computerized classification (e.g. cluster analysis and ordination procedures) they even are frequently omitted and only 'yes-or-no-occurrence' is used! Visual, 'subjective' classification with tables, punch cards, etc. can make a great deal of use of these cover, abundance and frequency figures. But also for these purposes, however, it is not always absolutely necessary to make a strict distinction between cover and abundance. The main aim is to see a degree of relative representation of each species. Therefore estimation systems have been developed in which abundance and cover are combined. The most widely used is the so-called 'combined estima-

tion scale' of Braun-Blanquet where in the lowest degrees abundance plays a role, and in the higher figures cover is the only criterion. Examples of estimation scales for abundance, cover and combined estimation are given in Appendix A. The most 'natural' scale and therefore the easiest, is the one estimating in 1/10 of cover with a base of abundance in terms of rare, few, abundant and many resp.: r,p,a,m. See page 66: recommended scale.

1.7. *Vitality and phenology*

Vitality is a somewhat ambiguous term, referring to a subjective estimation of the vigour of the plant. Usually it is only used to denote that a certain plant shows clearly reduced vitality (e.g. looks weak). This is usually denoted by a small 0 (zero): e.g. 4° next to the figure for cover, frequency etc. If the plant is dead, a † is put in the same place.
Phenology indicates the stage of growth of a plant, which gives important information not only about the actual stage but also about the trend of growth of the community. Vitality may partly be related to this.
The usual divisions are:
K (G) just germinated (incl. very young plant)
V   only vegetative (no flowers or fruits)
Vb bare (deciduous woody plants)
Fl   flowering
Fr   with fruits
†   dead
If only seed or subterranean parts are present, this can be noted as S (seeds) and R (roots, rhizomes, etc.).
In range ecology information on the stage of tillers might be important and can be indicated using T in addition to other symbols.

1.8. *Distribution and density of sampling*

The distribution of sample spots over the area to be surveyed, refers to the dilemma of random sampling or stratified sampling.

Random or grid sampling results in an unbalanced sampling of the different vegetation stands. Those with the largest extension will be sampled more frequently than those less represented. But when making a classification system, acreage of vegetation stands is of minor importance. Both for scientific and practical reasons often more rare stands or stands with a small total acreage, may be much more important or interesting. Such vegetation stands can even determine the aspect of the landscape (e.g. hedges, narrow tree belts, etc.). So there is no reason to spend more time in sampling the larger acreages than the smaller ones.

The defenders of random sampling, however, state that any other form of sampling would make the selection subjective because before one has sampled it, it would be impossible to judge whether individual vegetation stands are different or not. Theoretically this may be true, especially if only pure floristical parameters were used. However, landscape-ecological knowledge is sufficiently advanced for us to know that there is a certain relation between the cultural and natural landscape and the vegetation. Also in main lines there is some correlation between floristics and vegetation structures.

The solution is the aerial photograph. Photo-interpretation provides, in the first stage of survey, a rather objective analysis of the land. This results in delineation of physiographic (landscape) units. Landform and vegetation structure (including pattern) are main criteria in these delineations. The physiographic units form a reasonable basis for stratified sampling. Within these units indeed random sampling may be preferred. The problem of the requirement of homogeneity of vegetation stands that are used as sample spots so is practically solved, at least for practical purpose, sufficiently reduced.
The density (number) of the sample plots in a study area now mainly depends on the amount of vegetation units to be expected. That means it depends on the type of terrain and the type of classification one aims at. A rule of thumbs says that preferably each classification unit in the final classification should be represented by some ten examples (for statistical reasons). This means that type of classification and terrain and not *scale of survey* (mapping) determines the total number of sample points for classification.

1.9. *Sample (plot) size*

Sample size has two aspects:
a) how many samples have to be taken to have sufficiently statistical reliable data;
b) how large should one sample plot be in order to be representative for a classification unit (vegetation type).

Aspect a) is especially important for quantitative measurements like biomass, primary production, etc. In Van Gils and Zonneveld (1982/83) some data are given. In practice about ten 'relevés' per unit is quite good. More is often not possible (time constraints). The ranked set method (McIntyre, 1952) provides for mass determination a reasonable method to reduce number. The second aspect is especially a question about the minimum area. Arnolds

Table 1. Examples of empirical minimum area (= sample plot size), estimation for some main formations.*

Short and medium grass- and herbage land, tundra	$2 \times 2 = 4 \text{ m}^2$
Tall grass- and forb land dwarf shrubland	$5 \times 5 = 25 \text{ m}^2$
Shrubland (e.g. maquis) dense "steppe", arable field	$7 \times 7 = 50 \text{ m}^2$ (7.07 × 7.07)
Forest (tree + herb layer), "steppe" and herb layer of savannas	$20 \times 20 = 400 \text{ m}^2$
Tree layer woodlands – savanna woodland and herb (shrublayer) of tree-savanna and very open "steppe"	$31\frac{1}{2} \times 31\frac{1}{2} = 1\,000 \text{ m}^2$ (31.6 × 31.6)
Tree layer savanna – tree savanna	$50 \times 50 = 2\,500 \text{ m}^2$

* It is often advisable not to sample in a square, but to adapt to the terrain. In homogeneous terrain a circle is theoretically best, but practically less suitable:

$$\varnothing = 2 \sqrt{\frac{7 \times \text{area}}{22}}$$

PHYTOCENOLOGICAL RECORD No. _____

Location: _____

Height above sea level: _____.

Slope and exposure: _____.

Landscape: _____.

_____

Date: _____

Base map: _____

Aerial photograph No. _____

Type (transect, quadrat etc.) and size of

stand samples: _____

Structural Analysis

Height classes: / Life forms:	B broadleaf evergreen	D broadleaf deciduous	E needleleaf evergreen	N needleleaf deciduous	O aphyllous	S B+D: semi-deciduous	M D+E: mixed	G graminoids	H forbs	L lichens mosses	C climbers (lianas)	K stem succulents	T tuft plants	V bamboos	X epiphytes
8 = >35 meters															
7 = 20 - 35 "															
6 = 10 - 20 "															
5 = 5 - 10 "															
4 = 2 - 5 "															
3 = 0.5 - 2 "															
2 = 0.1 - 0.5 "															
1 = <0.1 "															

coverage: c = >75%; i = 51-75%; p = 26-50%; r = 6-25%; b = 1-5%; a = <1%.

leaves: h = hard (sclerophyll); w = soft; k = succulent; l = large (>400 cm²); s = small (<4 cm²).

Notes

Fig. 3. Phytocenological Record. This side of the printed form is used to record the structure of a phytocenose. Such data are recorded separately in each of the 20 sample plots located within a given stand.

Floristic Analysis

Stand samples:	1	2	3	4	5	6	7	8	9	10	11	12	13	14	15	16	17	18	19	20	F
Number of species:																					

Fig. 4. Phytocenological Record. This side of the printed form is used to record floristic data. Such data are recorded separately in each of the 20 sample plots located within a given stand. The letter F in the last column denotes frequency, i.e. the number of sample plots in which a given species occurs compared with the total number of plots within the stand. Frequency is expressed in percentages.

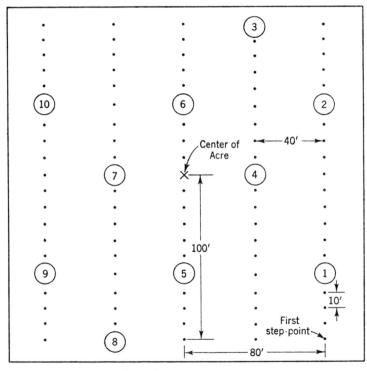

- • Step-point
- ① Square foot quadrat and step-point

Fig. 5. The step-point method. Type of acre plot for sampling level I. (From California Forest and Range Experiment Station, 1955b).

(1977) gives a summary of some methods. For practical surveying a reasonable result can be reached applying the data from Table 1.

2. Küchler's Phytocenological Record

This method always establishes all of the species including cryptogams (lichens, mosses, etc.) in the plant community using one form for many relevés. It is a method suitable in relatively detailed surveys where during field survey, walking in the field, one tries to delineate boundaries by assessing floristic differences using tabular comparison of relevés (which goes ahead of the final statistical treatment of all relevés leading to communities–associations, etc.).

The reverse side of the Phytocenological Record (Fig. 3) is used when the floristic composition of the plant community is to be determined (Fig. 4). The observations are thus recorded in tabular form. Following standard practice, the names of the species are written into the wide column at the left, and the coverage of each species is recorded in the column of each stand sample in which it occurs. If a species is found which does not occur in the first sample plot, its name is added to the list. The vegetation is not sufficiently homogeneous to be regarded as a type if a rather large number of new species must be added to the list with each new stand sample.

The mapper will find it useful to group the species according to some plan. For instance, in a meadow he can often distinguish such groups as grasses, sedges, legumes, other forbs, mosses, etc. If the community is multilayered, the species may be grouped by synusia. If the mapper uses such divisions on his tables, he must leave ample space after the species names of each division when analyzing the first plot. Additional species found in succeeding sample plots can then be listed in the appropriate divisions without difficulty. The stand samples, too, can be arranged in some pattern, for instance, according to wetness of the soil. The record taken in the field with pencil is carefully and neatly copied on clean sheets with a typewriter immediately after the day's field work is completed. At the time of preparing his clean copy, he can arrange the order of his sample plots in any fashion he chooses and number them consecutively. It is desirable to record the exact location of each quadrat on the base map.

Once a mapper has recorded his floristic observations in the table, he then holds a document which serves as the basis of all manipulations for establishing plant communities.

The classification procedure for vegetation tables is treated in Chapter 6.

3. The transect method*

The transect method has certain advantages which recommend it highly. Line transects are distinguished from belt transects and both may be used profitably. The line transect is the simplest and the fastest method to analyse the floristic composition of a plant community and therefore is particularly useful where time is an important element in the mapper's budget. The mapper simply runs a long measuring tape (preferable a metal tape) along a straight line through the community, and on his base map or in his notebook he draws the line to scale. He then observes the species that touch or shade the line and the distances between them, and records these on his map or in his notebook on both sides of the line just as they occur in the field. If the vegetation is homogeneous, the line transect will produce results quite as reliable as those of the quadrat and in less time. Instead of a series of sample plots, the mapper may prefer a few line transects and thus progress rapidly. The disadvantage of a line transect is that it is one-dimensional, which makes it sometimes more difficult to record the structure of the

---

* Not to be confused with line intercept methods, paragraph 1.3.2., p. 98.

vegetation and to establish subdivisions of the community. The mapper will find little difficulty, however, in recording the structure of the vegetation in his notebook in such a manner that the notes can be correlated with points along the line. For instance, points at given intervals can be numbered and the structure described at each point. Or else the structure at a given initial point is described in the notebook and assumed to remain essentially unchanged for some distance. At the point where the structure changes, a new entry is made in the notebook.

The line transect is frequently replaced by the belt transect, which combines some of the advantages of the line transect with those of the quadrats. In a belt transect, the mapper also runs a measuring tape through the vegetation but records the communities on narrow strips of even width on both sides of his line. Belt transects are therefore like quadrats except that one dimension is much longer than the other. The strips are narrower than the quadrats or else a belt transect would be no more than a number of quadrats put end to end. It is usual to observe the vegetation the full length of the belt transect but where this is very long, it is entirely feasible to record the vegetation only at regular intervals, say every other 20 m, and still obtain satisfactory results. In tropical forests where progress is difficult, the belt transect offers great advantages over other methods (Cain and Castro, 1959). A straight line is cut through the forest to allow the mapper to proceed, and from this line he can investigate the narrow strips on both sides with relative ease because the strips in such a case need not be wider than 5 m. This gives the mapper a better chance to progress rapidly and with satisfactory results than any other method of intensive mapping. Belt transects should be arranged in straight and parallel lines but where there is a distinct zonation of the vegetation, as at the sea shore or from rivers across their flood plains to surrounding uplands, the belt lines should be more or less at right angles to the vegetation zones. The width of the actual strips must be uniform throughout, but the decision as to the actual width depends, of course, on the character of the vegetation. In the Kansas prairie, for instance, 10 cm on either side to the line are adequate. The belt transects are recorded in the notebook where a drawing supplements the notes. Needless to say, the mapper must be careful to find the exact location and direction of the transect on the base map in order to assure the proper relation of the vegetation to the topography and other features of the landscape.

4. The Step-point method

The step-point method is used only in herbaceous vegetation. It was employed successfully in the grasslands of California and the California Forest and Range Experiment Station (1955b) has issued detailed instructions on how to analyze vegetation with it. The following paragraphs are based on these instructions but are slightly modified by using the metric system.

The step-point method of sampling consists of recording one herbaceous plant at each of 20 points, called step-points, along each of five transects, and of recording the herbaceous coverage estimates for ten quadrats of $1/4 \, m^2$ $(50 \times 50 \, cm)$ on a square plot. The sampling design is portrayed in Fig. 5. In addition to his usual equipment, the mapper needs a sample pin, 3–5 mm in diameter, and about 25 cm long, a loop at one end, and a sharpened point at the other. He also needs a collapsible frame of $1/4 \, m^2$ which is divided into four equal sections. Specially prepared forms or 'grassland data sheets' facilitate a rapid recording of the observations. On these forms, annual grasses, perennial grasses, and forbs are distinguished and each of these three groups is divided into desirable and undesirable species.

At each step-point, the mapper lowers his sample pin to the ground. The first herbaceous plant touched by the pin is recorded on the grassland data sheet. If the pin does not touch any herbaceous plant, the herbaceous plant

nearest to the pin in a forward direction (180°
arc) is recorded.

Herbaceous coverage is estimated using the
frame of 1/4 m², subdivided by crossbars into
four equal sections. At each of the designated
points, the mapper locates a quadrat by aligning
one of the crossbars of the frame with the line
transect of the step-points. He then estimates
the percentage of herbaceous cover in the area
bounded by the frame. All herbaceous material,
green or dry, of the current growing season is
considered herbaceous cover. The coverage estimate is then recorded on the grassland data
sheet as well. The method permits rapid progress and produces a reasonable, if generalized,
description of the sampled grassland.

5. The Curtis transect

Curtis and his collaborators in Wisconsin experimented much with various methods of
sampling forests. They found the following
technique to be relatively fast, simple, and statistically reliable.

First, a line of progress is selected along
which to make a transect through the forest.
Near the beginning of the line, one tree is selected and its surroundings divided into four
equal quadrants. The species of the tree nearest
to the center tree is determined in each quadrant. Trees must be over 10 cm (4 inches) in
diameter at 1.5 m (5 feet) above the ground in
order to be considered. This observation reveals, therefore, the generic and specific names
of five trees. The same observation is repeated
along the transect, i.e. the line of progress determined earlier. Observations should be spaced
evenly about 40 m apart or approximately
25 observations per kilometer. Marginal areas
of the forest should be avoided if disturbances
may be confusing.

To speed up the work, it should be done by
teams of two. The leader selects the center tree
and the other four trees. His helper measures
the distances of these four trees to the center
tree to determine density. He also measures the

circumference of all five trees to obtain the basal areas. The leader records names, distances,
and circumferences.

Lower synusias are recorded with line transects (as discussed under 1.3.2. or 3 of this
chapter) between the center tree and the nearest
observed tree in each of two opposite quadrants, for instance, the northeast and the
southwest quadrants.

6. Floristic versus structural analysis

The reader is referred to the scholarly and comprehensive treatise, Manual of Vegetation Analysis, by Cain and Castro (1959), and Knapp
(1983) for further information on the theoretical
and practical aspects of sampling phytocenoses.

It has often been said that a floristic analysis
is more detailed than a structural one. While
this is often true, there are many vegetation
maps where the legend items are described floristically but without much meaning. If, for instance, on a vegetation map of the United
States, a legend item reads 'oak forest' then this
says only that members of the genus *Quercus*
are likely to be dominant, or at least common.
Considering the very large number of oak species, 'oak forest' does not even reveal whether
a particular forest is deciduous or evergreen.

Species that are particularly common in a
phytocenose and give it character, especially in
the uppermost closed synusia, are called dominants. A plant community may have a single
dominant or it may have several. It cannot
have very many because in such a case each
species ceases to be dominant. It has to share
the available space with so many other species
that it can no longer be said that it dominates
the community. Five or six dominants is usually the maximum number and should be exceeded only in exceptional cases. The classical
example of a phytocenose without dominants is
the tropical rain forest. The large number of
species excludes dominants. However, this is
not always true, and there are numerous cases

where dominants can be recognized in a tropical rain forest, as for instance, the mora forests of northern South America.

Dominant species are obviously not the only ones in the community they dominate. The other species are sometimes referred to as companion species. Many vegetation mappers, when analyzing a phytocenose, list only the dominants and some of the more common companion species. They thereby obtain a relatively good description of the community without spending much time on an effort to list all species present. This is often but not always adequate for characterizing the community.

No matter how detailed and precise the structural and the floristic analyses of phytocenoses may be, it is important that both should be presented. Richards, Tansley and Watt, (as long ago as 1939), observed that 'vegetation should be primarily characterized by its own features, not by habitat, indispensable as is the study of habitat for the understanding of its nature and distribution. It is the structure and composition of a plant community that we must first ascertain and record as the secure basis of all subsequent knowledge.' Drude (1907) anticipated this development on his vegetation maps of part of the Ore Moutains. Dansereau (1961) said: 'The application of the physiognomic system to very different types of vegetation, from the humid tropics to the arctic, has convinced me that structure characterizes vegetation quite as significantly as does the floristic composition. I have never thought that one could do without the other.' The value of combining structure and floristic composition in an analysis of plant communities is so great that when doing one, the extra effort of doing the other is insignificant compared with the advantages the combined results have to offer.

The mapper must be aware of the fact that a structural vegetation unit does not necessarily correspond to a floristic one. A structural type can include various floristic subtypes, and a floristic type can include various structural subtypes. This means that units of vegetation with different floristic compositions may have identical structures and vice versa. As a result, a structural type of vegetation distinguished on an aerial photograph does not necessarily indicate a floristic vegetation type (Table 2).

Table 2. Structural types and floristic types of vegetation from a sample in Libya (after Hashim and Hermelink, 1981).

Structural types	Floristic types							
	1	2	3	4	5	6	7	8
Forest	7	1	1					
Woodland	3	3	3					
Shrubland			2	10	8	6	1	
Dwarf shrubland		1		1	1	2	13	
Grassland								20

APPENDIX A

Examples of sampling scales

1. Abundance scales
2. Cover scales
3. Combined scales

1. Examples of Abundance scales

Hanson and Love 1930 [1]
Class
    1 Very scarce
    2 Scarce
    3 Infrequent
    4 Frequent
    5 Abundant

Tansley and Chipp [1]
Class
    1 Rare
    2 Occasional
    3 Frequent
    4 Abundant
    5 Very abundant

[1] See Brown 1954 p. 34.

Braun-Blanquet 1932
Class
    1 Very sparse
    2 Sparse

    3 Not numerous
    4 Numerous
    5 Very numerous

Weber (Brüne, 1935)
Class
    1 Single
    2 Almost single
    3 Very few
    4 Few

    5 Rather few
    6 Fairly numerous
    7 Numerous
    8 Very numerous
    9 Dominant
   10 Dominant to the exclusion of other plants

Hanson [1] (See Brown 1953 p. 34)

Scarce:	1– 4	p. m^2
Infrequent:	5–14	p. m^2
Frequent:	15–29	p. m^2
Abundant:	30–99	p. m^2
Very abundant:	100 and more	p. m^2

Acocks 1950 [1] (See Brown 1953), *symplified.*

Very rare:	1 pl. per ha	100 m distance
Rare:	2 pl. per ha	70 m
Occasional:	40 pl. per ha	17 m
Fairly frequent:	450 pl. per ha	5 m
Frequent:	1 pl. per m^2	1 m
Common:	7 pl. per m^2	30 cm
Abundant:	40 pl. per m^2	15 cm
Very abundant:	150 pl. per m^2	75 cm
Extremely abundant:	1300 pl. per m^2	2,5 cm

Acocks 1950 (53) [1]

		Plant distance inch	cm	plants per morgen (2.1 acre = 0.9 ha)	plants per m^2
vvab	Extremely abundant	1	25	12.960000	1300
vab	Very abundant	3	75	14.40000	150
ab+		4.5		640000	70
ab	Abundant	6	15	360000	40
ab−		9		160000	20
c+		1	30	90000	10
c	Common	1.25		57600	6
c−		2		22500	4
f+		3	90	10000	$2\frac{1}{2}$
f	Frequent	3	90	10000	$1\frac{1}{2}$
f−		6		2500	0,3
ff+		12		625	0,1
ff	Fairly frequent	15	450	400	0,05
ff−		20		225	0,03
o+		30		100	0,012

[1]) See Brown 1954.

o	Occasional	50		36	0,004
o−		75		16	0,0018
r+		125		6	0,0007
r	Rare	200	6000	2	0,0002
vr	Very rare	300		1	0,0001

Barkman, Doing, Segal, 1964

					comp. Acocks
1	10	individuals per ha		10^9 mm^2	v.r.occ.
2	10^2	individuals per ha	1 ind. per 100 m^2	10^8	occ.
3	10^3	individuals per ha	1 ind. per 10 m^2	10^7	f. frequent
4	10^4	individuals per ha	1 ind. per m^2	10^6	frequent
5	10^5	individuals per ha	1 ind. per 10 dm^2	10^5	common
6	10^6	individuals per ha	1 ind. per dm^2	10^4	abundant
					very abundant
7	10^7	individuals per ha	1 ind. per 10 cm^2	10^3	extr. abundant
8	10^8	individuals per ha	1 ind. per cm^2	10^2	extr. abundant
9	10^9	individuals per ha	1 ind. per 10 mm^2	10^1	extr. abundant
10	10^{10}	individuals per ha	1 ind. per mm^2	10^0	extr. abundant

2. Examples of cover scales

Kylin and Samuelson
1: 0 – 8.3 %   0 –1/12
2: 8.3 – 16.7 %   1/12–1/6
3: 16.7 – 33.3 %   1/6 –1/3
4: 33.3 – 66.7 %   1/3 –2/3
5: 66.7 – 100.0 %   2/3 –1

*Sernander**
1: 0 – 5 %   0 –1/20
2: 5 – 10 %   1/20–1/10
3: 10 – 20 %   1/10–1/5
4: 20 – 50 %   1/5 –1/2
5: 40 – 100 %   1/2 –1

Braun-Blanquet 1932†
1: 0– 5 %   0 –1/20
2: 5 – 25 %   1/20–1/4
3: 25 – 40 %   1/4 –1/2
4: 50 – 75 %   1/2 –3/4
5: 75 – 100 %   3/4 –1

*Hult-Sernander**
1: 0 – 6¼ %   0 – 1/16
2: 6¼ – 12½ %   1/16 – 1/8
3: 12½ – 25 %   1/8 – 1/4
4: 25 – 50 %   1/4 – 1/2
5: 50 – 100 %   1/2 – 1

Doing
 0 – 2 %
00 2 – 5 %
01 5 – 15 %
02 15 – 25 %
03 25 – 35 %
04 35 – 45 %

05 45 – 55 %
06 55 – 65 %
07 65 – 75 %
08 75 – 85 %
09 85 – 95 %
10 95 – 100 %

3. Examples of combined scales

Codom method (Tansley 1946)

Dom (dominant) = one plant covers more area than all others together
cod (codominant) = a small group of plants (codominants) together cover more than ½ of the total area.
ab (abundant) = less cover than ½ but relatively high cover or very numerous
fr (frequent) = relatively low cover but rather frequent
sp (spase) = low cover, not numerous
r (rare) = only one or a few individuals
loc (locally) = can be combined with one of the other symbols but not with r.

Braun-Blanquet 1932

\+ = sparsely or very sparsely present, cover very low
1 = plentiful but poor cover value
2 = very numerous, or covering at least 5%, of the area, but less then 25%
3 = any number of individuals, covering 25–50%
4 = any number of individuals; covering 50–75%
5 = covering more than 75% of the area.

1) *Note.*
(Cover is often also a function of the plant itself. Tiny slender species may occur in high numbers but cover only a very small area. A few individuals of bulky species may cover almost the whole plot. Absolute figures are often not the most suitable for comparing different species)

Braun-Blanquet – Trepp (1950)

\+ = sparsely to very sparsely present or cover up to 1%
1 = plentiful but with cover   1% – 10%
2 = very numerous or   10% – 25%
3 = any number   25% – 50%
4 = any number   50% – 75%
5 = any number   75% – 100%

Domin (1923) modif. by Dahl (See Cain and Castro 1959)

\+ = Occurring as a single individual with reduced vigor; no measurable cover
 1 = Occurring as one or two individuals with normal vigor; no measurable cover
 2 = Occurring as several individuals, no measurable cover
 3 = Occurring as numerous individuals but with cover less than 4%
 4 = Cover 4–10% of total area
 5 = Cover 1/10 to 1/4 (11–25%)
 6 = Cover 1/4 to 1/3 (26–33%)
 7 = Cover 1/3 to 1/2 (34–50%)
 8 = Cover 1/2 to 3/4 (51–75%)
 9 = Cover 3/4 to 9/10 (76–90%)
10 = Cover 9/10 – complete (91–100%)

Domin (See Poore 1962)

1 one to two individuals (sprouts)
2 few individuals (sprouts)
3 rather many individuals cover 1/20
4 cover 1/20 – 1/5
5 cover 1/5 – 1/4
6 cover 1/4 – 1/3
7 cover 1/3 – 1/2
8 cover 1/2 – 3/4
9 cover 3/4 – 9/10
10 cover 9/10 – complete

Recommended:

Doing, Zonneveld, Leys	Compare Br. – Bl.
r cover < 5% rare	r
p cover < 5% few (paucus)	+
a cover < 5% rather numerous (abundant)	1
m cover < 5% very numerous (many)	2
1 about 10% (number irrelevant)	2
2 about 20% (number irrelevant)	2
3 about 30% (number irrelevant)	3
4 about 40% (number irrelevant)	3
5 about 50% (number irrelevant)	4
6 about 60% (number irrelevant)	4
7 about 70% (number irrelevant)	4
8 about 80% (number irrelevant)	5
9 about 90% (number irrelevant)	5
10 about 100% (number irrelevant)	5

6. The Classification of Vegetation

A.W. KÜCHLER

Every map that shows differences of vegetation implies a classification, and one might argue that every vegetation map is an application of the classification which a particular author happens to choose. This leads to the conclusion that the relation between classifying and mapping vegetation is that of a sequence in which the first item (classification) is arrived at more or less arbitrarily, whereas the second item (vegetation map) expresses the first one cartographically.

The scientific method requires that observed phenomena be described, classified, and explained. In the case of vegetation as an object of scientific investigation, this requirement has caused considerable difficulty; and with regard to classification, no agreement has been reached. Vegetation is heterogeneous, the approaches to its classification are manifold, and the personal attitudes of the phytocenologists perhaps equally varied. There are, therefore, different avenues to an orderly arrangement of the vegetational mosaic, and each of these avenues may lead to success in certain regions or for certain purposes, but may be less satisfactory under different circumstances. It is of fundamental importance that the vegetation mapper is aware of these aspects of classification because he must be sure to select the correct classification for any given region and purpose. He cannot ignore the point that the quality, and hence the value, of a vegetation map rest more heavily on the selected system of classification than on any other feature.

Classification means abstraction. Vegetation has a multitude of properties and features. From these, a relatively small number of readily recognized ones can be selected to be used as diagnostic characteristics. With the help of such characteristics, groups of particular populations can be identified. Such groups represent plant communities as units of a classification of vegetation.

A vegetation map presupposes that a number of vegetation categories can be established. These categories can be arrived at in two ways, both of which have been used frequently. One of these is to view the vegetation of the earth as a whole and consider a variety of methods to divide this total into meaningful parts, which may then be repeatedly subdivided until one arrives at units to small that further division becomes undesirable. The result of this process is a classification of vegetation, a hierarchy in which all classes are abstractions of the actually occurring vegetation types; a number of such classifications has been developed. This means that the classifying work has already been completed before the mapper ventures into the field, and he need only assign the observed vegetation units to the proper place in the hierarchy of the classification. This may sometimes cause problems but, at least in theory, the principle is simple. But the marvelous variety of forms which plant life assumes on all continents has been a major stumbling block to the development of a simple and universally accepted classification, and the problem of

A.W. Küchler & I.S. Zonneveld (eds.), Vegetation mapping. ISBN 90-6193-191-6.

classifying vegetation remains an unending one.

Many scholars have therefore taken to the field and proceeded on the basis of direct observation: one sets down on a map what one sees and thus accumulates data which can be sorted and organized. Because mapping implies observation, this other way has been used in the case of many vegetation maps. This a posteriori mode of classifying differs fundamentally from the first, or a priori, method by its direct approach and, at least theoretically, its freedom from preconceived notions. The results can be very graphic, as in the case of the Patuxent map (Stewart and Brainerd, 1945). Wagner (1961a) is concerned with large-scale vegetation mapping and observes that an important condition for the determination of local details consists first of all in renouncing all types of classificatory units of vegetation established elsewhere and by different methods. The vegetation mapper must begin with conditions just as they actually exist in his area. The entire vegetation is divided into broad physiognomic divisions such as forests, grasslands, and cultivated fields. Within each of these divisions, all stand samples are brought together in a single table. In Wagner's particular instance, the stand samples are arranged ecologically, from dry to wet, i.e. according to increasing soil moisture.

Similarly, Zonneveld (1963) reports on the vegetation mapping activities in the Biesbosch (Netherlands) and says: 'We deliberately chose not to use any existing system of classifying vegetation. This was done in order to recognize without prejudice those differences in the vegetation which are most significant in this landscape, i.e. which are directly related to the specific combination of environmental factors of this area.' The success of Wagner, Zonneveld, and others and the practical usefulness of their results demonstrate that mappers will find it worth their while to experiment more often with a posteriori classifications.

The a posteriori classifications offer great advantages: they fit the observed vegetation perfectly, they are highly flexible and adaptable and they can portray the vegetation more precisely and correctly than any other classification. They are particularly useful when the map scale is large, implying a minimum of generalization. Every detail can be recorded and utilized in elaborating and perfecting the description of the individual phytocenoses. A posteriori classifications therefore approach the ideal. However, the very characteristic detail of such a classification may also limit its use. A posteriori classifications are often applicable only in the mapped area or, at best, only short distances beyond the borders of the map. Only where the vegetation is simple over large areas as in high latitudes, can they be used extensively. Usually, a posteriori classifications prepared for large scale vegetation maps must be modified when applied to larger areas.

Most authors, however, reveal the powerful influence that existing classifications can exert. The postmapping classifications show in most instances a distinct affinity to the a priori classifications, more or less modified as local circumstances and the author's originality dictate. The a posteriori method can be highly descriptive. It is more flexible and, hence, can be more readily adapted to local conditions. But maps based on a posteriori classifications are sometimes difficult to compare because what is important in one area may be irrelevant in another. This is especially so when the regions to be compared are widely separated or ecologically quite unlike.

The classification of vegetation, the author's observations, and the purpose of the map combine to control the character of the map. Observations and purpose are the less debatable factors: the former can be checked in the field as to accuracy, and the latter is predetermined. On the other hand, when an author develops his own classification or employs one already in existence, with or without modifications, the result is always open to question, no matter how carefully devised, how clearly defined, or how well proved. As a result, vegetation maps show a great variety of classifications, of which the older ones sooners or later give way to more modern ones or are revised to meet changing standards.

Classification systems

For the vegetation mapper, as for all phytocenologists, it is imperative that the vegetation be studied in the field, thoroughly and without bias; such a study should include the plant communities, their evolutions, and the conditions under which they exist. Plant communities have so many characteristic features that it is not possible to use them all in devising a classification of vegetation. By being selective, however, authors necessarily emphasize some features and neglect others, according to their individual purposes.

Whenever a classification of vegetation is to be organized, it is important to first establish the guiding principles which depend on the purpose of the map. Secondly, it is important to clearly define the criteria which serve to select the appropriate diagnostic characteristics of the various vegetation types to be mapped. For instance, when UNESCO decided to prepare a physiognomic-structural classification of vegetation, only the growth form controlled structural divisions of the vegetation could be considered. Floristic considerations are irrelevant in such a classification. As plant communities are described and defined (a) by their structure, and (b) by their floristic composition, two types of vegetation maps result. These show the phytocenoses strictly from a structural point of view, or else they show the plant communities as defined by their floristic composition. Küchler (1956b) prepared both these types of maps of the same area and at the same scale. Their comparison reveals some interesting details.

It is, of course, possible to combine structural and floristic systems as for instance on the maps of Kansas and of California (Küchler, 1974, 1977). It is also possible to combine vegetation types with environmental features. Such maps are discussed in detail in Chapters 33 and 34.

The various classifications of vegetation have emphasized one or the other of these features. The best-known systems are the following:
1. physiognomic-structural systems;
2. floristic systems;
3. physiognomic-floristic systems;
4. dynamic-floristic systems;
5. the areal-geographical system;
6. 'ecological' systems.

Each of these systems has advantages and disadvantages. A classification based strictly on any one concept can be clearer and simpler than a classification based on a combination of concepts. If, on the other hand, more than one basic concept is introduced, the scope of the map can be extended a great deal. The mapper's best approach to vegetation is to be adaptably diversified in order to suit individual cases, and to avoid rigidity. It is desirable to begin in physiognomic-structural terms, without which a world-wide comparison is impossible. Floristic features gain in emphasis as the study of vegetation becomes areally more restricted and intensified. But in any case, the guiding principle should always be to base a vegetation classification directly on the facts as observed in the field.

Many of the best-known phytocenologists now recognize the importance of growth forms and physiognomy. Braun-Blanquet (1964) devotes ample space to growth forms and growth form communities, although later in his book he attempts to play down the significance of physiognomy. However, Aichinger (1954) states clearly: 'In erster Linie müssen wir auch in der pflanzensoziologischen Forschung vom Erscheinungsbild (Physiognomie) ausgehen und dieses floristisch, pflanzengeographisch, ökologisch und syngenetisch, unter besonderer Berücksichtigung der menschlichen Einflüsse untermauern.' This demand that phytocenological research start with physiognomy agrees well with the statements quoted in Chapter 3; but there are phytocenologists who are as yet too unaware of the significance of a careful physiognomic and structural analysis of the vegetation.

The modern vegetation mapper should make every effort to avoid this pitfall.

As plant communities can be identified by their structure and their floristic composition, it was inevirable that systematic classifications of vegetation should be either physiognomic or floristic. Actually, the number of criteria that can serve as a basis for classification of vegetation is large, and Ellenberg (1956) presented them in tabular form (Table 1). Whittaker (1962, 1978) discussed classifications of vegetation in detail.

Table 1. Criteria for the classification of phytocenoses.*

A. Features of plant communities
  I. Physiognomic criteria
    a. Life forms
      1. The dominant life form(s)
      2. The combination of life forms
    b. Structure
      1. Layers of different life forms and their height
      2. Density within layers
    c. Seasonal periodicity
  II. Floristic criteria
    a. An individual taxon (rarely 2-3 taxa)
      1. The dominant taxon (or taxa)
      2. The most frequent taxon (or taxa)
    b. Groups of taxa
      1. Statistically established groups
        α. Constant taxa
        β. Differential taxa
        γ. Character taxa
      2. Groups that are independent of statistics
        α. Taxa of similar ecological constitution
        β. Taxa of similar geographical distribution
        γ. Taxa of similar dynamic significance
  III. Numerical relations (community coefficients)
    a. Between different taxa
    b. Between different stands
B. Features outside the present plant communities
  I. Final state of vegetation evolution (climax)
    a. As a physiognomic unit
    b. As a floristic unit
  II. The site
    a. Geographical location of the site
    b. Individual site factors
      1. Physical factors
      2. Chemical factors
      3. The water factor
      4. The human factor
    c. The site as a whole

* After Ellenberg (1956), slightly modified.

Physiognomic-structural classifications

Vegetation maps based on physiognomy and/or structure are, in fact, based on their appearance. Physiognomy (from 'face') is defined as the general appearance of the vegetation and is used as a relatively broad term. It includes such items as broad-leaved deciduous forest, broad-leaved evergreen dwarf-shrub formation etc. Structure, on the other hand, goes into detail and reveals the degree of stratification of the growth forms in the plant communities as well as the height and coverage of the individual strata. Lianas, epiphytes and special growth forms tend to blur a simple stratification; they nevertheless form an integral part of structural descriptions and analyses.

A uniform classification of vegetation for world-wide mapping use must necessarily be physiognomic in character and principle. This cannot be said of floristic classification because the flora remains incompletely known in too many areas of the world. Furthermore, the number of species is too vast, i.e. impractical for a world-wide floristic classification that can be usefully employed on maps.

A priori physiognomic classifications of vegetation are nothing new. They go back to Humboldt (1807). Grisebach (1872) and Schimper (1898) were among the early pioneers who attempted world-wide classifications. Schimper's monumental work was revised and modernized by von Faber (Schimper and von Faber, 1935) and again by Walter (1968–1973). Champion (1936) successfully used physiognomy in classifying the forests of India and Burma, while Beard (1944) developed a similar approach for the vegetation of northern South America. He states categorically: 'The physiognomic classification of vegetation meets all the essential requirements for the treatment of tropical formations. First, structure and life forms are capable of exact measurement and record in the field and secondly, on the basis of actual types so recorded, structure and life form of any desired formation can be mathematically defined.' Today it is, in fact, difficult to be a competent phytocenologist without a thorough acquantance with at least some of the more recent physiognomic systems, e.g. Trochain (1961), Fosberg (1961c), Eiten (1968), Box (1981), in addition to the systems by Beard and Champion.

Eiten (1968) attempted to improve Fosberg's classification by making his system as morphometric as possible, using only citeria that can be readily observed. He avoids any terms that may imply an ecological bias. As a result, he avoids the use of leaf size, leaf type, sclerophylly because precise measurements are imparactical. Characteristically, he uses growth forms, their stratification and the height and coverage of the strata.

The most important and useful classification of vegetation among the more recent ones was published by UNESCO (Anonymous, 1973). It was originally designed for vegetation maps at the scale of 1 : 1,000,000 or less, but it has been shown to be equally useful at larger scales. Thus Williams et al. (1974) successfully applied the UNESCO classification on a map at the scale of 1 : 12,000. This classification has the great advantage that it is highly flexible. It is organized in the form of a hierarchy and is world-wide in its application. If any type of vegetation is observed anywhere which does not seem to fit well into the scheme, then it can be added conveniently by expanding the appropriate section of the classification. The UNESCO classification of vegetation is concerned with physiognomic-structural features of the vegetation. Occasionally, there appears an ecological elaboration, e.g. montane, ombrophilous, seasonally flooded, etc. The authors explain that such terms are used where they help in the identification of a given type of vegetation. The UNESCO classification of vegetation does not systematically relate the vegetation to one or more features of the environment (Appendix I).

Floristic classifications

Basically, it is not difficult to design a physiognomic classification of vegetation. Even though the number and variety of growth forms may be large, it is small when compared with the number of taxa. Even in structurally and floristically very simple communities, the number of taxa far exceeds the number of growth forms. This means that necessarily, the approach to classifying vegetation floristically is quite different from any physiognomic classification.

When a vegetation mapper plans a large-scale map, he may find the Phytocenological Record (see Chapter 5) a convenient way to establish the floristic conditions of a plant community. One side of this record is devoted to establishing the structure of the community while the reverse side of the record is for recording the floristic composition. This can be done in all the required detail, i.e. every individual species in every sample plot is listed separately with its degree of coverage. By using a sufficiently large number of plots, a relatively complete list of the present taxa can be assured.

Most botanists study vegetation by analyzing it floristically. The floristic composition becomes the basis of procedure and the units shown on the maps are plant communities characterized by certain species combinations. The great advantage of such a floristic approach is that it permits a high degree of detail and accuracy. It is therefore usually preferred on large-scale maps.

With the solid record of the floristic composition of a phytocenose, the mapper can proceed in a variety of ways. The record is there to be used but that does not mean that all present taxa are necessarily included in the description and identification of a plant community. A majority of mappers describe their mapped units by their dominant species, e.g. beech (*Fagus silvatica*) forest. Where the forest is dominated by two or more species, these can be listed as well, and there are many forests which are so poor in tree species that listing the dominant ones exhausts almost the entire list of trees. This is especially true of some forests in high latitudes. The richer a phytocenose is in species, the less descriptive is a dominant or even two or three dominants. For instance, Küchler (1964) describes the Mixed Mesophytic Forest of the eastern United States with seven dominants. In spite of this large number of cominants, the forest is not well described by them as indicated by the fact that Küchler then lists an additional 32 species which are 'prominent'.

One of the major limitations of floristic methods rests on the fact that the number of important species in a given area may be so large as to make it impossible for the mapper to convey an adequate idea of the vegetation. This is especially true of many tropical and subtropical regions where even very large map scales cannot overcome this difficulty. Seibert (1954), who accurately mapped a very small segment of the relatively simple vegetation in central Germany – and at the large scale of 1 : 15,000 – nevertheless shows as many as 134 different categories of vegetation! One can easily visualize the problems one would encounter in humid equatorial regions. According to Cain and Castro (1959) in their discussion of the tropical rainforest, Warming found 2,600 species of vascular plants on 3 square miles, Whitford observed 120 tree species over 3 m tall on only 0.12 ha, and they themselves recorded 173 tree species of more than 10 cm diameter on 2 ha. Not all tropical rainforests are so diverse; nevertheless, in many areas with a complex vegetation, a purely floristic method of mapping is not likely to produce the best results (see however also Chapter 6B).

Various Dutch vegetation scientists (Lindeman, 1953; Cleef, 1981; Hommel, 1986; i.p., Zonneveld and his student, pers. communication) concluded that the Braun-Blanquet method (= using total species combination) can be successfully applied in tropical rainforests. This means that those species which sufficiently frequently occur can be grouped in meaningful sociological groups (differential species, preferential species, exclusive species). It appears, however, that many species (often about 50%) have such a dispersed distribution that they appear with such a low frequency in the table that they have no sufficient statistical value to be grouped in sociological groups and hence have no diagnostic value. This, however, is not

a unique character of rainforests, it can be observed also in temperate zones (like the dune areas of Europe) and the savanna regions of Africa and South America where the Braun-Blanquet method is successfully in application for more than half a century. See also Chapters 6B and 29.

Richards (personal communication) working in the rainforests of northern South America found that none of the tree species occurring in a given sample plot of 1 ha also coccurred in the adjacent sample plot of the same size. So it was concluded that sample plots therefore must have an area of at least several hectares. As it is necessary to have a considerable number of plots in order to arrive at the floristic composition of a plant community with a reasonable degree of accuracy, one may conclude that such a method as the one of Braun-Blanquet may become too time consuming hence impractical (see, however, Chapter 6B 29 and Hommel 1987).

Poore (1963) observed that 'the situation can arise in which two exactly similar sites, close together and having the potential of developing identical vegetation, would be found to have forests which differed greatly in species composition. Conversely, it would not be possible to conclude that because two areas were covered with forests which differed from one another, there was necessarily any difference in habitat or potential.' This means that not everybody is convinced that it is feasible to employ certain floristic techniques in some regions of the world because they have proved to be valuable elsewhere (see, however, Chapter 6B).

A major problem arising from the floristic approach to mapping is the frequent lack of familiarity with the flora of many regions. Even well-trained systematists rarely know the species of more than a few limited areas, and the majority of the users of vegetation maps have little taxonomic training, if any. The assistance of local plant and tree connoisseurs can, if treated with caution, be of great help in certain cases (Zonneveld, Pers. Comm.). In addition, in many regions, especially in the tropics, knowledge of the flora is still far from complete. In all such cases, a detailed analysis of tropical vegetation, especially forests, should be based on the proposals by Richards, Tansley, and Watt (1940), or as outlined in Chapter 28 and 29 of this book.

Physiognomic-floristic systems

It has been pointed out repeatedly that phytocenoses can be characterized by their structure as well as their floristic composition. It is therefore not astonishing that many authors of vegetation maps present their plant communities in such a fashion that both these features are clearly indicated. Any author who wishes to proceed on this basis is likely to portray the vegetation in a useful manner.

It is, however, important to realize that both the structure and the floristic composition can be shown on a map in greater detail if they are used exclusively. As soon as both are to be used, the map is more apt to become crowded and hence more difficult to read, or else, the author will become more selective and choose only those features he considers essential to describe both structure and floristic composition. Such selectivity varies, of course, from one author to the next and it is not possible to determine a choice which is automatically the 'best' for all cases. It is more likely that the inevitable variations of such maps depend on the purpose of the map. The fact that one author makes one selection and another author makes a different selection does not at all imply that therefore one map is better than the other one. An author of such vegetation maps should always make a choice that best serves the purpose of his map.

Dynamic-floristic systems

Vegetation is dynamic, and it is not astonishing that some vegetation mappers should have attempted to present this dynamism on their maps. The first successful attempt to do this was made by Lüdi (1921) when he portrayed the dynamism of the vegetation in the Lauterbrunnental in Switzerland, calling it a Sukzessionskarte, i.e. map of succession. The dynamism of vegetation was explored in detail by Clements (1928) who succeeded in formulating the process from pioneer communities through a sere of unstable communities that follow one another in a prescribed sequence, to the climax vegetation which is considered stable. Aichinger (1954), too, was much concerned with succession, but Lüdi (1921) is perhaps the only one who mapped the vegetation for the purpose of illustrating succession. But their ideas fell on fertile ground and reappeared on many vegetation maps. The manner of their appearance, however, varied greatly.

On some maps, vegetation types were shown that where obviously unstable but they were not designated seral types. Only rarely did vegetation mappers indicate the seral nature of phytocenoses. Gaussen's vidid imagination presented series. A 'série' includes all phytocenoses that lead through their succession to a given climax, including the latter. It is therefore a climax community and all other communities that sooner or later evolve into that climax.

A map of the natural vegetation, be it actual or potential, does not show the dynamism of vegetation because the natural vegetation is assumed to be a stable climax vegetation. Vegetation that is not natural is, of course, subject to succession and hence implies instability. This may be expressed in the legend of a map but often is not. Dynamic vegetation maps are nearly always (but not necessarily) floristic in character.

The areal-geographical system

This system was introduced by Schmid (1948, 1949) who based a number of vegetation maps on it. It is discussed in detail in Chapter 32.

Ecological systems

There is a number of important and frequently used so-called 'ecological' systems. They are discussed in detail in Chapters 33 and 34.

Choosing a classification system

The vegetation mapper must be familiar with a variety of approaches and classifications. Only then he is in a position to make an intelligent choice. He may usefully remember that all types of vegetation throughout the world have only two basic features in common: structure and floristic composition. The vegetation mapper may well employ these two features together as the foundation for his work. He will find it particularly rewarding to experiment with two or more classifications. The results will clarify his thinking and, in addition, assure the best choice when he selects a classification for his particular area and purpose. Thus Ozenda (1963) compared the classifications by Gaussen and Braun-Blanquet for the purpose of mapping the vegetation of the French Alps; he also considered the classification by Schmid. Once he was thoroughly familiar with each of these three classifications and could express the observed vegetation in the terminology and categories of each, he chose the one by Gaussen as best adapted to his purpose. His work is highly instructive. The best developed and in Europe widely used system of classifying vegetation floristically is the one designed by Braun-Blanquet who in a personal communication insisted that his system is useful only for large scale vegetation maps and types of vegetation which are not too rich in species. This would mean that tropical rain forests, the Brazilian Campo Cerrado and other such tropical types must be classified in some other fashion.

As has been mentioned before, there are however good examples of application of this system (at least the principle of using the full floristic composition) in tropical rainforests and related vegetations. Braun-Blanquet's system was used primarily in Europe and Japan with great success and its principles are now being applied in other parts of the world with growing success as in the arid and semi-arid zones of Africa, Asia and South America and also in the humid tropical forests of these continents. See among many others: Duvigneaud (1960), Hey-

ligers (1963), Donselaar (1965), Meulen (1979), Cleef (1981), Van Wijngaarden (1984), Zonneveld (1980).

The system is floristic and identifies units of vegetation by diagnostic species, often distinguishing character species and differential species for associations and their subdivisions respectively. Ellenberg (1956) has given a useful presentation of how to establish such units, see also Chapter 6B.

Classical studies of the school show a characteristic hierarchy, indicating the level by an appropriate ending (Table 2).

Table 2. Nomenclature of the Braun-Blanquet system.

Rank	Ending	Example
Class	– etea	Querco-Fagetea
Order	– etalia	Fagetalia
Alliance	– ion	Fagion
Association	– etum	Fagetum
Subassociation	– etosum	Fagetum festucetosum

Whittaker (1962) has the following to say about the Braun-Blanquet classification of vegetation: 'When phytosociological study of an area is sufficiently advanced, its community types may be presented in a systematic outline. Community types may then be concisely defined by their diagnostic species and their subdivisions and relations to habitat, succession and management indicated. No other system of classification has made possible comparable classification of the vegetation of a region, both in broad outline and fine detail. For the initiate, familiar with the system and the vegetation, such classifications embody a wealth of information on species relations, community types and habitat relations such as can hardly be conveyed with equal efficiency .by any other system. No other system can claim successful application over an equally wide range of vegetation conditions and research purposes. The standardization of concepts and methods is also a major advantage; this standardization makes the work of one author directly interpretable by and useful to another, and permits the work of many authors to be integrated into a master scheme of the school's classification. These standardized methods have also notable merit because they tend to enforce careful work: (1) knowledge of the whole flora of the community is required; (2) detailed records from actual stands must be compiled as the basis of community description and interpretation; (3) definition of community types must further be made specifiable and unambiguous by designation of diagnostic spcies. Some of the most impressive successes of the system are in the area of applied phytosociology. In English language ecology there is no real equivalent of the extensive and effective phytosociological work in vegetation mapping, site indication and land management.'

Braun-Blanquet began his work on the classification in a purely theoretical way. His sole aim was to classify. It was under the vigorous and constructive leadership of Tüxen that the system assumed the practical character that distinguishes it today. But the system also has serious flaws as Whittaker (1962) reveals in his penetrating analysis of this famous work.

The mapper has therefore a variety of possibilities to characterize and classify the floristic composition of the plant communities he wishes to portray on his map. For large scale mapping, he should, if at all possible, record the entire flora. This is useful because it offers him the possibility to later exploit such species lists for a variety of often unforseeable purposes. In the end, however, it is left to the judgment of each vegetation mapper just how he whishes to proceed and what criteria he wants to use in defining the phytocenoses on his map.

If the mapper works in a thoroughly studied region like Western Europe, he may well select a strictly hierarchical classification like that by Braun-Blanquet (1964). Elsewhere, a non-hierarchical approach is more advantageous. Emberger of the Zürich-Montpellier (Braun-Blanquet) school of thought stresses that there is often no need to show plant communities as members of a hierarchy, as only some specialists are

interested in this. Usually, it is preferable to establish and map all plant communities as equals, without regard to rank in some hierarchical scheme, and the result will be more practicable for most purposes. Ellenberg (1956) and Wagner (1961a) have had the same experience. See also Zonneveld et al. (1979) and the I.T.C. approach (Chapter 29).

The relation between classifying and mapping vegetation is very intimate, and each strongly affects the other. Mapping is a method of portraying nature, and the classification must permit the mapper to approximate the true conditions as closely as possible. Actually, a vegetation map is the meeting ground of two poles: an author's systematic classification and nature's kaleidscopic arrangement of plants. The degree to which these poles meet depends on the imagination, insight, and skill of the mapper.

One of the most important aspects of considering classifications when selecting the most appropriate one for a particular vegetation map, is of course the purpose for which the map is to be prepared. Often it is best to map the vegetation on a strictly scientific basis. Such a map will lend itself to many purposes. However, such a map requires a great deal of detailed work. It may therefore be faster and less expensive to prepare a vegetation map for a given purpose, even though the map may not necessarily be useful for other purposes, too. In the following paragraphs, a number of examples are presented to illustrate how classifications and purposes are related. This should enable the mapper to make an appropriate choice when he is faced with such a situation.

Scharfetter pointed out as long ago as in 1932 that 'nothing contributes more to phytocenological cartography than the presentation of vegetation of one and the same area according to different points of view'. This implies that the mapper will find it useful to compare different types of vegetation maps before he decides on a method for his own map. He studies and compares a variety of vegetation maps and critically investigates their characteristics in the light of his own needs.

The comparative method is old and has been used many times. But attempts to compare vegetation maps are frustrated all too often. In fact, one of the lessons learned in making such comparisons is that many vegetation maps simply are not comparable. Differences in scale and resulting differences in generalization are disturbing. Frequently it is difficult to compare vegetation maps because authors disagree in their interpretation of vegetation, and even contradict one another. The differences in approach and interpretation usually find their basic expression in the classification, and the classification is therefore the critical feature of vegetation maps.

Küchler (1956b) sought to solve this problem by preparing a set of vegetation maps of southeastern Mount Desert Island, Maine, i.e. of the same area and at the same scale but each one based on a different classification of vegetation. On three different maps, he presented the structure of the vegetation, the floristic composition, and a combination of these. The first of these is based on the author's own method of recording the structure of vegetation. The second map is based on the floristic approach used by Hueck, one of Germany's most distinguished vegetation mappers, and the third map is based on the work of Wieslander. Wieslander's numerous large scale vegetation maps are affected by the need to serve forestry interests. Also, his method is devised for California. However, Wieslander has demonstrated a particularly keen insight into the problems of vegetation mapping and used a scientific approach to their solution. It is not difficult to adapt his method to other parts of the United States. The legends of the three maps (Appendix 2) will assist the reader in his appreciation of the problems involved and the selections made on the basis of comparison.

In the following paragraphs, the three vegetation maps of Mount Desert Island are used to determine an appropriate classification of vegetation by comparing the maps in the light of given purposes. The purpose of the map is always established first, followed by the criteria that serve such a purpose. Each of the three maps is then analysed with these criteria in mind which leads to a comparison and the most appropriate choice.

Example 1

Purpose. The Internal Revenue Service wishes to assess the value of extensive landholdings of lumber companies.

Criteria. The economic value of forest vegetation depends on several factors, notably the extent and condition of the forests and the floristic composition. Usually, only a relatively small number of species is, therefore, a prime consideration.

Analysis; physiognomic map. For a tax assessor a physiognomic map is useful because it tells him a great deal about the condition of the vegetation. He can see at a glance which land is forested and which is not, and he can easily distinguish different forest types. However, as there is no indiciation of the species, he cannot tell which forests are of value.

Floristic map. A floristic map reveals the distribution of species, which to the tax assessor means species that are economically valuable. But it tells nothing about the character of the vegetation. A plant may be a valuable species, but that does not mean it can be sold. The value of a spruce is potential rather than actual if it is only 1 m tall.

Physiognomic-floristic combination map. This map combines flora and physiognomy. But it omits many details of the physiognomic map, and the physiognomic and floristic units are grouped in a manner that must at once arouse the interest of the tax assessor. The map not only reveals which areas are wooded and which are not; it singles out the forests that have commercial value and gives the names of the species that compose these forests.

Comparison. Comparing the physiognomic, floristic, and combination maps, the revenue officer finds that all have merit but that the combination map serves his purpose best because it tells him clearly how much of the area is forested, which are the commercially most important trees, how they are distributed, and where they are concentrated. Thus he is able to evaluate the landholdings accurately.

Example 2

Purpose. A soil scientist has been assigned the task of mapping the soils of an extensive forested area at a large scale and as accurately as possible.

Criteria. Formerly some pedologists, even though not yet familiar with physiographic methods, nor remote sensing, made their maps by taking soil samples systematically every few feet in a grid pattern throughout their area. Presently the most common practice in soil survey is to use landscape features (including vegetation) for stratification for sampling (seel also Chapter 29, ITC approach).

The accuracy of their results will be in direct proportion to the density of the grid. But this is a tedious, slow, and expensive procedure. The pedologist knows that a close relation exists between soil types and vegetation types and that the vegetation changes with the soil. Aerial photographs may reveal mainly vegetation data and a detailed vegetation map thus becomes the most valuable tool.

Analysis; physiognomic map. The physiognomic map tells the pedologist where to find forests, and where the grasslands are, and such information is valuable; if some grass grows in wooded area, this knowledge, too, may be useful. But oaks, maples, beeches, and many others are all lumped together under the designation 'broadleaf deciduous trees', and the pedologist cannot find the kind of detail that his purpose requires.

Floristic map. The distribution of species in their various combinations is an excellent indicator of the distribution of soil types. A vegetation map that shows the extent of such plant communities can therefore tell the pedologist the extent of his soil types and the location of their boundaries. Usually one soil sample for each plant community or combination of species is then adequate, except perhaps where one community merges gradually into the next. The more accurately plant communities are portrayed, the more useful is the floristic map to the pedologist.

Physiognomic-floristic combination map. The physiognomic map with floristic overprints presents several advantages to the soil scientist. Although the physiognomic features are of relatively little value to him, the indication of the dominant species is useful. But the organization of the map, the use of colors for physiognomy instead of for the flora, and the difficulty in seeing at at glance the distribution of the floristic communities reduce the significance of this map for his purpose.

Comparison. Comparison reveals that the floristic map is the most useful of the three for the pedologist because it indicates the features of the vegetation that respond most directly to variations in the soil. He can correlate the floristic combinations more readily with soil types than any other aspect of the vegetation, and therefore he requires a map that depicts the distribution of plant species and their communities in the greatest possible detail.

Another solution is to procede here according to the system of a land unit survey. The procedure is almost identical. Floristic and soil sampling are done simultaneously and the result is a map with a vegetation and soil legend combined (see Chapter 29 and 34).

Example 3

Purpose. A geographical report with emphasis on the physical features of the landscape is to be prepared on a developing humid tropical area, and this report is to include a vegetation map. It is important to convey as graphic a picture of the vegetation as possible because the report is to serve as a basis for further investigations.

Criteria. The area is underdeveloped and not well known. The geographer requires descriptive categories of vegetation that he can identify in the landscape. He analyzes various vegetation-map types in order to determine what kind of map he can reasonably hope to prepare in his area.

Analysis; physiognomic map. The physiognomic map shows in reat detail many different vegetation types and their combinations. All these types are readily observable in the landscape and can be described easily and with a high degree of accuracy.

Floristic map. The floristic map presents major problems. It is not possible to describe the vegetation floristically unless the names of all or nearly all plant species are known. The geographer is not a botanist, and even most botanists would be blaffled by the array of species in an area with which they are not familiar.

Physiognomic-floristic combination map. Once again the geographer has a physiognomic map, at least as far as the colors are concerned. The floristic overprints are not so useful to him, and he may want to ignore them, but to do this would eliminte a significant part of the map and reduce its value accordingly.

Comparison. Comparing the three map types, the geographer soon concludes that the floristic map does not suit his purpose. The physiognomy of the vegetation is shown on both the physiognomic map and on the combination map. However, if on the combination map the floristic features are ignored, the physiognomic characteristics do not in any way match the distinctive and varied features of the physiognomic map. This latter, then, serves the purpose of the geographer best because it illustrates a method with which he can portray the vegetation of his area rapidly, accurately, and in detail.

Example 4

Purpose. An animal ecologist whishes to study to habitats of deer in order better to understand their life history; his interest is focused on their food and shelter habits.

Criteria. As deer are primarily browsing animals, their food and shelter come from the vegetation, and more especially the woody vegetation. The ecologist's attention is therefore directed primarily to the distribution of certain plant species that serve the deer as food and to the structure and density of the vegetation, because deer seek refuge from cold winds and carnivores in thickets and dense growth.

Analysis; physiognomic map. The physiognomic detail of this map reveals the height and density of the tree and shrub layers of the vegetation, and therefore the map is of considerable significance for the ecologist. Unfortuntately, the lack of floristic data reduces its value, since the distribution of food plants is not shown.

Floristic map. The distribution of plant species that serve the deer as food can be determined from this map, but the physiognomy and structure of the vegetation are ignored. Again, as in the case of the physiognomic map, the information is good in quality but inadequate in quantity.

Physiognomic-floristic combination map. This combination of physiognomy and flora presents all the features of interest to the ecologist, that is, the density and distribution of shrubs, thickets, and trees, and the dominant species of each vegetation type; moreover, there is an emphasis on woody plants.

Comparison. Comparison of the three maps reveals at once to the ecologist that the combination map best serves his particular purpose. True, the physiognomic map shows the structural details of the vegetation better than the combination map and the floristic map shows the distribution of the plant species and plant communities more clearly, but neither the physiognomic map nor the floristic map give both types of the needed information, and the combination of the two types on the third map more than compensates for the lesser detail.

Example 5

Purpose. A commercial beekeeper plans to expand his business, and a certain area has been suggested to him as being rich in plants that produce nectar. Before embarking on his expansion, he plans to investigate the prospects.

Criteria. As a first-class apiarist, he knows the needs of his bees, what plants they frequent, what plants are useless to them, and so on. Thus he is able to determine whether or not the vegetation of the new area holds out enough promise for an adequate profit.

Analysis; physiognomic map. This map of the physiognomy and structure of vegetation rveals, among other things, the occurrence of meadows and forests, whether there are layers of shrubs and dwarf shrubs in the forests. But bees are not wild animals that seek protection in thickets, and the apiarist finds little of interest on this map.

Floristic map. This map tells the beekeeper what species grow in the area, and he himself knows whether or not they are useful for his bees. It makes no difference to him whether a plant species occurs as a shrub or a dwarf shrub as long as it is nectar producing and is present in sufficient quantities.

Physiognomic-floristic combination map. Since the colors of this map refer primarily to the physiognomy of the vegetation, they are of only moderate value to the apiarist. The dominant species are shown by letter symbols, but it is difficult and tedious to discover the distribution of particular species because for each species it is necessary to examine the symbols of every vegetation type on the map.

Comparison. Comparison of the three maps reveals to the beekeeper that the floristic map serves his purpose best. It shows him what species grow where, and he can tell quickly and accurately which parts of the area (if any) are most promising for his expanded apiary, and how much they promise.

Example 6

Purpose. An army officer has to move men and their provisions through an area or occupy it, and this must be done in secrecy, that is, an enemy must not be able to observe the movement or occupation from the surrounding territory or from the air.

Criteria. The officer has orders to proceed in secrecy. His critical question, therefore, is what sort of vegetation map can show him best where and to what extent his troops can be concealed. He must discover whether the vegetation is dense enough to hide the troops and whether there is a continuous tree canopy under which the soldiers can find cover. He will want to know whether or not the vegetation is evergreen, since this may make a vital difference. The other important feature is the structure of the vegetation. A forest with many large trees per unit area and a thick shrub layer can seriously impede the movement of troops and their gear.

Analysis; physiognomic map. This shows the vegetation in the form of trees, shrubs, dwarf shrubs, grass, and so on. It shows combinations of these and such details as continuous layers of shrubs and patches of shrubs. It indicates which parts of the landscape to avoid because of open grasslands, dwarf shrubs, and similar low growth; troops would be detected here with ease. It shows where continuous tree cover exists, the proportion of evergreen to deciduous trees, of special importance during the winter and early spring, and where the closed forest gives way to scattered trees or patches of trees. An area with shrubs may offer excellent lateral concealment, but progress through it may not be easy, and cover from above may be wholly inadequate. Where evergreen and deciduous trees are mixed, the cover may be better than the proportion of evergreens would indicate, because the ground may appear sufficiently mottled when viewed from very high altitudes.

Floristic map. If the officer happens to be acquanted with the species of the area, and knows their growth forms, the floristic map may be helpful. But he cannot be relied upon to know them, and anyway the distinction between sugar maple and beech is not significant, since troops can hide under one as well as under the other. The map says nothing about the height, density, and other structural features of the vegetation, which, for cover, camouflage, concealment, and movement are of greater importance than the names of species.

Physiognomic-floristic combination map. This map distinguished between grasslands and forests, and so on, and therefore offers major advantages. But neither height nor density is revealed. Also, the distinction between woodland and commercial conifers is not useful; the officer is not selling timber, and the term 'woodland' includes both broadleaf deciduous and needleleaf evergeen trees. To know what the woodland really is like, he must know the specific names of the dominants, and this can hardly be expected of a military man.

Comparison. By comparing the three maps, the officer clearly sees that the physiognomic map serves his purpose best because it shows directly and in greater detail those features of the vegetation which interest him most.

6A. Establishing a Floristic Classification

I.S. ZONNEVELD

Introduction

Classification is in essence abstraction; that means to express the totality of the concrete reality by way of a limited number of certain diagnostic characteristics.

A floristic classification is defined by the fact that such abstracted diagnostic characteristics are (groups of) plant species (-taxa).

Any classification aims at ordering of the object. For a good ordering it should be a systematically one. To be systematic, guidelines are needed for the arrangements of units and for an hierarchy.

Guidelines for vegetation classification theoretically can be of various kinds. They can be purely structural (forest versus grasslands, etc., open vegetation versus closed vegetation, etc.).

They can also be purely statistical: let the computer according to a pure statistical algorithme on floristic or structural affinity make an ordination or clustering.

In most cases such a guiding principle is of 'ecological' nature. This means that the way the clusters of observed data (vegetation communities or types) are arranged and placed in a hierarchical order, is derived from envrionmental differences such as climatic differences, soil fertility or humidity, animal influence, etc. The diagnostic characteristics, however, are purely floristic, if the system is to be called a floristic system. If the environmental differences are directly used as diagnostic characteristics the classification is not anymore a pure vegetation classification. Compare the discussion on so-called 'ecological' maps.

A legend of a map is a kind of classification in which a chorological guideline is incorporated. Scale plays an important role in that case. Contrary a general floristic classification can be used on any scale. Only the relevé area should reach the minimum area, which practically ranges from a few m^2 (homogene grass swards) up to a few thousand m^2 (savannas). The legend can always be translated into a general classification by, if necessary, describing complex legend units in terms of a mosaic of floristic classification units.

Floristic classification can therefore be used at any map scale and in any vegetation type of the world. The ITC method of vegetation surveys combines the floristic methods of classification with areal photo-interpretation as a base, for practical vegetation survey in any existing vegetation type (see Chapter 29).

Characteristic for floristic classification is the statistical treatment of the data (relevés containing species list + environmental data at one spot of about the size of the minimum area).

It appears then that concrete boundaries that often can be observed in the field caused by geomorphological features or human interference, disappear in the statistical image. This is caused by the continuous character of the vegetation if seen in abstraction from the chorologic dimension. In Fig. 1 this is demonstrated. On the left hand is the concrete situation. A clear

A.W. Küchler & I.S. Zonneveld (eds.), Vegetation mapping. ISBN 90-6193-191-6.

SCHEME OF A CONTINUUM WITH TWO GRADIENTS

I CONCRETE SITUATION IN THE FIELD   II ABSTRACT MODEL

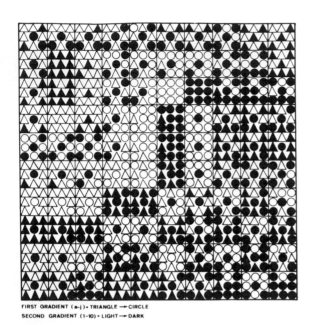

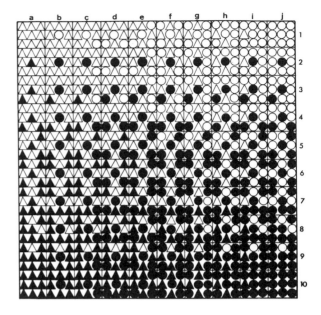

FIRST GRADIENT (a-j) · TRIANGLE ──► CIRCLE
SECOND GRADIENT (1-10) · LIGHT ──► DARK

vegetation mosaic can be seen. The vegetation is simplified by

a) making all stands of the same (square) size
b) by bringing only variations in color (black and white) and shape (circles and triangles).

The right hand figure shows a computer model (kind of ordination) of the same stands according to an affinity algorithme putting white by white, black by black, triangle by triangle and circle by circle. The result is a continum without boundaries. The researcher him (her)self can determine where the boundaries will be 'cut' (see Zonneveld, 1968 (1974)).

In reality the set of relevés (described stands) is seldom complete enough to achieve such a continuum. Often not all stands and certainly not all transitions are sampled and so 'natural' gaps (boundaries) appear in the classification continuum (right hand side of the figure). One can even adapt the sampling method to it, by avoiding transitional areas, guided e.g. by photo-interpretation. Nevertheless one should be aware of the basic continuum character of the vegetation, causing sometimes difficulties if

simple cluster analysis is used. Classification is, however, also possible in a continuum using the proper guidelines as is clear from our example.

Note: unfortunately statistical jargon uses the term 'classification' often as a synonym of 'clustering' which is not right. Any ordening (also 'ordination') can lead to classification.

Ordening in general

Before a classification into units can be made, the basic material (the relevés – records of concrete stands in the field) should be arranged appropriately. A list of present species is given per relevé as a minimum, preferably, however, with an indication per species about the occurrence of that species in terms of (estimated) cover, abundance or frequency within the relevé's sample area. Some identification of vitality or phaenology is also appreciated but too many kinds of data may hamper automatic computeration (see below). Quantitative data on biomass etc. are not useful for a general classification. Some environmental data should preferably always accompany the species list (see Chapter 17 (sampling)).

We have discovered before that in most cases an ideal arrangement will show a model with gradual changing properties in which only arbitrary boundaries can be made.

The larger the area under consideration and the more variation there is in habitats the less discontinuities that easily invite to 'cut' boundaries there are.

Usually the ecologist already has a certain idea of the units because of the physiographic situation. He is inclined to select his boundaries according to the concrete situation (discontinuities) in the field and to arrange (order) his concrete stands accordingly. (Compare left part of Fig. 1.)

From the explanation of the continuum concept above it is clear that this may give rise to difficulties because if he works in the upper part of the left of Fig. 1 his delimitation will differ from his arrangement if he only knows the lower part. It is because of this that classification gradually improves, the more data become available. It also causes the fight on boundaries in classification units between two surveyors working in different areas.

Ordering along known ecological gradients

A pure vegetation classification starts from a statistical ordering, based on pure morphometric properties of vegetation. If there is, however, a very clear and good measurable dominating environment factor, it may have sense to use, *within the process of trial and error,* a gradient of such a factor as a guiding principle in arranging the basic records directly. The result can be studied and may or may not show that this arrangement corresponds with morphometric differences in the vegetation. In the first case this knowledge can be used for the final classification and may save lots of men and computers time! (See e.g. Zonneveld, 1960; Van Gils/Zonneveld, 1982.)

(Statistical) ordering according to morphometric similarity

Sometimes the environment is so complex that it is difficult to known beforehand, which environmental factors (or interference between factors) will correlate with vegetation, or it is very difficult to measure or estimate, these environment factors.

In that case the comparison should start with nothing else than the morphometric* relation between the concrete records. Only afterwards knowledge of the relation with environmental factors can be used as guiding principle in 'cutting' the abstract continuum in two pieces or in hierarchical grouping of eventual aggregates. If a floristic description exists it is relatively easy to compare units more or less quantitatively (numerical). In this case we compare every record with every other record. So for every 100 records we have to make 10 000 comparisons. This can be done by way of estimation visually or by calculating indices of similarity and/or difference.

We need to take into account in both cases:
1. how many species do the two records have in common;
2. how many species do they each have that do not occur in the other record;
3. how many species are lacking in both records but are available within the total number of species in the whole material (see 4);
4. how many species are available in total of the records.

If not floristic but other morphometric properties are used that can be expressed in frequencies or other figures, in principle the same method can be followed.

---

* Due to species composition and structure.

THE TABLE METHOD FOR VEGETATION CLASSIFICATION

Relevées (floristic composition a – m + two environmental factors α and β):

1. α1, β7, a+, b+, c1, d3, e1
2. α1, β1, b+, f1, g2, h3, i4
3. α1, β3, a+, b1, c1, f2, g3, h2, i1, k1, l3
4. α1, β6, b+, c1, f4, l+, m6
5. α1, β7, c4, e3, m1
6. α1, β7, a2, c1, d+, e4, m2
7. α1, β8, a1, b+, c3, d2, e+, m2
8. α2, β2, c+, g+, h6, i2
9. α2, β4, a7, b+, c2, f3, g1, h4, i3, k4, l1
10. α2, β5, a1, c2, i2, l+, m1
11. α2, β9, a3, b2, d2, e1, i2
12. α3, β0, a1, b1, f+, g2, h3, i+
13. α3, β3, a6, d+, f2, g5, i2, k3, l2

14. α3, β4, a3, b1, c3, f+ h1, i+, l+, m3
15. α3, β6, a1, b4, c1, i1, m6
16. c7, d2, i1, m2, α4, β6
17. α5, β1, a1, b1, f+, h2, i+, k+
18. α6, β4, a2, b2, c5, m2
19. α7, β8, a3, b3, d+, e1, i1
20. α7, β3, a4, b3, e1, g2, k1
21. α7, β5, a8, b3, c3, d+, m4
22. α8, β2, a2, b4, f2, g1, h4, i+
23. α8, β10, a4, c+, d+, 32, i6
24. α9, β9, a2, b1, d1, e3, i1
25. α10, β9, a2, d1, e+, i2

a, b, c, etc. = plant taxa – α β = environment factors

I

Rough table (arranged according to environmental factor α)

Relevé nr	1	2	3	4	5	6	7	8	9	10	11	12	13	14	15	16	17	18	19	20	21	22	23	24	25
Envirm. factors α	0	1	1	1	1	1	1	1	2	2	2	2	3	3	3	3	4	5	6	7	7	7	8	8	9
β	7	1	3	6	7	7	8	2	4	5	9	0	3	4	6	6	1	4	8	3	5	2	10	9	9
a	+	.	+	.	.	2	1	3	7	1	3	1	6	3	1	.	1	2	3	4	1	2	4	2	2
b	+	+	1	+	.	.	+	.	+	.	2	1	.	1	4	.	1	2	3	2	3	+	.	1	.
c	1	.	1	1	4	1	3	+	2	2	.	.	.	3	1	7	.	5	.	.	3	.	+	.	.
d	3	.	.	.	.	+	2	.	.	.	3	.	+	.	.	2	.	.	+	.	+	.	+	1	1
e	1	.	.	.	3	4	+	.	.	.	1	.	.	.	.	.	.	.	1	1	.	.	2	3	+
f	.	1	2	+	.	.	.	.	3	.	.	+	2	+	.	.	+	.	.	.	.	2	.	.	+
g	.	2	3	.	.	.	.	+	1	.	.	2	5	.	.	.	.	.	.	.	2	.	1	.	.
h	.	3	2	.	.	.	.	6	4	.	.	3	.	1	.	.	2	.	.	.	4	.	.	.	
i	.	+	1	+	.	.	.	2	3	2	2	+	2	+	1	1	+	.	1	.	.	+	6	1	2
k	.	.	1	.	.	.	.	4	.	.	.	3	.	.	.	+	.	.	1	.	.	.	.	.	
l	.	.	3	.	.	.	.	.	1	+	.	2	4	.	.	.	.	.	.	.	.	.	.	.	
m	.	.	.	6	1	+	2	.	.	1	.	.	.	3	6	2	.	2	.	.	.	4	.	.	.

25 relevées

12 species

We will see that the classification process reduces these numbers to 5 communities and 5 sociological groups

II

Final table

Nr relevé	2	12	17	8	22	3	13	9	20	14	18	10	21	4	15	16	1	5	6	7	19	25	11	23
Envirm. factors α	1	3	5	2	8	1	3	2	7	3	6	2	7	1	3	4	0	1	1	1	7	10	2	8
β	1	0	1	2	2	3	3	4	3	4	4	5	5	6	6	6	7	7	7	8	8	9	9	10
1 f	1	+	+	.	2	2	2	3	.	+	.	.	.	.	+	.	.	.	.	.	.	+	.	.
1 g	2	1	.	+	1	3	5	1	2	.	.	.	.	.	.	.	.	.	.	.	.	.	.	.
1 h	3	.	2	6	4	2	.	4	.	1	.	.	.	.	.	.	.	.	.	.	.	.	.	.
2 k	.	.	+	.	.	1	3	4	1	.	.	.	.	.	.	.	.	.	.	.	.	.	.	.
2 l	.	.	.	.	.	3	2	1	.	4	.	+	.	.	.	.	.	.	.	.	.	.	.	.
3 a	.	2	1	3	2	+	6	7	4	3	2	1	1	.	1	.	.	+	.	2	.	3	2	3
3 i	+	1	+	2	+	.	1	2	3	.	+	.	2	.	.	1	1	.	.	.	.	1	2	2
3 b	+	.	+	.	+	1	.	+	2	1	2	.	3	.	+	.	.	+	.	.	.	+	3	.
4 c	.	.	.	+	.	1	.	2	.	3	5	2	3	6	1	7	1	4	1	3	.	.	.	+
4 m	.	.	.	.	.	.	3	2	1	4	7	6	2	.	1	+	2	.	.	.	.	.	.	.
5 d	.	+	.	.	.	.	+	.	.	.	.	.	+	.	.	2	3	.	+	2	+	1	3	+
5 e	.	.	.	.	.	.	2	.	.	.	.	.	.	.	1	3	4	+	1	+	1	2	3	

Final table

Sociological groups are formed (1 – 5)

These appear to be also ecological groups for factor β.

Group 3 is slightly negative indicative for factor α

III

Sociological Groups 1–5, Communities A B C D E

━━ = must occur
╍╍ = may occur

Final Classification

Classification Key (bar-table)

Comm. A = 1 + 3
B = 1 + 2 + 3 (+ 4)
C = 3 + 4
D = 4 + 5 (+ 3)
E = 3 + 5

IV

Table 1.

The hand-table method and the computer methods

The table method introduced by Braun Blanquet and clearly described by Ellenberg (1956; see also Shimwell, 1971, Mueller Dombois and Ellenberg, 1974), makes this comparison by way of visual estimation. In Table 1 the main principle is illustrated.

Mathematically treatments, using similarity indexes and/or correlation coefficients, are published by De Vries (1950), Greig Smith (1964), Curtis (1959), Bannink and Leys, Zonneveld (1973, 1975), Van der Maarel (1966); a recent compilation is given in Whittaker (1975). There is still progress in development of user-friendly computer programmes (e.g. Van Tongeren, 1986; see also Feoli et al., 1984., Van der Maarel (ed.), 1981, Gauch, 1985).

Calculation of correlation coefficients, but also similarity indices, is very time-consuming work that can only be done with the help of automatic computing machines. But even these appear to have their limits and are moreover requiring time-consuming preparations that sometimes take as much time as hand table making. In order to speed up the work and to make the material compatible for the computer often a wealth of data has been deleted.

Often only plain presence and absence were used; cover, vitality, phenology, etc. being neglected because the computer cannot swallow this if large numbers of relevés are involved. However, the newer computer generation and modern programmes tend to overcome these problems.

The time-consuming aspect of the table method is the re-writing by hand in the trial and error process (see below); we have, however, developed an instrument that facilitates this (Tabellation-table). Electronic systems to be applied on a normal computer exist in the meantime. The ideal integration of the visual table method and computer algorithmes will be achieved as soon as super-large monitors, able to show 500×500 matrixes, will become available.

A table resulting from the table method in which relevés (records) with their properties (e.g. species, or other morphometric description) are arranged can be multi-dimensional.

Although sometimes field data can be fed into the computer directly from the field description forms, usually it is better to make first (by hand) a tabel (matrix). This is necessary to 'clean' the material from impurities such as uncertainly determined species and/or infrequent species etc. A computer prints preliminary and only semi-final tables according to the used programme. The final touch, however, has to be done by visual iterative hand and brain work. The FLEX-CLUS method of Van Tongeren (1986) is a good example of a user-friendly iterative system.

Final arrangement and interpretation of tables

In the tables we can study the aggregation (discontinuity) or 'clining' of properties. (In an ideal continuum there is no aggregation but only clination that is increasing or decreasing qualitatively or quantitatively in a certain direction.) Moreover we can also study the relation with environmental factors as far as those are measured in conjunction with the descriptions of the vegetation stands.

On the base of this study we finally may select guiding principles and may decide which properties and combination of properties can be used as characteristics and where the boundaries between the units will have to be made.

In our example of the table method (Table 1) we see that various arrangements are possible. A hierarchical composition of the system enables us to combine various possibilities. The choice between these possibilities and the way of combination depends on the appreciation of the guiding principles. Each level in hierarchy will be governed by one and usually by different guiding principles. So in world-wide vegetation (also soil) systems, the climatic influences are often used as a guiding principle at the highest level and eutrophy often at one of the lowest levels. The main reason for this has been the wish to have as little difference as possible between the 'pure general classification' and the *small scale* (chorological) classification.

In fact it is so that the difference between both ways of classification in the beginning of the early stage of development of soil and vegetation science under influence of geography which did not classify by aggregation but only divided and subsidived, were not (and could not be) sufficiently recognized. The traditional system of hierarchy in classification still goes back to that early period. For many other practical purposes, however, especially the less small scale surveys, it would be much more appropriate to have eutrophy and/or soil hydrology as a guiding principle. There is no objection to do this, from classification point of view.

We already mentioned that before the selection of certain forming factors as guiding principle can be made, we should first investigate which factors correlate with the item to be classified (vegetation). This can be done by plotting values of measured or estimated environment factors into the model, respectively the tables.

In a mathematical model even a calculation of the significance (a regression calculation) can be made if exact figures are available. If there is a sufficient correlation the guiding principle can be used. If no correlation can be measured either the measured data are wrong, the calculation method was not appropriate, or indeed the guiding principle was not good, and there may be another factor than the supposed one.

Here a decision has to be made by the scientist taking into account his experience and knowledge of the subject. The first thing he has to consider is that distinct aggregations or clinations of morphological features are usually correlated with one or another factor. This may be an unknown one till now. So the use of the morphology as guiding principle (which is the purest way of typological classification!), which is done in the last-mentioned case, may help to reveal the existence of an unknown vegetation forming factor. Exactly because of this guiding character of morphology, it is so important to use an objective way of ordering.

If only factors that are already known are used as guiding principles we might miss the chance to detect and discover other forming ecological factors! However, for practical reasons this ideal way can only be approached and never wholly be reached.

Sociological groups

We have seen that the former method resulted into two arrangements:
1. arrangement of the relevés (records) (basic data) into classification units;
2. arrangements of the properties (in floristic system species) into groups of properties that can be used as (a set of) characteristics for the various classification units.

If floristic properties are used, the sets of characteristics are sets of species (taxa). A group of species (taxa) that are more or less similar in 'behaviour' in a classification system are called a '*sociological group*'. Instead of individual species, this group is used as a diagnostic classification characteristic. In the Braun-Blanquet system the group of so-called 'characteristic species' of an association (or alliance or higher unit) is such a sociological group.

The same is true for each group of 'differential species' of any kind in that classification system. If our classificiation system is constructed, we can determine any new concrete stand with the help of these sociological groups as characteristics. A presentation of these groups in a 'bar-table' is highly recommendable (see Fig. 2 - IV).

Ecological groups

If a table is arranged on the base of an environmental factor (water, nutrients, etc.) the group of species is called an '*ecological group*'. In the preceding text we have already mentioned the advantage and disadvantage of doing this. The ecological groups of Hejny (1960) with respect to the hydrological conditions are a good example (see also Iversen, 1936).

A special form of delineation of ecological groups is the 'Koinzidenz Methode' (coincidence method) of Tüxen (1954). Tüxen states that only within a narrow ecological range it is allowed to measure correlation between a single environmental factor and the occurrence of a single species. So only within the association or sub-association of his French-Swiss shaped classification system it would be allowed to arrange vegetation tables according to a measured factor (chemical soil data, hydrological data, etc.). The correlations found in this way Tüxen calls 'Koinzidenz' because one is not sure that coincidence also means causal correlation. Nevertheless the figures can be used to indicate and map the coinciding environmental conditions (see also Bannink, Leys and Zonneveld, 1973, 1974).

In the ideal case sociological and ecological groups should be identical. If there appears to be a difference this is due to the imperfectness of the methods of analysing the factors and methods of determination of correlation between species and between species and environment.

The problem of infrequent species

The arrangement of species into sociological groups depends on the distribution of species over clusters of relevés and the relative frequence within those clusters. One can distinguish species in relation to this frequence and distribution into the following categories (not taking into account transitional cases):
1. species occurring everywhere in reasonable numbers;
2. species occurring in some clusters in rather high frequence and in others not or rarely;
3. species occurring rather infrequent but only in certain clusters;

4. species occurring infrequently and dispersed without clear affinity to any cluster.

Category 2 represents the bulk of those plants that form clear diagnostic sociological groups (differential species, 'characteristic species').

Group 1 is, within the region concerned, not of diagnostic value, but may be very important for the vegetation description especially if it concerns plants with a high cover or high internal or external frequence or high biomass, determining the aspect or quality (e.g. fodder for animals, or timber or energy etc.) of the vegetation.

Groups 3 and 4 may create problems. If a very large number of relevés is available (many hundreds!) group 3 may still help in defining clusters. In the field, however, the diagnostic value is low, because usually the species are not present. Many of the better characteristic species of the Braun-Blanquet system tend to belong to this group. If the number of relevés is relatively low, groups 3 and 4 cannot be distinguished.

Group 4 represents plants with a dispersed distribution. Often they occur only a few times in the total table. According to our experience in temperate and tropical vegetations, natural as well as cultural, they may make up to 50% or even more of the total species composition. They have no diagnostic value for indication of regional differences. They may be of interest however in general plant geographical studies (see e.g. Hommel, 1987). Within the region, however, they cannot be included in a sociological group and so do not play a role in the classification. If these are tiny plants that never dominate in frequency nor in cover, nobody will have problems. Even if they dominate occasionally in an individual stand, Braun-Blanquet scholars, used to local 'phases' where one or another species dominates without changing the classification system, will see no problem. Others, however, appreciating dominance (as frequency or cover) as classification criterium will be confused by this.

The controversy on the possibilities of using total species combination for classification of tropical rainforest is for a great deal related to this problem. There the dispersed plants are often large trees dominating a reasonably sized sample area (several of 100–1000 m^2). The last category of scientists do not accept such differences in tree layer as tolerable. In the pure tradition of the Braun-Blanquet system the similarity between such stands indicated by the majority of small trees, tree seedlings, sedges and the relative few tropical rain forest herbs, is sufficient to keep them, however, in spite of a different tree layer, within one vegetation type. In this system, by definition, it is taken for granted that the dominant trees may vary within one vegetation type.

The influence of dispersity can be reduced somewhat by enlarging the sample plots. However, sample plots larger than a few thousands of square meters are not very practical (time-consuming). More important, however, is that they are seldom homogeneous enough. Even in the tropical rainforest the drainage system and other geomorphological features may cause deviation in site characteristics within areas of only several hectares. So the 'maximum area' put limits to the minimum area determined by the most disperse plants. It would go too far to discuss this problem of minimum and maximum area in detail. For practical purposes, however, it can be solved easily in forest and savannes by taking sample plots not larger than a few thousands of square meters and discard the dispersed species.

The dispersed character of species can only be shown after the treatment by table method by hand or computer. Only plants occurring less then about 2 (or 3) times in the total table can be left out of consideration. Plants with higher presence in the table may be united all in one cluster and then, even in low numbers, have high information value; so these should not be eliminated beforehand.

7. The Nature of the Environment

A.W. KÜCHLER

The phytocenose as well as the biotope are parts of the ecosystem, and the phytocenose is the integrated expression of the biotope. In order to appreciate the distribution of phytocenoses, and to enable the mapper to interpret his vegetation maps, it is necessary to focus attention on the main physical and chemical characteristics of the environment. The mapper is urged to acquaint himself with the excellent 'language analysis and the concept environment' (Mason and Langenheim, 1957) in order to assure clarity in thought and presentation when discussing an extremly complex concept.

A vegetation map is a map which shows the vegetation of some area. This has led some authors to the point of view that a vegetation map ought to show nothing but vegetation; they frown on vegetation maps with supplementary information on site qualities, such as soils, ground water, and climate. This attitude is easily defended. The vegetation of almost any area is so complex and so few features of the vegetation can be shown on any one map that an author of vegetation maps should give all available space to the vegetation. If he had planned to present certain characteristics of the vegetation and finds while preparing the map that he can show more, then he should show additional features of the vegetation rather than some aspects of the environment. In this manner, the strictly vegetational character of the map is maintained and all available space if well utilized.

However, even if nothing but vegetation is shown on the map, much environmental information is given by implication. If the vegetation of an area is to be presented on a map, the author obtains a base map with an outline of this area, the coastlines, if any, and the rivers, in addition to the grid of parallels and meridians giving latitude and longitude. All this is entirely legitimate and, indeed, quite necessary. But every phytocenologist knows, too, that the grid shows more than the geographical location, and that a change in latitude often results in vegetational changes also. Nobody would expect to find the same plant communities in Ohio and in Florida, or perhaps in Scotland and on the Riviera. The longitude often reveals the location of a place with regard to oceans and seas to the east or west, and as the great wind belts of our planet often extend west-east or east-west rather than along the meridians, an indication of the longitude immediately betrays significant climatic information. The courses of rivers and coastlines at once reveal a series of environmental features: the high atmospheric humidity along the sea shore, the high water table on flood plains, and others.

The complexity of the environment was understood early. In order to understand the processes involved, it was only natural to divide the environment into its component parts and thus seek a grasp of the whole. The major features of the environment, also called factors, were usually considered to be climate, soil, water and relief.

The climate was usually taken to consist of

temperature and precipitation. These were observed on a daily, monthly or annual basis. Gradually, instrumentation improved and the results were made comparable by placing the instruments into instrument shelters. Wind, atmospheric humidity, frostfree season and other features were gradually added to the basic observations. Geiger (1961) then showed that the climate near the ground, the so-called microclimate, can differ substantially from conditions above, the macroclimate, and so opened a whole new field of investigation.

Certainly, the seasonal aspects of the vegetation are basic in understanding the nature of the landscape. The preparation of phenological maps was therefore an important step forward in the direction toward site analysis. Gams (1918) pointed out that 'phenology must be employed where climatology always fails, i.e., in determining the total ecology of every site. Thereby phenology becomes a full-fledged member of ecological research.'

When ecology first became a field of study, ecologists hopefully studied the individual features of the environment, thinking thereby to solve the ecological problems (e.g. Livingston and Shreve, 1921). But as knowledge increased and information accumulated, the confidence waned. Tansley (1935) introduced the ecosystem. Egler (1951) insisted on holism and argued that splitting the environment into factors is misleading, and indeed, erroneous.

These ideas evolved simultaneously with the recognition of the need for more detailed knowledge of site qualities. As the world's population is growing rapidly, the most intelligent use of the land becomes ever more imperative. This implies that the land be devoted to those crops, pastures or forests which are most in demand at a given time. It implies that the productivity of the land be sustained indefinitely at the highest possible level without damaging the land. This requires that the capability of land be known in detail. Here the vegetation comes to the rescue and gives the needed clues because it permits, as nothing else, to recognize the critical features of a site.

Many vegetation maps have contours. This topographic information is so useful that it should be given on all maps of large or medium scale, possibly as far as 1 : 1,000,000. The altitude above sea level, and the exposure to light, heat, and wind, especially rain-bearing winds, are all features of fundamental significance in explaining the character of the vegetation. A splendid example is the vegetation map of the Nanga Parbat group in the northwest Himalaya Mountains (Troll, 1939), or the two Grossglockner maps (Gams, 1936; Friedel, 1956).

The importance of topography is revealed especially in the relation that exists between the vegetation on the one hand and the water economy of the soil and the features of the microclimate on the other. Convex surface features differ markedly from concave ones even though the contrast is ever so slight, and perhaps hardly noticeable at first sight. Even the slightest rise will occasion an increased runoff and an erosion of the finest soil particles. More pronounced elevations nearly always result in a localized microclimate with its own contrasts. In depressions, on the other hand, no matter how shallow, soil and water accumulate, promoting growth, but snow and cold air accumulate as well, retarding growth.

The topographic effect is well illustrated by the 'cove forests' of the Great Smoky Mountains. The narrow ravine-like valleys are protected against high winds; they are humid and relatively rich in soil, especially when compared with the exposed flanks and bluffs. The result is an unusually rich flora, the cove forests contrasting sharply with the much simpler plant communities on the less favorable sites. Scharfetter (1932, p. 147) stated the need for contours when he said: 'A good vegetation map clearly expresses the organic connection between a plant community and the local relief; the absence of contours results in an unmotivated side-by-side of color splashes.'

Relief is usually considered an environmental feature. Strictly speaking, however, it is not. Relief affects climate and soil which change with altitude, slope and exposure. As a result of the relief, climate and soil can change within very short distances. As the vegetation is directly affected by its environment, it, too, can change abruptly and within short distances. The resulting complexity is usually generalized on maps although it can be portrayed well if the map scale is large.

The study of the soil, pedology, was first placed on a scientific foundation by Dokuchaev and soon developed into an important science. The soil was found to consist of individual layers or horizons, each intimately connected with the history of the soil, i.e. the process of its formation. The chemical and the physical nature of the soil can vary greatly from one horizon to the next. Humus, the organic part of the soil, directly links its formation to the vegetation. The study of the nature of the soil was complicated by the discovery that it could not be handled properly without also considering the soil biota, the often microscopically small, even unicellular organisms.

Many ecologists consider water a factor, observing such features as precipitation, atmospheric humidity, saturation deficit, evapotranspiration, snow and the duration of its cover. The water-holding capacity of the soil, the depth of the water table and its fluctuations, the leaching effect of the water, especially in the A-horizon, and many other features became objects of study. It became clear, however, that many of these features of climate, soil and water were difficult to separate from one another.

It is important to realize the intimate relationship between vegetation and the site on which it grows because here is the key to a veritable treasure of information. The innumerable features of the environment are so intricately interwoven that it is quite impossible to unravel them, and in spite of the vast amount of ecological research we still lack the means to measure a site in all its complexities. Billings (1952) tentatively enumerated 71 factors of the environment, but it is not difficult to add to his list. In most ecological studies only a few of the more obvious features are singled out for investigation, but the inadequacy of this method is well known. As an example, it may be pointed out that in order to establish the hydrogen ion concentration (pH) of the soil, 'the' pH is measured by conventional means. Less often, the investigators remember that the pH may not be the same in all soil horizons; and that the pH may fluctuate appreciably in the course of the seasons (Fig. 1) is ignored all too often.

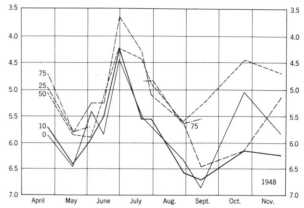

Fig. 1. Seasonal fluctuations of the pH values at different soil depths: at the surface, and 10, 25, 50 and 75 cm below the surface (after Ellenberg, 1950).

Another point in this connection is that many observers measure the environmental factors that come most readily to mind (e.g. soil acidity or precipitation) although the composition of the plant community is actually determined by quite different factors of which the observer is unaware. The following example from Ellenberg (1956) illustrates this point.

A comparison of the weed communities on fields of small grains and on fields of row crops revealed that the crop plants bore only a minor part of the responsibility for the consistent combinations of crops and weeds. It was found that the same weed communities can be obtained without the crop plants, if only the soil is cultivated at the same time and in the same manner as with small grains and with row crops. But even this cannot be the determining factor. Many weed species typical for row crops, e.g. *Solanum nigrum* and *Chenopodium polyspermum,* are more sensitive to cultivation (as it is commonly practiced with row crops) than

Matricaria chamomilla, Scleranthus annuus, and other small-grain weeds. The row crop weeds thrive best when the soil is cultivated only once and in the summer, when it is not too dry. If this one cultivation is shifted toward early spring or late fall, only small-grain weeds will germinate, besides the numerous indifferent species common to both crop types. It is the time of germination or, rather, the temperature prevailing at the time of germination that decides the composition of these weed communities. If quantitative measurements are to be made, they should therefore begin in this case with the germination temperatures.

Actually, the environment is dynamic and every change produces chain reactions which may be of great consequence. As an example of a chain effect, let us assume that, through a period of years, the precipiation at a given site is appreciably less than the average. This means first of all that less water is available. A dry soil heats and cools faster and the temperature fluctuations grow more extreme. If a moist soil helps to keep the relative humidity of the soil air and of the air layer immediately above the ground rather high, a drier soil means a drier air about it. The higher daytime soil temperatures combined with the reduced atmospheric humidity result in a more vigorous evaporation; this further aggravates the drought conditions. The increased evaporation at the soil surface draws more capillary soil water to the surface, enriching the surface horizons with solutes from below, possibly raising the pH. This is just the reverse process of what goes on during rainy years when the surplus water leaches the top soil and carries the solutes to lower horizons. It is easy to see that a change in precipitation, evaporation, temperature extremes, atmospheric humidity, and chemical composition of soil and soil water must necessarily affect the biota of the site. Many species, from vascular plants to soil bacteria, react more or less sharply to changes in any one of these factors. In the given example, the effect may well be that the production of dry matter is reduced; indeed, some species may vanish altogether. Such a disappearance may result in the invasion of the area by species from more arid regions. As a result, the competitive relations among the species are altered as well as the amount and kind of humus produced. This, in turn, again affects the biota and the microclimate. These chain reactions will continue until, over a period of years, everything is adjusted to the lowered precipitation and its effects.

The foregoing example of a chain reaction due to a change in one environmental factor (the amount of precipitation) may seem farfetched to readers who know only humid regions such as the northeastern United States or western Europe. Desert dwellers may have a similar reaction. But to readers in the great grasslands of the world, be they in Kansas or the Sudan, in the pampa of Argentina or the steppes of Russia, the given example is realistic because periodic droughts with the resulting profound effects on the character of the vegetation have been experienced all too often. Careful studies of the phenomenon have been published by Albertson and his collaborators (1957), Coupland (1959), and others. Striking illustrations were submitted also by Küchler (1964). But one of the most outstanding vegetation mappers of northwestern Europe with its equable climate noted the example above and commented that, undoubtedly, the vegetational fluctuations are mostly physiognomic. His remark is characteristic of many phytocenologists who fail to appreciate the extent of the floristic fluctuations that can occur in climax vegetation. Actually, the environment is not a constant and the annual deviations from a mean can be considerable. Rainfall fluctuations and their effects are usually at a minimum in very humid or quite arid regions but assume major proportions in subhumid and semiarid climates.

Küchler (1972) presented an example from the Kansas prairie which can be divided into three sections: the tall grass prairie in the east, the mixed prairie in the center and the short grass prairie in the west. The interesting feature here is that the mixed prairie lacks a fixed position: it oscillates back and forth from east to

Fig. 2. The oscillations of the mixed prairie in Kansas.

west and back to east (Fig. 2). As a drought intensifies, plants at the western edges of both the tall grass prairie and the mixed prairie are weakened or killed, opening the way for an invasion of western plants. Thus the borders of both the short grass prairie and the mixed prairie shift eastward. After some years, when the rains return, the process is reversed. A tight clay soil delays a westward movement as too much water runs off on the surface. A sandy soil delays the eastward movement as it can store more water at some depth. In the eastern and western border regions of the mixed prarie, the floristic composition can change dramatically and many species survive only in their seeds. These, however, have an extraordinary longevity, thus reassuring the return of the species when conditions improve.

The purpose of the foregoing example of an environmental change and (only some) of its consequences is to show how intimately all climatic, edaphic, and biotic features in the landscape are interrelated. At the same time, phytocenologists must be aware of the fact that theories evolved and tested in one region may not be practical in others.

Ecological research and site analysis

The observations concerning competition and the ranges of tolerance reveal clearly that one species is inadequate to portray the qualities of a site and that it is imperative to rely on the plant community as a whole. Vegetation is exposed to all ecological factors and their totality is more likely to be appreciated when studying phytocenoses rather than individual taxa. This is especially true because a given species will occur on a variety of site types (biotopes). On

the other hand, a given type of phytocenose will occur only on a given site type. This is particularly obvious in mountainous regions. In the United States, the indicator value of phytocenoses has been recognized and, indeed, employed for a long time, thanks to the observations of Shantz (1911, 1923), Shantz and Piemeisel (1924), Clements, (1928) and others.

If the indicator value of an entire plant community is employed, instead of an individual species, the results can be remarkably accurate. In Fig. 3, the solid lines may represent the ranges of tolerance of 18 species concerning the pH of the soil. Some species have a much narrower tolerance than others, but all have a range so wide as to preclude the accurate determination of the pH by a single indicator species. However, the degree of soil acidity must lie between the values of x and y. This becomes clear when the plant community is taken as a

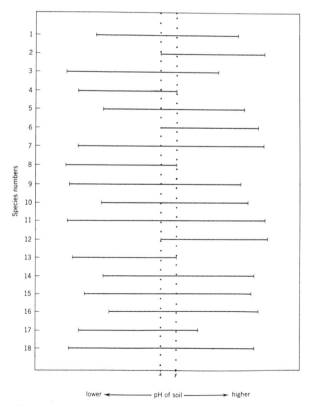

Fig. 3. The indicator value of a plant community as compared with the ranges of tolerance of the component species.

whole, because a higher pH would result in the disappearance of species no. 4, 8, and 13, whereas a lower pH would force out species no. 2, 6, and 12.

The ecological potential is relatively uniform throughout the area of a biotope. Walther (1960) concludes quite correctly that a comparison of biotopes therefore reveals qualitatively the controlling factors that are responsible for floristic differences among the phytocenoses. This also permits an insight into the intimate relationship between the phytocenoses and their sites.

'Site' is here understood to mean the total of all physical and chemical factors and influences of the environment, and biotopes are types of sites characterized by specific phytocenoses of the natural vegetation. If a given feature of the environment is not a part of the biotope, then the phytocenose will not indicate anything about it, as, for instance, in the case of rock strata deep below the surface and beyond the reach of all roots. On the other hand, the site analysis can be refined wherever measurable site qualities are of significance for the vegetation.

This sort of consideration applies to all site qualities, not just the pH of the soil or the depth of the water table. As time progresses, more and more such information will become available, but always it is the plant community rather than the individual species which can and must be used for a worthwhile analysis of site qualities. The community taken as an entity reveals the nature of a site more accurately and more comprehensively than anything else known to man. Detailed vegetation analyses have made it possible to relate several phytocenoses to one feature of the environment. As they react differently to such a feature, it is now feasible to calibrate biotopes. As a result, a series of phytocenoses can reveal the increase or decrease of an environmental feature or its effect with considerable accuracy.

The effects of the environmental influences are not limited to the floristic composition of the plant communities. They find their expres-sion also in the growth forms of the species and the resulting physiognomy and structure of the vegetation. The following observations by Troll (1951) illustrate this: 'For me it was one of the greatest scientific experiences to become acquainted with the tropical and subtropical vegetations of the Old and New Worlds with their convergent life forms in all variations from the desert to the rainforest and from the lowlands to the highest mountains. There were not only the well-known convergences of particular life forms (trunk succulent cacti and Euphorbias; leaf-succulent aloes, *Sansevieria* and agaves, bromeliads; trunk-forming Espeletias and Senecios; *Usnea barbata* and *Tillandsia usneoides,* etc.) but also the growth form communities of whole formations, e.g., the caatinga type of South America also in Africa, the African miombo type also in South America, the monte formations of South America in the karroo of South Africa, the tropical cloud forests of both continents, etc.'

Of course, it is well known that similar ecological conditions often result in vegetation types of strikingly similar physiognomy, and it is not difficult to add examples of such homologous types of vegetation to Troll's list: the California chaparral and the Corsican macchia, the Dakota prairies and the Russian steppes, the Nevada sagebrush and the Aaralo-Caspian wormwood formation, etc. The observation of homologous vegetation types logically led to the conclusion that particular combinations of ecological features find their expression in particular structural types of vegetation.

The convergent evolution of life forms under analogous environmental conditions must not lead to the conclusion that the physiognomy of vegetation by itself is an adequate indicator of these conditions. For instance, Wells (1962) shows in the mountains of the California Coast Range that 'both in pure form and in complex mixture, all physiognomic types often occur in close proximity on the same substratum, and on the same slope, and at the same elevation and distance from the ocean. Since these coincidences were encountered repeatedly on a wide variety of substrata, it is difficult to avoid the conclusion that the physiognomic end-members are not segregated primarily on the basis of climatic or edaphic factors in this area.' Later in the same paper, Wells says: 'Since grassland, shrubland and forest occur on almost all geological substrata in the San Luis Obispo area, it is obvious that the nature of the substratum has little direct influence on the physiognomy of vegetation under this range of climate, except when considered in connection with fire. But if the vegetation is regarded from the floristic standpoint, then the nature of the geological substratum has a pronounced effect.'

Wijngaarden (1985) observed in the Tsavo ecosystems of East Africa that large herbivores, especially elephants, can affect the structure of the vegetation, reducing it from woodland almost to steppe types. The reverse development occurs when the animal pressure is lifted. However, the floristic composition of the plant communities remains more constant.

In the case of the tall grass prairie, natural fires do not usually affect its floristic composition, provided these fires are natural, occurring on the average every four or five years in the summer. The story is different when ranchers set fires to the prairie every spring, a practice which is fatal to many forb species.

These are illustrations of the fundamental rule that the phytocenose must always be considered both physiognomically and floristically if it is to help in the interpretation of a site and its potential.

The study of vegetation types offers the added advantage of revealing long-term phenomena, such as are hidden from direct observation, e.g. fluctuations in the water table. Those trained to interpret the nature of the vegetation can therefore often tell at a glance what conclusions to draw from their observations. Not only is the vegetation the most sensitive site indicator, but the vegetation map is therefore the most comprehensive, the most reliable, and the simplest tool in analyzing the site qualities of the landscape (Krause, 1955, p. 11).

8. Aspects of Maps

A.W. KÜCHLER

The scale

The amount of detail that can be shown on a map is mostly a function of the scale; it diminishes with a diminishing scale, and the information on a small-scale map is usually more generalized than on a map with a large scale. As the scale controls the detail of the map, it also has a strong bearing on the type of material that can be shown. Broad formations (sensu Flahault and Schröter, 1910) are readily shown at small scales, whereas small plant communities usually require a large scale. Where phytocenoses gradually merge with others, an author may be tempted to show too much detail.

In compiling the data in the field, an author is quite easily impressed with unusual features of the vegetation, e.g. the occurrence of a rare species, or the relict of a formation which has long since migrated elsewhere. Such features, interesting as they may be, should not be shown unless the scale is sufficiently large to place them on the map without clutter, or if the organization of the material is worked out in such a manner that the unusual features can be fitted in as integral parts of the vegetation. In case of doubt, an author should always choose the simpler method and avoid complications. He must feel quite sure of the quality of his results before attempting to produce complex maps.

Even before the mapping work is started, the size and scale of the final version of the map should be determined, at least approximately. Whether the map is to be a wall map, appear on the page of a book or journal, or is produced as an individual sheet for research purposes, definitely affects the scale. To know the scale in advance of the mapping work may spell the difference between economy and waste in time, funds, and energy. On extensive plains with horizontal geological strata, e.g. in parts of Kansas, the vegetation can be remarkably monotonous for miles. In such an instance, a small scale is adequate. If, on the other hand, the vegetation to be mapped occurs in a region of alpine topography (e.g. the Olympic Peninsula) or in a region of great geological complexity (e.g. the Coast Ranges of northern California), there will be many significant plant communities that must be shown on the map and the scale must therefore be large. Wagner (1961a) has demonstrated this point in Austrian grasslands where a scale of 1:20,000 and even 1:10,000 may be inadequate to serve the purposes of the map. The vegetation maps of the Pasterze in the Alps (Friedel, 1956) and of a part of the Elbe Valley (Walther, 1957) are classic examples of vegetation mapping at very large scales.

Only large scales permit an accurate presentation of the geographical distribution of phytocenoses. Wherever a high degree of accuracy is required, as for instance in detailed land-use studies, the map scale must necessarily be large. In this, vegetation maps parallel soil maps and geological maps. But this only underlines the need of the vegetation mapper to select a map scale which is in harmony with the purposes that his map is to serve.

A.W. Küchler & I.S. Zonneveld (eds.), Vegetation mapping. ISBN 90-6193-191-6.

It is therefore quite essential that the vegetation mapper is thoroughly acquainted with the purpose and its exigencies. If the purpose is complex, it will require a large scale. If, on the other hand, the purpose is very limited or the vegetation is simple and uniform over large areas, the scale can be reduced.

Maps of small scale are usually derived from maps of larger scale. Such small-scale maps serve the interests of government agencies that are in charge of large areas. They serve as a basis for over-all directives. Once such directives are to be translated into practical application on the local scene, the small-scale maps must be replaced by vegetation maps at large scales. The vegetation mapper will appreciate by now that it is to his advantage to select an appropriate scale before his mapping activities are begun.

Some authors distinguish between large, medium, and small scales, but these terms have no fixed values and are entirely relative. Usually, a scale smaller than 1:1,000,000 is considered small. A scale from 1:100,000 to 1:1,000,000 may be called medium, and a scale over 1:100,000 may be termed large. Sochava (1961) reported on five scale classes used in the Soviet Union. Matuszkiewicz (1963) calls 1:37,500 a large scale whereas Doing Kraft (1963) speaks of a small-scale map at 1:25,000! In the United Nations Food and Agriculture Organization (FAO) it is common to distinguish four scale classes:
- exploratory scale: 1:1,000,000 or smaller;
- reconnaissance scale: 1:100,000 to 1,000,000;
- semi-detailed scale: 1:25,000 to 1:100,000;
- detailed scale: larger than 1:25,000.

In a small west-European country like the Netherlands, 1:50,000 may be considered a reconnaissance scale while 1:250,000 may be called a semi-detailed scale in Amazonia. It is usually simpler, more accurate, and more effective to mention scales by their actual values than to refer to them as large, medium, or small.

There are maximum and minimum scales, and above all there is an optimum scale. It is not always feasible to choose the optimum, but the author must be aware of the limits within which he is free to operate, if the scale is to be selected intelligently.

It is not feasible to state categorically which is the 'best' scale; nor is it possible even to suggest one optimum scale for vegetation maps. Where great detail is to be shown, a large scale becomes imperative; where broader vegetation units suffice, the scale should be correspondingly small. Ultimately, the selection of the scale depends on the purpose of the map; however, every purpose permits some flexibility.

The cost of production if often another controlling factor. If the map is to have several colors, a large scale may render the printing cost prohibitive. Unless there is no financial problem involved, the author must compromise between what is desirable and what is necessary.

It is customary to reduce a manuscript map to the desired size and scale of the printed map. Such a reduction is frequently influenced by financial considerations. The vegetation mapper must remember, however, that there are limits to which he can carry such a reduction. If he reduces his map below a minimum scale, he will sufficiently alter the character of the map so as no longer to serve its purpose. Where this minimum is reached depends largely on the organizational skill of the author and again, of course, on the purpose of the map. Whereas the minimum scale is often a critical problem, the maximum scale presents no difficulty. A map shows the areal differences of the vegetation and, in the case of great uniformity, the scale may well be of modest proportions as a larger scale will result in too few variations on the map. Too large a scale implies waste.

To illustrate this, the reader is invited to compare the vegetation map of Germany at 1:1,000,000 (Hueck, 1943) with the Major Forest Types of Georgia at the same scale (Anonymous, 1934). It is not necessary to study the two maps at length in order to realize that the map of Georgia could have been published at half its cale (reducing its area to one quarter of its present size) without sacrificing a single feature and without cluttering the map. Hueck, on the other hand, adapted his information to the given scale and used the available space to capacity. It would, therefore, not be feasible to reduce Hueck's map without changing its character.

Another disadvantage of an unduly large map scale is the difficulty of observing the distributional pattern of the phytocenoses. A reader wants to see the pattern of vegetation and to compare it with those of other maps and areas. However, if the map scale is too large, the vegetational pattern is obscured and the value of the map is diminished.

The optimum scale varies only with the purpose of the map and the skill of the author in organizing the map content, but it has nothing to do with production cost. Ideally, all vegetation maps should be prepared at their respective optimum scale; prospective authors are urged to seek enough financial support to permit the optimum scale for their maps. Such funds are usually well spent, as the best scale greatly enhances the value of the map. Vegetation maps tend to be complex, and the optimum scale is therefore particularly desirable.

The problem which every author must face is to determine the optimum scale for his particular purpose. This depends largely on the individual case, but the following points should be remembered. The smallest units which are to be shown on the map should have a diameter of at least 1 mm if round in shape, and if the map is printed in color. If the item is long and narrow, less than 1 mm is permissible. In the case of galeria forests, the space should be wide enough to show the river between two narrow strips of forest.

In order to determine the optimum scale, the mapper must consider the size of the area to be shown on the map and the kind of detail that is required by the purpose of the map. By correlating one with the other and by selecting a fixed value for the minimum size of any area shown on the map (usually at least 1 mm), the determination of the optimum scale becomes a matter of simple arithmetic. The mapper must

remember to base his calculation on the final version of his map as it will appear when it is printed. The manuscript map is usually done at a larger scale for a variety of compelling technical reasons. The accuracy and the legibility of a vegetation map depend inter alia on the dimensions of the smallest areas to be shown. If the map is to be printed in color and if the minimal

Table 1. Scale relations.

At the scale of	1 mm on the map equals	1 mm^2 on the map equals
1:1,000,000	1,000 m	10,000 ha (1 km^2)
1:500,000	500 m	25 ha
1:250,000	250 m	6.26 ha
1:125,000	125 m	1.5625 ha
1:100,000	100 m	1 ha (10,000 m^2)
1:50,000	50 m	2,500 m^2
1:25,000	25 m	625 m^2
1:20,000	20 m	400 m^2
1:10,000	10 m	100 m^2
1:5,000	5 m	25 m^2

At the scale of	1 mm on the map equal	1 mm^2 on the map equals (approx.)
1:1,000,000	3,280.90 feet	247.11 acres
1:500,000	1,640.45 feet	61.77 acres
1:250,000	820.23 feet	15.44 acres
1:125,000	410.11 feet	3.86 acres
1:63,360	207.88 feet	1.00 acre
1:62,500	205.05 feet	0.97 acre
1:31,680	103.94 feet	0.25 acre
1:24,000	78.74 feet	6,200.00 square feet
1:20,000	65.62 feet	4,306.00 square feet
1:15,840	51.97 feet	2,700.88 square feet
1:10,000	31.81 feet	1,076.50 square feet
1:7,920	25.98 feet	675.22 sqaure feet
1:4,800	15.75 feet	247.00 square feet

At the scale of	1 inch on the map equals	1 square inch on the map equals (approx.)
1:1,000,000	15.78 miles	249 square miles
1:633,600	10.00 miles	100 square miles
1:500,000	7.89 miles	62.25 square miles
1:250,000	3.95 miles	15.6 square miles
1:125,000	1.97 miles	3.88 square miles
1:63,360	5,280 feet (1 mile)	1.00 square mile (640 acres)
1:50,000	4,167 feet	160.62 acres
1:31,680	2,640 feet (½ mile)	160 acres
1:24,000	2,000 feet	91.82 acres
1:15,840	1,320 feet (¼ mile)	40 acres
1:10,000	833 feet	15.93 acres
1:7,920	660 feet	10 acres
1:4,800	400 feet	3.67 acres

area is to be at least 1 square millimeter, the mapper will find it useful to consult Table 1 for correlating the field work with the final map.

There are times when, for scientific reasons, very small areas ought to be shown on a vegetation map even thought their extent is well below the calculated minimum. In such cases it is permissible to exaggerate these small areas to such a degree that they can be drawn on the map. A small scale usually necessitates the suppression of many details to prevent the map from being cluttered. This procedure is a matter of expedience. But if it is justifiable at times to reduce certain items to nonexistence, it is equally justifiable to enlarge some items in order to assure their presence on the map.

Color permits the use of smaller areal units than do black and white patterns. In the latter case, the areas must be at least large enough to show the patterns clearly. When the vegetation types are shown by sets of letter and number symbols, it is important to keep the sale so large that the vegetation types of the smallest areas can still be identified.

One item the mapper must necessarily consider is the cost of producing his map. Such costs include the cost of printing. This is important because changes in the scale are directly reflected in the printing costs. The mapper must realize that a given increase in the scale implies a larger increase in the printing costs. For instance, if a scale is to be doubled, the printing costs will be substantially more than doubled. As a result, the mapper should consider what scale he would consider ideal, and which would be his absolute minimum scale. He can then consult with his printer and decide which is the largest scale that his budget will permit.

If any uncertainty exists concerning the scale, one should be careful to choose the largest feasible scale, at least for the manuscript map. This permits the maximum amount of detail to be shown and to be relatively accurate. If it is found later that the scale is too large, the map can be reduced. This can be done by diminishing the size of the map photographically either without making any changes in the map content

or by omitting some details, or by a certain amount of generalization, or both, but the relation of accuracy to scale can be maintained. The advantage of being able to make such adjustments is lost, however, if the original scale is found to be too small. The maps can, of course, be enlarged photographically, but one cannot thereby improve the refinement and accuracy of details which the larger scale calls for. For this, it is necessary to return to the field and map these vegetational details anew.

Rübel (1916) suggested for Switzerland that a scale of 1:25,000 is the best, that 1:50,000 is all right, but that 1:100,000 is too small. Such a rigid approach cannot lead to good results. A good scale is good only for a particular purpose and may be poor for another purpose. Most maps made under Tüxen's direction are necessarily of a scale larger that 1:20,000; Schmid's (1948) excellent map of Switzerland is at 1:200,000. Rübel's categorical statement is therefore without merit. Special purposes require special scales. For instance, Ellenberg (1950) discusses the scales for mapping weed communities and finds that the scale should be very large, somewhere between 1:5,000 and 1:10,000 but surely not less than 1:25,000. For the cadastral manner of vegetation mapping by Kuhnholtz-Lordat (1949), Ellenberg's scales would appear distinctly too small.

Base maps at scales of 1:20,000 to 1:25,000 are now available in many countries, and their number is increasing. These are the largest scales that can be used conveniently for mapping sizeable areas. The scale is so large as to show adequate detail, and the widespread availability of maps at such scales permits a country-wide or region-wide terminology and methodology. A larger scale, however, focuses the attention on such a small area as to be largely of local interest. This permits the introduction of ideas and methods that are of local use only, especially in ecological considerations. Wagner (1961a) gives some excellent examples of how he successfully modified the conventional procedures to serve local needs. His units of vegetation are ecologically controlled and their rank in the hierarchy of Braun-Blanquet is largely irrelevant.

The scale of accurate and detailed vegetation maps is also affected by the local relief. A mountainous terrain usually calls for a larger scale in order to assure an appreciation of the three-dimensional character of the distribution of the phytocenoses.

Molinier et al. (1951) demonstrated the relation between the scale and the usefulness of a vegetation map with the help of the maps of the Forest of Sainte Baume. This forest was mapped at five different scales, thus permitting an illustrative comparison. The forest is 2 km long and 800 m wide and of historic interest.

— At the scale of 1:2,000 the map of the forest covers an area 1 m long and 40 cm wide. At this large scale it is possible to show all details, even the exact location of some rare species.
— At the scale of 1:5,000 the map is only 40 by 16 cm. It is still possible to show most of the details that could be shown at 1:2,000 although it is necessary to use overprinted symbols to avoid cluttering.
— At the scale of 1:20,000 the forest measures only 10 by 4 cm. Only the major facies of the oak forest and the beech forest, can now be shown, without any indication of their density. But all essential features of the vegetation can still be shown.
— At the scale of 1:50,000 the forest is no more than 4 by 1.6 cm. Only the oak forests and the beech forests can still be distinguished, with perhaps some adjacent communities. The boundaries are only approximate.
— At the scale of 1:200,000 the forest is reduced to 10 by 4 mm. The boundaries of the two major vegetation groups are now quite inaccurate and the reader can get no more than a general idea of the vegetation.

Interestingly enough, Molinier then concludes that a general idea is inadequate for practical purposes and that the very largest scales (1:2,000) are the most useful for agricultural planning because conditions sometimes vary within a few meters. His conclusion reveals the extraordinary detail in which Europeans have studied their landscapes; it reveals also the constant pressure to improve the productivity of the land. In the end, however, and as a practical vegetation mapper, be adopts a scale of 1:20,000 for work on a nation-wide basis. This approaches the scale of 1:24,000 (1 inch = 2,000 feet) of the topographic sheets of the U.S. Geological Survey.

A problem arises if the map is to be done in a series of many sheets covering a large area. Unless the scale is predetermined the author should select the most complex area, often the one with the greatest relief, and establish the optimum scale for it. On the other sheets this scale may then appear larger than necessary but a smaller scale would clutter or oversimplify the more complex maps in the series.

Special caution should be exercised where manuscript maps are to be reduced to a scale which requires generalization of the vegetation features on the map. Again the purpose is the deciding factor. For instance, if the general appearance of the vegetation is wanted, the ground cover should be suppressed in favor of the forest trees. On the other hand, if a pasture map is desired, the ground cover must be emphasized, and the generalization should affect primarily the trees.

Another concern arises with reducing the manuscript map to the final scale of the printed map: the legibility of the names, terms, numbers, etc., written on the manuscript map. It happens all too often that this point is overlooked; as a result, it is practically impossible to read anything on the map without a magnifying glass. The value of such maps is unnecessarily low and care should be taken when preparing the manuscript map that all types of writing are large enough to permit the desired reduction without endangering the legibility of the printed page.

Giacomini has given more thought to scale problems than perhaps any other phytocenologist. In reviewing the types of maps that have been prepared at medium and large scales, he concluded that there are at least eight factors that help to determine the scale of a vegetation map and that they should be considered in deciding on the final scale whenever a vegetation map is planned (Giacomini, 1961).

In the past, good topographic maps were available for only a few areas. The new topographic maps at 1:24,000 of the U.S. Geological Survey, like those of other countries, are now so accurate that they can be enlarged and thus offer a base for a wide variety of scales. However, even today, there are vast regions where good base maps are either unavailable or of modest quality. Aerial photographs can help here a great deal to improve the less accurate base maps. Where only medium-scale maps are available, they will have to be enlarged photographically before they can be used for field work.

The various classifications, of course, permit a considerable range of scales: the latest maps by Gams attain their optimum at very large scales, for Schmid's method the scale need not be larger than 1:100,000, and for the Braun-Blanquet method the best scales remain at from 1:5,000 to 1:50,000 in spite of the wider range that has been employed successfully. Obviously, differences in scale affect not only the amount of detail but the classification of vegetation, too.

There are many different uses to which vegetation maps are put and sometimes a vegetation map is prepared for narrow and sharply focused purposes. The astonishing variety of purposes for which vegetation maps are made is illustrated in Section G of this book. But it is quite obvious that different uses require different scales, and the vegetation mapper must fully appreciate the purpose of his map in order to select a scale that is best fitted for it.

Large areas require, of course, a relatively small scale. However, if the mapping is done by a government agency so that many parts of the country can be done simultaneously and by a large staff, such a national effort permits a larger scale. In such cases, one uniform approach is needed. But if areas are more limited and local considerations increase in importance, the scale must be adjusted to these changing circumstances; it must be enlarged accordingly.

Not all scales can do equal justice to all phytocenoses. For instance, it may be that a multi-layered forest must be shown at a scale that is larger than that of grassland communities in order to properly reveal the internal structure of the vegetation. In other words, the scale is at least in part a function of the complexity of the structure.

A rather small scale can present the essential features of the vegetation where a region consists of extensive and relatively uniform plains, as in parts of the United States or the Soviet Union. Frequently, however, the vegetation grows increasingly complex as the local relief increases and the topography includes ever greater differences in altitude and steepness. In mountainous countries it is therefore often necessary to map at very large scales in order to do justice to the complex features of the vegetation in the narrow confines of steep-sides valleys and similar microlandscapes.

The choice between colors and black-and-white greatly affects the scale. The black-and-white maps require a scale two to three times as large as that of colored maps in order to show the details with an equal degree of clarity.

These observations are not to be taken as rules or directives. They rather reveal some of the problems the vegetation mapper must face. He will also discover that it is difficult to consider these factors in isolation because they affect one another. The proper choice of a scale implies therefore a delicate balance between a whole series of considerations.

Giacomini then suggest five different scales for different types of purposes:
— A scale of 1:200,000 to 1:250,000 for mapping the vegetation of a whole country.
— A scale of 1:100,000 for a synthesis of several small regions or for forestry maps on a nation-wide basis.
— A scale of 1:20,000 to 1:25,000 for synthesizing the natural vegetation of areas that can be selected for their scientific interest or their economic significance.

— A scale of about 1:10,000 for correlating phytocenoses with the characteristics of their sites. At this scale, the vegetation maps can present so much floristic detail that they can also portray ecological conditions more effectively.

— A scale of 1:5,000 or larger is preferable for all messicol vegetation as well as the structure of natural communities and their synusias. This seems reasonable because the features of cultivated crops and of synusias are apt to change and vary within very short distances.

Here again, as in the case of Molinier, the European origin of the proposals is evident, especially in the idea that a scale of 1:100,000 or 1:200,000 should be used for mapping an entire country. What a great step forward it would be, for instance, if the vegetation of the 50 states of the United States were mapped at 1:500,000 or even at 1:1,000,000! This is now feasible (Küchler, 1952) but also expensive because of the enormous areas involved. Recent, more highly developed remote sensing methods may reduce such costs considerably but the quality is likely to suffer. The lack of properly trained personnel presents another problem.

It is always of the greatest importance that the scale is clearly stated on the map, and it is equally important that the reader become aware of the scale before analyzing the map features.

There is more than one way to show the scale on a map (Fig. 1). One method is to show the scale in the form of a fraction (hence, the fractional scale), e.g. 1:200,000. It means that every distance measured on the map is 200,000 times longer in the landscape, or conversely, that any distance measured on the map is 1/200,000 of the corresponding distance in the landscape. The great advantage of giving the fractional scale is that it readily permits the reader to compare map scales quickly and without having to take any measurements. One glance will tell him which scale is larger and how much. A map should therefore always bear the scale in the form of a fraction.

Another method of showing scales is to draw a straight line and on it mark off the scale equivalents of kilometers, miles, feet, meters, etc. This is called a linear scale. It permits the reader to make direct measurements of distances on the map without any calculations.

Two additional points may become significant. First, the linear scale should always be expressed in terms of the metric system. In countries where other units for measuring distances are used, the scale should be repeated in such units. For instance, on a map of the United States or any part thereof, the scale should be given in kilometers and in miles.

The other point concerns maps that are being enlarged or reduced. When this happens, the scale changes as well. The mapper must then take special care to make sure that the final map will have the correct scale.

The grid

The map scale reveals only the dimensional proportions. For a true and accurate orientation of the reader, the map must also have a grid. This is a network in which lines drawn west-east (called 'parallels') indicate latitude, and lines drawn north-south (called 'meridians') indicate longitude, and between them give a frame of reference for every point on the map. These parallels and meridians are all the more necessary, as vegetation maps from all over the world are now becoming available and the reader is often quite unfamiliar with the names of places and areas on some of these maps. The grid will reveal the exact location at once even though the names of the areas portrayed may be Kwango, Riverina, or Thiès. Some vegetation

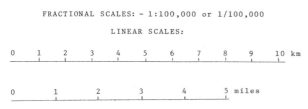

Fig. 1. Methods of showing map scales.

mappers and phytocenologists will, of course, recognize these names because of the important vegetation maps of these areas (Devred, 1955; Moore, 1953b; Gaussen et al., 1950).

Many phytocenologists prepare vegetation maps but have little experience in cartography. The result is that they occasionally omit to show either a scale or grid, or even both. If both are missing, it is impossible to make any measurements on the map, thereby reducing the value of the map seriously and unnecessarily. If the scale is missing but the map has a grid, this grid can serve as a scale because the distances between the parallels are known or can be determined. The reader need only remember that one degree of latitude very nearly equals 111.3 km (69.15 miles). Without a grid, a vegetation map cannot be compared with analogous maps because individual locations cannot be determined accurately. This is true even if the map is to be compared with other maps of the same area, if these are done on a different projection.

If an author objects to having a grid, he can do without it by marking the degrees of both latitude and longitude on the margin. He can, in addition, show the intersections of the parallels and meridians in the form of crosses of fine lines. Such a substitute is found on a number of maps but it is undesirable because it does make it more difficult to locate a point precisely.

Sometimes the scale is so large that the mapped area is very small, so small in fact as to fall between two parallels and between to meridians. This applies to maps of individual peat bogs and other vegetation types of small areal extent. But this is not an adequate justification for omitting a grid. Each degree of latitude and longitude is divided into 60 min, and each minute into 60 s. Topographic sheets will readily permit the accurate placing of a grid by minutes in such a manner that at least two parallels and two meridians can be shown on the map. There is, of course, no excuse for having no scale, no matter how small the mapped area is.

The Toulouse Colloquium of 1960 (Gaussen, 1961a) recommends in Resolution 11 that every vegetation map have a grid, the scale in both the fractional and the linear form, and also the name of the projection.

Many phytocenologists, unfamiliar with the nature of maps, may not appreciate the significance of the projection. Yet, it can be of major importance.

It is mathematically not feasible to transfer a curved surface like the surface of the earth onto a flat surface like a map. Any effort of this kind results in a distorted image. However, when the map scale is large and hence the mapped area is small, as for instance on the topographic sheets of the U.S. Geological Survey, then the degree of distortion is so small as to be entirely negligible. In fact, it is less than the normal expansion and contraction of the map paper due to changes in the atmospheric humidity. On large scale maps, it is therefore entirely feasible to make measurements with complete confidence that the results are accurate.

The situation is quite different on small-scale maps where large areas of the globe are presented such as the United States, Europe or Australia. The distortion now becomes a real factor. Measuring along a straight line no longer results in true values except along the central meridian. It is particularly difficult to transfer information from one small-scale map to another if the projection of the two maps is not the same. Mappers must then use extreme caution and care to obtain acceptable results.

An important aspect of mapping vegetation is the production of a base map. This is prepared with the help of aerial photographs and/or topographic maps. The base map is particularly useful because it lends itself to the production of a variety of vegetation and applied maps as dictated by different purposes. The following section gives some more details about those base maps.

9. Boundaries, Transitions and Continua

A.W. KÜCHLER

The chief task of a vegetation map is to show the areal distribution of different types of vegetation. This implies that unlike types will be contiguous and, on the map, separated by a boundary. The author of vegetation maps must therefore concern himself with the problem of boundaries.

A boundary separates two different types of vegetation. It is not enough to observe one type in the field and draw its boundary. The accuracy of the map is much greater, if the character of the surrounding communities has also been determined before the boundaries between the types are established.

When the vegetation mapper observes the vegetation, his task is to establish mappable vegetation units which fit the purposes of his map. His endeavour to identify the vegetation units inevitably leads to a search for their boundaries, and sometimes these are not easy to find. This may be due to the fact that a boundary is actually rather vague. It may also depend on the observer's point of view concerning vegetation types. For instance, the Wisconsin school of thought under the leadership of John Curtis developed the idea of the continuum, which implies in many instances a denial of the existence of vegetational boundaries. At the other extreme stands Tüxen, who draws boundaries with remarkable precision, based at times on the presence or absence of a single species. Where the area of occurrence of such a species ends there ends the vegetation type, too (see below).

The mapper's power of observation must be keen if he is not to misinterpret what he sees. An example will illustrate this. If a vegetation mapper works in high altitudes he will recognize timberline as one of the most important vegetation boundaries of his area. The location of timberline can easily be determined and often it is a very clear line, especially when viewed from a distance, as for instance, from one mountain to the next across a valley.

If the mapper asks himself (as he should) why the timberline is where be observes it, he may discover that he cannot necessarily show it on a map of the natural vegetation in the place where he found it in the field. The alpine meadows above the forests are often fine pastures; the forests below them are rarely as fine. Hence it has been the tendency of local herders to extend the alpine meadows downward at the expense of the forest. If the vegetation mapper plans a map of the natural vegetation and places the forest boundary where he observed it, then his natural forest vegetation will be more restricted than it should be, as the alpine meadows will be overextended.

The location of timberline can have other causes and need not be due to climate or to man. For instance, the forests on the eastern slopes of the Sierra Nevada in California along the Owens River Valley have two sharp boundaries: a climatic timberline forms the upper limit and another timberline terminates the forests against the desert below. However, it is not the lack of precipitation that enforces an end to

A.W. Küchler & I.S. Zonneveld (eds.), Vegetation mapping. ISBN 90-6193-191-6.
© 1988, Kluwer Academic Publishers, Dordrecht. Printed in the Netherlands.

tree growth. The steep and rocky slopes of that impressive mountain wall have been eroded through thousands of years and the debris in the form of coarse, loose rock and gravel has piled up at the foot of the mountains, slowly climbing higher as the millennia pass. This gravel has little ability to retain water from rain or melting snow, and trees therefore find it impossible to get established on it. There is, therefore, a sharp forest boundary coincident with the upper boundary of the loosely accumulated coarse gravel. The vegetation boundary is hence neither climatic nor cultural, but owes it existence to the abrupt change in the quality of the ground. This timberline is therefore a natural timberline. At this point, the vegetation boundary cannot be used to illustrate the effect of climate on vegetation as in the case of the upper (alpine) timberline; instead, it clearly indicates the boundary between two unlike physiographic features of the landscape.

Boundary lines – types and advantages

The technical problems of vegetation boundaries consist above all in having to decide how the boundaries are to be shown on the map. One way is to draw lines separating one vegetation type from another. The lines should always be quite thin so as not to affect the quality of the map and detract the reader's attention from more significant features; the space occupied by thick boundary lines may be more than is justifiable, especially on small-scale maps. Lines may be continuous for the major vegetation types and dashed or dotted for their divisions and subdivisions. Sometimes continuous lines are used where vegetational boundaries are well established while dashed or dotted lines imply that the location of the boundary is uncertain. In this manner, Hueck (1960) showed the reliability of his vegetation map of Venezuela by establishing three categories of boundaries: continuous lines for accurate boundaries, and dashed and dotted lines for fairly accurate and uncertain boundaries, respectively. In order to

render boundary lines unobtrusive they can be printed in grey.

Boundary lines have the advantage of making a map clear and of permitting precision to a very considerable degree. But some authors object to the sharp contrasts between vegetation types produced by continuous lines as being unrealistic; therefore they omit boundary lines altogether. The contrast between different colors or patterns is considered adequate to offset one vegetation type from another whereas the sharp lines are felt to exaggerate the differences. Gaussen's (1945) map of France or Seibert's (1954) map of Schlitz are good examples of maps without boundary lines. The absence of sharp contrasts is pleasing, indeed.

There are, however, other aspects to the problem of having boundary lines. One such aspect is that lines help to distinguish between similar though unlike colors or to offset similar black-and-white patterns from one another. A further aspect is the fact that the lines are a great aid in assuring better registration when printing a map with different colors. This is important because poor registration makes it difficult to read the map and distracts from more significant features. Putting boundaries is schematisation. Schematisation is the mother of science. So for maps which really have to be used and applied scientifically, clear boundaries are a must.

Locating boundaries

A boundary is a line between two vegetation types. It is not always easy to locate such a line in the field, and where good topographic maps or soil maps are available, these may be consulted to advantage. In this manner, a line which at first may have seemed rather arbitrary is given more justification. Aerial photographs of the area or other remote-sensing means will aid in locating the boundaries accurately and rapidly; the amount of time, energy, and funds saved thereby is very great and the map is much more reliable (see Chapter 28 and 29).

Theoretically, a boundary line on a vegetation map corresponds to a boundary in the field. But in practice, such a boundary may or may not exist. Scharfetter (1932) quotes Beck-Mannagetta, Cajander, du Rietz, Fries, Gleason, Osvald, Scharfetter, Tengvall, and Warming as authors who feel that boundaries, as a rule, are sharp and clear. They admit exceptions in the form of transitions, but it is felt that these only confirm the rule. Küchler (1955a) observed boundaries between plant communities that were remarkably sharp and clear: a wagon road led through a relatively undisturbed forest in the state of Maine. The turning wagon wheels and the action of the horses resulted in a herbaceous plant community that covered the road but never once penetrated into the surrounding forest. Indeed, the road community and the herbaceous community on the forest floor had not a single species in common. The species listed in column I of Table 1 occurred exclusively on the road, i.e. in or between the ruts. The species in column II grew in the surrounding forest – not one of them occurred on the road.

Table 1. Example of sharp difference between road and forest vegetation.

Column I	Column II
Agrostis alba	Clintonia borealis
Leontodon autumnalis	Cornus canadensis
Lycopus americanus	Gaultheria procumbens
Oxalis corniculata	Maianthemum canadense
Plantago major	Mitchella repens
Poa sp.	Osmunda claytoniana
Prunella vulgaris	Pteridium aquilinum var. latiusculum
Ranunculus acris	
Veronica officinalis	Vaccinium angustifolium
Viola blanda	

Perhaps the most detailed method of establishing sharp boundaries is given by Tüxen (1954b), whose work is based on the tenets of phytosociology sensu Braun-Blanquet. He says: 'Sites merge unobtrusively and gradually because neither the individual ecological factors nor their unlimited combinations change abruptly. However, phytocenoses do change at definite points quantitatively because their components are (generally indivisible) species. For this reason, too, a plant community can be defined and its extent can be established much more accurately than the site on which it depends. These one-species variations gain in significance the poorer the floristic composition is, or in other words, the more extreme certain ecological factors are (e.g., Salicornietum, Puccinellietum, etc.). Floristically rich communities on balanced heterogeneous sites usually differ from each other by several or many species. Abrupt and considerable site variations imply strong vegetational differences whereas small changes in site qualities result in slight floristic variations which, however, remain abrupt as long as one species disappears from the community or is added to it. Still more delicate changes in the character of the environmental complex find a fluid expression: the numbers and frequencies of the individual specimens of the species within the community fluctuate gradually and smoothly between various degrees of intensity'.

There is no doubt that sharp vegetation boundaries are common, due to more or less abrupt changes in the topography (relief), the soil, the geology, the climate, or the water economy of the substratum, and especially due to direct human interference (cultivation, engineering, etc.).

In contrast to these observations are those to the effect that sharp boundaries are rare, that transitions are usually more or less broad, that, indeed, the vegetation of a region may consist more of transitions than of clearly identifiable vegetation types. The most extreme view was perhaps held by Curtis, who believed that vegetation changes continuously, that boundaries are largely absent and, therefore, cannot be observed and recorded. Even Curtis, however, agreed that vegetation can change quite abruptly (Curtis and McIntosh, 1951; Curtis, 1959; see also Zonneveld, 1968, 1974).

Transition zones

In sifting the evidence, it becomes clear that boundaries exist indeed but are not universal. On the vegetation map, existing boundaries can be shown without much difficulty. Where it is desirable to show sharp boundaries when, in fact, they are blurred, the author must decide whether a transition zone should be shown on the map. The transition should certainly be shown if it is very wide, considering the scale of the map. Where it is narrow, it may, in some instances, be ignored. This implies that a vegetation type appears to be more uniform on a map than it is in fact because at its margins it may be merging gradually with the neighboring type.

Where a boundary is not clear because of gradual transitions, it can be shown as broken lines, zigzag lines, or by an interpenetration of two contiguous vegetation types (Fig. 1). Some transitions are so gradual that there may, in fact, be no place where a boundary line is justified. It is then necessary to decide arbitrarily how much of an admixture may be tolerated in an established type and to draw the boundary accordingly. Even if the place for the boundary in a transition zone has been selected very judiciously, field inspection may reveal little or no difference on the two sides of the boundary. An interpenetration, including its width, can be shown by alternating bars or arrow-like extensions. The degree of dominance of each of two vegetation types in a transition can be shown by the relative width of the color bars. In a transition zone, two vegetation types A and B merge; type A gradually gives way to type B. Therefore, in crossing the zone, the mapper usually finds at first type A, more prominent; hence the color bars representing it are wide compared with those of type B. But sooner or later the situation is reversed and type A remains but is poorly represented; its color bars are therefore shown distinctly narrower than those of type B before they vanish altogether.

Where symbols are used, the more prominent vegetation type is shown by its color whereas

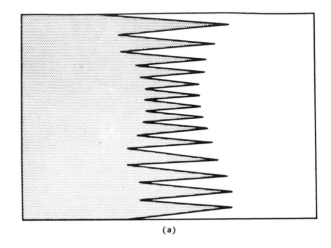

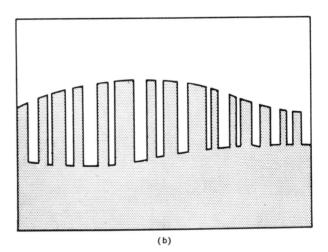

Fig. 1. Examples of showing transitions between two unlike vegetation types.

the competing type is shown in the form of symbols overprinted on the color. Near the opposite end of the transition zone, the case may be reversed: the second type is now shown in its color with the symbol of the first type overprinted on it.

In case of an invasion, the use of arrows can be quite effective. The arrows are shown in the color of the invading community, overprinted on the invaded one. The shape of the arrows is related to the degree of success of the invasion, thus: broad arrows imply a massive invasion, slender arrows an invasion of modest success at least up to the time of mapping, and the length of the arrows shows the depth of penetration.

The arrows can also be shown in grey and black. In that case, the mapper is advised to experiment amply so as to find just the right size and number of arrows, else the black arrows can easily become too prominent.

Ultimately, the use and manipulation of boundaries and transitions are left to each author. However, it is well to be acquainted with the existing possibilities and their variations in order to assure the most appropriate and the most effective choice, and thereby keep the quality of the vegetation map at its highest possible level.

While phytocenoses are the mappable units of vegetation on which the mapper must rely, the idea of the continuum challenges their very existence. But although this has led to some very spirited debates, it is now generally accepted even by such continuum authorities as Whittaker (1970) that phytocenoses and continua are by no means mutually exclusive, and that the existence of both must be accepted. From the point of view of the mapper, however, it seems desirable to distinguish between continua and transitions.

A transition is the population of taxa observable along a gradient between two unlike phytocenoses. Thus a beech-maple forest may grade into an elm-ash forest. The gradient between these two types may be steep or gentle as local conditions dictate, but the contrasting biotopes necessarily lead to contrasting species combinations. In a transition, therefore, one type ends, albeit gradually, and another one begins.

On the other hand, the term continuum implies something continuous. As vegetation types end in transitions, the application of continuum to a transition seems neither precise nor logical. A continuum may rather be said to result from changing conditions within a vegetation type which is here considered synonymous with formation sensu Whittaker (1970), even though this sounds almost like a contradiction in terms. For instance, the bluestem prairie of Iowa can be expected to differ from the bluestem prairie of Oklahoma simply because of a difference of eight degrees latitude. However, a

mapper is hard put in locating a boundary within this area as the gradient between north and south is very gentle, resulting from a gradually changing climate, and the bluestem prairie continues from one end to another. Some arbitrary boundaries can always be drawn, but they are debatable and the diagnostic taxa may not include the dominants because their wide range of tolerance permits them to maintain their dominant position throughout the area. Vegetation boundaries can be drawn only when a small section of the bluestem prairie is mapped at a large scale and local differentiation of structure and floristic composition results from unlike topographic and substratal conditions.

This floristic change within the boundaries of the bluestem prairie continuum differs therefore from the transition between the beech-maple forest and the elm-ash forest. As a result, a continuum is a population cline within a given vegetation type, at least a relatively broad one, to be differentiated from a transition between two unlike types. Used in this sense, the difference between transitions and continua is reflected in boundary considerations. The contrasts between the two ends of a transition is such as to permit a relatively clear and simple definition of boundaries, and they will be relatively acceptable, no matter how arbitrary they may be. In a continuum, however, there are no boundaries. Any boundary introduced into it has little meaning, and it is usually best omitted.

The vegetation mapper therefore focuses his attention on phytocenoses with their more or less distinct boundaries. It is not often practical to map continua and the mapper usually ignores them. He must nevertheless be aware of them because they will affect his interpretation of the observed communities.

Zonneveld (1968, 1974) has made it clear that one should distinguish between the concrete boundaries between phytocenoses in the field or on maps on the one hand, and boundaries in the abstract floristic classification typification systems on the other. He supposes that some confusion in the discussion on the subject of

SCHEME OF A CONTINUUM WITH TWO GRADIENTS

I CONCRETE SITUATION IN THE FIELD   **II** ABSTRACT MODEL

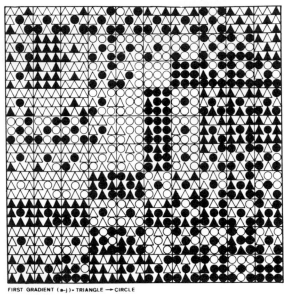 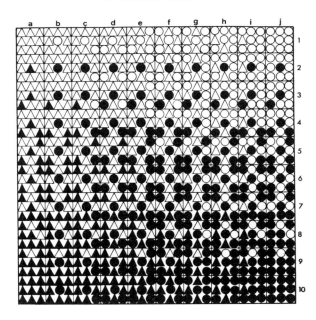

FIRST GRADIENT (a–j) • TRIANGLE → CIRCLE
SECOND GRADIENT (1–10) • LIGHT → DARK

Fig. 2. 6B1.

(dis) continuity of boundaries may be due to not making this difference.

What has been said before now deals with concrete boundaries. In the abstract, boundaries tend to be more continuous. Everybody who has experience with floristic vegetation classification using table and or statistical methods knows that the more complete his floristic relevé material is, the more transitions it contains, the less sharp are the boundaries between clusters. Fig. 2 shows an exaggeration of this phenomenon. Here all 100 sites are different. In the left image, however, they form a clear observable pattern, partly with more contrast, partly with less. In the right picture, however, they are arranged in an abstract two-dimensional model. Virtually no boundaries exist. It is one clear continuous row. Further valuable discussions on boundaries, concrete as well as abstract, can be found in the proceedings of the symposium. 'Tatsachen und Probleme der Grenzen in der Vegetation', Herausgeber R. Tüxen. Verl. Cramer 3301 Lehre, 1974.

10. Patterns, Colors and Symbols

A.W. KÜCHLER

The method of presenting vegetation and its units and divisions is an important feature of the vegetation map because the clarity and legibility of the map depend on it. The primary purpose for using black-and-white patterns or color is, of course, to distinguish between areas of different types of vegetation.

Letters and numbers

In the case of black-and-white patterns, there are three basic approaches, each of which permits a considerable number of variations. The first method employs letters, numbers, and combinations of these. Letters can be used in the form of capitals and lower case letters, italics, etc. Numbers can be shown as Roman or Arabic numerals, and the combinations of all these make for a vast variety. The possibilities are further enlarged by arranging the letters and numbers as straight sequences, fractions, or in some other form. The place of a number in such a fraction can then be given a definite meaning as well. This is illustrated (Fig. 1) by the method developed by A.E. Wieslander (1949).

The method of using letters or numbers to distinguish different types of vegetation permits the production of vegetation maps at a very low cost. The black color is used for everything: vegetation boundaries, symbols, grid, frame, title, legend, etc., and hence only one plate is required for printing.

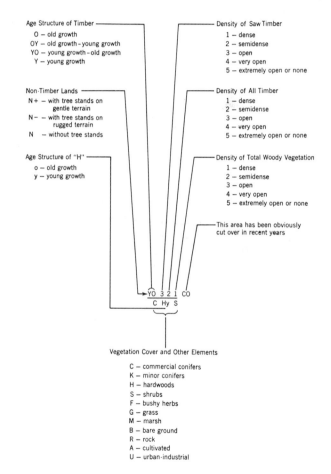

Fig. 1. The classification of a vegetation type with letter and number symbols, arranged as a formula. Note that the position of each letter and number within the formula is significant (after Gardner and Wieslander, 1957).

The chief disadvantage of the method is that it fails to show at a glance the distribution of the various vegetation types. True, it is easy

A.W. Küchler & I.S. Zonneveld (eds.), Vegetation mapping. ISBN 90-6193-191-6.

112

enough to establish the character of the vegetation at any given point. But in order to become aware of the distribution of any particular vegetation type, it is necessary to read all the symbols on the map. The reader is obliged to color by hand those types that are important to him, or indeed all of them, and that is time-consuming (Fig. 2).

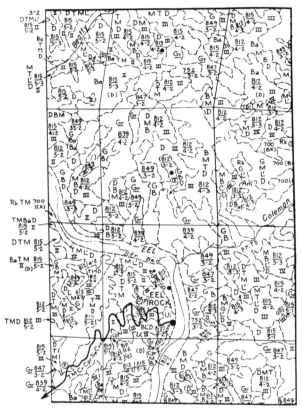

Fig. 2. Soil-Vegetation map (California Vegetation-Soil Survey).

Shading and patterns

Another method to distinguish different types of vegetation is by using different shades of grey. These may range from black to white and as they, too, require only one printing plate, the cost is not high. The shades of grey simulate different colors and can be very effective. The areal distribution of vegetation types is quite clear. However, the number of possibilities is relatively small because the various shades of grey must differ enough to allow a real contrast between them. Otherwise it is very difficult to read the map, and misinterpretations are made easily. Jenks and Knos (1961) have studied this problem in detail. Contrasts result from the varying features of the patterns, especially darkness, texture (coarseness) and the design of the pattern units (Fig. 3 A, B and C).

A map can be made more meaningful by using the shades of grey in a given sequence. Rather than at random, they can be arranged in the legend in such a manner that an advance from dark to light coincides with a progression in one or more vegetational features, e.g. density, xericity, etc. It is then possible to see at a glance how such features are distributed, and this aspect makes the various shades of grey more valuable.

Finally, as a third possibility to show the vegetation only in black on white, there is the large variety of patterns. These consist of dots of various sizes, arranged regularly or irregularly, thick and thin dashes, fine lines and broad stripes running horizontally, vertically, diagonally, and in combinations of these. In addition, there are crosses, stars, triangles, squares, and a host of symbols of every description.

Unless the patterns are selected very judiciously, the map will be a bewildering jumble and the character of the vegetation cannot be determined readily. If at all possible, here, too, there should be a sequence from dark to light, meaningful in terms of some aspects of the vegetation, but such a sequence is difficult to achieve when many types of symbols are used.

Another method may be mentioned parenthetically. This is to print the names of the vegetation types directly on the map, as was done by Louis (1939) on his map of Turkey; he uses no patterns of any kind and neither colors nor symbols. At first sight, this methods looks interesting enough, but it is not very practical and has rarely been adopted.

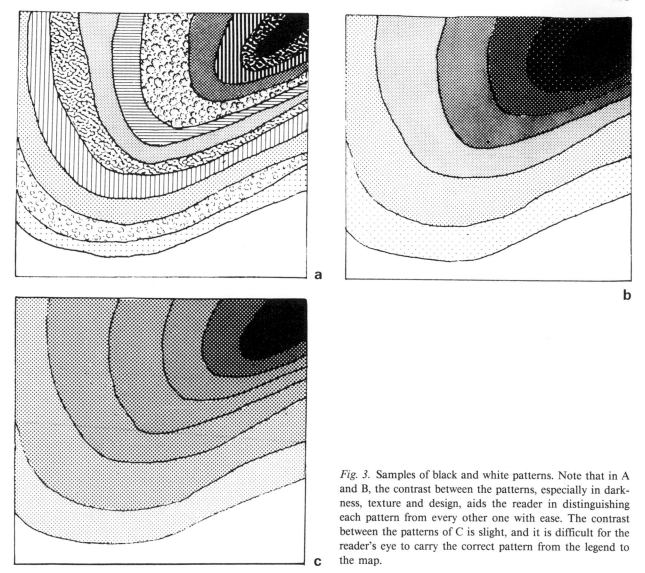

Fig. 3. Samples of black and white patterns. Note that in A and B, the contrast between the patterns, especially in darkness, texture and design, aids the reader in distinguishing each pattern from every other one with ease. The contrast between the patterns of C is slight, and it is difficult for the reader's eye to carry the correct pattern from the legend to the map.

Color

A new world is opened up by the introduction of color. It is as if a composer of piano sonatas had suddenly discovered the dazzling world of symphonies. Colors occur in great variety, are readily distinguishable, and can be applied flat or in patterns, in combinations, or as over-printed symbols. They can be mixed and super-imposed. Often they can be made even more effective by simultaneous use of black or co-lored patterns and symbols and by a judicious use of white (Gaussen, 1945).

Just as there are various possibilities of applying black-and-white patterns, there are also several approaches to the use of color. By far the most common use of color is simply to select arbitrarily a variety of contrasting colors in order to facilitate the distinction of one vegetation type from the next (Drude, 1905; Wangerin, 1915; Rübel, 1916; Wagner, 1948). This is a rather primitive method but also the simplest. Indeed, it is not easy to raise the use of color to a higher level, and Scharfetter (1932, p. 157) repeats Rübel's (1916) advice urging prospective authors to follow the early leaders in using their color scale, possibly with slight modifications, rather than introducing new scales, even though these colors were selected arbitrarily, not to say accidentally. Their continued use implies no more than a 'tradition' which lacks a scientific foundation. Rübel (1916) proposed eleven colors for the vegetation of Switzerland. Of these, five are for meadows and bogs, three for needleleaf forests, and one each for deciduous forests, dwarf shrubs, and cultivated land. This scheme was not generally adopted because of its obvious limitations.

Emberger and Molinier (1955), on the Carte des groupements végétaux de la France, tried to give their color selections a stronger foundation by basing them on a recommendation by the 6th International Botanical Congress (Amsterdam, 1935), to the effect that floristically related plant communities are shown in related colors. But they circumnavigate this reef by stating that such must not be done at the expense of common sense. Thereby, they are once again free to use colors as they see fit. Emberger and Molinier use the classification of vegetation by Braun-Blanquet exclusively. Hence they assign one color to each order that appears on the map, and the various alliances are shown in different hues of the color of that order to which they belong. Smaller units, such as associations, are shown by the color of their respective alliances with the initials of the association overprinted in black, for example, 'Q.i.' for Quercetum ilicis and 'Q.p.' for Quercetum pubescentis. Where an alliance is represented by a single association, the overprinted letter symbols are omitted. Within this arrangement, the darkest colors are assigned to alliances of the smallest areal extent. This permits the reader to see these items quickly and clearly even though they occupy only a small area. But at times the authors are guided by the dynamics of the vegetation and give the darkest colors to those communities that are closest to the climax. A Quercetum

ilicis climax is therefore shown by a darker green than a Quercetum cocciferae, a seral phase. In a vague way, the colors are assigned to the alliances according to the degree of xericity of their sites, but this idea is not applied rigidly and experimentation continues. Broadly speaking, the use of colors is as follows: blue and purple for more or less hydrophilous (grassland) communities and aquatic associations; red, orange, and yellow for xeric communities; and green for forests.

Transitions between two alliances are shown with alternating bars of the colors of the respective alliances; these bars are relatively broad. In a mosaic-like irregular distribution, the color of the most extensive alliance is used as a base and the less extensive alliance is shown in its own color in overprinted circles. Pioneer communities are shown by narrow vertical bars. Regressive communities (e.g. due to erosion) are indicated by narrow horizontal bars.

Emberger (1961) showed that the Service de la carte des groupements végétaux in Montpellier has changed its approach to the use of color. The influence of Gaussen in nearby Toulouse resulted in adopting color schemes based on ecological rather than floristic criteria.

Thus, vegetation units on biotopes with similar environmental conditions receive similar colors. This offers the advantage that ecological relations and affinities become readily evident, and the vegetation map thereby becomes more useful. This approach to the use of color permits a high degree of accuracy in interpreting the vegetation if the scale is very large and, hence, the area involved is correspondingly small.

The large scales used in Montpellier (1 : 5,000 to 1 : 20,000) lend themselves well to such exact analyses. However, as the scale shrinks this accuracy fades. Obviously, a given color cannot show the degree of xericity on a vegetation map of a single Kansas county. At the same time, the reader must always be aware of the fact that a yellow hue in Oregon does not signify the same xericity as in Texas, even though the vegetation maps of Oregon nd Texas may be at the same scale. The meaning of colors changes when ecological conditions of one region differ from those of another.

Schmid feels free to select his colors at random, but being artistically inclined, he is careful first to establish color harmonies, so as to avoid any clashing colors. As a result, his map of Switzerland (Schmid, 1948) is restful to the eye and easy to read.

Sometimes a given color system is best limited to a given range of map scales. Beyond this range, the system may be less successful because the change in detail alters the problems of presentation. If, for example, such a system is found to be well suited to scales from 1:200,000 to 1:1,000,000, as for instance the system proposed by Gaussen (1949), then it cannot be expected to do as well on the map of the Elbe Valley at 1:5,000 (Walther, 1957) because the entire map would have but a single color.

Colors come in a variety of tints. For instance, red occurs in carmine, rose, pink, lake, vemilion, and others, and these tints, in turn, can be varied by using them flat, ruled, stippled, or in some other fashion. In this manner, it is possible to correlate colors with vegetation features by varying their character. For example, various tints of green may be employed for deciduous forests according to what species dominate the forest (Küchler, 1956b).

Rübel (1916) suggested that colors be selected in such a manner that they correspond to the actual colors of the vegetation. This same idea was much later tried again by Burke and Sheldon (1953). But such an approach is rarely feasible because it ignores the seasonal color changes of the vegetation as well as a large number of details.

Lüdi (1921) introduced an interesting idea by correlating colors with succession. Pioneer stages are shown in shades of grey, seral communities in shades of dull green, and only the various climax types have their own different and uncoordinated colors. Anthropogenic vegetations, as illustrated by certain types of meadows, are shown on the map in a very vivid green, thereby indicating that they are neither seral nor climax communities and basically different from natural phases of succession. Ideas such as Lüdi's can be developed to meet a great variety of needs.

Kuhnholtz-Lordat, in his historical approach to vegetation mapping at very large scales, has devised his own color scheme, although he was strongly influenced by the work of Gaussen, who works at much smaller scales (Marres, 1952). A local color scheme relating the various phytocenoses to their respective substrata was developed by Duvigneaud (1961). Applied only on maps at 1:50,000 and in a limited region, the number of phytocenoses is relatively restricted, yet the author was able to adapt Gaussen's ecological ideas to his area with a considerable degree of success.

It is always best to select light colors. They are usually more pleasing, make a map more legible, and are especially useful in connection with overprinted symbols. The latter are difficult to recognize if they are printed on a dark color.

Color-number combinations

If the number of colors on a map is large, it becomes difficult to distinguish between them, especially when the eye must go back and forth between the legend and the map. In such a case, it is best to number the legend items and repeat these numbers in the respective areas on the map. Any doubt about the color can then quickly be dispelled by referring to the number. Such numbers are not symbols in the strict sense, but they compete with the symbols for space and attention. Küchler (1953a) presented a vegetation map of the United States on which he used colors for the physiognomy of the vegetation, and black overprinted number symbols to indicate the dominant genera. In this manner the physiognomic and floristic aspects of all vegetation types were kept strictly separated and could everywhere be identified with ease.

In a later publication, Küchler (1964) again combines various unrelated colors with numbers. However, this time, he is noticeably influenced by Gaussen. While insisting on a free choice, he admits that Gaussenian ideas have affected his selection of colors. On this, his second map of the United States, the colors have therefore technically no other purpose than to assist the reader in distinguishing one vegetation type from the next. Supposedly, it is incidental that there are parallels between Küchler's choice of colors and Gaussen's system. The influence of the French master is nevertheless striking.

Similar principles have been adopted by Soviet vegetation mappers on their small-scale maps (Lavrenko and Sochava, 1956; Lavrenko and Rodin, 1956).

Gaussen's use of color

The most advanced and refined method of selecting colors yet devised is that by Gaussen. His first masterpiece was his map of France in four sheets at 1:1,000,000 (Gaussen, 1945); the delicacy of the colors and the manner of their arrangement make it a joy to behold this map, especially if the four sheets are mounted together. But Gaussen is probably known best (and rightly so) for his map series at 1:200,000 of France and some areas overseas. The originality of this cartographic method consists in giving ecological values to the colors: blue represents moisture, black implies shade, etc. By superposition of well-chosen colors it becomes possible to present on paper a certain picture of the complexity of the ecological factors. When two factors have analogous physiological effects they are shown by similar colors. If the preponderance of one ecological factor becomes manifest in a much stronger color, the strength of the other colors (representing other factors) may be decreased correspondingly.

Gaussen (1949, 1961b) has given detailed instructions on the use of colors. By employing colors for physiologically effective environmental factors, he established a series of what might be termed 'plant climates' with implied natural vegetation. The color shows the more permanent aspect of the habitat, and the physiognomic character of the vegetation is shown by the manner in which a color is applied: flat for forest, rules for heath and other shrub, stippled for grassland, etc. The colors, therefore, have a triple function: they set off one type of vegetation from another; they indicate climatic conditions; and they show the physiognomy of the actual vegetation. Similar colors imply similar climatic conditions. The dominants are shown as letter symbols in black where the background is light and in white where the background is dark.

In transitions between two or more vegetation types, the colors of each type are given. The colors are arranged in alternating vertical bars, and the relative width of each such bar indicates the degree of dominance of one vegetation type over another.

Gaussen leaves all cultivated areas white, and on this white background, the messicol vegetation is shown in the form of symbols. These are usually letter symbols but not always. Letters are here used to great advantage, especially as they are the initials of the represented items. Their size and shape imply statistical values, and their color reveals the ecological affinities and ranges of the crops.

Gaussen therefore employs colors for multiple purposes, i.e. for the natural vegetation and its substitute communities, for physigonomy and floristic dominants, the cultivated crops and their ecological optima. Gaussen's imagination and genius for employing and organizing colors have never been matched, and authors of vegetation maps are urged to study Gaussen's instructions and his maps at 1:200,000 and 1:1,000,000 with the greatest care before embarking on their own color scheme.

Choosing appropriate color schemes

The significance of colors and their standardization or lack of it was well illustrated at the exhibition of vegetation maps held in connection with the International Botanical Congress at Montreal, Canada, in 1959. Küchler (1960a) made the following observations:

On many vegetation maps, the colors have no purpose other than to distinguish one vegetation type from the next, as on the maps of the Congo, e.g. the vegetation map of Kwango (Devred, 1955). Sometimes a certain reasoning can be detected in the selection and use of the colors, as for example on the three vegetation maps of southeastern Mount Desert Island, Maine (Küchler, 1956a), but here, too, the choice was essentially arbitrary. In contrast to his, Gaussen's maps of France reveal a carefully developed method of selecting colors so as to show more: the selection and application of colors inform the reader on the ecology and physiognomy of the vegetation. Thus these vegetation maps were greatly enriched.

The following examples will illustrate some of the contrasts in the use of colors on the exhibited vegetation maps.

1. The map of the Soviet Union (Lavrenko and Sochava, 1956) and of the Labrador Peninsula (Hare, 1959) both show the vegetation of the Far North with tundra and types with sparse plant growth amidst deserts of rock and gravel. On the Russian map, blue colors have been chosen for these vegetation types, whereas on the map of Labrador the analogous types are shown in red and yellow.

2. The tropical rain forest is done in blue on the vegetation map of Africa (Keay, 1959); thus it has the same color as parts of the arctic tundra on the Soviet maps! Hueck (1960) presents the tropical rain forest of Venezuela in green whereas the color scheme of Gaussen requires it to be in purple.

A discussion of the choice of colors makes it quite clear that methods like that of Gaussen offer great advantages on maps at medium scales, even though it may be necessary to adjust them to different regions. In contrast to this

are the vegetation maps at large scales, where an analogous choice of colors is no longer feasible. Under the influence of local conditions and circumstances it becomes inevitable that the choice of colors and their use must differ more or less from the world-wide systems.

In general, it has so far not been feasible to develop a color scheme for vegetation maps that can be applied in all regions and at all scales. Indeed, such a scheme is not feasible because some maps show only vegetation whereas on others it may be linked to a host of unrelated features.

Insistence on one color scheme that must be applied to all vegetation maps is therefore futile and, in fact, quite undesirable. There is, after all, only one spectrum, and it must be applied to all the maps listed above and to all others. It is therefore quite inevitable that a given color will have different implications on different maps. This was demonstrated years ago at the Bundesstelle für Vegetationskartierung in Germany.

It is best that authors remain free to use colors as they wish. The Toulouse Colloquium (Gaussen, 1961a) urges them, no matter what plan they may wish to follow, to devise a method which is logical with regard to the distribution of colors as well as their intensities. The Toulouse Colloquium also recommended (Resolution no. 10) that on maps at medium and small scales the height and density of the vegetation be correlated with the colors. Accordingly, the colors should be darker, the taller and denser the phytocenoses are, and lighter, the lower and more open they are. Colored symbols on a white background are considered specially effective for all forms of messicol vegetation.

Whatever an author of vegetation maps may think of these recommendations, the following observations seem to hold true anywhere and for all types of vegetation maps: A systematic and well-reasoned approach to the use of colors is most desirable, and as far as is feasible, an author should keep his colors light. He should feel encouraged to enrich his colored maps with symbols but use them sparingly. Beyond this, he should permit himself plenty of time to experiment with colors and color schemes; he will find this both revealing and rewarding.

The geologists have solved the color problem for their maps by using a one-dimensional system as implied in the time sequence of the geological formations. Such an approach is not possible in vegetation mapping because phytocenoses cannot be arranged in a one-dimensional system.

Use of symbols

There have been several references to the use of symbols. As the name implies, these signs of different character symbolize what can be shown by other means only in a less satisfactory way, if at all. Symbols may indicate individual taxa, structural features or entire phytocenoses. What a symbol represents usually covers such a small area that it is not feasible to portray its exact outline. Often these areas are so small that they should be suppressed altogether. Whether a feature is shown by a symbol or in some other fashion is thus a question of scale. A symbol, of course, covers a certain area on the map, but it also implies that its true area stands in no relation to that on the map. It is therefore 'off-scale' and cannot serve as a basis for measurements.

The overprinting of black symbols on black-and-white patterns should be avoided as the reader finds it difficult to read and hence to use such a map. On a colored map, symbols may be black or in color. Black symbols are difficult to see on dark colors, and colored symbols are not easily distinguished from surrounding colors if the color contrast is weak. The author must therefore be very careful with the selection and placement of his symbols.

Gaussen (1936), Rübel (1916), Trochain (1961) and others have introduced black symbols of given shapes, and assigned one each to the most common species of their regions. Symbols of this kind are also used extensively on the Soviet vegetation maps by Lavrenko and Sochava (1956) and others. This idea would be even more useful if the phytocenological authorities had agreed on what symbols to adopt for which species, but such general agreement has not been reached so far. A map on which the vegetation is shown by symbols only is difficult to read. If, on the other hand, the vegetation is shown in color, a judicious use of black symbols can greatly enrich the map without lowering its legibility. Wieslander (1937), on his colored vegetation maps of California, uses black letter symbols for species. He thus shows the dominant species of every plant community, using letter combinations of capitals and lower case letters and thin lines above or below the letters to cope with the wealth of species. Colored symbols have been employed with great success by Gaussen (1948) but they are expensive. Küchler (1956a) and Troll (1939) have also employed colored symbols in order to portray the geographical distribution of some growth forms or taxa more graphically.

The spacing of symbols can be done in two ways. It may be irregular, possibly implying that the position of a symbol on the map corresponds to the actual location in the field. Where numerous stands of a plant community permeate an extensive vegetation type, it is clear and simpler to have the symbols spaced evenly thoughout the area involved. In neither case may the symbols be so crowded as to affect the legibility of the map adversely.

Symbols can be given statistical values by varying their character, size, and density (number per unit area). The Canadian Forest Service

Table 1. Informational use of colors and symbols on Canadian forestry maps.

Shown in color		Shown as overprinted symbols	
		Crown cover	Height
Softwood	above 30 feet		
Hardwood	above 30 feet	A: 10– 30%	1: up to 30 feet
Mixed	above 30 feet	B: 30– 50%	2: 30 to 50 feet
Softwood	below 30 feet	C: 50– 70%	3: 50 to 70 feet
Hardwood	below 30 feet	D: 70–100%	4: 70 to 100 feet
Mixed	below 30 feet		

(Anonymous, 1948a) and Gaussen (1948) have offered good examples of this method (Table 1).

Symbols may represent taxa which occur sparsely while colors show these same taxa when they form the bulk of the vegetation. For instance, a spruce forest may be shown by a green color, but in addition, the legend can show spruce by a symbol. If, then, an area is covered with a spruce forest, it appears on the map in green. If, on the other hand, a beech forest has a significant admixture of spruce, then the beech forest is shown by its own color with spruce symbols overprinted on it. Of course, the use of symbols is by no means limited to taxa; symbols may just as well be employed to indicate the occurrence of plant communities, especially those which cover small areas but are scattered through a larger formation at relatively high frequency. In this manner, Küchler (1977) showed the location of the individual groves of *Sequoiadendron giganteum* on the western slopes of the Sierra Nevada in California.

To assure the legibility of the map, care must be taken to use symbols sparingly so as not to overload the map or to overshadow the colors on which they are printed. A cluttered map is difficult to read and its usefulness is correspondingly impaired. But a vegetation map can be greatly enriched by symbols, especially when these are well selected and sparingly used. A good example for a judicious use of symbols is the map of Central Asia (Lavrenko and Rodin, 1956).

In a very real sense, patterns and colors determine the character of the vegetation map. In a way, they 'make' the map. On a colored map, the colors are the most obvious feature of the map. With such prominence, colors should be used for the most prominent items to be portrayed. Patterns and symbols are secondary and should therefore be employed to designate items of secondary importance.

When an author must decide on a color scheme for a new vegetation map, he will find it wise to study first the use of patterns, colors, and symbols on other vegetation maps. He will thus become aware of a great variety of possibilities in applying them. But not every vegetation map is a model. The author can therefore learn from past mistakes in judgment and execution, too, and so, perhaps, avoid these on his own map.

11. The Legend: Organizing the Map Content

A.W. KÜCHLER

Forms of legends

The organization of the map content implies an orderly arrangement of the various types of vegetation and their characteristics. This arrangement is presented in the legend of the map. A legend portrays the content of the vegetation map, which reveals the location, extent and geographical distribution of the individual types of vegetation horizontally and often also vertically.

A vegetation map cannot portray all aspects of plant life in the landscape. An author must therefore select one or several features of the vegetation, e.g. certain physiognomic aspects or perhaps the dominant species. The author will present the selected features throughout the map, preferably to the exclusion of all others.

In order to organize the map content, the vegetation mapper begins by establishing just what sort of items are to be shown on the map. He makes a complete list of all items, reviews them carefully, and then proceeds to group them meaningfully.

For instance, if the natural vegetation is to be portrayed, the mapper can establish a number of physiognomic groups such as broad-leaved deciduous forests, needle-leaved evergreen forests, various types of grasslands, shrub formations, etc. Each group consists of one or more communities that may be described floristically. By reviewing his list of vegetation units critically, the mapper quickly discovers what groups can be established, and into which of these he must place every individual unit of vegetation.

Depending on the purpose of the map, the same material may be organized in different ways. For instance, ecological factors may be the main concern, and therefore the major groups are established with this in mind. However, where environmental features are considered secondary in importance to other features, they are relegated to subdivisions of the major groups. It is possible, for instance, to establish a few major groups, each described as a forest type, a grassland type, etc., and within these groups arrange the items ecologically, requiring more or less heat, water, etc., as the case may be. In other circumstances, the major groups may be distinguished on the basis of some ecological factor, e.g. water. In such cases, the major groups range from lake and river vegetation via swamp vegetation and communities periodically flooded, to those with a permanently high water table, a medium high water table, and finally a low water table or one subject to great seasonal fluctuations. The different communities of the region on the mapper's list are then placed in the appropriate ecological group, and the vegetation of the entire area has been presented in an orderly fashion, and according to criteria established a priori. Thus it is possible to show major features of the vegetation in a manner that permits the reader at once to grasp the significance of each group and, within the group, of each community. Zonneveld's (1960)

A.W. Küchler & I.S. Zonneveld (eds.), Vegetation mapping. ISBN 90-6193-191-6.

ecological diagram applies this principle in a visual manner.

Some legends are mere lists of a number of items denoting types of vegetation. This is the simplest form of composing a legend. If the legend is short, then this is undoubtedly the most convenient and adequate form of a legend. It is employed frequently.

When a legend is longer, it becomes increasingly desirable to arrange the legend items in a

Table 1. The natural vegetation of Kansas.

A. Prairie
   1. Short grass prairies
      a) Northern grama-buffalograss prairie (*Bouteloua-Buchloe*)
      b) Southern grama-buffalograss prairie (*Bouteloua-Buchloe*)
   2. Mixed Prairies
      a) Bluestem-grama prairie (*Andropogon-Bouteloua*)
      b) Chalkflat prairie (*Andropogon-Bouteloua-Distichlis*)
      c) Alkali sacaton prairie (*Agropyron-Distichlis-Sporobolus*)
      d) Salt marsh (*Distichlis-Suaeda*)
      e) Cedar Hilis prairie (*Andropogon-Bouteloua-Juniperus*)
   3. Tall grass prairie
      a) Bluestem prairie (*Andropogon-Panicum-Sorghastrum*)
   4. Sand prairies
      a) Sandsage prairie (*Andropogon-Artemisia-Calamovilfa*)
      b) Sand prairie (*Andropogon-Calamovilfa*)
B. Forest
   5. Oak-hickory forests
      a) Oak-Hickory forest (*Quercus-Carya*)
      b) Ozark forest (*Quercus-Carya*)
      c) Cross Timbers (*Quercus-Andropogon*)
   6. Floodplain vegetation
      a) Floodplain forest and savanna (*Populus-Salix*) incl. freshwater marsh (*Spartina*)
C. Mosaics, transitions and boundaries
   7. Transition between No. 2a and No. 3a
   8. Mosaic of No. 3a and No. 5a
   9. Boundaries
      a) Western red line: western boundary of No. 1a with western wheatgrass (*Agropyron smithii*) as a codominant
      b) Eastern red line: western boundary of prairie with significant forest islands
D. Lakes and reservoirs

manner that permits the reader to understand more readily which vegetation types belong together, which are divisions or subdivisions of others, etc. For example, Küchler (1974), on his vegetation map of Kansas, arranged the various types, their divisions and subdivisions in this manner (Table 1). Each unit is then described in greater detail in an accompanying text.

On this basis, map legends can become complex, listing certain items which are then divided; the divisions are further divided, and even such divisions can be further divided and subdivided. An appropriate organization of the legend becomes therefore a guide for the reader, and the more clearly the legend is organized, the easier it will be for him to find his way through the vegetation of the mapped area. Excellent examples of vegetation maps with complex legends are published in the Documents de Cartographie Ecologique, e.g. Dobremez et al. (1976), Richard (1978), Pautou et al. (1979) and many others.

Complexity

The preparation of a good vegetation map costs much time, effort and funds. Mappers therefore strive to make their maps as useful as possible. This means that they often attempt to show as much as the scale permits. This may be laudable but it requires that the mapper knows exactly what to present on his map and how to organize it. Otherwize, the map is overloaded with detail, cluttering up the map content, and making the map difficult to read. This has a discouraging effect on the prospective readers who will soon lay the map aside. Complex detail at the cost of legibility can sharply reduce the value of a vegetation map. Mappers must therefore experiment with their manuscript map in order to discover how far they may go in presenting complex details. There is no sharp boundary between legible and cluttered maps, and the mapper has to use his judgment if he wants to produce the best effects.

The permissible degree of complexity of vegetation maps depends only in part on the scale, and very considerably on the ability of the author to organize his material. In addition to skill, the organization requires an appropriate terminology and a clear grasp of the purpose of the map. If these prerequisites are not met, irrelevant material will needlessly complicate the map. The various prerequisites for good organization are easily underrated or ignored and the value of the map thereby depreciated unnecessarily. The need for thought and experience can hardly be overemphasized, and authors should always limit the scope of their offerings to such an extent that whatever they show is readily seen and appreciated by the reader of the map.

The degree of complexity can vary enormously on maps of similar scale; this is best illustrated by two examples:

1. The Distribution of Forests in the Upper Peninsula of Michigan (1:250,000; Anonymous, 1941b) is a very simple map. The forests are shown in the following manner:
1. northern hardwoods, saw timber;
2. northern hardwoods, young growth;
3. mixed hardwoods and softwoods;
4. pine;
5. conifer swamps;
6. aspen brush;
7. agricultural land.

In the legend, a number of dominant tree genera follow each forest type parenthetically. No variations in the floristic composition within these groups are shown. Seven different flat colors appear on the map, each one representing a type. As the description of the forest types is limited to the mention of a few genera, there exists no particular problem in showing each type on the map. All one has to do in mapping the vegetation is to name the forests according to one of the established categories. Greater simplicity can hardly be expected of a vegetation map at this scale.

2. The second example, showing the highest degree of complexity, is the Carte de la végétation de la France (1:200,000), feuille Perpignan (Gaussen, 1948). The following information appears on this map: the natural vegetation is shown in 41 categories; floristic distinctions are made both by genera and by species; growth forms from trees to dwarf shrubs appear in nine different height classes, in addition to herbaceous and aquatic vegetation. All these are grouped into series of dynamic character with facies as subdivisions; several related series are combined into zones (étages) which are climatic in character. The cultural vegetation (land-use) is shown in 42 categories ranging from different types of afforestation to truck farms and individual crops. In every canton, land-use is shown in four major types, each of which is indicated according to the percentage of the area it occupies; there are five percentage classes. Nine kinds of fruit trees are shown in groups of 1,000, 10,000 and 100,000 each; ten other crops appear, in five percentage classes, and the ecological affinities are given for both orchard and other crops. Vineyards are indicated separately by symbols representing 10 ha each. Finally, there are 16 individual 'botanical curiosities and significant sites'. In addition, the nature of the environment is shown, based above all on temperature and precipitation.

It is obvious that such a bewildering amount of information cannot be shown on one map of this scale without rendering the map completely illegible, unless it is subjected to the strictest organization. Thus the natural vegetation is shown in variously screened colors and overprints whereas the cultural vegetation is shown by symbols which have statistical values and each of which has a significant size, shape, and color. To this may be added the number of symbols per unit area. Even then, it required the keenest imagination and skill to keep the map on a level of usefulness. That we have in the Perpignan map proof that this can be done so well is, of course, gratifying and propitious. Yet, we must understand that to reach such heights of achievement is the exception rather than the rule, even as not every painter can be a Rembrandt. On the other hand, the way has now been shown, and it may be followed by less imaginative minds, too.

There are, of course, other vegetation maps which reveal a very carefully planned organization of the map content. Models among these are the maps emanating from the Bundesstelle für Vegetationkartierung in Germany, and the map of Leonberg by Ellenberg and Zeller (1951).

Godron et al. (1964) proceed schematically as do Ellenberg and Zeller, but their legend is much more complex. They also show how the various units of vegetation are related to moisture and the effects of man.

Consistency

Organization implies consistency. A consistent organization of the map content places all parts of the map on the same basis and permits comparisons, both within the scope of the map and on maps of other areas with a homologous vegetation, and even with maps showing climate, topography, soils, or geology.

Once the mapper has selected an approach to mapping and decided on the features of the vegetation he wants to portray on the map, he must then express his observations accordingly. If the dominant species are to be mentioned, they must be mentioned in every plant community on the map. It is not appropriate to list the dominants for some communities and then elsewhere on the map include items like meadow, deciduous forests, chaparral, etc. Similarly, if the structure of the vegetation is to be shown on the map, it must be shown throughout the area covered by the map. Only then can one part of the map, or one community, be compared usefully with other parts or communities so that the similarities and contrasts really become clearly evident.

Sometimes it is desirable to show items which clearly differ in character from all others on the map. For instance, on his well-organized map of Perpignan, Gaussen shows a number of 'botanical curiosities'. Items of this kind should be shown in such a manner that the reader is immediately aware of their exception-al character. In the legend, these items can be placed in a group by themselves, preferable at the end, and on the map they should appear as a boldly different overprint. In any case, the number of such exceptional items should be kept to a minimum.

It may not be possible or even desirable to be entirely consistent at all times, but in general it is true that greater consistency implies a greater usefulness of the vegetation map.

Some authors may find these remarks superfluous because they take consistency for granted. Examples of maps which are entirely consistent are those by Gaussen (1948), Küchler (1953a), Schmid (1948), Tüxen (1956a) or Wieslander's blue-line series of California (e.g. Cushman and Lusk, 1949). On these maps, consistency implies that all parts of a given map are comparable. If a certain aspect of the vegetation is not shown in a section of a map, then the reader knows that it has not been shown because it is absent, and not because, although present, another aspect was shown instead, for whatever reason. In other words, all vegetational features to be considered appear on the map wherever they occur.

There are numerous vegetation maps on which a consistent organization of the map content is lacking, and the character of the map is thereby changed completely. If the material is not organized with any degree of consistency, different parts of the map are not comparable and the interpretation of the map may lead to erroneous conclusions. Two examples will suffice to illustrate this point.

1. The map of the United States (Shantz and Zon, 1923) is not consistent in that the vegetation is shown sometimes by species, at other times by genera; elsewhere on the same map the vegetation is shown by growth forms, regional affinities, or according to ecological controls. This approach can be very descriptive but makes a comparison of any one part of the map with another impossible.

2. Another more recent example of a lack of consistency is found on some Swedish vegetation maps where the 'classification system for

vegetation mapping' is applicable in the central and southern parts of Sweden (Ihse et al., 1975). In this classification, there are four physiognomic classes plus one of urban environments and agricultural lands. The latter is not comparable with the others and stands by itself. The four physiognomic classes are divided into 27 vegetation units which are further subdivided into 42 lower categories. While the four major physiognomic classes are consistent, this consistency is wanting in the divisions and subdivisions, where the items are described by genera, by species, physiognominally or ecologically. Some items have little or no meaning, e.g. 'other trypes', 'other species', or 'other deciduous species'. Deciduous is used in contrast to floristic identification or physiognomic descriptions, leaving the reader to wonder whether these types are evergreen (which they are not). Rich fens are distinguished from poor fens without indication of what richness means, etc.

The classification of vegetation as applied on these vegetation maps is based on aerial photography. It gives the impression that the ground truth was not established, or else the classification could have been more explicit. This also explains such items as 'clear felled areas'. The fact that trees have been cut is of passing interest as such conditions are usually temporary, but it would have been significant to indicate what type of forest was felled. One is reminded of the vegetation maps of Oregon and Washington where 'burns' only indicate that there has been a fire without saying what type of forest was destroyed. Aerial photography is so advantageous that one can hardly do without it. But only after establishing the ground truth does it really become meaningful. Long (1974) emphatically expressed a similar opinion. The vegetation may remain inadequately known in spite of extrapolation when the possibilities for establishing the ground truth are too limited. The mapper should indicate this state of affairs wherever applicable.

Standardization

Standardization implies that a method has been found which can serve the many purposes of vegetation mapping, which is applicable to many types of vegetation, and which permits an ideal degree of standardization on all continents. Such a standardization is highly desirable because, in a sense, all phytocenologists will be speaking the same language, and semantic confusion, one of the most difficult problems of our times, will cease to exist. For example, Braun-Blanquet's method of classifying vegetation is successfully applied in various parts of Europe, has been introduced successfully in Japan and has been applied, though with less success, in other parts of the world. The advocates of this classification appear to feel that it is only a matter of time before all phytocenologists and vegetation mappers throughout the world will employ this method of describing and mapping vegetation.

However, an objective appraisal of many vegetation maps leads one to believe that the ideal degree of standardization is not likely to be achieved. If we compare the vegetation maps by Braun-Blanquet and his followers, it becomes at once evident that this strict standardization used in classifying vegetation is not applied on vegetation maps. Men like Emberger, Molinier, Lüdi, Tüxen, Zonneveld and even Hueck, the very leaders of European vegetation mapping all profess to be followers of Braun-Blanquet. And yet, their maps are not at all alike. One need only hold side by side such maps as that of Montpellier by Braun-Blanquet and Tchou (1947), of Aix by Molinier (1952), of Schlitz by Seibert (1954), of Baltrum by Tüxen (1956a), and of the Riesengebirge by Hueck (1939). It would be difficult to speak of any degree of standardization.

It is not necessary here to make a detailed study of the differences. But it is important to observe that some of the finest examples of vegetation maps that emerged from the school of Braun-Blanquet are quite unlike in character, leading to the conclusion that standardization on vegetation maps is not likely to be achieved simply because the intellectual approach to the classification of vegetation is the same.

Standardization aids communication and makes vegetation maps comparable. It should be promoted. But it should not be adopted to the exclusion of different non-standardized methods. A new and original approach to mapping can reveal new problems and thus become highly stimulating. Anything that will aid in research and lead to new ways of progress must be given a fair chance to evolve. Ozenda (1961a, 1961b) has expressed similar thoughts. The general trend of thinking among the world's leading vegetation mappers seems to be toward standardization. It may therefore seem regrettable that the vegetation maps of our time show no more uniformity than they do, i.e. practically none at all. Even one and the same author will employ different methods to map vegetation at different times or under different circumstances.

In fact, however, such divergence is to be welcomed rather than regretted. If the limitations of our methods to classify and map vegetation are liabilities, let us turn them into assets. It may not be quite so desirable to attain the degree of standardization which at first appeared so attractive because different approaches and different methods stimulate the imagination. They allow a greater degree of originality; they also permit a greater flexibility with which to adjust a vegetation map to varying local conditions.

Authors of vegetation maps are encouraged to become thoroughly familiar with all major methods of mapping vegetation, and then review critically their particular needs as dictated by the purpose of their vegetation maps. In the light of such considerations, they will be able to study the vegetation of their area and devise a method of mapping it which will do a maximum of justice to the readers of their maps as well as to the vegetation and to their own originality. This may be done by utilizing one of the existing methods, perhaps with modifications which local conditions make desirable. Or else,

it is possible to develop a new and different method incorporating the ideas of several other authors, and of course, the door is always open for an entirely fresh and original approach.

In order to prepare a vegetation map, the mapper should consider all the means at his disposal to portray the vegetation in such a manner that the map will best serve its purpose. Although the ways in which he can present the vegetation on the map are limited, they are nevertheless so numerous, especially if used in combination, that he must make his selection with great care. A variety of mapping methods is discussed elsewhere in this book, but all methods rest on certain basic features of the vegetation, and it is these that must be considered, and the extent of their use weighed judiciously, before actual mapping can begin.

There are six such basic features or groups of features but the mapper will be quick to realize that each of these features is more or less complex, that parts of one may be combined with parts of another, and that therefore his means of presenting the vegetation must be flexible. This flexibility permits him to adapt his map to any one purpose or to certain groups of purposes. The degree to which the mapper succeeds in serving his purpose depends to a considerable degree on his organizational skill.

The first of the basic approaches to a description of the vegetation is to use its physiognomy and structure. The physiognomic approach can range from the broadest descriptions: geographical terms such as heath, meadow, savanna, etc., to the most detailed structural analysis of growth forms, their height, density, and layering (Küchler, 1956b, map I).

The second basic feature, the floristic composition is in part a matter of detail (scale), and to a considerable degree a matter of judgment. One approach lists the dominant species, and this may be contrasted with the community approach. Species may also be grouped according to their areal affinities or in some other manner. The mapper must be familiar with the existing possibilities. To insist on one method because he has always used it or because he has not thoroughly acquainted himself with others implies a limitation that can only be detrimental.

The third basic approach to mapping the vegetation is to present it as the natural vegetation, be this actual or potential. This is usually described physiognomically or floristically, or preferably both. For many purposes, a map of the natural vegetation is particularly valuable, and the most recent methods permit the construction of such a map for certain areas with a reasonable accuracy. This accuracy, however, depends somewhat on the regions, on the experience of the mapper, and on the time and information at his disposal. In the light of these arguments, the mapper must decide whether under given circumstances he can prepare a map of the natural vegetation with such a degree of accuracy that his effort will be worthwhile, and the result of his endeavor will meet all requirements.

The fourth basic means to show vegetation is to present the cultural vegetation types or replacement communities, i.e. the actual vegetation. These may be considered the derivatives and counterparts of the natural vegetation. They can be shown in a great variety of ways and may include statistical values. The purpose of the map is here the determining factor, as usual. Maps of such substitute communities are usually made for 'practical' people, such as silviculturists, range managers, planners, etc. However, this should not detract from the fact that maps of the cultural vegetation can also be valuable for purely academic purposes, especially as a basis for research.

The fifth basic means to show vegetation on a map is to employ the phytocenoses as expressions of the environment. Some vegetation mappers object to this approach because it includes features that are not vegetation. However, these 'ecological' vegetation maps have been found useful and popular (Küchler, 1984). It is, of course, not possible to combine vegetation with all environmental factors, and the mapper is forced to be selective. He usually has two choices. One is to introduce on his map

those environmental features that are of particular significance for a given type of vegetation. In the case of marshes the dominant environmental feature is the high water table, in the case of deserts it is lack of rainfall, on alpine meadows it is the short growing season, and so on. Such maps are often highly descriptive. But it must not be overlooked that many environments are not so overwhelmingly dominated by a single factor. References to the environment are then usually omitted. To ignore the environment when elsewhere it was stressed may leave the reader wondering. But the omission at least implies that the site qualities are relatively

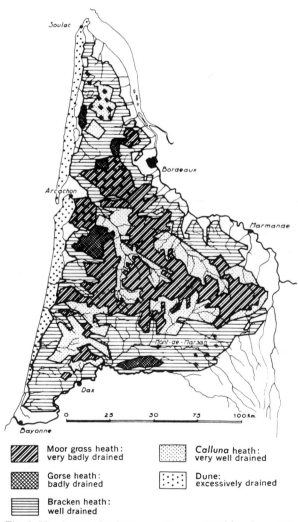

Fig. 1. Heath types in Gascogne, France, resulting from differences in drainage (after Emberger, Gaussen and Rey, 1955).

balanced in their effects on the plant communities.

The other choice for the vegetation mapper lies in selecting one particular environmental feature and relating all plant communities to it. This method is particularly useful with regard to various aspects of water, be that precipitation, drainage (Fig. 1), depth and fluctuation of the water table, duration of snow cover (in mountains), or others. Although such maps are limited in their uses, they can be very valuable. See Chapters 33 and 34 for further comments on 'ecological vegetation maps'.

The sixth and last feature of the vegetation the mapper must consider is its dynamism. It is, of course, much simpler and less time-consuming to ignore the dynamic qualities of vegetation altogether and to map only what can actually be observed in the field at the time of mapping. This is the usual procedure and can be entirely adequate and satisfactory. There are, however, instances where information on the dynamic aspects of vegetation is desirable, even necessary. It is not difficult to organize the vegetation on the basis of its dynamism because all plant communities are either climax communities, or members of a sere evolving toward a climax. Assignment of a plant community to its proper sere, and possibly to its place in the sere, is the chief problem. Although the problems are not numerous, they are not therefore solved easily. Mapping the successional aspects of vegetation is discussed in Chapter 23.

The six complex approaches to showing vegetation on a map permit a very complete treatment of the subject. But it is not feasible to show all features of vegetation and it is important that the right selection be made. The information must be available and it must serve the purpose of the map. How the features are combined on the map is left to the imagination of the mapper. Sometimes it happens that he finds it difficult to organize his material so as to get it all on the map. In that case he may want to consider printing part of his material on transparent paper which can be overlaid on the map.

Alternative approaches

Dansereau (1961) illustrates five different approaches to mapping vegetation (Fig. 2A–E). On each of these maps all phytocenoses are expressed and shown according to one method. If, for instance, the mapper wants to show the vegetation floristically and as members of some formation, he will want to consult Fig. 2A and C. His solution might be to show the formations in six different colors and overprint 30 types of symbols in the appropriate places to show the floristic features. Dansereau uses dynamic, i.e. successional, phases of formation to

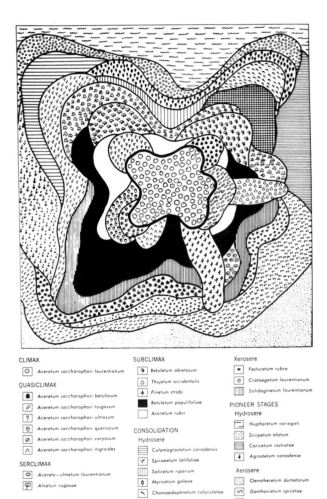

CLIMAX
☐ Aceretum saccharophori laurentianum

QUASICLIMAX
☐ Aceretum saccharophori betulosum
☐ Aceretum saccharophori tsugosum
☐ Aceretum saccharophori ulmosum
☐ Aceretum saccharophori quercosum
☐ Aceretum saccharophori coryosum
☐ Aceretum saccharophori nigroides

SERCLIMAX
☐ Acereto – ulmetum laurentianum
☐ Alnetum rugosae

SUBCLIMAX
☐ Betuletum abietosum
☐ Thujetum occidentalis
☐ Pinetum strobi
☐ Betuletum populifoliae
☐ Aceretum rubri

CONSOLIDATION
Hydrosere
☐ Calamagrostetum canadensis
☐ Spiraeetum latifoliae
☐ Salicetum riparium
☐ Myricetum galeae
☐ Chamaedaphnetum calyculatae
☐ Piceetum ericaceum

Xerosere
☐ Festucetum rubra
☐ Crataegetum laurentianum
☐ Solidaginetum laurentianum

PIONEER STAGES
Hydrosere
☐ Nupharetum variegati
☐ Scirpetum elatum
☐ Caricetum rostratae
☐ Agrostetum canadense

Xerosere
☐ Oenotheretum dumetorum
☐ Danthonietum spicatae
☐ Trifolietum repentis

Fig. 2A. The thirty most characteristic plant associations of the Montreal plain, assembled in a model that shows their actual contacts (after Dansereau, 1961).

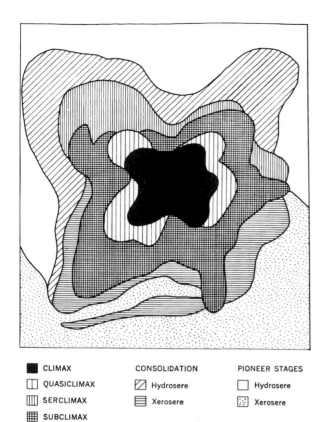

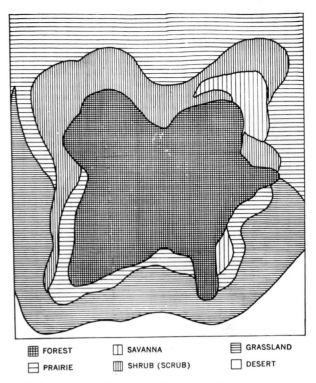

■ CLIMAX	CONSOLIDATION	PIONEER STAGES
☐ QUASICLIMAX	▨ Hydrosere	☐ Hydrosere
▥ SERCLIMAX	▤ Xerosere	⊡ Xerosere
▦ SUBCLIMAX		

Fig. 2B. The associations of Fig. 2A, classified according to their dynamic position (after Dansereau, 1961).

▤ FOREST	☐ SAVANNA	▤ GRASSLAND
⊟ PRAIRIE	▥ SHRUB (SCRUB)	☐ DESERT

Fig. 2C. The associations of Fig. 2A, classified according to the type of formation to which they belong (after Danserau, 1961).

establish his major vegetation types. As a result, his map content is strictly organized: every phytocenose is shown both floristically and in its proper formation, which was the goal. But his map says nothing about ecological controls, or other aspects. To introduce any one of these would have added irrelevant material that might have reduced the legibility of the map or detracted in some other way. The mapper can also combine parts A and E, or C and E, or select some other combination. But if he decides to combine three approaches, he must experiment patiently in order to obtain a clear and unequivocal map.

On maps at very small scales, the number of features basic to the organization should be kept small, preferable not more than half a dozen or so. But on large-scale maps, an extraordinary variety of detail can be shown if the map content is well organized.

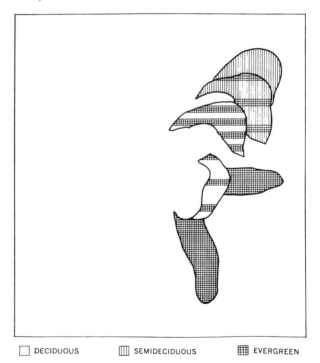

☐ DECIDUOUS	▥ SEMIDECIDUOUS	▦ EVERGREEN

Fig. 2D. The associations of Fig. 2A, classified according to their periodicity (after Dansereau, 1961).

130

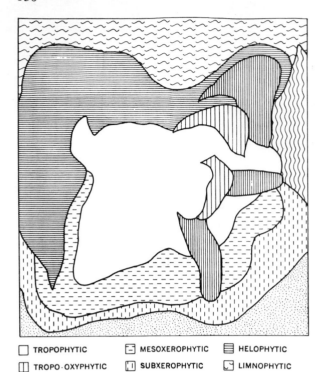

☐ TROPOPHYTIC	☐ MESOXEROPHYTIC	☐ HELOPHYTIC
☐ TROPO·OXYPHYTIC	☐ SUBXEROPHYTIC	☐ LIMNOPHYTIC
☐ OXYPHYTIC	☐ PSAMMOPHYTIC	☐ OXYHYDROPHYTIC

Fig. 2E. The associations of Fig. 2A, classified according to the dominant ecological control (after Dansereau, 1961).

A strict organization of the map content should result in a map on which the reader can see at once not only that there are different types of vegetation, but also which of these are related and which are not. This may best be accomplished by a judicious use of colors. Thus, several related communities, perhaps all within a given group, are shown by various shades and hues of the same color, whereas a very different and quite unrelated group will be shown correspondingly in a different color. If a vegetation mapper wishes to show both the natural and the cultural vegetation on one map, the most effective solution is to show each of the various phytocenoses of the potential natural vegetation with its own color. Usually there are several types of actual vegetation for each type of potential natural phytocenose. The different types of actual vegetation are then shown in the form of symbols or line patterns overprinted on the potential natural vegetation. This method permits the mapper to show both natural and cultural together. In further reveals which types of cultural vegetation are tied to particular natural phytocenoses, and with what degree of consistency this is so. The natural vegetation is actual rather than potential where a color has no overprint (e.g. Lauer and Klink, 1973). Kalkhoven (1976) prepared a map of the potential natural vegetation of the Netherlands at 1:200,000 using colors for the units of the potential natural vegetation. These units are then subdivided according to the actual vegetation types.

Constructing legends

The legend is the key to the map. It gives the map its meaning, reveals the number and variety of vegetation types and their mutual relationships (if any). It is always an integral part of the map and therefore requires much thought and care. The mapper will find it useful to study a large variety of legends; this will assist him in devising the best possible legend for his own map. The legends, of course, present the map content. It has already been pointed out that the legend should reveal how the map content is organized and that it must consist of terms which are clear and unequivocal to the reader. There is, in addition, the possibility of manipulating the legend items in such a manner that the author can thereby affect the organization, the use of colors, the length of the legend, etc. An example will illustrate some possibilities.

In a landscape to be mapped there occurs a grassland type of vegetation, pure stands of junipers, and grassland sections over which junipers are thinly scattered. There are here three types of vegetation and they may be shown as such (Fig. 3A). It is also possible to select two patterns that may be combined. In that case, one pattern is used for the grassland, one for the junipers, and where the junipers occur in the grassland, one pattern is superimposed over the other (Fig. 3B). In this instance, it would be quite feasible to omit all boundaries and there-

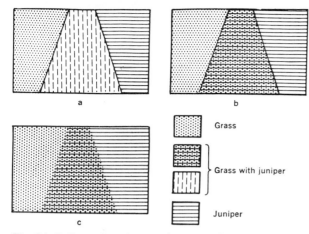

Fig. 3A–C. Examples of presenting vegetation on a map.

by produce an effect of a more gradual change from one type to the next (Fig. 3C).

If the junipers occur not only as a pure stand and in the grassland but also in other types of vegetation, e.g. sagebrush, it may be better to add a symbol for juniper to the legend. The juniper is then shown by its pattern where it occurs as pure stands and as a symbol where it grows in a matrix of grass or sage.

It is also possible to show the junipers only as a symbol, even where they form pure stands. In this last case, there are only three basic types and hence patterns: (1) grasslands, (2) sagebrush, and (3) juniper. But in fact, there are five types, and the additional two are (4) grassland with junipers and (5) sagebrush with junipers. However, the last two types do not require separate patterns. The most logical arrangement of these items in the legend would be as follows:

1. junipers
2. grassland
3. grassland with junipers
4. sagebrush
5. sagebrush with junipers

If the mapper prefers to show the close relationship between the grasslands with and without junipers, etc., his arrangement of the legend should be rather as follows:

1. junipers
2. grassland
   a) same with junipers
3. sagebrush
   a) same with junipers

Correspondingly, the mapper can separate all types on his maps by boundaries. Or else, he can omit those between items 2 and 2a and between 3 and 3a. The latter way will be the most pleasing and also appear most 'natural' and hence convincing.

The term 'grassland' is physiognomic whereas 'junipers' is floristic. To be consistent, the grassland should be described floristically, too, and the genera of the grasses should be given, for example 'wheatgrass and needlegrass' or else 'grassland of *Agropyron* and *Stipa*'.

If junipers, grass and sagebrush occur together, it may be that they represent a transition. If this is the case and the author wishes to further refine his map, he can divide the transition according to dominance: grass with sagebrush and sagebrush with grass, etc. In such a case, the map legend should read as follows:

1. *Junipers* woods
   1a) same, with *Agropyron* and *Stipa*
   1b) same, with *Artemisia*
   1c) same, with *Agropyron, Stipa,* and *Artemisia*
2. *Agropyron-Stipa* grassland
   2a) same, with *Juniperus*
   2b) same, with *Artemisia*
   2c) same, with *Junipers* and *Artemisia*
3. *Artemisia* shrubs (sagebrush)
   3a) same, with *Agropyron* and *Stipa*
   3b) same, with *Juniperus*
   3c) same, with *Agropyron, Stipa,* and *Juniperus*

In such a detailed legend, it is best to have one color each and also one symbol each for numbers 1, 2, and 3. Item 2b will then be printed in the color of no. 2 with the symbol of no. 3 overprinted. In this manner, items like 1a and 2a cannot be confused. The difference between 1a and 2a is clear: 1a represents a forest, perhaps rather open, with some grass between the trees. The over-all impression of this type is that of a forest. In item 2a, on the other hand, the trees are scattered in a matrix of grass. It is the latter which dominates the landscape and the junipers seem supplementary.

The author must establish criteria to determine the degree of density or coverage of each of the basic items (grass, juniper, sagebrush); only then can be hope to assign the various combinations to their appropriate legend category.

If his mapped area shows all these entities, his legend will correspond to the complete table. But it may well be that some items do not occur in the mapped area. In such a case, the author will find it wise to set up such a table anyway, on a separate sheet of paper. This will assure him of a logical and clear arrangement of the possibilities. From these he can select the items that occur in his area and present them in their proper sequence.

The legend of a vegetation map may take the form of a diagram if it is to express the relationship between some major features of the vegetation with some features of the environment. Such a diagram may improve the readability and hence the appreciation of the map content on the part of the reader. In this manner, Zonneveld (1960) related the vegetation of the Biesbosch in the Netherlands with geomorphologically determined soils, with soil-hydrological factors and the tides. Similarly, vegetation maps issuing from the Tropical Science Center in San Joseé, Costa Rica (Holdridge et al., 1971) relate vegetation diagrammatically to mean annual biotemperature, mean annual precipitation, elevation and latitude with the help of a diagram (Fig. 4).

Some authors use cross sections or profiles to illustrate the relations between vegetation and the relief, soils, geology etc. Perhaps the most comprehensive attempt comes to us from Klinka (1977) who successfully portrays the rela-

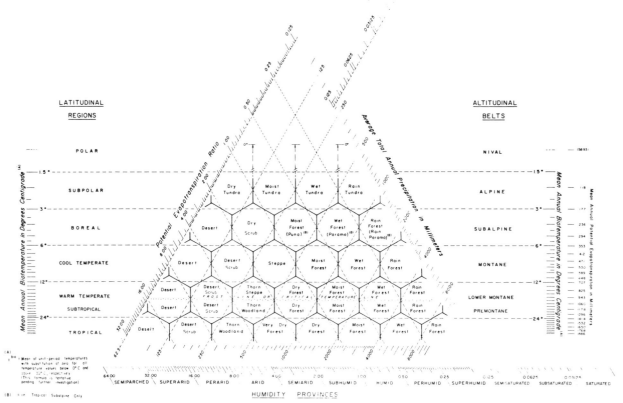

Fig. 4. Life zones, showing the relation of vegetation to climate (after Holdrige et al., 1971).

tionships of the vegetation with elevation above sea level, exposure, 18 different soil types, each with all its horizons and their respective thickness, 23 climatic features plus the evapotranspiration for nine individual months (in tabular form), and finally forest productivity.

It takes both imagination and a major effort to design a diagrammatic approach to a legend but if the result is a success then it is also gratifying to the author and helpful to the reader. But the author should in any and all cases supply a 'regular' tabularly arranged legend. In the following pages, Zonneveld presents six different approaches to the construction of a diagrammatic legend.

11A. Examples of Vegetation Maps, their Legends and Ecological Diagrams

I.S. ZONNEVELD

Introduction

In this handbook a number of vegetation maps are reproduced by printing a fragment of them including the relevant part of their legend. This gives us the opportunity to make a realistic comparison. Also some legends c.q. the ecological schemes belonging to them are added to this chapter. For a good understanding the reader is referred to Chapter 11 and especially to Chapter 29 and its appendices dealing with the compilation of a legend as part of the landscape approach in vegetation survey (mapping) using aerial photo- (or other remote sensing image) interpretation.

In this book the following maps are presented.

Real vegetation maps

— A detailed map of the 'Marlemontse Plaat', fragment of the 'Verdronken Land van Saeftinghe' (1971) in the Schelde estuary, the Netherlands. The classification used for the legend is according to Braun-Blanquet principles. Due to the large scale, most units are pure communities of low rank. Original scale 1:10,000, Fig. 1 (see colour section).
— A semi-detailed map of the 'Verdronken Land van Saeftinghe' (1967–1971) (Schelde estuary, the Netherlands), at the same time an example of sequential vegetation mapping using old and new aerial photography. The classification is adapted to the special type of mapping (interpretation of old photography). Original scale 1:40,000, Fig. 2 (see colour section).
— A reconnaissance vegetation map of the Kaarta region Mali (fragment). The classification used is a classical floristic one according to Braun-Blanquet principles using the full species composition. Due to the scale all units have a complex character. Original scale 1:200,000, Fig. 3 (see colour section).
— A reconnaissance vegetation map of Corsica (France) (fragment) with a classification system specially made for this map. Original scale 1:200,000, Fig. 3, Chapter 31 (see colour section).
— A reconnaissance vegetation map of Southern Sumatra (Indonesia) (fragment) with classification specially made for the legend of this map. Original scale 1:1,000,000, Fig. 5, Chapter 31 (see colour section).

Potential natural vegetation maps

— The potential natural vegetation map of Germany. Classification used 'Braun-Blanquet principle'. Original scale 1:200,000, Fig. 1, Chapter 26 (see colour section).
— The potential natural vegetation map of the Netherlands (fragment). Classification already existing according to Braun-Blanquet principles; resulting map units described as complexes of associations and alliance level including also percentage of actual land-use. Original scale 1:200,000, Fig. 2, Chapter 26 (see colour section).

A.W. Küchler & I.S. Zonneveld (eds.), Vegetation mapping. ISBN 90-6193-191-6.

VEGETATION MAP 1971

STRAND

100 0 100 200 300 400 500 m

SLIK EN LAAG SCHOR
MUD FLAT AND LOW SALT MARSH

 gemeenschap van Zeekraal
community of Salicornia europaea
(Salicornietum europaeae)

 gemeenschap van Zeeaster, lage rozetten
community of Aster tripolium, vegetative

 gemeenschap van Zeeaster met Engels slijkgras
community of Aster tripolium with Spartina anglica

 gemeenschap van Zeeaster, hoog en bloeiend
community of Aster tripolium, flowering

 gemeenschappen van Zeeaster en van Engels slijkgras
communities of Aster tripolium and of Spartina anglica
(Spartinetum anglicae)

NAT TOT DROGE KOMMEN
WET TO DRY BACK SWAMPS

 gemeenschap van Engels slijkgras
community of Spartina anglica
(Spartinetum anglicae)

 gemeenschap van Engels slijkgras met veldjes Heen*
community of Spartina anglica (Spartinetum anglicae)
with small clusters of Scirpus maritimus

gemeenschap van Engels slijkgras, dichte velden
community of Spartina anglica (Spartinetum anglicae), dense clusters

gemeenschap van Heen, dichte velden
community of Scirpus maritimus (Scirpetum maritimi), dense clusters

 gemeenschappen van Heen en van Engels slijkgras
communities of Scirpus maritimus (Scirpetum maritimi) and of Spartina anglica
(Spartinetum anglicae)

 gemeenschappen van Engels slijkgras, van Heen en van Zeeaster
communities of Spartina anglica (Spartinetum anglicae), of Scirpus maritimus
(Scirpetum maritimi) and of Aster tripolium (sociation of Aster tripolium)

 gemeenschappen van Engels slijkgras, van Heen en van Schorrezoutgras
communities of Spartina anglica (Spartinetum anglicae), of Scirpus
maritimus (Scirpetum maritimi) and of Triglochin maritima (sociation of Triglochin
maritima)

gemeenschap van Riet
community of Phragmites australis
(sociation of Phragmites australis)

DROGE KOMMEN EN OEVERWALLEN
DRY BACK SWAMPS AND NATURAL LEVEES

16 gemeenschappen van Gewoon kweldergras, van Spiesmelde en van Zeeaster
communities of Puccinellia maritima (Puccinellietum maritimae), of Atriplex hastata
(Atriplicetum hastatae), and of Aster tripolium (sociation of Aster tripolium)

18 gemeenschap van Spiesmelde
community of Atriplex hastata (Atriplicetum hastatae)

19 gemeenschap van Spiesmelde en Strandkweek
community of Atriplex hastata and Elytrigia pungens (Atriplici-Elytrigietum
pungentis)

20 gemeenschap van Gewoon kweldergras met Echt lepelblad e.a.
community of Puccinellia maritima (Puccinellietum maritimae) with Cochlearia
officinalis et al.

Compiled by: Drs. J. Leemans en Drs. B. Verspaandonk under supervision of Dr. Ir.
I. S. Zonneveld, I.T.C. Enschede in cooperation with Dr. Ir. W. G. Beeftink, Delta
Institute, Yerseke.
Published by the Foundation ,,Het Zeeuwsch Landschap'', Heinkenszand,
supported by the Prins Bernhardfonds.
Cartography : S. de Silva, Intermap B.V., Enschede 1979.

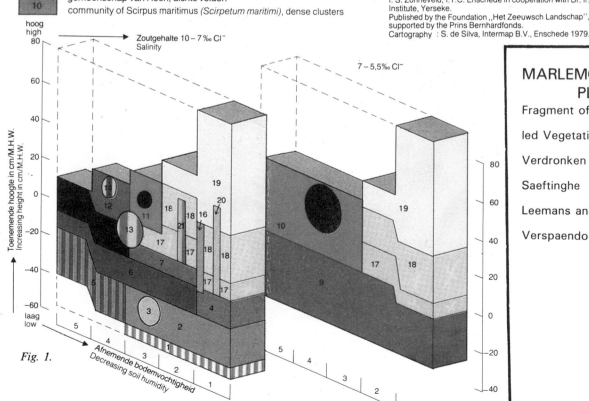

Fig. 1.

MARLEMONTSE PLAAT
Fragment of the Detailed Vegetation Map
Verdronken Land van
Saeftinghe
Leemans and
Verspaendonk

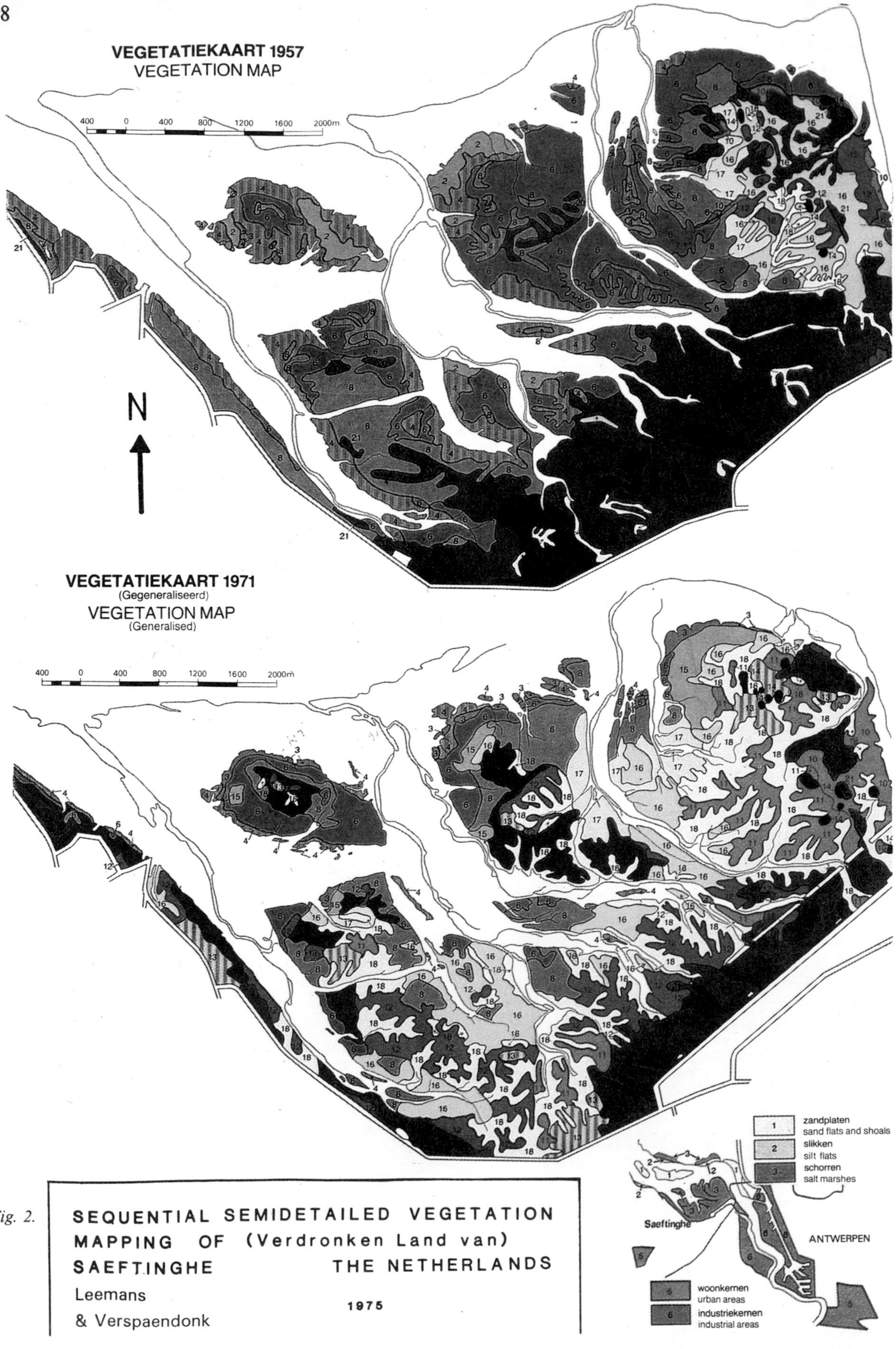

138

VEGETATIEKAART 1957
VEGETATION MAP

VEGETATIEKAART 1971
(Gegeneraliseerd)
VEGETATION MAP
(Generalised)

Fig. 2. SEQUENTIAL SEMIDETAILED VEGETATION
MAPPING OF (Verdronken Land van)
SAEFTINGHE THE NETHERLANDS
Leemans 1975
& Verspaendonk

zandplaten
sand flats and shoals
slikken
silt flats
schorren
salt marshes

woonkernen
urban areas
industriekernen
industrial areas

Saeftinghe

ANTWERPEN

SLIK EN LAAG SCHOR
MUD FLAT AND LOW SALT MARSH

1 aanslibbende schorrand met ijle begroeiing van Zeeaster en Zeekraal
salt marsh shore in accretion with thin vegetation of Aster tripolium and Salicornia europaea

2 opslibbing met Zeeaster
silting up with Aster tripolium

3 hogere plekken in 2 met hoge bloeiende Zeeaster
higher spots in 2 with high flowering Aster tripolium

4 aanslibbing met Zeeaster en Engels slijkgras
accretion with Aster tripolium and Spartina anglica

▲ zich vestigend Engels slijkgras
settling Spartina anglica

GEBIEDEN MET NAT TOT DROGE KOMMEN
AREAS WITH WET TO DRY BACK SWAMPS

5 Zeeaster en Gewoon kweldergras met zich vestigend Engels slijkgras
Aster tripolium and Puccinellia maritima with settling Spartina anglica

6 lage kommen met Engels slijkgras. Lage oeverwalfen met Gewoon kweldergras en Spiesmelde
low back swamps with Spartina anglica. Low natural levees with Puccinellia maritima and Atriplex hastata

7 kommen met Gewoon kweldergras, Zeeaster, Spiesmelde en velden Heen.* Oeverwallen met Spiesmelde
back swamps with Puccinellia maritima, Aster tripolium, Atriplex hastata and clusters of Scirpus maritimus. Natural levees with Atriplex hastata

8 hogere kommen met Engels slijkgras, Spiesmelde en veldjes Heen. Oeverwallen met Spiesmelde (en Strandkweek)
higher back swamps with Spartina anglica, Atriplex hastata and clusters of Scirpus maritimus. Natural levees with Atriplex hastata (and Elytrigia pungens)

9 vrij hoge natte kommen met dichte velden Engels slijkgras. Oeverwallen met Strandkweek
rather high wet back swamps with dense clusters of Spartina anglica. Natural levees with Elytrigia pungens

10 kommen met dichte velden Heen. Oeverwallen met Strandkweek
back swamps with dense clusters of Scirpus maritimus. Natural levees with Elytrigia pungens

11 hoge kommen met velden Heen. Hoge oeverwallen met Strandkweek
high back swamps with clusters of Scirpus maritimus. High natural levees with Elytrigia pungens

12 hoge kommen met velden Heen en velden Engels slijkgras. Hoge oeverwallen met Strandkweek
high back swamps with clusters of Scirpus maritimus and of Spartina anglica. High natural levees with Elytrigia pungens

13 hoge kommen met mozaïek van Heen, Engels slijkgras, Gewoon kweldergras en Schorrezoutgras. Hoge oeverwallen met Strandkweek
high back swamps with mosaic of Scirpus maritimus, Spartina anglica, Puccinelli maritima and Triglochin maritima. High natural levees with Elytrigia pungens

14 rietvelden
stands of Phragmites australis

GEBIEDEN MET DROGE KOMMEN, OEVERWALLEN
AREAS WITH DRY BACK SWAMPS, NATURAL LEVEES

15 hoge kommen en lage oeverwallen met Zeeaster en Spiesmelde. Invloed van vloedmerk
high back swamps and low natural levees with Aster tripolium and Atriplex hastata. Influence of tide mark

16 hoge kommen en lage oeverwallen met Spiesmelde, Gewoon kweldergras, Zeeaster en veldjes Heen. In laagste kommen Engels slijkgras
high back swamps and low natural levees with Atriplex hastata, Puccinellia maritima, Aster tripolium and small clusters of Scirpus maritimus. In lowest back swamps Spartina anglica

17 brede oeverwallen en hoge kommen met Spiesmelde
broad natural levees and high back swamps with Atriplex hastata

18 brede oeverwallen met Strandkweek
broad natural levees with Elytrigia pungens

GEBIEDEN MET BEWEIDING
AREAS WITH GRAZING

 19 vrij hoge natte kommen met dichte velden Engels slijkgras. Oeverwallen met Gewoon kweldergras, Spiesmelde
rather high wet back swamps with dense clusters of Spartina anglica. Natural levees with Puccinellia maritima, Atriplex hastata

20 overgang tussen 3 en 19
transition between legend unit 3 and 19

 21 grasmat van Gewoon kweldergras in de kommen en van Zilt rood zwenkgras op de oeverwallen. In het oostelijk deel van Saeftinghe met Zilt fioringras
sward of Puccinellia maritima in the back swamps and sward of Festuca rubra f. litoralis at the natural levees. In the eastern part of Saeftinghe with Agrostis stolonifera subvar. salina

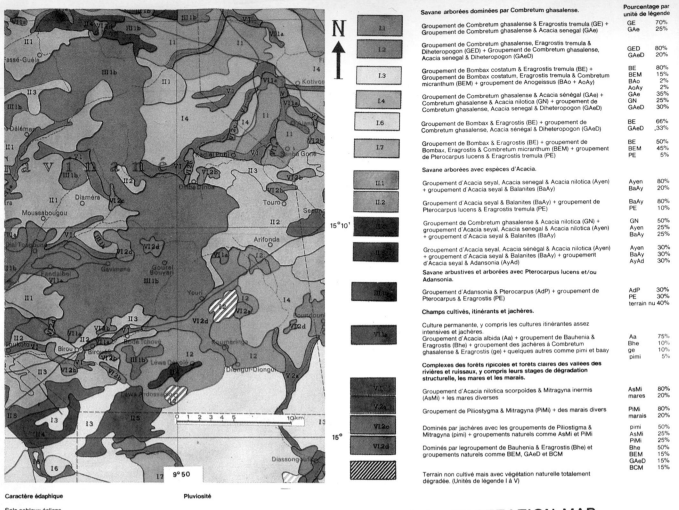

Savane arborées dominées par Combretum ghasalense.

		Pourcentage par unité de légende
I.1	Groupement de Combretum ghasalense & Eragrostis tremula (GE) + Groupement de Combretum ghasalense & Acacia senegal (GAe)	GE 70% GAe 25%
I.2	Groupement de Combretum ghasalense, Eragrostis tremula & Diheteropogon (GED) + Groupement de Combretum ghasalense, Acacia senegal & Diheteropogon (GAeD)	GED 80% GAeD 20%
I.3	Groupement de Bombax costatum & Eragrostis tremula (BE) + Groupement de Bombax costatum, Eragrostis tremula & Combretum micranthum (BEM) + groupement de Anogeissus (BAo + AoAy)	BE 80% BEM 15% BAo 2% AoAy 2%
I.4	Groupement de Combretum ghasalense & Acacia sénégal (GAe) + Combretum ghasalense & Acacia nilotica (GN) + groupement de Combretum ghasalense, Acacia senegal & Diheteropogon (GAeD)	GAe 35% GN 25% GAeD 30%
I.6	Groupement de Bombax & Eragrostis (BE) + groupement de Combretum ghasalense, Acacia sénégal & Diheteropogon (GAeD)	BE 66% GAeD 33%
I.7	Groupement de Bombax & Eragrostis (BE) + groupement de Bombax, Eragrostis & Combretum micranthum (BEM) + groupement de Pterocarpus lucens & Eragrostis tremula (PE)	BE 50% BEM 45% PE 5%

Savane arborées avec espèces d'Acacia.

II.1	Groupement d'Acacia seyal, Acacia senegal & Acacia nilotica (Ayen) + groupement d'Acacia seyal & Balanites (BaAy)	Ayen 80% BaAy 20%
II.2	Groupement d'Acacia seyal & Balanites (BaAy) + groupement de Pterocarpus lucens & Eragrostis tremula (PE)	BaAy 80% PE 10%
II.3	Groupement de Combretum ghasalense & Acacia nilotica (GN) + groupement d'Acacia seyal, Acacia senegal & Acacia nilotica (Ayen) + groupement d'Acacia seyal & Balanites (BaAy)	GN 50% Ayen 25% BaAy 25%
II.3a	Groupement d'Acacia seyal, Acacia senegal & Acacia nilotica (Ayen) + groupement d'Acacia seyal & Balanites (BaAy) + groupement d'Acacia seyal & Adansonia (AyAd)	Ayen 30% BaAy 30% AyAd 30%

Savane arbustives et arborées avec Pterocarpus lucens et/ou Adansonia.

III.1a	Groupement d'Adansonia & Pterocarpus (AdP) + groupement de Pterocarpus & Eragrostis (PE)	AdP 30% PE 30% terrain nu 40%

Champs cultivés, itinérantes et jachères.

VI.1a	Culture permanente, y compris les cultures itinérantes assez intensives et jachères. Groupement d'Acacia albida (Aa) + groupement de Bauhenia & Eragrostis (Bhe) + groupement des jachères à Combretum ghasalense & Eragrostis (ge) + quelques autres comme pimi et baay	Aa 75% Bhe 10% ge 10% pimi 5%

Complexes des forêts ripicoles et forêts claires des vallées des rivières et ruisseaux, y compris leurs stages de dégradation structurelle, les mares et les marais.

V.1	Groupement d'Acacia nilotica scorpoïdes & Mitragyna inermis (AsMi) + les mares diverses	AsMi 80% mares 20%
V.2a	Groupement de Piliostigma & Mitragyna (PiMi) + des marais divers	PiMi 80% marais 20%
VI.2c	Dominés par jachères avec les groupements de Piliostigma & Mitragyna (pimi) + groupements naturels comme AsMi et PiMi	pimi 50% AsMi 25% PiMi 25%
VI.2d	Dominés par le groupement de Bauhenia & Eragrostis (Bhe) et groupements naturels comme BEM, GAeD et BCM	Bhe 50% BEM 15% GAeD 15% BCM 15%
	Terrain non cultivé mais avec végétation naturelle totalement dégradée. (Unités de légende I à V)	

Caractère édaphique	Pluviosité
Sols sableux éoliens	
Sols sableux profonds sur des dunes longitudinales, GAe dans les dépressions interdunaires	moins de 600 mm
Sols sableux sur des dunes ondulées avec GAeD dans les dépressions interdunaires	plus de 600 mm
Sols sableux profonds sur des dunes longitudinales basses avec BEM dans les dépressions interdunaires et BAo et AoAy dans les mêmes dépressions et le long des ruisseaux	600 mm – ± 700 mm
Des sols sableux dans les dépressions interdunaires des dunes ondulées	moins de ± 700 mm
Des sols sableux peu profonds sur sous-sols limoneux ou argileux	600 mm – ± 700 mm
Des sols limoneux très peu profonds et sols limoneux sableux sur sous-sols argileux ou rocheux	600 mm – ± 700 mm
Des sols limoneux durs à argileux et localement sur sols limoneux-sableux humides.	
Des sols argileux-limoneux sur les pédiments et les pentes de vallées	moins de ± 700 mm
Des sols argileux plats fortement érodés	moins de ± 700 mm
Des sols sableux pas profonds, localement sols argileux-limoneux sur les dépressions dans les pédiments	moins de ± 700 mm
Comme II.1 et II.3, spécialement sur les champs de cultivation abandonnés	moins de ± 700 mm
Sur roches doléritiques escarpées	
Des sols sableux divers dans les dunes, les vallées, les dépressions et sur les terraces	moins de ± 700 mm
Des sols sédimentaires limoneux-argileux	moins de ± 700 mm
Id.	moins de 700 à 800 mm
Sols limoneux-argileux dans les vallées	
Sols sableux bas et sols limoneux	

Fig. 3.

Fragment of the **VEGETATION MAP**
of the **KAARTA–REGION** in **MALI**

ITC 1977

Projet de développement rural intégré de la Région Kaarta, République du Mali.
Phase de reconnaissance du volet VIII: Cartographie
© ITC-Enschede. Pays Bas. 1977

Auteurs responsables: Ir. A. de Gelder
Drs. H. A. M. J. van Gils
Dr. I. S. Zonneveld
A. Vreugdenhil

L'Ecologie de la végétation de la région du Kaarta

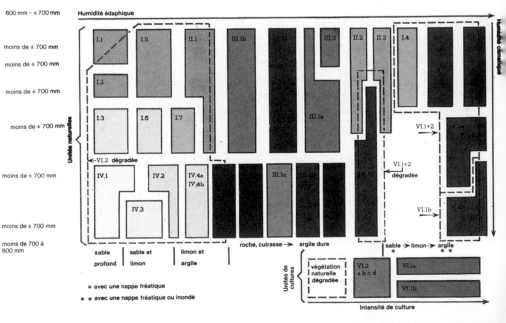

Legend and classification

A legend of a map can be considered as a kind of classification in the sense of a scientific tool to store data in a logic and conveniently arranged way, usually with some hierarchy which makes the 'look up' process easier. Beside the map legends, however, also other vegetation classification systems do exist, e.g. syn-taxonomic systems (in Europe and Japan) based on the Braun-Blanquet way of species composition, structural and physiognomic systems as the Unesco system and other systems mentioned in the former chapter. There are legends which use such existing classification systems or systems that are designed according the principles of such systems (see ITC method, Chapter 29). One can say that these legends are adapted to such type of classification systems (usually floristic ones, see maps 'Marlemontse Plaat', Kaarta*, Potential. Nat. Veg. of the Netherlands). Other legends, however, are descriptive classifications at their own. There the classification is adapted to the legend (the map of Corse, Sumatra and the semi-detailed map of Saeftinghe).

The main differences between a general classification (syn-taxonomic) compared to a legend is the chorological character of the latter. Area means for a general syn-taxonomy only that the minimum area requirement should be fulfilled, which means that one considers a sample spot of several square meters (in case of fine grained grass swards) up to several thousands of square meters in case of savanne. On a map those areas are usually neglectable. The hierarchy in a general syn-taxonomy is fully independent of the area occupied and especially of the location and the way concrete vegetations concerned have common boundaries. In legends, however, the location in relation to other types and the area of the different vegetations plays an important role.

First of all the unit should have a sufficient size in relation to the map scale to be represented. Areas that would appear in patches of less than a few millimeter on the final maps will be neglected as individuals and either not mentioned at all, or combined with others in a so-called mosaic, a complex unit, which will be described as a combination of several classification units. These mosaics may have a specific pattern that could only be depicted as such on much larger scales. In such mosaics, units may be combined which never would appear together in a general syn-taxonomy, because they need not to be ecologically related. Also the hierarchy on a higher level may follow quite different guidelines and be according to different characteristics as does the general vegetation taxonomy (typification) that is not bound to location.

So are the complex units I.1 and I.2 of the reconnaissance vegetation map of Kaarta Mali (Fig. 3, see colour section) characteristic (different) dune pattern with Combretem ghasalense savanne community on the dunes and Acacia savanne communities in the valleys; both too narrow to be mapped on the used scale. Comparison of the detailed map and the semi-detailed map of the 'Verdronken Land van Saeftinghe' (Figs. 1 and 2, see colour section) shows clearly the tidal creek levee pattern of the complex units of the semi-detailed map. Here, however, the mapper has tried to make a generalization in such a way that the main character of estuarine for-land, characteristically intersected by creecks, is still present in the generalized map. This is done by some exaggeration of creek levees.

It would have been possible to map only complexes of creek levees and backswamps and so have a map with only the major creek levees separate and hardly any creek pattern left. In the same way the dune valleys fully disappeared from the Kaarta Mali map (Fig. 3, see colour section), except the unit II.4, which is in fact a large interdunal valley. In these examples

---

* This does not mean that a general taxonomy for the Kaarta region already existed. The classification is made during the survey according the principles and procedure described in Chapter 29, which is not a mere description of legend units, but a general classification based on a sampling guided by preliminary legend units.

LEVELS OF SYNTHESIS IN VEGETATION SURVEY

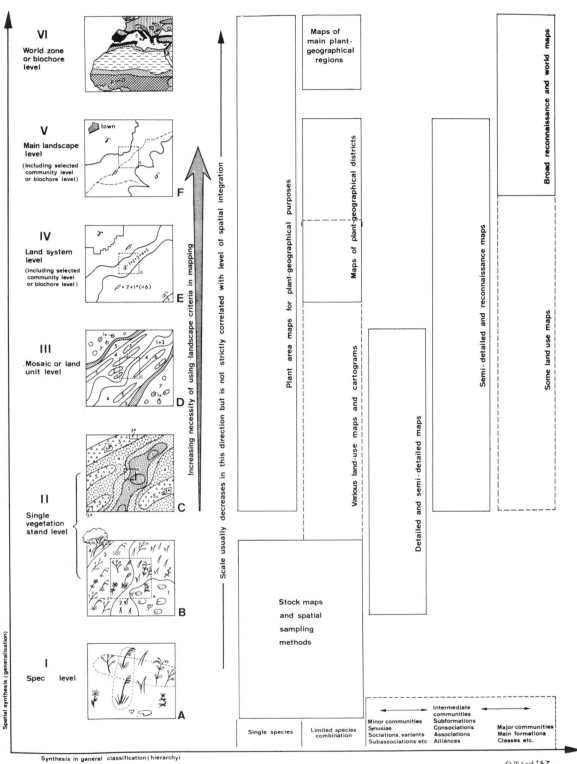

Fig. 4.

the vegetation communities combined in one map unit are quite different in composition. They belong at least to different associations. There are examples of even greater differences (alliances, classes). On the other hand in syn-taxonomic systems very much related units (variants of one sub-association) may occur in different groupings of map units, which is the case in the Mali map, but cannot be demonstrated because of lack of space. So there is no strict relation between the hierarchical level in map legends and general syn-taxonomic systems. This means that as an example it *is not so that* one could map on a reconnaissance map the orders of the Braun-Blanquet system, on a semi-detailed the alliances, and on a detailed map (1:10,000) only associations. This is shown in Fig. 4.

There, from botton to top, the same area appears as part of a larger area on a smaller scale map. It can be seen that it is possible to make very small scale world maps of the distribution of a single species. On the other hand it does occur that on very detailed maps it may be feasible to map only alliances, if one would be satisfied with a simple legend, for a detailed map image.

Contrary the reconnaissance map of Kaarta (Mali) distinguishes also at reconnaissance scale units that are similar in floristic composition belonging virtually to the same syn-taxonomic units, but that only differ in structure and form of the pattern due to intensive human activity (degradation). The units on the reconnaissance scale of the Mali map, but also the two potential natural vegetation maps of Germany and the Netherlands, using a general floristic classification, are complexes of plant communities which would have the rank of an association that may even belong to different alliances in a Braun-Blanquet system.

On detailed maps of scale 1:10,000 and larger complex units are usually rare. The map of Saeftinghe (Fig. 1, see colour section) is a good example. Most units are rather pure and belong either to a community with the rank of an association or a lower level (sub-association, var-iant). Some units refer to a very pioneer stage where side by side different communities occur on very small areas. In that case even on this scale complexes occur. They are indicated by the two colours used for the pure communities elsewhere in this case. In the semi-detailed maps of the Scheld estuarium (Fig. 2, see colour section), the reconnaissance map of Corse (Fig. 3, Chapter 31, see colour section) and of South Sumatra (Fig. 5, Chapter 31, see colour section) the occurrence of complexes is not so pronounced in the legend description. This is because here a general syn-taxonomy has not been used. Contrary one has tried to find descriptions that may give a straightforward non-complex description of the smallest vegetation areas that could be mapped.

The 'Verdronken Land van Saeftinghe' map (1971) (Fig. 2, see colour section), is a generalization of the detailed vegetation map of which the 'Marlemontse Plaat' of Fig. 1 (see colour section) is a fragment. This map could have been expressed in complexes of (sub)-associations, but the map of 1957 has been derived via aerial photo-interpretation of old photos, using today's knowledge of vegetation patterns on photographs. In that case the strict sampling in the field for determining the full plant composition was not possible (see Chapter 29) and therefore a more descriptive way was chosen, using mainly the dominant species.

The vegetation map of France (Fig.3, Chapter 31, see colour section) does not use a full vegetation classification such as a Braun-Blanquet system as a base, but also tries to describe the units not as complexes. One of the characteristics of this map is that for the scale (1:200,000) the map pattern is extremely complex, so that reading almost requires a magnifying glass. The map pattern in contrast to the written legend description has therefore almost a semi-detailed character if compared to several other reconnaissance maps. Comparison with the potential natural vegetation map of the same scale shows that the detail in patches is mainly caused by the fact that on the map of Corse not only the potential, but also the other stages in the series

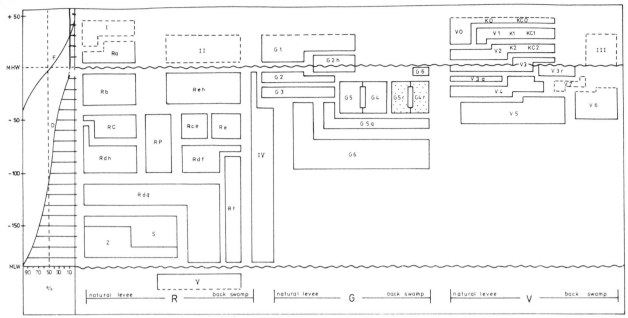

Fig. 5. Legend of the vegetation map, scale 1:10,000, of the Brabantse Biesbosch, a fresh water tidal delta (Zonneveld, 1960). F = frequency of overtopping; MHW = mean high water; D = duration of overtopping; MLW = mean low water; R = rush and forb vegetation; G = Phragmites reed; V = tidal forest (cultivated).

are mapped separately. The main colour indicates something about the potential vegetation.

The map of Sumatra (Fig. 5, Chapter 31, see colour section) has a similar character as far as the avoidance of complex units is concerned. Here the classification is also adapted to the legend as much as possible. For the differences of these two maps (one is using the succession series principle, the other does not) you are referred to Chapter 31.

The groupings of the map units in the legends of the mentioned maps are done according to (mesological) ecological principles. The Scheld estuarium map legends are grouped according to sequence of accretion which is at the same time a sequence in soil moisture regime, salinity and flooding frequency. The Mali map shows groups related to climate and soil, also expressed in a rough structural division of the natural vegetation such as tree savanne (sav. arboré) versus, shrub savanne, (savanne arbustive) forest and (not included in the fragment) savanne woodland. A dominant species (group) is also mentioned at that level.

The maps of Corse and Sumatra have a main hierarchical level straightforward based on altitude and climate. The potential vegetation map of the Netherlands (Fig. 5) does not show hierarchical subdivision; the legend, however, is arranged more or less according to a continuous ecological gradient. From dry and poor towards moist and mineral rich.

The legends of the two Scheld estuarium maps and the map of Kaarta (Mali) show a special ecological diagram, a two- or three-dimensional graphic represention of the ecological relations. In Figs. 5–8 some other examples are added. Fig. 1, the oldest of these diagrams, was in fact primarily made to design a colour system on. By representing it graphically one can get easily an expression about the visual effect of the colour selection. But then it appeared useful to have the legend in this diagram form also on the map, in addition to the list of units and there description.

The diagram of Fig. 5 shows the main gradients in that area: the soil moisture regime related to the gradient levee-backswamp from left to right and the flooding intensity and fre-

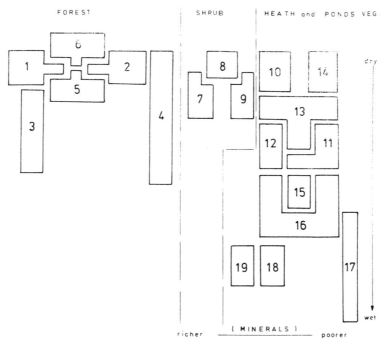

FOREST   SHRUB   HEATH and PONDS VEG.

Fig. 6. Legend of the vegetation map, scale 1:5,000, of the Buurser Zand in the Netherlands by Y. Kaplan, thesis ITC (non publ.).

quency from bottom to top. There is also a division in three structural units. The herbaceous vegetations (rushes and forbs, mainly symbols R) are on the left, the pure Phragmitis stands (partly under cultivation, symbol G) are in the middle and the tidal (willow) forests with symbol V are on the right. The measures in the diagram are connected with realistically measured elevations, flood frequency measurements and aeration depth determination in the soil. Fig. 6 represents the scheme on a students' map of an heathland area. Here the vertical dimension is the moisture regime of the soil. Horizontal is the trophy degree (rich left, poor right). Also here the structure groups the whole diagram in three: forest, shrubland, heath and other low vegetations. Fig. 7 depicts the legend diagram of a map of Spanish wooded grazing lands ('Dehesa'). It is three-dimensional, moisture is again vertical and mineral trophy degree (fertility) is from left to right, the structure gradient directs backwards. By combining the structure gradient with the two edaphic gradients the interrelation between the various gradients could be shown. The differences in structure are here mainly due

RANKING OF GRAZING LANDS IN RELATION TO THE FACTORS SUBSTRATUM FERTILITY[1], MOISTURE[2] AND TIME[3]

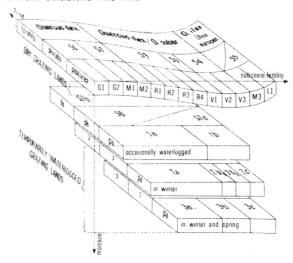

1. Horizontal ranking deduced from observations of texture and pH of the topsoil and additional soil characteristics

2. Vertical ranking deduced from estimated internal and external drainage, climatic conditions

3. Ranking in perspective: successive development of formations from pasture to climax vegetation

Fig. 7. Legend of the vegetation map, scale 1:50,000, of 'Dehesa' near Merida, Estramadura, Spain (Spiers, 1978).

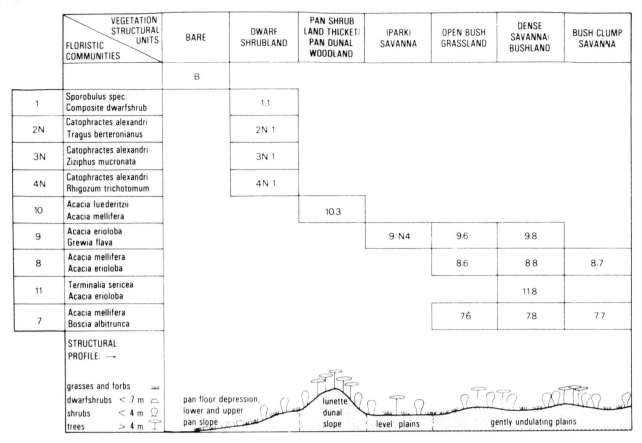

FLORISTIC COMMUNITIES		BARE	DWARF SHRUBLAND	PAN SHRUB LAND THICKET/ PAN DUNAL WOODLAND	(PARK) SAVANNA	OPEN BUSH GRASSLAND	DENSE SAVANNA/ BUSHLAND	BUSH CLUMP SAVANNA
		B						
1	Sporobulus spec. Composite dwarfshrub		1.1					
2N	Catophractes alexandri Tragus berteronianus		2N 1					
3N	Catophractes alexandri Ziziphus mucronata		3N 1					
4N	Catophractes alexandri Rhigozum trichotomum		4N 1					
10	Acacia luederitzii Acacia mellifera			10.3				
9	Acacia erioloba Grewia flava				9/N4	9.6	9.8	
8	Acacia mellifera Acacia erioloba					8.6	8.8	8.7
11	Terminalia sericea Acacia erioloba						11.8	
7	Acacia mellifera Boscia albitrunca					7.6	7.8	7.7

STRUCTURAL PROFILE: →

grasses and forbs
dwarfshrubs < .7 m
shrubs < 4 m
trees > 4 m

pan floor depression, lower and upper pan slope — lunette dunal slope — level plains — gently undulating plains

Fig. 8. Vegetation structure-floristics-landform relationships (C.A.R.A.P., 1980).

to the grazing intensity. The little bend in the top layer is an artist's expression of the fact that gradients are never pure and straight. The units around V2 used to be a bit moister than the others in this layer which is in general rather dry. Fig. 8 shows a somewhat different diagram. Here a profile is sketched showing the main landforms in the horizontal direction.

The vegetation structure is depicted in addition in the horizontal direction. Sometimes only one structural type, sometimes more than one, occurs in one landform. At the vertical axis the floristic types described, following the (computerized) table method of Braun-Blanquet, are indicated. From this diagram it is directly clear that in this case of the Kalahari area, structure and floristic composition are not strictly correlated. This is often the case in savanne-like vegetation types, because of temporal influence of animal grazing or destruction, fire,

etc. This type of diagram added to a legend gives a good impression of the reality. An interesting aspect of this diagram is that it was made as a reaction on the differences in opinion between the Anglosaxion-vegetationschool-educated commissioner, who wanted a map strictly with a structural legend and the concerning vegetation surveyors who were convinced that structure only would give the map a too temporal character and by far not enough indicative value for the environment. So during the survey due attention was paid to both aspects.

The diagram of the Kaarta map (Fig. 3, see colour section) is again a two-dimensional graph with global climate and edaphic gradients as axes. Human influence such as agriculture or (over)grazing are also indicated which could be read as a third dimension. For more information on the diagrams is referred to the original publications.

As main advantage of these diagrams it can be stated that it makes maps, which will be used for environmental indication and management, easier to read. Without the translation of (complicated) vegetation typology, the user can see at a glance which kind of ecology can be expected. The construction of these diagrams is moreover a healthy exercise for the vegetation surveyor, which stimulates a better understanding in a semi-quantitative way of the relations between vegetation and the various environmental gradients. The usefulness for designing the (so important) colour 'language' has already been mentioned.

12. Language of Map Text

A.W. KÜCHLER

Terminology

It appears that terminology is a greater problem than many mappers realize. It is therefore important that they pay particular attention to the manner in which they express the units of vegetation and their organization.

A clear and unequivocal terminology is imperative on vegetation maps. Many examples can be cited where authors have disregarded this basic rule, apparently without concern for the reader, who is left baffled by the obscurity of the terms. Mangenot (1955) discussed the analysis and mapping of tropical rainforests and urged that the results of our researches be presented in a generally comprehensible manner. The confusion in phytocenological terminology has been pointed out many times since long ago. Küchler (1947) gives a series of good examples of this perplexing problem and agrees with Barnhart (1930) that 'the rules of terminology ought to be simple and based on reasons clear enough and strong enough that everyone understands them and is disposed to accept them'. Similar pleas have been made again and again (Küchler, 1947, 1950, 1954b).

A rigorous organization requires a set of terms each of which produces the desired picture in the mind of the reader. Strictness in terminology is even more necessary than in organization because a poorly organized map makes reading difficult whereas poorly chosen terms may prevent any appreciation of the information which the author wishes to convey to his readers.

The terms must conform to the purpose of the map. To illustrate: on a vegetation map which is to show the floristic associations of a given area, all terms should be floristic names. To introduce occasional ecological or physiognomic terms instead of association names only weakens the organisation of the map. Such procedure undermines the unity of the map and leads to confusion because it requires a change in thought as one passes from one item to the next and in most cases this is not practical.

Usually an author is very familiar with the area on his map and, as a result, he may be inclined to use terms that are based on local features which he knows so well that to him they become self-evident. He easily forgets that the reader lacks this intimate acquaintance, is not familiar with the local terminology, and is therefore unable to interpret the meaning of the map. For example, on the map of Arkansas (Anonymous, 1948b) there appears the term 'Crowley's Ridge Vegetation'. One can gather from this term that there is a topographic feature named Crowley's Ridge, and that it bears a vegetation distinct from the surrounding types. But the reader has no means to discover what sort of vegetation is growing on Crowley's Ridge. It may be a forest or perhaps a grassland, it may be one plant community or several. Unless the reader happens to be familiar with Crowley's Ridge (most people are not), this term is so obscure as to be useless. The map

A.W. Küchler & I.S. Zonneveld (eds.), Vegetation mapping. ISBN 90-6193-191-6.

itself is thus useless throughout the area of Crowley's Ridge, and if more such terms are used on one map, the corresponding amount of the time, effort, and money spent on it have been wasted.

Vegetation is usually so complex that only a few characteristics can be mapped directly. For this reason, many authors establish categories with given names, and these names are then expected to act as symbols and imply the sum total of all features of such a vegetation category. For instance, the name 'Caricetum curvulae' indicates only that there is a vegetation type in which *Carex curvula* may be assumed to be prominent. The name reveals neither the physiognomy of the community nor its structure, neither its floristic composition nor the qualities of the site on which it occurs. But to a phytocenologist acquainted with Braun-Blanquet's method of classifying vegetation and familiar with the vegetation of the Alps, this name implies a certain combination of species that form a community, and in addition, it indicates a series of environmental features. This is not necessarily the most desirable way to describe vegetation, because most of the implications are lost to most readers.

It happens often that authors use the names of taxa to characterize vegetation types. This is, of course, perfectly legitimate and often the most useful method. The quality of the map is, however, influenced by the manner in which these names are used.

The clearest and therefore the best way is to use both the local names and the scientific names, as for instance 'tall grass prairie of big bluestem (*Andropogon gerardi*) and Indian grass (*Sorghastrum nutans*)'. Such a statement is descriptive, explicit, and clear. Where it is not feasible to employ such lengthy terms, the author will omit part of the information. In so doing, he must be careful to remain consistent. Throughout his legend, the plants should be named uniformly, either in the vernacular or in scientific terms. If the latter, all names should be uniformly on the same level, i.e. all names should be specific or else all names should be generic. It is not good practice to mention only the genus here, give the species there, use the vernacular elsewhere, and perhaps no plant names at all in a fourth instance.

A different kind of problem was described by Aubréville (1961). The large evergreen and semi-deciduous forest type which stretches from Guinea to the Congo Basin is called Uapacetalia by one author in Ivory Coast. Lophiretalia by another, and Strombosio-Parinatea with a subdivision of Piptadenia-Celtideltalia by a third author in the Congo. However, *Strombosia, Parinari, Piptadenia,* and *Celtis* are just as characteristic of this forest type in Ivory Coast as *Lophira* and *Uapaca* in the Congo Basin. These names do not, therefore, correspond to the floristic composition of this forest type. They can give the wrong impression, namely, that the species mentioned are particularly frequent, constant, or abundant in these forests, which is not so at all. Their distribution is irregular and discontinuous like that of the other species and to name the forest by yet another species would in no way improve the matter.

Where maps with this kind of terminology are prepared for some practical application, they must therefore be accompanied by a lengthly statement so that the reader may obtain the full benefit from his map. On the other hand, it is quite feasible to describe on a map, by other methods, the most significant features of the vegetation, and, if desired, of the environment as well. What is significant is left to the discretion of the author, who is guided by the purpose of his map.

Tüxen (1956a), one of the most prominent members of the Braun-Blanquet school of thought, has sought to overcome the terminological difficulties of his map readers by describing the vegetation in nontechnical terms. On his exemplary map of Baltrum he presents items such as grasslands of silver grass, dune shrub, crowberry heaths, salt meadows, etc., each of which is followed by the technical (Latin) name of the community in parentheses. In this manner, the map is rendered useful to a wider public and hence more valuable, solving an important problem without sacrificing anything. Tüxen's success is based solely on his manipulation of the legend.

The author who sets out to draw a vegetation map must critically analyze every term he proposes to use in order to discover whether it fulfills the basic rule of being clear and unequivocal to his readers. Wherever there is any doubt, he should discard the term and replace it by one which leaves no doubt in the mind of anyone. To be ambiguous is unscientific. The more a reader can be in doubt about the meaning of a term, the weaker is the map as a whole. Nor is it useful to select an esoteric phraseology. Here as elsewhere simplicity is a virtue, and the more readily a term produces the correct picture in the mind of the reader the more valuable is the term. Cryptic terms, on the other hand, are often useless or can lead to misinterpretation.

This attention to the quality of the terms must not be limited to the legend items. It must be trained on all terms used anywhere on the map, including the title. If, for instance, a map is entitled Vegetation Map of Jefferson County then the logical conclusion is that it portrays the vegetation. If, however, the effects of climate, soil, water, topography, or other environmental features are introduced, then the map ceases to be a vegetation map sensu stricto. As the author does not want to mislead his readers, he should state in the title what he really attempts to show. If the title does not lend itself to an adequate description, a subtitle can be used to elaborate the title and thus lead the reader in the right direction. As a result of Fosberg's observations (Gaussen, 1961a), Resolution no. 3 of the International Colloquium on Vegetation Mapping recommends that vegetation maps be based strictly on features of plant communities. These are to be distinguished in their title from maps which include or are based on ecological facts or ecosystems, describing both vegetation and environment. For example, the Vegetation Map of Tunis (Gaussen and Vernet, 1958) might be entitled more adequately Vegetation Map of Tunis with Colors of Ecological Values.

The terms must also be as true as possible. Who, indeed, would choose terms that are untrue? But no map can be quite true unless its scale approaches 1 : 1, which is obviously absurd. The truthfulness of a map then becomes a function of the scale: the smaller the scale, the greater the generalization, i.e. the greater the departure from the truth. Boundaries become less meaningful as the scale shrinks, the areal content emphasizes, more and more, one feature at the expense of all others and, in the end, the departure from the truth on small world maps is so great that the maps can give, at best, no more than broad hints.

A map of the province of Ontario may show muskegs scattered through the forests of that region. On a small-scale map this is permissible in a variety of cases. But if these types of the vegetation are to be shown at a large scale, then the term 'muskeg' becomes ambiguous to the extent of being meaningless, because both floristic composition and the growth forms may vary greatly from one muskeg to the next. Ambiguity and obscurity are the greatest enemies of good vegetation maps and must be avoided at all cost.

There are of course a considerable number of terms that are well established in phytogeography. Every phytogeographer is familiar with paramo, prairie, heath, taiga, tundra, savanna, maquis, steppe, and many others. Many of these terms are variously defined. Tundra, taiga, steppe, and a few others are used on more than one continent, but the great majority of these terms refer one to a definite area. The terms are frequently used and should be employed as in the past, although they should perhaps be more clearly defined wherever this is possible or desirable. For instance, the term tundra describes the vegetation on the poleward side of the arctic timberline. It originated in N. Europe and is now accepted and employed for analogous phytocenoses in Siberia, Canada, Alaska, etc. This extension in the use of the term tundra is logical and practical.

The matter becomes more problematical when the same term is applied to high altitude vegetation in regions like Colorado and Utah. Different conditions prevail in these lower latitudes, and significant differences may exist between the vegetation of Utah summits and that of the plains around Great Bear Lake. The use of the term tundra should therefore be avoided for high altitude vegetation in the middle latitudes, or else additional details must be given to avoid ambiguity or misinterpretation. The mere presence of a small number of true tundra species far south of the arctic timberline does not justify the extension of the use of the term beyond its range. It simply means that the species involved have a broad range of tolerance for some environmental conditions.

In an altogether different instance, profound changes like burning or clear-cutting a forest can result in mapping problems. No such problems exist if the potential natural vegetation is to be shown, unless severe soil erosion must be anticipated, and hence an altered substratum. Otherwise, the vegetation will promptly return, perhaps at first in the form of herbaceous communities, surrounding and engulfing the stumps of the departed trees. These stumps have their own significance, affecting the soil, the fungal population, etc., and may therefore be of interest to the mapper. It is not necessary to have a separate color, pattern, or symbol for the stumps. All that is needed is to mention them as part of the community in which they occur. If for instance, this is a herbaceous community of *Epilobium,* etc., the mapper will elaborate his legend item by saying '*Epilobium* community on recently logged spruce forest sites'. The stumps are then implied. If this seems inadequate, a symbol may be added, such as fine black stippling, to be overprinted on the logged sections. The U.S. Forest Service published maps showing 'old burns' and 'recent burns'. This procedure is unsatisfactory insofar as it says nothing about what was burned nor what existed at the time of mapping. Such legend items constitute a severe but unnecessary reduction in the value of the map.

It is necessary to name the phytocenoses that have been established in an area or that are used in a key because these names will be employed in the legends of the vegetation maps. Such a name should describe the plant community so completely that a person unfamiliar with it can visualize it with a resonable degree of accuracy. On many vegetation maps, such names could have been improved, if their authors had given them more thought.

Once for instance, on an excursion, a vegetation mapper made the following experiment. He selected a phytocenose and then asked his unsuspecting colleagues individually to describe how they would map it. Every vegetation mapper consulted had a different answer, of which the two extremes may be given here. One colleague spoke vaguely of a bog. When he was reminded that there are different kinds of bog, he concluded by dividing the plant community into a Callunetum and a Sphagnetum. Unfortunately, the Spagnetum contained much more *Calluna* than the Callunetum. The trees were not mentioned at all. The other colleague described the vegetation well as follows: a boggy dwarfshrub heath of *Calluna* and *Sphagnum* with scattered *Pinus uncinata* (Küchler, 1961).

Diagrams expressing the relations between plant communities and their environment also imply a type of map language. The value and success of such diagrams depend on the author's ability to convey to the reader what effect individual environmental features have on different phytocenoses.

When the mapper has composed his legend, he should go over every individual item and analyze the terms he has used in describing the vegetation. He should focus his attention on the ease with which a reader can appreciate what he means and wishes to convey. The mapper's publications may be prepared for a limited circle of users in his particular area. In fact, however, publications are distributed world-wide. Readers everywhere become aware of them, and if the publications are related to the field of interest of such readers, the publications will be distributed to all parts of the globe. Readers notice publications by direct reference, citations in bibliographies and otherwise. The clear and unequivocal expression of the mapper's ideas and observations assumes therefore a singular importance.

Supplemental texts

It is well known that the vegetation of an area can consist of phytocenoses, which are so complex in their structure and composition that it is difficult, if not impossible, to express them concisely. Many authors then resort to the method of coining a conveniently brief term and elaborating on the details in a text which accompanies the map, or which, at times, is printed on an enlarged margin of the map.

Occasionally, a map consists of a number of more or less independent sheets, like the vegetation map of France (1 : 200,000) or the vegetation map of Western Australia (1 : 1,000,000). In such a case, the terminology of all sheets must be coordinated so that, in a sense, the entire series has but one legend. However, this does not prevent the mapper of a given sheet from presenting all justifiable details even though they may not occur on any other sheet of the series. The main consideration remains that the map must be in harmony with its purpose and must be designed so as best to serve the reader.

As a general rule, a good map is independent of a text. This does not mean that a text cannot greatly elaborate and supplement the information of the map. It does mean, however, that even without such a text and with no more than a well-chosen legend, a good map is a complete unit, capable of conveying the information which the author wishes to convey.

Sometimes it is desirable to have a text which brings additional information, as it is not always possible to place all vegetational features on a map, although they can easily be described in a text. Excellent examples of this approach have been presented by Puig (1976), Klinka (1977), Pickard (1983), and many others. At times, however, authors have put the cart before the horse: a complete text is written first, describing the vegetation. Then a vegetation map is attached, almost as an afterthought, if at all. This procedure is dangerous. If somebody wishes to describe the vegetation of an area and also have a map, let him prepare the map first. This is the best procedure because the cartographic representation of the vegetation forces the author to resort to an exact and comprehensive observation. The map then becomes an excellent guide to the formation and organization of the ideas on which the textual description of the vegetation rests.

It has already been pointed out that a good vegetation map is strictly organized and uses clear and unambiguous terminology. It is difficult to do this unless the author spends a good deal of thought on just these items; if he succeeds, his map is likely to be good.

But if he writes his text first, the situation is different because this pointed clarity of the map may then be absent. A beautiful and eloquent description may be highly readable, and each individual aspect, thus described, can be visualized by the reader. But the usual subjective approach emphasizes one aspect here, another there, and if this information is then, after completing the description, placed on a map, the author can show only what he has already put into words. He has thus limited himself, and his map will be weak in its terminology because he uses the terms of his subjective and possibly inconsistent descriptions rather than purely objective ones. Of course, this need not be so, and after a good description, no matter how subjective, it is still possible to prepare a good map. But most authors are by then so steeped in the text's approach that a new and independent terminology is impossible for most of them, largely because it simply does not occur to them. It is easy enough for a mind to get into a groove, but it is difficult to get from there into another one.

One might ask here, parenthetically, whether observations that are embodied in a description must be subjective. If the author plans primarily a description and later decides to attach a vegetation map, then his activity is influenced by what he sees subjectively. For instance, in describing the physiognomy of the vegetation, he may be impressed by the unusual height of the trees in a given area. He will certainly mention this in his description. Elsewhere, not being equally impressed, he will not mention the height of the trees at all, even though local differences in the height of trees may be more significant. A good example is the vegetation map of Tanganyika (Gillman, 1949) with an excellent descriptive text. Nevertheless, the result is not a map of superior quality because Gillman stresses here the density of vegetation, there the height, in a third place other features, so that the chances of analyzing the vegetation of Tanganyika on the basis of this map are reduced, and different parts of it cannot be compared.

If, on the other hand, an experienced author of good vegetation maps plans to describe the vegetation of an area, he will draw his map first and will express his data in a manner which embodies all necessary prerequisites for a good map. In other words, he is sufficiently aware of his needs for a vegetation map that his original observation will be both objective and comprehensive to the extent of expressing every feature in terms that can be accepted on the map. If he collects more information than he can place on the map, he will have to select those items which can conveniently be shown; this becomes a matter of organization. But the subjectivity which creeps in so easily can be avoided more readily. Even though, in his text, the author should assemble his data in the form of a poetic essay with a style and beauty all its own, his map will yet be one of a high quality in a scientific sense.

Some authors realize that a vegetation map should be a unit, complete in itself. But they are also aware of the fact that the value of the map would be increased if the legend information on the map could be supplemented by a text. In order to permit the map to be independent of any book or article and yet supply desirable additional information, they place a well-chosen, concentrated text directly on the map, i.e. on an enlarged margin. Gaussen (1948), Braun-Blanquet and Tchou (1947), and Wieslander (1937) may be cited as examples; that their maps have gained through the addition of the marginal texts is beyond doubt.

Gaussen, on his map of France at 1:200,000, brings a fourfold elaboration for each major legend item: (1) the habitat conditions, (2) the natural vegetation (climax), (3) the evolution of the vegetation, and (4) the land-use. This sort of text is enlightening not only because of the information it contains. It also illustrates Gaussen's outlook: though technically a botanist, he is a geographer at heart, and the colors on his vegetation maps refer in reality to climatic regions.

Gaussen then further elaborates his material by adding six inset maps of the same area at 1:1,250,000; these show (1) climax vegetation, (2) soils, (3) land-use, (4) agriculture, (5) climate (precipitation and temperature), and (6) agricultural hazards.

Braun-Blanquet and Tchou (1947) group their plant communities at 1:20,000 into alliances of one or more associations each. For each of these, they report on the soil, list the more important species, and hint at the possible land-use.

They add two inset maps (about 1:78,000) of (1) the natural vegetation (climax), here showing only two associations and 'aquatic vegetation', and (2) soils (six soil types).

Gaussen is therefore more elaborate than Braun-Blanquet and Tchou in spite of his smaller scale. It might be argued that Braun-Blanquet does not require such elaborations because the very nature of his associations implies certain environmental features. But this is not a strong argument because, if taken at face value, it would make the entire supplementary text unnecessary.

Wieslander's (1937) major legend items are physiognomic. On an enlarged margin, he lists the most prominent species of each vegetation type and comments on the height of the formation, fire hazard (extremely important in California), effects on runoff, economic aspects, and other features. He has no inset maps. But Wieslander then adds a cross-section or profile to illustrate the relation of the vegetation to altitude and exposure.

Other authors, too, have used profiles. Their purpose is to relate the vegetation as portrayed on the map to the elevation above sea level, to topography (exposure, etc.), to the substrate (soil, underlying geological strata, etc.), etc. Good examples have been presented by Haffner (1968), Klinka (1977), and others. It is, however, important to realize the profiles, useful as they may be, cannot replace a map. For instance, Schweinfurth (1966) presents the vegeta-tion of New Zealand and illustrates his text with a number of profiles. The profiles are well done, but they cannot take the place of a vege-tation map because they apply only along the line of the cross-section and do not reveal the areal distribution of the different phytoce-noses.

Once an author has decided to have a text on his map margins, he can organize his material in such a manner that the text and the map become an integrated unit, each one supple-menting the other.

13. Other Technicalities

A.W. KÜCHLER

Editing and layout

A completed manuscript may have different forms and the author must be aware of this because the printer depends entirely on the manuscript as it is submitted to him.

The manuscript is not complete when all vegetation patterns have been applied correctly. There must be a title which reveals what the author intends to show and what the reader may hope to find. It is most important that the title be explicit. If the author wants to show the potential natural vegetation or the dynamic features of the vegetation or its structure or its relations to some ecological factor, then the title should say more than 'Vegetation Map' and include a reference to the particular character of the map. For instance, a title may read 'Potential Natural Vegetation of Humboldt County' or 'The Structure of the Vegetation in the Humboldt Reserve'. It is desirable to print the title in large and bold letters so as to make it easily identifiable.

The map must have a grid showing latitude and longitude, and where feasible, the projection should be given, too. There is no absolute need for these where the area involved is very small, such as a bog, but they should be given if at all possible. A grid does become necessary as soon as the mapped area exceeds 2 km in diameter. Accurate data on latitude and longitude can be gleaned from topographic maps as published by the U.S. Geological Survey. The grid lines may indicate minutes of degrees and their spacing must conform to the needs of the map.

The map must have a frame. Such a frame is most effective when it consists of two lines, a thin inner line and a heavy outer line. The numbers which indicate the degrees of latitude and longitude may conveniently be placed between such two lines of the frame.

The map must have a scale. This is more essential than some authors seem to realize, for there are vegetation maps without a scale of any kind and such maps are much less useful than they might have been. It is always best to have both a fractional scale and also a linear scale. The first type readily permits comparisons with other maps and the second allows the reader direct measurements on the map. To have only one these may be adequate, but there is no problem involved in having both and therefore both should be given. The author must take care to give on his manuscript the scale that the final printed version is to have, and not the scale of the manuscript. This applies only to the fractional scale. A linear scale is reduced with the manuscript, hence the proportions do not change. But the fractional scale on the printed map must be known in advance and placed on the manuscript even though the scale is quite different, i.e. larger.

The name of the author or authors must be given on the map. Brevity is best; there is no need to say more than 'by John Smith' if that covers it. Sometimes it is desirable to list individually the names of the persons who did the

A.W. Küchler & I.S. Zonneveld (eds.), Vegetation mapping. ISBN 90-6193-191-6.

field work, the compilation, the drafting, etc., if the author did not handle these matters himself. But the author's name should stand out boldly among the others as the one who is essentially responsible for the map content and its organization. The map is not the place to reveal the reason for making the map. For instance, the map may have been prepared as part of a thesis in partial fulfillment of the requirements for the degree of Doctor of Philosophy. If it is necessary to state this, then it should be done in a paper that may accompany the map but not on the face of the map itself.

The map should also show the year or years during which the map was prepared. Considering the dynamic character of the vegetation, this can be a crucial item and should not be omitted. The year when the field work was done may not coincide with the year of publication; the latter is a useful thing to know, especially for bibliographical reasons; both should appear on the map together. From a scientific point of view, the year of the field work is the more significant. On the other hand, only the year of publication should be shown if the map is compiled from source maps and is not based on field work. This is the case with most maps of small scale.

Title, author, scale, etc., and the legend must be arranged and distributed on the map so as to serve their purpose best. This is referred to as the layout of the map. Printing costs are high and therefore the layout should be such as to avoid a waste of space. A waste of space on the map implies larger printing plates and this raises the cost substantially and unnecessarily. On the other hand, the various items should not be squeezed together too tightly, as this will spoil the appearance of the map. If an author is at all uncertain about his layout, he will find it most useful to observe the work of others. The layout is a technicality and has nothing to do with the map content. The author may therefore consult not only other vegetation maps but also geological maps, soils maps, and others. Their problems are the same.

The layout includes the map itself, the title and possibly a subtitle, fractional and linear scale, the projection of the map, year of preparation and publication of the map, authorship, and complete legend. If the map is part of a journal article, appropriate bibliographic data should be added.

The mapper should make a complete list of items to be shown and then experiment with their distribution and with the sizes of the letters, which may be different in each line. If one or more inset maps are used, their location on the main map, and their own layout and letter sizes must be considered separately. Letters printed on gummed transparent material may be purchased in a wide variety of sizes.

The mapper should not hesitate to reject his layout if he is not entirely satisfied, and improve it. This may take a great deal more time than an inexperienced mapper may anticipate but undue hurry at this stage must not be permitted. Vegetation mappers will find patience a virtue of singular value.

Reduction

It is always best to prepare the manuscript of a vegetation map at a scale appreciably larger than that planned for the final, printed version. Drafting is easier, and changes and corrections can be made more readily without affecting other parts of the map. Eventually, however, the map must be reduced, and it is most important that this reduction be kept in mind during the preparation of the manuscript. Patterns, symbols, and words on the manuscript map must be so large that they remain plainly legible after their reduction. The proper choice of the lettering size, etc., implies therefore that the author knows in advance the approximate scale of the printed map.

Many times a manuscript looked most pleasing to its author but turned into a disappointment when the reduced version was printed. The names may be difficult to read without a magnifying glass if they are legible at all. Line and dot patterns may be ruined similarly. "Theoretically, reproduction of shading patterns should not affect the value of the tones, even at reduced scale. Many cartographers, however, have experienced failures where patterns have either 'dropped-out' or 'closed-up' in the reproduction process. This is a vexing problem since a pattern may be used satisfactorily at one reduction or with one printing technique and give poor results with another. The problem becomes especially acute when dot style patterns of fine texture are used" (Jenks and Knos, 1961). The reduction may therefore lower the utility of the vegetation map or even destroy it altogether. Obviously, the author must calculate the probable degree of reduction before he starts his manuscript and then select his patterns and his lettering accordingly.

Some authors find it convenient to draw their manuscript maps with inks of different colors. They must remember that the usual photographic reduction not only eliminates color differences but affects the colors unevenly. For instance, red in the original turns black on a photograph, whereas blue appears very pale and, indeed, may vanish altogether. The mapper can use this reaction of the colors to his advantage. For example, he may obtain a base map of his area on which such items as county lines, etc., are printed in pale blue. These blue lines will disappear and only the vegetation types remain.

Some mappers keep map and legend independent of one another during the drafting stage. This is entirely legitimate and is indeed an advantage because one may be in the way while the other is being done. But it is important that the reduction for printing applies equally to the map and the legend so that all patterns and symbols in the body of the map and in the legend are of the same size. If this is ignored, the symbols may be appreciably larger in the legend than on the map and the result can be confusing and make the reading of the map more difficult (cf. Zohary, 1947).

Reproduction and printing

If the map is to be printed in black-and-white, the vegetation types will best be distinguished by a variety of dot and line patterns. The author can place such patterns on his manuscript map by using Zipatone, Craftone, or similar materials which may be obtained commercially. A vast variety of such patterns is available. The patterns are printed on transparent material which is easily cut. The author need only lay it on his manuscript and cut out the areas of the different types of vegetation. He places it in the exact position wanted and presses it down. It will then stick to the paper of the manuscript map.

The matter is more delicate if the map is to be printed in color. The author should consult the printer on available colors and discuss the matter with him. A great deal of research has been done on color printing (e.g. Laclavère and Dejeumont, 1961) and the author will find it worth his while to acquaint himself with the latest developments in this field. The author must know the form in which it is to be applied, i.e. as a flat color, in the form of lines, dots, or some other pattern.

A good printer usually has some kind of a color chart which the author may study. There are three basic colors, red, yellow, and blue, and of course, black. The latter is used for title, frame, grid, etc. The three basic dolors can be combined in a variety of ways and thus greatly enlarge the number of available colors. The printer's inks may be transparent or opaque. Two colors will combine only if the inks are transparent. If large dots are printed with blue ink on a yellow background, the dots will be green. If the author wants blue dots on yellow, then the blue dots must be printed with opaque ink. This means two different printing plates for blue and the author must be sure to discuss with the printer the cost of printing, its relation to the number of color plates, the need for more than four plates, etc.

Most printers can identify their colors by a number. This is a great aid to the author. All legend items can be numbered and these same numbers can be entered on the map in the appropriate areas. For example, if an area on the map bears the number '8', then this refers

160

to legend item 8. Wherever legend item 8 occurs on the map, the 8 should appear written plainly. Every area on the map, regardless of size, must have a number that relates it to a legend item. Once this is done, the author can make a list of his legend numbers and after each number give the printer's number of the color that is to be used. This will be a simple and clear guide for the printer and will greatly aid in proofreading the color proof.

When all the details have been worked out, the printer can make an accurate estimate of the cost of printing the map. Authors should insist on a color proof. If they find on such a proof that some colors are wrong or too bright or too dark or are unsatisfactory for any reason, then a different ink can be selected for the final printing. A color proof must also be proofread with regard to the accuracy of the color selection and application for every vegetation type on the map and in the color boxes of the legend.

The cost of printing will vary according to the quality of the paper. The printer will have samples and the author should be guided in his choice by the use of the map. If the map will be used much, possibly even in the field, the paper should be relatively heavy and sturdy. If the map serves only to illustrate an article in a journal, the paper may be lighter. It also makes a difference whether the map on publication is to be folded or rolled. It is always desirable to have strong paper for maps because they tear easily when much used. But heavy paper does not fold easily and the author must compromise accordingly. One solution to this set of problems, while maintaining reasonable costs, is to print on a lighter weight of paper and have those copies to be used in the field, etc., laminated. The printer can 'sandwich' the map between pieces of plastic in the same way that he does a restaurant menu.

A minor cost feature is the width of the margin of the map. The margin is the strip of paper which surrounds the frame. There again the author must use his judgment: too wide a margin is wasteful and too narrow a margin spoils the appearance of the map. The author can find adequate guidance by studying other maps that correspond to his in size.

The printer must know how many maps he is to print. The mapper determines this number by consulting his publisher. If the map is to be published in a book, then the publisher should tell the mapper what size of edition is planned, e.g. 5,000. This means that 5,000 books will be printed, each of which will contain one copy of the map. If the vegetation map is to appear in a journal, then the mapper consults the journal editor concerning the circulation (number or subscriptions) of the journal, e.g. 2,000. The mapper will want a generous number of reprints, and the editor will want extra numbers of this journal because later orders for back numbers will have to be filled at undetermined dates. All these must be added to the basic 2,000 maps.

Finally, the author will want to obtain an adequate number of free copies of the map for his own use and for distribution among his colleagues. Most authors do not publish colored vegetation maps very often. But a colored vegetation map is an impressive piece of work and an adequate number should be kept on hand because requests for reprints have a tendency to arrive for long periods after publication.

The mapper must discuss the possible folding of the map with the printer. If the folded map will be inserted in a pocket at the end of a book or journal number, then the mapper must consult the journal publisher with regard to the size of the pocket. The map must then be folded so as to fit comfortably into such a pocket. There are different ways of folding a map, and the mapper, when discussing this with the printer, should insist on having the map folded 'accordion fashion' as this will reduce the wear and tear to a minimum.

Proofreading

To err is human – even the most careful mapper cannot avoid making mistakes. All vegetation maps must therefore be proofread. However, authors are warned that proofreading a vegetation map is very different from proofreading an article. It is more difficult and more tiring for the eyes and very time-consuming. Just because it is difficult and tiring, the proofreader himself is apt to make mistakes, i.e. he overlooks or fails to recognize the mistakes on the map. Therefore proofreading should be done at least twice, and preferably more often. It is always best to have the proofreading done by more than one person.

Whatever the mapper places on his map must be proofread. This excludes the items he did not place on his map. For instance, if the author uses a standard base map with grid, cities, rivers, etc., then there is no need to proofread these. On the other hand, if he places rivers, cities, or anything else on his vegetation map, he must proofread these, too.

Proofreading a vegetation map includes above all the proofreading of the vegetation types, and their boundaries. If the map was compiled from source maps or aerial photographs, the proofreading consits of comparing the new vegetation map with the source maps or the photographs. This is done most efficiently by cutting a rectangular hole into a white sheet of paper. The paper may be standard letter size (8 1/2 by 11 inches) and the hole should approximate 3 by 4 inches. Sometimes it is convenient to adjust the hole to the grid. For instance, on a map at 1 : 1,000,000, the hole may correspond to one degree of latitude and longitude. This should be considered the maximum size for the hole because a larger size reduces its effectiveness.

The white sheet of paper with the hole is placed on the manuscript map. This focuses the eye on the part of the map that can be seen through the hole whereas the remainder of the map is excluded. The paper is laid on the manuscript map so that the northwest corner of the map is seen through the hole; the left and upper edges of the hole should coincide with the western and northern margins of the map in the northwestern corner. The author then compares the vegetation features and everything else that must be proofread within the hole with the corresponding part of the source maps or aerial photographs. The paper may not be shifted until everything within the hole is completely proofread. The author then moves the paper eastward (to the right) just enough so that the left margin of the hole comes to rest where the right margin was before. The latitude is not changed. The new area is now proofread just as carefully, and when this is completed, the paper is shifted eastward again. In this manner, the author works his way methodically across the map from the left margin to the right one. When he has arrived at the right margin of the manuscript map he has finished only a strip across the northern part of the map.

The white paper is now shifted back to the left margin of the manuscript map. This time the hole is placed below the original starting position so that the upper edge of the hole comes to lie where the lower edge used to be at the beginning. From here, the author proceeds eastward again in the same manner until he arrives at the right margin. Then he proceeds to the third strip, still farther south, and so on until the entire map is done.

If the author must interrupt his work, he should note on a paper the exact latitude and longitude of the position of the hole. He can then return to this position on resuming his work without having to fear duplication or omissions.

Everything must be proofread: the colors, the patterns, the symbols, the boundaries and their location, the title, the scale, the legend, etc. If the legend items are numbered and the numbers are repeated on the map, then the numbers must be proofread, too. The author must check whether the number on the map portrays the correct item and is properly placed on the map, i.e. into the area of the correct type.

If a variety of features must be proofread, it is sometimes more efficient to proofread just one at a time through the entire map, and then proofread the whole map again with regard to another feature.

When the proofreader finds a mistake, he records it. This is better and less time-consuming than to correct every mistake as soon as it is discovered. There are two ways of recording mistakes. If the map is small and simple, the proofreader will draw a straightline from where the mistake is located to the margin. At the margin end of the line he records the distance to the mistake and its character, as, for instance: '14 cm, change number to 28', '4 inches, blue missing', '10 cm, remove triangular symbol', etc.

Another way of recording mistakes is useful on larger and more complex maps. It is done on a separate sheet of paper, not on the margin of the map. On the separate sheet, the location of the mistake is indicated by latitude and longitude; the character of the mistake is described as above.

Proofreading means moving one's eyes from the hole to the source and back to the hole, back and forth, back and forth, dozens of times, hundreds of times. This is the tiring feature. The efficiency of the proofreader usually does not last long. As soon as he observes that his eyes are tired, he should rest them for 10 or 15 minutes. Even so, many proofreaders find that their concentration suffers after 2 or 3 hours. It is a waste of time and effort to proofread a map when tired.

The reader will appreciate now why proofreading vegetation maps is so time-consuming. This should not induce him to rush the job. Hurry is fatal. In planning his mapping project, the author must therefore allow ample time for proofreading every individual step at least twice. Proofreading is not likely to be overdone.

A more recently developed important technique is to print vegetation maps with a computer. The information on the manuscript map can be fed directly into the computer where it is digitally stored. This permits a variety of comparisons, analyses and syntheses. However, proof-reading remains as important as ever and the aesthetic quality of computer made maps is usually low. The nature of electronic methods will be discussed in the following chapter.

14. Automated Cartography and Electronic Geographic Information Systems

D. VAN DER ZEE and H. HUIZING

Introduction

Until some decennia ago the exclusive means to store and present the geographic distribution of land attributes like vegetation was as a classical map: a two-dimensional piece of paper with colours and/or symbols. Since then it became possible to store such data in a computer memory digitalized and ready for retrieval in map form at any moment (see Tomlinson, 1968). There are three more important aspects of this development:

— A map is a final document that cannot be converted or amended after it has been printed. Data in an electronic memory, however, can be altered anytime depending on the system used. The geographical data stored can be converted and amended whenever needed or wanted.

— The electronics do not only give storage possibilities, but also make it possible to study coincidences to bring several data of land together in their mutual relationship.

— It is possible to generalize or transform data in a special way or format wanted by the user, like combinations of attributes (land evaluation) or exclusive retrieval of special aspects, in the course of special studies.

Especially these two latter aspects have given rise to the concept: geographic information systems. Such a system can be a most valuable tool in bringing vegetation mapping data in better contact with the various applications.

Concepts of geographic information system

Geographic information is information about entities that have both attributes and a position in space with respect to a geographical coordinate system (Burrough 1984a). Or to phrase it in a different way: Geographic information is information about the state of affairs (conditions, circumstances), the properties and the mutual relations of factors related to a geographic area (Berg). A third characteristic is that of time. The attributes may change character over time but retain the same spatial location or the attributes remain the same but the location changes (Dangermond). See Fig. 1.

The *processing* of such information embraces the compilation, storage, retrieval, transformation, integration, analysis and display or presentation of the information. A *system* that allows this processing is called a *Geographic Information System (GIS)*, but they may also be called Geo-Information Systems, Spatial Information Systems or Land Information Systems. See Fig. 2.

Most of the work associated with geographic information systems can be done with traditional tools: pencil, paper, ink, colourprinting, photography (Burrough, 1984a). Graphics in the form of maps have provided and still provide us with a useful medium for record keeping, conceiving of ideas, analysing concepts, predicting the future, developing decisions about geography and communicating spatial concepts to others (Dangermond). Still, at present the term geographic information system commonly ref-

A.W. Küchler & I.S. Zonneveld (eds.), Vegetation mapping. ISBN 90-6193-191-6.

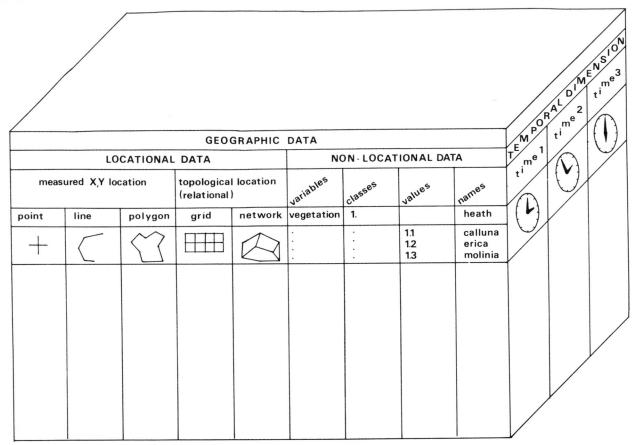

Fig. 1. The three components of geographic information (after Dangermond).

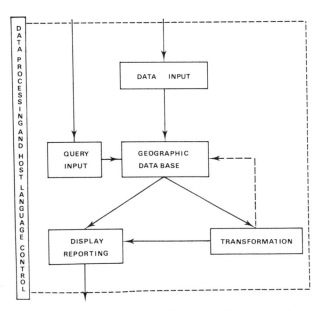

Fig. 2. A simplified overview of a geographic information system (source: Burrough, 1984a).

ers to the rapidly evolving technology of automated tools for the efficient storage, analysis and presentation of geographic data. Instead of thematic atlases computerized data bases are used.

A computerized geographic information system consists of a *databank* or *database* containing the data related to the geographic area (*geocoded data*) and a package of computer procedures for processing the data (Berg).

In this context it is useful to distinguish the concepts *data* and *information*. Data consist of observations, measurements, recorded properties of objects, obtained without taking possible applications into account, and in themselves may have no meaning (Berg). From these data information may be derived either directly or indirectly (after processing of the data). Information is knowledge on an object system that can be used in the context of a decision-making process or a communication need.

In the same way a distinction can be made between a *database* and an *information system.* A database is no more than a collection of data, which may be meaningless without further processing. In addition to a database an information system also contains the tools necessary for extracting information from the data. In computerized data systems the tools consist of algorithms translated into computer programmes.

Why to use automated geographic information systems

The uses of the available technology with respect to geographic information sysems are vast and are cutting across a large number of professions. By passing the question why geographic information is needed, the question arises if and when geographic information is needed, why should it be in an automated computerized system? For this various reasons can be mentioned (Burrough, 1984a).

One group of reasons is related to the quicker and cheaper production of maps with a minimum of skilled staff, to the possibility of experimenting with different graphical representations of the same information and to the possibilities for easy updating of maps. These reasons lie mainly in the domain of *cartography.*

Another group of reasons is focused on the possibilities to analyse data, to compare the information of different maps, to subject these maps to statistical analysis and to compile new maps out of a combination of information on existing maps. These reasons therefore lie in the domain of *analysis* and *synthesis.*

A third reason is that a geographic information system may make it possible to create a dynamic model of part of the earth's surface on which ideas about future land use, the impact of planning decisions and consequences of natural and man-controlled processes can be tested. This reason lies in the domain of *model making.*

All these reasons for using a geographic information system have in common that without automation and computerization many activities would take too much time and labour and with the traditional technology would not be carried out.

What then are the *advantages* of automated systems (Dangermond).
— Data are maintained in a physically compact format:
  a small section of a magnetic disc or tape may replace a large number of map drawers or bookshelves;
— data can be maintained and extracted at a lower cost per unit of data handled (once the use of the system has become a routine);
— data can be retrieved with much greater speed;
— various computerized tools allow for a variety of types of manipulation with a minimum of time and effort;
— graphic and non-graphic (i.e. attribute) information can be merged and manipulated simultaneously in a 'related' manner;
— rapid and repeated analytic testing of conceptual models with a spatial (= geographical) context can be performed;
— change analysis can be efficiently performed for two or more different time periods;
— interactive graphic design and automated drafting tools can be applied to cartographic design and production;
— certain forms of analysis can be performed cost effectively that when done manually simply could not be done efficiently.

As a result of all these advantages there is a tendency to integrate data collection, spatial analysis and decision-making processes into one common information flow context.

Geographic information systems also have some *disadvantages:*
— converting existing geographic records into an automated file by digitizing, scanning etc. is still a time-consuming labour intensive activity associated with technical problems;

— in general the input of data into the system remains the bottleneck, once that is done manipulation and output are easy;
— a large amount of technical as well as financial overhead is required to maintain automated files, i.e. a computer in a special air-conditioned space with reliable electricity supply, skilled technicians, software, maintenance, etc.; it will make a great difference whether a GIS can be put up as a mere addition or extension of an already existing overhead or a completely new overhead has to be created;
— for the initial acquisition of systems high costs are involved;
— for certain application areas the cost-benefit ratios are just marginal;
— existing thematic data sets are now mostly handled by specialized agencies; output formats are often not compatible and the integration or mutual exchange of datasets therefore is difficult.

Whether or not to use an automated geographic information system and if yes what type, will depend on the balance between advantages and disadvangages and on what group of reasons the emphasis is put.

A factor of importance that should be included in the considerations is the decay rate of information, i.e. the speed with which information becomes obsolete. The following list of subjects is ranked according to an increase in the decay rate of information (which more or less coincides with an increase of human influence):
— geology;
— geomorphology;
— soils;
— forestry;
— vegetation/rangelands;
— agricultural land use;
— human geography;
— socio-economic aspects;
— urban environment.

It may be clear that for information with a high rate of decay the advantages of use of an automated geographic information system will be greater than for information with a lower rate of decay.

It is here that the usefulness of geo-information system technology for vegetation mapping becomes evident. Vegetation and land-use both tend to change much more rapidly due to natural or anthropological factors than soils, geomorphology and geological structures. Indeed most vegetation maps are already obsolete in certain details when they are printed. In an electronic system they may be updated easily.

Another question to be considered is that of the rate of turnover of information: how frequently is there a demand for (updated) information; how frequently are data collected and put into the automated files for updating?

Types of geographic data and their spatial representation in GIS

In general there are three basic notations used for representing the spatial location of geographic phenomena:
— *points:* solitary occurrences of single objects, observations of soil or groundwater at borings, sites of vegetation samples, etc.;
— *lines:* transportation network, drainage pattern, administrative boundaries, etc.;
— *polygons* or *areas:* mapping units of vegetation, soil, land-use, terrain; administrative units; enumeration districts, planning zones, etc.

There are two fundamental ways of representing such topological data (Burrough):
— *vector representation:* points, lines, and polygons are defined by (sets of) coordinates in a coordinate system (x and y, longitude and latitude) and linked to given attributes;
— *raster or grid representation:* a set of cells contained in a grid, their location being defined in the x and y coordinates of the grid-system. Each cell is independently addressed with the value of an attribute.

In most geographic information systems these two ways can be interconnected. Both ways will

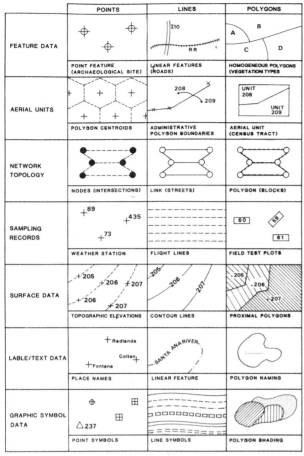

	POINTS	LINES	POLYGONS
FEATURE DATA	POINT FEATURE (ARCHAEOLOGICAL SITE)	LINEAR FEATURES (ROADS)	HOMOGENEOUS POLYGONS (VEGETATION TYPES)
AERIAL UNITS	POLYGON CENTROIDS	ADMINISTRATIVE POLYGON BOUNDARIES	AERIAL UNIT (CENSUS TRACT)
NETWORK TOPOLOGY	NODES (INTERSECTIONS)	LINK (STREETS)	POLYGON (BLOCKS)
SAMPLING RECORDS	WEATHER STATION	FLIGHT LINES	FIELD TEST PLOTS
SURFACE DATA	TOPOGRAPHIC ELEVATIONS	CONTOUR LINES	PROXIMAL POLYGONS
LABLE/TEXT DATA	PLACE NAMES	LINEAR FEATURE	POLYGON NAMING
GRAPHIC SYMBOL DATA	POINT SYMBOLS	LINE SYMBOLS	POLYGON SHADING

Fig. 3. The different types of geographic data and methods of representation (source: Dangermond).

be discussed in somewhat more detail now (see also Fig. 3).

Vector data structures

The vector representation of an object is an attempt to represent the object as exactly as possible. The coordinate space is assumed to be continuous and not quantized as with the raster space.

Point entities. Point entities are all geographical and graphical entities that are positioned by a single x,y coordinate pair. In addition to these coordinates other data should be stored to indicate the attribute, i.e. what kind of point it is,

and other associated information, e.g. type of symbol to use for its presentation or display.
Line entities. Line entities are linear features built up of straight line segments. The simplest line requires the storage of a start and an end point (two x, y coordinate pairs) plus an indication of the attribute and the display symbol. A *chain* or *string* is a set of n x, y coordinate pairs describing a continuous complex line composed of n−1 straight line segments. The shorter the line segments, the closer the chain will approximate a complex curve but also the larger the number of coordinate pairs to be stored. See Fig. 4.

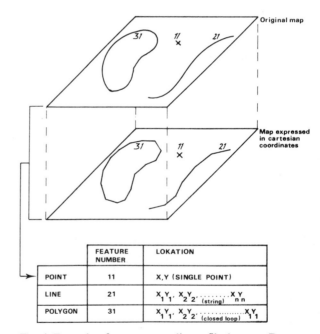

	FEATURE NUMBER	LOKATION
POINT	11	X,Y (SINGLE POINT)
LINE	21	X_1Y_1, X_2Y_2, ... X_nY_n (string)
POLYGON	31	X_1Y_1, X_2Y_2, ... X_1Y_1 (closed loop)

Fig. 4. Example of an x, y coordinate file (source: Dangermond).

Area entities. Areas or *polygons* can be represented in various ways in a vector database.

The simplest way to represent a polygon is an extension of the simple chain. This means to represent each polygon as a set of x, y coordinates on its boundary. See Fig. 4. A disadvantage is that lines between adjacent polygons should be digitized and stored twice, with all the risks of errors and gaps. Also island polygons may create a problem. These problems are serious if the polygons also have to be displayed as polygons for the final presentation. If however, polygons are only digitized in this way to be converted to a raster format, these problems can rather easily be overcome.

A more complex way of digitizing and storing polygons is in the form of a segment or chain file in which each chain is stored as a list of x, y coordinate pairs and two pointers which refer to the attributes of the areas on both sides of the line. Each line has to be digitized only once, thus avoiding gaps. This data structure allows the production of simple derivative maps from the basic polygon network and display and presentation as polygon map is very well possible. See Fig. 5.

Raster data structures

Simple raster data structures consist of an array of gridcells (= pixels or picture cells). Each cell is referenced by a row and column number and contains a number representing the type or value of the attribute being mapped. In raster structure a point is mapped as a single gridcell, a line as a number of neighbouring gridcells and an area as an agglomeration of neighbouring gridcells. See Fig. 6.

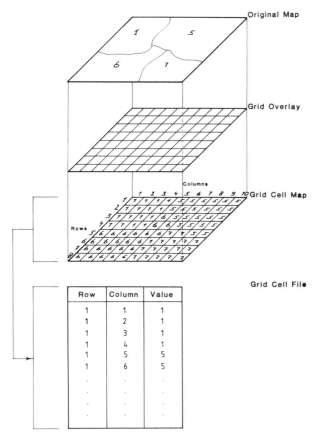

Fig. 6. Example of a grid file (source: Dangermond).

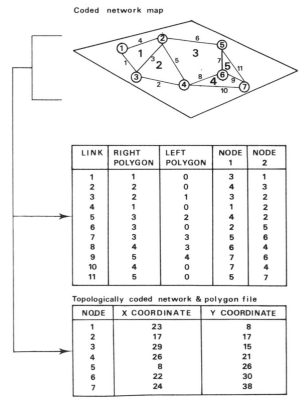

Coded network map

LINK	RIGHT POLYGON	LEFT POLYGON	NODE 1	NODE 2
1	1	0	3	1
2	2	0	4	3
3	2	1	3	2
4	1	0	1	2
5	3	2	4	2
6	3	0	2	5
7	3	3	5	6
8	4	3	6	4
9	5	4	7	6
10	4	0	7	4
11	5	0	5	7

Topologically coded network & polygon file

NODE	X COORDINATE	Y COORDINATE
1	23	8
2	17	17
3	29	15
4	26	21
5	8	26
6	22	30
7	24	38

Fig. 5. Example of an x, y coordinate node file (source: Dangermond).

This type of data is easy to handle in a computer. The two-dimensional surface, upon which the geographic data are represented, is not continuous but quantized. This can have an important effect on the estimation of lengths and areas when gridcell sizes are large with respect to the features being represented. The *resolution* of the raster system is the relation between the cellsize in the database and the size of the corresponding area on the ground. The smaller the gridcell, the higher the resolution, but also the larger the storage space needed in the computer. Also in this respect a proper balance has to be found between efficiency and economy on the one hand and the accuracy wanted on the other hand. With the selection of a proper cell size the inaccuracies with respect to e.g. area measurements will remain in an order that is similar to that experienced when generalizing maps for scale reduction.

Because each cell in a two-dimensional array can only hold one number, different geographical or graphical attributes should be represented by separate sets of arrays or *overlays*. See Fig. 7.

Vector data structures are more commonly used in the GIS aimed at automated cartography, whereas raster data structures seem more appropriate for analysis and synthesis as well as model-making.

Input of data into a GIS

Data input is the operation of encoding the data and writing them to the database.

Two aspects of the data need to be considered for a GIS:
— the positional or geographic data, the *where*;
— the associated attribute (in this case vegetation classification) categories or graphic data, the *what* (Burrough, 1984a).

The procedure of data input into a GIS can best be distinguished in three different activities:
— the entering of the spatial data or digitizing;
— the entering of the non-spatial associated attributes;
— the linking of the spatial to the non-spatial data (Burrough, 1984a).

The way in which these activities are interrelated will be different from one programme to the other. Some programmes call for cumber-

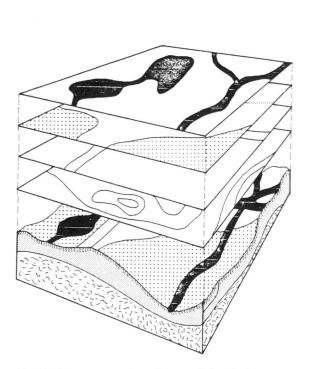
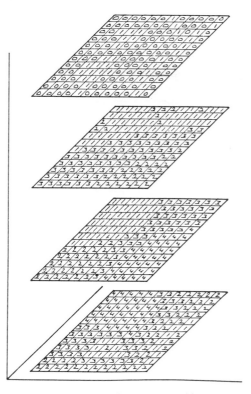

Fig. 7. The representation of mapped date in layers or overlays, both conceptually (left) and in the computer (right) (after: Burrough, 1984a).

170

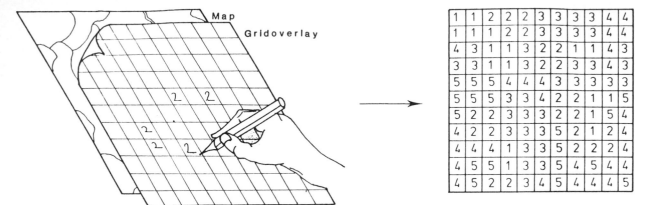

Fig. 8. The manual cell encoding method and the resulting grid cell file.

some procedures, other programmes are relatively easy and userfriendly. Some procedures will be reviewed now, in which emphasis will be given to those that are not specific for automated cartography.

Manual input of data

For manual input into a vector system a reference grid is put on the source map. The source map is a vegetation or land-use map produced according to the method described in the various other Chapters of this book. It need not be a printed map, a corrected manuscript map may do as well.

The coordinate pairs of points and/or line segments are obtained from that grid and typed into a file or as input to a programme. Depending on the programme the attribute codes can be included in the same procedure or have to be attached in a later stage.

For manual input into a raster format on the source map is put a transparent overlay with the selected gridcell size adapted to the mapscale. The value of a single map attribute is then written down on an encoding sheet and later typed on punchcards or directly into a computerfile for each single cell. This is a very tedious and time-consuming procedure. See Fig. 8. The procedure can be streamlined if the programme allows to put adjacent cells of the same value as *runlenghts*. Especially for simple

maps this can have advantages (Burrough, 1984a).

Because of the enormous amount of time and labour involved in these input methods most advanced geographic information systems now include the possibility of digitizing.

Digitizing

A digitizer is an electronic (electromagnetic, electrostatic or electromechanical) device, consisting of a table or tablet upon which the source map can be placed. See Fig. 9. Embedded in the table or located directly under it is a sensing device that can accurately locate the centre of a cursor. The *cursor* or *point locator* is a small device that can be moved over the table and is connected to the same digitizing system.

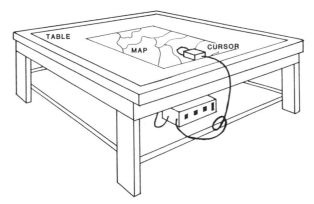

Fig. 9. A digitizing table for manual digitizing.

When the cursor is placed over a point on the map and a button on it is pressed, an electrical signal is sent directly to the computer, indicating the coordinates of the cursor with respect to the digitizers frame of reference. These coordinates can then be converted to absolute map coordinates by simple scaling routines. The main aim of the digitizer is to input quickly and accurately the coordinates of points and lines (Burrough, 1984a). In many digitizing systems the associated attribute codes can be attached in the same procedure.

There are two types of programmes for digitizing lines. In one type a line is digitized by recording individual points along it with the cursor. The operator has to decide himself about the spacing of these points as a function of the complexity of the line and the rate of generalization allowed. In the second type of programmes the cursor is placed at the beginning of the line and, after a start command is sent to the computer, the operator moves the cursor along the line until at the end a stop command is given to the computer. In between start and stop the computer records coordinates at either equal time intervals or at equal intervals in x or y direction (Burrough, 1984a).

Raster databases can be created by computer programmes that transform the coordinate input of digitized point, line or polygon databases into space filling arrays of cells.

The *digitizing accuracy* is limited by the resolution of the digitizer itself and by the skill of the operator. What accuracy is required also depends on the topologic accuracy of the source material and on the objectives for using the data in the geographic information system. Digitizing is time-consuming and enervating work and the main bottleneck in using a GIS. It may take nearly as long to digitize a map accurately as it takes to draw it by hand. But once a map has been digitized it can be stored indefinitely on magnetic disc or tape for future use.

Automatic entry devices. Various technologies and instruments exist that directly capture automated spatial location at the time when non-locational geographic data are recorded, e.g. in earth observation satellites such as Landsat and related image processing technology (Dangermond). This may be related to automated cartography in order to e.g. provide satellite images with texts or with linear features like roads or administrative boundaries.

Other advanced entry services include laser driven automated line following devices and automated scanner technology. This technology is still rather new, only in use at few locations and have not yet proven to be overly economically (Dangermond). They will therefore not belong to the standard equipment of most of the available geographic information systems.

Processing or editing of input data. All of the input procedures mentioned may involve a more or less substantial amount of processing and/or editing subsequent to the initial data input. In vector data structures lines may have to be removed or added, gaps in polygons closed, attribute codes revised or added. In raster data structures empty cells may have to be filled, cells with wrong attribute codes have to be recoded. For this type of editing various programmes exist. Sometimes this editing may be as time-consuming and labour intensive as the whole digitizing phase, but cannot be missed if one wants to start manipulations with an accurate database.

Output of data from a GIS

Data output is the operation of presenting the results of data manipulation in a form that is understandable to a user or in a form that allows data transfer to another computer system. *People compatible output* comes in the form of maps, graphs and tables. They can be ephemeral displays on electronic screens or permanent images (hard copies) on paper or other material. *Computer compatible output* may be in the form of a magnetic tape that can be read into another system or it may involve some form of electronic data transmission over telephone lines or radio links (Burrougn, 1984a).

Corresponding to the distinction between vector and raster data structures is the division between vector and raster display devices.

Vector data can be displayed by plotters or on screens. *Penplotters* are robot draughtsmen and essentially consist of a plane surface on which paper or other material can be laid and a penholder that can be moved independently in the x and y direction. Penplotters are available in very simple design up to the most advanced cartographic drawing tables. With *vector screens* or *storage displays* the pen is replaced by a beam of electrons and the paper by a green phosphor screen. Once the image has been written on the screen it remains visible until the whole screen is cleared. A quick semi-permanent paper copy of the image on the screen can be obtained from a *hardcopy device.* This is a simple thermal printer linked to the display generator of the screen, that prints the information on an A4 size sheet of specially treated paper in about 10 seconds (Burrough 1984a).

Rasterdata can be displayed by printers, plotters or screens. The simplest raster display is by means of the *lineprinter* or *printing terminal* and the earliest computer mapping systems were designed to use these crude but simple and fast tools for output. For these systems the attribute codes have to be translated into the alphanumerical symbols of the keyboard. With these same symbols the maps can be displayed on black and white (television) screens or on the green phosphor screens. A disadvantage is the non-square character size. Somewhat more advanced are the *matrixprinters* which can print fine dotrasters in varying density to create varying greytones or print greytones by way of several dot and linepatterns. Some matrixprinters can be provided with a three colourribbon to facilitate colour printing. Penplotters may also be used to present rasterinformation after different line- or dot-hatching symbols have been assigned to the attribute codes. At present also colour screens are available for the display of raster data.

Manipulation of data in a GIS

Many geographic information systems seem to have been developed for automated cartography and graphical display and the manipulation capabilities are focused on that. The data analysis capabilities of those systems often leave much to be desired. But fortunately also systems have been developed with a larger analytical power, although they sometimes may give less satisfactory cartographic results.

Structure of the database

In order to discuss some of the manipulation processes that are possible first the structure of the database has to be explained.

A *database* could be compared to a thematic atlas with maps on the same scale of the same area drawn in a way that they can be easily compared. A database also contains a series of *maps* or *overlays* that are defined by the same space in x and y in either vector or rasterformat and by the type of coding it can store: two digit, three digit or real figure data. Each map or overlay can then be fed with one attribute or set of attributes for the area concerned.

An *attribute* is a value of a geographical variable associated with a given topographic position. These attributes need not exclusively be vegetation categories but may include soil, geology, hydrology etc. (see also Fig. 7).

Transformation programmes

The attribute values can be transformed using appropriate logical and mathematical procedures. For study of correlation of vegetation and land-use with other land attributes these transformations are highly important. In theory it does not matter whether the data are in vectorformat or in rasterformat, but in practice map transformations are much easier implemented when using pixels rather than polygon networks.

Each map or overlay is stored separately on a storage device. For the implementation of a

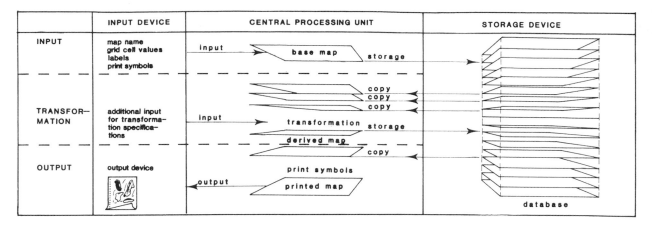

Fig. 10. Schematic overview of data manipulation in a geographic information system (source: Berg).

given transformation the necessary overlays are copied to the processor. There they are subject to an appropriate algorithm supplied by the user and converted to a new output overlay. This new overlay then can be written to the storage device as an additional overlay or as a replacement of one of the transformed overlays. See Fig. 10.

Simple transformations are the recoding of single maps, e.g. if maps have to be generalized or simplified or if maps have to be cleaned or wiped out to make them ready to receive new attribute codes. In such cases the new overlay usually replaces the old one on the storage device. Other, still rather simple, transformation programmes can combine the codes from two or more overlays into a set of codes on a new overlay or into a crosstable. These simple programmes only look at individual pixels or points and ignore their surroundings.

Somewhat more elaborate transformation programmes can also take the surroundings of pixels into account and can respond to the instruction: map all areas with code *a* within a radius *r* from an area with code *b*. In still more complex programmes also dynamic aspects, e.g. flows of matter or energy from or towards points, may be captured. Distance and the occurrence of obstructions may be included as factors that influence the intensity and way of the flow.

Especially in model-making the element of repetition over time may become important. If a process is repeating itself or is continuous the output of a first cycle of the process can be taken as the input to the second cycle and so on.

Data retrieval

Before transformation programmes can be implemented it may be necessary first to find out exactly what data are stored on which overlay. Of course a print-out of each overlay could be processed and inspected, but the printing will be rather time-consuming especially when the data files are rather large. And since in many cases it can be established at a glance whether a certain overlay is of relevance for the particular objective at hand it will also be a waste of paper. In that case the procedure of *browsing* is more appropriate.

Browsing is the visual scanning of files on a displayscreen (Dangermond). Going through complete files when one is only interested in a particular subarea of the map can be short-cut by a *windowing* technique that allows the user to specify certain windows in x and y or certain restrictions that can be taken from other overlays, e.g. only data for a specific administrative unit (Dangermond).

174

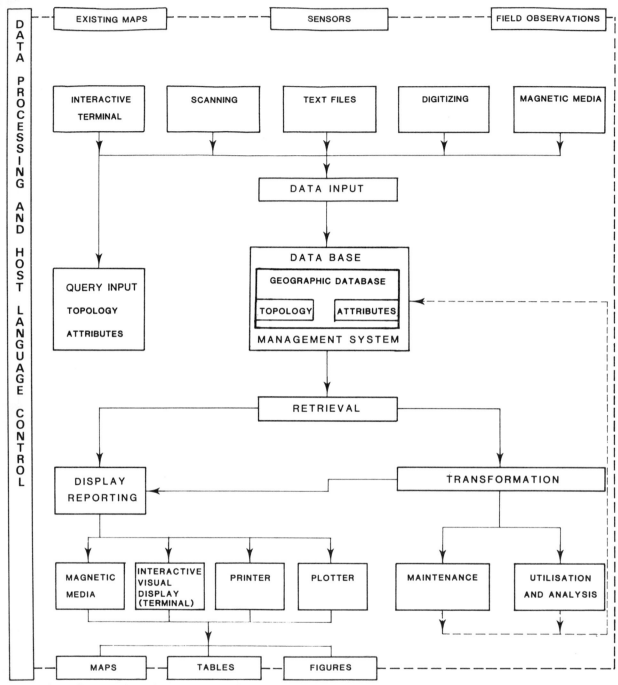

Fig. 11. A detailed overview of a geographic information system (source: Burrough, 1984a).

Measurements

Apart from retrieval and transformation in many cases measurements or countings are wanted. Therefore various programmes have

been developed in order to carry out such measurements in a GIS.

Point data can be counted in total or for a specific polygon unit. Lines can be measured from point to point or along a curve. Areas or perimeters of polygons can be measured which is a very useful application taking the place of the old fashioned time-consuming planimeter or dotgrid counting. Although it should be stated that at present there are digital planimeters attached to mini-computers with which planimetering takes about the same time as digitizing the map. Even volumes can be measured through cross-section techniques or through overlays of multiple surfaces (Dangermond).

Advantages and disadvantages of GIS modeling

Whether the geographic information system is used only for simple transformations or for more complex model-making or *modeling* a number of advantages has to be balanced against a number of disadvantages in order to make out whether using the GIS indeed is the most appropriate method to achieve a given aim or solve a certain problem. A GIS should be used as a *tool*, rather than a *toy*.

Advantages

— The user is forced – maybe more than in other methodologies – to define the problem clearly, to decide which data are required, and if necessary to collect them. Then he has to develop a logical flowchart using well-defined basic spatial operations that can be linked using computer (= algebraic) language.
— The methodology allows a very large number of options or scenarios to be examined in a relatively short time and with relatively little labour input, i.e. after the basic data have been put in by digitizing or other methods, thus allowing optimal use of the data.
— The analysis is much easier in rastermodes than in vectormodes (Burrough, 1984a).
— The gridcell (= raster) technique for map manipulation is typically much more efficient both in data storage and in the operation of analytical tasks (Dangermond).

Disadvantages

— The algebraic algorithms used in the transformation programmes only work with the assumption that all source maps are absolutely correct and that there are no important sources of error. A lot of confusion or mistakes can be created when this assumption in reality is not true.
— Another working assumption is that the spatial surfaces representing the variations of topography, cost or other variables are smooth and differentiable. But especially in the case of *environmental data* it is impossible to say with certainty what the value of the attribute at a point is, let alone to give an exact value for the whole area. The source data therefore should be regarded as no more than averages that are bounded by a degree of uncertainty that of course is carried forward or propagated by the operations of the algebraic transformation (Burrough, 1984a). Using exact procedures on inexact data does not make the resulting data more exact.

The use of a GIS for vegetation surveys and related environmental aspects

In the context of the use of GIS for environmental surveys two questions are central:
1. Does a GIS offer substantial advantages over the conventional methods?
2. If so, what type of GIS will be most efficient?
From the preceeding chapters it may have become clear that the use of a geographical information system will become interesting if comparison of many different attributes is required. That is when the interrelations or correlations of *many* different aspects of an area have to be analysed or when the same aspect has to be analysed for its change through time. In this

176

case a raster or grid-based GIS will be the most appropriate one. The ITC *USEMAP* system is a useful example.

Another case in which the use of a GIS seems promising is that in which model-making is involved, e.g. in the context of land evaluation or regional planning. Using a GIS many more alternatives and variations can be considered and calculated than with conventional means.

Also in such a case the gridformat is the most suitable one. If, however, the turnout of variations and alternatives only involves the input by digitizing of a complex landunit map and the output of suitability maps by assigning different values to the units for different alternatives the problem is mainly a cartographic one and could be solved as well by a vector-format GIS.

After the more general explanation of what geographic information systems are and how they can be used, it is time to discuss some practical experiences and applications at the hand of some case studies. These case studies are centered around the use of one specific GIS at use at the ITC, i.e. the USEMAP system. The USEMAP system is essentially gridcell based, but maps can be digitized as polygons. Also pointdata can be digitized but no single line elements. Output can be printed in alpha-numerical symbols but also in greytone symbols and even in colour. Since for vegetation and land ecological mapping we often are interested in rather large areas with a still reasonable resolution the USEMAP programmes were tested on some results of a photo-interpretation and fieldwork project in the Moneragala district of Sri Lanka. Landcover/use maps of different data (1956; 1970/72/75; 1982) were digitized as well as a landscape ecological map and a land unit map.

Resolution

A gridcell size corresponding to 1 ha was selected in order not to loose too many of the

Table 1. Comparison of the same map in different gridcell sizes.

	1	2		3		4		5	
		Gridcell of 4 ha		Difference between 1 and 2		Gridcell of 25 ha		Difference between 1 and 4	
Land-use codes*	Gridcell of 1 ha (no. of cells = no. of ha)	No.of cells	No. of ha	In ha	In %	No. of cells	No. of ha	In ha	In %
11	445	115	460	+15	3.4	18	450	+5	+1.1
12	10 182	2 510	10 040	−142	−1.4	387	9 675	−507	−5.0
21	463	118	472	+9	1.9	9	225	−238	−51.4
22	1 061	254	1 016	−45	−4.2	36	900	−161	−15.2
23	984	245	980	−4	−0.4	38	950	−34	−3.5
27	15	4	16	+1	6.7	0	0	−15	−100.0
28	199	47	188	−11	−5.5	1	25	−174	−87.4
34	31 573	7 869	31 476	−97	−0.3	1 294	32 350	+777	+2.5
35	12 858	3 177	12 708	−150	−1.2	535	13 375	+517	+4.0
41	8 793	2 173	8 692	−101	−1.1	334	8 350	−443	−5.0
42	11 274	2 816	11 264	−10	−0.1	444	11 100	−174	−1.5
43	520	129	516	−4	−0.8	22	550	+30	+5.8
44	3 695	935	3 740	+45	1.2	148	3 700	+5	+0.1
45	405	105	420	+15	3.7	18	450	+45	+11.1
46	3 212	801	3 204	−8	−0.2	141	3 525	+313	+9.7
47	2 564	625	2 500	−64	−2.5	93	2 325	−239	−9.3
51	914	224	896	−18	−2.0	38	950	+36	+3.9
53	536	139	556	+20	3.7	20	500	−36	−6.7
Total	89 693	22 286	89 144	−549	−0.6	3 576	89 400	−293	−0.3

* The meaning of the codes is not essential in this example.
Source: Berg, J. van der (1985).

categories that occur in small areas only. With a map covering an area of 37.5×45 km the total number of cells is so large that many of the programmes take quite some time to be executed and some programmes cannot be used at all. Subdividing the map in smaller portions and adding them together later might be a remedy for that, but will also mean that a lot of the advantages of the use of a GIS are lost.

A test with gridcells of different sizes showed that gridcells of 4 ha also would have given reasonable results whereas gridcells of 25 ha already give a considerable loss of information (see Table 1).

Source material

Digitizing was done from the maps that were made as the final products of the interpretation and fieldwork project. It could as well have been done from corrected and amended preliminary maps.

Digitizing of interpretation data directly from the photograph was not done because of various considerations:
— In the map compilation elements of generalization and abstraction are incorporated, resulting in some complex categories. Trying to come to a similar abstraction in the GIS after the digitizing will be very difficult as experiences with automatic classification of Landsat images show.
— In many of the photographs too few points can be identified that correspond with points on the topographical map and therefore no planimetric control can be obtained.
— Distortions caused by tilt and relief displacement cannot be corrected when directly digitizing from the photograph.
So for all these reasons it was preferred to digitize from maps.

Limitations of output

Printing out maps of this format in alphanumerical symbols is very impractical, giving piles of paper that have to be assembled again. With the greytone and colourprint, the output size of the gridcells can be adjusted and the map can be made to fit on one width of computerpaper. However, this is at the cost of the number of greytone or colour densities that are in principle available in the programme. Add to that that some colour-density combinations give only a very low contrast and it is clear that the computer colourprint is no match for conventional cartographic products, not even for lightprinted and manually coloured maps. The advantages of USEMAP have to be found in the analytical sphere rather than in the cartographic sphere.

Comparison of sequential maps

When maps of different dates are made of the same area the wish to analyse the changes is obvious. One method is to superimpose a clear transparent copy of one map on the coloured copy of the other map and to analyse visually what changes occur and in what place. When the maps are as complex as the ones of the Moneragala area, either the main trends of change are only given in descriptive terms or with much more time and effort a very complex land-use change map can be compiled which in itself needs to be interpreted again. Elaborate manipulation of the (digital) planimeter has to be carried out to measure the areas of different types of change. It is for this type of analysis that a GIS may be very useful.

Of course a comparison of all categories in one map with all other categories of the other map will also result in a change. But if producing a new map can be done by simply typing a command on the computer terminal one can easily display comparisons and analyse the changes category by category.

Fig. 12 (see colour section) shows the comparison in three colours of the forested areas in 1956 and in 1982. Areas of increase and decrease can be easily detected. If for one of these categories the landcover/use is printed out an easy overview can be obtained of which categories have replaced the original forest or which categories are replaced by the new forest. Relating these changes to the information comprised in the landscape ecological or landunit map of the area may give information about the factors behind such changes. See Fig. 13.

Crosstables relating these change maps to areal extents can be easily made. They are expressed in numbers of gridcells, but it is very simple to change them into hectares. If one gridcell already equals 1 ha it is no problem at all. The making of one large crosstable to relate the changes of all categories can give the general overview of the change pattern.

The results of such map comparison should not be considered too absolute, since there are various causes for inaccuracy. First of all many interpretation lines cannot be considered as real sharp boundaries since especially with respect to the (semi)natural vegetation transitions are often gradual. A slight shift in boundary therefore may not always necessarily correspond to a real change. Also during digitizing very small shifts in the position of lines may occur resulting in inclusion or exclusion of numerous gridcells in or from a category. Some comparisons may even reveal some quite illogical changes, in which case, interpretation or mapping mistakes are the source of errors. But for relative comparison of two sequential interpretations GIS can be a very helpful tool. A condition for doing so is, however, that both interpretations are carried out with consistent use of the same criterion. A comparison of two digitized interpretations of the same photographs by two different teams revealed about 50% differences which can only be explained by the use of different criteria to outline the same categories.

Landscape ecological and land unit maps

Landscape ecological and landunit maps both contain similar information, the main difference being that the terrain form and geology is the more dominant aspect in the landunit map and the vegetation type and structure in the landscape ecological map. Both types of maps are rather comprehensive maps containing information about soil, terrainconfiguration, slope, rocktype, vegetationcover, land-use, etc.

In principle such landscape ecological or landunit maps can be compiled by digitizing individual existing aspect maps. The problem then is that these maps are seldom available on the same scale with identical levels of generalization and that most of the boundaries are rather arbitrarily drawn to indicate gradual transitions instead of representing real sharp boundaries and therefore will seldom coincide exactly when various aspect maps are superimposed. Therefore it has many advantages to compile first a comprehensive map using existing source maps and airphoto interpretation in which the boundaries which logically represent the same type of transition in the various aspects can be made to coincide, and then digitize such a map. Once they are digitized, each of these elements can be easily separated into individual aspect maps which then can be compared and related to each other and to maps containing other information, e.g. the landcover/use changes.

Fig. 14 (see colour section) shows the landscape ecological map of the central part of Moneragala district in which the selection of the colours is made in such a way that the transition from wet to dry via an intermediate zone is reflected. This subdivision in three categories can also be symplified as in Fig. 15 (see colour section).

A relevant application of the combination of landscape ecological or landunit maps with a GIS is land evaluation.

Land evaluation

In land evaluation the characteristics of the different landunits (LU) as they appear on the landunit or landscape ecological map are confronted with the requirements of the various land utilisation types (LUT) which are considered to be relevant in order to assess the suitability of each LU for each LUT.

Fig. 12.

FOREST APPEARED   FOREST DISAPPEAR   FOREST REMAINING

LANDUSE 1982 THAT REPLACED FOREST SINCE 1956

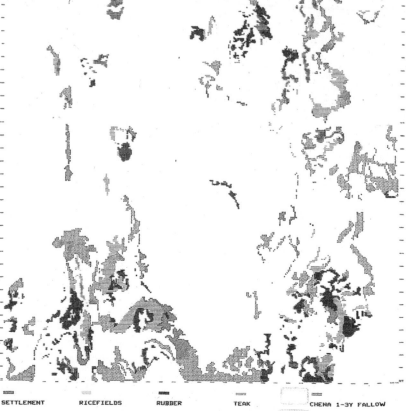

Fig. 13.

SETTLEMENT   RICEFIELDS   RUBBER   TEAK   CHENA 1-3Y FALLOW

CHENA 4-7Y FALLOW SAVANNA   WOODLANDS   TANK   ROCKOUTCROP

180

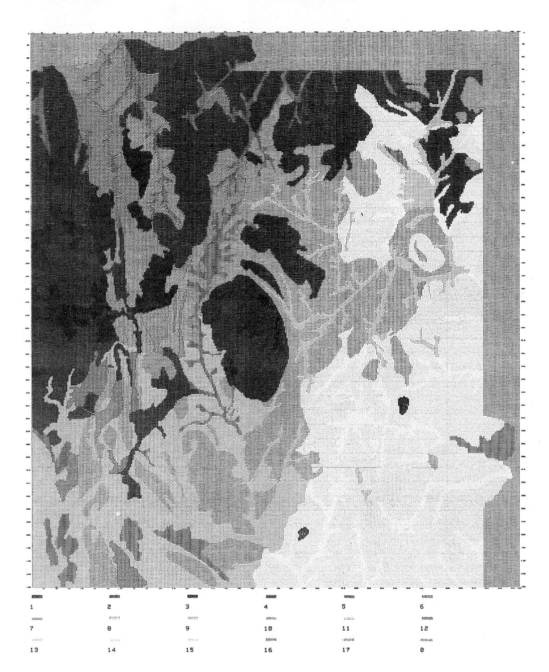

Fig. 14.

The use of a GIS makes it possible to test many more variables and alternatives than with the conventional means. Some experiences will be illustrated with examples from a case study carried out in a test area around Lemele in the Netherlands.

Case study. The Lemele area was chosen for the case study because a great variety of landscapes and land cover types occurs there at relatively short distances. These landscapes and land cover types are representative for large parts of the eastern Netherlands. In addition, adequate data

VEGETATION ZONES IN MONARAGALA-DISTRICT.   19-JAN-84 14:23:01

SURROUNDING AREA   DRY ZONE   INTERMEDIATE   MOIST ZONE

Fig. 15.

182

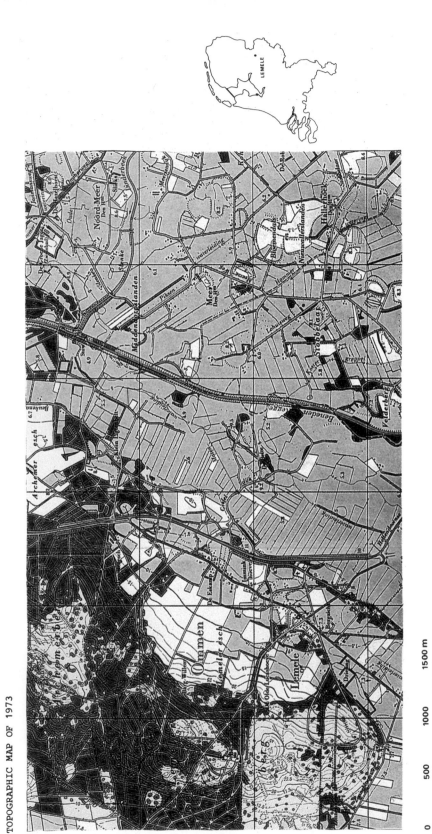

Fig. 16.

THE LEMELE AREA: LOCATION, TOPOGRAPHY, POLYGON
MAP OF LAND UNITS AND TABLE OF LAND ATTRIBUTES

TOPOGRAPHIC MAP OF 1973

1500 m

1000

500

0

Attribute column labels (read top to bottom of the list):

- soil type
- depth surface soil
- texture surface soil
- groundwater table class
- organic matter surface soil
- dense/compact subsoil
- subsoil texture
- presence high Fe₂O₃ contents
- relief
- presence peat(y) layers
- groundwater supply (50 cm)
- groundwater supply (90 cm)
- av. water capacity (50 cm)
- av. water capacity (90 cm)
- root development (maize)
- root development (s. grains)
- root development (grass)

polygon nr.	codes for land attribute classes
2	1,1,4,9,2,6,1,4,5,0,0,5,5,5,
3	1,2,4,2,5,3,0,1,1,0,0,1,3,2,4,
4	1,2,4,3,5,3,0,1,1,0,0,1,2,4,2,
5	1,2,4,5,5,3,0,1,1,0,0,1,2,4,1,4,
6	1,2,4,6,5,3,0,1,1,0,0,1,2,4,1,2,
7	1,2,4,7,5,3,0,1,1,0,0,5,4,1,2,
8	1,2,4,8,5,3,0,1,1,0,0,5,4,1,1,
9	1,2,4,9,5,3,0,1,1,0,0,5,4,1,1,
10	1,2,5,7,5,3,0,1,1,0,0,5,4,1,1,
11	1,3,4,9,4,3,0,1,1,0,0,4,4,1,1,
12	1,3,5,2,4,3,0,1,1,0,0,4,4,1,1,
13	1,3,5,8,4,3,0,1,1,0,0,5,4,1,1,
14	1,3,5,7,4,3,0,1,1,2,0,5,4,3,2,4,
15	2,2,8,2,7,4,3,1,1,5,0,5,4,1,1,
16	2,2,8,2,7,4,4,0,1,1,0,5,4,1,1,
17	3,4,5,7,4,4,0,1,1,0,0,4,4,1,1,
18	3,4,5,8,4,4,0,1,1,0,0,4,4,1,1,
19	3,4,5,9,4,4,0,1,1,2,0,4,4,1,1,r,
20	3,5,5,6,5,6,0,2,2,0,0,0,4,3,2,2,
21	3,5,5,7,5,6,0,2,2,0,0,0,4,3,2,2,
22	3,5,5,6,5,6,0,2,2,0,0,0,4,3,2,2,
23	3,5,5,7,5,6,0,2,2,0,0,0,4,3,2,2,
24	3,5,5,8,5,6,0,2,2,0,0,0,4,3,2,24,
25	3,5,5,9,5,6,0,2,2,0,0,0,4,3,2,2,
26	3,6,4,8,5,4,0,1,1,0,0,0,4,3,1,1,

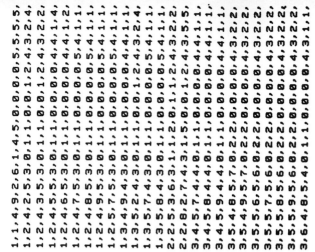

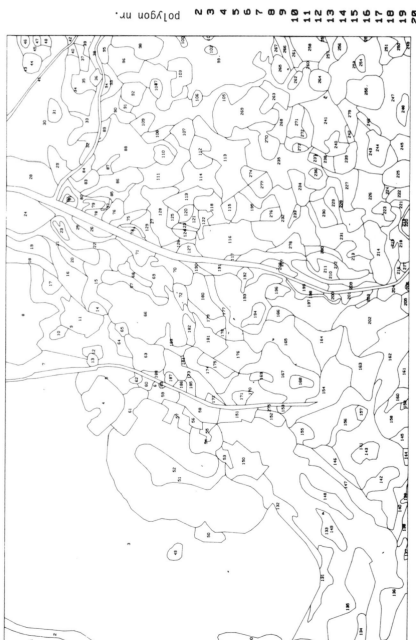

DIGITISED POLYGONS OF LAND UNIT MAP

SECTION OF TABLE WITH POLYGON NUMBERS OF LAND UNIT MAP AND CODES FOR LAND ATTRIBUTE CLASSES

on the geomorphology, soils and groundwater (mapped at a scale of 1:25,000) are available. Different generations of topographic maps, also at a 1:25,000 scale, show the land cover/land use situations during the last 80 years (see Fig. 16, left). In the area, a rapid change has taken place from a labour-intensive, mixed farming system based on food crop cultivation associated with extensive grazing by sheep in the late 19th century to a capital-intensive dairy farming system at present in which arable land is mainly used for fodder crops.

The ITC-USEMAP system was used in the case study. Data on landform, soil and groundwater was combined into a land unit map. This map was digitized (see Fig. 16, right) and converted to a data base of 13 200 cells covering an area of about 2 000 ha. Class values of land attributes (slope, relief, soil depth, groundwater depth in winter/summer, etc.) were stored in table format. Attribute values, or combinations of these, could be related directly to the mapping units of the digitized map so that, whenever required, printed maps of these could be obtained. Land cover data of 1903, 1933 and 1973 were also digitized and stored in the same data base.

The procedure of the land evaluation is illustrated in Figs. 17 to 21 (for Fig. 21, see also colour section). The land evaluation was carried out for three land utilisation types (LUTs):
— maize for fodder production;
— grassland for pasture or fodder production;
— winter rye for food production.

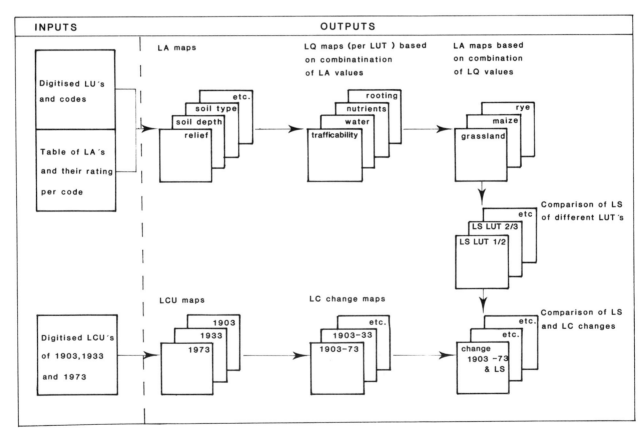

LU = land unit
LA = land attribute
LQ = land quality
LS = land suitability

LCU = land cover unit
LC = land cover

Fig. 17.

Present management and technology (i.e. the use of tractors, large amounts of fertilizers and mechanized harvesters) is assumed for the first two LUTs. Past technology (i.e. horse traction, heath sods and stable manure as fertilizer and manual harvesting) is assumed for the third LUT. Winter rye was the main arable crop in the area in the first part of the 20th century.

For each LUT, relevant land qualities were assessed on the basis of the class values of several land attributes. The land qualities were rated as high, medium, low and very low. These ratings take into account the expected yield levels under the management considered for the LUT. Four land qualities were assessed:
— trafficability (for fodder maize and grassland);
— available moisture (for all LUTs);
— limitations for root development (for all LUTs);
— soil fertility (for winter rye only).
Fig. 18 (see colour section) shows the assessment of the land quality 'available moisture' and the decision rules applied in this assessment.

The assessment of the physical suitability is based on the assumption that the least favourable land quality determines the suitability class. Fig. 19 (see colour section) shows computer-produced land suitability maps for the three LUT's. Land suitability as shown on these maps is based on yield and input levels, but does not take into account prices of outputs and inputs.

Fig. 20 (see colour section) is an example of the results of the analysis of land cover changes.

There are many possibilities to further utilize and/or expand this data base. One example of the use of the data base is given in Fig. 21 (see colour section). In this example, the following question was asked: is the considerable change from arable land to grassland between 1903 and 1973 (see Fig. 20, colour section) in agreement with the suitability of land for arable crops and grass? For this purpose, the land suitability maps for arable and grassland were combined in one map (see Fig. 21, colour section). This map was compared with the area that changed from arable to grassland. A new map was produced on the basis of this comparison. This map (see Fig. 21, colour section) shows that most of the changed area is equally suited for arable and grassland uses; relatively smaller parts are better suited for grassland than for arable land.

The case study demonstrates that the USE-MAP system can be a valuable aid for the analysis of sequential and multi-thematic data. Major advantages of the use of such a system are:
— the quick retrieval of data inputs in the form of coloured grid-cell maps;
— the quick production of new maps based on combinations of different data according to selected decision rules;
— the possibility of quick tests to determine the effect of (slight) changes in decision rules on the spatial patterns of land qualities and land suitabilities.
Disadvantages of the system used are:
— the rather time-consuming data input which may only be worthwhile in practice when a regular use of the data base for problem analysis and solving is envisaged;
— the limited topographic information that is available for orientation in the computer-produced maps;
— the relatively small cell size that is needed to reduce boundary errors (the choice of a small cell size was not a limitation in the case study, but will considerably increase the time required for data processing when a larger data base is used).
The examples given in this case study all refer to agriculture. Expansion of the data base would be possible by adding spatial data on semi-natural vegetation composition, value of habitats for birds, scenic value of land areas, information on farm size, rural settlements, infrastructure, etc. Incorporation of such data in the data base would allow the analysis of, for instance, beneficial or adverse effects of land use changes in terms of (agricultural) production, conservation, recreation and socio-economic conditions.

186

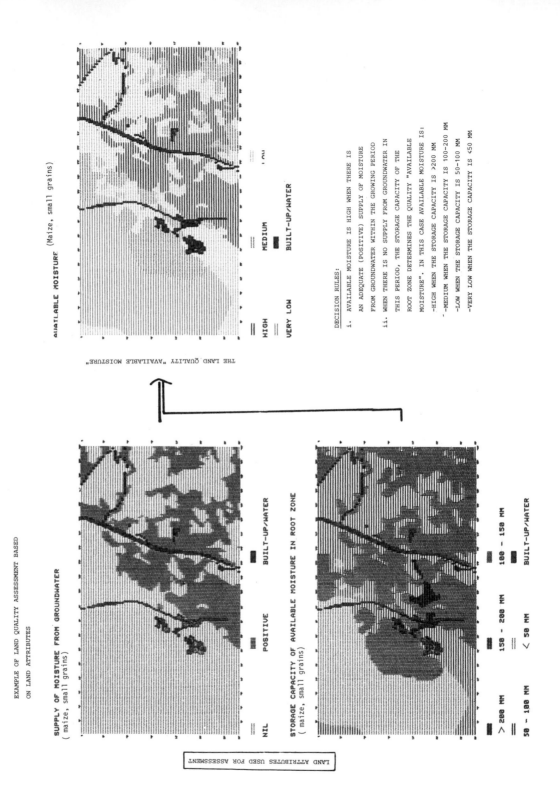

EXAMPLE OF LAND QUALITY ASSESSMENT BASED
ON LAND ATTRIBUTES

Fig. 18.

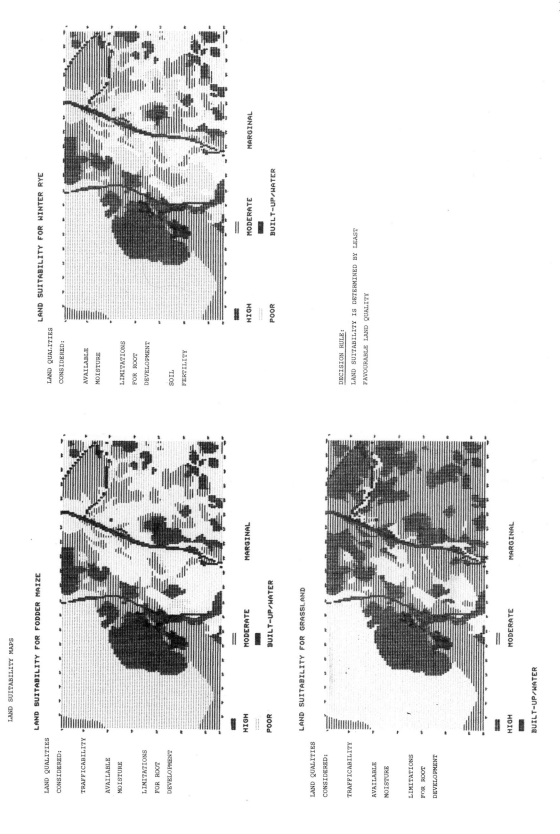

Fig. 19.

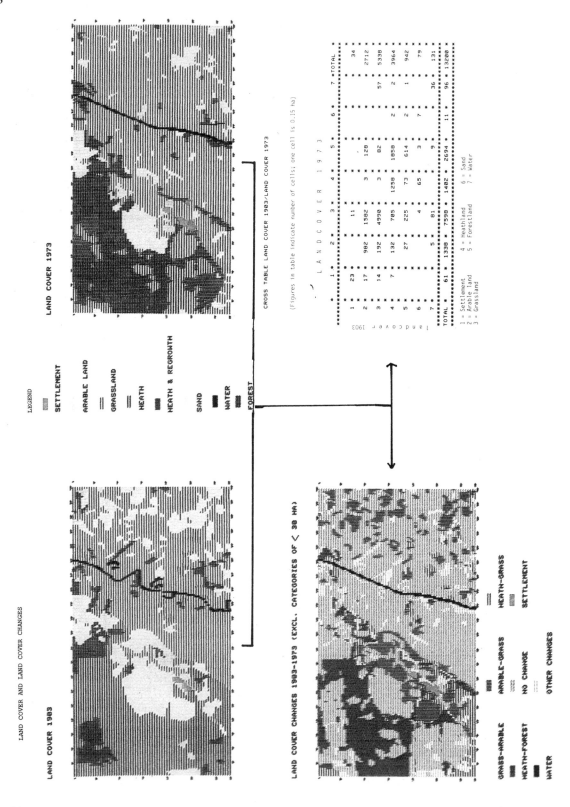

Fig. 20.

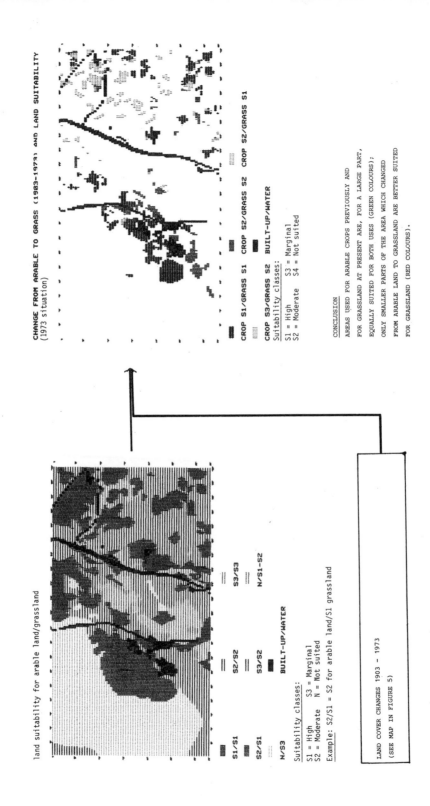

Fig. 21.

CONVERSION OF ARABLE LAND (1903) TO GRASSLAND
(1973) IN RELATION TO SUITABILITY OF LAND

land suitability for arable land/grassland

S1/S1   S2/S2   S3/S3

S2/S1   S3/S2   N/S1-S2

N/S3   BUILT-UP/WATER

Suitability classes:

S1 = High   S3 = Marginal
S2 = Moderate   N = Not suited

Example: S2/S1 = S2 for arable land/S1 grassland

LAND COVER CHANGES 1903 - 1973
(SEE MAP IN FIGURE 5)

CHANGE FROM ARABLE TO GRASS (1903-1973) AND LAND SUITABILITY
(1973 situation)

CROP S1/GRASS S1   CROP S2/GRASS S2   CROP S2/GRASS S1

CROP S3/GRASS S2   BUILT-UP/WATER

Suitability classes:

S1 = High   S3 = Marginal
S2 = Moderate   S4 = Not suited

CONCLUSION

AREAS USED FOR ARABLE CROPS PREVIOUSLY AND
FOR GRASSLAND AT PRESENT ARE, FOR A LARGE PART,
EQUALLY SUITED FOR BOTH USES (GREEN COLOURS);
ONLY SMALLER PARTS OF THE AREA WHICH CHANGED
FROM ARABLE LAND TO GRASSLAND ARE BETTER SUITED
FOR GRASSLAND (RED COLOURS).

15. General Considerations

A.W. KÜCHLER

The most accurate portrayal of plant communities is obtained by mapping them. This can be done by a variety of methods, each of which has its own peculiar merits and imperfections. It happens occasionally that an inspired author develops a new method of mapping vegetation. In the course of many years, and heeding the advice and criticism of his colleagues, he gradually improves his method until it reaches a high degree of perfection. The author is then convinced that this method is superior to all others. Frequently, he also believes that his method can be employed everywhere with the same degree of success and produce equally good results. It is now well established that such authors are carried away by their enthusiasm.

There is no single method of mapping vegetation that will do justice to all the divers purposes for which vegetation maps are now prepared, and it was quite inevitable that different methods should evolve in order to serve the whole gamut of purposes in widely different regions (Küchler, 1960a). For instance, maps like those of Malaysia by van Steenis (1958) and of the Pasterze in Austria by Friedel (1956) are not only quite unlike but required quite different methods. In a region like Malaysia, the vegetation is known only in its broader aspects and not even that is some parts of the area. In Austria, on the other hand, a great deal of skill and talent has been focused for decades on a vegetation already well known and understood in its details. The two maps could hardly have been prepared by the same method and still

be meaningful. Of course, these are maps of widely differing types of vegetation and scale. But even within a given area, one mapping method can never do what two or more methods can achieve in presenting the character of the vegetation. Scharfetter (1932) even went so far as to say that nothing will advance vegetation mapping as much as the portrayal of the vegetation of a given area according to different points of view. This is not only scientifically sound, but has been shown to have considerable practical value as well (cf. Küchler, 1956a and b).

The methods of mapping vary according to the classification of vegetation. Some of these classifications are better adapted to very large scales, as illustrated by the map of the Brabantse Biesbosch (Zonneveld, 1960), whereas others fit smaller scales much better als illustrated by Hueck's map of Venezuela (1960) and Zonneveld's (1977) map of the Kaarta region in Mali.

The need for different methods is therefore well established, and it is useful to discuss several such methods. Some of these are widely used; others are not but nevertheless rest on interesting ideas. A cross section of such methods will permit the vegetation mapper to choose whatever is best adapted to his purpose. It will also allow him to make comparisons and perhaps to modify one or more methods through simplifications, amplifications, or combinations, thus serving his particular needs even better.

The purpose of a vegetation map determines

A.W. Küchler & I.S. Zonneveld (eds.), Vegetation mapping. ISBN 90-6193-191-6.
© 1988, Kluwer Academic Publishers, Dordrecht. Printed in the Netherlands.

the content and character of the map and these, in turn, determine the method to be used in preparing the map. When the purpose of the map is clearly established and the desired content and character of the map have been determined, then, and only then, can the mapper select an appropriate mapping method. For instance: a vegetation map is to serve as a basis for the establishment of site classes. As the site class is equivalent to the biotope, a vegetation map will be useful only if it shows the plant communities as completely as possible. The purpose of the map therefore demands that the scale be large and that both structure and floristic characteristics be shown in combination, directly or implied. A physiognomic classification like Rübel's (1930) is therefore inadequate, because it ignores the flora. On the other hand, methods like those developed by Ellenberg and Zeller (1951), Gaussen (1948), or Küchler (1955) permit a satisfactory combination of the necessary details.

Basically, there are two types of methods: mapping by field observation and mapping by compilation. The latter is treated fully in Chapter 27; the following material deals therefore only with mapping methods based on observations of the actual vegetation in the field by direct observation and/or by remote sensing.

Planning the field work

Before the mapper begins to map, he must plan and organize his work. This is the only way in which to complete the work without waste of time and funds and to assure the highest quality. Such planning includes first of all a knowledge of the relevant literature. This will acquaint the mapper with the various approaches used in the past, the types of vegetation, and their interpretation. A study of the literature includes a study of maps. If large-scale vegetation maps of the area have been prepared before, they should be scrutinized with great care. The author may for instance, discover some old large-scale maps prepared by early surveyors or by railroad companies, and he is likely to find one or more maps of small scale. He should not scorn any of these; rather should he collect them, treasure them, and use them to understand and appreciate the character of the vegetation. If there are no large-scale maps, there may nevertheless be a variety of small-scale maps which will reveal general relations and affinities of the phytocenoses.

It is also useful to study all available topographic, geological, climatic, and soil maps. They help the mapper appreciate the problems in interpreting the actual vegetation and he will understand better which environmental features are dominant.

Finally, the mapper should become thoroughly familiar with the flora of the region. He should be able to identify the species on sight, including grasses when they are not in bloom. If the mapper does not know the flora adequately, he must obtain the assistance of a professional taxonomist (systematist).

Having made these preparations, the mapper will review and sift the information as it relates to the purpose of his map. He will then know in what types of categories the vegetation is to be expressed.

The inductive approach to vegetation mapping is particularly recommended. Whatever method the mapper selects, he must avoid premature generalizations and must not project into his observations concepts that he or others evolved elsewhere and at other times. The field work should always begin with establishing the vegetational units as they occur in that area at the time of mapping. In other words, the mapper should always map first what he observes: the actual vegetation. This basic mapping must not be complicated by the introduction of ideas on affinities, dynamisms, etc. Every one of these may be important but must be considered separately after the vegetation is mapped in its present state. Only experienced mappers who are thoroughly acquainted with the character and history of the vegetation of their area may successfully combine the phases of (1) mapping the actual vegetation and (2) interpreting it. However, the mapper must always be equipped with a notebook in which he records observations on dynamic and other features of the vegetation. Such a record will be of great value later, when the time has come to interpret his map.

As the mapper walks through the area to be mapped and carefully observes and distinguishes the types of vegetation, he concentrates exclusively on what he sees before him. The resulting freedom from preconceived notions will permit him to test the validity and usefulness of existing concepts and points of view. This is particularly relevant to maps of large scale, because they force the mapper to observe in great detail and very accurately.

By definition, vegetation consists of mappable units. It is therefore the first task of the mapper to establish such units. Under optimum conditions, these are homogeneous with regard to structure, floristic composition, and site qualities, but such optima may not always prevail. Indeed, the very concept of homogeneity is not entirely clear; discussions by Gounot (1956) and Augarde (1957) do not reveal any general agreement on the subject. As a first step, it may therefore be wise to single out all types of vegetation that can be clearly identified in one manner or another. What lies between these can be established more readily as further types, transitions, subtypes or variations, etc.

Of course, the eventual vegetation map is to be useful and, even during the preliminary inspection, the mapper must therefore endeavor constantly to see the vegetation and all its units as a feature of the landscape. In Sukachev's terminology (1954, 1960), the mapper should observe and record biogeocenoses and not only phytocenoses. In mountainous terrain, the three-dimensional character of the distribution of vegetation types must also find expression on the map. Troll's (1939) observations will prove useful in this respect. In part, the landscape determines the character of the vegetation, and in part the character of the landscape can be deduced from the vegetation, for the interrela-

tions between the two are most intimate. If the vegetation changes, something else must have changed in the landscape. Just what it is that has changed must then be determined by ecologists through careful investigation. Such a comprehensive approach to mapping vegetation will help the reader of the map to interpret the vegetation as an integral part of the landscape; it results in the most meaningful vegetation maps and permits their fullest exploitation.

It is desirable to map an area as quickly as possible and also in the greatest possible detail. Therefore some preliminary considerations are necessary. According to the size of the area, the available time, and the state of knowledge on the vegetation to be mapped, four types of mapping may be distinguished:
1. exploratory;
2. reconnaissance;
3. extensive;
4. intensive.

Exploratory mapping

This type of mapping applies to areas where the vegetation is not well known, if at all. Often it is difficult or impossible to identify the species, as in large areas in the tropics where the vegetation has not been studied adequately. Many people in non-tropical countries picture the tropical vegetation as a bewildering jungle with thousands of species arranged without order or patterns over vast areas. It seems hopeless to map such an area for lack of distinct vegetation types and it is simpler to refer to the vegetation of the whole area with some general term like 'tropical rainforest'. One may realize that within the tropical rainforest different types do exist, but they merge imperceptibly and cannot be identified.

There are indeed cases to exemplify such difficult conditions but in many regions these examples are the exception rather than the rule. Many of the forests, especially in regions where the rainfall is less extreme, are simpler and can be described and classified without the seemingly insurmountable difficulties presented by some rainforests of tropical lowlands. Where the forests give way to more open communities like the Brazilian campo cerrado or some of the African savannas, the possibilities of identifying particular vegetation types improve further still.

One way to overcome some of the obstacles to mapping an unknown vegetation is not to think in fixed terms, or in the terms of any one particular classification of vegetation (Küchler, 1960b). It happens often that one classification will succeed where another fails, and frequently it is possible to combine certain features of one classification with those of another. Men like Champion (1936), Burtt-Davy (1938), Richards, Tansley and Watt (1940), Beard (1944), and others have prepared very useful ways to classify the physiognomy and structure of tropical vegetation down to the finest detail. This includes the height and density of every individual synusia, the character, amount, and location of epiphytes and lianas, and even the size and character of the leaves and tree trunks. It may not be necessary to always use all characteristics listed in such classifications, but wherever any one of them can contribute to a better and more accurate description of vegetation types, it should be used. This is particularly important where such a feature permits a mapper to distinguish one vegetation type from another.

The mapper can go one of two ways and should make it clear which one he selects: (1) he may prefer to mention only those features which contribute directly to the characterization of the community. This means that while the community has many other features, these contribute nothing to set it apart from the neighboring communities. (2) Or else he carefully prepares a list of all features he deems significant, and then consistently describes all of them in every stand throughout the area. The features may at times be of very little significance but the uniformity of analysis is useful later on, when the mapper, working in his laboratory, compares his data on the various units of vegetation. On the other hand, the second way is distinctly more time-consuming than the

first, and this may make a difference to the mapper.

Therefore, while the resources to analyze the physiognomy and structure of the vegetation permit the mapper to indicate any differences that may distinguish two types from each other, the use of physiognomic data is not at all the only means at hand. The vegetational analysis can be greatly enriched and refined by floristic information. Of course, this is the very approach which, in the humid tropics, is so forbidding that it forced the evolution of advanced physiognomic classifications. But even though the number of species may seem overwhelming, there are yet certain floristic groups which differ so much from the rest that they can be treated as separate items, useful in identifying vegetation types. Such groups include especially the palms, bamboos, tree ferns, cacti, and others. Among these, however, the number of species is often relatively small in any one vegetation type. This can help materially in refining the details established physiognomically.

In addition to the physiognomic and structural analysis, supplemented and refined by floristic data, a third type of information can give additional help in establishing individual plant communities. This consists of a large variety of environmental features which affect or even control the extent and distribution of certain vegetation types. The task of the vegetation mapper is therefore to recognize the various biotopes of the landscape. Both the physiognomic and the floristic features of the vegetation react usually to changes in the environment, especially where such changes are relatively abrupt. A few examples may illustrate this. Phytocenoses on river flood plains which are periodically inundated can often be distinguished from those on low terraces rarely flooded and from those on still higher land, always out of reach of even the highest floods. Or compare the flood plains of rivers with the steep sides of the valleys, and possibly with less steep uplands: each is likely to have its own peculiar type of vegetation. The sequence of salt water, brackish water, and fresh water displayed along the seashore reveals a zonation of the vegetation, and even the beach alone can possess two, three, or more striplike vegetation zones, based on the amount of spray blown by the wind from the waves to the shore. Changes in the substratum are often reflected in the vegetation with remarkable clarity, and limestone, sandstone, granite, lava, sand, and organic accumulations all produce their characteristic plant communities. Varying degrees of exposure to drying or rainbearing winds can result in noticeable differences in the vegetation of mountainous regions as can many other microclimatic features.

Such physiognomic, floristic, and ecological features permit the mapper to identify a host of plant communities even if he should be mapping the vegetation of an area with which he is quite unfamiliar. It is therefore entirely feasible to map an unexplored vegetation and thereby lay the foundation for future research. The example offered by Wyatt-Smith (1962) illustrates this case.

Another way of exploratory mapping was demonstrated by Hueck, who had the task of mapping Venezuela, an area larger than Texas and Oklahoma combined or about as large as France, West Germany, Switzerland, the Netherlands, Belgium, and Luxemburg put together. Hueck flew back and forth across the country, mapping the vegetation from the air. He marked the start and direction of a given flight on graph paper and then noted the character of the vegetation at regular intervals and, in addition, at all clearly identifiable points in the landscape. Sometimes he was equipped with base maps but often there were none. Even those available proved so unreliable that occasionally galeria forests had to be moved for miles in order to accompany their rivers! Yet Hueck was able to produce a remarkably fine physiognomic vegetation map of Venezuela (Hueck, 1960) which can now serve as a guide for more detailed studies.

A vegetation mapper must be thoroughly acquainted with all possible approaches to the description of the vegetation. He can then use a wide variety of resources and distinguish units of vegetation that might well have been overlooked. The chief purpose of such a broad approach is to permit the actual mapping of vegetation, especially where highly organized systems cannot readily be applied. What is important here is to have the vegetation mapped at all. Once different types are shown on a map, the more important ones can be singled out for further study.

Reconnaissance mapping

Reconnaissance vegetation mapping is in a sense the same as exploratory mapping, but the term is commonly employed in regions where the vegetation is already better known than in unexplored regions. Whereas exploratory vegetation mapping usually serves the purpose of discovering the general nature of the vegetation or of the biogeocenoses, reconnaissance mapping is mostly a preliminary step to more intensive mapping. It fits best into large parts of the United States and Canada, the Soviet Union, much of Africa, Australia, Argentina, etc. Smaller areas permit greater detail but the chief function of reconnaissance mapping remains the same; to give a general impression of the vegetation as a prelude to work at larger scales.

Reconnaissance maps imply the availability of some kind of base map on which the field observations can be entered. If the area to be mapped is very large, the base map must be of relatively small scale. But the danger of misjudging an appropriate scale for the base map is quite real. For instance, Tansley and Chipp (1926, p. 36) state that 'except in totally unexplored regions, some sort of survey map will usually exist, and if it does not, it is clear that a topographical survey must be made either before or concurrently with the recording of vegetation. A map on the scale of 1:100,000 or thereabouts can be used as a field map for reconnaissance work (though this is on the small side) but the largest scale map obtainable (up to, say, 1:25,000) should be used.' For Englishmen whose country has been completely mapped at very large scales, it may be difficult to appreciate that even in a country like the United States, technologically among the most advanced nations on earth, there are large areas for which maps 1:100,000 or larger do not exist, even several decades after Tansley and Chipp made their statement. This is, of course, even more true of many other countries.

Fortunately, map sheets at 1:1,000,000 are now available for practically all parts of the world. When during the process of mapping, a 1:1,000,000 sheet is found unreliable in some of its features, detailed notes on the errors should be taken during the field work so as to permit the eventual vegetation map to be more accurate than the original base map. Even for very large areas and for general surveys, the scale of 1:1,000,000 should be the smalles scale to be considered for base maps. In the great majority of cases, the areas to be mapped at any one time are small enough to justify base maps of scales larger than 1:1,000,000.

In the old days, explorers mapped the vegetation along their routes of travel through unknown lands. But their days have passed. With the availability of modern base maps and possibly of aerial photographs, such 'route maps' have slipped into history and can now be justified only in rare cases. The need today is for maps that show areas rather than lines and while in the days of the early explorers this was not feasible, modern technology has reached a level where that need can be satisfied. As a matter of fact, line surveys are still used but they are placed so closely together that they enable the mapper to use such lines as a basis for a vegetation map of the entire area. In a sense, the lines become part of a grid. When the lines of such a grid are spaced closely the vegetation of the entire area will be included.

Where modern topographical maps exist, the observed vegetation can be shown directly on the map. The use of color crayons is advisable, but care must be taken that every color and every symbol is explained on the margin or the back of the map. In countries where communi-cations are good, as in the United States and in western Europe, much reconnaissance work can be done with the help of an automobile, but the mapper should not be the driver of the car. The journey may lead for many miles in one direction but a return journey should be planned on roads parallel to the outward trip, so as to cover different, though perhaps similar, terrain. Several such round-trip journeys will cover a large area in a relatively short time. Whenever the mapper is in doubt about the dominant species, he should stop and ascertain them or collect samples, mindful to number them so that he can later, i.e. upon identification, properly relate the species to the areas where he collected them. If at all possible, the identification of the species should be done upon return from the day's drive or, at the latest, immediately after the end of the trip, if this lasted more than a day. On such trips, the mapper must also be careful to map only the vegetation in the proximity of the highway and confine his observations of more distant terrain to the notebook. If he nevertheless wants to record his observations directly on the base map, he should distinguish between vegetation clearly seen near the road and distant types for which he can vouch with less assurance. In a sense, this is the modern version of the ancient route maps but there is an important difference: on the modern reconnaissance map the spaces between the routes can easily be filled in with the help of cross journeys and aerial photographs. Remembering that he is confining himself to reconnaissance work, the mapper should avoid spending much time on investigations of soil or geology. Such matters are better left to more intensive work. In mountainous terrain it is often possible, with the help of powerful binoculars, to recognize vegetation types of one mountain side from another one across a valley. This saves much time as progress in mountains is often slow. The general accessibility of the landscape in the United States leaves the recognition of the species as the major problem. If the mapper works in an area where he does not know the flora well, he will find it less expensive to hire a taxonomist who is well acquainted with the local flora than to do without one.

In many parts of the world the landscape is less accessible than in the United States or in western Europe. One must therefore rely on other means of locomotion. Trains, busses, horseback, and flying are all possibilities, but the latter is useful only if the plane remains at low altitudes. A helicopter can come close to being ideal. Trains and busses have the disadvantage that they do not stop when this may be desirable and travel on foot is very slow and seriously retards progress.

The actual mapping in the field is affected by a variety of features. The time available to the mapper is a consideration only when it is severely limited. In that case, experience will soon teach the mapper how much territory he must cover every day in order to complete his task. However, while the mapper will want to complete his field work in the shortest period possible, his objective should not be record speed, but high quality. Unforeseen delays are almost inevitable and it is better to return from the field with a good map of part of an area rather than a poor map of the whole area.

Usually the mapper can estimate in advance the size and scale of the planned vegetation map in its published form. This enables him to calculate the minimum size of an area he can show. However, remembering that he is engaged in reconnaissance work, he will find it more practical to record only areas which are very much larger than the minimum area. This will speed up his work considerably and keep his map from being cluttered.

A reconnaissance map will give its reader a good though broadly generalized view of the vegetation of a region. It can reveal clearly where problem areas exist or are likely to arise, thus guiding the reader in his planning of detailed mapping. One reason why even a reconnaissance map should always be of the highest possible quality is the very fact that the sections to be mapped in detail may not be contiguous and can therefore be related to one another only on the reconnaissance map. The latter is the only link between all parts of the map or between all types of vegetation in the mapped area.

Gaussen and Vernet (1958) called their vegetation map of Tunisia the Tunis-Sfax sheet of the International Vegetation Map at 1:1,000,000; it is essentially a reconnaissance map. The implication of the title is that the vegetation of the world is to be mapped at this scale. Such a development should receive every possible encouragement. The International Map of the World at 1:1,000,000 can serve as a uniform base, and progress could be rapid if every vegetation mapper mapped the vegetation on the sheets that cover his country or at least the sheet with the area in which he lives.

Extensive mapping

There is no clear boundary between reconnaissance and extensive mapping, nor is there one between extensive and intensive mapping. For exploratory to intensive mapping there is, therefore, a gradual change, and one should not insist on placing a map in any one category. As extensive and intensive mapping merge imperceptibly, the differences become a matter of degree: intensive mapping is mapping at a high degree of intensity and extensive mapping implies a lower degree of intensity.

Terms like 'intensive' and 'extensive' are therefore relative, depending primarily on scale and map content. For a mapper who always works at scales of 1:5,000 or larger, a scale of 1:25,000 will seem extensive. But this same scale will imply intensive mapping to one who usually maps at 1:250,000 or 1:1,000,000. Another distinction can be made on the basis of the map content. In the case of extensive mapping, the different types of vegetational landscapes occupy the center of attention whereas the result of intensive mapping reveals the individual vegetational features of a given landscape, resulting from the microclimate and other similar details. The vegetational map of the Nanga Parbat (Troll, 1939) is intensive at at scale of 1:50,000. Similarly, the Pflanzenstandorskarte (1:50,000) by Ellenberg and Zeller (1951) in intensive. On both maps the correlations between vegetation and environment are shown in a detail that permits an intimate insight into the character of the vegetation and of the sites on which it evolved. Larger scales would not appreciably alter the character of these maps or the methods of their production, in spite of the possibility of showing more detail.

Gaussen on his map series of France at 1:200,000 also establishes relations between the vegetation and the environment, especially the climate. But at this scale, a regional survey has replaced the details on the other maps and the work has become extensive. Instead of analyzing individual phytocenoses, the broader vegetation types are shown and thus the character of these maps is different, too. Indeed, the difference between intensive and extensive mapping may be based on the individual plant communities as the criterion. The mapping may be thought of as intensive if the individual plant community is a matter of major concern; if this is not so, then the mapping may be considered extensive.

It is usually the purpose of the map which dictates the mapping method. If the mapper bases his choice of an appropriate mapping method on the character of the vegetation, he should begin by focusing his attention on the most complex part of the area. Once he has satisfied himself that the method of his choice can be relied upon to produce the desired results where the vegetation is least homogeneous, he will find no serious difficulty in the remainder of his area. If, on the other hand, he begins in the simplest section, the procedure is reversed and he may find himself confronted with a series of increasing difficulties, leading to continual revisions of his method. It may also imply that the mapper will have to retrace his steps and review the work already done in order to keep it in harmony with the section done last.

In portraying the vegetation on the base map, the physiognomy is recorded first. A method of analyzing the physiognomy and structure of the vegetation is described in Chapter 4; it is adapted to mapping at any degree of intensity. In extensive mapping, it is not necessary to show every individual synusia; the mapper will prefer to limit himself to the uppermost stratum of the vegetation or to those two strata which seem to him the most significant. Likewise, the dominant species should be listed in every established type but there is no need for detailed floristic analyses. However, the mapper may find it worthwhile to mention the dominant species of each synusia he includes in his physiognomic description. It will be useful to view the recorded plant communities in their landscape setting and to link them broadly, where possible, to altitude, exposure, rock, water, etc. This is largely a matter of appropriately formulating the legend items, as has been discussed in Chapter 11.

Aerial photographs are most useful. They give an amount of detail that fits extensive mapping very well and frequently suffice for the preparation of the map.

Intensive mapping

Intensive mapping is, just as in the case of extensive methods, a generalization of the facts, although the mapper attempts to reduce the degree of generalization to a minimum. Every plant community is mapped individually as far as the scale will permit. The mapper now records every detail of the structure and all species instead of the dominant or most abundant ones only. Intensive mapping usually implies a small area, but this may require the same amount of time as a much larger area mapped extensively. It is now possible to reach an unprecedented degree of accuracy in intensive mapping, based on an advanced technology which results in superior base maps of any desired scale, and on high-quality aerial photographs which permit the precise mapping of vegetational boundaries. The mapper, however, must not permit himself to be seduced by beautiful photographs. Granted that scale and quality can now approach the ideal, the photographs do not give adequate information on all features of the vegetation. In forests, the upper synusia often covers or blurs the lower synusias to the effect that a careful study of every individual stand remains indispensable. In grassland communities, aerial photographs frequently fail to show the details the vegetation mapper wants to record. On the other hand, the aerial photograph is the ideal means for mapping the outlines and boundaries of the communities. The quality of large-scale vegetation maps can now be assured as far as the tools are concerned.

During the past decades, the mapping methods have been greatly improved, too. However, much discussion continues on which method is the best, the most reliable, the fastest, the cheapest, and so on. Some authors, believing they are objective and scientific, use statistical methods, but a lot of numbers is not science, and some statistical methods lead to poor results. Objectivity is, of course, desirable, but it is not always attainable; there are instances when subjectivity seems preferable, as for instance in the choice of sample plots. Such a choice must always be left to the mapper, and his good judgment and experience are often better guides in selecting a sample plot than some statistical method.

From the mapper's point of view, the important first step is the recognition of mappable vegetation units. How he determines such units may vary from one occasion to the next. When his work in intensive, it should always include the most detailed description feasible of (1) the physiognomy and structure of every individual phytocenose, and (2) the floristic composition and its characteristics. Recording these features on the Phytocenological Record will facilitate and speed up this work and assure the desired and needed completeness of information. The vegetation is then adequately described as it is at the time of observation. In his notebook, the mapper should collect further information on the vegetation, such as dynamic features, land use, etc.

The recognition of plant communities is based on the experience that a given combination of growth forms and taxa will occur in an area where given environmental conditions prevail. Such an area may be large or small; but the assumption is always that, if its character is relatively uniform, the plant community will also be relatively uniform. Where environmental conditions are quite unfavorable to plant growth, as in deserts, on salt pans, in high latitudes and in other extreme conditions, the vegetation may indeed be remarkably uniform with regard to both physiognomy and floristic composition. But as living conditions improve, it becomes increasingly difficult to speak of real uniformity. The vegetation is then perhaps 're-latively' uniform, i.e. it is uniform in its broader features and its minor local variations may be ignored under given circumstances. What constitutes minor local variations depends on the purpose and the scale of the vegetation map. For instance, Friedel (1956) mapped the Pasterze area in Austria at a scale at 1:5,000 and showed much fine detail. But a large amount of this detail would have been suppressed, i.e. incorporated in broader units, if the scale had been 1:20,000, which is the standard scale of the Carte des groupements végétaux de la France, of the Carte de la végétation de la Belgique, and others. All these, however, are considered intensive mapping. On the other hand, Friedel's Pasterze map might have been different, had it been published at a scale of 1:1,000, although it may be argued that a larger scale would not have changed Friedel's map because the extreme environmental conditions above the alpine timberline result in a rather simple vegetation which can be shown in all its details at 1:5,000; a larger scale would therefore mean nothing but an enlargement of the map without leading to a more refined content.

Phytocenoses are not usually uniform. The distribution of growth forms and species within the community is irregular and often leads to more or less distinct subdivisions without boundaries. Such a random distribution makes

200

it necessary to establish an abstract phytocenose, i.e. one that incorporates the various
characteristics observed in the field, even
though they may not all occur on the same sample plot. It is this abstraction that is then recorded on the map. During the field work, the
mapper may very well map individual stands,
give each one its own name, and treat them as
individual units even though they are essentially alike. The abstraction is best made in the
laboratory when the nature of all mapped communities can be compared.

The mapper should always move about on
foot in his area to acquaint himself thoroughly
with the general features of the landscape and
its vegetation prior to beginning his mapping
activity. This aids the mapper materially in recognizing what vegetation types are most characteristic in his area and on what sites they find
their optimal living conditions. The first observation in the field is directed toward the physiognomy and the structure of the vegetation. It
happens often that two communities differ in
both physiognomy and floristic composition so
that the more readily observed physiognomic
units imply floristic units as well. For instance,
a landscape may be covered with a needleleaf
evergreen forest of spruce (*Picea rubens*) which,
in places, is enriched with birch (*Betula papyrifera*) and in other parts with birch and fir (*Abies
balsamea*). A closer look reveals no undergrowth in some parts, a few widely scattered
forbs like bunch berry (*Cornus canadensis*) in
others, and in a third section bracken (*Pterid-
ium aquilinum* var. *latiusculum*) and dwarf
shrubs like blueberries (*Vaccinium angustifol-
ium*) seem rather common. These distinctions
are physiognomic; they are also floristic. Frequently, a physiognomic unit consists of two or
more floristic divisions. This means that several floristically distinct communities are physiognomically alike. The opposite happens, too,
on rare occasions.

Physiognomic and structural features on one
hand and floristic characteristics on the other
are the basic criteria in recognizing a plant community. They allow a detailed analysis and also
comparisons between the different communities. The mapper in the field, wishing to establish units that he can record on his map, must
therefore base his observations on these criteria.
They have the great advantage that they are
common to all types of vegetation anywhere in
the world.

In intensive vegetation mapping, an analysis
of the site is necessary too, for at large scales the
vegetation should not only be described but
also interpreted. An interpretation, however, is
not feasible without an analysis of the sites.
Mapping vegetation, strictly speaking, means
showing the vegetation on a map. But the value
of such a map, especially if it is done at a large
scale, is greatly increased if the various types of
vegetation are shown in their relation to the
landscape. Certain landscape features exercise a
controlling influence on the vegetation pattern
and a vegetation map is more useful to its reader if he can see the relations between the vegetation and the environmental characteristics
that are responsible for its distribution in his
area.

Relief, slope and exposure are noted with relative ease and should never be ignored. Where
geological maps are available, they should be
consulted; and even without them, the mapper
will often be in a position to distinguish the
general geological character of the landscape.
Many mappers ignore the geology of their area,
thereby weakening their maps unnecessarily.
This is not to say that the geological features
should be shown on a vegetation map. In fact,
they should definitely not be shown. But the
map is greatly enriched if, in its legend, vegetation types on limestone can be distinguished
from vegetation types on granite, etc.

A major feature of the environment is the
soil, which is in part the result of the vegetation
and in part derived from local bedrock. In some
areas, wind, water, and ice deposit loess, silt,
and till; the soils are then independent of the
local bedrock, at least if these deposits are so
thick that the roots of the vegetation do not
reach through them. Soil maps are most valuable tools for the vegetation mapper and he

should obtain them if at all possible. Where no soil maps are available, the necessary soil samples can be obtained with a soil augur. However, it is better to dig a pit, about 1 m^2 and with walls smoothed out carefully. This permits very close observation of the soil and its profile, but field data must be supplemented with laboratory tests. Soil observations and analyses require special training and the mapper will find it most useful to get such training or else enlist the cooperation of a professional pedologist.

Another feature of the environment which can have a decisive effect on the distribution pattern of the plant communities is water. Frequently there are distinct contrasts between the windward and the leeward sides of mountains, as for instance in trade wind regions or in the Sierra Nevada of California. The amount of available water is sometimes a matter of altitude so that increasing elevation means more water. The mountain ranges in the arid west of the United States supply good examples and Shreve (1915) has made excellent studies of these phenomena . The available soil water is also of major significance. In the vegetation-water relations it is not only the precipitation falling on the ground that is important, but also the water-holding capacity of the soil, the depth of the water table and its fluctuations, the rate of flow of underground water, and its chemical nature. In the Flint Hills of Kansas, horizontal geological strata outcrop on the slopes and the differences in their water-holding capacity are faithfully reflected in the different plant communities on each outcrop. Some of these water relations can be observed only by specialists and are not of immediate concern to the vegetation mapper. But he should consult with such a specialist whenever he wishes to both map and interpret the vegetation and finds the distribution pattern of the plant communities baffling.

Methods of describing the physiognomy and structure are discussed in Chapter 4; the reader is referred to these for further information.

The detailed floristic analysis of plant communities records far more than the dominant species to which the mapper confines himself when mapping extensively. The detailed floristic analysis is not simply a more intensive version of the extensive approach. Besides recording all species (cf. Chapter 5), it is now important to indicate the role they play in the community. This is one reason why it is always preferable to precede the floristic analysis with a structural analysis. The observed species are then studied in each individual synusia separately and so recorded.

Minimal area

Mappers should always be concerned with the minimal area. This is the smallest area that can be shown on the map, or perhaps that should be shown. A first class printer can print remarkably small areas. Assuming then that he prints all such tiny areas that the mapper may compile, the resulting map may be illegible, and hence worthless. The mapper must therefore use his best judgment and decide how much detail he wants to show so as to present as many characteristics of the vegetation as possible without cluttering the map. He must also understand that the minimal area on a black-and-white map is substantially larger than on a map printed in color. On a black-and-white map, differences in the vegetation are usually shown by using different patterns. In order to make sure that the reader understands which pattern is shown, the entire pattern must occur within an area. For instance, if the pattern consists of dots spaced a certain distance apart, then obviously the pattern is not revealed if in a small area only one dot appears. There should be at least four dots. On the other hand, on a map printed in color, a small dot printed in red will stand out sharply if surrounded by other colors. It can therefore be small.

The smallest colored area that a mapper should consider is a circle with a diameter of one millimeter. This statement may be modified when applied to areas which are very long and narrow, e.g. galeria forests. In such cases the length of the vegetation type somewhat compensates for the width, and the latter may therefore be somewhat less than 1 mm. The vegetation map of the United States (Küchler, 1964) exemplifies this approach.

Sometimes, the mapper wishes to portray a vegetation type which, for whatever reason, seems important but the area of which is so small that it is less than the minimal area allowable at a given scale. The mapper can then choose between two ways: either he enlarges the area. He will justify this with the argument that if he is justified to suppress units of vegetation when the scale is too small, he can also enlarge them enough so they can be shown on the map. Obviously, this argument is useful only in exceptional cases. The other way is not to show the area in its proper outline and extent but instead replace it with a symbol printed in a bold color like red. The symbol then indicates that at the location of the symbol, the type of vegetation occurs but its area is too small to be shown.

There is, of course, a close relationship between the minimal area and the scale. One square millimeter on the map corresponds to an area in the field of $1\,km^2$ at the scale of $1:1,000,000$, but at the scale of $1:10,000$ this same square millimeter will correspond to $100\,m^2$. When the mapper is engaged in field work, this consideration will be useful. If he knows the approximate scale of the map when it is published, then he can calculate the minimal area and use it in the field when he considers the finer details of the vegetation.

Photography

Photography is extremely useful for vegetation mappers. Aerial photography is essential in modern mapping; it is discussed elsewhere in this book. But there is also photography on the ground, and this is far more important than some mappers seem to realize. The best practice in photographing vegetation on the ground is to take at least one, preferably several photographs of every plant community that the mapper records and intends to present on his map. If he is mapping at a large scale and has a topographic map at his disposal, then he should record the exact spot where the photograph is taken and also the direction in which he was facing while taking the photograph. This is done by marking the letter V on the map in such a manner that the vertex of the letter is located at the point indicating the mapper's position while the two arms of the V include the area recorded by the camera. This means that, on the map, the V opens in the direction in which the mapper pointed his camera. The number of the photograph is then written on the map as closely to the vertex of the V as possible.

Having taken the photograph and recorded it on the map, the mapper records in his notebook first the number of the photograph, then the plant community which he photographed or the number of the Phytocenological Record (if he used it), and finally the particular features of the phytocenose which he considers significant in selecting this particular view of this particular aspect of the plant community. The photographs will later assist him materially in recalling what the observed and, above all, in interpreting it. If at all possible, he should have his films developed at once. If it turns out that some of them are not so good as expected or desired, he may be in a position to repeat his efforts with more success.

16. Survey Approaches

I.S. ZONNEVELD

Introduction

The choice of survey approaches for vegetation survey depends on aim and means available or scale required and kind of vegetation to be mapped. Lack of aerial photography or other remote sensing means will force the surveyor to apply old-fashioned field survey methods using topographic maps as a base. Some semantic information may also help him in guiding his survey.

Are teledetection means present then at a very large scale (detailed mapping) the 'photo-guided field survey' will be appropriate. If the classification (legend) will be simple (land-use or main formations) the 'photo-key method' may be feasible. In most cases using an inter-mediate to small scale, the (ITC-) landscape approach using teledetection (photo-) means will have to be used.

We have already stated before that other re-mote sensing records such as radar, MSS* and to a certain extent also thermal infrared, can be transferred into photograph-like images that can be interpreted along the same principles as photographs. Even if they are reprocessed into other images (computer line prints) the inter-pretation follows the same methodology as landscape-guided, or key photo-interpretation or even the photo-guided field survey.

The systematic reconnaissance flight using small aircraft and direct observation is used for

monitoring vegetation changes and is mainly developed for surveying moving targets like domestic and wild animals.

For experienced surveyors of various types (soil, geomorpology, vegetation, geology) it will be clear that almost every practical approach makes use of the principles of each of the above methods. In a clear key-type interpretation sur-vey some landscape aspects may be included. A pure landscape type survey always has some similar classification procedures as in keys. In a pure key interpretation the ground truth collec-tion will be guided mostly by the photo image, and in the systematic landscape approach it will not always be possible to have a strict land(mapping) unit classification and hierarchy ready before the sampling stage starts, which means that a check of various plots belonging to the same mapping type will be necessary. If one has the chance it is always useful to fly in a small aircraft over the area or at least to have a view from a high point. It contributes to the understanding of the whole. Nevertheless one of the methodological principles mentioned will usually dominate, depending on aim, scale, type of area and experience of the surveyor. The approaches will be described separately below.

Field survey without aerial photos or other teledetection means

In the 20th century it is ill-advised to do vege-tation surveys not using aerial photo-interpreta-

* MSS = Multi Spectral Scanning.

A.W. Küchler & I.S. Zonneveld (eds.), Vegetation mapping. ISBN 90-6193-191-6.
© 1988, Kluwer Academic Publishers, Dordrecht. Printed in the Netherlands.

tion. Still, even when aerial photographs are available there may arise a need to survey without the help of photo-interpretation. This may be so in the following cases.

— In detailed and semi-detailed surveys in intensively cultivated areas, where delineated boundaries between vegetations that are not shown on photographs are required (weed communities under crop cover, etc.).

— Check areas in reconnaissance surveys. If possible photo-guided field survey should be applied.

— A third need for field survey is where for some reason no aerial photographs or other remote sensing images are available. This should not occur but sometimes it does.

In all these cases the existence of a base map with sufficient topographic details to enable accurate orientation is essential. Topographic maps are usually provided by the land survey offices. In case they don't exist the surveyor should make such base maps himself. For this mapping refer to geodetic handbooks.

The type of mapping procedures not using aerial photos or remote sensing images are mainly characterised by the pattern of the observation points, and depend mainly on the possibilities of orientation in the field, and in relation to this, on scale. One can distinguish among: transect method, grid method and 'free' method. In the transect method (or base line surveying, cf. Steur, 1961) the surveyor walks through the area in a straight line that has been plotted previously on the base map and is chosen so that it runs as much as possible in a direction where a rapid succession of vegetation types can be expected, e.g. perpendicular to a slope, a river etc. Within these lines the surveyor makes notes wherever the vegetation changes (boundaries). If he has already a well-worked-out vegetation classification, in between these boundaries he notes only the vegetation type. If he still has to make a classification he should make occasionally a more detailed description. The boundaries should be drawn on the map as far as he can see them from the line where he is walking. The next line will be paral-

lel to the previous one and the same procedure will be repeated. The lines will be selected so that from each line about half the distance between those lines can be seen. In reconnaissance surveys this is often not possible; there, the course of the boundaries in between the unseen area will be drawn by estimation. In a mountanous or hilly area one usually can see over longer distances than in flat terrain. A telescope or pair of binoculars is essential in these surveys. In detailed surveys in crop lands and dense forest it is often necessary to estimate parts of the boundaries in between the lines, because there only a very small area of several meters can be seen.

When the base map is not very accurate the lines should be rather rigid and the surveyor should use a compass and measure or estimate the distance he walks rather accurately, although he never needs to aim at photogrammetric accuracy which is a waste of time! If a good base map (or photo) with sufficient orientation possibilities is available he may deviate more from the rigid line. The extreme of such deviation is the 'free' survey. Usually a very accurate base map with a wealth of orientation marks on it is required to make a 'free' survey feasible.

The transect or base line method is in most cases the most useful and least fatiguing one. The opposite of the free survey is the grid survey. This is in fact a transect method where the base lines may form a grid of squares. For very detailed surveys on scale 1:100 and similar surveys this is the best method. For such large scales accurate maps and orientation points never exist. A carefully measured grid is then the only basis for an accurate orientation. Such mapping, however, is more 'sampling of patterns' than a mapping procedure, and moreover tends to sampling methods similar to point measurements.

In cultural areas with a regular pattern of farm or forest lots (parcels) the practise of semi-detailed to reconnaissance survey (ca 1:50,000) might be to visit each lot and note the vegetation type. A grid or base lines are not necessary because all the lots are on the map and form a kind of grid already.

In detailed survey it is not advisable to use the parcels as units. Often various communities may occur in one parcel and it is just one of the aims of the vegetation survey to show the diversity that may exist in one parcel. Also without aerial photographs one can (and should) make a kind of interpretation of the physiography of the terrain with the help of existing topographical maps and sample (and plan base lines, etc.) according to the units that can be derived from such a map interpretation. Old maps (giving land-use patterns of the past, usually well adapted to natural conditions in those days, may also be very useful for this purpose.

Between mapping purely on a photo-interpretation basis and mapping not using aerial photographs at all, many transitions may exist as

will be clear from the foregoing. Photo-guided field survey of Küchler (see Chapter 28), is such a transitional type.

Also Viktorov, Vostokova and Vyshivkin (1964) describe various mapping systems used in Russia belonging to the above-metnioned types and the combination of these mapping systems with aerial photography.

Photo-guided field survey (see Fig. 1)

The most elementary and simple way of using aerial photos or other remote sensing means is using the image as a field map. Orientation on a good photo or other image is usually at least as easy as on a map. On aerial photos various details may be much clearer and orientation much more efficiently done. Especially vegeta-

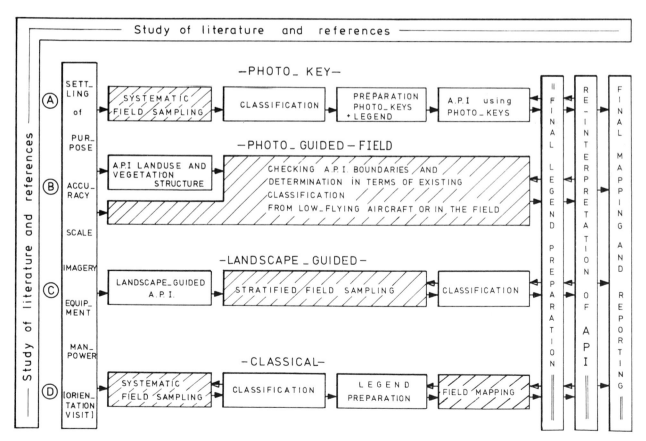

Fig. 1. Scheme of four methods of vegetation survey. A = photo-key; B = photo-guided field; C = landscape-guided; D = classical.

tion boundaries, which are not on the map, may be directly visible on the photo. On the present satellite imagery it is usually necessary to plot the main roads and orientation points on the image via an existing topographical map. The procedure is further the same as with the field survey, with this difference that during the orientation in the field, the map (in this case the photo or remote sensing image) is also interpreted while walking or observing in the field. Transitional to the landscape method is the method described by Küchler. Here, before going into the field, the whole area is delineated into mapping units but, contrary to the landscape approach, without hierarchy or classification. This means that boundaries are drawn only on the main vegetation cover differences obvious on the photo. After that, each plot is visited in the field and described according to field observation. It is comparable with a grid system survey in which the rigid grid has been replaced by a more flexible one formed by the boundaries already predesigned. It has the advantage that under-sampling of small and over-sampling of large areas is prevented.

Key method

The key method starts with observation in the field resulting in annotations of photos, with data on 'ground truth'.

Each subject to be surveyed is compared with the image it shows on the photo. In this way keys are prepared which are used to determine each feature on the photo during the systematic interpretation. The final key can comprise a series of photos (cuts from the original airphotos) or also be a dichotome (forked) determination key (as in floras, soil classification books etc.).

In this case the criteria of division are grey tone or colour (hue, intensity = value and saturation = chroma)* texture/structure (horizontal pattern) and vertical size and shape (stereoscop-

---

* These terms refer to the Munsell colour chart.

ic) of the image. In the classification of the image convergence of evidence plays a part, but only on vegetation (land-use) details, not on land features as a whole. The interpreter moreover concentrates during his work on features 'point by point' in the same way as the terrestrial surveyor is in a ground survey. After the interpretation, some field checking is done for correction and elucidating of doubtful interpretations.

The systematic landscape approach using aerial photos and other teledetection means

The landscape (or 'physiographic', or landunit) approach, developed at ITC, was born out of the so-called 'physiographic soil survey'. which already before applying aerial photo-interpretation, used landscape features as mapping tools. Soil surveyors often use the word physiographic. It seems, however, that in certain scientific languages it is synonymous with geomorphologic which is too narrow! Therefore, we use the term landscape approach. It starts with a preliminary interpretation of mapping units. The interpreter concentrates on the photo-image as a whole. Just as a photograph of a person reveals to a certain extent the identity of this person, and not only details of his skin or the form of his nose and colour of her eyes, a photograph of the land gives an impression of the total character of the land in a holistic way. Convergence of evidence of all kinds of land features (not only strict vegetation features) thus plays an important part in the delineation of boundaries and classification of the content. The result is a map showing units, and a preliminary legend showing a certain hierarchy. Similar units of land are grouped in the same, different land units appear in separate legend classes. The content is, however, not yet known in detail. The interpreter makes use of his general landscape ecological knowledge in which geomorphology especially, and natural and cultural vegetation and other land-use aspects play an important role, but the relation between these land attributes and climate and geology are also taken into account. The more he knows already of the particular area under survey, the more he may suspect of the content of the units. However, in principle the fieldwork after the photo-interpretation is the appropriate stage at which the units are translated into vegetation terms. The fieldwork is in fact not a check, it is a sampling stage.

A most important aspect is that the photo-interpretation provides a basis for stratified sampling. This is the great advantage of survey using photographs over surveys without photographs. There are survey schools teaching that sampling in the field should be done either randomly or in a rigid grid, or on lines. However, for reconnaissance type surveys especially such a system is a waste of time because large mapping units will be far oversampled and/or the smaller units undersampled. On the other hand if one works without photos, and selects sampling points from the field, a great danger of bias exists. This is especially true for vegetation surveys. If stratification is done on photos the great advantage is that in most cases the details of the subject (single plants and species) cannot be distinguished on the photo. Within homogeneous photo-interpretation units the sampling should be done at random, which prevents the surveyor from being biased in the terrain by too obvious, but on the whole less interesting, local vegetation differences. It regularly occurs that differences in vegetation do not appear on the photo because they are either not able to influence the photographic process (only slight colour differences not shown up in grey tones) or because the scale is too small, or because there is a dominant layer hiding the undergrowth (certain forests, most crops), or the photo quality is too poor (often heterogeneous!). If there is, however, another physiographic or landscape feature (relief, macro-pattern) correlated with the vegetation difference, the unit will be distinguished and can be included in the stratified sampling. It is clear that this advantage is more evident in small-scale surveys and in more natural areas and less important in large-scale and strongly man-influenced vegetation where distinct vegetation differences may often not be correlated by any geomorphological or hydrological land property.

The difference with the classical method not using aerial photographs is clear by comparing Fig. 1C with Fig. 1D. In Chapter 29 (the ITC approach!) the various steps in landscape-guided vegetation survey are described in more detail.

Systematic reconnaissance flight (SRF) and related approaches

Systematic reconnaissance flight has been developed, especially in East Africa. In two congresses the methodology has been described and discussed quite extensively. A general review will be given below. Elsewhere the direct observation is practiced. The author has some experience with the so-called 'survoles' practiced by the RADAM scheme in Amazonia, where from small aircrafts observatios were made, and by hand indicated on radar images used as basemaps. Observations are completed by photos from hand-held cameras.

Viktorov, Vostokova and Vyshivkin (1964) describe the Russian 'aerovisual survey'. This can be done with or without the help of aerial photography. This type of mapping is done by drawing the vegetation boundaries from an aeroplane and classifying the difference directly. A special device is made that provides a drawing table with a mechanically moving base map. Thus speed of movement can be synchronised with the actual ground speed of the aeroplane. Ground checks are required in order to have a proper vegetation classification. This method is practiced in tundra areas in Russia on reconnaissance level. The most developed systems, however, are operational in Africa and carried out by Wattson, Northon-Griffiths, and others. See next chapter.

16A. Low-level Aerial Survey Techniques

W. van WYNGAARDEN

Low-level aerial survey techniques are used for gathering information on the natural resources from low-flying small aircraft and is mainly used in surveys of semi-arid and arid rangeland areas (covering in general several 10,000 km^2). Sometimes it is referred to as SRF (systematic reconnaissance flights), but this are low-level aerial surveys with a specific sample design.

Although developed originally to study animal populations (E. Afr. Agric. For. J. 1969), it was soon recognized that also several aspects of the animal habitat could be studied in this way. Especially in east and southern Africa a large number of low-level aerial surveys have been carried out. There is nowadays a general consensus on the methods of how to survey animal populations, but less so on vegetation aspects (see e.g. ILCA, 1981; Norton-Griffiths, 1978; Western and Grimsdell, 1979). However, Viktorov et al. (1964) describe already an 'aerovisual survey' whereby the vegetation of the tundra areas of the USSR was mapped directly from a low-flying airplane. Therefore a number of aspects of the vegetation which can be surveyed from their air will be discussed and some examples given.

Structure and phenology

The vegetation of rangelands shows often considerable variation in cover by trees, shrubs and grasses. Also the phenology of the vegetation varies strongly over the year due to the strong seasonality of the climate. Both these aspects can be easily observed from the low flying aircraft. These data were initially mainly collected to explain the distribution of animals in relation to their environment, but with an appropriate sampling design they can be used for mapping purposes sec also. Examples can be found e.g. in Cobb (1976), Western and Grimsdell (1979). One of the major problems is the calibration of the aerial estimates on cover of the various strata of the vegetation with the actual cover. Data from the Tsavo ecosystem (Kenya) suggest that aerial visual estimates tend to overestimate the cover (Table 1). Combining aerial estimates with ground measurements can,

Table 1. Comparison of aerial and ground estimates of vegetation structure in the Tsavo ecosystem (Kenya).

	From the air	On the ground
Tree cover	32.5% +/−1.1	max. 45% aver. 18%
Shrub cover	51.6% +/−1.2	max. 50% aver. 24%
Grass cover	47.0% +/−0.9	max. 80% aver. 35%
Method	visual estimate from low-flying aircraft (100 m)	estimated at the ground; checked with measured values; average calculated from vegetation map
Source	Cobb 1976	Wyngaarden and Engelen 1985

A.W. Küchler & I.S. Zonneveld (eds.), Vegetation mapping. ISBN 90-6193-191-6.

210

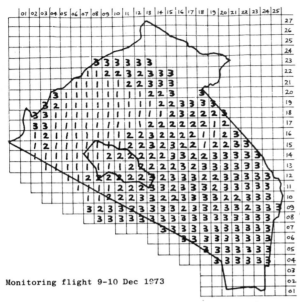

Monitoring flight 9-10 Dec 1973

Fig. 1. Grass greenness in the Amboseli ecosystem (Western and Grimsdell, 1979). 1 = dry (< 10% green); 2 = slight green (10–20% green); 3 = very green (> 20% green).

however, easily solve this discrepancy. The phenology of the vegetation is often expressed in terms of greenness of the grass and/or woody stratum (Fig. 1). Both aspects are often used in explaining the distribution of animal populations (Table 2).

Table 2. An example of a multiple regression analysis between environmental factors and zebra density distribution in the dry season in Tsavo National Park West (Kenya) (Cobb, 1976).

Significant variable	*r*	% variance accounted for
Grass condition (greenness)	0.36	13.1
Damage to woody vegetation	−0.32	10.1
Rainfall in 1973/1974	−0.17	5.0
Grasscover	0.23	4.1
Rainfall in previous 2 months	0.08	4.7
Topography	−0.22	2.9
Rainfall in previous month	0.25	2.5
multiple *r* =	0.65	total = 42.4

r = correlation coefficient; dependent variable = zebra density distribution; independent variable = environmental factors.

Biomass

An estimation of the biomass of the vegetation is an indispensable part of rangeland surveys. Two ways have been tried to do this from low-flying aircraft. Western and Grimsdell (1979), found a good correlation between estimated cover and height of grasses and standing biomass of grasses. Both these aspects were estimated from the air, and so grass biomass was mapped for the Amboseli ecosystem (Kenya).

A more sophisticated technique was used by McNaughton (1979). He established a good relation between biomass and vegetation reflectance ratio at 8000 A and 6750 A. Using photometers fitted in a small aircraft he was able to map green biomass for the Serengeti ecosystem (Tanzania) (Fig. 2). However, this method seems to work only in areas with a rather homogeneous vegetation, strongly dominated by one species. Other authors found it often difficult to establish simple relationships between reflectance measurements and biomass in more varied vegetation (see e.g. Esselink and Van Gils, 1985).

Timber inventory

Watson and Beckett (1976) describe a method for an aerial inventory of the timber resources in southern Sudan. They propose a two-stage sampling technique. First calibration plots are established from which vertical small-scale aerial photographs are taken. The timber resources are inventorized in the calibration plots on the ground. Remarkable was that after some training more than 80% of all timber species in various size classes could be correctly identified on 1:1000 small-scale aerial photographs.

Then the whole survey area is stratified and (photographic) survey samples are taken from each stratum. Together with the data of the calibration plots an assessment of the timber resources is made of all the strata in the survey area. The final result is a listing of timber in m^3/km^2 per stratum.

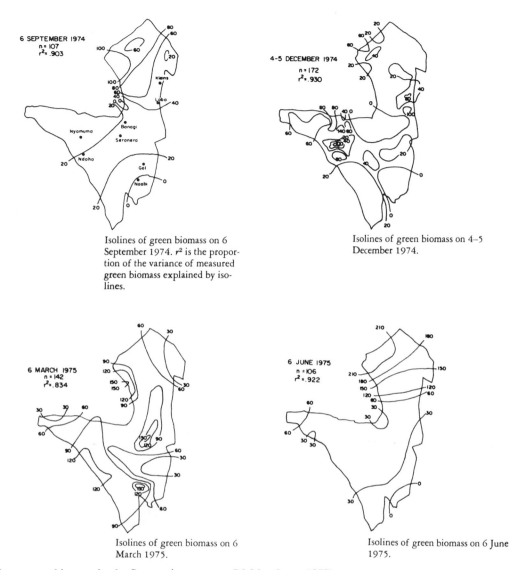

Isolines of green biomass on 6 September 1974. r^2 is the proportion of the variance of measured green biomass explained by isolines.

Isolines of green biomass on 4–5 December 1974.

Isolines of green biomass on 6 March 1975.

Isolines of green biomass on 6 June 1975.

Fig. 2. Green grass biomass in the Serengeti ecosystem (McMaughton, 1979).

Land-use

Using a line intercept, different categories of land-use can be recorded along the flight line over an area. The percentage of land in each category is then taken as proportional to the length of the line intercepting land in that category. Fig. 3 gives an example from Mali (West Africa). Combining visual observations with smallformat large-scale aerial photography was found to be very useful in mapping land-use in Kenya (Norton-Griffiths et al., 1983).

Monotoring

Aerial photography is in general too expensive to be repeated at short-time intervals. Satellite images can provide sequential scenes over the year, but for most areas the result is often disappointing because of cloudcover, especially during the rainy season. For monitoring various aspects of the vegetation, low-level survey techniques can be a very useful tool, as can be seen e.g. in Fig. 2.

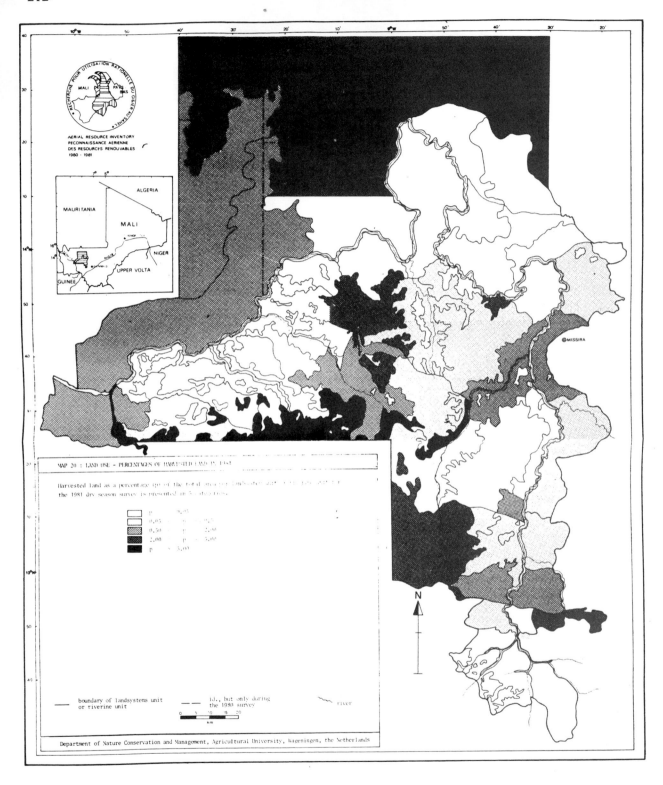

212

Fig. 3. Example of a land-use survey by low-level aerial survey (Bie and Kessler, 1981).

Conclusion

So far collecting data on vegetation and land-use by low-level aerial survey techniques, has been mainly a secondary activity in surveys of animal populations. However, very promising results have been obtained in recording and monitoring various aspects of the vegetation in a relative simple and inexpensive way. Problems in accuracy and sample design are probably different then in surveying animal populations and should be looked into in more detail. Ground truth observations are indispensable, but when properly designed it can result in high quality data (see e.g. Watson and Beckett, 1976).

17. Sampling

I.S. ZONNEVELD

Type of sample points

In vegetation survey, sampling is the description, quantitatively and/or qualitatively, of examples of vegetation stands. It is irrelevant whether or not material is taken home or to the laboratory for measurement, determination, weighing etc. As far as measuring is concerned this may be done accurately or by estimation. Sampling has two purposes:

a) to collect data by close observation in order to provide knowledge of the object. This knowledge is used for classification and for evaluation in a scientific and practical (applied) sense ('*data points*');

b) to supply points to serve as fixed references for the delineations on the map ('*hang points*').

In surveys not using aerial photographs both aims of sampling are usually separated in time; (a) is done first and (b) follows later in the mapping stage. In that case (a) is always more intensive and (b) has more a character of a quick check. Data observed in (a) are always accurately recorded but the data of (b) may be just noted in a symbol form on a map or not even noted at all but expressed only in the position of a line on a map.

In surveys using aerial photo-interpretation a part of the mapping is already done beforehand, sampling of type (b) is sometimes not necessary or much more restricted. In that case most of the sample spots will be observed with (a) as main objective, hence are 'data points'.

There are many methods of sampling. They depend on the purpose and type of survey and also on the school of the vegetation scientist carrying it out. For these methods refer to Chapters 4, 5 and 6.

First some remarks will be made about the relation of sampling, accuracy and mapping scale, the distribution of sample spots over the mapping area, the demand for recording data of mesological ecology, the difference and relation between sampling and mapping.

The relation between sampling details and mapping scale

It is often assumed that there should be a relation between the amount of detail collected on a sample spot and the detail of the map. In general this is, however, not strictly true. Also a broad reconnaissance survey needs a clear legend. To describe a good legend there should be a good general classification. It is true that many of the units seen in the field cannot be mapped separately, but nevertheless they appear in the legend description of the complex legend units and are used for delineation of the complexes! On the contrary there are examples of detailed surveys where sample spot descriptions may be restricted to 'single values'. The extreme is a detailed stock map giving numbers and volumes of individual tree species for forestry purposes. The individual observations are, in that case, very simple. The survey in that case is usually very detailed (large-scale).

People who are doing reconnaissance may be in a hurry. Therefore, it can be psychologically explained that they tend to neglect accuracy in sampling also about data points. However, many reconnaissance maps would gain much in value if sampling at data points was done in more detail on each sample spot. A commonly used reason for doing less accurate sampling is lack of floristics knowledge. Detailed sampling on flora would increase the time required so much that a reconnaissance survey would be impossible. However, structure measurements can be done in some detail there and the use of the 'quick herbarium method' may help. Moreover in reconnaissance survey, contrary to detailed surveys, sampling time is minor compared with traveling time, so reduction of sampling time does relatively not give much gain.

The balance between time for sampling and total time of the survey should be chosen for each project individually. Indeed it is possible that very detailed sampling may not be feasible, but for classification a minimum of data per sample unit is required. However, this is as much true for detailed surveys as for reconnaissance. Therefore it is good to start with the rule that *scale and sampling time are not necessarily interrelated!*

The selection of sample plots and distribution over the mapping areas

The distribution of sample plots over the area to be surveyed refers to the dilemma of random or stratified sampling. Random and rigid-grid sampling would result in an unbalanced sampling of the different vegetation stands. Those with the largest extension would be sampled more frequently than those less represented. But when making a classification system, acreage of vegetation stands is of minor importance. Both for scientific and practical reasons often more rare stands or stands with a small total acreage, may be much more important or

interesting. Such vegetation stands can even determine the aspect of the landscape (e.g. hedges, narrow tree belts, etc.) or the (grazing) value (limited wadis in deserts, etc.). So there is usually no reason to spend more time in sampling the larger acreages than the smaller ones.

The defenders of random sampling, state that any other form of sampling would make the selection subjective because before one has sampled it, it would be impossible to judge whether individual vegetation stands are different or not. Theoretically this may be true, especially if only pure floristical parameters were used. However, ecological knowledge is sufficiently advanced for us to know that there is a certain relation between the (cultural and natural) landscape and the vegetation.

The solution is the aerial photograph. As can be seen in Chapters 16 and 29, in the first stage of a 'landscape-guided' survey, a rather objective analysis of the photo-image should be made. This results in delineation of land(scape) units. Landform and vegetation structure (including pattern) are the main criteria in these delineations. These landscape units form a reasonable basis for stratified sampling. Within these units indeed, random sampling is preferred. The problem of the requirement of homogeneity of vegetation stands that are used as sample spots is not fully solved, but for practical purposes is sufficiently reduced by the above-mentioned use of photographs.

In reconnaissance surveys especially, a considerable time is spent in moving from one sample plot to the other, so it is advisable to group them in clusters in such a way that a minimum distance exists between a group of samples. The cluster should be situated so that it is as accessible as possible, for instance by road, river or air. It is clear that the arrangement of units determines the possibility of clusters. This clustering, however, may not lead to dominance of one region that may not clearly differ in airphoto (or other remote sensing) image, but still may have a difference in climate. It is wise to distribute the clusters rather regularly over the image, not all in the north, east or south, but a few in the middle, north, west, south, etc.

The samples should also be regularly distributed over map patches of one type, not too many in just one patch. So the distribution of samples is:
— stratified (acc. to photo-interpretation);
— random (within the units);
— clustered (making use of clustered units);
— regular (equal distribution of clusters as far as is allowed by the photo-image).

For sampling, procedures using the 'key' method and the photo-guided field survey, the principle is the same – see therefore Chapter 4 and 5. For discussion about the related problem of the minimum area of the sample spot refer to Chapter 4 and 5 and to the metioned handbooks. Theoretically this is a very difficult problem. In practice the simplest solutions (see Braun-Blanquet; Cain and Castro; Van Gils and Zonneveld; Zonneveld, Thalen and Van Gils) are recommended and give no problems in most cases.
— *Hang point observation* should never be done at random. There are three ways.
— A kind of grid system is useful in detailed work, in terrain that is difficult to overview and where airphotos do not give enough orientation.
— A landscape-guided approach, either with airphotos or without (see Chapter 16 and 16A and bibliography) can be applied in terrain where landforms and covertypes can easily be observed and oriented on field-maps and/or photos.
— Arrangement in the form transects, preferably perpendicular to the vegetation zonation.

Mesological-ecological sampling

Pavillard (1936) made a useful distinction in *mesological ecology:* the study of relations between organisms and environment by analysis of the environment and *ethological ecology* which is the study of relations by analysis of the behaviour of the organism.

It is advisable to estimate or measure not only vegetation data at the sample spot. For the explanation of the map picture and especially for evaluation for scientific or practical reasons, many types of *mesological,* synecological data are necessary. Time may be a limitation here. But there are many simple estimations and measurements that can be done during vegetation sampling. A simple soil auger to estimate texture, consistency and colour, a little spade to reveal structure (and other data of at least the upper soil strata), a colorimetric pH tester, an altimeter, a compass and a slope meter are instruments that can be easily and quickly handled.

The most favourable situation is found in those types of 'integrated', multidisciplinary, surveys where the survey team includes representatives of various disciplines. Here, field sampling for soil survey and for vegetation can be combined usually quite well. Not all disciplines can work at the same sample spot because the required data and the time necessary for description are different, e.g. geology and vegetation (see for details Zonneveld, 1979).

The relation and difference between mapping sensu stricto and sampling

The principal character of sampling is that there is a restriction to small parts of the vegetation cover ('sample plots') which are considered (or hoped) to be representative for much larger parts of that cover. On the contrary mapping aims at the 'description' (in the form of a map) of the total. From these characters of sampling and mapping it may be clear that there is no strict separation between them. Sampling may include a detailed description of the spatial horizontal structure of a sampling plot. This may result in a real micro-map (compare stock maps). On the other hand a rather detailed map may be compiled solely from sampling data taken e.g. in a grid system. In that case the only additional mapping process is the location of the sample points on a base map, the compilation of a legend system and the drawing of boundaries in between the dots in the map without further observations in the field or from the photograph. (Indeed there are scientists who do not know a different procedure.)

However, besides both extremes there is a wide range of observation types that may be used for mapping but cannot be considered as sampling. Compared with sampling in its narrow sense, some of those usually have a less exact, less basic character. Because of that they require, unlike sampling, real skill and experience in the mapping discipline tending in certain aspects even to an 'art'.

At the same time, however, the latter remarks point to the necessity of accurate sampling beforehand, to provide a sound and objective basis for classification and legend compilation on which arbitrary observations constantly have to be tested and checked. (Here a similarity with soil survey may be mentioned. There the word 'sampling' is frequently used in the more limited sense of taking soil samples to the laboratory to be tested. Detailed profile description, however, can also be considered as sampling.)

The relevé

General

In the vegetation schools using species composition ('floristics') as main classification criterion, sampling is usually called doing a 'relevé'. The core of a 'relevé' (Aufnahme, opname, etc.) is a list of plant taxa (species) usually with an indication of abundance, cover, etc. per taxa. It makes sense to add beside the obligatory administrative data like sequence number, photo number, data, name observer, locality and/or coordinates, also 'meso-ecological data' such as: general geological and geomorphological data (at least relief, exposure, slope), soil data, altitude above sea level, etc.

The amount of data should not be too large.

If one has a team of two people a 'relevé' may as a general rule last no longer than 1 hour and preferably shorter.

Certainly one should not dig a profile pit on each sample area. With a good soil auger one makes good observations in a much shorter period of time.

Nor should one try to identify all species in the field. The more one knows the better, but still unknown species can be taken in a 'quick herbarium' (using no more time for collection than 1 minute (31 seconds!) per plant (see section 'The quick herbarium').

In the developing countries many epople still have a keen knowledge of plants (in our developed countries this knowledge is usually lost). One should use this knowledge expressed in the vernacular names, which later on via the quick herbarium can be translated in scientific names. Usually one can employ 'local botanists' with much advantage. For a classical herbarium, if wanted, one should make special expeditions in the most favourable season and stratified according to experience, gained in the field before. Also quantitative vegetation determination (primary production) measurement cannot readily be combined with collection of 'relevés' for classification. If one does this at each 'relevé' there is a great risk that one does not get enough 'relevés' for good statistical treatment, which is required for classification.

Contrary, the quantitative sampling should be done on special field trips, preferably after the general survey has been done and stratified according the classification and final mapping results. If this is not possible one should take at regular distances (e.g. each 3rd or 5th 'relevé', etc.) or stratified occurring best professional judgement, from time to time a more intensive 'relevé' where an hour more is spent for more detailed measurements, determinations, etc.

The 'relevé' sheet

For systematic sampling it is good to have a preprinted form that (1) can serve as a checklist and (2) has all specific information always at the same spot, which enables quick perusal during the classification stage.

In areas where the flora is known beforehand, one can even print the plant names in advance. The latter is a great advantage in the field as well as in the processing (table making) stage. If the flora is not well known this is not possible, because one has to use nick-names and numbers which are documented by herbarium material (see next paragraph). In the ITC approach a 'relevé' sheet has been developed as a synthesis between the wish to have much information and the requirement to have a handy format that is useful in the field as well as in the office. Too often one sees elaborate sheets that are not used in practice because it takes (often too) much time to be fully filled in. On the other hand, small forms may be handy in the pocket, but they do not have enough space.

Fig. 1a gives the ITC 'relevé' sheet front side and Fig. 1b gives the reverse. The sheet is developed for a rather complete ecotope description and can also be used by geomorphologists and soil scientists if they are not too demanding. All administrative date, and the main aspects of vegetation structure and aspects are on the front side. The reverse (which can be used by folding the paper over a line running vertical over the paper, so one does not need to turn the paper, nor tear it out of the book) gives space for plant taxa and some columns for abundance cover, phenology, etc.

One should also note the administrative number at that side. The rest of that side can be used for making a sketch of the sample plot if wanted, and for other notes.

One can use the space behind the species names for making a line intercept where this is relevant. The same part of the form can be used for any other information. The form can be used for quick descriptions as well as for detailed ones. Note that the front side makes a separation between data for the sample spot (the 'relevé' sensu stricto being 2×2 to 100×100 m) and the area around it. The latter can be the mapping unit as appearing on the preliminary photo-interpretation. This may be the ecotope in large-scale maps, in small-scale maps rather higher categories of landscapes may be described here. The space at top-right is meant for final classification later in the office, so there should be nothing filled in the field. Each form filled in at daytime, should be rewritten every night. So a double set of 'relevé' sheets grows to be filed and stored separately for safety reasons.

Material sampling

The type of sampling in a 'relevé' is mainly descriptive, one does not always take material home. Nevertheless this will be necessary from time to time. It is useful to take some rock samples, some soil samples of selected profiles and some vegetation samples (also of selected areas, see above) in which determination of texture, dry weight and various other (e.g. nutritive) parameters is done in the laboratory. For the way of sampling and for parameters to be determined see Chapter 4, 5 and 6.

A most important requirement is to take specimen of those plants that are not yet known. This means making a herbarium.

A warning should be given here. Those who try to make a classical herbarium in a practical survey during the relevé-field survey have experienced that either the survey, or the herbarium, usually both, will suffer from such a combination, unless time and money (field personnel and equipment) are not at the usual restricting level.

The best solution, found after long experience, is the 'quick herbarium' as practiced in ITC-fieldwork (practical consulting work as well as students fieldwork). See following section.

The quick herbarium

The quick herbarium is made by sticking plant parts quickly, *not* on loose, but on fixed paper sheets like pages in a book, using masking tape, on sheets (one plant per sheet) that are filed in spiral bindings, or any other handy file cover in

ITC RELEVÉ   DATA SHEET   ITC Vegetation and Agricultural Land Use Survey

PHOTO No. (run, type etc.)		Area, Country, etc.	DATE / /	No.
			day month year	

PRELIMINARY LEGEND SYMBOL:	SIZE SAMPLE m²	ALTITUDE IN m:	FINAL CLASSIFICATION (not for fielduse)
			FINAL MAP SYMBOL

PRELIMINARY LANDUNIT NAME:	OBSERVER(S):

TERRAIN DATA

SITE/ELEMENT	MAPPING UNIT	LANDFORM - (HYDROLOGY)	SYMBOL
ROCK LITHOLOGY:	ROCK LITHOLOGY/GEOLOGICAL FORMATION	SOIL	SYMBOL
LANDFORM/TOPOGRAPHIC POSITION WITHIN MAPPING UNIT	LANDFORM	VEGETATION	SYMBOL
		structure	
SLOPE TYPE straight, concave, convex, irregular	RELIEF TYPE	composition	SYMBOL

RELIEF TYPE
- Very flat
- Almost flat (< 2%)
- Undulating (3 - 7%)
- Rolling (8 -13%)
- Hilly (14-20%)
- Steeply dissected (21-55%)
- Mountaineous (> 55%)

SLOPE :	EXPOSURE N.NE.E.SE.S.SW.W.NW.	RELIEF INTENSITY	LAND USE	SYMBOL
LENGTH:				
MICRO-MESORELIEF				

SOIL AND WATER DATA

EROSION			SOIL DRAINAGE	WATER SOURCE	RUNOFF		FLOODING/PONDING			
TYPE	RATE	AREA AFFECTED					DEPTH	AGENT	FREQUENCY	DURATION
none	very low	<25%	excessive mod.well	rain	very rapid	slow		none	...per	days
sheet	low	20-50%	somewhat imperfect	run-on				rain	.....years	weeks
rill	moderate	50-75%	excessive poor	aquifer	rapid	very		run-on	month	months
gully	strong	>75%	well very poor	irrigation	medium	slow		river	week	
eolic	severe							lake		
								sea		

HORIZON symbol	DEPTH cm	TEXTURE	COLOUR	PH	HCL	MOTTLING	CONSISTENCE	SURFACE SEAL	SURFACE STONINESS/ ROCK OUTCROPS (%)	REMARKS
									effective soil depth (cm)	groundwater depth (cm) .....

PRELIMINARY SOIL CLASSIFICATION

LAND COVER/USE DATA (semi-)natural or planted

SAMPLE PLOT				SITE/ELEMENT	MAPPING UNIT
STRATA	HEIGHT	COVER %	DOMINANT SPECIES (for details p.t.o.)	general cover/use type (if complex, estimate % cover of each type)	general cover/use type (if comple , estimate % cover of each type)
trees					
shrubs					
herbs					

total real cover %

PRELIMINARY COVER CLASSIFICATION:
    form:
    composition:

LAND USE
    type:
    field size/shape:

Fig. 1a.

(repeat number here) No.	LIST OF SPECIES						OBSERVATIONS/INTERVIEWS ON VEGETATION, CROPS, ANIMALS AND MANAGEMENT ASPECTS (SAMPLE SITE)
							(SEMI-)NATURAL VEGETATION
							- burning
							- fuel wood collection
							- range condition
							- grazing traces
							- fences
							- watering points
							-dropping/footmarks/ tracks
							CROPS
							-planting distance -stage -height -crop condition
							Timing
							-date of planting -date of harvesting
							-rotation
							Mechanisation
							Input use
							-fertilizer -pesticides
							Yield
							(expected) yield in year of survey
							-normal yield range
							ANIMALS
							-type/breed
							-number
							-estimated body weight
							-age/sex
							-condition

SAMPLE SIZE:	TOTAL NO. OF TAXA:					

Fresh WEIGHT DATA   g/m2   dry

1	2	3	4	5	a. $\xi \frac{1 \to n}{n}$	c sub.sample
6	7	8	9	10	b. sub.sample	c/b.a=g/m2

Fig. 1b. Example of a ITC relevée data sheet.

which papers are *fixed*. Usually A4 format is best. If one has to carry everything on one's own back the system is more tiresome, unless in the tundra-like vegetations where one has tiny plants which can be filed easily in small pocket notebooks,

Usually, however, one has transport, in less than a few hours' walking distance, which ena-bles one to carry several spiral bindings (files). The difference from a classical herbarium is shown in Table 1.

As said before: it would be good to spend also some time for making a special real herbarium, but in doing so one needs to make special trips or to have a special expedition member for it, it is a precious document for the future.

The quick herbarium, however, is less beautiful but more complete. From time to time one files together the plants that are apparently similar and so gradually a systematic order grows in the books. Therefore one needs to file in spiral bindings or so and not in books with permanently fixed sheets.

The more one knows, the better the systematic order starts to become according to the current plant taxonomic system. Any possibility is used to increase knowledge. So any visiting plant taxonomist or local-vernacular flora specialist can be asked to look through the books and assist with the identification.

If one has sufficient transportation means one will have the books in the field as much as possible. One can then compare specimen of unknown plants with the still not identified specimen in the quick herbarium book and give the same 'nick-name'. Then it is not necessary to collect all unidentified species in each 'relevé'. So the total volume of quick herbarium books can be reduced. Besides a nick-name each unidentified taxon will get a number, consisting of the relevé number by which it was observed at first, followed by a letter (or figure) indicating the sequence in that 'relevé'.

Table 1. Comparison classical and Quick herbarium methods

Classical herbarium	Quick herbarium
– Large sheets, the sheets are loose	– Sheets are in a book
– If there is wind, rain, dust or any other usual disagreable field situation, it doesn't work handily, one postpones to 'press' the plants till the evening, when one is tired or someting else has priority (rewriting 'relevés', food preparation) and one postpones once more, etc., the specimen gets lost eventually	– The whole process of putting the specimen takes about 1 minute or less ('preferably 31 seconds')
– One has to label the plants in a special way, which takes extra time	– Wind, rain, etc. do not disturb much
– One is inclined to look for good specimen, which make a good impression in the herbarium, etc.	– The 'relevé' number and the sequence letter is written on the masking tape
– Even if everything goes well, it takes many minutes per plant to be filed	– The books are dried (if necessary) every evening
– If ready one will be so careful with it that it will not be in the field where one needs it	– In dry climates the drying proceeds during the field work
	– The books are daily in the field

18. Reflection, Absorption and Transmission of Light and Infrared Radiation through Plant Tissues

I.S. ZONNEVELD

Introduction

Radiation emitted by the sun is not only the basis of life but also of interpretation of photos and several other remote sensing images. A good understanding of the pathway of the radiation through the canopy of plants (the green mantle) is necessary for a good explanation and interpretation of the differences that can be observed. The studies on this subject, making use of spectrometers both in the laboratory and the field, reveal that in publications by practical users of different types of film and filter combinations, unproven hypothesis have too often been perpetuated, especially where use of infrared such as in false colour film, was concerned (remote science fiction). It would be too great an effort to discuss (or even to mention) here all the publications dealing with the use of different film and filter combinations and the supposed influence of leaves of different plants, different development stages, differences in health conditions, etc. A limited review only will be given based on publications, in which apparently careful research has been carried out using sophisticated modern instruments: e.g. Gates, Keegan, Schleter and Weidner (1965), Weidner (1965), Hoffer and Johannes (1969), Howard (1971), Knipling (1969), (Rosetti Ch. Kowaliski at Have, 1967), Bunnik (1978) and from the authors not published experience (see for details the Manual of Remote Sensing, 1983).

The following major factors/vegetation properties, are influencing the radiation by vegetation as recorded on airphotos, graphs or other remote sensing images. To explain a certain intensity of reflected light of a vegetation stand one needs to judge at least the following items:

1. total structure of the part of vegetation under consideration also in relation to angle of incidence and recording
   a) density of the foliage in the horizontal and vertical directions,
   b) inclination of the leaves;
2. reflection within the plant tissues;
3. absorption within the plant tissues;
4. transmission through the plant tissues;
5. reflection, absorption and transmission by cell walls;
6. reflection, absorption and transmission by pigments (chlorophyl, etc.).

For a good understanding of the following a basic knowledge of colour theory is required. Items such as complementary colours, colour formation by addition (television screen system), colour formation by subtraction (dia system) should be known as a routine. Handbooks or a good encyclopedia can be used for recapitulation.

In the following part of this chapter we will describe subsequently the radiation spectrum (its composition and intensity at the earth's surface; the plant tissue and vegetation structure and components; the pathway of radiation (reflection, transmission and absorption).

A.W. Küchler & I.S. Zonneveld (eds.), Vegetation mapping. ISBN 90-6193-191-6.

The spectrum and energy of solar radiation

The dominant part of the radiation spectrum of the sun reaching the earth is usually divided into three parts: ultra violet, light and infrared, (X-rays and 'micro-waves' originated by the sun, hardly play a role at the earth's surface). In fig. 2 can be seen that the optimum energy is delivered to the earth by the near infrared (wavelengths of between 700 and 1500 nm) and that beyond 300 and 2,000 nm hardly any important energy amounts reach the earth.

This means that the optimum energy in the radiation occurs on the transition line between 'light' (visible radiation) and 'heat' (thermal radiation). Plants use mainly parts of the visible

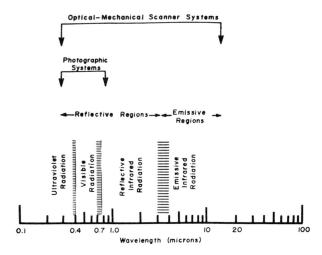

Fig. 1. The electromagnetic spectum as far as relevant for remote sensing.

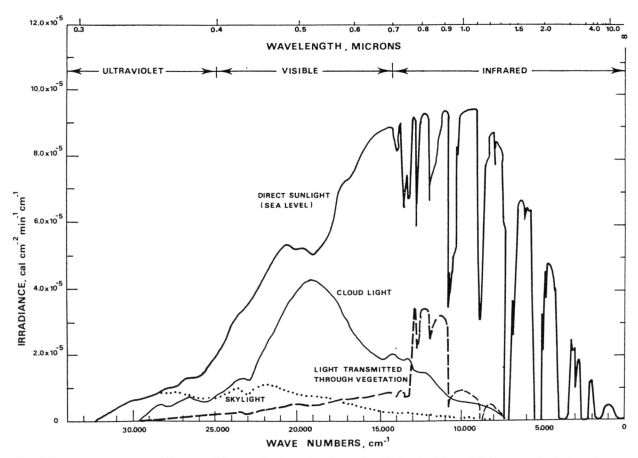

Fig. 2. Spectral distribution of direct sunlight on a horizontal surface, cloud light, skylight and light transmitted through vegetation (after Applied Optics 13/vol. 4, no. 1, 1965).

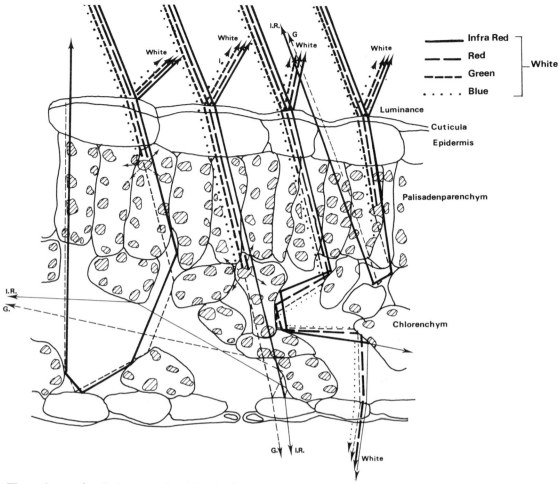

Fig. 3. The pathway of radiation through a plant leaf.

radiation, for the most part blue (0.4–0.4 μ) and red (0.6–0.7 μ) and also a considerable part of green (0.5–0.6 μ) for photosynthesis, as will be seen later. (See Fig. 3 and 4). This means that all the infrared, if absorbed by vegetation, would have to be transformed into heat. This would require a very intensive 'cooling' system. In Figs. 2 and 5 it can be seen that in the range of maximum energy incidence (near infrared) the transmittance of the vegetation is also at its maximum. We should see later that an approximately comparable amount of energy is reflected. The remaining energy warms up the plants, and excess of heat not necessary for the growing processes is reduced by transpiration. This is regulated by a highly sophisticated cybernetic system. Reflectance and transmittance decrease strongly with increasing wave length (towards the pure thermal infrared). However, the total energy input in this part is also strongly decreasing.

Note in the figure that the areas of strong absorption are mainly in the infrared at several well-defined wavelengths. This is mainly due to absorption of water in the atmosphere. We shall see later that plant tissues also show strong absorption at these wavelengths. Fig. 2 shows, however, that the intensity of these readily absorbed wavelengths is reduced in the atmosphere before it reaches the plants; only at high altitudes are special plant adaptations necessary (special structures reducing radiation incidence, better heat regulation, etc.). We see that the total amount of radiation, the absorption activity of the atmosphere, and the properties of plants in relation to the use of and the defence against radiation are well-balanced. If this were not the case plants (and man) would not exist.

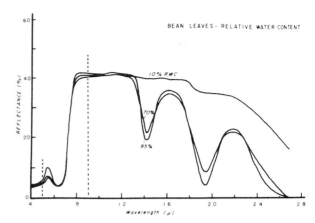

Fig. 6. The effect of leaf dehydration on the spectral reflectance of bean leaves. The dashed lines delineate the spectral sensitivity range of Ektachrome Infrared Aero film (from Kipling, 1969).

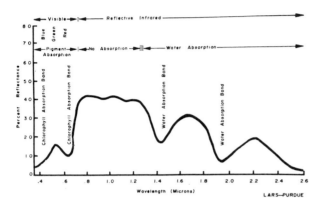

Fig. 4. Three primary regions of response in relation to leaf reflectance. Water absorption accounts for the strong decrease in reflectance in the two remaining wavelength bands, located at approximately 1.45 and 1.95 μ (from Hoffer and Johansen, 1969).

Plant tissue and structure
of vegetation cover in relation to reflection, absorption and transmission of radiation

Plants are made up of cells arranged to form the various tissues. Most important for our purpose are the herbaceous parts of plants and especially the leaves or leaf-like ('green') organs. Wood and bark will not be treated in detail. These are opaque for most of the solar radiation which reaches the earth. Only some far infrared and longer micro-waves (radar, etc.) may pass through. Leaves have a cuticle on the outside which is often glossy in appearance. Hair or wax may influence the reflection of radiation. The cuticle usually contains suberin. The cuticle is closed, and permits direct contact with the outer air only by stomata. The inner part of the leaf (mesophyll) is made up of parenchyma, a tissue of cells with cavities in between, giving it a spongy character. The cells are filled with cell moisture in which the pigments are present in discrete, usually globular particles. These pigments are: chlorophyll a and b (2/3 of total), xantophylls (about 1/3) and some carotenes. The cavities are normally filled with air which is usually saturated with water vapour. On the upper side most leaves have a layer of paren-

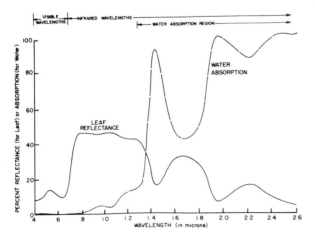

Fig. 5. Relationship between leaf reflectance and water absorption in the 0.4–2.6 μ wavelength region (from Hoffer and Johansen, 1969).

chyma cells elongated perpendicularly to the surface of the blade (see Fig. 3). This type of parenchyma is called 'palysade', and differs from the spongy parenchyma both in the form of the cells and in their arrangement. This paly-sade parenchyma may occur on both sides of the leaf (e.g. eucalyptus) or on the upper side only (most broad-leaved species of temperate and humid zones).*

Reflection and transmission of radiation is only partly influenced by the plant tissues. The *structure of the vegetation cover* also plays a role. If the vegetation cover is less than 100%, part of the solar radiation reaches the substrate directly. The reflection, absorption and trans-mission of the radiation then depend upon the properties of the soil or water surface. Even when vegetation cover is complete, but thin, the substratum may reflect radiation which has been transmitted through the leaves. This re-flected radiation is then partly transmitted a second time, on the way up.

Hence the total thickness of the leaf tissue in a vertical direction plays a role. Reflection from the surfaces of leaves and the bark of branches and stems is moreover strongly influenced by the angle of incidence. The latter is determined by the position of the surface, which may vary from vertical to horizontal, and by the position and type of radiation source. Moreover at least part of the radiation is coming from all direc-tions. In order to make a correct interpretation of recorded reflection, as much as possible must be known about the structural features describ-ed above. So to explain the final effect, it is far

* It has been suggested that there is a difference in reflection between the spongy parenchyma and the palysade parenchy-ma, the latter reflecting more infrared. Nowadays, however, this seems unlikely (refer to Knipling, 1969 Howard, 1971 and their cited literature). Most probably all the cell walls of the parenchyma reflect radiation. There is no reason to assume that palysade cell walls would differ in this respect from these of the spongy parenchyma. Palysade parenchy-ma, however, is usually richer in pigments than the spongy parenchyma, and this may affect the passage of light.

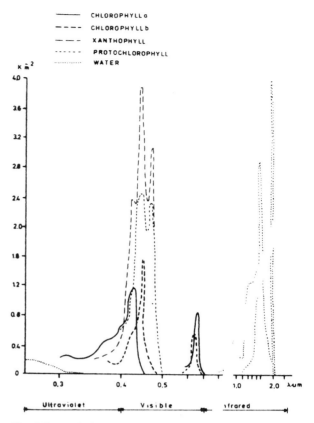

Fig. 7. Spectral absorption of leaf pigments and water (from Gates et al., 1965).

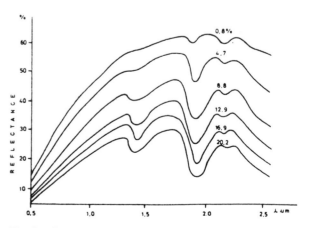

Fig. 8. Effect of moisture content on spectral reflectance of Newtonia silt loam (from Bowers and Hanks, 1965).

from sufficient merely to determine the properties of *leaves* only with respect to radiation.

The path of radiation through vegetation

Terms

Radiation hitting the vegetation element, e.g. a leaf surface, usually is partly reflected, partly transmitted and partly absorbed. The reflected part contributes to the formation of a photo or other remote sensing recorded image. The absorbed part is converted into energy and contributes to assimilation or heat formation and related processes in plant life. The transmitted part reaches lower levels of the vegetation or the substratum and it may again be reflected, absorbed or transmitted. Thus transmitted radiation may subsequently follow a path diverging from, or even opposite to, the original angle of incidence.

In the substratum reflection and absorption are common, transmission is rare and only holds good for microwaves and longer electromagnetic waves. In the substratum (but also in plant tissues) radiation is often transformed into another kind of radiation and re-emitted. This happens with some of the short-wave radiation (visible and near infrared), which is converted into heat energy, and may be re-emitted as active far infrared.

Reflection is a sending back of radiation without change of the frequency of various monochromatic components. Reflection can be subdivided into three types:
1. *specular reflection,* the direction is only towards the source if the angle of incidence is 90° (e.g. mirror);
2. *diffuse reflection* reflects in all directions more or less equally (e.g. non-specular surfaces);
3. *reflex reflection* is reflection into the direction of the radiation source (certain 'reflectors' of cars, etc., a mirror at right angles to the source).
The term *luminance (or brightness)* is used to

refer to the radiation flux per unit area of surface as observed from a given point. A perfect diffusing surface gives equal luminance in all directions. A mirror gives luminance only when viewed from one direction. It is useful to distinguish between pure reflection, which takes place at a two-dimensional surface, and reflection which is the sum of reflection, absorption and transmission through an entire three-dimensional body. The term *reflectance* is used to describe the reflecting properties of a surface (e.g. a leaf surface) and *reflectivity* for a body (e.g. several leaf strata taken together, total vegetation cover, etc.) (Howard, 1971). Good examples of spectral reflectance curves are given by Hoffer and Johansen (Fig. 8–11).

Absorption means that the radiation is always transformed into other forms of energy (chemical), or heat, which may later be again transformed into radiation of another frequency.

Albedo is a term used to indicate the total amount of sunlight sent back by reflection into the cosmos.

The combined effect of absorption, reflection and transmittance

It is known that the various frequencies of radiation are not absorbed in equal amounts. This is clear from Fig. 2, where it can be seen, especially in the infrared parts of the spectrum, that certain frequencies are absorbed before they reach the earth's surface. This absorption has been carried out by water, in the form of vapour or liquid.

Fig. 7 shows that leaf pigments absorb strongly in the 0.4–0.5 µ wavelength (ultraviolet to blue) and the 600 to 700 nm wavelength (red). It is well-known that wavelenghts play a dominant role in carbon dioxide assimilation, the basis of the life process in green plants. In this case the absorbed radiation is converted into chemical energy. Most of the green, 0.5–0.6 µ, is also absorbed. However, plant pigments are completely transparent to near infrared (Fig. 7).

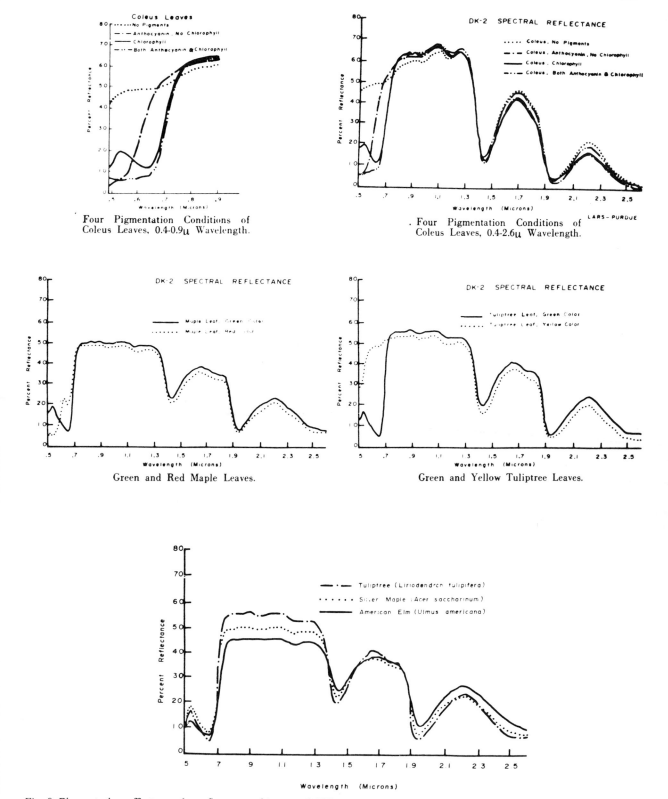

Four Pigmentation Conditions of Coleus Leaves, 0.4-0.9μ Wavelength.

. Four Pigmentation Conditions of Coleus Leaves, 0.4-2.6μ Wavelength.

Green and Red Maple Leaves.

Green and Yellow Tuliptree Leaves.

Fig. 9. Pigmentation effects on the reflectance of leaves of different plant species (from Hoffer and Johansen, 1969).

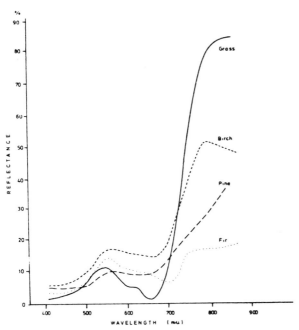

Fig. 10. Spectral reflectance curves of various typical foliage types (from Fritz, 1967).

moist soils. These figures show that the total reflection of even those parts of the spectrum, apparently not absorbed by plant pigments or by water, is only about half of the original amount. It follows that half the radiation must pass through the vegetation.

We may compare the mesophyll to a thin layer of snow. In snow light is scattered from crystal to crystal in all directions. The final result is that half the light is reflected, and the other half is transmitted. The snow layer appears bright white when viewed from above or below. This is also the case with the near infrared radiation in the mesophyll of a plant leaf. The cell walls reflect and refract the radiation in a similar way as the snow crystals (see Fig. 3).

It might be expected that this would also happen with the visible part of the spectrum. However, the radiation encounters many pigment cells on its path through the leaf, and these absorb the frequencies which are suitable for photosynthesis. These frequencies do not, of course contribute to the final reflectivity of the leaf. The radiation, which passes the leaf eventually, is called its transmittance. The type of pigments present determines the final colour of the leaf.

The leaf reflects and transmits relatively more green than red and blue, i.e. if chlorophyll dominates. A dead leaf, however, will give relatively more red. The total amount of reflected infrared is, however, strikingly constant in most

Absorption by pigments is the main reason for the differences in final reflectivity of the leaves between the various parts of the spectrum in the visible and nearest infrared. The dips in the various graphs 4–6 and 8–11 in the infrared have the same explanation as those in Fig. 2 there water is responsible for the absorption, from Fig. 8 it can be seen that the same dips occur in the spectral reflectance curves of

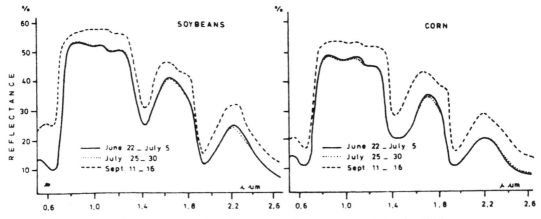

Fig. 11. Changes of reflectance of corn and soya bean leaves over the season (from Sinclair, 1971).

cases. This holds good when we compare various species and also in maturity, old and even sick or dead leaves (Knipling), although slight yet important differences may occur.*

Refraction is deviation of the direction of radiation. This is a well-known property, used in optics (lenses and prisms), and occurs when a ray passes from one medium into another. Refraction decreases with increasing wavelength and hence is specific for each frequency. However, because in most situations there is some scattering, refraction has no influence on the 'colour' of the reflected or transmitted light because the final mixture remains the same. (In detail, the various types of monochromatic light might be seen as being the result of diffraction prism effect of raindrops, irization in plant

cells, etc. However, this effect is usually unimportant for the interpretation of vegetation images). Diffraction can also be generated by gratings and certain surface structures of plants may diffract light in this way. As was mentioned before, most of the radiation reaching the earth is in the visible and near infrared wavelength. We have already seen that light radiation (particularly blue and red) is intensively used for photosynthesis. The near infrared is of less use to plants and could be harmful if absorbed and converted into heat. It is therefore easy to understand why it is almost completely reflected and transmitted. Much of those parts of the infrared that can be absorbed by water is already absorbed in the atmospheric water vapour before reaching the biosphere, thus reducing the possibility of harmful overheating, whereas sufficient energy still remains for the transpiration and respiration of plants. Reflection diminishes strongly in the direction of the far infrared, the true heat radiation. The total intensity of the latter is much lower than that of light and near infrared radiation.

A surface perpendicular to the radiation source (the sun) receives a maximum of energy. Any deviation from this angle reduces energy reception. At the same time it reduces the amount of radiation reflected back towards the source, especially on glossy surfaces. The position of the scanner, etc., in relation to the sun and to the reflecting target should be taken into account when interpreting aerial photographs or other remote sensing means because the luminance of surfaces is influenced by their relative positions.

---

* From this it is clear that the so-called 'chlorophyll effect on infrared radiation', which has been mentioned in many older publications, does not exist. Theoretically it is still possible for two green leaves to absorb different amounts of green radiation. It is not the total amount of one monochromatic component that makes the colour, but the additive mixture of all the colours together (blue + green + red). If the ratio between these remains constant the colour remains the same, even when the total level changes. However, if infrared plays a part in the colour impression this would change the picture. If red, green and blue have a constant ratio but have e.g. in one leaf a higher level than in another, the ratio between those colours and infrared differs!

This most probably is one of the most favourable aspects of false colour photography. The colour differences on false colour are often erroneously attributed to unexplained and often probably non-existant fluctuations in the infrared radiation. Only on the transition between red and infrared (about 700 nm), so in the 'dark-red', some influence of chlorophyll absorption may be apparent.

19. Observation Means and Platforms

I.S. ZONNEVELD (with a contribution by A. Kannegieter)

Introduction

Science starts with observation. Observation is primarily done with the help of our natural human sense organs. The vegetation scientist needs all five of them. To identify a plant he has to *feel* with his fingers the roughness and consistency of the leaves and other plant organs. The *scent* of rubbed plant material is often an essential way of recognizing a species (e.g. various eucalyptus species, many herbs); even from a distance the odour may be useful (e.g. various fungi). The *taste* of plant material (sweetness, salinity, bitterness) plays a similar role. An ecologist may even use his *ears* in sensing from a distance the occurrence of unseen plant species, e.g. Aspen (Populus tremula) and others.

However, by far the most important sensing medium is electro-magnetic radiation. *Vision* with natural *eyes* is the most common way of using this medium. Vision can be applied not only for close observation, but also for more remote sensing and in both cases can be improved by incorporating simple optical instruments. Both a magnifying glass and a pair of binoculars are indispensable tools for a vegetation surveyor in the field. The wavelength (λ) used is 0.4–0.8 μm (see Fig. 1).

Technical sensing devices extend the possibilities of *using visible radiation. Photography* (a result of optical and chemical technology) enables us to record visible images in *two-dimensional* and (with use of stereoscopic vision)

three-dimensional pictures. The unavoidable loss of accuracy and flexibility is compensated for by the possibilities of viewing the object *from positions where natural viewing is difficult.* Moreover, photography takes over the difficult task of transferring the visual image to a lasting document on paper. The *mapping procedure* especially, *is simplified* greatly by this. In fact photographs, whether taken by a camera hanging on a kite and showing a few square meters, or taken by a camera carried by a satellite, and showing almost the whole earth are kinds of maps.

Another advantage of technical devices for sensing and recording electromagnetic radiation is that *parts of the invisible radiation can also be recorded* (Fig. 1). Photography is able to visualise the near infrared radiation (wavelengths untill about ca. 1 μm). Radiation of lower frequency cannot be sensed through optical means. But other instruments for sensing and recording exist. So *heat radiation* (far infrared) can be recorded with the help of the 'infrared or thermal scanner'. Here a rotating scanning mirror takes over the task of the optical system. An electric device converts the signals into a kind of television image (in lines with varying light intensity) that can be recorded on a normal photographic film, or directly on computer compatible tape and/or on a television screen. Wavelengths between 0.3 and 20 μm can be recorded by the existing apparatus that are at present available for civil use. So also light and near infrared can be recorded in this way (MSS, multi-spectral sensing).

A.W. Küchler & I.S. Zonneveld (eds.), Vegetation mapping. ISBN 90-6193-191-6.

234

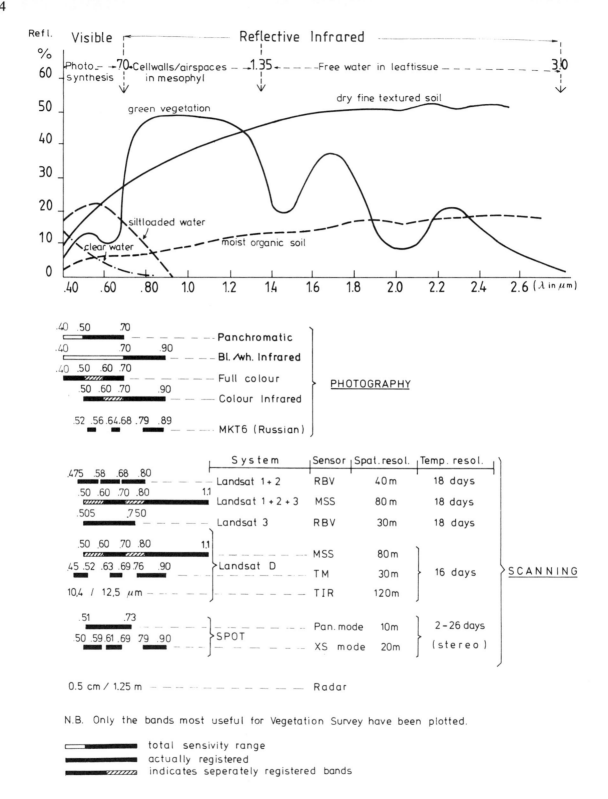

Fig. 1. Reflectance patterns/sensor bands.

Radiation with wavelengths of millimeters (*known as 'micro-waves'*) is used in radar systems. *'Side looking radar',* using waves of 8 mm and 3 cm, 12 cm, recently also larger, is a very useful way of remote sensing on a reconnaissance scale. It is a common type of radar as used in navigation, but in this case using nonrotating antennae mounted on an aircraft. The returning signals are electronically converted into images on photographic paper. The result is a photograph-like picture that can be used in a similar way as classical aerial photography can be.

Radar technology is rapidly developing and will increasingly be a valuable tool for vegetation mapping. For the types of reflective and emissive radiation mentioned above it holds that it is possible to store digitally the signals of the bands that seem to be useful for recording a certain aspect of vegetation or its environment.

The value of and possibilities for vegetation survey of the various methods will be dealt with below.

The possibilities of direct observation (sensing) of morphometric vegetation properties of the vegetation itself are very important in connection with this. This may sound self-evident and therefore a superfluous remark. However, we shall see in the next paragraph that it is impossible to avoid using indirect means (i.e. landscape features) instead of direct observation of the vegetation itself in order to compile a vegetation map. Before judging the usefulness of various sensing means the radiation itself and the influence of the vegetation cover thereupon should be discussed (see Chapter 20).

The *platforms* of observation depend on the recording means. For taste and touch observation, one should stay in the vegetation; visual observation of course can be done from the same place in the vegetation viewing in any direction or very classically, from a hill-top or another high landmark. It can also be done from moving objects such as horseback, camel, cars, boats and also aeroplanes or balloons, rockets and satellites. The ground observation is commonly done in the sampling phase. Real mapping observations in this modern age will be (unless the survey is very detailed) done from flying remote platforms as mentioned above.

The Systematic Reconnaissance Flight (SRF) system uses direct eye information from aeroplanes (see Chapter 16a). An important aspect of flying platforms is that one can make vertical observations with the cartographic advantages mentioned already above. Photos from high fixed points have necessarily to be oblique. Still this may give useful results (see Zonneveld, 1975). In special cases if monitoring and not cartography is the aim, oblique photos from aircrafts can serve the same special purposes. In general, however, vertical photography and other imagery has to be preferred, although the user has to learn to get used to the (for a human observer) unnatural vertical view, compared with the natural horizontal or oblique way of looking at the environment.

The term remote sensing or teledetection is used for all types of sensing from platforms at a certain distance from the object. Aerial photography from any type of aircraft is usually included in the terms teledection or remote sensing. However, in certain circles, a distinction is being made between aerial photography and remote sensing (teledetection); the latter being any sensing means other than photography from aeroplanes.

Eye observation

The human eye is a very delicate and ingenious sensing device that can observe an object or area at the same time by keeping it fixed on the retina and yet can combine this observation by scanning it through moving the field of view over it. The recording system is situated in the human brain and memory with all advantages and disadvantages of that. Direct observation to be used for cartographic use must be (even by the man with the most ideal cartographic memory) directly laid down on paper by hand-drawing.

The eye is able to distinguish between various wavelengths of the electromagnetic radiation through a system of four types of sensors. One type ('rods') which occurs predominately at the fringes of the retina is sensitive for the total wavelength range between 0.4 and 0.7 µm (light!). Three other, cone-shaped, sensor types have each a smaller range of sensitivity. One for the wavelength around blue (ca. 0.4 µm), one for about green (ca. 0.5–0.6 µm) and one for red (about 0.6–0.7 µm). This means that we do not have a possibility to separate the spectrum in more than three bands.

All colours other than the ones mentioned are formed by combination of the three bands (blue, green and red). Blue + green = bluegreen (cyan); green + red = yellow; more red to it gives orange, etc.; blue + red = purple (magenta). An equal amount of the three main colours (green, red, blue) makes white (or grey if low intensity). The colour circle (Fig. 2) is the easiest means to help in finding out how an observed colour should be interpreted. In fact colour photography and MSS recording as is done by current satellites do in principle what the eye does. The restriction to the vision of only the three main colours one can make clear for one-self, by looking at a low quality 'snowy' black and white television image from such a distance that the individual points of which the screen exists are pictured on your retina in a similar size as the reception cones.

For an average television apparatus this is a distance of about 2 m. If the light is not too strong or too weak, one sees on that black and white television a mixture of blue, green and red points in a kind of 'Browns' movement. This can be explained by the fact that quite a number of times one cone is hit by the image of one point and then signals its own colour to the brain for that point alone (being either red, blue or green).

Usually, however, more than one cone is hit by the image of one image element. Then both signal their colour to the brain's visual centre. The reader should be reminded in this context also of the techniques of the 'pointilistic painters' who use colour dots closely put on the canvas instead of colour strokes and so just as on a television screen, create colour effects using only the three basic colours, blue, green and red. Also nature uses this principle depending on scale! So the combination of (many) red flowers in green grass will in any scale that is too small to depict individual flowers, show a yellow field!

Even more than for the form, the colour pattern observed is difficult to remember in detail and painters of all ages have tried to make the impression permanent by using pencil and paint to picture it down; also cartographers. Another means is the photographic image or another geo-information system. In those cases one can use another observation means like photo cameras, radar or other scanning devices for electromagnetic radiation. The advantages of a vertically taken photo are, being beside an observation record immediately also a cartographic representation: in fact a 'map', as has been mentioned already before. Still, direct 'eye observation' is practiced especially for moving targets but also quick multi-temporal observation as in monitoring (see Chapters 16 and 24).

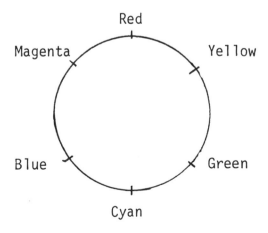

Fig. 2. The colour circle.

Aerial photography

The camera is in a sense a copy of the human eye using the same system of diffraction of light to form an image on a sensitive plate. An important difference is that the result is more static. No scanning can be done. The advantage is the direct formation of a map as is already mentioned before.

It will be assumed that the reader is familiar with the basic principles of photography and aerial recording. The use of five types of photographic plates will be dealt with briefly. These are: panchromatic, orthochromatic, infrared, true colour and false colour.

The first three have only one emulsion layer, the last two are made up more usually of three layers.

Panchromatic film

Panchromatic film is more or less equally sensitive to light radiation and the near ultraviolet. The image gives, if no filters are used, a sum of the intensities of the whole spectrum range to which it is sensitive. Filters have the property of transmitting only a narrow range of frequencies. Note that this does not mean that they are completely opaque to all other frequencies and a slight influence of other radiation should always be taken into account. This is the most widely used type of film. It is commonly used with a yellow filter (= minus blue filter) to eliminate the ultra-violet and blue radiation which is often scattered by the atmosphere by reflection and refraction (haze). With a flying height above about a thousand meters this factor starts to diminish the resolution of the image. Other filters may increase the value for special use. So a green filter is applied for sensing into water (not too deep submerged water plants!) and reducing other influences at water surface. A blue filter has the opposite result.

Orthofilm

Orthofilm is mainly sensitive to blue and blue-green (not to red) and has a limited application.

Black and white infrared

The so-called black and white infrared film is sensitive to the near ultraviolet and blue range and in the near infrared and darkest red part. It is mainly used with a dark red filter, (0.7 µm filter), to register mainly the infrared radiation. Occasionally a 0.5 µm filter (modified infrared photography) is also used, when green and red should contribute to the picture. Indeed differences can be seen between the use of the two types of filters. The contribution of red and green, should be counted for because of the sensitivity of the film also in these frequencies. The film can also be used for ultraviolet and blue photography, by using a blue filter (= minus green + yellow + red + IR). This, however, is only done for special purposes and usually not for vegetation. Because of the scattering of blue light mentioned above, it can only be done on large photo scale (low altitude) flights.

Infrared photography is more widely used. Its advantages are that the near infrared has a high energy in the sun spectrum. Although absorption of this wavelength does take place in the atmosphere, there is less scattering than with light (especially blue, see above), hence no haze effect at all. The highest resolutions are often obtained in infra-red photography. Moisture conditions at the earth's surface are sharply recorded because of the 'dips' as previously shown in figure 1. Hence drainage patterns, etc. are also clearly recorded. Pure water almost fully absorbs IR. In forestry inventory infrared and the so-called modified infrared photography have a rather wide application, because coniferous trees usually appear much darker on infrared than on panchromatic films, whereas broadleafed species appear in relatively light value on both types of film. The explanation of this is that the arrangement of the needles of conifers permits more radiation to reach the soil, and there are more (micro) shadowy areas than with most broadleaf crowns. Quite a lot of

238

the IR radiation is absorbed by the soil but also less radiation is returned from the shadowy areas between the leaves. The (visible) light, especially green and blue, is also more scattered in the shadowy parts (compare the shadowy parts on panfilm, which are usually less dark than on infrared). The many shadowy parts between the needles cannot be seen individually, but contribute to the darkness of the whole. (Former theories that the internal leaf anatomy was responsible are most probably incorrect, see Rosetti, Knipling, Howard.)*

Besides crown density, the type of undergrowth and forest floor may also play a role in determining total reflectivity of a forest. For distinguishing between broadleaf trees and all kinds of herbacious vegetation, IR film is inferior to panfilm. This would follow directly from the remarks made about the universal tendency of plants to get rid of infrared radiation by reflection and transmission (see Chapter 18). This has been confirmed by the author's own experience on several occasions. So IR photography also has a rather limited and specialised application, in spite of the fact that it has been stated in some handbooks and publications, that IR photography is very useful in vegetation survey in general.

True colour film

True colour photography depends on the principle that, with the help of three coloured layers, all colours can be made visible. The process makes use of colour formation by subtraction. The colours used for the three layers are all complementary (mixed) colours: magenta (= red + blue), cyan (= green + blue) and yellow (= green + red). Note that each of these layershasonecolourincommonwithanotherlayer. At the end of the colour photography process, there is a variation in the intensity of the colour in each of the three layers. For instance,

at a certain place on the film, let us say an image of an homogeneous meadow, the magenta layer is fully removed but the cyan and the yellow layer allows only the blue and green to pass, the red is 'filtered out'. Then the remaining radiation hits the yellow filter. This transmits only green and red. There the blue is filtered out. Red was already left behind in the cyan filter, so only green light remains. Hence the picture appears green as a meadow ought to be. Thus to record a green image the magenta layer has to be removed. The magenta-coloured layer is therefore made sensitive to green light. In the same way it can be reasoned that the cyan layer has to be sensitive for the light that passes through a magenta and yellow filter, which is red. The yellow bearing layer is made sensitive to blue. Intermediate hues and chromas are formed by partly removing the dyes in the layer. Because the layers are also sensitive to ultraviolet, a UV filter should be used, especially at higher altitudes. The above-mentioned situation holds for diapositive photography. If we work with negatives from which positives can be made, either on film or on paper, the sensitivity and colour combinations should be different, and complementary colours used. For details about this, refer to photo handbooks. However, it is sufficient to have the above principles in mind explaining the colour formation on the photo especially when one sees deviation from reality caused by filtering (purposely done or by the atmosphere), or caused by wrong development or alteration of the colour film. It is very important to make use of this reasoning when false colour film is used. In the case of false colour film all colours deviate from reality, and only by reasoning based on the colour 'play' described above the user can understand the results. Such reasoning will also enable him intelligently to plan his use of this type of film, and to select the appropriate filters for use during recording and processing.

Warning. The information on the photographic technology given here, however, is only to inform the reader who may have to judge the

---

* Only Wolff (1967) shows figures probably showing some influence of leaf anatomy which, however, needs careful rechecking.

quality and possible deviation of the material. In doing *photo-interpretation,* however, one should also continually be aware of the meaning of the colour, and one should try to explain what one sees. In that case *one should not try to explain the colours according to the photographic process of colour-formation by subtraction, but use the principles of colour-formation by addition (like a television principle).*

The colour circle (see Fig. 2) shows clearly how with the three main colours (green, blue and red) all other colours are formed by addition. For semiquantitative measuring of the contribution of the several wavelength bands one could make use of colourcharts, e.g. the one's used for the colour printing process for selection of the saturation (grids intensity) of the composing colours. An improved version of such a map is published by A. Brown (1982) and available at ITC, Enschede, The Netherlands. See also what is remarked above on colour theory and the principles of the human eye.

False colour film

The most commonly used false colour film deviates from true colour film in having one layer sensitive to near infrared, therefore it is also indicated as 'colour infrared film'. The dyes used are exactly the same as in true colour film: cyan, yellow and magenta. The differences are that the blue-sensitive layer is replaced by an infrared sensitive layer. The corresponding dyes are shifted in such a way that (in the diapositive process) the layer sensitive to *green* is connected with the yellow layer and that the one to red is connected with the magenta dye. This means that a *green* object gives a blue colour and the magenta and cyan layer stay intact. (See above description of what happens with true colour.) The layer sensitive to *red,* connected with the magenta dye, shows an object which is actually red, according to this process *green.* The layer sensitive to *infrared* is connected with the cyan dye, which results in the recording of pure infrared radiation as *red.* Colour mixtures

are intermediate. As mentioned already, to work with it easily one can better think in terms of colour formation by addition of final colours (red + yellow = orange; green + blue = cyan, etc.). Equal distribution of all these frequencies in the final image means white light. On a true colour film this occurs only if the object really is white. On a false colour film a white object only appears white if the IR radiation is similar in intensity to the green and red. This may for white subjects often be the case but is not necessarily so.

In Fig. 3 a scheme is given (kindly supplied by KLM Aerocarto) of the process of colour formation in false colour diapositive.*

A healthy leaf appears bluish red (magenta, purple) on false colour because the recorded radiation consists of green and IR, giving respectively blue and red. When both the red and green radiation increase, the balance of the colour mixture moves to white, because the green (= blue on the photo), and the red (= green on the photo) tend to become equal to the IR (= red). On the photo this means that the colour shifts from magenta through orange to whitish colours. This is what happens in diseased plant tissues. This is mainly due to decreased absorption of red and green, and consequently more reflection of these colours. The position (inclination) of the diseased leaves may also influence the balance between the reflected radiation as was the case in diseased Ulmus trees in Amsterdam (Vries de and Aldewege, 1970). Some of the theories about detection of diseased trees, etc. are hypotheses and speculations. Often the features can also be seen on true colour photographs, but more clearly on

---

* It should be noted that the layers sensitive for one spectral band may also be slightly sensitive for the others. Especially the layer sensitive for IR is also nearly equally sensitive for the other colours (blue, green and red). Hence also these colours contribute to the image formation on this layer. A false colour film therefore cannot be fully analysed as if it were a multispectral record of completely separate spectral bands and should be interpreted accordingly. In practice, however, by filtering of false colour films one can reasonably well simulate pan, infrared photos, etc.

240

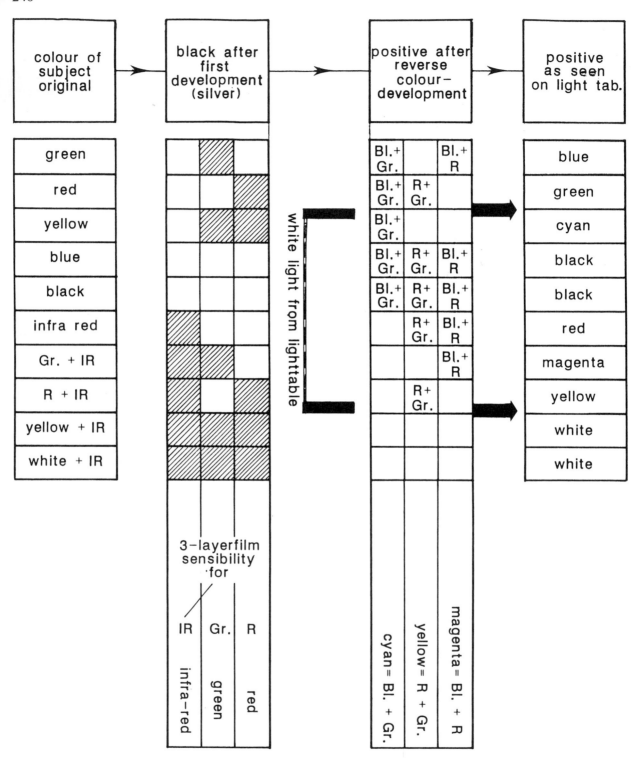

Fig. 3. Colour presentation by Kodak ektachrome false colour (transparent Film-posture).

false colour. This may be partly due to the colour balance described above, but also partly because the human eye may be more sensitive to slight differences in the red, orange and blue, than in the green and yellow colours of nature. A 'chlorophyll effect on infrared radiation' which has already been stated before, does not exist. However, a 'chlorophyll effect' might exist in the sense of a reduction of absorption capacity (increase in reflectance) of the chlorophyll for red and green. However, if water enters into the cavities of leaves, this might absorb IR radiation, thus changing the normal colour balance, as probably is the case in the very young leaf shown in Howard (1971) and which holds for any decaying plant material. There is then no question of a 'chlorophyll effect', it is purely a case of the water absorption effect of green leaves. Another more simple example of the colour balance is given by photo-images of sediments or plants or any suspension in water. These items always appear blue on false colour photos. Water absorbs IR and red to a much greater extent than blue and green. If these colours are reflected they can be recorded. Blue is filtered away by yellow filter, hence green remains giving a blue image on the photo. Uneven water surfaces (waves, etc.) give much scattered radiation reflection of all frequencies, so much so that all layers are developed and the result is white! Pure water reflects almost nothing if viewed perpendicularly, hence appears black.

Multi-spectral photography

For certain purposes the use of separate wavelengths (colour) may be advantageous, this often is done via multi-spectral photography. This may be produced by one camera with more than one (usually four) lenses, or with more than one simultaneously operating camera. The camera(s) are loaded with films of the same or different sensitivity (e.g. pan and IR) and different filters are used. The results are a number of photos, each representing a certain wavelength as determined by film-filter combination. These photos can be used separately or used in special additive viewers. For routine surveys of large areas any increase in the number of photos to be handled (and compared) increases time and cost and therefore here, multispectral photography is less suited.

Multispectral scanning, airborne and satellite images

So far photography is treated as an observation and recording means. A very different type of recording is scanning. For the electronic principle refer to handbooks. The result is an image in lines, comparable with a television screen. It can be easily stored on tape, analogous or in digital form. The wavelengths recorded vary from visible radiation to the microwave. Scanning can be done from aeroplanes as well as from satellites. The use from aeroplanes has not proven to give more practical results than photography. Photography is superior to any scanning as far as resolution and planimetric accuracy is concerned. Moreover the possibility for stereo application and stereo vision is half of the classical interpretation success! Also the multi-wavelength character of scanning can be equalised in practice at least by multispectral photography (MSP). This is so because the greater variations of wavelength bands that can be used in Multiband Scanning (MSS) appear in practice not very useful compared with the much higher cost of airborn scanning. It is very different, however, with satellite MSS – here large applications are possible especially because of the sequential aspects of satellites in orbit. The recording on an electromagnetic way (on tape) here is a great advantage. The data can be stored and easily combined; especially multitemporal combination of images is a useful tool (see Chapter 25). A great advantage is the synoptic view on the present small-scale satellite imagery.

While airborn MSS will only be used for experimental purposes, there is hope that satellites, who are still also mainly experimental, may become operational for civil use in vegetation survey and monitoring. At present the American satellites already provide material that can be used for semi-operational purposes (Landsat I to IV). In the near future the French satellite SPOT, may add to the opportunities (see also Chapters 29 and 31).

The Landsat MSS operates in four spectral bands (see Table 1). The data are available in various forms that allow conventional imagery interpretation.

Table 1. Landsat spectral bands.

NASA code	Wavelength (μm)	Real "colour"	'False' colour composite (4/5/7) image colour
4	0.5–0.6	green	blue
5	0.6–0.7	red	green
6	0.7–0.8	near IR	—
7	0.8–1.1	IR	red

The data are available in various forms that allow conventional imagery interpretation.

The available Landsat products are:
— black and white positive paper prints of single bands;
— black and white negative film of single bands to produce paper prints yourself;
— black and white positive film to produce colour composites of different bands with the colour-additive viewer or by using diazos (the 'poor man's colour composite');
— false colour paper prints; the layers sensible for band 4,5 and 7 are converted in respectively yellow, magenta and cyan superimposed;
— besides that computer compatible tapes are available that enable the user to produce himself special imagery for specific goals.
In chapter 21 the various possibilities and methods are treated. Important applications are multitemporal images, 'biomas' (cover) indication, enhanced false colour images. For those who cannot afford to buy tape and use computer facilities the products mostly used in vegetation surveys are the black and white images of

one band for a quick look and the diazo and paper false colour composite image. The 'poor man's colour composite' is a superposition of diazos in the complementary colours, which gives often remarkable results (see below).

Beside MSS other types of images are also produced using satellites. So return beam video recording (RBV) was tried in Landsat, being an optical system transmitted by radio waves. From manned satellites the existing photography is well-known. The near future will teach us what real operational images will be used. Resolution and scale are becoming already more detailed. For the scale this is partly a disadvantage (the synoptic view being one of the advances of satellite imagery). The larger resolution will increase the use possibilities, however, tremendously. The present Landsat allows no larger accuracy than in average that of vegetation maps of scales 1:250,000 at most. The smallest element (pixel) being about an acre. The thematic mapper (the newest Landsat experiments) comes already down to 30×30 meter. The Spot will have higher resolutions (see Chapter 31).

The MSS system provides absolute reflection intensity data in digital form which can be computer processed for the purpose of identifying and classifying vegetation/land-use date on the basis of their reflectance patterns. The framesize (image produced by Landstat) is $185 \, km \times 185 \, km = 34225 \, km^2$, which is imaged in 30 seconds. Such a large 'instantaneous' coverage is very useful for reconnaissance survey purposes. The usual scale of the images (one black and white image per band) is 1:1,000,000, but good quality images can be photographically blown up to a scale of around 250,000 before the image-constituting elements (picture elements or 'pixels') become disturbingly individually visible. The radiometric resolution is such that 2^6 intensity levels are distinguished. This is interesting in the case of computer-processing of the data; since the human eye can only distinguish around 2^3 greytones, this advantage gets lost when standard B & W images are used.

Producing false colour composites e.g. by colour coding the reflection values in band 4 (the green band) in blue, band 5 (the red band) in green and band 7 (the IR band) in red and superimposing the transparencies under careful spatial registration, compresses the data from three bands into one image. It means at the same time a form of colour enhancement, particularly the presentation of vegetation. Like in false colour photography the range of composite colours utilized to present the information to the eye is restricted, due to the fact that the colours blue and green are hardly used for discrimination within 'vegetation', since reflectance values in bands 4 and 5, respectively, are so low.

A strong point of satellite remote sensing is the repeated passage over the same area, which moreover is always at the same time of day (same sun-angle). The temporal resolution of the system (the revisit interval), however, was 18 days (for Landsat-D 16 days). For many parts of the world, this temporal resolution is far too coarse for the study of dynamic aspects, of which land use is one. The chance that cloudcover or other unfavourable atmospheric conditions will prevent the gathering of (good quality) data within the critical period(s) for successful land-use/crop identification, is there very low (humid tropics, Western Europe, etc.). The spatial resolution of the MSS system is only some 80 m on the ground (net pixels measure in reality 56 m $\times$ 79 m). Within these, roughly $\frac{1}{2}$ ha rectangular 'pixels' on the ground, the radiation reflected by (whatever variety of individual) objects, is simply compounded into one value per pixel for each band: no finer spatial detail. Unless the reflection of one pixel is strongly dominated by a landcover type with very different reflection from that of surrounding pixels, it will not be individually detected. Usually a group of at least 3×3 pixels is needed for a land-use unit to be observed in the image. Narrow linear features with strongly contrasting reflectance characteristics (roads in band 5, canals in band 7) are often easily detectable.

The one-band B & W MSS images resemble small-scale panchromatic (band 4 and 5) and B & W IR (bands 6 and 7) airphotos. Bands 6 and 7 are often practically mutually interchangeable for the information they contain. The superimposing of the colour-coded images of bands 4, 5 and 7 (see before) to form false colour images can be done electronically but also very well 'mechanically'.

One way is to project blue, green and red light through the B & W transparencies (diapositives or transparent positives) of bands 4, 5 and 7 images, respectively, and registering these colour-projections on a screen. The resulting false colour image can thus be studied and interpreted on the screen or a hardcopy photograph can be made of this screen-image for the same purpose. Special equipment is required for doing all this colour added viewer (I_2S). An even simpler and less costly way is the "poor man's composite" mentioned earlier: transparent images on yellow, magenta and cyan forming diazo-material, respectively for bands 4, 5 and 7. For the exposure of the sheets, even sunlight can be employed instead of the UV-lamp whereas the developing of the exposed diazo sheets is done with ammonia vapour. In doing so we produce the equivalents of the three emulsion layers of the false colour diapositive, each acting as a filter when viewing on the lighttable (or even against the window!). These approaches to data-compression/image enhancement, particularly the latter, are often looked upon with scorn by the 'electronics people' but may constitute the only successful way within reach of land-use interpreters/Landsat users in the developing world. Admittedly a lot of precise and absolute information on reflection-characteristics contained in the original digital data is lost this way, and the colours are not reproduceable, etc., but a useful colour-enhanced sinoptic view over a large area is available for use, where no information was (and would be) available otherwise.

If a simple photographic laboratory is available, photographic enlargements in transparent positive can even be made from the original

1:1,000,000 Landsat B & W material in band 5, 5 and 7, up to a scale of 1:250,000 (for a selected part of subframe); these enlargements can again be used to produce a false colour composite at this larger scale.

The MSS images are very useful for a first stratification into main landform/land-use regions or strata: large areas under one land-use (forest, estate plantations) can be easily delineated. It may also be possible to delineate areas of complex, small-scale land-use which are homogeneous in their composition. Those latter areas are classified as complex land-use categories. Areas with too many 'mixed pixels' because each pixel on the ground contains several land-use types/crops, defy further analysis. In this case the MSS imagery serves to determine where aerial photography with (the always necessary) fieldwork will have to unravel the land-use picture more in detail if so required. The spatial resolution of Landsat MSS data is generally too coarse to detect any of the spatial elements employed so prominently in airphoto-interpretation for land-use mapping: pattern, texture, shape, shadow and associated features. The Landsat system neither provides stereo vision to be used in land-use interpretation.* The future generation of satellites, however, will have higher resolution and so may have increased applicability for land or sea.

The Landsat *return beam vidicon* (RBV) sensor system provided a finer resolution with pixels of 40 m on the ground. The image could therefore be blown up to as much as roughly 1:150,000, thus affording much more detail than the MSS data. Landsat RBV band 1 registered the radiance values in 0.45–0,60 µm, band 2 in 0.57–0,68 µm and band 3 in 0.66–0.82 µm (following on this RBV band coverage are bands 4, 5, 6 and 7 of the MSS sensor). The RBV sensor in Landsat 3 registered the total reflection intensities in the range of 0.505–0.750 µm (panchromatic); thanks to this wider spectral range, a spatial resolution of 30 m could be achieved.

It is now realized that practically all spectral band ranges of the Landsat 1–3 sensor systems were poorly located for giving information on important vegetation/crop parameters such as green leaf area index (GLAI) or green biomass index (GBI).

Landsat 3 was also to register in the thermal band (10.4–12.6 µm). The spatial resolution of 237 m was, however, too coarse for crop (transpiration) studies.

Landsat 4 and 5, the only satellites of the Landsat series, at the moment operational, also carry a conventional MSS sensor, for continuity sake. However, they further carry the thematic mapper (TM), a sensor system which collects reflection intensity data in seven spectral channels. Of these, particularly bands 1 (bluegreen), 3 (red) and 4 (near IR) are well placed for crop surveys, because of the close relation between reflection values in these bands and single plant (growth) parameters.

— Band 1 (0.45–0.52 µm) values are closely related with carotenoids and chlorophyll absorption; they should give a good discrimination of conifers.

— Band 3 (0.63–0.69 µm) values are closely related to chlorophyll absorption and give information on cover percentage and degree of maturity of crop canopies.

— Band 4 (0.76–0.90 µm) values are closely related to leaf area index and biomass.

The *spatial resolution* of the TM data in the reflective (sun-spectrum) range is 30 m. Imagery can thus be produced at scales around 1:100,000 scale. This relatively high spatial resolution improves the ratio between pure 'infield' pixels and mixed border pixels which improves detail and precision in interpretation, mapping and per-class area determination. Also cultural patterns and features (roads, canals, etc.) are better discernable which improves orientation and proper location e.g. of training site areas for supervised computer-aided classification of land-use.

---

* Only if fringes overlap some stero is allowed which, however, only may be of use for detecting strong terrain relief features like mountain ridges, etc. See however SPOT, below.

The radiometric resolution is supposed to be 2^8 (256 intensity levels distinguished in each band); this helps in making even finer distinctions between classes, if at least the in-class variations are not so great that classes overlap.

There is also a thermal band (10.4–12.5 μm) which, again, has a coarse resolution of 120 m. The temporal resolution of Landsat 4 and 5 is 16 days; again too large for the study of dynamic processes as land-use, in many areas of the world where cloudyness is a problem during the agricultural season. Landsats 4 and 5 neither provide stereo-viewing capacity useful for vegetation and land-use surveys.

The French SPOT satellite

This became operational recently and carries a sensor-system which offers the choice of recording a 60 km × 60 km scene in
— the panchromatic mode (0.51–0.73 μm) with 10 m spatial resolution;
— the XS mode, i.e. in three spectral bands, with 20 m resolution; band locations and widths have been particularly defined for vegetation/crop studies:
— band 1 (0.50–0.59 μm);
— band 2 (0.61–0.69 μm): chlorophyll absorption;
— band 3 (0.79–0.90 μm): green biomass, green vegetation density!
The 10 m resolution in the pan mode means that imagery can be 'blow up' to around 1:40,000 scale and even somewhat larger, the XS mode images to around 1:80,000. There are two identical sensor-systems on board the SPOT satellite, each of which can be aimed sideways over 0–27° from the vertical; each can be used in the pan or in the XS mode.

It might be useful in land-use studies to employ the XS mode for identification/classification and the pan mode for exact location, cultural features, field boundaries. This may even obviate the need for flying separate aerial photography (for e.g. the location of training sites); aerial photography will then be limited to real problem areas in (semi-) detailed mapping.

The side-look capability offers flexibility in temporal resolution; the vertical overpass interval is 26 days; by looking sideways, review-intervals of as low as 2 days may be possible! This greatly enhances the probability of successful data collection within the critical time periods for land-use/crop survey.

The slant-view capability also offers the possibility of combining two images from different view angles into a stereo model; thus, 20 m height resolution appears to be within reach, good enough to get an impression of landform in its relation to land-use.

The smaller frame-size of SPOT (60 km × 60 km) imagery means less area overview but this is amply compensated for by the considerable gain in detail for even semi-detailed land-use surveys, for greater precision in land-use identification, mapping and per class area determination.

With the spatial resolution of SPOT it should be better possible to unravel successfully the complex land-use and cropping of small-scale farming areas like in Western Europe and the developing world.

The stereo-view capability should also prove useful for automatic topographic base map production and updating at the semi-detailed level which is very time-consuming and costly when it has to be done in the traditional (optical) way.

The CCD pushbroom scanner of SPOT anyway provides data of much higher geographic fidelity than those produced by the transverse-sweep type of scanner used in Landsat.

In summary we can say that remote sensing from space has unique characteristics and benefits in that it offers:
— large area uniformity of perspective by nearly ortho-viewing the earth surface;
— repetitive coverage;
— the facility of visual presentation and communication, e.g. to demonstrate to authorities responsable for land-use policy how devastation of vegetation cover is proceeding and where vegetation and land-use surveys measures are urgently required;

— the possibility to overlay multi-temperal data to minotor change in vegetation cover and use the different character of change in the aspect of the land;

— the possibility to integrate spatially defined data from various sources;

— the possibility to combine the advantages of manual and digital analysis, combining the visual image elements of shape, texture, pattern and location with the intensive use of the spectral characteristics ('signature').

Some notes on thermal radiation

The main difference between the recorded radiation by thermal recording devices is that it uses emissive radiation transmitted through a few 'windows' in the atmosphere mainly about $10-12\frac{1}{2}\mu$ and almost no reflection. So it can be done during the night as well as by day. Even if the geometric quality of the image is all-right, what in principle may be achieved, the application value for most vegetation mapping is minor, but it can be of use in land-use studies and special land ecological studies.

The main advantages are the following. It can be done during the night. It records direct ecological features that may be of interest for advanced aerial ecological studies. Occurrence of frosty valleys and other micro- and mesoclimatic features (depending on and so indicating soil moisture, etc.). On not too small scale images, animals can be surveyed provided the right time of the day is chosen in relation to contrast between animal body and environment temperature and provided the animals will not hide under trees, etc. Fire survey and forest fire watch schemes can benefit from it. In fact the latter application is the one most widely used, already on a routine base. Especially from satellites thermal imagery will be able to give a better insight into the 'fire ecology' of the savanna areas of the world.

Disadvantages for vegetation survey, however, is the inconsistency of the recorded information. Emissive infrared contrary to reflective radiation is influenced by a number of factors strongly varying in kind and time. It is very difficult to predict (and explain) each feature. One can observe a pattern but it is very complicated to do the necessary groundtruth measurements in an efficient way. Recorded is the emitted radiation from the top of the vegetation surface (or if lacking) of the soil or water. This radiation is influenced beside the emission co-efficient of the vegetation material by transpiration effects and by the amount of energy transported from lower levels.

This amount depends on: total amount of energy stored, transmissivity of the materials and also on disturbance due to wind action. In areas with moist soil, by daytime the top of the vegetation may be warmer than the top of the soil or water surfaces. The image then has much in common with reflective infrared. During the night, water and moist soil may appear 'warmer' than the vegetation, rough surface may have stronger (cooling) transpiration than smooth surfaces. These differences could be of interest for pure vegetation mapping because vegetation structure may be indicated by this. Slight wind influence may, however, disturb the image, which makes it an especially vulnerable method. Most important is that one image is not enough. A sequential set of images during 24 hours is usually required, which makes the application, already very costly, and often too expensive. Therefore it will remain a marginal means, only useful for special purposes.

Some notes on radar

Contrary to thermal infrared, radar is a very useful means for direct vegetation recording.

The difference with the other means is that it represents an active sensing by emitting its own radiation (usually of micro-wavelengths of 8 nm, 3 cm and around 12 cm) of which the reflection is sensed and converted into an image very much like a photograph. Modern developments enable also stereoscopic study. It enables us to record directly differences in vegetation structure. As for any electromagnetic radiation, structure elements of about the size of the wavelength cause resonance so that they can be well distinguished from coarser or finer elements. (So colour is 'made' by structure dimensions in the order of magnitude of the wavelength of light, 0.4–0.8 µm, which are intercellular structures in plant tissues. Distribution, orientation and also the moisture content of the structure elements influence the reflection.)

Radar is able mutatus mutandus to record structure elements of the order of mm to cm which correspond with leaf size. The finest radar (8 mm) therefore is still influenced by atmospheric bodies (dust, hail, raindrops); 3 cm radar is very suitable for recording vegetation differences and can 'look through the clouds'. Larger microwave radar even looks through the vegetation a little bit into the soil, and is not suitable for vegetation survey. For surveys in the humid tropics 3 cm radar has been applied with great success. Polarisation permits enhancement of structures in various directions.

Smooth water surfaces, drysmooth soil surfaces, road surfaces and fine textured homogeneous vegetation canopies reflect the incoming radiation away from the antenna, and thus appear in black in the image. Also radar shadows, caused by relief features, are black. However, buildings and constructions such as bridges, roadbanks, canalsides, but also forest-edges and plantrows (when facing the antenna) strongly reflect and appear in very light greytones. Open forest and crop canopies give a mixture of vegetation canopy and soil reflection, resulting in a speckled texture in their image.

The backscatter of relatively shorter wavelength radar will be determined by the roughness and leaf-orientation of (dry) vegetation covers; the longer wavelength radar will give more information on the soil under such a dry vegetation canopy. Succulent foliage, even when fine textured, such as a grasscover, however, because of its high dielectric constant gives a strong backscatter in the direction of the antenna and thus appears in light greytone in the image; so does moist soil. Problems arise in areas of strong relief: the slopes facing the antenna produce higher overall reflection intensities than those facing away and this affects strongly the comparability of greytones for land-use elements/crops over the entire area, which upsets the consistency in their interpretation. On the other hand, on small scales such as 1:250,000, the same origin causes minor differences in vegetation. Bushes of only several meters high can, for instance, clearly be seen in relief by 'shadow' signature.

Disadvantages of radar are the small acquisition scale (in the order of 1:250,000 to 1:1000,000), the relatively poor resolution of the data in comparison with aerial photography in the order of 20 m, though the civilian SAR (synthetic aperture radar) goes down to as little as 3 m. Furthermore, only large area missions are economical. Special skill is also required for the interpretation of radar imagery.

In the interpretation of radar images for land-use the problem of interpreting tonal variation is often that we do not know whether it is caused by differences in vegetation density, ground material or minor changes in landform. Mostly, the approach is that of integrated landscape analysis: broad physiographic units are delineated.

Integration with Landsat imagery and aerial photography of selected representative areas will improve detail and accuracy of the interpretation considerably.

Using different polarizations (in emitting and in the receiving), the simultaneous use of different wavelengths (multi-frequency or 'multiband'), using different angles, different flight-directions, and particularly multi-temporal radar coverage, all offer possibilities for improving the discrimination capability of radar for land-use/crop surveys.

Geometrical registration, required for superimposing radar date for different dates and with e.g. Landsat data, to detect change or for the multi-temporal approach, is rather problematic.

Multi-data radar for studying the changes in the landcover backscatter, in close relation with the agricultural calender for the area covered, has given very encouraging results in crop classification in some cases.

The certainty of obtaining high quality data (all weather capacity of radar) at the optimal periods for crop discrimination/identification is of great importance in this.

In the early 1960s and 1970s large parts of continuously cloud-covered Latin America have been surveyed and mapped with the help of airborn radar for the first time. The RADAM project in Brasil involved 8.5 million km^2, Colombia followed with the PRORADAM project in 1972, Venezuela with the CODESUR project. Also in Ecuador, Peru, Bolivia, blind spots in the map were finally filled in this way.

In Nigeria, a nationwide reconnaissance survey of land-use and forest was realized with the help of radar; 932,000 km^2 were covered in 3 months. Covering this area by (medium-scale) photography would at least have taken 3 years, utilizing all periods of suitable weather.

The radar strips were flown, 'looking' into two directions, and with side overlap to obtain stereo-views.

The scale of the imagery was 1:250,000, the resolution of 15–50 m (near and far range).

In 1981–1983 the European Space Agency (ESA) conducted its SAR580 airborn Synthetic Aperture Radar test flights.

In 1978 *spaceborn* imaging radar came into operation for earth-observation: the essentially sea-observation satellite SEASAT with its 25 m × 25 m resolution Synthetic Aperture Radar also produced images over land which proved very useful also for land-use study, in conjunction with Landsat imagery.

In Nov. 1981 the (Space) Shuttle Imaging Radar (SIR-A) produced 1:500,000 scale, 40 m × 40 m resolution data.

In 8 hours' sensor activity it 'imaged' digitally 10 million km^2.

A SIR-B flight, planned for late 1984, will carry a SAR with variable look-angle and will offer stereo-imagery possibility.

A future SIR-C mission is to incorporate multi-frequency and multi-polarization capability with 15 m × 30 m resolution.

All these research-missions are trying to find which combinations of λ's, angles, polarization, timing, etc., may be most suitable for, amongst other things, land-use/crop/vegetation classification and mapping. The latter (mapping), may take the form of geometrically defined digital data input into comprehensive geo-data bands to be combined with spatial information derived from other sources and serve as a basis for inventory and planning purposes.

19A. A Practical Application of Radar Imagery for Tropical Rain Forest Vegetation Mapping

G. SICCO SMIT

Introduction

The humid tropical forest is up to now the major area in which practical experience has been obtained with radar images for vegetation mapping. We have therefore chosen tropical forest mapping as a realistic example to demonstrate the application problems and possibilities.

For the practical application of radar imagery in forest vegetation mapping, several requirements have to be fulfilled, both by the user and the imagery. The user should have a broad knowledge of the characteristics of the forest type to be mapped and how these types might be imaged on radar. This information can be obtained by practical training and research in combination with a theoretical background of the principles of radar. The image should be simple in handling. The purpose of interpretation and deliniation of the forest types should be according to a classification system. A broad frame for the system is generally available, but in reality adjustments have to be made in the sub-classification of the forest types of previously unmapped areas. In this respect a visual interpretation with a built-in training aspect can be more practical in application than a rigid computerized classification system.

The accuracy of the deliniation and classification has to be checked in restricted pilot areas for its limitations by using other imagery – aerial photography – if available, reconnaissance flights and a fieldcontrol. The latter is always a necessity if forest survey information of the forest types has to be obtained which is not visible or deducible from the radar imagery, such as the species composition and relevant timber volume of the tropical rainforest.

The product, the forest type map, should be comparable or even better in accuracy and cost (including the 'time' factor) in relation to a map obtained of the same region by other remote sensing techniques.

The practical application of radar imagery also depends on the requirements of the map. There are differences in requirements between a small-scale map of a large tropical rain forest area in the Amazon region with a limited classification into wet- and dryland forest types as a base for a reconnaissance forest inventory, and a large-scale map of small areas of plantation forest in the temperate region of Western Europe with a detailed classification into species, age classes and volume assessment as a base for a 10-year management plan. In the first case radar imagery is a proven tool; in the second case the application of this imagery is under investigation.

The research possibilities are great; inherent in the radar systems, combinations can be obtained with different polarizations, bands and depression look under controlled and known conditions. The influences on the reflection of the radar beam by the terrain conditions as slope and soil moisture content, wetness of the vegetation, wind and seasonal differences of the forest vegetation are more difficult to obtain.

A.W. Küchler & I.S. Zonneveld (eds.), Vegetation mapping. ISBN 90-6193-191-6.

Apart from the visual interpretation of a radar imagery the computerized classification may become important in view of the satellite imagery. There are two types of radar systems: non-coherent or real aperture (brute force) used for instance by Westhinghouse (Ka-band; 0.86 cm) and coherent or synthetic aperture radar (SAR) used for instance by Goodyear Aeroservice (X-band; 3.12 cm).

Use of radar imagery for
forest type mapping in general

The main advantage of radar is that for large areas, where climatical conditions hamper a coverage with small-scale aerial photography or Landsat MSS, a total coverage of the terrain can be obtained in a short time. For this purpose the radar is installed in an aircraft as a side-looking airborn radar (SLAR) system. The antenna emits a narrow beam of radio waves in the selected band (X-, C- or L-band) and polarization (horizontal -H or vertical -V). The returning echoes are recorded in time sequence and provide the cross-track component of the image; the along-track dimension is the forward motion of the aircraft in the flightline direction. SLAR produces a continuous strip of which the scale is independent (between certain limits) of the flying height.

For a flying height of 6 or 12 km the image is usually on scale 1:200,000 to 1:400,000 and covers a width of 20 to 40 km and up to 1,000 km length in the terrain.

The returning radar pulse is initially received as a function of the slant range distance from the aircraft and will be transformed to ground range. The SLAR image has a line-perspective geometry. Linear features as drainage pattern, forest type limits with an abrupt difference in the height of the vegetation and topographical differences will be observed more clearly in the direction parallel to the flightline than perpendicular to it. A high depression angle covers the near range and a low one in the far range of the image. The radarshadow of objects of the same height are larger in the far range than in the near range.

SLAR strips can be interpreted in monovision. The radar shadows should fall preferably towards the interpreter, otherwise the impression may occur that rivers and valleys are higher than the hills: the so-called 'pseudo three-dimensional vision'. Also real stereoscopic vision is possible. A technique is to take the SLAR strips with a 60% overlap in such a way that the terrain, imaged in the near range and in the far range, can be observed under a stereoscope.

The radar shadows for stereoscopic vision are preferable to the left direction. Because of movements of the aircraft and instrumental errors it may be difficult to estimate or measure the absolute height differences and even a horizontal terrain configuration may result in stereoscopic vision as a rolling surface with the level of rivers going up and down. However, SLAR gives excellent information on the relative topographic height differences and on abrupt differences in the height of vegetation, it may be even better than on aerial photographs at the same scale.

The interpretation elements tone and texture should be used with the utmost care. The tone will be lighter if the reflection towards the antenna is larger and will depend on the depression angle, the angle of incidence of the slope towards the beam ('aspects') and the dielectric constant of the object. A hillslope with an incidence angle of nearly 0° towards the radar beam will have a light tone and near 90° a dark tone; with more than 90° radar shadow will occur. This influence of slope may obscure great differences in vegetation types. Water will have a dark tone, the heterogeneous canopy of a forest a medium tone and the limit of a smooth low vegetation and a tall forest will have a light-toned line in the along track direction. This white line and the radar shadow explains the clear distinction between agricultural plots within the forest and the problem of differentiation of a secondary forest from an undisturbed forest in shifting cultivation.

The element texture is mostly combined with volume. It is true that a shrubforest with a smooth texture has a lower volume than the heterogeneous aspect of a hill forest, but the volume of this forest type can be lower than the smooth texture of a dense homogeneous plantation forest.

As a base map for the forest type mapping, the images can be used to compile a mosaic with an accuracy of 1% in the cross- and along-track directions.

Use of radar imagery for forest type mapping in the tropics

Side looking airborne radar (SLAR) is a well-known, tried and tested system of obtaining small-scale images of vast tropical rain forest regions. Projects like RADAM in the Brazilian Amazon and PRORADAM in the Colombian Amazon with a synthetic aperture radar system in the X-band (Goodyear Aeroservice), in Nigeria (Motorola – X-band) and of the Pacific coast region of Colombia (Westinghouse Ka-band) with a real aperture radar system have shown the importance of both systems in mapping forest regions with radar imagery on a worldwide basis.

Forest type mapping in the tropics started in 1968 with the radar mosaic of the Darien area of Panama and Colombia. The legend consisted of four dryland forest and four wetland forest types and three types of human influences. This means that the very important delineation can be made between dry- and wetland forest types with a sub-delineation according to vegetation and topography. On the other hand it seems that some information was superimposed over the radar images because the areas with timber exploitation could not be identified directly on the radar images.

For the interpretation of SLAR images for forestry purposes, the interpreter may have additional material such as aerial photography, topographic maps, 'obsolete' forest type maps and at his disposal information about the vege-tation from forest inventories or other sources. Logically these data are a 'must' if the interpreter is not yet well trained.

However, even a well trained SLAR interpreter will have problems adjusting an existing forest type classification to SLAR interpretation with its properties and limitations. Moreover, a SLAR project is generally not carried out for one single purpose such as forestry alone, but as a multidisciplinary tool which may be used for a variety of purposes including thematic mapping for geology, geomorphology, soil survey and land-use. With good interdisciplinary coordination one may take advantage of the possibilities of interchange of valuable data and a better directed field control.

The importance of a field control is demonstrated by the following example. In June 1969 radar images according to the Westinghouse real aperture system AP/AP-97 (Ka-band, 0.86 cm) were taken at the scale of 1:220,000 of an area of more than 10 million ha of the Pacific coastal and mountain area of Colombia. About 70% of this area consists of tropical rain forest of which large parts have never been mapped before.

A company suggested that on 1:50,000 scale enlargements the following two forest types could be delineated without field verification or collateral sources:

— type 2.1: trees 20 to 25 m tall, heavy underbrush, spacing approximately 8 m, bole diameter 0.2–0.5 m;
— type 2.2: trees 30 to 40 m tall, continuous canopy, spacing approximately 13 m, bole diameter 0.5–1.0 m.

This delineation suggests that without ground-truth data it would be possible to measure or to estimate the tree height, the under growth, closure of the tree canopy, the number of trees and the diameter of the on radar images treebole; in other words the volume of standing timber. Such a delineation system would be more applicable for a homogeneous even aged forest in the temperate zone than for the heterogeneous tropical rain forest types of the dryland region. Based on photo interpretation of existing

1 : 60,000 aerial photographs, reconnaissance flights and field visits it was found indeed that there was no relationship between the above-mentioned two forest types as interpreted on radar images and the existing forest types. For the practical application, however, three of the elements visible on SLAR images, namely *drainage patterns, human influence* and *physiographic features,* are particularly useful for forestry mapping purposes.

Drainage patterns

Large bodies of water like the sea, estuaries, lakes and main rivers are easily detected by their black tone and smooth texture. Swamps, former river courses and oxbows in the wet land, border areas dividing wet- and dryland, the pattern of gallery forest along streamlets in grassland and strongly curved rivers are also easily detected.

The smaller rivers and streams with an 'along-track' alignment in the dryland, with small topographic height differences and in the low hill regions are more easily detected than the rivers and streams with a 'cross-track' alignment. For the latter the general course can be predicted but an accurate delineation is difficult. In the high hill region, and even more pronounced in the mountainous region, the pattern of the drainage is difficult to establish. The presence of radar shadows and the high reflection of the slopes towards the radar beam prevents the delineation of a reliable pattern of the partly visible drainage.

In places it may be difficult to distinguish between small lakes or recently burned areas. A road with many curved stretches can give the same impression as a river.

Human influences

The straight line pattern of human influences is clearly visible if tonal differences are present and if the abrupt change in height between lower vegetation and high forest is towards the radar beam (thin line of high reflection) or away from the radar beam (thin dark line of radar shadow). A city has an overall light tone; other areas with a light tone will not be mistaken for a city if no roads are visible nearby. Villages cannot be detected if their location is not already known. The main roads with their straight line configuration can generally be delineated accurately; however, in some parts, even with an 'along-track' alignment, too many changes in the natural vegetation may prevent delineation of the main roads. Smaller roads and tracks are not visible. In the dryland region of terraces and low hills the plantations, permanent agriculture or grassland can be delineated if the area is large enough and its limits are straight lines. Tonal differences within these areas may indicate the type of crop or agriculture but the aspect of hillslopes in relation to the radar beam can prevent a high degree of accuracy. In the high hill and mountainous regions the delineation becomes more difficult and because of radar shadows or high reflectance of the hillslopes towards the radar beam differences in vegetation due to human influence will be unpredictable.

The biggest problem is the detection of the small areas of agriculture classified as shifting cultivation and a possible differentiation between secondary forest and natural low forest types of the dryland regions. I.S. Zonneveld (pers. comm.) mentions the possibility in the east Amazon region to map (cocoa) plantations in the forest areas.

Physiographic features

Usually the tropical rainforest is heterogeneous in species composition and in diameter distribution. Of all the species in the heterogeneous forest types only a few have an actual commercial value. The presence of these commercial species can either be dispersed throughout the forest or more or less grouped according to special growing conditions. These conditions are mostly related to topography, stagnating rainwater, impermeable or sandy soils and the influence of salt water near the coast. As on small-scale SLAR images with a resolution of 15 m the individual tree is not visible, it is not recommended to use interpretation systems based on the delineation according to species, tree height, crown closure or crown diameter or even volume. The topographic height differences under a dense tropical rain forest are better reflected on SLAR images than on aerial photographs.

The delineation between flooded (wetland) and non-flooded (dryland) areas is fairly easy to delineate accurately on SLAR and is important in view of species composition and exploitation methods. The delineation of the dryland forest into classes of ruggedness of the terrain may have a correlation with species composition and volume; even when there are no significant differences in species composition or volume according to inventory data, this delineation into ruggedness classes gives important information for the exploitation of the forest (see also Chapter 20).

The text and pictures of radar surveys in tropical regions will be given in the following examples.

Proradam – Colombian Amazon (1973–1979)

The 'Projecto Radargrammetrico del Amazonas' in Colombia is an excellent example of the practical application of radar for forest type mapping of a vast tropical rain forest area of which no accurate maps were available. This area of 38,000,000 ha is covered mainly by dense natural rain forest with the exception of savannas and savanna forest types in the northern part.

The terrain consists of broad floodplains along the main rivers, a flat to low hill configuration of the dryland and in the centre a zone with higher hills and low mountains. The human influence of shifting cultivation, secondary

forest and permanent grassland or agriculture is rather small in extent.

The project was multi-disciplinary incorporating forestry, geology and pedology. The total duration was 93 months and the amount of work was in total 6,500 man/months.

In 1973 the area was flown with SLAR. The synthetic aperture radar (SAR) system of Aeroservice Corporation in the X-band (3.13 cm) was used with a depression angle between 15°–39° and a flying height of 12,500 m. The flight line was N-S, always with a west-look direction. A total of 104 radar strips were flown with a 60% overlap having a coverage in the near and in the far range. The radar strip had a width of 37 km and a length up to more than 1000 km. The original scale was 1:400,000 with an enlargement to 1:200,000. A total of 69 semi-controlled radar mosaics on scale 1:200,000 was compiled. If cloud coverage permitted it, additional false colour aerial photographs on scale 1:80,000 were taken.

The interpretation system was based on the differentiation between wet- and dryland types, the physiographic features like topography and drainage patterns and the differentiation between dense forest and the savanna types. The classification of the main forest- and vegetation types (total area in percentage) is as follows:
— Region A: forest and vegetation types of the floodplain (8.95%); four sub-types and six combinations based on inundation characteristics;
— Region B: forest and vegetation types of the dryland (terraces and low hills):
  1. dense and heterogeneous forest (69.20%); six sub-types and six combinations based on topography;
  2. savanna and savannaforest (15.05%); five sub-types based on vegetation differences and topography;
— Region C: forest and vegetation types on high hills (3.85%);
— Special types: human influences and non-forest (2.95%).
The 1:200,000 scale semi-controlled radar mosaics were used as base maps for the delineation

and classification of the relevant forest and vegetation types. The final 11 maps were reduced to scale 1:500,000 and published in colour. To verify the accuracy of the interpretation and to obtain information – not visible on radar images – related to the species composition and volume of the different forest and vegetation types, a reconnaissance forest inventory was carried out.

The inventory consisted of 874 one hectare sample plots in 30 pilot areas and 174 one hectare individual sample plots mainly along the rivers (sampling density 0.0036%). For the statistical calculations the forest types were grouped according to the volume of the standing timber into exploitable and non-exploitable types. An overall map on scale 1:1,000,000 (two parts) gave this information of standing timber volume in combination with the ruggedness of the terrain.

Example 1a

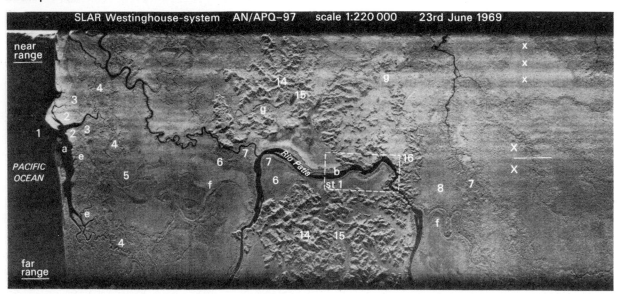

Example 1b

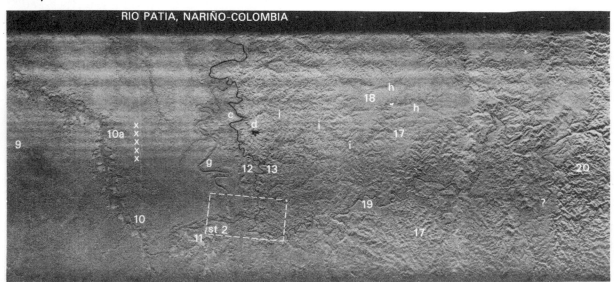

Example 1c

Fig. 1A–C. SLAR Rio Patia region, Nariño Colombia, scale 1:250,000.

Example images

Example 1 (Fig. 1A–E)

Example of a radar image tropical rain forest SLAR, Westinghouse system AN/APQ-97, scale 1:220,000.

23rd June 1969, Rio Patia region, Nariño, Colombia.

In this example covering a distance of more than 100 km, the special features of the coastal area of the Pacific Ocean, the swampy conditions of the lowland, the differences in ruggedness in the hilly zone and the mountains of the 'Cordillera Occidental' are all clearly visible.

All water bodies have the same dark tone, whether it was saltwater of freshwater rivers with silt or clear standing water in lakes; the depth of water is not reflected, the sandbars near the estuary were above the waterlevel when the image was taken.

a) saltwater
b) river with silt
c) river with clear water
d) lake

In the lowland area the drainage pattern is clearly visible: a new bifurcation of the river, old meanders and former river courses can be identified. In the hilly area the direction of the streamlets is less clear and it is sometimes impossible to delineate them correctly because of the 'shade' due to the backwardslope of the hill being bigger than the depression angle of the radar beam. This influence is more pronounced in the far range part of the image.

e) tidal creek with typical pattern
f) old meander
g) former river course
h) drainage clearly visible
i) drainage less visible
j) drainage not visible

The difference in tone depends on topographical differences related to vegetation types. In addition, strips with tonal differences can be observed parallel to the flightline, which are sharp and large X/X or less sharp and smaller x/x without any relation to topography. In the mountain area the topographic height differences can be so large that in the near range part the slope towards the radar beam is forshortened and in the far part the backward slope radar shadow 'blackens' a considerable area.

The location of the most important forest types can be deduced by the concept of 'confidence of evidence' between the tonal differences and 'groundtruth data'. In this example the forest types are indicated and not delineated in order to prevent biased impression in forest type mapping from radar images. The combination of drainage pattern and topographical differences makes it possible to divide the area into three main zones:

— coastal zone with influence of saltwater (estuary, tidal creek, ecotonic swamps)
  1. bank above waterlevel
  2. mangrove forest
  3. swamps, salt or brackish water
  4. limit between mangrove vegetation and freshwater forest
— wetland forests of the floodplain and swamps (rivers, old meanders, former river courses, low topography)
  5. swamp with a herbaceous vegetation
  6. low swampforest with 'sajo' (Campnosperma panamensis)
  7. forest on levee
  8. swamp with palms
  9. swampforest with cuangare (Irianthera joruensis)
  10. levee forest with shifting cultivation not clearly distinguishable (10a) from swampforest or more clearly (10b) due to straight borders
  11. mixture of levee forest and swampforest
  12. levee and inundated terrace forest with shifting cultivation which because of its 'texture' can be interpreted as forest of dissected terrace
— dryland forest with differences in topography (drainage pattern, ruggedness)
  13. forest of dissected terrace
  14. forest of dissected terrace and lower hills
  15. inundated valleys with swampy conditions

  16. forest of undulating terrain with swamps in between
  17. forest on lower hills
  18. forest on undulating terrain at higher level
  19. forest on floodplain of small river
  20. forest on high hills and rugged terrain
  21. forest on undulating terrain, highest level
  22. mountain forest

This classification can be useful in excluding from an inventory those areas which are too wet or too rugged for an economic exploitation. The species composition and volume determination can be executed by inventory and according to these broad delineations of forest types.

In addition two stereograms of aerial photographs on scale 1:50,000 are given as sample areas to indicate the differences in tone and textures of the rain forest and vegetation types between these two techniques of remote sensing. The stereograms have been constructed for stereoscopic observation with a pocket stereoscope (Fig. 1D and E).

Example 2 (Fig. 2)

Nigeria; 1:1,250,000 (original 1:250,000); SLAR mosaic, NB 31–05 (Benin region), Motorola Aerial Remote Sensing, Inc., Federal Department of Forestry, Nigeria, 1976–1977.

This mosaic, covering 500,000 ha, is constructed of 'South-look' imagery – this giving a better 3-dimensional impression of topographic differences than the 'North-look' imagery also available.

Three of the elements visible on SLAR mosaics, namely drainage patterns, human influences and physiographic features, are particularly useful for forestry purposes.

Drainage patterns. Large bodies of water like the sea, an estuary, lakes and main rivers are easily detected by their black tone and smooth texture. Swamps, former river courses and ox-bows in the wet land, border areas dividing wet and dry land, the pattern of gallery forest along

Stereogram 1

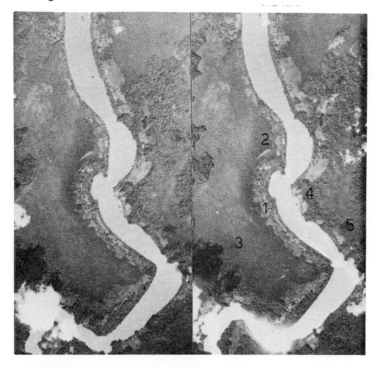

1. levee with
   shifting cultivation

2. swamp forest
   (alluvial plain)

3. swamp
   (alluvial plain)

4. hill forest

5. swamp forest
   (in former valley)

Stereogram 2

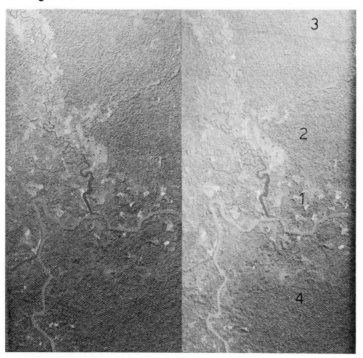

1. shifting cultivation

2. swamp forest
   (Cuàngare)

3. swamp forest
   (Sajo)

4. dry land forest on
   disected terrace

Fig. 1D and E. Stereograms Rio Patia region, Nariño Colombia, scale 1:50,000. (D) 1 = levee with shifting cultivation; 2 = swamp forest (alluvial plain); 3 = swamp (alluvial plain); 4 = hill forest; 5 = swamp forest (in former valley). (E) 1 = shifting cultivation; 2 = swamp forest (Cuàngare); 3 = swamp forest (Sajo); 4 = dryland forest on dissected terrace.

258

Example 2

SLAR mosaic, NIGERIA

NIGERIA

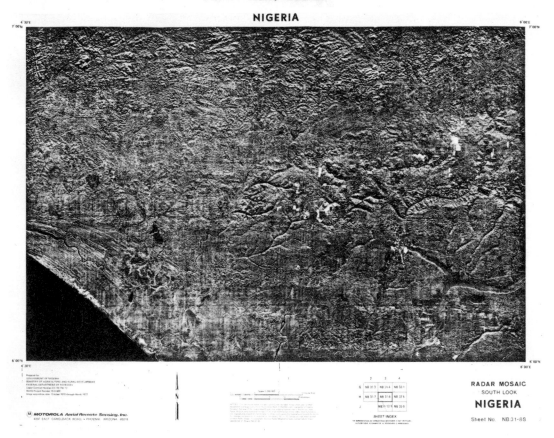

Fig. 2. SLAR mosaic Nigeria (Benin region), scale 1:1,250,000 (original 1:250,000).

streamlets in grassland and strongly curved rivers are also easily detected. The smaller rivers and streams with an east-west alignment in the dryland with small topographic height differences and in the low hill regions are more easily detected than the rivers and streams with a north-south alignment. For the latter only the general course can be predicted. In the high hill region, the pattern of the drainage is difficult to establish. The presence of radar shadows and the high reflection of the slopes towards the radar beam prevents the delineation of a reliable drainage pattern as it is only partly visible. In places it is difficult to distinguish between small lakes or recently burned areas.

Human influences. The straight line patterns associated with human influences are clearly visible if tonal differences are present and if lines indication abrupt changes in height between low vegetation and high forest, towards the radar beam (thin line of high reflection) or away from the radar beam (thin dark line of radar shadow).

Generally the alignments in the east-west direction are more easily detected than those in the north-south direction. The city of Benin has an overall light tone; other areas with a light tone will not be mistaken for a city if no roads are visible nearby. Villages cannot be detected if their approximate location is not already known. The main roads with their straight line configuration can generally be delineated accurately; however, in some parts, even with an east-west alignment, too many changes in the natural vegetation may prevent accurate delineation of the main roads.

Smaller roads and tracks are not visible on this SLAR material. In the dryland region with low hills, the areas with forest plantations, permanent agriculture, and grassland are delineable if the area is large enough and its limits are straight lined. Tonal differences within these areas may indicate the type of forest plantations or agriculture but the aspect of hillslopes in relation to the radar beam influences the accuracy of interpretation. In the high hill and mountainous regions the delineation becomes more difficult because of radar shadows or high reflectance of the hillslopes. A problem is the detection of small areas of agriculture and the differentiation between secondary forest and natural low forest types of the dryland regions.

Physiographic features (zones). Normally the tropical rain forest is heterogeneous in species composition. Of all the species in the heterogeneous forest types usually only a few have actual commercial value. These commercial species can either be dispersed throughout the forest or be grouped as a result of special growing conditions. These conditions are mainly related to topography, soils (waterlogged or nutrient deficient), human activities (shifting cultivation) or the influence of saline water in coastal regions. Individual trees are not visible on small-scale SLAR and it is therefore not recommended to use integration systems based on delineation by species, height crown closure, crown diameter or volume.

The differences in ground elevation under a dense cover of tropical rain forest are usually better detected on SLAR images than on aerial photographs.

The boundary between flooded (wetland) and non-flooded (dryland) areas is fairly easy to delineate on SLAR, as is the division of terrain into ruggedness classes. This is important since such distinctions often relate to species composition and exploitation methods.

Example 3 (Fig. 3)

Example of a satellite radar image, Rio Guaviare, Colombia, SIR-A.

The epoc of spaceborne microwave surveying started with the launching of Seasat by the USA in 1978 (L-band, SAR spatial resolution 25 m, 100 km swath width). Because of the necessity of real time transmission of the tropical forest areas only those of Central America were covered by the Florida Station. The shuttle imaging radar programme started in November 1981 with the launching of SIR-A (L-band, SAR spatial resolution 40 m, 50 km swath width, 43° depression angle) and over 10 million km^2 were imaged during the 8 hours of sensor activity covering in single strips some of the tropical rain forest areas in Central and South-America, Western Africa, India and the Far East and Papua New Guinea/North Australia.

In october 1984 the SIR-B was launched, but still in January 1985 no images were received for interpretation. Other programmes for satellite radar images are in advanced stages of planning such as those of Europe and Japan.

Description. Clearly visible is the natural grassland region of the 'Llanos' (I). The black-toned grass (Ia) shows no indications of recent burnings; and the light-toned gallery forest along the streamlets (Ib) indicates the high response of the forest vegetation.

The floodplain region (II) of the river Guaviare is detected by the dark-toned river (IIa), the oxbows (IIb) with stagnating water and the light-toned reflection of the swamp vegetation (IIc) in former oxbows of the river. However, the limit between the floodplain and hill regions (IId) can only be presumed. In the forest region (III) the gray-toned dryland forest of the hill region (IIIa) can be differentiated with difficulty from the river forest (IIIb) by tone and texture.

This satellite radar image does not indicate the important feature of topographic differences as is common in SLAR taken from an aircraft.

260

Example 3

SATELLITE RADAR IMAGE
Rio Guaviare, Colombia

SIR-A

I. natural grassland
  "Llanos"
a. grassland
b. gallery forest

II. floodplain
  Rio Guaviare
a. river
b. oxbow
c. swamp vegetation in
  former oxbow
d. presumed limit
  between floodplain
  and hillregion

III. tropical rainforest
  of the Amazone
a. dryland forest
b. river forest

Scale 1:500,000

N

location center circa
3°N – 71°30′ W

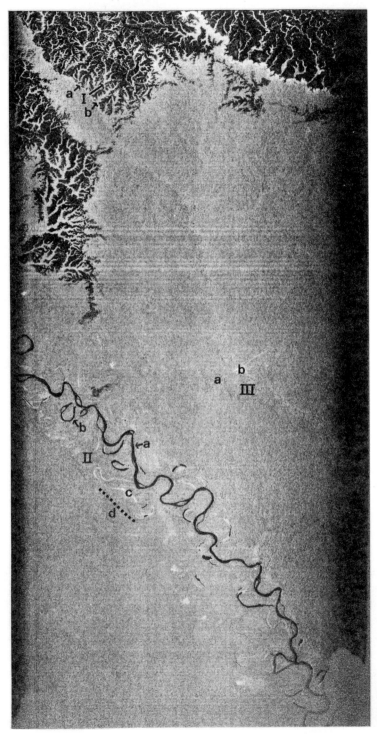

Fig. 3. Satellite radar image, Rio Guaviare, Colombia, SIR-A, scale 1:500,000.

Example 4

RADAM PROJECT—BRASIL
RIO ARAGUAIA
SB 22—2—B Sc

Scale 1:250,000

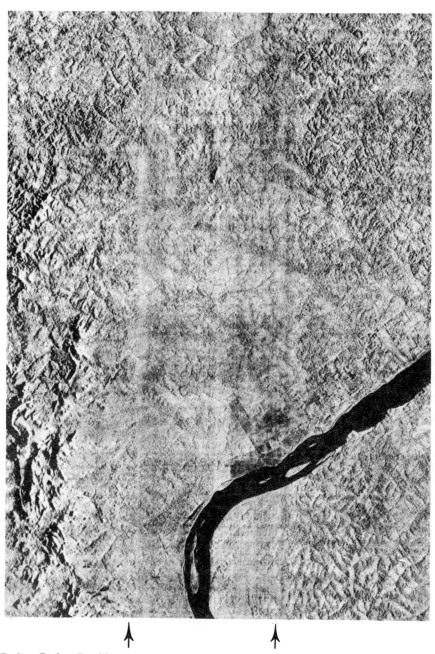

Fig. 4. Radam Project Brasilian Amazon, Rio Araguaia, SB 22-2-B Sc, scale 1:250,000.

262

Example 4 (Fig. 4)

Example of the Radam Project Brasilian Amazon, Rio Araguaia.

This part of a semi-controlled mosaic, made from individual radar strips (X-band, SAR), flown in 1971/1972, clearly shows the relative topography. Along the river Araguaia the straight-lined limits of large areas of farm grassland can be observed; in the central part with dark-toned recent burned patches and to the right still in the process of shifting cultivation. The original forest is of lighter-tone. To the left the mountainous region – with its radar shadows falling to the left – is differentiable from the more undiluted region.

Example 5

COLOMBIAN AMAZON
Rio Putumayo

near   far

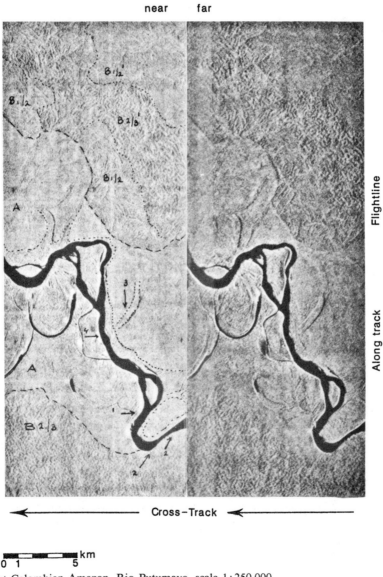

Cross–Track

0 1   5 km

Fig. 5. Proradam Project Colombian Amazon, Rio Putumayo, scale 1:250,000.

The forest area is the transition zone of the broad type of tropical woodland forest (Floresta alberta) and the tropical submountain forest, with sub-deliniations according to relief and crown closure.

Example 5 (Fig. 5)

The Colombian Amazon was flown with side looking synthetic aperture radar in the X-band, scale 1:400,000 (1973). The flightline was always in the north or south direction with the side look towards the west and an overlap of 60% between the strips.

This overlap enables observations of the terrain in the combination of near and far range under a stereoscope (mounted for a pocket one with enlarged scale of 1:250,000).

Clearly to distinguish is the limit between the actual and former flood plain zone (A) and the zone of dissected terraces and low hills (B) subdivided into classes according to the ruggedness.

In zone A can be observed that near point 1 the vegetation with a gradual increase in height from the river land inwards has a broader light-toned reflection in the near range than in the far range; the same is observed on the island in the river. The human settlements near point 2 are not detectable as such. The former oxbox at point 3 with a low swamp vegetation is detectable by the thin black line of the radar shadow and the thin white line of the limit of low (swamp) and tall vegetation (forest); this phenomenon is better detectable in the far range.

The forest vegetations along the river in the along track alignment, near point 4, is visible especially in the far range.

20. Interpretation of Remote Sensing Images

I.S. ZONNEVELD

General

The use of aerial photographs in mapping has two aspects:
— the photogrammetric and cartographic aspect and
— the interpretation aspect.
We shall deal in detail here only with the last one. For the photogrammetric and cartographic aspects of using aerial photography refer to other handbooks (see also Chapters 7, 8, 9 and 10). First of all we have to discuss some common concepts: recognition, analysis and interpretation. These words are used as far as possible here according to the common use of these words, which may be slightly different from their meanings in some handbooks about photo-interpretation. Finally modern remote sensing systems, but even also classical airphotography allow or even demand a certain form of preprocessing of the image before interpretation may start. This is treated in Chapter 21. The place and type of photo-interpretation depend also on the survey approaches used. This is treated in Chapters 16 and 29.

Recognition, analysis and interpretation

A photograph shows an image that may be composed of silver grains on paper or also of patches of one or another coloured chemical compound. Nevertheless, we may 'recognize' at a first glance the face of a person or a kind of flower or animal or ... on a vertical photo, a landscape, which are immortalized with the help of the 'camera obscura'.

We can only *recognize things we already know.* The process of recognition from the image of a photograph is an unconscious 'interpretation' of the illusion created by a certain combination of silver grains or paint dots on a piece of paper.

Using aerial photography for vegetation survey will be more effective the more the surveyor can *recognize,* hence the more he already 'knows' about the object of survey. This means he should be a good field observer in the first place. From this simple conclusion follows the main law of photo-interpretation. *Photo-interpretation is the task of the field surveyor.*

A photo-interpretation expert (whatever that may be) who does the interpretation, sending somebody else to do the field sampling, cannot exist. Although there are a great many features that can be recognized by an experienced vegetation scientist who is acquainted with photo-interpretation, there are many properties of vegetation that cannot be seen directly. With *conscious interpretation* the surveyor may guess what these may be. Such a guess, however, needs a check in the field. So photo-interpretation is an iterative process of photo study and field check in immediate alternation. For a simple survey approach such as the photo-guided field survey (see Chapter 16) this is indeed the procedure. For the key-method approach (see Chapter 16) field and photo knowledge are

A.W. Küchler & I.S. Zonneveld (eds.), Vegetation mapping. ISBN 90-6193-191-6.

266

brought together in the beginning of the survey
by direct comparison of remote sensing image
and the reality by near observation in the field.
For the most important approach, however, the
landscape-guided one, photo-interpretation
plays a crucial role in the stratified sampling as
base for final typification. There it is important
that the preliminary classification is made as
objective and consequent as possible, because
the check (better called field description or sampling in this case) is done much later and never
over the whole area. The conscious interpretation of the photo-image in terms of features
known in the field is a dangerous tool for such a
classification! It is too easy for imagination
(however, essential this may be) to run wild and
obscure the desired objectivity ('hinein inter-
pretieren'). Moreover unknown aspects may be
overlooked in this way. There is a way to increase the objectivity as much as possible and
that is through 'analysis of the photo-image features' and subsequent classification of the various parts of the image. In doing this it is necessary to be detached from what is already
'known' or considered to be known.

The piece of paper with its pattern in tones
and colours, preferably seen stereoscopically, is
for the time being the object of study, not the
reality behind it.

Such an analysis could theoretically be done
by automatic recording instruments to a certain
level but is usually done without. The features
looked upon are:
— tone (colour = hue, value, chroma);
— texture (smooth, fine, rough);
— size in horizontal and vertical (stereo) direction;
— shape also horizontally and vertically.
Classification of the image according to these
features is helped by finding 'convergence of
evidence' (clustering of classification characteristics). These features help the interpreter who
is doing a direct comparison in the field very
much in recognizing his subject on the photo
(see key method and photo-guided field survey,
Chapters 16 and 29). In doing an objective study of features and trying to classify them objec-

tively in a pre-photo (image) interpretation
there is a dilemma, however.

Efficiency of interpretation and image classification can be enhanced considerably by using
knowledge about the subject although we stated
above that dangers exist of subjectivity. As we
shall see, especially the 'systematic landscape-
guided approach' (see Chapters 16 and 29)
makes use of landscape features such as landform and spatial and temporal *relations* between patterns applying landscape ecological
knowledge. Contrary to the earlier mentioned
features automatons will never be able to do
interpretation about such types of features.
These involve photo-analysis on:
— situation or location in relation to other elements or units;
— temporal change (if multi-temporal images
are available) which gives data about seasonal and successional features. Convergence of evidence with these aspects increases the interpretation results considerably. Vegetation cover types and geomorphology (terrain analysis) represent the necessary knowledge for using all these (especially the latter) data properly in classification of the image.
The objectivity of the interpretations is not
much threatened by this latter pre-knowledge
because the aspect of cover type that can be
observed at photos and images is usually simple
(see Chapter 4). The geomorphological knowledge relies very much on stereoscopically analyzed terrain forms that can be judged quite
objectively.

The photo-interpretation, executed for those
survey approaches where the interpretation is
done simultaneously or after the field work, will
result in terms of the final mapping legend.

The pre-photo-interpretation that preceeds
the systematic landscape survey approach, however, leads to a legend description in the form
of cover type, type of patterns and terrain form
(geomorphological landforms) including drainage patterns (see Chapter 29).

All the image features should, however, always be mentioned in the legend because:

— this stimulates a systematic analysis of all image features of each pattern element;
— it facilitates the systematic comparison of field data with the preliminary photo (image-) interpretation map.

As we shall see in Chapter 20, the preliminary interpretation maps serve as base for stratified sampling.

The study of the sampling results (and the final classification based thereupon) in relation with the photo-interpretation units, will increase the interpreters knowledge greatly and gradually also his unconscious knowledge (= intuition). Now he is at the stage to do the reinterpretation of the photographs incorporating his full knowledge and intuition into the final mapping result. The new knowledge will help him to interpret photographic features, including landscape features (see above) in terms of vegetation communities that cannot be recognized directly. The increase of 'unconscious knowledge' (intuition) helps him in direct recognition (also in future surveys) of more objects and with more accuracy than was possible before. This is the main reason that also in survey approaches where pre-interpretation is done, photo-interpretation and field sampling should be done by the same person.

Preprocessing of aerial photographs and other remote sensing means as a part of the interpretation procedure

Results of optical sensing appear directly as a photograph. Development of the negative and printing processes are kinds of preprocessing that do not need to be treated here. If, however, filters are used or different intensive development of different layers, this is done to enhance certain features. This type of preprocessing is done with a certain aim of interpretation in mind. If one filters a (false) colour film with a well saturated red filter one wants apparently to use only the red-labeled radiation. One will only do this if one's interpretation aim demands this. Still the result is a not very

much deterministic image for which general interpretation methods will be applied. It is different if one wants to concentrate on very special features. In that case one has to prepare first a kind of 'key', e.g. one wants an automatic interpretation of a certain vegetation type say forest or pasture or subdivisions of these. Suppose we have a multispectral image either on photos (analogue image) or on digital tape and the computer line outprint of that tape for the various recorded spectral bands. It is possible then to investigate whether the vegetation type has a unique combination of intensities in reflection of several bands. This can be done on analogue images by comparing several colour composites. There are machines (colour additive viewers) which enable us to vary colour and intensity of that colour of each band. In essence it is a combination of a series of projectors that project the image of each band on each other with different filters determining the colour and with a possibility to make a variation in light intensity for each projector separately. One can look by trial and error, guided by some knowledge of the area and concentrated knowledge in some key areas for those combinations that give the clearest contrast and discrimination. As soon as one knows this combination one systematically passes through all the material in which now the wanted features are strongly enhanced compared with the not preprocessed material in single bands or general colour composites (such as common pan, true colour or false colour photos). So preprocessing and interpretation proceed here simultaneously. Especially for small-scale imagery (satellite), where not many photos exist, this is a feasible method. By methods of colour slicing or density slicing a machine could be trained in certain cases to pick out objectivily without further human subjectivity, the wanted discrimination.

For normal photography of large survey areas for which a surveyor has to deal with hundreds or even thousands of photos such procedures are too complicated and too time-consuming. It is clear that this type of preprocessing has more application in key type of surveys than in landscape-guided type of interpretation, although the landscape-guided survey can also benefit from it especially because it remains a trial and error process. It can be applied in mapping the pattern of landscape elements of which the character still is unknown and not only to extrapolate data about key areas with well-known details.

The above mentioned procedures represent a kind of automatization, which, however, only works in limited cases.

Other examples of such procedures are laser preprocessing which have hardly any practical application for normal vegetation survey especially because laser needs repetitive geometric line structures which usually are rare in vegetation. Automation is more prospective if digital data exist. This can be supplied by digitizing photos (Astroscan, etc.) or if the data are available on tape, which is usually the case if a scanning sensor is used (MSS from satellite or aircraft platforms). In this case statistical manipulation can be applied to enhance special features. Cluster analysis, various kinds of polyfactor or principle component analysis can be used to find characteristic combination of wavelengths per pixel (resolution unit). An 'unsupervized clustering' resulting in a line outprint with various classes is comparable with a common photo (see Chapter 21). The interpretation is done according to the same methodology as a common photo-interpretation and is rather suitable for a landscape-guided approach. The main disadvantage, if compared with normal good photography, is the usual lack of the height dimension (stereoscopy). So it is useless to apply this on scales where much easier cheap common optical photo-images can be made (false colour, true colour or common pan or optical photo-multispectral sensing). It is, however, useful for small-scale work such as satellite imagery where the third dimension is less important and moreover the units are limited in number and of a general character.*

More attention to special information can be paid by applying supervised statistical classification, using principal component analysis including transformations of the co-ordinate systems (see Chapter 21). In key areas the ground-truth is collected and via existing computer programmes that may be also adapted to special aims, rather detailed automatic interpretation can be done by letting the computer do the extrapolation of the knowledge over the whole area. Still the application of this is limited according to two reasons:

— the human brain can make correlations of 'converging evidence' of very many items; the brain of a computer is in relation to the human brain extremely poor, although a bit more consequent and objective within the scope of its limitations;
— the computerized interpretation requires special skill and especially expensive apparatus; in Chapter 24 (see also Chapter 14) this is treated in more detail.

---

* The present Landsat imagery at the not too low latitudes show some overlap that makes stereovision possible. The new system of SPOT with much higher resolution (see Chapter 31) foresees in full overlapping stereo images.

21. Digital Image Processing, Computer-aided Classification and Mapping

N.J. MULDER

Introduction

The purpose of vegetation mapping is to store information about vegetation in a symbolic way in a format which allows easy, accurate and timely retrieval of that information. Initially the information extraction was performed on photographic remote sensing data by means of visual interpretation. The information storage format was defined by the limitations of paper and pencil. The updating of maps and the combination of maps was at best tedious and often not possible at all in operational environments.

The availability of digital remote sensing data has enhanced the possibilities of information extraction by the use of computerised classification methods and the possibilities of image enhancement before visual interpretation. Visual interpretation still plays an important role but the format of the stored maps has now become digital too. Modern image processing techniques applied to cartography allow interactive map design and map updating and even the generation of movies of the change in vegetation properties with time.

The common format of digital map data and remote sensing data has the added advantage of providing multiple source data to the interpretation process. It is common practice now e.g. to estimate a vegetation index from remote sensing data which with the estimated vegetation class and some ground based measurements then leads to a biomass estimation.

This in turn can be combined with elevation and slope data from digital elevation models leading to sun exposure and sensitivity for erosion. Climatological maps and actual meteorological data from weather satellites with soil maps leads to the practical possibilities for timely environmental impact predictions and the monitoring of dynamic processes taking place on the face of the earth. The only limitation on the use of these new digital techniques is the speed of learning and the acquisition of a sufficient level of 'computer litteracy'.

In the end a good balance has to be found between the capabilities designed into the human brain for thinking and on the other hand the fast application of simple decision rules and data transformations and combinations by digital processors. The popular field of artificial intelligence (AI) is better renamed as pseudo intelligence (PI). Nevertheless in AI there are very usefull concepts for knowledge engineering as e.g. implemented in expert systems for various fields of expertise and in due time the interpretation rules from pattern recognition will be complemented by 'fuzzy' rules derived from human expertise.

One way to present the subject of information extraction from remote sensing data is the bottom up approach as followed here. We start with the basic units of remotely sensed data and gradually go to a higher level of symbolic information, contained in data about objects and collections of objects in scenes.

Hence we start with image processing. The

A.W. Küchler & I.S. Zonneveld (eds.), Vegetation mapping. ISBN 90-6193-191-6.

270

purpose of image processing is to present the information contained in numerical remote sensing data as images. The presentation must be optimized for visual interpretation. In pre-processing errors are removed and the data are transformed to a common georeferenced format. The usual format is a square grid of cells on the earth's surface, called scene elements.

The purpose of pattern recognition is to automate the classification of remote sensing data and to produce a map-like display or print. In practice the classification is performed interactively with a trained user rather than 'automatically'.

Proper interpretation and classification requires at least the right interpretation of the meaning of the numbers which code the remote sensing information in the multispectral, the thermal and the microwave domain. Further it requires the correct application of domain (vegetation) knowledge to the data as represented in images and quantitative maps.

For the correct interpretation of the basic remote sensing measurements it is essential to consider the following. Multispectral data (reflected sunlight) and thermal infrared (emissive) data are generated by photon sensors. The numerical data represent the number of photons detected over a short period of (electronic) exposure time. In these sensors each photon will generate one or more electrons. A stream of photons will thus be converted into a current of electrons which is further manipulated by electronic devices and finally converted into a digital number.

For each scene element and for each spectral band (filter) one number is generated which is proportional to the photon count for that scene element and that spectral band. For each scene element for each spectral band there is a corresponding number in a numerical image. The numerical image is stored as one two-dimensional array for each spectral band. Mapping each element of the numerical image into a (coloured or grey) picture element (= 'pixel') generates a picture. Picture elements are usually squares of the order of 0.1 mm. Pixels should

not be confused with scene elements the size of which is expressed in meters rather than fractions of a millimeter (see Fig. 1).

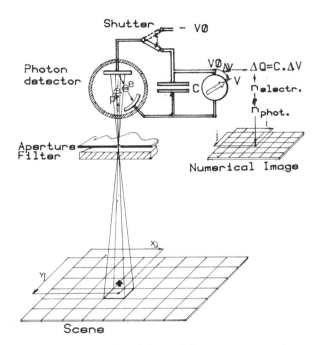

Fig. 1. Multispectral and thermal detectors are transducers based on the photo-electric effect (transducing photon energy into electron energy). A photon hitting the metal kathode may generate an electron. The electron is moved to the anode by the electrical field between anode and kathode. A stream of photons is detected as a current. The current of electrons is proportional to the stream (flux) of photons falling onto the sensor.

Microwave sensors form the other class of sensors. For microwaves the theory of electro magnetic waves is appropriate as the sensor actually produces a signal which is proportional to the average electrical field strength (power) over a limited sample time. Scene elements are defined by range resolution and antenna opening. The data are stored in a two-dimensional numerical image (Fig. 2).

The next level of interpretation requires a good understanding of the interaction of radiation with matter. Multispectral data carry information about the surface spectral reflectance, the illuminating spectral photon flux and the

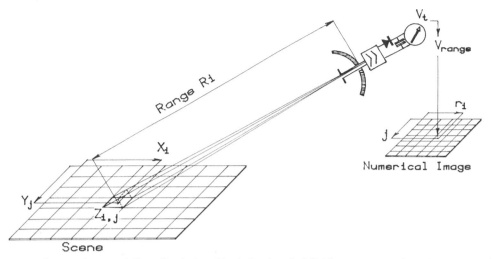

Fig. 2. A microwave detector is essentially a dipole 'catching' the electrical field component of an electromagnetic wave. The oscillating voltage is rectified and the average rectified voltage as a function of time (range to the reflecting object) is digitized and constitutes the numerical radar data. The function of the antenna is to direct and focus the radar waves. With side looking airborn radar the length of the antenna determines the resolution of the radar system in the foreward direction.

angle between the normal [1] to the surface of a scene element and the sun.

Thermal (emissive) infrared data contains always a mixture of information about the spectral emissivity, the temperature and the angle between a normal to the surface and the sensor direction.

Active microwave data contain mostly information about the structure of the reflecting scene element and some information about the dielectric constant which in turn depends on surface composition.

A further level of interpretation requires extensive knowledge of the processes occurring at the Earth's surface and the relation between processes. As an example take the cyclic changes of vegetation over the seasons and the dependance of the yield of crops on weather, soil and human activities. Such knowledge has to guide the choice of processing algorithms and processsing parameters. A further development is that heuristic [2] image interpretation rules are being embedded in classification software called expert systems.

Presently available data are generated by imaging weather satellites such as Meteosat, NOAA satellites, specific remote sensing satellite systems such as Landsat MSS, RBV and TM [3] and SPOT and occasional experimental systems.

Operational airborn systems are rare. Large areas have been covered by radar (Amazonas, Indonesia) in cases where clouds do not permit a sufficiently timely photo coverage. Airborne multispectral scanners are not really operational because the rapid aircraft attitude changes necessitate extensive geometric corrections. This correction effort is not economically compensated by improved information extraction from the multispectral data. Airborn charge coupled devices, push broom scanners (linear multispectral arrays), have the possibility to be completely attitude stabilised when fixed to an inertial platform but apparently there is not yet a suffi-

---

[1] Normal: arrow directed perpendicular to a surface.
[2] Heuristic: from heureca!; 'I found it!' (Archimede), finding the solution to a problem by insight and intuition.

[3] MSS: multi spectral scanner; RBV: return beam vidicon; TM: thematic mapper.

ciently large market for such systems to be developed for commercial use.

The digitizing of false colour (infrared) film or multispectral camera film gives data with high spatial resolution but low radiometric accuracy. A further problem with film digitizing is that commercial systems do not produce the equivalent of a photon count but the equivalent of film density. Image processing applied to film densities is one of the prevailing sources of blunders in processing.

In the near future it is to be expected that there will be data with a high spatial resolution ($= < 10$ m) a few well choosen spectral filters (< 4) and some form of on board data compression/feature extraction. At least one radar satellite will be operational. Meteorological satelites will continue to evolve in time resolution and space resolution. If a market can be identified then airborn CCD[4] cameras will be developed with high resolution (> 6000 elements in a linear array) and complete stabilisation of attitude. Film will probably be processed by hybrid analog/digital machines which will perform all necessary geometric corrections (ortho images) and radiometric corrections as well as image enhancement and partial classification (thematic ortho image maps).

Digital terrain models, digitised topo maps, digitized soil maps, etc., are already used as supporting data for classification algorithms. Any type of image or map data can be used in the classification process assuming that the relations between classes and data are understood and properly formulated. These relations are ordered and stored in a data base of decision rules (rule base).

Digital image processing is the activity of transforming two-dimensional data arrays (numbers) into images or into geo data bases and maps. Scene elements are mapped into picture elements (pixels).

Digital processing of remote sensing data became popular after the launch of the ERTS

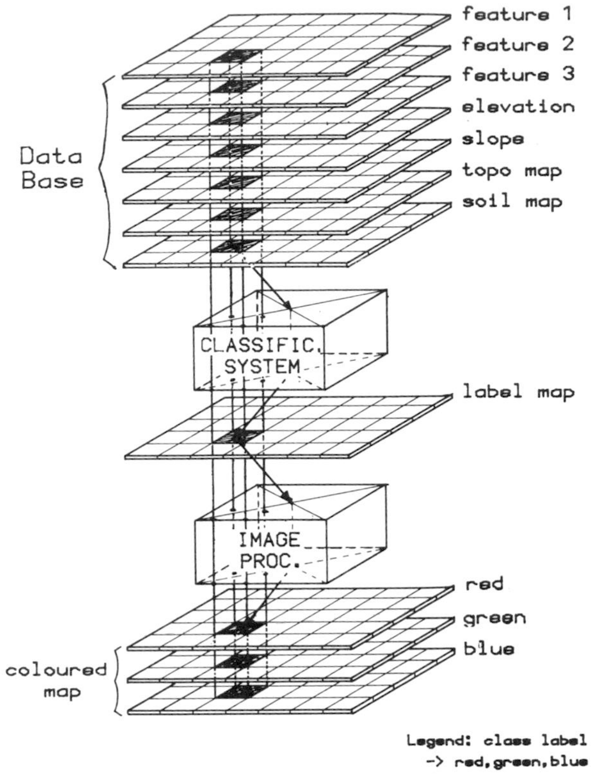

[4] CCD: charge coupled device.

Fig. 3. The classification of multiple source data.

(now Landsat) multispectral scanner (MSS) and return beam vidicon (RBV) sensors which provided digital data at low prices. Remote sensing systems produce arrays of numbers referenced to a grid of samples (scene elements) on the Earth's surface. Special hardware and software are available for the efficient correction, information extraction and display of arrays of remote sensing data and other geo referenced data in raster format. Non-raster data such as maps can be converted to raster format by digitizing lines and polygons followed by software rasterising (scanning and sampling). Information extracted from remote sensing and rastered map information is stored in a geo data base for further information extraction and information retrieval (see also Chapter 14).

Developments in the fields or artificial intelligence (AI) and pattern recognition (PR) lead to so-called expert systems which consist of a data base combined with a rule base. The expert system translates a user request to the application of a set of rules from the rule base (knowledge base) to the database (Fig. 4).

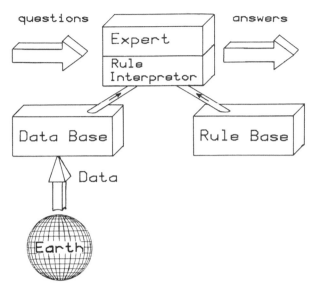

Fig. 4. The division of a classification system (expert system) into a data base a rule base and a rule interpretor. A request for information is analysed, the rule interpretor selects the appropriate rules from the rule base and applies the rules to the data base contents, producing the requested information with a quantitative measure of certainty Cf.

As all image processing systems must have interactive displays and image printers, image processing systems are highly suitable for the generation of most types of maps. Presently the exception is formed by maps which require extensive special symbol generation. The solution is in that case to do all the colour coding (colour separations) with the image processing system and to generate a separate overlay for the (hand drawn) special symbols.

In data classification and data base applications it is often necessary to report on statistics or other forms of data compilation such as area covered by a certain class of landuse. As each pixel is the image of a scene element of known area, counting 'pixels' with a certain label or attribute is equivalent to area determination of a class of fields or the area with a certain attribute.

The depth of treatment of the subject image processing and related subjects in this handbook is based on the assumption that the general reader wants mostly to be oriented in the field. An increasing number of standard enhanced images becomes available which can be used without an exact knowledge of all the mathematics involved. On the other hand it is very important to have an intuitive understanding of the subject. This understanding can be developed by always carefully studying the assumptions made in the development of any method of processing and by trying to visualise the effects of transforms. Therefore standards products will be shown, do's and dont's will be given without extensive proof. Examples are chosen in the context of vegetation surveying and monitoring. For those who only want to use pre-processed image a general perusal of this chapter will suffice, others may study the following chapters more extensively.

For more information on digital image processing and the practical use of an image processing system in combination with land information systems (LIS) and geo-information systems (GIS) the reader is advised to follow a short introductory course (minimum three months) in image processing and pattern recognition with a large amount of hands-on work and integration of remote sensing information extraction with LIS and GIS data handling.

In due time concepts of artificial intelligence, especially expert systems will develop to a practical level where the reader may want to update his knowledge of automated classification and interpretation methods by attending workshops or courses on the subject.

Pre-processing and corrections

General remarks

In applications of multispectral and thermal remote sensing the sensor output is proportional to the input photon flux (number of photons per second) at the sensor aperture. The output signal is determined by the wanted signal (reflection or emission), but also by unwanted effects such as variation in sensor gain [5] and sensor offset [6] between sensors, and by atmospheric effects such as scattering and absorption of photons.

Difference in sensor gain and offset between sensors is the cause of 'striping' in uncorrected and not properly corrected data. With the increase of the number of sensors in the same spectral band from 6 in Landsat MSS to 6000 in the new CCD scanners proper sensor correction is becoming a non-trivial problem.

Atmospheric effects vary with time and location. The strongest time varying effect is that of haze. Haze is caused by the scattering of photons coming directly from the sun or reflected from local scene elements into the field of view of the sensor. The effect is that photons are added to the photons coming directly from the scene element in the instantanous field of view of each sensor (IFOV). For airborn sensors there is an added complication that the scan angles are rather wide and hence the optical path from sensor to scene element is varying. Therefore the number of scattered photons is also dependant on the scan angle. A less important atmospheric effect is the effect of sunangle. With low sunangles the illumination of the scene is spectrally changing from white to more red and the total intensity of the scene illumination goes down for two reasons. One reason is the increased atmospheric absorption with lower sunangles. The other reason is the decreasing number of photons per second falling on a horizontal surface with lower sunangles.

The sensor scanning pattern does usually not coincide with an ideal square grid on an ideal Earth surface. This means that the data has to be rescanned and geometrically corrected. After the geometrical correction the data is either geo-referenced or referenced to a choosen scene in a time series of scenes. With multi-source data it is often necessary to resample the data in order to bring it to the same raster grid size; e.g. combining 50 m digital terrain elevation data with 79 m Landsat MSS data, 30 m Landsat TM data and 25 m Seasat SAR data.

Radiometric corrections

Radiometric corrections [7] should be done before any geometric correction or geometric transform. The reason for this rule is that in the raw data there is a direct relation between the 'scanline' of each sensor and its position in the sensor array. After e.g. rotation of the data values this direct relation is lost.

If the data are 'raw' data
Then start with applying sensor correction.

If the data have been corrected by e.g. a central facility then check the quality of the data using e.g. a colour display unit. If you still detect 'striping' in the display then re-do the sensor correction.

Else check whether atmospheric corrections are needed.

Correction of sensor effects

Gain and offset correction. In multispectral and infrared sensor systems the input of the sensor consists of a stream of photons (photon flux) flowing through a conical field of view of each sensor. The sensor electronics integrate the photon flux during an electronic 'exposure' time resulting in a photon count. The output of the sensor is a number which is a scaled value of this photon count. Various sensors in the same photon energy band have different gains and offsets; e.g. Landsat MSS has four spectral bands with six sensors in each band. The aim of the correction procedures is to get the same effective gain from all sensors in a spectral band and to get a signal with zero sensor offset.

---

[5] Gain = difference in output divided by corresponding difference in input.
[6] Offset = output at zero input.

[7] Radiometric correction: correction of sensor effects, correction of atmospheric effects.

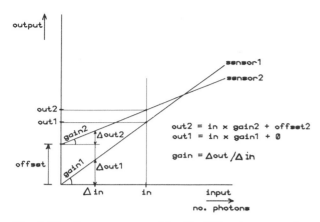

Fig. 5. The linear sensor model with different sensor gain and offset between sensors. Gain = increase in output/increase in input; offset = output at zero input.

Sensor correction in multi sensor systems is based on the assumption that sensors differ only in sensor gain and sensor offset (linear sensor model, see Fig. 5). Two methods of sensor correction are used:
— calibration table inversion;
— histogram matching.

Calibration table inversion. A calibration table is generated (for each sensor) e.g. during the retrace time of the scanning mirror of Landsat MSS. The sensors are exposed to a photon flux of known intensity and the numerical output is recorded in a table. What is needed, however, is a relation between a measured output and the corresponding correct input. To achieve this the table has to be inverted viz. a new table will store input as function of output. As the output is in integer [8] values the output is used as an index to the 'inverted' table. The contents of the table at address 'output value' is the correct input of the sensor concerned.

An interpolation algorithm is used to calculate an input value for those cases where there was no input $-$> output pair in the calibration data.

[8] Integer: whole number e.g. 0, 1, 2, 3, ...

The transform (table) relating actual sensor output to ideal sensor output:

N-table(s) (old-value(s))

for sensor s, apply to all sensors a transform according to:

new-value(s) = N-table(s) (old-value(s))

As an example: let sensor s = 1 and the table be:

index :	0	1	2	3	4	5	6	7	8	..
N-table:	0	0	0	0	1	2	3	4	5	..

N-table implements the relation: new-value = old-value $-$ 3; if old-value < 4 then new-value = 0. So if e.g. old-value = index = 7 then

new-value: = N-table(7) = 4.

The above-mentioned use of tables for the transformation of data is very common in image processing. Often the term look-up tables or table look-up is used to indicate the mapping of an input value into an output value using the (integer) input value as the address (index) in the table with the output at the contents of the table.

output: = look-up table (input)

Calibration data may not be available or may be of bad quality. In such cases radiometric corrections are based on the statistical method of histogram matching.

Histogram matching. The basic assumption in histogram matching is that a scene or part of a scene (window) has a spatial uniform distribution of sensor input (photon flux per sample). As the histogram of the input for each sensor is the same for the same part of the scene, the histograms of the output values must also be the same for ideal sensors. The cumulative histograms per sensor must also be the same (allowing for some noise). If the cumulative histograms do not coincide, then this indicates a different gain and/or offset for the sensors.

Fig. 6A shows a hypothetical input photon

flux histogram transformed into two output histograms by sensors with different gain and offset. The two different sensors produce values with histograms with different mean and standard deviation. Fig. 6B shows the corresponding cumulative output histograms.

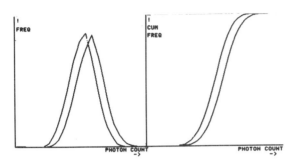

Fig. 6. (A) output histograms of two sensors; (B) output cumulative histograms of two sensors.

A simple method of histogram matching is to take the histogram of the sensor with the highest gain as a reference and map the others sensor outputs onto this sensors cumulative histogram using histogram mean and standard deviation:

$$new(s): = \frac{old(s) - old\ mean(s)) * old\ sd(r)}{old\text{-}sd(s)} + old\text{-}mean(r)$$

s = sensor;
r = reference sensor;
sd = standard deviation.

Instead of performing the above transform on all old data a look-up-table is generated with old(s) as address and new(s) as contents.

In case the linear sensor model is not applicable (not just gain and offset) it is better to use a method which directly maps old(s) values

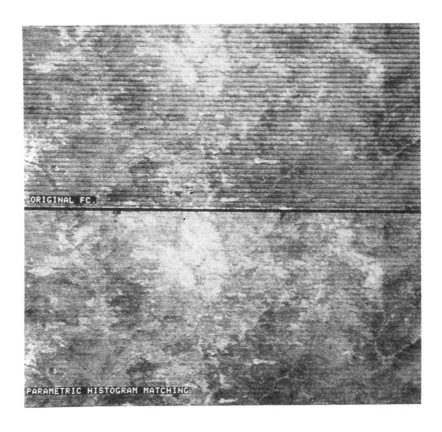

Fig. 7.

through the cumulative histogram s and cumulative histogram r into new(s).

Fig. 7A (see colour section) displays uncorrected Landsat MSS data with six sensors per spectral band. The effect of different sensor offset and gain is so-called six line 'striping'. Fig. 7B (see colour section) displays the same data corrected by histogram matching. The effect is the removal of the 'stripes'. Some manufacturers call the sensor correction program a 'destriping' program.

Corrections of atmospheric effects

Haze correction. Haze is the scattering of photons by the atmosphere into the field of view of the sensor. It depends on the local concentration of water vapour, dust or other particles in the atmosphere. The haze effect is in first approximation a linear additive effect. Correction requires that the local haze contribution is estimated from 'black' ground samples or from minima in the histograms of the various spectral bands (global estimation). The haze contribution is strongest in the high energy (short wavelength) bands of multispectral sensors. In the near infrared such as Landsat MSS band-7 the haze contribution is often neglegible.

Fig. 8 shows cumulative histograms of Landsat MSS bands 4 to 7 with estimated haze values per band.

A general rule of thumb is: take the minimum value of the histogram for a spectral band, substract 1 or 2 and apply this according to:

$$new(b) := old(b) - haze\text{-}corr(b).$$

b is a band index, haze-corr, is the haze correction factor.

It is important to apply haze correction before any other transform such as e.g. leaf-area-index or biomass-index estimation involving ratios. Many authors ignore haze correction, viz. Van Dijk (1984).

Sunangle, spectral illumination correction. The spectral contents of the scene illumination will change with atmospheric conditions and sunangle (pathlength through the atmosphere). As scene illumination is a multiplicative factor the correction must also be multiplicative (per spectral band).

Using multiseason data or trying to extend a method for biomass estimation to other places or other dates it is important to realise that the spectral illumination changes with sunangle and atmospheric conditions (red sunset, white sky, etc.). Correction for these effects can only be performed if a reference reflecting area is available. If such a reference has been found the band gains are adjusted per band b for reference spectral signature REF(b, t) at dates t0 and t1 according to:

$$new(b, t1) := \frac{old(b, t1) * REF(b, t0)}{REF(b, t1)}$$

b = spectral band b;
t0 = time 0;
t1 = time 1;
REF = reference area or field.

Suggestions for REF are: large areas of concrete, quarries, dry bare sand areas, etc.

If   the data are raw data
Then   perform skew correction next.
Else   check whether geometric transforms
       are needed and apply those needed.

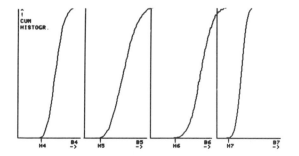

Fig. 8. Cumulative frequency histograms of MSS bands b4, b5, b6 and b7 with haze contribution h4, h5, h6, h7.

278

Geometric corrections, resampling

The raster of picture elements of a picture has to be a square raster because film writers write square picture elements. The numeric image of the scene can also be considered as a regular raster of numbers. The raster of scene elements is not such an ideal square raster. It is determined by the scanning motion of the sensor system, motion (six degrees of freedom) of the sensor platform, earth curvature and earth rotation and the effect of perspective projection of a three-dimensional scene onto a two-dimensional numerical image and a two-dimensional picture.

To correct for the non-ideal raster of scene elements values have to be calculated at the position of the ideal raster points centered at ideal scene elements. The calculation of values at ideal rasterpoints involves the interpolation of neighbour scene values. The process of the calculation of values at the position of an ideal raster from a given raster of scene elements is called resampling.

For resampling it is necessary to be able to specify the geometric raster distortion model and the interpolation model for the values.

Earth rotation, skew correction

Satellite data have to be corrected for the effect of earth rotation. This correction is often already applied at the receiving station. It is also commonly applied the wrong way. This sort of errors in the central facilities are irrecoverable.

The effect of the earth rotating to the east and satellite rotation to the south is that each new set of e.g. six scanlines projected onto the moving earth will start west of the previous set.

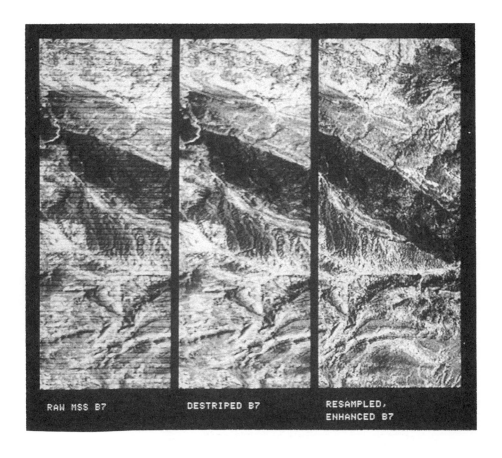

Fig. 9.

The matrix of scene elements on the surface of the earth is skewed (in e.g. blocks of six in the case of Landsat MSS) while the data matrix is stored as a square array. Mapping the array of scene elements into a pixel array (pixel = picture element) would give a wrong correspondence between image and scene. For a correct correspondence between scene geometry and picture geometry the information contents has to be shifted left per block of e.g. six scanlines for consecutive blocks. The processing error in the central (receiving) stations is to apply this shift without taking account of the block structure.

Correcting the aspect ratio of Landsat MSS

Landsat MSS was designed with a sampling raster of 57 m by 79 m on the ground. This design was based on the criterion that MSS output would be printed mostly on lineprinters (scale 1:25,000). Because of this outdated user requirement and the fact that image output and map output is generated with square picture elements the data have to be resampled. Resampling means that the data along each scanline is interpolated and data values are determined at e.g. 79 m intervals to make a square 79 m × 79 m scanraster. Square scene elements then correspond to square picture elements (pixels). Image scale is determined by:

scale = pixel size / scene-element size

A typical pixel size for optimum visual perception is 0.1 mm. For Landsat this leads to an optimum scale of 0.1 mm/79 m = 790,000.

Fig. 9 (see colour section) shows first (A) raw Landsat band-7 data displayed with square pixels. The effect is a stretch of the data in scan direction with a factor 79/57 and skew to the left. The B image is the destriped version of A. The C image is corrected for skew and resampled to a square sample raster of 79 m × 79 m. Fig. 10 shows the sample raster of Landsat MSS as projected on the eastward moving earth. Mark the staircase effect caused by the typical scanning with six sensors per band at the same time.

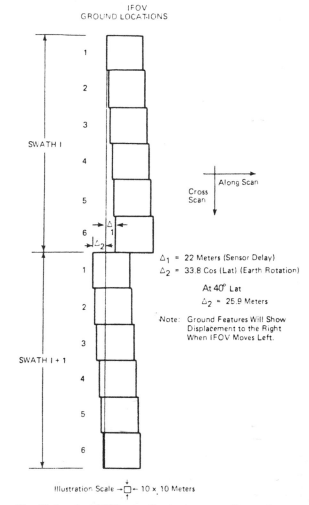

Fig. 10. Landsat MSS sampling raster geo-referenced.

Skew correction and all other geometric corrections involve resampling of the data.

Resampling

Conceptually data samples are thought of as functions having a value only at the sample point. The sample point being the centre of the field of view (FOV) at the time the scene element in the FOV was imaged. By drawing a smooth curve through the measured data points the original intensity profile over e.g. a scanline is estimated. Taking the values at new intervals along a scanline is equivalent to resampling by

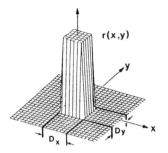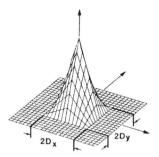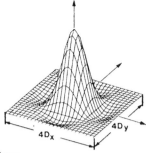

Fig. 11. Interpolation functions; nearest neighbour, linear, and cubic interpolation weight functions.

interpolation to the nearest old values from the new sampling positions.

Usually one has the choice of nearest neighbour-, bilinear- or (bi)cubic interpolation. The corresponding continuous weight functions (2-dim.) are depicted in Fig. 11.

Choosing an interpolation method:

If data is thematic data
Then choose nearest neighbour interpolation.

If data is remote sensing data
Then
    If emphasis is on minimum distortion of radiometric data
    Then use bilinear interpolation
    Else use bicubic interpolation.

Explanation:
— nearest neighbour; copies the nearest neighbour value of label. With map data any other interpolation of map units would be meaningless. For remote sensing data nearest neighbour interpolation is too coarse, it introduces positional errors in the data;
— bilinear   ; for remote sensing data, fast but smoothing effect, no artefacts at edges and lines;
— (bi)cubic   ; for remote sensing data, has some enhancement effect; sharpens edges and lines (artefacts).

Unfortunately most Landsat processing stations do use nearest neighbour resampling for a first order geometric correction thereby spoiling the data for good because a reconstruction of the original 'raw' data is not possible.

Correction of e.g. Landsat MSS data should be done by shifting blocks of six scanlines at a time. The amount of shift depends on latitude. Fig. 10 depicts the proper sampling geometry for Landsat MSS. Data pre-'corrected' in regional processing centres using nearest neighbour interpolation do not take into account the six line staircase effect and are therefore unnecessarily distorted.

Deconvolution

The smearing and smoothing effects of a scanning and moving sensor with finite electronic exposure time and non-ideal optics are described as a 'convolution' of the scene radiance distribution with a point response function. A point response function (PRF) is the function generated by the system when a single light point moves through the field of view of the sensor. Correcting the effect of the smoothing of the data contents is called deconvolution, the opposite of convolution. Deconvolution has the effect of image enhancement (Fig. 9, see colour section).

Fig. 12A shows the effect of a simulated PRF of (0.2, 0.6, 0.2), taking the weighted average over three scene elements. The dotted curve is the result of convolving f(x) with the PRF. The overall effect is a blurring of sharp features.

Deconvolution of the scanner's point response function (PRF) should occur in the same step as resampling to a square grid. The smoothing effect in Landsat MSS data extends over five scene elements in the scan direction and a fraction of a scene element in the cross scan direction. As Landsat MSS is severely oversampled (2 times) no information is lost when going from e.g. 57 m sample distance to 79 m sample distance. The point response function tells us how much signal of the neighbour samples is contained in the momentarily central sample as in Fig. 12A. Deconvolution takes the central sample value and substracts a fraction of the neighbour values in order to compensate for the overflow of value (smoothing effect) as in Fig. 12B.

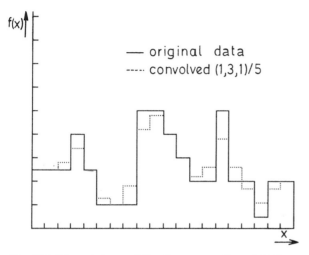

Fig. 12A. The point spread function acts as the weight function int he process of taking the average weighted value at a sample position.

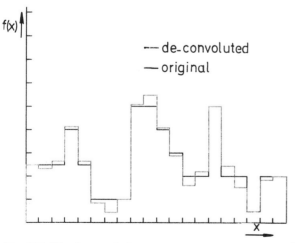

Fig. 12B. The deconvolution operator compensates most of the smoothing effect caused in the scanning process and characterised by the point response function (PRF).

In Fig. 12B the effect is shown of countering the effect of the PRF (0.2, 0.6, 0.2) by a deconvolution operator (−0.5, 2.0, −0.5). The combined effect of both operators is (−0.1, 0.1, 10.0, 0.1, −0.1). In many places the original signal is reconstructed, but there will always be artefacts near prominent features. However, the human observer will appreciate the result of the deconvolution as image enhancement.

If no options are available for in-line resampling and deconvolution in the software provided, then use resampling with cubic convolution and new sample size of e.g. 75 m × 75 m.

Resampling, skew correction and deconvolution can be combined into a single operation (Mulder, 1982) with user defined interpolation operator.

Image scale

Image scale is determined by the ratio of display pixel size to scene element size.

Example: pixel to raster =
75 μm : 75 m = 1 : 1,000,000

For easy viewing a pixel size of 0.1 mm is optimum. For visual interpretation involving splitting of pixels (e.g. interpolation of jagged boundary lines) one could go up to pixel size of 0.3 mm or 0.4 mm.

Table 1. Optimum display scale (250 mm viewing distance) for a standard eye resolution of 0.1 mm.

raster	0.1 mm pixel	0.25 mm pixel
80 m	1 : 800,000	1 : 320,000
40 m	1 : 400,000	1 : 160,000
20 m	1 : 200,000	1 : 80,000
10 m	1 : 100,000	1 : 40,000
4 m	1 : 40,000	1 : 16,000
2 m	1 : 20,000	1 : 8,000
1 m	1 : 10,000	1 : 4,000

Registration

Image registration is the process of fitting the data of one 'image' exactly on top of a reference image or reference map. Because it is very unlikely that the scene elements of one image coincide with the scene elements of another image it is necessary to resample the data. The difference between the geometry of different data is modeled as a distorted rubber sheet. Control points are used to determine the parameters used in the corrective 'rubber stretch' which brings both images into registration.

Image to image registration is essential in the use of multi-temporal and multi-source data. The user has to specify control points in both images and he has to specify the interpolation method for resampling. Fig. 13 (see colour section) shows a number of pass-points on a split screen display (one screen is used to display two separate images).

Image to map registration can usually be done with the same program as image to image registration. The only problem is that pass-points have to be specified in sample (pixel) coordinates which have to be derived from map coordinates.

Map to image geometric transformation including e.g. UTM grid (universal transverse mercator) is simple if the map data have been digitized, sampled and rastered to the same raster as the image data. Otherwise resampling with nearest neighbour interpolation has to be performed on the map. Fig. 14 (see colour section) shows the overlay of digitized land-use map and a topo map onto a Landsat MSS band-7 image.

Colour coding and spectral feature extraction

Introduction

Remote sensing data and other data (geo information) can be presented to the user/interpretor as colour images or colour 'maps'. The colour output of an image processing system is defined in terms of additive colours red, green and blue. For each picture element (pixel) the intensity of the basic colours must be specified. The transformation of data to colour is called colour coding. The user can specify the transformation from data to colour in such a way that a maximum colour difference exists between classes of interest and for an easily remembered colour key or colour legend.

Feature extraction is the transformation of e.g. raw multispectral data to less data but keeping the essentials needed for classification or interpretation. Feature extraction implies data reduction (e.g. fewer spectral channels).

A problem common to colour coding and spectral feature extraction is that the spectral data are redundant. In both cases information

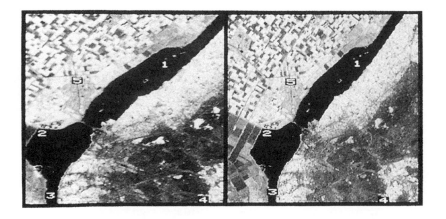

Fig. 13.

Fig. 14.

extraction will lead to a reduction of the number of data channels. In the case of colour coding any number of spectral bands have to be mapped into three colour channels containing the relevant information for the specific application.

Information here means anything which answers a question, or anything that is used in the evaluation of a hypothesis.

In the case of feature extraction a decision model or classification scheme will indicate a minimum sufficient subset of spectral bands or linear combination thereof for the separation of classes as defined by the user.

Example 1: in Landsat MSS four spectral bands have to be mapped into three colour components; and often two spectral features are sufficient for the classification of land-use.

Example 2: in airborn scanner MSS there may be as many as ten spectral channels which must be mapped into three colour components for visual interpretation. The number of relevant spectral features is usually three.

Feature extraction leads to information extraction and data reduction (the inverse statement is often not true).

Proper colour coding is based on the physics of photon absorption and photon reflection and the physics of colour perception. Data intensity has to be mapped into image intensity and data spectral reflectance must be mapped into equivalent image (colour print) reflectance.

Proper feature extraction is based on understanding the relation between reflected spectral intensity for the classes which should be separated in the classification process. This understanding can be formalized in reflection models such as canopy reflection models for vegetation with leaf area index and leaf structure as important parameters. Feature extraction is in that case equivalent to the estimation of the model parameters from the spectral data. As model parameters have a more direct relation with class separation, model inversion is an optimum method of feature extraction.

In colour coding the aim is to optimize the colour contrast (hue and saturation) and the intensity contrast for the visible separation of user defined classes.

In feature extraction it is important to extract those features that allow the best numerical separation between user defined classes (supervised classification).

Proper treatment of colour coding and spectral feature extraction in literature is rare. The following section therefore goes into more numerical detail than the other sections.

Colour coding and notation

Colour output. The elementary unit of output of an image processing system is a picture element (= pixel). The size of the pixel is defined by the output aperture of a filmwriter and subsequent photographic or cartographic printing/reproduction. The colour of each pixel is (additive colour generation) defined by three numbers specifying the intensity in red, green and blue at the time of display.

Any nonlinear effects in the chain from digital output value to displayed value can be corrected by the use of output correction look-up tables, one table for each colour and specific hardcopy material.

Notation. A colour image is numerically specified by an array of numbers:

$$c(i, j, l) = (r(i, j), g(i, j), b(i, j)) \qquad (1)$$

indices i, j are position indices, l is a colour channel index with:
$l = 1 \rightarrow$ red, $l = 2 \rightarrow$ green, $l = 3 \rightarrow$ blue.

The 3-dimensional data matrix can also be considered to consist of three 2-dimensional data matrices; $r(i, j)$, $g(i, j)$, $b(i, j)$. A red channel, a green channel and a blue channel as depicted in Fig. 15.

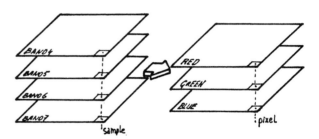

Fig. 15. The data matrix $b(i, j, k)$ mapped into the colour data matrix $c(i, j, l)$, $k = 1$ to N MSS bands, $l = 1$ to three colour components; red, green and blue.

For a single pixel at position (index) i, j the display colour is defined by:

$$c(l) \qquad l = 1 \text{ to } 3 \qquad (2)$$

With the same assignment of l as in (1). The 1-dimensional array $c(l)$ is equivalent with three colour coordinates defining a colour by its position in a 3-dimensional colour space as in Fig. 15. The numerical range of $c(l)$ is usually 0 to 255.

We will also use the equivalent vector notation $\vec{c}$ which has components: $c(1)$, $c(2)$, $c(3)$ or:

$$\vec{c} = \begin{bmatrix} \text{red} \\ \text{green} \\ \text{blue} \end{bmatrix} = \begin{bmatrix} c(1) \\ c(2) \\ c(3) \end{bmatrix}$$

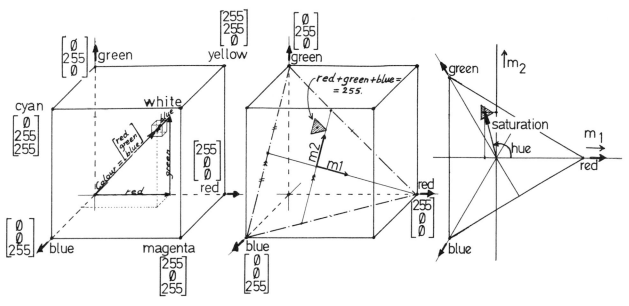

Fig. 16. (A) Colour space = colour cube, the colour of each pixel is defined by its R, G, B (red, green, blue) components; (B) colour cube with triangle plane with intensity = $R' + G' + B'$ = constant = 255, $R' = R*255/\text{int.}$ $G' = G*255/\text{int.}$ $B' = B*255/\text{int.}$, R', G', B' are normalized colour components of colour vectors with tips in the int. = 255 plane; (C) normalized colour triangle with intensity = 255, hue (angle) and saturation (radius) in m_2, m_2 coordinate system.

For the human mind it is easier to specify colours in terms of the total (red + green + blue) intensity (brightness), the colour of the rainbow → hue and the pureness of the colour ∑ saturation (the Munsel system uses the terms value, hue and chroma respectively).

Intensity, hue and saturation are equivalent with conical coordinates in the colour cube. Planes of constant intensity intersect the planes of the colour cube producing triangles and hexagonals as in Fig. 17 which shows planes of constant intensity with the corresponding colour values for each plane. Hue and saturation are usually properties of reflecting materials and must be independent of the (illuminating) intensity and are equivalent to polar coordinates in a normalised colour triangle. The hue changes with the angle (pointer to the corresponding rainbow colour) and the saturation increases from zero in the neutral grey point to a maximum at the border of the colour triangle as defined by the primary colours.

The normalised colour triangle (Fig. 18, see colour section) is defined by coordinates:

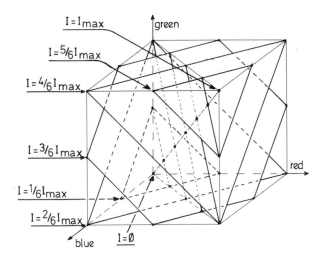

Five planes of constant intensity,
I = red + green + blue.

Fig. 17. The colour cube and intersections (triangles and hexagonals) of planes of constant intensity with the colour cube.

$$c'(l) = (c(l)/i)*255 \qquad (3)$$

where $i = c(1) + c(2) + c(3)$ the total intensity per scene element. For numerical reasons the values are scaled by 255.

286

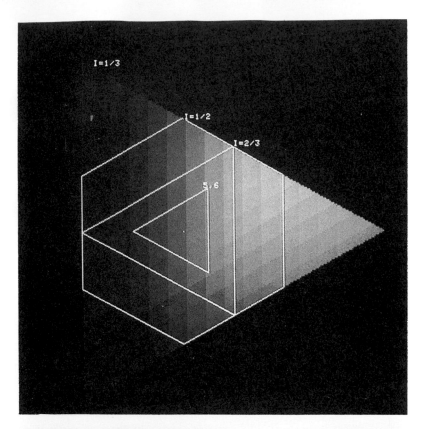

Fig. 18.

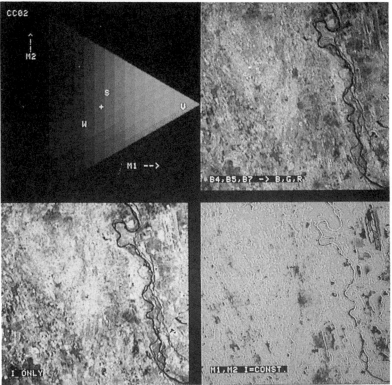

Fig. 19.

Fig. 19 (see colour section) shows a Landsat MSS false colour image (band-4 → blue, band-5 → green, band-5 → red) with the data points of vegetation ($\vec{v}$), soil ($\vec{s}$) and water ($\vec{w}$) plotted in the normalized colour triangle. A separate image shows the intensity component of each pixel in black and white and the fourth image shows the corresponding normalized c' (l) colour image with $i =$ constant $= 255$. The last-mentioned picture is equivalent to the hue and saturation picture; because only the variation in hue and saturation is shown, the intensity is constant (numerical 255).

The colour coordinates as represented by position in the normalized $R'G'B'$ colour (mixture) triangle can also be specified by two rectangular coordinates m1 and m2 as in Fig. 16. A hue transformation is equivalent to a rotation of the data in the colour triangle. A saturation transformation is equivalent to a scaling (stretch of the data triangle) in m1, m2 space. A change in white point (colour balance) is implemented by a shift of origin as depicted in Fig. 20.

The relation between conceptual coordinates m1, m2, i and display coordinates r, g, b (red, green, blue; l = 1, 2, 3) is given by:

$$\begin{bmatrix} m1 \\ m2 \\ 1 \end{bmatrix} = \frac{1}{i} M \begin{bmatrix} r \\ g \\ b \end{bmatrix} \qquad (4)$$

with

$$M = \begin{bmatrix} .817 & -.408 & -.408 \\ .0 & .707 & -.707 \\ .577 & .577 & .577 \end{bmatrix} \qquad (5)$$

and $\quad i = M(3) \begin{bmatrix} r \\ g \\ b \end{bmatrix} \qquad (6)$

and the inverse transform:

$$\begin{bmatrix} r \\ g \\ b \end{bmatrix} = i \times M' \begin{bmatrix} m1 \\ m2 \\ 1 \end{bmatrix} \qquad (7)$$

with

$$M' = \begin{bmatrix} .817 & .000 & .577 \\ -.408 & .707 & .577 \\ -.408 & -.707 & .577 \end{bmatrix} \qquad (8)$$

For a numerical implementation on data channels with byte words (0,255 range) scaling of

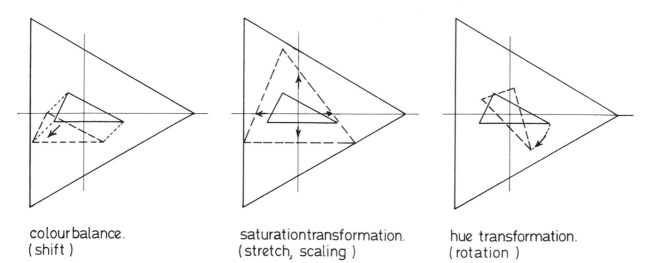

colour balance.
(shift)

saturation transformation.
(stretch, scaling)

hue transformation.
(rotation)

Fig. 20. Colour enhancement in the normalized colour triangle with m_1, m_2 coordinates, implemented as data shift, stretch and rotation.

288

m1, m2, i to the signed byte range of $-127,127$ is necessary.

Colour enhancement

Colour enhancement transforms are specified in terms of intensity hue and saturation enhancement but numerically implemented as matrix transforms in the domain of colour output, calculating new colour coordinates r', g', b':

$$\begin{bmatrix} r' \\ g' \\ b' \end{bmatrix} = M' A M \begin{bmatrix} r \\ g \\ b \end{bmatrix} \tag{9}$$

with $(M' A M)$ being a 3×3 matrix with A specifying the colour enhancement in the hue, saturation (m1, m2) domain

Change in white/grey reference. Assuming that the MSS data has already been mapped into red green and blue colour coordinates, it is often necessary to shift the data triangle over the colour triangle with:

$$\begin{bmatrix} m1' \\ m2' \\ 1 \end{bmatrix} = \begin{bmatrix} 1 & 0 & sh1 \\ 0 & 1 & sh2 \\ 0 & 0 & 1 \end{bmatrix} \begin{bmatrix} m1 \\ m2 \\ 1 \end{bmatrix} \tag{10}$$

$$A(sh1, sh2) = \begin{bmatrix} 1 & 0 & sh1 \\ 0 & 1 & sh2 \\ 0 & 0 & 1 \end{bmatrix} \tag{11}$$

positive sh1, sh2 values will shift the data right or up in the colour triangle of Fig. 20.

Saturation enhancement. The effect of saturation enhancement is to stretch the data triangle (v, s, w) in the colour triangle. For maximum data stretch the data is first shifted to the centre

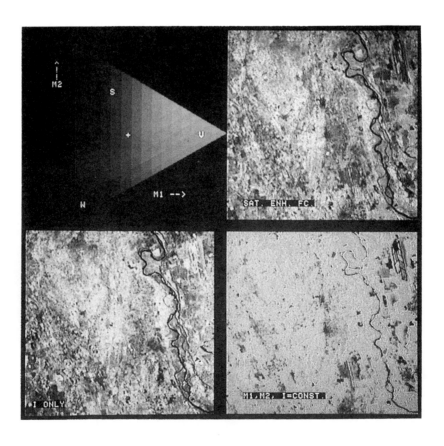

Fig. 21.

of the colour triangle using A(sh1, sh2). The colour stretch is implemented with:

$$A(s1, s2) = \begin{bmatrix} s1 & 0 & 0 \\ 0 & s2 & 0 \\ 0 & 0 & 1 \end{bmatrix} \quad (12)$$

Fig. 21 (see colour section) depicts the effect of saturation enhancement applied to the data of false colour image (Fig. 19, see colour section) (Novara, N. Italy).

Hue transformation. Hue transformation is equivalent to a rotation of the data over the colour triangle, useful for putting the maximum hue contrast between classes to be separated. Data rotation in m1, m2 space is implemented with:

$$A(a) = \begin{bmatrix} \cos a & -\sin a & 0 \\ \sin a & \cos a & 0 \\ 0 & 0 & 1 \end{bmatrix} \quad (13)$$

Fig. 22 (see colour section) shows the effect of a 120° rotation of the data triangle and interchanging red and blue, putting vegetation near the green corner, water remains blue, bare soils are coloured cyan/red (Novara, N. Italy).

Affine colour transformation. Combination of the above-mentioned shift, stretch and rotation transforms; A(sh1, sh2), A(s1, s2), A(a) allows the mapping of any data triangle onto the colour triangle. A more direct way of accomplishing this is by using one A matrix, implementing a so-called affine transform from a given data triangle to the corners of the colour triangle. As an example specify that v (characteristic vegetation point) → green, water w → blue, bare soil s → red. A data matrix in m1, m2 space is calculated from data space samples v(k), s(k), w(k), (k = 1 to 3, → r, g, b) using the M transform resulting in a data matrix D:

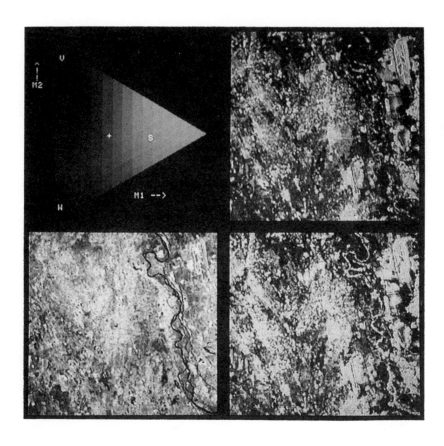

Fig. 22.

$$D = M/i \begin{bmatrix} sr & vr & wr \\ sg & vg & wg \\ sb & vb & wb \end{bmatrix} \qquad (14)$$

$$D = \begin{bmatrix} sm1 & vm1 & wm1 \\ sm2 & vm2 & wm2 \\ 1 & 1 & 1 \end{bmatrix} \qquad (15)$$

To be transformed into corresponding m1, m2 coordinates matrix C:

$$C = \begin{bmatrix} rm1 & gm1 & bm1 \\ rm2 & gm2 & bm2 \\ 1 & 1 & 1 \end{bmatrix} \qquad (16)$$

The affine transform mapping D to C is calculated from:

$$A\,(aff) = C\,D' \qquad (17)$$

with D′ being the inverse of D. The complete transform is then:

$$\begin{bmatrix} r' \\ g' \\ b' \end{bmatrix} = (M'\ C\,D'\ M) \begin{bmatrix} r \\ g \\ b \end{bmatrix} \qquad (18)$$

Fig. 23 (see colour section) shows the affine mapping of characteristic data samples $\vec{s}$, $\vec{v}$, $\vec{w}$, onto red, green, blue in the colour triangle and corresponding pseudo natural colour image.

Spectral correlation

Spectral correlation is based on the idea that any spectral signature can be decomposed (described on a basis of) a number of standard spectral signatures. Early work on principal components showed that the total variance in multispectral data can be described on the bases

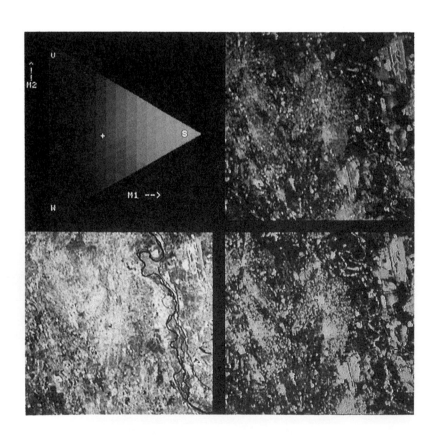

Fig. 23.

of the spectral signatures of water, vegetation and bare soil (Mulder, 1974 and 1981). The spectral variation can also be decomposed into intensity variation and two mixing factors, the spectral mixing of bare soil and vegetation and the equivalence of mixing water and bare soil spectral reflectance (equivalent to the variation found in three principal components).

Spectral correlation is equally useful for colour coding as it is for feature extraction in the case of general land-use and vegetation classes. It is based on the observation that most MSS data can be described as a spectral mixture of the reflection of a sample of dense green vegetation ($\vec{v}$), a sample of deep pure water ($\vec{w}$) and a well chosen sample of bare soil ($\vec{s}$). Even with ten channel MSS data very little information is lost if all the spectral values are projected onto the base of $\vec{v}$, $\vec{w}$, $\vec{s}$.

Two approaches can be followed in expressing a datum in the (sum) normalised feature space in terms of a mixture of vegetation, water and soil reflectance:

1. The oldest approach projects data on the lines connecting the average:
$$\vec{cg}' = (\vec{v}' + \vec{s}' + \vec{w}')/3$$
with the vectors $\vec{v}'$, $\vec{s}'$ and $\vec{w}'$. This is the 'centre of gravity' method;

2. a better method is to project the data on lines which are perpendicular to the opposite sides of the $\vec{v}'$, $\vec{s}'$, $\vec{w}'$ data triangle in the sum normed feature space.

   For example the green axis is defined by the line perpendicular to the difference of $\vec{s}'$ and $\vec{w}'$, and pointing towards $\vec{v}'$. Projecting $\vec{v}'$ on the green axis gives the maximum value in the data set. Projecting $\vec{s}'$ and $\vec{w}'$ on the green axis gives the minimum (equated to zero) value for both $\vec{s}'$ and $\vec{v}'$. The same reasoning is true when we substitute:
   $$\vec{v}' \rightarrow \vec{s}', \quad \vec{s}' \rightarrow \vec{w}', \quad \vec{w}' \ \vec{w}' \rightarrow \vec{w}'$$
   in the above given argument, with green axis → red axis. Another round of circular resubstitution defines the blue axis.

Of the methods presented in Fig. 24, method B is preferred because it maps the data triangle $\vec{v}' \vec{s}' \vec{w}'$ exactly into the colour triangle with

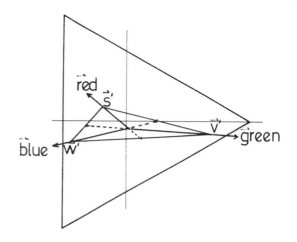

a Centre of gravity.

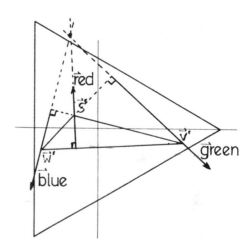

b. Intersection of perpendiculars.

Fig. 24. Spectral correlation on MSS bands b4, b5, b7 in the m_1, m_2 colour triangle space. Reference vectors are $\vec{v}'$, $\vec{s}'$, $\vec{w}'$ for vegetation, (bare) soil and water. (A) Spectral correlation with the centre of gravity of the $\vec{v}' \vec{s}' \vec{w}'$ triangle as origin; (B) Spectral correlation with the intersection of perpendiculars as origin. This is the preferred method because if e.g. red is maximum for $\vec{s}'$ then both $\vec{w}'$ and $\vec{v}'$ have zero red at the same time.

$\vec{v}' \rightarrow$ green, $\vec{s}' \rightarrow$ red, $\vec{w}' \rightarrow$ blue. Method B is worked out numerically in the next section.

This method works on more than three spectral bands. It is a generalisation of the method of mapping the data triangle into the colour triangle with an affine transform in the m1, m2 space.

Method: spectral correlation in a normalised subspace with new colour axes perpendicular to opposite sides of the data triangle:

a) let b(k), k = 1 to N be the spectral data matrix (spectral vector) of a data sample (scene element). Using a program for supervised training select three samples v(k), w(k), s(k) for green vegetation, water and soil which are the most extreme in distance from the average of the general cluster and have the maximum saturated colour in a three channel colour display.

b) normalise the data (to a (N−1)-dimensional subspace) for all (i, j) data as well as for the characteristic samples b(k) → b′(k) k = 1 to N.

$$b'(k) = b(k)/i$$
$$i = sum\,(b(k))/\text{square-root}\,(N), \quad k = 1, N$$

c) calculate the sides (difference vectors d1, d2, d3) of the data triangle defined by normalised characteristic data vectors $\vec{v}'$, $\vec{s}'$, $\vec{w}'$:

$$\overrightarrow{d1} = \vec{0} + \vec{s} - \vec{w}'$$
$$\overrightarrow{d2} = -\vec{v}' + \vec{0} + \vec{w}'$$
$$\overrightarrow{d3} = \vec{v}' - \vec{s}' + \vec{0}$$

As only the direction of the d vectors is important they are next normalised to unit 'length'. The normalised vectors d are all in one plane, the plane of the data triangle. In this two-dimensional subspace of the N−1-dimensional subspace of sumnormed spectral data we can now calculate vectors perpendicular to the d vectors:

$$\vec{g} = -(\overrightarrow{d3}.\overrightarrow{d1})\overrightarrow{d1} + \vec{0} + \overrightarrow{d3}$$
$$\vec{r} = \overrightarrow{d1} - (\overrightarrow{d1}.\overrightarrow{d2})\overrightarrow{d2} + \vec{0}$$
$$\vec{b} = \vec{0} + \overrightarrow{d2} - (\overrightarrow{d2}.\overrightarrow{d3})\overrightarrow{d3}$$

the vectors $\vec{g}$, $\vec{r}$, $\vec{b}$ are the new colour axes defining green, red and blue, in N-dimensional space. For good measure they are also normalized to unit 'length'. $\vec{g}$, $\vec{r}$ and $\vec{b}$ have as a property that the sum of their components is zero. This means that the new colour axes are orthogonal to the intensity axes.

The colour components of sumnormed data $\vec{b}'$ are now calculates from:

$$green = \vec{g} . \vec{b}'$$
$$red = \vec{r} . \vec{b}'$$
$$blue = \vec{b} . \vec{b}'$$

with colour vector:

$$\vec{c} = \begin{bmatrix} red \\ green \\ blue \end{bmatrix} \text{ and}$$

$$COL = \begin{bmatrix} r1 & r2 & r3 & r4 & .. & rk \\ g1 & g2 & g3 & g4 & .. & gk \\ b1 & b2 & b3 & b4 & .. & bk \end{bmatrix}$$

the relation between the colour vector c and the sumnormed data vectors b′ is now given by:

$$\vec{c} = COL\,\vec{b}' \tag{19}$$

The proper scaling factors on the red, green and blue axes are found by substituting s′, v′ and w′ for b′ giving the following table (Table 2):

Table 2. Scale minima and maxima for red, green and blue before scaling.

	Soil	Veg.	Water
Red	rs	rv	rw
Green	gs	gv	gw
Blue	bs	bv	bw

The scaling factors to be applied to the result of (19) are such that rs, gv and bw end up with the maximum value, e.g. 255, and the off-diagonal elements reduce to zero.

After the scaling the colour/data matrix is as follows (Table 3):

Table 3. Scale minima and maxima after scaling.

	Soil	Veg.	Water
Red	255	0	0
Green	0	255	0
Blue	0	0	255

The transformation as described above from normalised data space to normalised colour space orthogonalises the base vectors $\vec{v}, \vec{s}$ and $\vec{w}$. The colour picture does not contain the intensity factor, it could be called a hue + saturation picture, mapping the data 'reflectances'.

The original intensity i can be printed as a black & white picture or can be recombined with the 'hue, saturation' picture:

$$\begin{bmatrix} r \\ g \\ b \end{bmatrix} = i \times \begin{bmatrix} r' \\ g' \\ b' \end{bmatrix} \qquad (20)$$

For Landsat MSS and Landsat TM the method is worked out in a numerical example. The samples taken for vegetation and water are rather universally applicable, the reference for bare soil is more dependant on local circumstances. We are using a very red soil from Spain for the TM reference. The program used to produce the numerical example, runs on the microcomputers used for the exercises of our students. It consists of about seven simple subroutines.

Table 4. Matrix of weight coefficients and scaling parameters for the four bands of Landsat MSS

data	$\overrightarrow{\text{VEG.}}$	$\overrightarrow{\text{SOIL}}$	$\overrightarrow{\text{WATER}}$
B4	12.000	36.000	20.000
B5	9.000	66.000	6.000
B6	54.000	58.000	3.000
B7	68.000	64.000	0.000

diff.	$\vec{S}'-\vec{W}'$	$\vec{W}'-\vec{V}'$	$\vec{V}'-\vec{S}'$
B4	−0.232	−0.843	0.730
B5	−0.700	0.140	0.173
B6	0.358	0.248	−0.330
B7	0.573	0.456	−0.573

$\vec{B}-\vec{C}$	$\overrightarrow{\text{GREEN}}$	$\overrightarrow{\text{RED}}$	$\overrightarrow{\text{BLUE}}$
B4	0.163	−0.453	0.827
B5	−0.852	0.846	−0.507
B6	0.277	−0.170	−0.098
B7	0.413	−0.223	−0.223

	VEG.	SOIL	WATER
Green	0.261	−0.035	−0.035
Red	−0.155	0.069	−0.155
Blue	−0.105	−0.105	0.455

Table 5. Matric of height coefficients and scaling parameters for the six reflective bands of Landsat TM

data	$\overrightarrow{\text{VEG.}}$	$\overrightarrow{\text{SOIL}}$	$\overrightarrow{\text{WATER}}$
B1	11.000	75.000	17.000
B2	9.000	·52.000	10.000
B3	7.000	82.000	10.000
B4	112.000	95.000	4.000
B5	70.000	212.000	6.000
B7	21.000	146.000	4.000

diff.	$\vec{S}'-\vec{W}'$	$\vec{W}'-\vec{V}'$	$\vec{V}'-\vec{S}'$
B1	−0.169	−0.604	0.493
B2	−0.102	−0.322	0.271
B3	−0.242	−0.198	0.286
B4	0.888	0.178	−0.705
B5	−0.041	0.556	−0.322
B7	−0.334	0.390	−0.022

$\vec{B}-\vec{C}$	$\overrightarrow{\text{GREEN}}$	$\overrightarrow{\text{RED}}$	$\overrightarrow{\text{BLUE}}$
B1	−0.057	−0.353	0.582
B2	−0.042	−0.180	0.309
B3	−0.208	0.027	0.155
B4	0.870	−0.540	−0.011
B5	−0.148	0.476	−0.574
B7	−0.415	0.569	−0.461

	VEG.	SOIL	WATER
Green	0.330	−0.050	−0.050
Red	−0.089	0.150	−0.089
Blue	−0.178	−0.178	0.180

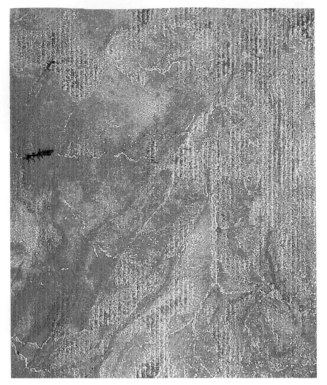

Fig. 25A.

Fig. 25B.

Fig. 25 (see solour section) shows the result of the 'spectral correlation' method applied to Landsat MSS data with a mountainous scene.

The b & w image contains the structural information; geomorphology (relief), hills, valleys mainly through shadow and shades. The colour picture is 'flat' it contains only 'spectral reflectance' information which depends only on surface materials and vegetation leaf area index and not on changes in illumination or shadows. It is therefore a prime image for land-use interpretation and classification.

Spectral feature extraction

Spectral features. The purpose of spectral feature extraction is to transform the spectral data into spectral features which form a better description basis for spectral 'signatures'. Feature extraction preceeds classification of e.g. land-use and vegetation classes. The original spectral data contain so-called spectral signatures, a term meaning that each class of data has a characteristic curve of spectral reflection as a function of wavelength. Spectral reflection data are generated by multispectral scanners with rather arbitrary sampling (spectral bands) of the spectral range. In reflection data the illumination effects, slope shading and shadow are all included with the spectral reflectance data in a multiplicatively way:

$$R(k) = I(k) \times \cos(\text{theta}) \times r(k) \times dt \qquad (21)$$

$R(k)$ = reflection data (photon count); k = central wavelength of the spectral band; $I(k)$ = illuminating intensity in band k (photon flux); theta = angle between the normal to the illuminated surface and the sun direction; $r(k)$ = spectral reflectance in band k which contains the spectral information about the reflecting surface; dt = electronic shutter time of the sensor (seconds).

The data in various spectral bands are highly correlated and therefore redundant. Physical factor analysis (common sense, rather than statistics) leads to proper methods of feature extraction:

A first principle is to separate the illumination and slope effects from the spectral reflectance as in the above formula. This is performed by calculation of the total photon flux over all spectral bands (intensity) and then normalizing all spectral bands by dividing by that photon count.

Canopy reflection models, provide a guide to spectral feature extraction. Model inversion means that the model parameters are estimated from the spectral reflection data.

Spectral feature extraction has as first step the separation of the intensity factor and the spectral reflectance factors as already demonstrated (21). Bands normalized for intensity:

$$b'(k) = b(k)/i \qquad (22)$$

k = spectral band index; b(k) = reflection (photon count) in band k; b'(k) = reflection in band k, normalized by intensity.

$$i = \sum_{k=1}^{N} (b(k)) \qquad (23)$$

Substituting (22) in (21)

$$b'(k) = \frac{I(k) \times r(k)}{\sum_{k=1}^{N} (I(k) \times r(k))} \qquad (24)$$

r(k) = spectral reflectance in band k; I(k) = spectral sun irradiance in band k, (photon flux in band k).

I(k) is constant in space and time; b'(k) is a constant for a given r(k) and proportional to r(k).

The b'(k) values are interpreted as weighted spectral reflectance values only depending on the spectral mixture of reflecting materials in the field of view of the sensor.

Green leaf area estimation. Spectral features of vegetation are best derived from vegetation canopy models. The most important parameters of these models are the green leaf area index (GLAI) and the vertical over height ratio (V/H). The GLAI is the area of horizontally projected green leaf area per square meter, V/H is the ratio of vertically projected leaf area over horizontally projected leaf area per unit volume of

canopy and is equivalent to the tangent of the average leaf angle.

Model inversion relates MSS data to the GLAI and the V/H parameters. The GLAI is estimated from the spectral mixture of vegetation v'(k) and the underlying soil s'(k).

— v'(k) = vegetation spectral reflectance in band k according to (24);
— s'(k) = bare soil spectral reflectance in band k according to (24);
— x'(k) = unknown mixed spectral reflectance as function of k, see (24).

The relation between the GLAI and the mixing of (interpolation between) $\vec{s}'$ and $\vec{v}'$ is shown in Fig. 26. The relative distance from a leaf/soil mixture reflectance x'(k) to bare soil reflectance s'(k) is in first approximation proportional to the inverse exponential of the GLAI.

$$x'(k) = s'(k) + gvi \times (v'(k) - s'(k)) \qquad (25)$$

$$gvi \sim (1 - \exp(GLAI)) \qquad (26)$$

gvi = a 'green vegetation index' indicating the spectral mixture of green vegetation and bare soil.

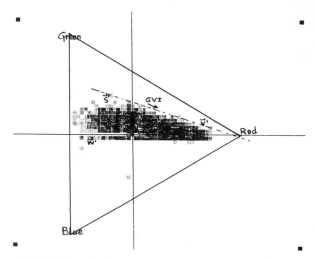

Fig. 26A. The relation between GVI (green vegetation index) and the spectral mixture of normalized dense vegetation v' and bare soil s'. All possible mixtures of vegetation and soil reflectance are on a line between $\vec{v}'$ and $\vec{s}'$. GLAI is related to GVI on a non-linear scale.

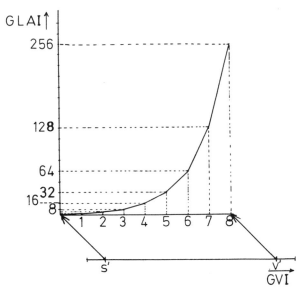

Fig. 26B. The transformation from GVI green vegetation index to GLAI estimator, using a piecewise linear function in order to compensate for the non-linear relation between the GLAI and the relative distance between $\vec{s'}$ and $\vec{v'}$ (= GVI).

This first-order approximation is based on a random leaf distribution in a plant canopy and the derivation of the formula is simple. Non-trivial second-order effects can be found in literature (Bunnink, 1977; Suits, 1972). However, the more involved models are plant type specific which means that the vegetation class must be known, which is not the case at the time of feature extraction. So GLAI and biomass[9] estimation should be performed hierarchically;

— first: use first order models to determine the vegetation class,

— next: use the second order, class specific model for a more accurate estimate of the green leaf area index and the related biomass.

One should also remember that the inverse exponential model assumes a homogeneous canopy over whole scene elements. If the cover consists of scattered plant then for the plant distribution a linear spectral reflectance model should be used.

[9] Biomass: proportional to green leaf area and leaf thickness (kg/m^2).

GLAI estimation from MSS data: from formula (25) it follows that:

$$\text{gvi} = \text{const.}\, 1 \times \sum_k (x'(k) \times (v'(k) - s'(k)) - \text{const.}\, 2$$

const.1 and const.2 are determined experimentally. $v'(k)$ and $s'(k)$ are determined by hand selection of typical samples of green vegetation and bare soil.

a) determine the spectral reflectance mixture line as $v'(k) - s'(k)$, where $\vec{v'}$ is a sample of normalized spectral reflectance with the highest GLAI in the scene and $s'(k)$ is a typical bare soil sample for k = 1 to N spectral bands; For example Landsat MSS data:

$$v'(k) - s'(k) = v'(4) - s'(4) + v'(5) - s'(5) + v'(6) - s'(6) + v'(7) - s'(7)$$

b) project all intensity normalized data $x'(k)$ onto the mixture line, giving a GVI (green vegetation index):

$$\text{GVI} = \sum_{k=1}^{N} (v'(k) - s'(k)) \times x'(k)) \quad (27)$$

For example Landsat MSS data: $b'k = x'(k)$

$$\text{GVI} = (v'(4) - s'(4) \times b'(4) + (v'(5) - s'(5) \times b'(5) + \dots \text{ etc.}$$

c) determine the value of $v'(k)$ (maximum GLAI e.g. = 4) and $s'(k)$ (GLAI = 0) on the GVI scale and curve fit formula (26) in this range, expanding the range over the interval (0,255). The result is approximately linear with the GLAI and therefore a good relative estimator of the average GLAI over the field of view of the sensor and hence a good feature for the classification of vegetation classes.

Warning. The use of a global vegetation reference $\vec{v'}$ for a given sensor does not present problems but the bare soil reference $\vec{s'}$ may not be a constant over the area of the image file. A varying $\vec{s'}$ has severe effect on low to very low GLAIs. Also the GLAI scale on the GVI axis must be determined experimentally for each vegetation type separately. The green biomass should be related to the GVI scale experimentally (in the field). The V/H parameter as estimated by the intensity is not very accurate because many other local effects influence the intensity. It can be used to separate e.g. grass from trees.

For areas where the GLAI is low (< 0.5) the variation in soil spectral reflectance will have a non-neglegible influence on the zero point for estimated GLAI on the GVI scale. The following procedure can be used for adaptation to a local bare soil spectral reference s' loc (k).

If estimated GLAI is low (< 0.5)

And there is considerable variation in soil colour

Then apply Procedure for local soil reference.

Procedure: digitize polygons of constant soil spectral reflectance areas. Link the polygon attribute to the typical bare soil signature per zone (polygon). Proceed to estimate GLAI from the GVI but with the local s' as the zero point on the mixture axis (GVI) and GLAI scale.

Fig. 27 (see colour section) shows the implementation of the transformation of GVI to GLAI using a piecewise linear function (loop-up table). The GVI was calculated using al four bands of Landsat May data Novara Italy. The direction of the spectral mixture axis was found to be:

$$\vec{e}\,(v'\,(k) - s'\,(k)) = \begin{bmatrix} 0.730 \\ 0.173 \\ -0.330 \\ -0.573 \end{bmatrix}$$

Which is practically parallel to $\vec{v}' - \vec{s}'$ in a band-7, band-5 subspace where

$$\vec{e}\,(v'\,(k) - s'\,(k)) = \frac{-b'(5) + b'(7)}{\sqrt{2}} \tag{28}$$

For the above reason the trick of:

$$gvi = \frac{b(7) - b(5)}{b(7) + b(5)} \tag{29}$$

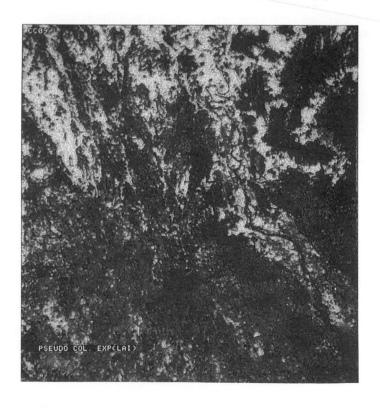

Fig. 27.

gives about the same results as the properly defined BVI of (27).

Band ratios. The trick of taking e.g. the ratio band-7/band-5 or any other ratio of two bands has no scientific rationale.

A number of other vegetation indices exist, often called biomass indices. These indices are mostly based on a statistical approach, correlating biomass with functions of Landsat MSS reflection values. Biomass estimation should, however, be estimated in a separate step after calculating a GVI.

Biomass estimation.
a) Use intensity (i) and green leaf area index (GLAI or GVI) to classify the data into various vegetation classes (supervised classification or digitized manual classification);
b) use available (ground data or database information such as classification result previous season(s)) information about plant density and plant pattern to link the GLAI (or GVI) to the biomass for each vegetation class separately and for each plant density zone separately. The expected relation between GLAI and biomass is: biomass is proportional to GLAI to the power 3/2 as GLAI is an area and biomass is proportional to a volume.

The above method of biomass estimation is relatively accurate for GLAI from 0.2 to 1.2. Above and below this range the spectral variation in soils and the overlap of projected leaves cause to many additional inaccuracies.

Other vegetation indices as found in literature are:
— normalized difference vegetation index VI:

If only two spectral bands are used in Landsat MSS data
Then the above introduced GVI based on the axis $(\vec{v}' - \vec{s}')$ and GVI $= (\vec{v}' - \vec{s}') . \vec{b}'$ reduces to gvi, VI = gvi:

$$VI = \frac{b(7) - b(5)}{b(7) + b(5)} \qquad (29)$$

— transformed vegetation index TVI:
The TVI is a variation in the VI (29):
$$TVI = \sqrt{(VI + 0.5)} \qquad (30)$$

It is essentially a numerical trick to get rid of negative numbers, the reason for the square root is probably to get a better correlation with biomass.

Some use:

$$VI = \frac{b(6) - b(5)}{b(6) + b(5)} \qquad (31)$$

and corresponding TVI.

Soil brightness index (SBI) and green vegetation index ('GVI). Knauth and Thomas (1976) use two indices derived from four Landsat MSS bands but do not use reflectance data $(b'(k) = b(k)/i)$, but reflection (photon count) data. Because these indices are not normalised for intensity any variation in intensity will cause variation in the indices.

$$SBI = .43 \times b(4) + .63 \times b(5) + .59 \times b(6) + .26 \times (7) \qquad (32)$$

'soil brightness index' and:

$$'GVI = .29 \times b(4) - .56 \times b(5) + .60 \times \times b(6) + .49 \times b(7) \qquad (33)$$

'green vegetation index'. Both indices are not normalised for intensity and are thus not proper indices.

Principal components (PC), Karhunen-Loeve transform (KL). In classical pattern recognition PC (or KL) transforms were used for statistical feature extraction along the lines of statistical factor analyses.

Principal components are sometimes used for colour coding, violating the principle that 'data intensity must be mapped into colour intensity and data reflectance into colour hue and saturation'.

Principal component analyses can be used to order the data (statistical factor analyses) after all the known physical factors have been found and removed from the data set. Like: first find the data intensity (sum of the b(k)) remove this factor by dividing (b'(k) = b(k)/i), remove next the vegetation to soil mixing factor, next the water to soil mixing factor and then apply a principal component analysis to the remainder for ordering the spectral data on variance, interpret the result looking for a physical model explaining the variance of the residual data.

The Karhunen-Loeve or principal component transform was an early technique for feature extraction by data reduction. It is based on the property of the K-L transform that with a given number of principal components one retains a maximum of statistical variance. The weakness of variance based methods is that variance is not equivalent with (real) information. Multiplicative factors like that of the intensity component are not properly taken into account in the standard way principal components are applied to multispectral data. A common mistake is to use principal components for colour coding. This violates the principle that one cannot add apples and horses, which in this case means adding spectral reflectances to shadows and shades.

In the following figure (Fig. 28, see colour section) Landsat TM data of the Netherlands has been used. The known multiplicative intensity factor has been removed by first calculating the intensity over the six reflective bands (total photon count):

$$i = \sum_{k=1}^{N} b(k)$$

and normalizing the bands to weighted reflectances b'(k):

$$b'(k) = b(k)/i$$

In the domain of reflectances b'(k) a principal component analysis is applied resulting in six eigen vector axes. The b'(k) data is then projected on the eigen vectors by:

$$pc(e) = \sum_{k=1}^{N} ev(k) \times b'(k)$$

pc(e) = principal component for eigen value e; ev(k) = weight factor relating eigenvector $\vec{e}$ to spectral band k; b'(k) = weighted spectral reflectance in band k.

Resulting in six image files, each file is displayed after multiplication with a constant illuminating intensity factor resulting in six images pc(1) to pc(6). pc(6) should be zero because one degree of freedom was removed from the original six degrees of freedom by the condition:

$$\sum_k b'(k) = \sum_k \frac{b(k)}{i} = 1$$

The pc's are ordered according to their variance (eigen value) starting with the largest variance and ending with the smallest one. pc(6) will not be completely zero because of round off effects in the calculations.

Multi-temporal features

With multi-temporal (multi-season) data the first step is to register the data of different seasons exactly to the same geo-reference grid of scene elements as discussed under geometric corrections and geometric transforms.

Satellite remote sensing offers the possibility to collect data over short intervals. The interval length is called the time resolution of the system. Between various systems there is a trade off between spatial resolution and time (temporal) resolution. NOAA weather satellites (polar orbits) have a spatial resolution (scene element) of the order of 1 km but a time resolution of 1 day. The change of spectral signature with time may be indicative of the health (yield) of vegetation or the succession of e.g. crops. It is also possible to follow other dynamic processes such as deforestation, desertification and the effects of increasing salinity of soils. Certain stress factors in vegetation can only be detected under specific conditions over a limited time span. In such a case the temporal feature extraction amounts to a selection of one or two frames of data from a time series.

Multi-temporal (multi-season) features are dependent on spectral features. Spectral feature detection is performed first followed by an analysis of the change of that feature with time.

300

Fig. 28.

If there is a model for the change of vegetation cover with time such as a crop calendar then that should be used as a first basis for a multi-temporal analysis. Such an analysis compares the predicted spectral features with the actual ones and selects those features which show a significant deviation from the predicted value(s).

The change of the green leaf area index (GLAI) with time is often characteristic for certain vegetation types. Multi-temporal feature extraction is thus mainly a matter of selecting the right dates of data acquisition. The further procedure is to transform each MSS data set into GLAI and intensity and to use the multi-temporal set of e.g. three GLAIs for classification or interpretation. For three GLAI values per season it is possible to show each component of the set in a primary colour e.g. Spring-GLAI in red, Summer-GLAI in green and Autumn-GLAI in blue.

Fig. 29 (see colour section) shows three date colour coded GLAI. The GLAIs are coded for May red, for July green, for October blue. Display intensity indicates average GLAI over three seasons, display hue indicates the time of maximum GLAI and display saturation the magnitude of change in GLAI. The scene is from Landsat MSS of Novara, Northern Italy. The data acquisition times were selected from the local crop calendar and knowledge about change in GLAI with time for each crop class. A problem may be the lack of availability of data due to cloud cover, other meteorological conditions or organisational problems. Good Landsat data are seldom available with a time resolution of better than a (few) month(s). If the scene element size (1 km × 1 km) of NOAA satellites is not a too serious problem then one should consider the acquisition of digital visible and reflective infrared data for every 1 or 2 days in a critical season. This allows for the estimation of the dynamical aspects of vegetation at low spatial resolution. This can be supplemented by data of opportunity at higher resolution.

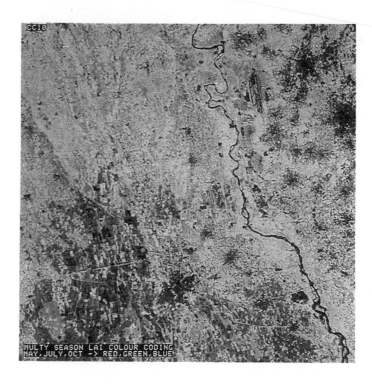

Fig. 29.

Spatial features, spatial filters and operators

Spatial feature extraction. In the extraction of spatial features from an image we determine the building blocks for structural pattern recognition. Well known building blocks are area features, edge features and line features. Point features are often treated as 'noise' but can be part of a texture feature.

Spatial filters will only pass spatial information which fulfills certain conditions.

Edge detection and line detection. Edge features can be detected by using an edge filter. The assumption is that the edge of objects (e.g. fields) coincide with 'edges' (strong gradients) in image intensity.

Line features and point features are detected by filters which pass only data if a pixel value is significantly different from the average of its surrounding pixel values (see Fig. 30).

Area segmentation. The 'computer' recognition of spatial patterns is based on numerical or symbolic measures for the spatial/spectral relationship of a pixel with its neighbours.

In vegetation surveys the emphasis is mostly on features related to extended areas (more than e.g. 15 scene elements). Topographic features are more often line features and edge features. For this reason the further treatment of spatial feature extraction in the following section concentrates on first finding coherent areas (segments). This can be followed by e.g. a shape analysis based on the edge features of these segments.

Segmentation is a first step in structural pattern recognition. It brings together (by (re)labeling) pixels with comparable spectral signatures such that all pixels of one segment wil display one colour. A segment property table contains features of the combination of groups of pixels rather than individual pixel features. This in turn allows the use of shape and texture feature for the classification of segments.

Shape recognition is based on features describing the contour of segments. A shape factor

Fig. 30. Spatial (boxcar, convolution) filters used in feature detection, applied to Landsat thematic mapper data channel 3 (TM3). (A) original TM3 in black and white. (B) TM3 after edge detection, shown in pseudo grey scale (abs (gradient (tm₃))). White pixels indicate locations with a high image intensity gradient. Edges are passed as single white lines. Line features are shown as double white lines (two edges with opposit sign). (C) TM3 after line and point detection, shown in pseudo grey scale (abs (Laplacian (tm3))). Lines and points are passed. Lines are shown as white single lines. Edges are shown as double lines because the line filter is a second derivative filter (Laplacian). As point features are also passed the image is 'noisy'.

is e.g. the ratio of the area of a segment to the edge length squared (for a circle pi. radius squared/(2,pi.radius) squared = 1/4 pi).

Texture features are derived after segmentation as a statistical variance of the original values within a segment boundary. Segments are linked together by merging to objects such as fields or land units. The decision rule for merging segments is based on both spatial and spectral features (properties) of the segments.

Fig. 31 (see colour section) shows the effect of segmentation on data of 30 m ground resolution (thematic mapper TM channel 3). The lack of coherence in the original data is made visible by coding the data in pseudo colour. Segmentation (into primitive segments) is achieved by a new method (Mulder, 1984) of conditional rankorder filtering.

Conditional rankorder filter. For each pixel in an image that pixel and the surrounding eight nearest neighbours are subjected to a rankorder sort on the value of these nine pixels. If the new pixel value is equal to the rank 5 of the sorted values the filter acts as a median filter. A median filter passes the median values and removes the extreme values and hence it removes also 'noise' values. On edges there are always about six pixels with more or less the same value thus replacement by rank 5 will preserve edges. With line features the number of values which is the same is three or four and the median filter will replace line values by values from the non-line elements which are in the majority. In order to keep the line features, a test is performed for the detection of line elements. If the test is positive then at that location the original (central) value is copied into the new file. Else the median value is taken.

Example condition median:

subimage
1 7 9
2 9 8   rank 5 value = 8 and feature is edge feature →
1 9 8   new central value = 8 (was 9)

subimage
2 8 1
1 9 2   rank 5 value = 2 but feature is line feature →
0 7 1   new central value = 9 (was 9)

On the condition that the central pixel is not part of a line it is replaced with the median value in the 3 × 3 subimage surrounding the central pixel. Repeated application of this edge and line preserving 'smoothing' operator results in segments with the same local median value. This new method has a faster convergence rate than the older method of taking the conditional local average repeatedly.

Conditional mean filter. The old method calculated the average value only over elements with

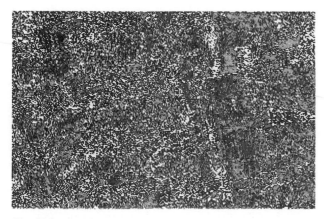

Fig. 31A.

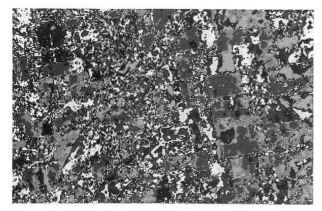

Fig. 31B.

values which did not deviate more than a specified amount from the central value.

Example conditional mean threshold = 3:

```
             2 1 0
subimage     0 6 7   mean of
             1 7 8

. . .
. 6 7   is 7 → new central value = 7 (was 6)
. 7 8
```

The effect of the described conditional filters is a segmentation of the image into natural mapping units.

Texture is a property of a segment. Texture features can be derived from segment statistics, e.g. the average absolute value of the differences between the segment median value and the original values.

Segmentation techniques based on repeated conditional filtering (= pixel merging) work also well on radar data if certain precautions are taken first. In radar return signals the noise is multiplicative which means that if the signal increases, the noise will increase proportionally. In conditional smoothing the decision to include a neighbour pixel in the smoothing operation or not is defined by a threshold. It is desirable to make the threshold just larger than the noise amplitude. For radar this would mean that the threshold has to be proportional to the value of the central pixel. In order to get a constant threshold for the whole image one could take the logarithm of the (power) signal. This makes the noise additive so that the condition for smoothing or not smoothing becomes a simple thresholding condition.

As radar contains much coherent noise because it is based on sensing with a coherent source, it is quite critical to find the proper threshold for segmentation. Otherwise pattern recognition on radar data will fail. The classical filters give very little improvement. One has to use segmentation first followed by the extraction of average signal and e.g. variance calculation per segment as segment features (field properties).

Classification and decision-making

General remarks

Image processing systems have two types of output; image data and 'map' data. Map data are generated by classification algorithms which in turn implement decision rules applied to e.g. spectral data (features). Usually map elements (pixels) represent the class label as a colour → colour legend. This type of maps can be produced in an extremely short period because film writers, which produce e.g. colour separations of maps, are much faster than classical drafting machines. The combination of fast classification algorithms, up to date remote sensing data and fast output devices allows the production of up to date maps. By storing the result of the classification in a database, it is possible to do change detection by comparing the most recent classification with a previous one.

In pattern recognition (PR) there are two main streams; statistical PR and structural PR. Statistical pattern recognitition is based on decisions rules which make use of the statistical relations between observations (data) and classes defined by 'users'. Structural pattern recognition is much more concerned with structure as a composition of basic elements, regularity in spatial relationships and topology.

At the present state of classification of remote sensing data both approaches are used in a complimentary way. The application of statistical decision rules to individual scene elements will merge elements into small segments. Segments have shape, structure, size. They form the basis of a structural classification by merging segments into objects. Objects are then classified on the basis of statistical spectral features of objects and structural features of objects.

Classification is the assignment of labels or names (nominal data) to samples (scene elements) or aggregations of samples (segments or objects).

Decision rules map ordinal (numerical) data into nominal data. Decision rules developed in statistical pattern recognition are equivalent to decision rules in business decision-making. This is specially clear if a cost is assigned to each wrong classification and benefit to each right classification. The class of Bayes maximum likelyhood (MLHD) decision functions can be programmed to be minimum cost/benefit decision rules by specifying the costs of misclassification relative to a benefit for right classification. A problem is that the user cannot easily specify proper cost and benefit factors.

Supervised classification

Supervised (statistical) classification requires the user to specify in advance the class definitions (labels) and to provide a training set for the determination of the decision functions or decision rules. A training set is constructed by either carefully selecting fields in the image (on the display) which do not contain a mixture of classes and for which the class is known or to do an exhaustive interpretation on a representative part of the image and store the interpretation result as a map in the database. The last-mentioned approach can only be used if the average field (patch) size in the scene is larger than 25 resolution elements. The reason for this rule is the high number of mixture elements on field boundaries relative to pure elements in the middle of small fields. The function of the training set is to determine the statistical relationships between classes and data.

In unsupervised classification the assumption is that the user cannot supply meaningful class labels or class samples. All clustering programs are essentially based on finding maxima in a N-dimensional frequency distribution. They will define decision functions, grouping the data under a predefined number of arbitrary labels which have no known relation with the actual application. In the case of a single spectral band, the frequency distribution is depicted as a histogram [10] and clustering is equivalent to peak detection in the histogram.

The main disadvantage of unsupervised clustering is that no knowledge about the relation RS-data with classes is used. The second disadvantage is that the users of unsupervised clustering do not understand that the detection of a peak in a histogram may be completely irrelevant to the classification task at hand.

Fig. 32 indicates the weakness of the unsupervised clustering appraoch: in a 2-dimension MSS band-5, band-7 (b5, b7) histogram the clusters are not found for meaningful class definitions such as dense vegetation (veg), bare soil (soil) or water but rather for the statistically dominating mixtures of special signatures. The same effect is seen in two-dimensional histograms on intensity and GLAI (see Fig. 33).

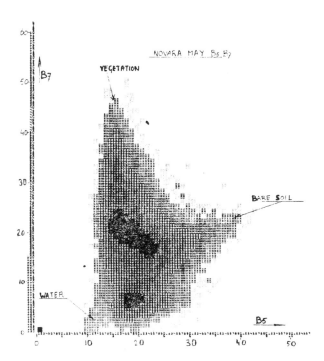

Fig. 32. Two-dimensional histogram on Landsat MSS b5,b7. Density (black) is proportional to the number of pixels with the combined value b5,b7. Clusters are found in parts of the feature space where a mixture (overlap) of class clusters occurs.

[10] Histogram: a histogram is the display of the frequency of occurrance of a datum or combination of data values in a file. E.g. freq(b7, b5) contains the number of scene elements which have a certain (unique) combination of b7 and b5.

306

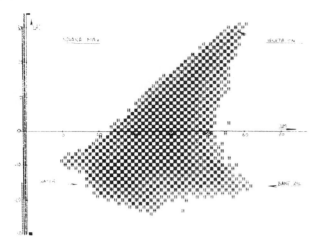

Fig. 33. Two-dimensional histogram on features intensity i = b7 + b5) and gvi (GLAI) vegetation index vgi = (b7 − b5)/(b7 + b5). The checkerboard effect is caused by numerical side effects of the transformation to i and gvi and the rounding off to integer values.

The sequence of steps for the classification of multispectral data is:
— radiometric correction of the data;
— extraction of spectral features, data reduction;
— correction of geometric errors, registration, resampling;
— extraction of multitemporal/multispectral features;
— extraction of spatial features;
— collection of field data guided by available images and maps;
— training of the classifier, calculation of decision parameters, decision rules;
— run the classifier, apply the decision rule to the data;
— calculate statistics of e.g. area per class;
— compare the class label map with digitized reference area or with the test set, produce a confusion table (omission/commission table); [11]

[11] Omission/commission table, confusion table: a 2-dimensional table giving the statistical relation between the test data and the final classification result in terms of the frequency of the occurrance of test labels and classification labels. On the diagonal of such a matrix one finds the frequency of commission, off the diagonal the omission frequencies are found. Other name: confusion table.

— use context masks to reassign labels in context; [12]
— define a colour table linking class-lable to legend and colour of the thematic map or display;
— overlay useful existing map data (data base, e.g. topography) information onto the classification result;
— store new classification result and/or print it as a map.
The critical steps in a classification sequence are:
— the extraction of features;
— the definition of a training set or test area map for training the classifier;
— the definition of context masks and rules for context reclassification.

Decision functions and decision rules

General remarks. A decision function transforms numerical data values into logic values (class labels) by inverting the relation between classes and measurements. E.g. the function fw defined on Landsat MMS band-7 transforms an actual band-7 value (b7) into a decision for water or non-water:

$$fw = b7 - 6 \tag{34}$$

If fw < 0
Then class is water
Else class is non-water

In order to go from numerical to logical (nominal) data fw is evaluated and compared with the value 0. The number 6 plays the role of a decision threshold. The logical part of the decision function is an If ... Then ... decision rule.

The choice of a specific decision function is not too critical for the accuracy of the classification result in multispectral data classification. The effect of a decision function is to partition

[12] Context masking: a context map defined by an expert or derived from a map is used to change the class labels of those elements which cannot exist in the specified context, to a context compatible class label.

a feature space into areas containing the label that the datum will get at that specific feature space coordinate. With one or two features the partitioning of the feature space can be pre-calculated and stored in a one- or two-dimensional decision table.

For example a two-dimensional decision look-up table on two features i and gvi produces for each datum (i, gvi) a classification result l (a label from the set of predefined class labels):

$$l = \text{decision-table } (i, gvi) \qquad (35)$$

l = class label; i = intensity feature; gvi = green vegetation index feature.

i and gvi are integer numbers which are used as indices in the decision-table. This is the principle of associative memory, the unknown data values address the class label of that data. For each scene element with data pair (i, gvi) in the file, the decision-table produces a label l in the label (thematic data) file. A classification deci-

sion (look-up) table can be compared to a two-dimensional array of pigeon holes. The feature i adresses the column address, the vgi feature addresses the row number of a pigeon hole. The contents of each pigeon hole is a class label which is copied into the same position in the output map file.

Fig. 34 shows the feature space partitioning of four decision rules applied to features: i = (b7 + b5) and gvi = (b7 − b5)/(b7 + b5) for bands 7 (b7) and 5 (b5) of Landsat MSS. The decision functions in Fig. 34 are subdivided into two nonparametric maximum likelihood (mlhd) methods; sigma and k-nearest neighbours and the methods; (parametric) maximum likelyhood and handpainting.

Maximum likelihood, minimum cost. Maximum likelihood decision rules are based on a formula by reverend Bayes. It linkes a priori (in advance) class probability to posterior (afterwards) class probability (likelihood).

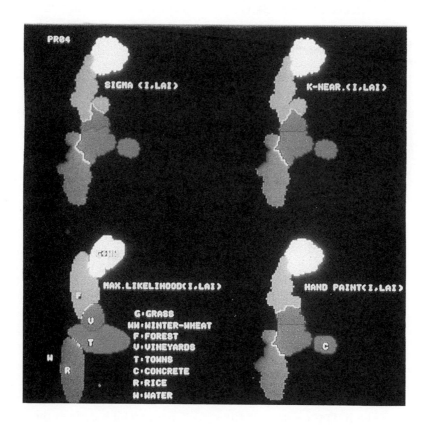

Fig. 34.

For each scene element with value b, we want to know the most likely class label. The likelihood that a scene element belongs to class l given the observation of feature value b (b can be a scalar or a vector) is:

$$p(l/b) = p(l) \times p(b/l)/p(b) \qquad (36)$$

l = class label; b = value in band (or feature) b; p(l/b) = probability that the class label is label l (from a set of class labels) given the value of b, posterior probability = likelihood; p(l) = the probability that in the classification result class l will occur, the prior probability of class l; p(b/l) = the probability that for a given class l, value b will occur; p(b) = the probability that value b will occur without bothering about the class of the scene element.

p(b) is estimated from the relative frequency (histogram) of spectral band values, p(l) is estimated from the number of elements with label l over total number of elements (a priori class probability which we do not know well enough!), p(b/l) is estimated from the relative frequency (normalised histogram) of spectral band values for class l only, as defined by the training set.

The maximum likelihood rule means for one observation b we calculate the likelihood that the scene element under consideration belong to one of the classes 11, 12, 13, 14, ..., the classes li with the largest likelihood p(li/b) is choosen as being the class of the scene element. The original statistical relation class-to-measurement as represented in p(b/l) is inverted to give the relation measurement-to-class as max-i (p(li/b)).

If the benefit of a right classification is the same as the cost of a wrong classification then the maximum likelihood rule is also a minimum cost rule.

The catch of all Bayes decision rules is that if we knew the a priori class probability we would know the number of scene elements per class and hence the area covered by that class. If we already know the area we need not classify at all in many applications! The prior class probability is also dependent on the choice of scene. In practice most people make p(1) the same for all classes which defeats the Bayes rule completely and maximum likelihood rules degenerate in fact to maximum class relative frequency rules. This is one of the cases where statistical theory provides only a smoke screen for the unwary user.

Although there exists only one Bayes rule there are various ways of estimating the probability element p(b/l) in the Bayes rule.

The conditional class probability p(b/l), which is the chance of value b given the class is l can be estimated from the training set in a parametric or non-parametric way.

Parametric estimation. Parametric estimation of p(b/l) means that the distribution of the data is assumed to be Gaussian and is characterised by mean (vector) and variance (matrix). With multispectral data the Gaussian assumption is basically wrong and should for that reason not be used. When Gaussian MLHD is applied, decision boundaries in feature space will be ellipsoids. Overlapping clusters will generate second-order decision boundaries between classes.

Non-parametric estimation. Non-parametric estimation is equivalent to estimation of P(b/l) from the relative frequency per class in the training set, without making an assumption about the distribution function of p(b/l). There is hence no need to calculate the parameters of such a distribution function. As the number of training samples is often low relative to the volume of the feature space, gaps in the N-dimensional frequency distribution (histogram) are filled up by the use of a smoothing algorithm.

— Sigma is the name of a Fortran programme for the non-parametric estimation of the data probability for a given class p(b/l). It uses a narrow Gaussian function with width sigma, centered at each sample point in the feature space, effecting a smoothing of the sometimes 'spiky' multi-dimensional histogram. The name of the algorithm comes from the smoothing parameter of the same name which defines the width of a Gaussian smoothing function. Other cheaper smoothing functions (potential functions) perform as well and may cost much less computer time.

— k-Nearest neighbours is another non-parametric maximum likelihood rule. It is a zero order interpolation function, equivalent to class histogram smoothing by piecewise constant smoothing. The width of the smoothing function is determined by a search radius in feature space which is extended until k-neighbours are found. k-NN decision-making is implemented by asking k (e.g. = 3 or 5) nearest neighbours in feature space to vote for the class membership of the data in question. It is a 'democratic' decision rule (decision function) which will supress local minority group samples in feature space.

— Handpainting relies on a visual estimation of a realistic $p(b/l)$ distribution in 2-dimensional feature space and the definition of the decision table by interactive means. Starting with the display of the training samples of all classes the decision class areas in the table are painted in. This is effectively applying smoothing and determination of the decision boundaries at the same time. The big advantage of this method is that it is implemented to work as a real time classifier using the colour look-up table of the colour display as a classification table for the data displayed on the screen. Any change in painted decision boundaries will have immediate effect on the classification results. The spatial pattern of classified themes provide the most direct feedback to the user for improving the classication rule or for trading of one type of error against another type, depending on which type of error is easier to edit (see Fig. 34).

The number of features plays an important role in classification and decision-making. Proper feature extraction will limit the number of spectral features to two or three per season. With more than four features the training set will be too insignificant as a cluster of data points in a high volume feature space (e.g. a 5-dimensional feature space with 256 level data has a volume of $2**40 = 1,000,000,000,000$ cells). The solution to this problem of a large number of features (usually in multi-temporal data sets) is to either do a classification per date first and combine classifications later or to update the previous classification with the new data removing uncertainty (label: unknown) from the set of pixels with each new date.

Hierarchical decision-making. Hierarchical decision-making is also called layered decision-making. As an example: it is often quite possible to classify the data into vegetation and non-vegetation, in a next step the vegetation can be split up into sub-classes. In the case where there are too many features the dimension of the feature space becomes too large and a sequential approach is indicated.

The procedure of hierarchical classification is as follows:
First classify the data in large classes e.g. vegetation, non-vegetation, then edit the classification result using context information and visible pattern. Proceed to split up vegetation into e.g. forest, non-forest, edit, etc.

Context masking, reclassification

The human interpretor has a very good feeling for which classes to expect where. His approach is often to start with large units and to look for detail only when necessary. Man is also selective in the collection of evidence (data). There are all sorts of contexts in which a certain class of data is very unlikely to occur. As it is very time-consuming to specify the context regions (e.g. geomorphological units) beforehand a method was developed to provide the context after a preliminary classification by the computer. The person looking at the classification result quickly spots regions where the classification result is contradictory to context. The problem areas are quickly outlined as polygons and a reclassification rule is defined (Mulder, 1983).

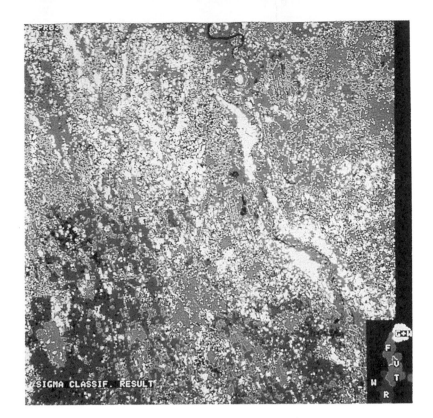

Fig. 35.

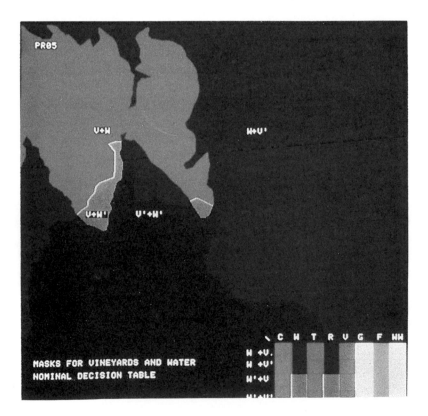

Fig. 36.

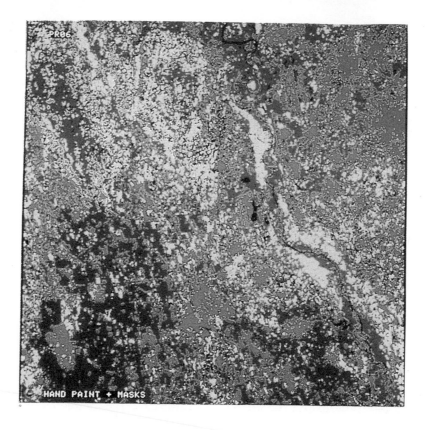

Fig. 37.

Context masking is a post-classification technique for bringing the class label into accordance with the spatial context. Spatial context is put into the system using a data tablet for digitizing context polygons (zones, areas) or the context areas can be derived from existing data base information such as a digital elevation model (height, slope classes, sun exposure, etc.). Some systems allow prior context information to be used in setting the prior class probabilities for each zone in a file. The classification program will have to calculate the decision function differently for each zone using the local prior class probabilities. Prior context information must be defined more precisely than posterior context information as prior information only need to be provided in areas where errors (in context) actually occur (see Figs. 35–37).

Expert systems

Human experts use a large body of mostly heuristic rules in their decision-making. Heuristics comes from heureca which means 'I have found it!'. Systems are being developed now in the field of artificial intelligence which show a pseudo-intelligent behaviour. A set of rules which relate to a narrow field of experience is collected in expert systems.

An expert system consists of three parts:
1. a rule base, which contains the collected and ordered knowledge about a narrow field of application;
2. a rule interpretor, which applies the rules from the rule base, according to a certain strategy, to the data;
3. a data base, which stores the data and allows intelligent retrieval of the data. Some data

312

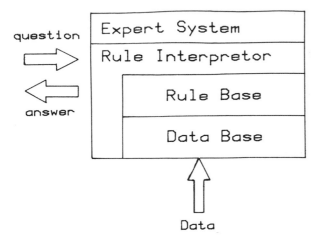

Fig. 38. The general structure of an expert system.

are provided directly by the user (see Fig. 38).

The approach of artificial intelligence (AI) to building expert systems can be beneficial in the design of automated interpretors and classifiers. An expert system can provide a common frame work for the integration of remote sensing data and features and heuristic rules about e.g. geomorphological, economical and even political context.

Two strategies of rule interpretation can be distinguished:

1. bottom up, data driven, forward chaining decision-making;
2. top down, hypothesis driven, backward chaining decision-making.

Operational systems are relatively strong in the application of relative small sets of heuristic rules. They are generally weak in the implementation of the existing body of rules in pattern recognition and image processing. 'The more exact knowledge there is, the less you need an expert!'

Systems for computer aided remote sensing data interpretation under development at the moment, aim to integrate the bottom up approach of the technical data base with the top down rules in human intuition.

A next step in the development of expert sys-

tems is to build hierarchies of specialised experts. One example is a system being built now (in the authors lab.) in which there are two slave experts: one for line finding and one for the detection of fields. The conclusions of the slave experts are evaluated by a master expert which has a more general and a more wide view.

Hardware and software

General remarks

Developments in hardware have been quite dramatic in the period of 1965 till now (1985). Microcomputers of today have more processing power and more memory than the most powerful minicomputers of two decades ago. Developments in hardware are very much market driven. If a high demand can be created then the technology will be developed.

Initially the demands for image processing and pattern recognition were mainly coming from the military and the medical applications. Today there is a high demand for computer-aided design and computer-aided manufacturing systems (CAD and CAM). CAD systems were initially automated drafting systems with vector screens. Vector screens record lines very much like pencil lines are recorded on paper. Selective erasure is not possible, a change in design requires complete erasure of the screen followed by rewriting. Raster screens record e.g. line information as a string of little squares (pixels). Each pixel can be individually changed (random access). Because raster screens have such advantages and moreover allow colour shading, all new CAD/CAM systems are using raster screens.

Graphics raster screens started off with only one bit (on or off) per pixel, but with large numbers of pixels like 2048×2048.

Image processing systems started off with more information per pixel, typically 12 bit to 16 bit per pixel, but smaller numbers of pixels, like 512×512.

Graphics systems used in cartography and image processing systems used in remote sensing are both evolving towards high resolution raster systems. A typical design for a new system would be a screen with $1280 \times 1024 \times 16$ bit resolution and a refresh memory of $4096 \times 4096 \times 16$ bit. This last option means that the user sees at any time a window of e.g. 1280×1024 of the data in memory with a roam and zoom option. The screen acts as a magnifying glass peeking into the digital data planes.

General purpose computers are too slow for operational image processing and pattern recognition. The role of minicomputers in image processing systems is mainly that of transport and administration of data. For the actual implementation of the algorithms special hardware processors are needed.

A generic type of processor is the array processor also called pipeline processor. It is good at performing single instructions (operations) very fast. Floating point multiplication and addition are typically executed at speeds of more than 10 million per second.

Modern image processing systems are designed with multiple specialised processors, many of which are pipeline processors. Some typical processors are:
- input processors, e.g. processors for reading keyboard input or coordinates and attributes from x,y-tablets;
- video processors, accessing video refresh memory at the speed of the screen scanning process, performing in real time the transformation of digital data into coloured picture elements on the screen;
- multiplication and addition pipeline processors, important for multispectral transforms and spatial filters;
- sorting processors, e.g. used in rankorder filters;
- geometric transformation processors, some times called warping processors;
- list processors, used for interacting with the user (language understanding), and the processing of e.g. string of coordinates and attributes for vector data processing;
- memory management processors for the efficient access of arrays or blocks of data;
- look-up tables, used when there is a fixed relation between input and output. The address of the table serves as input, the contents of the table serves as output.

The output of an image processing (pattern recognition and graphics) system is either a coloured image (map) or new data planes in the data base.

Enhanced images require high quality film writers in order not to loose the high quality of the processed data at the image generation step. Such filmwriters are often based on a rotating drum which carries the film to be exposed. A glow crater tube with colour filter or tuneable laser is used to expose the film per pixel per scanline and per colour per frame. Frame sizes are typically of the order of 250 mm. The spatial resolution of such systems start with 12.5 µm and upward. This allows eight times magnification of the film negative to the resolution of the eye (100 µm at 250 mm viewing distance).

If only classification results or maps need to be printed then the radiometric requirements are much less severe (16 levels per colour). The result can be directly printed at eye resolution (100 µm) but the size of the map must often be large (e.g. 30 cm × 40 cm to 1 m × 1.4 m). If much higher quality map reproduction has to be done then it is desirable to have a plotter with digitally screened output (dot size modulation) with selectable screen resolution (e.g. 60 lp/cm) and selectable screen angles or random dot generation to avoid Moire' effects in the overlay of colour separations.

Image processing system hardware

Starting at the input side:
- magnetic tape unit: needed for reading the data from computer compatible tape (CCT). Alternatively low cost systems could read prepared data from floppy disc;
- mini computer: will take care of transporting and administrating the data;

314

— disc: for massive data storage. The data is reformatted to the internal format of the system data base;
— interactive colour display processor and colour screen: plays an important role in training the system for classification tasks or in setting up parameters for image processing. The user can interact with the system through a:
  — keyboard or preferably through an x,y-tablet;
  — x, y-tablet: takes as an input a cursor position or the position of an electronic pencil, as output it gives the position of pencil or cursor as a pair of x,y coordinates. Attribute information is specified through one or more buttons on the cursor or through the general keyboard. A tablet can be used for the selection of parameters or processes from a menu.

In digitizing polygons a special fill operator is called to fill closed polygons with e.g. legend colours from a palette of typically > 4096 colours. Digitizing polygons with a colour screen has many advantages over the classical digitizing with lines only. A first advantage is that the user gets an immediate feedback on the closedness of the polygon. If there is a break in the polygon then the colour will leak out of the polygon. An inverse fill (erase) will withdraw the colour and the polygon can be closed.

— Graphics raster screen: some times added to the system as a specialisation for cartographic data entry and editing. The specialisation is in high screen (memory) resolution e.g. 2048×2048 but low colour resolution e.g. 16 colours = 4 bit. Such subsystems are also equipped with a (large) x,y-tablet. Vector to raster conversion is always done by special processors (or firmware);
— array processors are either part of the interactive colour display system, operating on the local video memory or they are general purpose array processors. General purpose array processors are fed by the minicomputer with data from the disc and the results of the array processor are usually fed back to the disc;

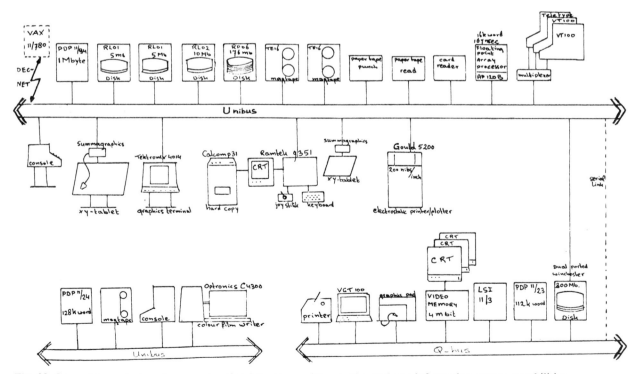

Fig. 39. General lay-out of an image processing laboratory with graphics and geo-information system capabilities.

- quick look cameras are relatively cheep devices which take the video signal from the colour screen and feed it to a separate flat CRT screen. The CRT screen is used with a colour filter wheel to expose 35 mm camera film, or Polaroid film for immediate use;
- electrostatic printers / plotters are used for fast but low resolution (200 to 400 lines/inch) output of graphics, grey tones or symbols;
- film writer: high quality colour film output device. Radiometric resolution 256 levels in red green and blue, optionally 256 levels in white. Spatial resolution selectable 12.5, 25, 50 or 100 µm. Film usually $10'' \times 10''$ colour dia positives or colour negatives. Optional b & w colour separations;
- Microcomputer based systems: it is possible to perform remote sensing data classification, some pattern recognition and a small geo-information system with a system costing less than $ 3000. In the authors lab we have ten such systems put together in a network (100 kbaud) with a file server with 30 Mbyte Winchester disc. The systems consist of a microcomputer with network device, an x,y-tablet for digitizing and menu

selection, a high resolution colour monitor and an ink jet (colour) plotter. All the processing is done locally, only the 30 Mbyte mass memory is shared. The main purpose of this system is to provide students with access to modern surveying tools. But it is quite possible to use the same system operationally e.g. in a local county office;
- mini survey: in one application photographs taken by a micro-light aircraft are interpreted individually or as stereo pairs, the interpretation overlay or the annotated photograph is then put onto the x,y-tablet. Four passpoints (map references) are entered and the lines and polygons are digitized. In the digitization process the image coordinates are automatically rectified to common map coordinates without need for the rectification of images. Various layers of information can be overlayed and combined in a geo-information system;
- colour ink jet plotter: the colour output device of the microcomputer based system. The display resolution is 640 dots over the width of A4 size paper. It is used to make a colour hardcopy of the data displayed from video memory or to generate the colour

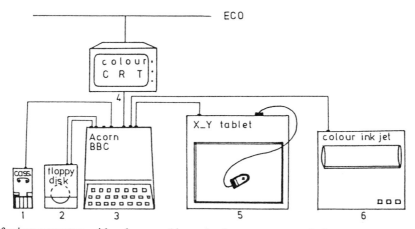

Fig. 40. Network of microcomputers with colour graphics and colour map output. Software for processing remote sensing data and the setting up of a geo-information system. ECO = Local area network; 1 = cassette recorder; 2 = floppy disk drive; 3 = microcomputer BBC, Acorn; 4 = colour cathode ray tube (CRT), 12″ colour display; 5 = XY-tablet, Calcomp 2000, USA; 6 = colour ink jet printer, Tandy.

map directly from mass memory. With the last option the size of the output map is not limited by the screen size ($\Rightarrow$ video memory).

Software

Hardware can be bought tested, evaluated and accepted, or rejected. Software is much more untouchable but certainly as expensive and as important as the hardware is.

To the user the system will present itself through the 'user interface':

— user interface: software for the communication between the system and the user. The quality of the user interface and the phylosophy behind it will largely determine whether the user will like the system or not. If the user likes the system he is motivated to make effective use of the system. The software is called 'user friendly' in that case. If the user does not like the system the user will soon be irritated and find all sorts of faults with the system and return very soon to working with transparencies and colour pencils;

— menu driven: the available options at a certain stage of the processing are presented to the user as a menu. The selection of the required option is by entering e.g. a two-letter selector using the key board, pointing with the cursor to the selected option on a menu screen or pointing to a menu on the x,y-tablet. Users with little experience or working irregularly with image processing systems have a strong preference for menu driven software. One of the many options is (should be) the help option. The help key will provide explanation on the items on the actual menu page;

— command driven software: engineers, software programmers and sophisticated users prefer to drive the system directly by giving (typing) commands. Commands are given in a command language which has a well defined syntax. This language must be learned and be remembered by the system user. There is no standard command language for image processing, pattern recognition or geo-information systems;

— application software: the actual transformation and classification algorithms together form the application software. The choice of methods and algorithms is defined by the manufacturer, his cultural background and that of his well to do clients. Too many algorithms are still coming from signal processing where the notion of 'pattern or structure or context' is alien.

It is not possible to give good advice in this limited space on the specification of application software. In the worst case still 60% of the software delivered will be of some use for everyday applications.

It takes about 4 to 5 years before good ideas perculate from the research laboratories to commercially available programmes. A good advice is to make arrangements for the updating of the software through contracts with the manufacturer and through informal circuits (e.g. user clubs).

22. Mapping Herbaceous Vegetation

A.W. KÜCHLER

One of the most important resources of many parts of the world is their grassland vegetation. Of course, plant life is a primary resource in any landscape, but in mapping it, the herbaceous vegetation reveals significant features which require special consideration. It becomes evident that in mapping herbaceous vegetation, a uniform approach to all vegetation types may not produce the best results.

The great majority of the older vegetation maps treat woody vegetation in greater detail than herbaceous vegetation. Trees are often enumerated by genera or species on maps which refer to the herbaceous vegetation with no more than 'grassland'; at times there is a broad distinction between tall and short grass, meadows, bunch grass, marsh grass and others.

These observations may lead to the conclusion that herbaceous vegetation is of slight importance. This conclusion is not sound, because in an area like the North American prairie or the pampa of South America, herbaceous vegetation is pratically the only kind of any significance. Furthermore, in various parts of the world, forests are used almost exclusively for grazing purposes, which gives the herbaceous ground cover more importance than the tree synusia. Also, herbaceous plants are primary indicators of soil conditions and changes, which explains their value in land management. This list of arguments can be enlarged considerably.

Why, then, has the herbaceous vegetation been treated like a stepchild on so many vegetation maps? Among the possible answers, three have perhaps special significance.

The first, rather intangible one, is the attitude of the authors of vegetation maps. The vast majority of these authors are forest-minded because they come from regions where forests are the natural vegetation. Their eyes react quickly to changes in the floristic composition of forests, but they are rarely trained to be equally sensitive to the features of grasslands. Malin (1947) pointed out that forest-minded people analyze all land from a forest point of view and, therefore, often speak of 'treeless' areas when referring to the great grasslands of the earth; had man build his civilizations above all on grasslands, he might well have referred to the great forest regions of the earth as 'grassless'. Vegetation maps themselves support this argument; the contrast is striking when vegetation maps by authors from forest lands are compared with maps by grassland authors.

The second answer is probably more important than the first one; it lies in the very nature of grasslands. The individual specimen of a herbaceous plant occupies little space. Within a short distance, therefore, it is possible to meet a considerable number of species; in proportion to their size, these species may occupy no less space than trees. To show the same detail on a map of herbaceous vegetation as on a forest map would, therefore, automatically increase the scale of the former. A map at $1:100,000$ showing a forest 20 m high would correspond to a map of grasslands at $1:5,000$ where the vege-

A.W. Küchler & I.S. Zonneveld (eds.), Vegetation mapping. ISBN 90-6193-191-6.

tation is 1 m high. Obviously, if the same details are required, herbaceous vegetation is either mapped on restricted areas only or many more maps must be produced. One is reminded of the problems of mapping tropical rain forests where a very large number of species per unit area makes any form of detailed mapping exceedingly difficult. In general, however, the scale problem of the contrast between woody and herbaceous vegetation disappears when the determination of phytocenoses is based on a more detailed analysis of the floristic composition, or when only dominants are used to characterize the plant communities.

A third troublesome aspect of mapping herbaceous vegetation is the impermanence of the constituent species. The species of woody vegetation can be identified without difficulty in any season. Herbaceous vegetation, however, consists of a considerable number of annuals, and also of perennials which are geophytic or hemicryptophytic in character. This precludes their identification during part of the year. Even within the course of a single growing season the floristic composition, as far as it is readily observable, may change almost completely, although some species will remain recognizable throughout the summer and, indeed, most or all of the year. The floristic combinations of the vernal and the autumnal aspects of a given herbaceous vegetation type belong, of course, to the same plant community. A detailed knowledge of the floristic composition of such communities is required before they can be mapped satisfactorily, and in most parts of the world such knowledge is only now being assembled.

In semi-arid regions of the world, where the vegetation is often predominantly herbaceous, the rainfall may be considerable for years, only to shrink to a fraction of its average amount for the ensuing years. The result is a more or less radical change in the floristic composition of the vegetation. Finally, changes in the floristic composition of herbaceous phytocenoses due to human activities are rapid and far-reaching indeed.

In the deserts of Arizona, California and Turkestan as in the steppes and savannas of the Sahel in Africa and elsewhere, a very short-lived herbaceous vegetation of ephemerals will develop after the rare but heavy rains. Any one who maps such a region when the herbaceous vegetation is fully developed is profoundly impressed by its spectacular masses and colors. It would seem inconceivable to ignore such an important part of the native vegetation. But should it be mapped if it lasts only 3 weeks out of 52, and does not even appear every year?

Such features seem to have discouraged authors from preparing maps of herbaceous vegetation. Is there then no prospect of preparing satisfactory maps of herbaceous vegetation?

Large-scale mapping

The Bundesstelle für Vegetationskartierung under the direction of Reinhold Tüxen has published numerous large-scale maps on which herbaceous vegetation is shown (e.g. Seibert, 1954). These are maps of forest regions, but the herbaceous vegetation that is characteristic of each forest community is included in all its variations. Thus, the forest types are differentiated into subtypes on the basis of the herbaceous ground cover synusias, and at the same time, the total vegetation is well presented. There seems, therefore, no major problem involved in mapping herbaceous vegetation at very large scales. It must be remembered, however, that Tüxen and numerous other phytocenologists worked in western Europe, a region with a humid, equable climate. Tüxen and Preising (1951) have given directions for mapping herbaceous vegetation in their area, and their fine maps bear witness to the validity of their arguments. But it is also quite obvious to those who live in the world's great semi-arid grasslands, that Tüxen's ideas cannot be applied there without important modifications, for it could easily turn out that, within 2 or 3 years after the area was mapped, a map would no longer correspond at all to the facts. This emphasizes the point that it may be necessary to develop different mapping techniques for different regions.

Small-scale mapping

As the map scale shrinks, generalizations become a serious factor. In this, there is no difference between woody and herbaceous vegetation. If an area is to be mapped at a small scale, which implies much generalization, some material must be suppressed. If the map is to be done in a forest region, the usual procedure is to suppress the herbaceous synusias first, as on the maps by Davis (1943) and Moore (1953). This may be justifiable from various points of view, but there are important aspects of land-use and management which require a different procedure because the herbaceous vegetation may be quite as important as the trees, if not more so.

The map by grassland authors show a different attitude. Costin (1954), on his vegetation map of the Monaro region, distinguishes a large number of herbaceous phytocenoses, but many of these are grouped together because the scale is too small to show them individually. One of the finest small-scale examples is the vegetation map of the Ukraine at 1:1,000,000 (Kleopov and Lavrenko, 1942) where the herbaceous vegetation is shown to a modest degree in forest lands, but in detail on the steppes. Forests and steppes are characterized by about the same number of dominants.

At still smaller scales, Küchler (1953a, 1964) attempts consistency on his maps of the United States at 1:3,168,000 and 1:14,000,000 by showing the major physiognomic divisions of the vegetation, each with its floristic subdivisions. This includes, of course, the great prairie. In spite of the condensed legends, the various aspects of the prairie are clearly discernible.

Possible solutions to mapping problems

Various possibilities exist to meet the difficulties of mapping herbaceous vegetation, and in given cases it may be useful to combine some of these possibilities on one map. Such possibilities are:

1. to select a 'normal' year, i.e. one which most closely approaches the averages in precipitation and length of growing season, and to map the herbaceous vegetation that persists longest in the course of such a year;
2. where there are distinct differences between the early and the late parts of the growing season, both the early and the late aspects can be shown separately in the legend for any 'normal' year as vernal and autumnal phases;
3. where the character of the vegetation is apt to change, due to more violent fluctuations of the climate, the 'normal' vegetation can be shown with indications of 'humid' and 'dry' phases.

Seasonal aspects and fluctuations over the years therefore invite an elaboration of the legend items. It is permissible to show the various aspects as individual items, but it is clearer and therefore preferable to group them. Thus, for instance, legend item 1 gives the 'normal' version of the phytocenose, with 1a and 1b giving early and late seasonal aspects. The same technique can be used to describe the character of the phytocenoses during dry and wet periods. Vegetational changes from wet decades to dry ones, and back, represent a form of dynamism, but the various aspects must not be confused with seral changes. Changes due to climatic fluctuations are not an evolution toward a climax because it is the climax itself which fluctuates in harmony with the prairie climate.

When the mapper wishes to record the physiognomy of herbaceous vegetation, he should select the period during which grasses and forbs attain the peak of their development. This usually begins with the flowering phase. Where vernal and autumnal aspects are mapped, the vegetation must be investigated several times during the growing season.

If the mapper has aerial photographs specially made for his area, he should insist on having them taken during the flowering phase. This

320

will emphasize color contrasts that might not be observable during other seasons. The photographs of herbaceous vegetation (i.e. their negatives) should be of the largest possible scale. It is easy to make enlargements from small-scale negatives but they do not always show the important details with the same clarity as photographs obtained from large negatives. Photography approaches the ideal when it is repeated every season, or at least twice in the course of one growing season. Such repetition is particularly useful when the herbaceous vegetation is to be mapped at large scales.

It is now well established that herbaceous vegetation can be shown in appropriate detail at all scales. It would seem important, for the sake of scientific accuracy, to devote more attention to herbaceous vegetation than in the past. But it is also clear that mapping herbaceous vegetation presents peculiar problems and in many instances requires its own specialized techniques. At any rate, the herbaceous phytocenoses must always be presented with the same care and detail as the woody communities if the map is to have a high scientific value.

23. Mapping Dynamic Vegetation

A.W. KÜCHLER

Vegetation is dynamic, i.e. it is involved in a continuous evolution. The rate of evolution is at its minimum in climax conditions, but in all other conditions it can range from very slow to fast. It is essential in phytocenological investigations to understand the processes involved in the evolution of vegetation, but such processes may be too slow for one man to observe during his lifetime. For this reason, authors compare various evolutionary stages and endeavor to conclude from such observations what processes are involved, their direction, and their effects. Such a procedure is, however, speculative and the results may remain debatable. Vegetation maps solve this problem especially if done with great care and accuracy and in great detail. Some time after mapping, measurable changes will have occurred and, even though a new observer may map and study the vegetation long after the first vegetation map was made, he will find it entirely feasible to compare the old and the new situations. Then, indeed, the exact rate and direction of change can be ascertained and these, in turn, throw light on the nature of the processes involved in this evolution. The appearance and spread of some species or the disappearance of others can be observed and followed, and measured with a degree of accuracy that would be unthinkable without such vegetation maps. The causes of the changes can then be established.

The recognition of the dynamic features of the vegetation is very useful. It not only permits an appreciation of what is going on in the landscape and where a situation is headed; it also permits the phytocenologist to avoid mistakes that might have serious consequences if applied to some aspects of land use.

To use one of Gaussen's (1959) graphic illustrations: the vegetation evolves from the pioneer phases of a sere through a number of phases to the final or climax phase. One such sere on a north-facing slope may have phases that we may characterize as 1, 2, 3, 4, 5, 6, 7, and 8 (climax) and the phases of another sere on a south-facing slope of the same mountains may be indicated by a, b, c, d, e, f, g, and h (climax). In either case, it is quite possible that the floristic composition of the early phases has little or nothing in common with that of the climax. On the other hand, it is likely that a number of species occur simultaneously in the early phases of both seres. Thus h is derived from b and hence is related to it floristically even though b and h may not have a single species in common. On the other hand, b and 2 may have some species in common even though they are unrelated.

The evolution of vegetation may be progressive toward a climax or regressive, away from a climax. The latter case is particularly common as a result of human activities such as lumbering, herding, burning, cultivating, mowing, etc. The prevailing condition of the vegetation is therefore by no means necessarily headed for a climax, at least not so long as man continues to interfere in the natural evolution of the vegetation. It is obvious that the correct

A.W. Küchler & I.S. Zonneveld (eds.), Vegetation mapping. ISBN 90-6193-191-6.

interpretation of the vegetation, particularly its stage in the evolution, is of basic importance if a phytocenological study is to result in an improvement of the prevailing conditions.

Vegetation maps have been compared with photographs, especially since the introduction of aerial photography. Both record the vegetation as it can be observed at a given moment. This gives vegetation maps a static character, but phytocenologists know their subject to be dynamic indeed. The vegetation mapper usually finds it wise to map the actual vegetation before he embarks on interpretive features such as dynamics, climax, yields or others. This is always the safest and soundest beginning, but he must realize, of course, that the actual vegetation is often quite unstable whether it is man-induced or not. His map, therefore, may soon be out of date. This is well illustrated in the United States by some of the so-called cover-type maps, which are maps of the actual vegetation. For instance, the map of Minnesota (Cunningham and Moser, 1940) shows a pin cherry type which is transitory and soon gives way to the next seral phase. Fenton (1947) made similar observations in Great Britain.

The vegetation mapper endeavors to map the plant communities of his region in the most meaningful way. He will, therefore, try to present not just the various species combinations, or physigonomic types, or ecological units. He will include, where possible, his observations on the stability of the vegetation, on its changes, rate and direction of change, and also on the meaning of the change. In so doing, he produces a dynamic vegetation map rather than a static one.

Causes and characteristics of dynamism

It is necessary to consider the character of vegetation dynamism before the techniques of mapping can be discussed meaningfully. What makes the vegetation dynamic? The causes of this dynamism must be understood if the interpretation of the vegetation is to be sound, and

only then can the dynamic vegetation map be expected to retain its usefulness over long periods.

The environment affects the vegetation and its geographical distribution. But what is significant here is not so much the relation between a given vegetation type and a set of environmental conditions; it is rather the stability of these conditions. Secular variations, such as periods of glaciation and long-term fluctuations of the climate, have long been known. Other than that, the environment was considered stable except where man rendered it unstable. Today we know that the environment is not stable, that it changes and fluctuates, that there are many causes underlying environmental changes, and that the rate, extent, and direction of these changes may vary from place to place. Perhaps it is difficult to accept the instability of our environment, but it can no longer be denied. At best, the rate of change is so slow at some times and places that for our immediate purposes we may speak of stability. But this should be done only with the clear understanding that we are dealing in fact with slow changes, even though these may be too slow to affect our needs.

Vegetational dynamism is therefore caused above all by the instability of the environment. But it is not enough to simply observe this instability, because different conditions will result in different types of vegetation. Therefore the mapper finds it useful to observe vegetational changes as well as their causes (Küchler, 1961).

Natural causes may be distinguished from anthropogenic causes, and among the former the regular or periodic ones may be separated from the irregular or aperiodic ones. The regular changes include the secular changes. For instance, the post-Pleistocene plant migrations are still in progress, at least in some parts of the world. Then there are cyclic changes covering a varying number of years for each cycle. A good example occurred on the High Plains in the heart of North America: the 20s of this century were characterized by ample rainfall; the 30s were dry; the 40s were wet; and the 50s again suffered from drought. Albertson (1957) and his collaborators in Kansas have described how the vegetation changed both physiognomically and floristically under the influence of these cycles. Cyclic fluctuations are known from most semi-arid regions of the world, for it is here that they are experienced most acutely.

The regular changes of the environment include also the seasons, resulting in the seasonal aspects of the vegetation (Küchler, 1954a). These are observed primarily in the herbaceous communities and range from the desert ephemerals, which may not even appear every year, to the grasslands and forest floor synusias. These seasonal aspects may be few or numerous. For instance in Gujarat, India, Saxton (1924) described eight such seasonal aspects of the vegetation which develop successively in the course of the year, and this series recurs each year. It ranges from marsh to desert and back again due to the heavy monsoon rains and the subsequent drying up of the soil. The seasonal series of plankton communities offer further examples of this type.

In contrast to the periodic changes in the environment stand the aperiodic ones. These include the formation of new land on growing deltas, by landslides, melting glaciers, volcanic eruptions, shifting sand dunes, the temporary and partial destruction of vegetation by grazing animals, insects or diseases, changes brought about by windstorms, floods, fire, etc. In the case of fire, the changes depend on the vegetation type that was burned, the intensity of the fire (heat, speed, wind, etc.), the effects of the fire on the soil, especially the humus layer, and subsequent erosion. Accordingly, the succeeding plant communities may be strongly affected by the varying degree of recovery of the survivors.

For instance, in sections of the Bar Harbor fire of 1947 on Mount Desert Island, Maine, the soil was affected only slightly; the fire burned the oaks but they survived and sprouted vigorously from their undamaged roots; an almost impenetrable thicket of oak shrubs replaced the forest of tall oaks. Elsewhere, the spruces on the mountain sides were burned and utterly destroyed; there were no survivors. The dry needles on the forest floor burned, too. What soil remained was soon eroded down to bedrock, and the charred stumps rose grotesquely on their mutilated roots high above the naked granite. Five years later a few grasses and forbs and some blueberry bushes were trying valiantly to recolonize the destroyed landscape, starting in cracks and shallow depressions in the granite where soil was beginning to collect once more (Küchler, 1956b). Fire alone can obviously not explain the difference between these two types of vegetation, both of which rose on a surface that seemed uniformly barren at the end of the holocaust.

There is little difference between a fire started by lightning and one started accidentally, i.e. involuntarily by man. If, however, man deliberately burns the vegetation repeatedly, and more or less regularly, perhaps even annually, then the effect on the vegetation is totally different. Fire thus becomes one of the many tools man uses in rendering the land more useful to him.

The vegetation mapper must also consider other features of land-use, such as selective cutting of forests, clear-cutting and reforestation (possibly with exotics), mowing of grasslands, and grazing of both grasslands and forests. The latter may eventually vanish from the landscape as a result of grazing because the animals eat the seedling trees and thus prevent any regrowth. Californian oak woods provide good examples for this. In contrast to such anthropogenous but semipermanent vegetation types is the deliberate removal of the entire vegetation and its replacement by cultivated crops and their associated weed communities. The great significance of all these substitute communities has been well established and the vegetation mapper can ill afford to ignore them.

In short, all vegetation changes continuously, although the rates of change vary from abrupt to extremely slow. The changes brought about naturally or by man usually imply changes in the quality of the vegetation and hence of its usefulness to man. Therefore they deserve close attention by the vegetation mapper.

The term 'dynamic' as applied to vegetation is here understood very broadly, implying simply that the vegetation is undergoing changes of any kind. On the map, it may be desirable to show a change that has taken place in the past; but it is particularly useful to indicate changes in progress at the time of mapping. The techniques are essentially the same in both cases.

Successive mapping

The most accurate and reliable manner of showing changes in the vegetation is to make detailed vegetation maps of permanently established sample plot and to repeat such mapping periodically. A comparison of the maps made at different times will then reveal the changes that have taken place. The time involved in mapping the vegetation repeatedly is usually rather short. For instance, Ellenberg allowed 7 years to elapse between his two maps of the grassland vegetation along the Seitenkanal (Ellenberg, 1952). Even though there are examples of longer periods, they are all short in the light of phytocenological evolution. One of the great problems is, of course, that vegetation mapping as a scientific field is so young. It would be most interesting to have some agency, such as a state or national academy, select one or more areas both disturbed and undisturbed, and map their vegetation every 20 years in great detail. The maps might well reveal many an unsuspected development.

Longer periods can be encompassed only through an historical approach [1]. Krause (1950) and Molinier (1952) give some interesting examples of this kind, but Schwickerath's treatment of the area around Stolberg remains the model achievement (Schwickerath, 1954). Mapping at the scale of 1:25,000 and using his own direct observations as well as historical records and documentary evidence, Schwickerath presents his landscape on three individual colored maps, each with a transparent, black-and-white overlay. These three maps of the same area reveal the evolution of the landscape and its vegetation with great clarity, and if, as the saying goes, one picture is worth a thousand words, then this is even more true of Schwickerath's maps with their overlays.

The first map shows the original vegetation during the 'beech forest period', 2000 – 1 B.C., and the overlay reveals the indications of settlements during prehistoric and Roman times, as well as the changes in the landscape from early Franconian times until the year 1800. The second map shows the vegetation as it prevailed in 1800 while the overlay presents the changes that took place between 1800 and 1940. Finally, the third map shows the vegetation as Schwickerath observed and mapped it in 1940. The accompagnying overlay indicated the smallest units of the physical landscape: the biotopes. This permits a close correlation between the individual phytocenoses and their physical environment, revealing a distributional pattern that man was able to modify in many ways but never to obliterate. One is reminded of Tüxen's examples of the potential natural vegetation types and their various substitute communities: the spatial pattern of the potential natural phytocenoses, neatly fitted into the mosaic of biotopes, remains the dominant feature of the landscape even through centuries of human occupation.

If on a vegetation map or its legend there is nowhere any indication of vegetational dynamism, then it may be assumed that the vegetation is relatively stable or else that its dynamic features are not important for the purpose of the map. But there are many vegetation maps where changes are shown clearly although the legends contain no particular reference to dynamism. One of the best examples is, of course, the wonderful vegetation map of Perpignan by Henri Gaussen (1948). For on this map it is possible to observe the potential natural vegeta-

[1] Leemans and Verspaendonk, 1980 used old arial photographs to map former vegetation and so showed vegetation succession. See also fig. 2 chapt. 11A.

325

tion as well as the various changes that have taken place, leading to a great variety of relatively stable substitute communities. Ozenda (1963) believes this map series is of such great interest to so many users primarily because of its dynamic aspects. It is important that the reader be fully aware of all the information and its implications given on a vegetation map, and it is a part of the mapper's task to present his material in such a manner as to help his reader find all that has been shown.

Scale

Not all vegetation maps are equally well adapted to include dynamic features of the vegetation. One major limitation is the scale. In general, maps at larger scales permit more information on vegetational dynamism than smaller-scale maps. For instance, a vegetation map of a continent as commonly found in atlases cannot be expected to include the rich information on the detailed maps by Kuhnholtz-Lordat (1949). Yet, an American 19th century atlas contains a simple vegetation map of North America which shows a large part of the interior grasslands as such but with the note that they would turn into forest if protected from fire (Sargent, 1880). This map is therefore remarkable for two reasons: its author introduced ideas of stability and dynamism on a small-scale map, and that at a time when our ideas on vegetational dynamism were still being formulated.

The scale alone is inadequate as a criterion for the number of dynamic features to be shown because the method of mapping, the classification of vegetation, and the organization of the map content also affect what information can be shown on a map. Nevertheless, where the dynamic features of the vegetation are important, the mapper should select the largest scale that time and funds permit. The more detailed and penetrating his information as revealed in the legend, the greater is the ultimate usefulness of his vegetation map.

Legend

The legend is the key to the map; it offers several possibilities to change a static vegetation map into a dynamic one. A vegetation map can present the plant communities more meaningfully by relating them to the vegetation changes. This requires information on the environmental processes active at the time of mapping, and their effect on the vegetation. Usually, there is not a simple reaction but a set of complex chain reactions. This necessitates the determination of the most effective factors producing the most significant changes in the vegetation, and a careful organization of all features. Where succession is in progress, a sequence of plant communities can usually be established. At one end of the sequence will be the type of vegetation that prevailed when the change began. At the other end will be the vegetation type that will have taken over when the change has run its course. Between these terminal points of the chain there are one or more links. When their character and sequence have been established, the vegetation map can portray the vegetation in a dynamic fashion by presenting the vegetation types not only as they are at the time of mapping but also with regard to their place in the succession. Of course, it should not be expected that the entire series of plant communities from pioneer stage to climax is present. But the recognition of a transitory type and its place in the sequence is significant.

On large-scale maps, it may be practicable to introduce an entire chain of successional phases as part of the legend. This can be done in the form of a series of rectangles or boxes, with the initial community shown in the first box and the stabilized one at the end. Each box is followed by a description of the plant community it represents. Only part of the series may occur in the mapped area and only those rectangles are shown in color which indicate plant communities on the map. In this manner, it is possible to see at a glance what series is involved, which phases of the series are present, their proximity to the final or climax stage, and their distribution in the landscape (Fig. 1).

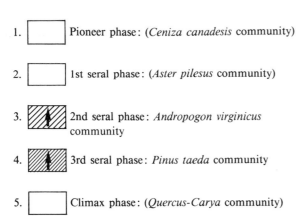

1. ☐ Pioneer phase: (*Ceniza canadesis* community)

2. ☐ 1st seral phase: (*Aster pilesus* community)

3. ▨ 2nd seral phase: *Andropogon virginicus* community

4. ▨ 3rd seral phase: *Pinus taeda* community

5. ☐ Climax phase: (*Quercus-Carya* community)

Fig. 1. Series 1: Oak-Hickory series.

In Fig. 1, the series is called 'Series 1', but there is no need to number it unless two or more series appear on the same map. Instead of numbering the series, it is also quite acceptable to name it. In such a case, the series is usually given the name of the dominant species of the climax community. In this instance, series 1 would be called '*Oak-hickory* series'. Gaussen uses this method on his map of France at 1:200,000. In Fig. 1, three phases of the series do not appear on the map and hence are left blank. The color of the two present communities is the same, but the individual hues range from light to dark, the final and most stable phase in the succession being the darkest. The communities differ from one another but are shown as different phases of the same series. (The arrows are added only when the relative quality of the phases is to be indicated; see next section, below.)

It may be that the potential natural vegetation or the climax does not occur anywhere in the mapped area. The mapper will nevertheless find it useful to show it in the legend in order to enlighten the reader on the affinities of the substitute communities. Usually, a legend item is suppressed when it is not represented on the map. In this particular instance, it is valuable to retain it, but it should be followed by a parenthetic note that it is not shown on the map unless the absence of color serves the same purpose. For example, such a legend item might read: '*Andropogon gerardi-A. scoparius* community (not on map)'. Thus the legend space devoted to seral plant communities which are not on the map need by no means be wasted.

Qualitative changes

Vegetation maps are very useful when they show improvement or deterioration in the quality of the vegetation resulting from man's use – or abuse – of the land and its cover. Not all forms of land use produce desirable results. In the United States, there are millions of acres of annual cheat grass (*Bromus secalinus*) where once perennial grasses of far greater value prevailed, and much ponderosa pine (*Pinus ponderosa*) has been replaced by worthless brush. But not all changes are bad: the valuable loblolly pine (*Pinus taeda*) in the southeastern part of this country is replacing the less desirable longleaf pine (*P. palustris*) in many areas.

In order to show such qualitative changes, Aichinger (1954) has proposed to list the entire successional sequence and then connect the individual stages by arrows. Ascending arrows (↑) indicate changes toward a richer, more demanding, and more valuable community (progressive vegetational evolution); descending arrows (↓) indicate a change toward a poorer, less demanding, and less valuable stage (regressive vegetational evolution); finally, cut descending arrows (↯) indicate such regression after clearcutting or burning a plant community.

For example, a spruce forest, rich in *Oxalis,* is located in the dolomitic areas of the conifer belt in the Alps on a shady slope where much snow occurs. When this forest is clear-cut, the top soil erodes. Snow slides, avalanches, and sheet erosion during heavy rains alter the environment to such an extent that the spruce can no longer regenerate in it. *Calamagrostis varia* spreads and enables *Pinus mugo* to invade because this pine is better able to endure the snow pressure and avalanches on the steep slopes deprived of part of their topsoil. Only slowly, under the protection of the Pinetum mugi, does a larch forest get established which, in turn, permits the return of the spruce (Aichinger, 1954, p. 23). The sequence is shown thus: Piceetum oxalidosum ↓ Calamagrostidetum varia ↑ Pinetum mugi ↑ LARICETUM DECIDUAE ↑ Piceetum. The capitalized stage is the one present at the time of mapping. If, for example, a map of this vegetation showed a Laricetum deciduae, a Piceetum, a Pinetum mugi, and a Calamagrostidetum side by side, with no further information, it would be impossible to observe the relations between them. But if the legend is organized so as to show the dynamism of the vegetation, the whole story is told at once and the reader can see at a glance that all these plant communities are parts of the same sere, most of them being unstable. It is quite possible that different stages of the same sere occur side by side because different sections of the spruce forest may have been logged at different times.

Some problems in mapping successions

Succession may, at times, follow a linear progression, and at other times the evolution toward the climax is very complex. Schwickerath (1954) developed 'association rings' to illustrate the successional complexities of the landscape. Sometimes, it may be possibile to reduce very simple 'association rings' to linear forms, just as a smaller map scale calls for a generalization of the portrayed plant communities. But where this is not feasible without altering or destroying the basic meaning of the 'association ring', the vegetation mapper must limit himself to showing the existing communities as actual or potential natural phytocenoses and their seral phases of undetermined rank. Various 'earlier' phases may possibly be distinguished from 'later' phases but this requires detailed information which, in may areas, is just being gathered or is not available at all.

Fosberg (1961c, p. 200) points out that mapping dynamism on a single map rather than on a series of successive maps, introduces into the map an element of interpretation. In addition to recording what the mapper observes, he records what he thinks is going to happen or has happened. This is very well, and may be desirable, but it should be made very clear in the explanation what part was observed and what is interpretation. Aichinger solved this problem by capitalizing the actual vegetation.

Much has been learned about the dynamics of vegetation since the pioneering work of the early ecologists. Many North American birch and aspen forests are now recognized as transitional phases, usually following fires. The vegetation of the western ranges can be described in degrees of overgrazing, with the floristic features of the various phases well established. Some aspects of succession are particularly interesting from the point of view of the vegetation mapper. Friedel (1956) observed in the Tirol, Austria, that succession need not proceed evenly through the various stages and their transitions. Under given circumstances, such successional phases can have a remarkable degree of stability. But when, in the course of their evolution, a certain threshold value in the environment is passed, one stage changes rapidly into the next one. He also demonstrated with the help of phytocenoses how the landscape evolved where the Pasterze glacier had retreated and thus exposed sizeable areas: the areas released between 1856, 1890, 1910, and 1933 can be recognized by the evolutionary stages of their vegetation (Friedel, 1934). On the other hand, Tansley and Chipp (1926) found that the various successional stages may not be so different from one another as they appear to be, and that the seedlings of the latest stages may already be present in the early stages, even though they are well hidden among the early dominants which grow faster and taller. Much later, this observation was confirmed by Daubenmire (1943) in the Rocky Mountains, and by Egler (1954) on abandoned fields in New England.

The dynamics of vegetation can also be mapped where one type invades another, e.g. a forest invading a grassland. Such invasions can be indicated with the help of arrows in the color of the invading community. Instead of colored arrows, black ones will do sometimes, but they become problematical on black-and-white maps. The length and width of the arrows can indicate the degree of invasion: broad arrows imply a massive invasion, thin arrows a modest one; the length of the arrows shows the depth to which the invasion has progressed.

Sometimes the problem consists in showing two or more stages of a fluctuation, as for instance the spring and summer aspects of a grassland, or perhaps its major phases during wet and dry cycles. The problem is solved by manipulating the legend items. The two stages are given together, one following the other, possibly the second one in parentheses. But each item must be clearly shown for what it is, as 'spring phase' or 'drought phase', etc. In this manner, the information of the map can be greatly enriched, increasing its value considerably.

Use of color

Where the dynamism of the vegetation is an important part of the information to be mapped, it can be shown with the help of colors. Lüdi (1921) made a 'succession map' of the Lauterbrunnental in Switzerland. He used one color for all pioneer communities, one other color for all transitional communities, and finally, one color each for the different climax communities. In addition, he has one more color for messicol vegetation, i.e. cultivated fields. This was probably the first cartographic attempt to portray dynamics as a major feature of a vegetation map. Much experience has been gained since Lüdi's pioneering work and it is now possible to show vegetational dynamism with the help of various color schemes.

When a given type of the potential natural vegetation has been established, it is presented in a flat color, e.g. sky-blue. All substitute communities of this type are shown in various patterns of the same sky-blue color, i.e. blue lines, blue dots, blue dashes, etc., thereby indicating which communities belong together without indicating any successional sequence. This is desirable when many substitute communities do not follow one another in a sere, especially when planted by man. Gaussen uses this idea in his system of color manipulation.

Some authors show the potential natural vegetation in flat colors throughout the area. Where substitute communities take its place, an overprinted pattern of a different color is used. The extent of the various potential natural phytocenoses is thereby shown, as well as the degree to which they have been replaced and by what. As the number and variety of overprinted symbols may become very large, this latter method is more useful on maps of a very large scale. Some authors use this method on small-scale maps, too; the overprinted symbols must necessarily be reduced in number in proportion to the area covered, and they must be shown in vivid colors to assure an adequate contrast with the more subdued colors of the potential natural communities. For instance, the overprinted symbols may be shown in bright red while the potential natural vegetation forms a background of soft variations of tan, green, and blue. Often the symbols are limited to the messicol vegetation and are used only where the crops assume a major economic significance in the landscape.

Where the vegetation is going through several phases of successional series, a different color may be assigned to each series. The climax is given a flat color and each stage receives a carefully selected hue of the same color. Running through the series from pioneer stage to climax, the hues should be arranged consecutively from light to dark, so that the lightest hue is closest to or at the pioneer stage and the darkest hue is at the final phase or nearest to it.

It does not seem difficult to relate the substitute communities to the potential natural vegetation with the help of colors, but this possibility seems to have escaped many vegetations mappers. For instance, Maack (1950) on his vegetation map of Paraná set aside a whole section of his legend in which he relates the substitute communities to the potential natural vegetation by appropriate statements, but each item is given a different and unrelated color.

Care must be taken to retain a maximum number of chances for using the colors. For instance, a dark color can be made lighter by using lighter tints or else by breaking a flat color into lines and dots of varying thickness and diameter. The effect may be nearly the same. But these are two different methods of using color and they should not be confused. Only one is necessary to show succession, and the other one may then be employed to indicate other features of the vegetation. Where both methods are to be used on the same map, care must be taken to keep the line and dot patterns sufficiently coarse so that they can readily be distinguished from flat colors of light hue.

As time progresses and more research is done in the field of vegetation mapping, new ideas will no doubt be presented to map the dynamism of vegetation. But some methods have already developed and these should be employed more frequently. In a sense, this will combine experimentation with the efforts to make vegetation maps more useful, as different authors use different approaches. It is precisely this sort of applied experimentation that promises the greatest and the most rapid advances of an aspect of vegetation mapping which so far has received little emphasis. Quite independent of such considerations is the fact that every carefully prepared vegetation map, showing the character and distribution of the phytocenoses in detail, will serve as a basis for later observations, whenever the evolution of vegetation is the object of study.

24. Monitoring Vegetation and Surveying Dynamics

I.S. ZONNEVELD

Introduction

The dynamic aspect of vegetation can be considered as a pattern in the fourth dimension. Study of it is a must for well understanding the green mantle. In the applied side of vegetation science it is also very important to know the dynamics, especially for the future. Man's interest for this goes back to the dawn of history. The herdsman since historical time has to carry his sheep to the 'pastures green' (Psalm 23). For this he should not only know where these pastures are, but also at what time they are expected to be green so that he may arrive there in time. Prediction is exlusively done by extrapolating knowledge of past and present into the future (see Table 1).

Modern application of this old herdsman's wisdom is done not only in the field of grazing but in any aspect of land from polution to commercial interest in expected yields and then is called 'monitoring'.

This term is derived from 'monitor' = an older pupil that should watch the younger one about his study and warn if something goes wrong. Later (since World War II) this word is generally used for military warning systems and also for medical systems, even for an aparatur (a television screen). The original word in Latin: 'monere' = to warn. But the aspect of watching (necessary base for warning and admonishing) became dominant in literature and is the part that has given rise to development of modern techniques and instruments.

Table 1. The process of monitoring and its aim – original meaning (monere) – monitoring sensu lato (watching and warning) – monitoring sensu stricto (watching).

To watch	To evaluate	To warn	To delliberate, to plan	Execute action
I. The herdsman sees a sheep deviating from the herd	He judges that the animal goes too far	He trows a stone by his sling or shepherd stick or sends the dog after the sheep	The sheep feels the stone and reaction starts	The sheep returns to the herd
II. The ITC trained official observes change in land-use or farm pattern or crop growth or range trend	He (she) evaluates this in terms of possible shortage of food or fodder in near future in certain regions	A warning is given to the local administrator and/or farmers or range manager	Discussion starts resulting in planning and or legislation or proposed measures	Plan, or measures are executed
	Feedback to watching the results of action			

A.W. Küchler & I.S. Zonneveld (eds.), Vegetation mapping. ISBN 90-6193-191-6.

Ways of monitoring are executed at different scales from different platforms and by different means of observation and as it is as old as life, because animals and man depend in their struggle for life on watching and warning (each other and themselves), they vary from classical to sophisticated technical methods.

Scale in space and time

The scale and platform vary from the herdsman's eye observation, standing in the middle of his sheep, to the weather satellite using low resolution MSS imagery. The warning can be immediately (and directly followed by throwing a stone via the herdsman's tools or sending the dog after a sheep going astray) to deliberately produced periodic reports after long study of the observed process (via sequential patterns study) with all kinds of sophisticated means, and taking into account political deliberation (e.g. the monitoring of degradation resulting from (over)population and warning about it with all its religious and social and political implications).

The monitoring of vegetation as attribute of land with modern means can be focused on:
— short-term (seasonal) aspects in order to guide short-term measures (agronomic, range management);
— long-term aspects; development of agriculture; general conservation aspects; range administration focused on rangeland improvement and conservation, legistation;
— short-term as well as long-term aspects are involved in monitoring for prevention of catastrophic results of climatic variations causing temporal famine among people and animals. (As any positive tool the latter mentioned application can be dangerous if only such catastrophies are temporally prevented and no structural improvement is done in balance between number of people and animals with the resources. In that case a good warning system will speed up the devastation by excluding a natural sound negative feed back factor);

— the basic unit of the sequential observation (watching) is the cell. This can be a *grid* cell unit of a computerized geo-information system. It could, however, also be a land-use parcel or a (multidisciplinary) land unit or a administrative unit. This all depends on aim and scale of the monitoring system. The way the data will be interpreted is important too. The pattern in the field (on the map), such as extentions and reduction of vegetation cover, intensity per grid cell, giving indirect information on production, is the main matter of interest.

The means

Monitoring can be done administratively by requesting e.g. farmers to register the land-use yearly and send these data to a central place. It can be done by special observers travelling on foot or any other means through the field, using their eyes as main observation means and existing topo-maps and files as recording means. They can do sequential observation on fixed sample areas and so produce tables and graphs depicting change.

The observation place may be a high place (hill or tower) or a moving aeroplain from where with naked eye observations are done (systematic reconnaissance flight, see Chapter 16A).

The recording means might also be any remote sensing (aerial photograph, satellite imagery, radar, thermal, sonar, etc.) that can answer the questions: what, where and when. So far monitoring is being done in rather detail by ground observation, by making photos of permanent sample plots to watch range condition.

Systematic reconnaissance flights is a common means for qualitative as well as quantitative recording (see Andere, 1981; Croze and Gwynne, 1978; Croze et al., 1978; Gwynne and Croze, 1975a, 1975b; Gems, 1979; Huizing and Zonneveld, 1980; Chapter 16A this book).

Experiments with plant cover estimation by

infrared/red ratio of radiation, directly measured from (small) aircrafts are being done (see Peden, 1985, and references; see also Vane et al.).

Satellite imagery is in spite of low resolution promising, because of relatively easy repetition possibilities. An extreme example is the use of NOAH (weather satellite) images which have a coarse spatial resolution of about 1 km^2, but are daily available, giving possibilities for interesting process analyses (Hielkema, Tucker). Irrespective which recording means or image is used, it should be interpreted or at least 'processed' and brought in a final form. This may be a qualitative image or set of images, or these images complemented with qualitative data, or tables, or graphs, or a combination of all these.

In certain cases it is possible to read dynamics from a single image. So a satellite image of a part of Mali (see Zonneveld 1979 and 1980) shows clearly the progressive penetration of savanne woodland devastation from above left → below right. The line where the devastation stops is about the Tsetse boundary. This single image clearly shows the dynamics of the process that is watched and it invites the warning what to do about it and what not (Tsetse control??). Also land-use maps often show trends even if still only one stage is present. It is clear that a second image or map produced somewhat later, may increase the visibility of the process and so increase the possibility of prediction of what may happen and what to do for steering the process.

If only three stages (images, maps) are available it is possible to make one image of it, provided they have each only *one* type of information in a (semi-) quantitative gradient, e.g. a vegetation cover parameter. In that case we can use the well-known colour print technique in such a way that e.g. the first stage is produced in red, the second in green and the last one in blue.

If the value is high (does not change) in all stages, the area then will be white; if there is a change so that the highest cover is in the last stage the colour will tend to blue. The opposite case will produce a colour more reddish etc. For executing this in diazo by colour formation by substraction, the colours should be taken as complementary colours, see further Chapter 21. If more than three stages have to be depicted a series of maps should be made. If there are many, a kind of motion picture can be the result, or at least a narrative strip can be produced (see Zonneveld, 1975).

Quantitative comparison requires exact 'registration' of the images, that means that they have to fit exactly on one another.

The two main errors then are:

1. orientation errors (in the field, or due to distortion of image, or basemap, etc.);
2. classification/interpretation errors.

Correlations can be calculated using a dot grid.

The noise according to the two mentioned errors can be estimated quantitatively by careful qualitative study of the images. It can also be estimated by calculating the correlation of the two stages. Van Heusden (1983) calculated a 'Correlation value for reciprocal sequence'. He calculates, using a dot grid, the coincidence between a unit in the old stage with the same unit in the new stage as a percentage of the unit in the old stage. Then he repeated such a calculation, but then as a percentage of the unit in the new stage. He added these two percentages and considered this sum as a measure for correlation. The maximum is 200, the minimum is 0. Values above 100 he considered as clear correlation and could so prove a clear trend in development of a Calluna heath vegetation into a grass vegetation.

The more recent development is the electronic automated cartography and geographical information systems (geo-information system or land information system). These systems combine the capacity to store data of land according to the quantitative character (what) in their geographical coordinates (where) at different intervals of time (when).

The system provides retrieval about these 'what', 'where' and 'when' functions in the form of automated mapping and in the form of graphs, tables, etc., in various combinations. For details see Chapter 14.

This monitoring, using pictures, should always be accompanied (as any remote sensing interpretation) by collecting ground observation of the kind mentioned in the beginning. So vegetation monitoring should preferably be accompanied by observation on so-called 'permanent quadrates'.

The change can then be expressed in table form or graphs or as mentioned before by detailed photographs taken on short distance, vertically or oblique. Also destructive or non-destructive sampling of cover or phyto-mass can add here to the completeness of the information (see Chapter 19).

Starting and executing a monitoring schema

Large monitoring schemes on vegetation (and other aspects) exist in the form of the KREMU scheme in Kenya and the UNEP sponsered scheme in Senegal (see Andere, 1971; Van Praet).

For details is referred to such schemes (see GEMS/PAC, 1979; Gwynne and Croze, 1975a, 1975b). Base line observations on the ground are an important aspect. They should be well designed and monitored in the desired detail, not too little, but certainly not too much, an evil that sometimes occurs.

A start should always involve a 'backwards' monitoring too, studying old maps and especially (if available) old photos.

Good examples of retrogressive monitoring using old photographs interpreted with recent ones as a 'key', are given by Leemans and Verspaendonk (1975); (Saeftinghe) and W. van Heusden (1983) for heath land vegetation in the Netherlands. This means that one need not wait until after several years a certain trend shows up; one can directly start to learn from the dynamic features of the past ('flying start'). Monitoring methodology and administration depends largely on aims and context.

The following deliberations are important in planning such schemes:

— what are the users' needs about vegetation and other related land items to be monitored;
— which land (and water and vegetation) properties are suitable (and indicative) to be monitored;
— what will be the method of observation and recording: personal interpretation of images, or (semi-)automated image processing of parameters suitable for that;
— over how long a period needs watching (= sequential observation) to be extended in order to be able to assess a trend which is sufficiently reliable to base warning on;
— which watching means are most suitable for the special case (satellite, small aircraft, photos, MSS, etc.);
— what will be the frequency (temporal resolution in days, weeks, months, years);
— what moment of the day or the year (season) is the most feasible and effective one for the aim;
— what kind of groundtruth is needed and when is it most feasible to collect it and how;
— what land data will be used as a starting point and a basis for indicating the change (a land unit map (?), a land-use map, a vegetation map, etc.);
— if data are obtained how can we process and express them in such a way that we can study the process of change optimally (simple visual comparison—→ automatic computerized geo-information system);
— how do we express the data to be able to execute the warning as effectively as possible; definition of users' needs is here also important;
— specific studies on interpretation of the results in terms of converting temporal data on pattern into process;
— the methodology of land evaluation on temporal data;
— last but not least, how does one organize monitoring systems in the local administration context.

25. Mapping Land-use

A. KANNEGIETER

Introduction

The relation between land-use and vegetation survey

By land-use we mean in this context the 'use' which is made of the surface of the earth. This can be expressed in 'formal' terms: the cover types or objects occupying the earth's surface such as buildings, crops, grassland, forest, bare soil, waterbodies. It can also be expressed in 'functional' terms: the use to which the different parts of the earth's surface are being put: residential/industrial area, dairy farming, timber production, conservation forest, recreation area, etc.

It is the expression in formal terms that justifies directly the incorporation of land-use survey in a handbook on vegetation mapping. By far the largest surface of formal land-use types are 'green' types hence parts of the vegetation cover: crops, grassland, forests. A vegetation survey in inhabited areas cannot neglect these and their classification. But also consideration about the functional aspects is of high value for the vegetation mapper, who is concerned about the explanation of the vegetation cover pattern and about application of the survey results.

Objectives of land-use surveys

A land-use survey often forms an integral part of a survey project to provide the information base for development planning.

Any planning for development which implies change in land-use should be based on a thorough knowledge and understanding of the *present* situation of that land-use and of the reasons and causes underlying this. The reasons/causes that land-use is less than optimal may be of a physical nature (terrain condition; soil type, depth, fertility status; drainage problems, etc.), but also social-economic character (land tenure situation, deficient infrastructure, incidence of diseases such as malaria, riverblindness, 'tsetse', lack of technical knowledge or capital, etc.). Land-use information has therefore to be complemented by information provided by other surveys, on landform, soil, hydrology, social-economic conditions. The comparison between the results of a land suitability study which shows potential land use alternatives, and present land use, will reveal the scope for development and which bottlenecks in the physical or socio-economic (political!) sphere will have to be overcome.

Objectives of landuse surveys may further be:
— the determination of the areas under particular crops;
— the determination of crop condition and monitoring crop development; crop acreage in conjunction with crop development and meteorologic information may enable us to make crop production forecasts;
— the monitoring of developments in land-use as they result from certain management/conservation policies or the lack

A.W. Küchler & I.S. Zonneveld (eds.), Vegetation mapping. ISBN 90-6193-191-6.

thereof, such as urbanization, deforestation (e.g. by shifting cultivation), erosion, salination, desertification; but also land reform, reforestation and impact of introducing new crop types/varieties, management practices;

— the determination of the need for detection of effectiveness of irrigation water supply, losses by leakage or illegal tapping;

— monitoring for early detection and watching the development of crop diseases, pest-attacks, damage by other agents; the inventory of the ultimate damage caused, also by floods, fires, etc.; a special and very important application is the monitoring for the localization of locust breeding places in desert areas;

— the location of illegal dumping sites of industrial waste material, threatening the environment.

This list of applications is long but by no means complete.

The aims and objectives of a land-use survey determine the extent and the degree of detail of information required with regard to the following aspects of the land-use in the study area:

1. What? (identification, classification);
2. Where? (location, delineation);
3. When? (processes such as crop or crop/fallow rotations, land-use changes);
4. How? (the technology employed in agricultural production);
5. How much? (the area occupied by various land use types, yields); and also:
6. Why? (the reasons, causes of the phenomena observed).

The information supplied by the surveyor may be of little value to the user if the information demand as expressed by the user does not reflect the specific information needs.

Close contact between surveyor and user before, and during the survey will obviate conflicts between information supply, demands and needs and will promote the actual and effective use of the information produced.

Presentation

The form in which the information is presented to the user should be geared to the user's capability and experience in handling it. Land-use information can be presented in the form of

— statistics, stating how many ha each of the various categories occupies in a certain (administrative, or naturally determined area);

— in the form of a map, constituting a permanent record of land-uses in their geographic location and extent. Such a map can be used for measuring the areas under each land use type. The land-use situation can be studied spatially in conjunction with other maps: on geology, geomorphology, soils, hydrology, population, etc. to study relationships between land-use and these aspects. It can also by compared with land-use maps of earlier or later land-use surveys for the area to study what changes may have taken place as a result of a certain policy (or the lack thereof);

— more recently, computer-aided systems of geo-data base compilation have been developed. In this approach, spatial information on land-use is generated, coded and stored on a geographic gridcell basis, together with a wealth of other information on land. Special computer programmes make it possible to correlate the different types of information for a better understanding of land-use in its spatial relationships with the physical and socio-economical environment and even make predictions about the effect of management/policy measures. In Chapter 14 this type of geo-information systems is treated more in detail.

Special dynamic character of land-use

Land-use is not a static condition but rather a dynamic process! That is what significantly distinguishes a land-use survey similar to vegetation mapping from a survey of e.g. geology or soils. Land-use information, particularly at the more detailed level, rapidly outdates. Land-use/cover maps should therefore show the date of the survey on which they are based.

Approaches to a land-use survey

A land-use survey may take the form of
— a pure field survey, if possible with the aid of a recent topographic map;
— a field survey with the aid of aerial photography as a base for orientation and recording. Advantage of the airphoto over the topographic map is that in the former, objects and features such as field boundaries are seen in their natural aspect which facilitates orientation and accurate delineation of land-use units. Enlarged prints and diazo copies may be used to advantage in this work;
— a survey based on airphoto or other remote sensing image interpretation, supported by a minimum of (costly in time and manpower) fieldwork. It is particularly the possibility of studying (in stereo-vision) the land-use in relation to its physical environment, which considerably adds to the information on and understanding of land-use which can be derived from the airphotos;
— for any approach to land-use survey, one needs a classification system. The following paragraphs deal with this subject.

The classification of land-use

Introduction

Land-use can be expressed in terms of 'formal' use or 'cover' on the land surface, e.g. buildings, crops, grassland, forest, bare rock. It can, however, also be expressed in terms of the functional use of the areas concerned: e.g. residential area, foodcrop rotations, dairy production area, conservation forest, recreation area.

The latter classification is a more subjective approach to classification than the former which is based on direct observation of facts.

Most of the land-use surveys are based on aerial photography or other remote sensing data/imagery, which usually do not give much direct information on functional aspects of land-use such as crops, soil or water management techniques and levels, yields, land tenure and property conditions, etc. which constitute important information to evaluate resource use and formulate development policy. For these objectives, much additional information from other sources and fieldobservation, is required, which is difficult to present in map-form. Most land-use surveys therefore result in maps showing the various *covertypes* in their areal extent and geographic distribution, to be compared with mapped information on physical and socio-economic aspects of the land and land capability.

It also should allow the comparison with maps of land-use in other areas, and of other points in time, for the monitoring of land-use change. Memoirs or 'reports' should accompany the land-use map to give an accurate description of the categories appearing in the legend and give additional information relevant to the aims and objectives of the survey.

The world land-use classification

At its international congress at Lisbon in 1949, the International Geographical Union decided to install a Commission to study the possibilities for a *World Land-use (WLU) Survey*. Factual knowledge on the present position of land-use (l.u.) and the understanding of the causes and reasons behind this, was badly needed for many parts of the world, particularly the newly independent nations, as a basis for planning improvement and development of their land resource use.

The increasing use of aerial photography provided the possibility of studying in stereo-view the land-use in its relation to its physical environment. The emphasis was on maps in which *actual land-use* was shown in its areal extent and distribution. Those maps were to be accompanied by explanatory memoirs. The basic survey had to be accurate and record facts, not merely opinions.

The WLU Commission was asked to stimulate and supervise the WLU mapping project to safeguard consistency and quality in the mapped information, it would advise the national committees and other bodies involved in the publishing of the master map sheets at the scale of 1 : 1,000,000.

To secure uniformity in land-use classification for this WLU map, a *master key* was devised. This could be expanded to suit regional and local conditions and other map scales but the legends had to remain correlated with the WLU master key. This master key contained *nine major categories* of land-use with a number of subcategories for each. Also, the colours to be used to graphically display were prescribed symbols. The categories are:

1. *settlement and associated non-agricultural land-use* (dark and light red);
2. *horticulture* (deep purple): all intensive cultivation of vegetables and small fruits (not fruit trees!). Where vegetables were grown in rotation with farm crops the areas should be classified under 4 'cropland';
3. *tree and other perennial crops* (light purple). The crops concerned must be named or indicated by means of symbols; groves of cork oaks, resin, turpentine or gum-producing trees are included. Further such 'permanent' crops/cultivations grown without rotation such as sisal, manila hemp. NB.: Sugarcane and alfalfa, though grown year after year on the same area, are not regarded as 'perennial' and are put under 4 'cropland';
4. *cropland* (brown colours). Subdivided into
   a) continual (rice, sugarcane, wheat and maize in monoculture) and rotation cropping (crops grown in a fixed or variable rotation, incl. fodder grass, clover, alfalfa, which may occupy land for 2–3 years); included are also 'current', i.e. < 3 year fallows;
   b) land-rotation: areas cultivated from fixed farms/settlements as the dominant occupation, but where cultivation is carried on for a few years followed by a per-

haps considerable resting period under scrub or grass with little to no economic importance. NB.: This in contrast to 'forest with subsidiary cultivation' (7f);

5. *improved permanent pasture* (managed and enclosed; light green);
6. *unimproved grazing land* (orange for 'used', yellow for 'non-used'). This category concerns the extensive native pasture or rangelands; this may have been enclosed but then dominantly in large units, it may be periodically burnt over as the sole management measure. The type of vegetation concerned should be described (in maps or notes), e.g. savanna, tropical grassland, steppe, dry pampas, heather moorlands, etc.;
7. *woodlands* (in shades of green). To be distinguished on the basis of the morphological character of the forest:
   a) dense; b) open; c) scrub; d) swamp forest; e) cut over or burnt over, not yet fully recovered;
   f) forest under subsidiary cultivation:
      i) 'shifting cultivation': patches re-cleared from time to time by wandering tribes;
      ii) forest crop economy: woodland holdings where some cultivation is carried out on a subsidiary basis to the working and management by replanting of the forest land. The types of forest can usually be distinguished by symbols as: evergreen broad-leaved, semi-deciduous, deciduous, coniferous, mixed, etc.;
8. *swamp and marshes* (in blue colours): fresh and salt water, non-forested;
9. *'unproductive' land*: appearing bare: bare mountains, rocky and sandy deserts, moving sanddunes, salt-flats, icefields.

Where the land falls into two categories (e.g. olive groves with cultivation of wheat underneath) it is to be indicated by combination of colours in alternative bands). Clearly, the WLU classification is based partly on cover and partly on functional land-use.

Many national/nation-wide land-use surveys

have been carried out using this classification framework as a basis with smaller or greater modifications to suit the local needs/conditions. A good example is the survey of land-use as part of an integrated resources survey, of the entire island of Ceylon (now Sri Lanka), based on aerial photography at the scale of 1:40,000 (map scale 1":1 mile = 1:63,360), started in 1956. The WLU framework was adhered to with only minor modifications.

An outstanding example is the Japanese comprehensive land-use survey with its legend used in the 1:50,000 mapping, started in 1951 and providing the most comprehensive record of land-use in the world to serve as a basis for development planning in the very intensive and small-scale agricultural land-use. In this map 32 types of land-use are distinguished by colours and symbols in provincial maps of great refinement. In paddy fields, for instance, four types are distinguished: (1) triple cropped; (2) double cropped; (3) single cropped; (4) perennial crops. In upland fields there are five subdivisions according to type and intensity of cropping, etc.

Modified world land-use classification in Sudan

The comprehensive *land-use map of the Sudan* by Lebon illustrates well the need for modification and the intricate combination of symbols necessary to suit the specific regional character of land use in those parts of the world. Lebon describes the problems of land classification and mapping involved. The survey was begun in 1956 and resulted in a set of maps at the scale of 1:1,000,000 and a reduction to a single sheet map of Sudan at 1:4,000,000 (in 1963), accompanied by a memoir.

The modifications/adaptations of the WLU classification and its expansion were justified as follows:

— In the first place, the chief modes of l.u. in Sudan are primitive, diverse and extensive: mainly 'land rotation' (4b) and 'used unimproved grazing' (6a). Neither mode of land-use metamorphoses or obliterates the original natural vegetation, only modifies it more or less, via burning or overgrazing. On the fallow lands, natural vegetation is apparent in its secondary stage of degeneration. This leads to class 7 (forest) being appended after most subtypes.

— Secondly the adapted classification conforms to actual disparities between economies:
 —in the Central Sand Zone (Qoz), open woodland savanna or low savanna, characterized by compact villages and widespread cultivation in rectangular plots;
 —in the Southern Clay Plains a greater variation of huts, cattle shelters and much smaller cultivation plots on highlands, intermittent land-use for grazing and often more or less intensive swamp grazing with cattle shelters and huts;
 A requirement was further that all types of the classification had to be readily and consistently identified on the particular types of aerial photography available.

These conditions necessitated the subdivision of WLU types, but also the creation of *combined categories,* particularly 4b–6a, i.e. 'land rotation' with 'used unimproved grazing', land rotation being the commonest mode of land-use among sedentary and semi-sedentary tribes and the grazing being of the nomadic herding type so widely prevalent in Sudan.

The grazing is actually *dominant in time:* active land rotation areas are only used from May to January, each aera thus being only 24 months in 10 years under crop-cultivation (4b), since the arable patches are only 2–4 year cultivated, followed by 10–15 years of fallow-rest. After the cropping, pastoral tribes utilize the land under 'unimproved grazing' whereas the sedentary people themselves practice some year-round grazing with their own livestock. Thus the land is used 120 months per 10-year period, under unimproved grazing (6a).

Only five of the nine WLU categories could be mapped in these Sudan maps: 4 (a + b), 6 (a + b), 7 (a, b, c and f), 8 and 9. Classes 2 (horticulture) and 3 (tree and other perennial crops) were of too little areal importance and thus included in 4 (cropland). A separate class 'water areas' was added. Very lengthy notations had to be used to accurately define and indicate important specific land-use types consisting of compound categories, e.g.

— 4 (a')b^{vi} — 6a (7c) = land rotation (Wadi Azum type), with grazing, in woodland savanna;
— 4b^{viii} — 6a (7b, 7c, 7') = land rotation (Nilotic type) with grazing in (low) woodland savanna or on 'high' land of Southern Clay plain.

The specific land-use conditions in the Sudan are so varied and special, that comparison with land-use in other parts of the world on the basis of a common legend has become theoretical. It is actually based on agricultural systems rather than on covertypes of crops in fixed fields. Only the order of classification is maintained but the classes are nearly all compound classes.

Modified world land-use classification in Sri Lanka

Sridas, reporting on *'Rural Land-use Mapping in Ceylon'* on the basis of 1:40,000 aerial photography, resulting into 1:64,000 mapsheets, reported the need for only slight modification of the WLU classification framework. Class 6 'unimproved grazing land' became 'Grassland and scrub' since the grazing use was not widespread and its consistent interpretation impossible. The class 'unproductive land' was retermed 'unused land', since the production-capability of tracts of land may be realized if new uses or technologies arise.

Subcategories were added as far as photo-scale demanded and the skill and experience of interpreters allowed to go into more and useful detail. Letters were used for this and as sufficed to the main class symbols, e.g.

— 2H for homestead garding;
— 2R for rubber plantations;
— 4P for paddy cultivation;
— 6V for Villu (riverbasin) grazing land, etc.

Wikkramatileke, however, writing on *problems of land-use mapping in the tropics,* states in 1959 that the WLU legend presented considerable difficulties when applied to Ceylon. In practice, the distinction between land rotation (4b) and shifting cultivation (7f) was, even in the field not only difficult, but of doubtful value: *'chena'* cultivation* in Ceylon's dry zone is practiced both by settlers from established villages and by a growing number of itinerants, with little or no visibly different impact on the landscape.

He remarks that 'most mapping techniques and schedules have been evolved in and applied to occidental mid-latitude areas that have an orderliness of land-use patterns, defined agricultural cycles and a high degree of agricultural stability'. Annual changes there tend to occur within the design of a permanent framework and even long-term shifts maintain recognizable patterns. In the tropics on the other hand, agricultural stability and relative permanence of land-use patterns can be associated more often than not only with the growing of irrigated rice and plantation crops. Subsidiary crop production in general is ephemeral both in respect of areal distribution and crop cycles. Crop patterns are largely conditioned by rainfall which is a highly variable factor in many tropical areas.

With respect to the SE dry zone region, the difficulties that attend the mapping of crop types and distribution or the location of the dwellings in domestic gardens ('homestead gardens') are many and varied: on the units there are many coconuts and bananas and other perennial tree crops together with annual food-crops under casual cultivation. The crop areas are minute, crops mixed and there is much bare ground and scrub. Most holdings are so tiny that field boundaries cannot even be shown. Terms of 'orchard', 'horticulture' or 'market gardens' just do not apply. Classification and

---

* Name used for rainfed agriculture as opposed to paddy, ranging from pure shifting cultivation to almost permanent cropping with regular fallow.

mapping must use complexes preferably using regionally significant concepts. Complex units of a particular composition are, however, very difficult to define spatially, their boundaries are rather obscure, they are determined by equally vague tenure conditions in constant state of flux.

The boundaries, the degree of permanency and the economic function (subsistence or commercial crop cultivation) of recently cleared areas are hard to determine from their one time appearance on the aerial photograph. They are strongly determined by rainfall and social-economic factors like tenure. And yet a distinction between 'chena' and 'fixed dryland' agriculture is important. This may mean the distinction between 'abandoned chena' and 'current fallow' in a deliberate crop rotation. Within areas of active 'chena' cultivation, transitions between cleared land, and scrub, jungle, forest and pasture are chaotic with variable plant-associations.

The classification of vegetation types on unoccupied ground can be considered even more subjective than the occupied categories of land-use. In fact the only category of land-use, where the question of boundary definitions and transitions does not arise, is the paddy fields. Here, the mode of irrigation water supply determining single or double cropping would be valuable. Between crops, paddy land forms an important grazing resource!

The *tenancy status* of paddy land is of much socio-economic value, but could only be mapped on very large-scale maps (tiny holdings!) and after extensive field inquiry.

Wikkramatileke's conclusions are that mid-latitude techniques, criteria, classifications, categories and nomenclatures do not fit tropical conditions and do not produce effective land-use maps. This is particularly true for areas in which traditional agricultural systems have long been developed through adaptation to very real ecological variation in environmental situations within short horizontal distances and over successive periods of changing annual conditions. The changes in these traditional economics themselves complicate the problem of land-use mapping.

Generalization of detailed information, because of cartographic problems and economy, in presentation on a reduced scale, often renders new categories, virtually meaningless.

Furthermore, the inherent characteristics of much of tropical agriculture, particularly the impermanence of many of its facets, make for qualitative rather than quantitative assessment and the need for very frequent revision. 'There is a great need for many experimental mapping programmes which can sample quite varied local environments in an effort to discern the techniques, classifications and nomenclatures that can be applied to the whole of the tropics with a more uniform degree of success than now comes out of the application of standard middle-latitude procedures.'

Land-use mapping in Latin America

The *Casebook of OAS Field Experience in Latin America* (1969) in its section D on 'Problems of Land-use Reconnaissance Mapping in the American Tropics and Subtropics' presents observations in the same tenure with respect to the applicability of the WLU classification framework. The surveys concern the portrayal on a small-scale map varying from 1 : 100,000 to 1 : 500,000 of the predominant land-use types and their classification and description based on the use of small-scale photography: 'It is worth mentioning that the techniques of surveying land-use..., by use of aerial photographs have been developed largely in temperate zone countries'. Here the land-use patterns generally reflect relatively advanced agricultural and urban development, conditioned by environmental, historical, economic and cultural influences, which has few parallels in Latin America. The resulting classifications reflect these temperate zone patterns.

Perhaps the greatest differences in land-use are found in the Latin American tropics and subtropics. In these climate zones, temperatures favour perennial plant growth and in areas without moisture deficits, farming is a year-round activity. Also the number and variety and physiognomic types of useful or economic plants are far greater than in temperate climates. Some outstanding features of land-use in these parts of Latin America which are not common to temperate zones, are: large total areas dedicated to tree crops, farming at high altitudes, considerable shifting cultivation, minifundia, cultivation on steeply sloping lands, considerable diversity of land-use in relatively small areas, owing to local variations in zonal climatic conditions. These and other land-use characteristics pose special problems in the description, classification and cartographic portrayal of present land-use in the tropical and subtropical regions of the Americas. Problems cited are e.g. those of cocoa and coffee under shade cover; the same problem is met in Africa. *'Mixed' land-use,* the cultivation of an assortment of crops and pasture together in the same, or side by side in adjacent small fields in complexes of small-scale family farming should rather be classfied as a discrete land-use type, providing the proportions of the different uses are more or less constant.

Grazing lands in tropical and subtropical areas often present special problems to reconnaissance mapping. In humid zones trees and brushy vegetation is often found in improved and unimproved pastures. These may have been purposely planted and may camouflage the pasturage in the photo-image. In subhumid or arid zones, 'grazing land' may present the aspect of low, scrub forest. Actually, the combined grazing-browsing-firewood and charcoal production form of land-use is so common throughout these regions that it should be considered as a distinct category of land-use. This presence of vegetation other than grass in tropical and subtropical grazing areas limits the feasibility of photo-interpretation techniques in classification and mapping this type of land-use.

The areas of 'discontinuous' or *shifting cultivation* present the same problem as mentioned by Wikkramatileke for Sri Lanka. Large total areas of the American tropics and subtropics are farmed on a discontinuous basis, primarily for short-cycle foodcrops by small subsistence farmers, producing characteristic mosaic patterns on the aerial photo-image.

In moist to wet tropical and subtropical zones, soil fertility degeneration will dictate only 2–3 years of crop-production to be followed by several years of fallow under natural regrowth vegetation. In wet zones, cleared plots may have to be abandoned due to weed-invasion requiring years of natural regrowth of bush/forest for its suppression. In the subhumid zone, areas are cleared or abandoned in response to erratic year-to-year rainfall distribution. New areas are opened up in the expectation of good rains, to be anbandoned when these expectations do not materialize.

In transition zones with forest, these patterns may signify transition to permanent cultivation/settlement but also conversion to pasture land. The resulting land-use patterns represent a mixture of cultivation, pasture, low and natural forest.

A *regional ecologic interpretation of agroclimatic* conditions is recommended to try and classify the intensity of agricultural land-use.

Land-use mapping in Mediterranean and (semi-) arid zones

With respect to land-use mapping in *arid zones,* Dudley Stamp, in 1962 emphasized the importance of a separation into 'irrigated' and 'non-irrigated' and the differentiation between 'perennially cultivated' and 'intermittently cultivated'. He further agreed that it is not always possible to distinguish within category 6 (unimproved grazing) the 'used' and 'unused'; 'each country should make its own divisions based on vegetation, natural or semi-natural, linked wherever possible with stocking intensity and types of livestock'.

Karmon, in his article on *'Land-use Survey in Mediterranean Countries,* Cartographical Suggestions' (1962), reviews the special conditions of land-use in semi-arid regions and especially the Mediterranean realms and points out the very distinct features which are the main concern of any planning or agricultural authority and which should stand out clearly on any land-use map:

1. the amount of irrigated/partly irrigated land as one of the most important denominators for the intensity of agricultural land-use;
2. the economic importance of fruit-orchards, demanding a distinct marking of the various species;
3. the intercropping of trees and arable crops as a very common feature;
4. grazing is predominantly by sheep and goats on mountain slopes and crop stubble land;
5. terracing is a common feature in parts of the region;
6. subsistence economy is still the rule in large parts of the region: scattered holdings and even patch-cultivation;
7. forest occurs mainly in degenerate forms of maquis or garrigue, except where modern afforestation has taken place. Their main use is not 'forestry' but grazing;
8. soil erosion has created barren soils or badlands in large areas; differentiation in types of 'unused' lands seems necessary;
9. a comprehensive survey of land-use should not confine itself to the mapping of agricultural lands; special attention should be given to recreational or touristic land-use which is of great importance in the region and which provides a much larger income from land than agriculture.

Karmon therefore proposes that the following land-use types should find expression in mapping in which the division into nine major categories, as in the WLU framework, is maintained:

1. *settlements and associated:*
   a) dense housing;
   b) dispersed housing;
   c) industry and transport;
   d) domestic gardens and recreation areas;
2. *horticulture:* should here only include irrigated intensive vegetable or flower gardening in plots between 0.4 and 1 ha;
3. *tree and other perennial crops:* of the utmost

importance here and thus to be as detailed as possible. Main distinctions: irrigated and non-irrigated. Intercropping indicated by a special symbol. Thus: three categories:
   a) non-irrigated fruit trees, incl. grapes;
   b) irrigated fruit trees, incl. grapes;
   c) intercropped fruit trees with arable crops;
4. *croplands:* again the main distinction: irrigated and non-irrigated. Karmon also wants to indicate the *purpose* of the crop: for food, fodder or industrial purposes. This seems unrealistic since it would require much inquiry whereas destinations may change during and after crop-development. In hilly regions, terracing should be indicated. Examples of the groupings are given in his paper;
5. *improved permanent pasture:* plays only a small part in semi-arid countries and thus should not appear as a separate category, but as a subcategory to rough grazing;
6. *unimproved grazing:* actually, every area, except perennial crop plantations, is used for grazing in the rainy season and after the harvest. This being understood, it needs not be specially indicated but mentioned in the accompanying memoir/report. The category should include only those areas where year-long grazing occurs as the only land-use. Transhumance as the prevailing type of economy, should be made a special subcategory;
7. *woodlands* are to be differentiated into 'afforested' and 'natural forest', marking:
   a) afforested land;
   b) natural forest;
   c) maquis a.o. types of degenerate forest (may well be identical with 'improved grazing');
8. *waterbodies:*
   a) lakes or reservoirs;
   b) fishponds;
   c) swamps;
   d) saltpans and playas (if utilized for the production of salt, with a special indication of industrial activity);

9. *unproductive land:* the main reason for non-productivity to be indicated, e.g.:
   a) sand dunes;
   b) badlands;
   c) barren rock;
   d) ruins (tourist value to be indicated).

Where the World Land-use Commission ruled that (1) main land-use categories should be marked by a certain combination of colours and (2) subcategories by the use of symbols, the colourscheme prescribed did not seem to be quite satisfactory and has not been widely accepted.

Karmon proposed that colours used should resemble *natural colours,* thus suggesting the type of land-use:
— the green colour should be reserved for the representation of perennial treecover, both in fruittree-cultivation and forests;
— for fieldcrops, the natural colour of ripe grain: yellow. This category has the largest number of subcategories, to be indicated by symbols. The yellow serves well as a background to show up these symbols;
— for water in the form of 'waterbodies' or to indicate irrigation: blue;
— red colours are usually reserved for representation of settlements as in any geographical map;
— purple to be used for industrial land-use and communications;
— grazing land could be indicated by brown colours (representing the general aspect of the semi-arid landscape in the dry season).

Further: no *letters* should be used as symbols as this would create difficulties of language.

In accordance with the aboven, Karmon presents the following scheme:
1. urban land-use and housing: red components;
2. horticulture: orange (as a combination of red for urban activity and yellow for fieldcrops);
3. cropland: yellow colours;
4. treecrops: green colours;
5. woodlands and grazing land: brown colours;
6. waterbodies: blue colours;
7. unproductive land: grey colours.

Actually Karmon proposes the use of colour, also in symbols, lines, dots, etc., in indicating subcategories. This, however, presents technical problems and tends to confuse the picture.

Mediterranean conditions in land-use can hardly be understood nor properly classified without knowing the very complicated sociological basis. The only way to bring out the general picture of land-use and the complex structure in details would be the combination of a small-scale map (e.g. 1 : 200,000) with an inset map on a large scale showing the land-use of small subareas representative for particular forms of complex land-use.

Land-use classification in Eastern Europe

Kastrowicki, in his contribution to World Land-use Survey Occasional Papers No. 9 (1970) titled *'Data Requirements for Land-use Survey Maps',* argues that in densely populated countries where the present pattern of land-use as the result of long centuries of man's pressure on land eventually leads to the most intensive use of every parcel, it is more important to map, not only the distribution of the major forms of land-use (covertypes as derived from aerial photography), but also how and what they are actually used for. He quotes as an example the Japanese land-use survey in which e.g. various cropping intensity levels are mapped.

He argues that the WLU framework is flexible enough to incorporate such refinements/modifications to suit local needs and conditions and maintains that 'if there are still major differences between particular classifications, these are due more to excessive individualism of the authors than to real needs or requirements'.

The same, he says, is true regarding the recommended notation of master key symbols and colours. In many land-use surveys they are not adhered to, not so much because there is a real need to change them but in most cases simply because some people prefer different colours.

He regrets that as a result, the original idea of Sir Dudley Stamp (the driving force behind the WLU Survey on a common basis of classification and mapping) that any land-use map (like any geological map) should be read easily without looking each time at the key of symbols, has been wasted.

Concerning the influence of the scale of mapping on the subdivision of the main classification categories, the Regional Subcommission Meeting for East Central Europe reviewed the question of maintaining comparability also in larger scale land-use mapping. It distinguished four levels of detail/scales in land-use mapping:

— large-scale 1:2,000–1:10,000, showing individual cropfields and patches of other 'uses' (covers) and their 'utilizations' (functions), at the individual farm or settlement level;

— 'detailed' land-use maps, at 1:25,000–1:100,000 scale showing main land-use categories and subdivisions to suit local needs and possibilities;

— 'simplified' land-use maps, at 1:200,000–1:500,000 scale showing main land-use categories with limited subdivision;

— 'general' land-use maps, at 1:>1,000,000 scale, the scale at which the WLU Survey aimed.

It is hard to reconcile these terms and scale limits with those defined by e.g. the O.A.S. in its surveys in South America.

The Regional Subcommission for E.C. Europe of the I.G.U. then reviews the experience in using the WLU classification:

1. *settlement:* subcategories within this class will depend on the purpose of the study; both large scale aerial photography and topographic maps (if of recent date! sic.) can provide information for this;

2. *horticulture:* small homegardens to be included in class 1. Settlement: 'settlement and homegarden'. For the rest, horticultural crop cultivation is often in rotation with 'arable crops' and should therefore not be mapped separately, except when occurring in *large* permanent complexes (e.g. greenhouses);

3. *tree and other perennial crops* (sometimes indicated as 'orchards'). Kastrovicki observes that the distribution of the areas covered by perennial crops (orchards, vineyards, tree plantations) can easily be defined from aerial photographs and most topographic maps. It is, however, impossible to interpret the kind of orchard fruit from aerial photographs. Only some semi-perennial crops such as cotton, sugarcane and strawberries would be less easily recognizable. The above statements may be true for particular areas and production systems, but are certainly not true in general.

In 'intercropping', dominance is not possible to determine, whereas in 'intercalary' cultivation of annuals it is still more difficult. In these cases it would be better to distinguish blocks with more or less uniform proportions of individual crop types and in more detailed surveys 'mixed crop' categories should be mapped, indicating individual crop proportions;

4. *cropland,* referred to as 'arable land'. The WLU subdivision into only (a) continual and rotation cropping and (b) land rotation, is not regarded as sufficient. Even in more general surveys (see Lebon on the Sudan Survey), as many as 13 forms of land rotation systems had to be distinguished. The mapping of only 'actual crops' in more detailed surveys is still more unsatisfactory. (1) It is only possible for very large scale agriculture and (2) in a given season/year the croptypes will change on account of croprotation.

In E.C. Europe (and elsewhere) therefore (standard) crop combinations should be presented on the map i.s.o. single crops on individual fields. The commission proposes to map 'crop-groups' based on their common agronomic properties (soil requirements, position in crop rotation, etc.) in view of the numerous crop types involved in small-scale unspecialized agriculture:

a) intensive or intensifying crops': those requiring much labour and fertilizer, finally becoming good fore-crops (most of the rootcrops, maize and some industrial crops);

b) 'structure forming' crops or 'restorative' crops, mostly legumes. These are also good fore-crops but require neither much labour nor fertilizer;

c) 'extractive' or 'extracting' crops: mostly cereals, that do not require much labour or fertilizer but deplete the soil most, thus being poor fore-crops.

It is difficult to see, however, how this grouping would alleviate the problem of mapping the spatial and temporal complexity in cultivation of a large number of crop types;

5./6. *grassland:* the IGU subdivisions into 'improved permanent' and 'unimproved' grazing land and into 'used' and 'unused' is recognized by the subcommission as too much based on West European (British) conditions. It does not apply to the reality in many other parts of the world. Also its colour code of yellow and pale green, where dark green is used for 7. (Forest) is regarded not logical/systematic; it leads to problems when several greens have to be used to mark different categories of grassland and/or forest.

Grass types and their productivity and nutritional values should be used as discrimination criterion (!) but this may not be easy in (semi-) natural grasslands. Therefore the mapping of 'grassland complexes' is suggested, giving the % share of individual vegetation types (airphoto-interpretation and field observation).

How individual grasslands are being used (cut for fodder or hay or grazed, or both) needs to be established by inquiry; it will not appear on maps of smaller scales.

Also the occurrence of shrubs or trees in grasslands presents classification problems (we met this before, e.g. in the OAS survey report on South American tropical and subtropical grazing lands). 'Land improvement measures' (permanent irrigation, drainage, terracing) would only have to appear on very detailed maps.

Livestock: difficult to establish and map numbers and types; their influence on the use, quality and composition of grasslands would, however, warrant mapping;

7. *forest:* in more detailed surveys it is general practice to indicate forest land by dominant species and/or degree of treecover. Much information can be obtained through the interpretation of density, age classes, dominant species, in aerial photographs.

In Europe, where most forests are government managed, data on management, density and species can be easily obtained.

'Utilization' of forest is not easily seen, however (timber, polewood, firewood, charcoal, recreation, hunting, conservation/reserve, watershed protection; turpentine, resins and even mushroom production! Sic.).

Aerial photography also supplies most of the information needed about mixed categories of forest and arable land (shifting cultivation/land rotation) or grassland;

8. *'water':* 'open water', which can serve many purposes, is not represented in the WLU framework! And finally:

9. *'idle' land or 'unproductive' land:* differentiation from grassland may be problematic. In many countries large areas of 'grassland' which are fit but *not used* for grazing, should be classified under 6b. A problem is the transition between grassland and bare soil surface. Areas reserved for urban or industrial development or so-called 'social fallow' are also difficult to classify either as 'grassland' or as 'unproductive'.

Areas which cannot be reclaimed without transformation and investment (such as various types of swamp, rocks, steep slopes, man-made elements like slag-heaps, holes resulting from surface exploitation of minerals), abandoned quarries and heavily eroded 'badlands' typically should be classified as 9 'unproductive' land.

Conclusions of the East Central Europe Sub-committee are that data requirements for a land-use survey differ according to the aim, type and scale of the map to be produced.

The ITC land-use classification

During a number of years the staff of the Rural and Land Ecology Survey group at the ITC (International Institute for Aerospace Survey and Earth Sciences at Enschede, the Netherlands) has used the World Land-use Survey framework as a basis for classification and legend construction.

In the light of experience in the Mediterranean region and with respect to land-use in tropical areas, this classification and the instructions for graphical representation of land-use in maps, have gradually been modified, incorporating elements of e.g. Karmon's proposals.

The surveys were mainly based on interpretation of aerial photography of scales between 1 : 10,000 and 1 : 50,000, with supporting field-work.

This land use classification in its final version was as follows:

1. *settlement and associated non-agricultural areas;*

    1a *settlement, residential:* only areas large enough to be mapped as area features (at least $0,25 \, cm^2$ in the map), should be indicated; the rest will appear as topographic features in the basemap;

    1b *associated non-agricultural:* this may include industrial areas, railway yards, airports, quarries, etc. which are only mapped when occupying large enough areas. If not, they will be represented in the basemap in the form of symbols: linear symbols for transport lines, pictorial for other features.

    If relevant, industrial areas, railway yards, etc., can be shown separately as 1b, 1c. This is preferable to mapping them as subclasses of 1b ('associated non-agricultural areas') such as 1b1, 1b2, etc., which symbols would occupy too much space for these relatively small mapping units. Also areas under use for recreation are mapped as a subclass of 1.

    If there are no other than 'residential areas' under class 1, the classification will be:
    —settlements and associated non-agricultural areas: settlements, residential.

2. *horticulture:* only areas of intensive cultivation of vegetables, small fruits and ornamentals in small fields should be classified as such. If this type of land-use does not occur in the study area, this is indicated in the legend as '2. horticulture: not applicable'.

    This category of land-use is on a global scale occurring so little, that it does not warrant the status of being one of the nine main categories! In regions where it is important enough to be mapped separately, this could be done as a subclass under 'cropland'. This would, however, disrupt the numerical order and thus the comparability of map legends.

    In tropical areas, 'homestead gardens' with fruittrees, vegetables and spices but also foodcrops/fibrecrops for home-use, are often spatially so closely associated with settlement, that separation becomes impossible. In such cases the combination is to be mapped as a subclass of class 1, e.g.: '1a = settlement and homestead gardens';

3. *treecrops and other perennial crops:* as far as required by the purpose of the survey and possible with regard to photo-scale, photo-quality and the complexity of the land-use, individual types or else functional groups of treecrops should be distinguished: 3a, 3b, 3c, in order of areal extent. If applicable, treecrops should be distinguished from shrub-crops and the other types of perennial crops (tea, coffee, grapes, sisal, hop, etc.), starting with treecrops, f.b. shrub-crops/vines.

Example:

3a olives
3b peaches
3c citrus
3d grapes

if they occur in this order of areal importance and can be consistently distinguished throughout the photo-coverage.

or

3a olives
3b other treecrops
3c grapes

if olives are important enough to be singled out and can be consistently distinguished, whereas the other treecrops cannot be and need not be separated.

4. *arable crops:* Depending on the scale of the survey, etc., individual crops or crop-groups can be distinguished: 4a, 4b, 4c, etc., e.g.

4a cereals;
4b foddercrops;
4c fallow, or

4a wheat;
4b barley;
4c other cereals (maize, rye, oats);
4d fodder crops (sorghum, alfalfa, ...);
4e other crops (horsebeans, ...), and also
4f fallow

If several arable crops occur on *complexes of fields* which are too small to be mapped individually, special subcategories could be used such as:

4b wheat and chickpea rotation;
4c wheat and sunflower rotation, but also
4d wheat and fallow rotation.

Also 'land rotation' or 'shifting cultivation' or 'semi-permanent' arable cropland can be mapped as a separate subclass of 4.
For irrigation, see below;

5. *improved grazing land:* depending on the scale of the survey etc., subcategories can be distinguished on the basis of type of improvement;

6. *unimproved grazing land:* without going into the question of 'used' or 'non-used', the following subcategories could be distinguished:

6a open grassland, dense cover;
6b open grassland, sparse cover;
6c grassland with thinly scattered trees, dense grass cover;
6d grassland with thinly scattered trees, sparse grass cover;
6e grassland with thinly scattered shrubs, dense grass cover;
6f grassland with thinly scattered shrubs, sparse grass cover;
6g open shrub vegetation (i.e. without grass cover).

An other addition could be 'with rock outcrops'.

Only those subcategories which do occur in the study area appear in the legend and are consecutively numbered in order of decreasing grazing/browsing value.

NB: This distinction of unimproved grazing lands on the basis of their grazing quality is preferable to using complicated notions as (6a + 7c + 9a) for grassland + shrubs + rock outcrops. It preserves the clarity of the map. It is, however, to a large degree subjective;

7. *woodlands:* this category may be subdivided into:

7a dense forest, natural;
7b idem, planted;
7c open forest;
7d dense shrub cover (open shrub cover will appear in a combination under 6);
7e swamp forest;

8. *waterbodies:* if the areas are large enough to be mapped, the following subclasses may be distinguished:

8a lakes;
8b reservoirs;
8c swamps and marshes, the latter if definitely unsuitable for grazing; otherwise they will appear as a subclass of 6: 'seasonally' or 'permanently grazed marshland';
8d fishponds.

Waterbodies too small to be mapped, if important in the context of the survey or for orientation purposes, will appear as linear (rivers, canals) or pictorial (basins, ponds) symbols in the topographic basemap;

9. *presently unproductive land:* this category may be subdivided into:
9a sandy beaches, riverbeds, sanddunes;
9b mudflats;
9c bare rock;
9d 'badlands' (term to be reserved for terrain that is so badly dissected that flat areas are almost non-existent and vegetation is so sparse that only very extensive grazing might be possible);
9e areas of active mass movement;
9f snow or ice covered areas.

Notes. If one of the *main* categories does not occur in the study area, this has to be indicated in the legend by the indication 'not applicable' behind the category title. This will ensure that the relationship between category numbers and category contents, as in the WLU framework, is maintained to preserve comparability of maps of different areas and periods.

Example:
1. settlements and associated non-agricultural areas;
2. horticulture: not applicable;
3. treecrops and other perennial crops, etc.

If a certain *sub*category does not occur in the total study area, however, this should be simply left out. On the subcategory level there is no fixed complement/order of subclasses with their numbers.

If a study area is worked by a team, the members of which cover different parts of the entire survey area, then there should be a common legend in which a subcategory only appearing in one of the sub-areas, should appear, even if it is 'not applicable' in all other sub-area legends.

'Mixed' and 'complex' land-use. A land-use is *'mixed'* when two or more land-use/crop types appear in the same field, e.g. olives with wheat underneath, or olives with grapes, or coconuts with grazing underneath.

In this case a mixed (.../...) symbol should be used, e.g.:
3a/4b = olives as the main crop, with wheat underneath;
4b/3a = wheat field with some olives;
3a/3b = olives + grapes.

The order indicates which is the dominant crop in the mixture. If both crops are of equal importance a notation *3a/4b* is suggested.

A unit is *'complex'* if each crop occurs in pure cultivation on its own field but when neighbouring fields carry different crops and fields are too small to be mapped individually.

In this case a 'complex' notation (... + ...) could be used, e.g.:
3a + 4a + 4b = olives + wheat + fodder crops, or
4a + 3a + 4b = wheat + olives + fodder-crops in small fields forming a complex.

Again, the order indicates the relative dominance of each crop involved. If two crops are of equal area-importance, the following notation can be used: 3a + 4a = olives and wheat, each comprising about 40–60% of the area, whereas in the case of 3a + 4a, the 3a fields will occupy ⩾60% of the total area in the complex. These combinations may, however, for the sake of simplicity in annotation be represented by a subclass-notation: e.g.:
3c = mixture (olives and wheat);
3d = complex of olives (dominant) and grapes.

Irrigation. In the symbols of '2. horticulture', '3. treecrops and other perennial crops', and '4. arable crops', irrigation may be indicated by suffixing an accent, e.g.:
3a = treecrop (e.g. peaches);
3a' = irrigated treecrop (peaches).

A better solution, avoiding this more complicated symbol, would be:
3. *treecrops and other perennials:*
    irrigated:

3a peaches;
3b apples;
3c vines; and then:
dry land cultivation:
3d olives;
3e grapes;
3f almonds.

Colours. Each main category that is applicable (!) should be assigned a basic colour, preferably its 'natural' colour (Karmon). Subcategories are to be indicated by different shades of this basic colour, e.g.:
3. *treecrops and other perennial crops:* green:
  3a olives = olive green;
  3b citrus = bright green;
  3c vineyards = pale green.
Thus:
— class 1 'settlements .....' in red colours;
— class 2 'horticulture' in orange;
— class 3 'treecrops .....' in green colours;
— class 4 'arable crops' in yellow colours;
— class 5 'improved grazing' in dark brown colours;
— class 7 'woodlands' in dark green colours;
— class 8 'waterbodies' in blue colours;
— class 9 'presently unproductive' in grey-tones.
If one main category has many subcategories, hatchings may be used to extend the range. Another solution is to reserve the colour-shades for logical subcategory grouping within which the individual subcategories are distinguished by their symbols. Subcategory symbols must anyway appear in every single mapping unit. This is not only to guide the colouring process but also to be able to verify in case subcategory colours should look alike. In the case of only few copies having to be made of a land-use map, the cost of colour printing/reproduction becomes too high per map-copy. Then only a few maps for direct use will be hand-coloured, the rest can then still be used without colouring or coloured at a later stage, as long as the land-use of each mapping unit has been clearly indicated by the appropriate (sub)class-symbol.
Irrigation may be indicated by horizontal hatching for gravity irrigation, by broken vertical lines for sprinkler irrigation.
Terracing can be indicated by diagonal lines crossed by widely spaced horizontal lines; in small areas by tracing generalized terrace contour lines.
Complex categories may be indicated by alternative colour-stripes in the colours of each of the crop components, the widths of the stripes indicating the relative importance of the land-use/crop concerned. Because of the labour involved, and for greater clarity of the map, it is to be preferred to regard the complex as a subcategory in itself and apportion one of the shades of the main class colour to it.
Mixed land-use should be indicated with a colour shade derived from (one of) the main composing (sub)-categories.
NB: Palest colour-(shades) should preferably be assigned to those (sub)-categories occupying the largest area. Economically important enough to be mapped (sub)categories occupying only small areas will then be shown in more aggressive colours in order not to be overlooked.

Preliminary map. Only in the preliminary map, for convenience during the interpretation and fieldwork stage, land-use/croptypes may be indicated by notations representing abbreviations of those (sub)categories, e.g.:
— O1 = olives;
— V = grapes (vines);
— Ce = cereals;
— Ct = citrus;
— Co = coconut, etc.
In the final mapping, these symbols are then translated into the 'official' symbols made up of letter and number combinations, e.g.:
4. *Arable crops:*
  — *irrigated:*
    4a rice;
    4b1 maize (graintype);
    4b2 maize (for forage);
  — *dryland:*
    4c barley;
    4d oats, etc.

The latest ITC world land cover and land-use classification

Only recently, a number of arguments have lead to the adoption on an experimental basis of a completely new classification scheme, in which *covertype* and *function* are separated (ITC Land Ecology Group).

Like foregoing schemes, it has to serve as a checklist from which the actual map-legend is to be composed. The arguments are the following:

In the interpretation of imagery, what is seen is the *cover* on the earth surface, in most cases the functional use of this cover can only be verified by extensive observation on the ground/inquiry. In project-oriented surveys of a specific area, it often happens that only a few of the nine categories of the WLU framework occur. On the map due mention was always made of the fact that 'the legend used is based on the WLU framework' and those WLU categories which did not occur in the survey area were duely indicated as 'non-applicable'. These non-applicable categories were mostly '2. horticulture' (on a global basis this type of land-use is anyway too unimportant to warrant its status as one of the nine main categories). Improved grazing is also on a global basis hardly important enough to be a separate main class.

Further the term 'grazing' in '6. unimproved grazing land' is not regarded realistic. In the first place no evidence of 'grazing' by domestic stock may be visible in the aerial photographs and it would be very hard to consistently verify the 'used' versus 'non-used' status in the vast areas concerned. 'Grazing' is in some regions an important 'use' of woodland types: pigs and sheep in oak forest, browsing by goats in shrub-vegetation. Arable land after the harvest (stubble) and during the fallow period is in some regions also important for grazing.

The function of woodland may further be anything like timber production, firewood, charcoal, resin production, mushroom production, watershed protection, wildlife reserve, hunting, recreation, soil regeneration (bushfallow). Often from a distance there is no indication as to such use.

Waterbodies may be used for shipping, irrigation, drinking water supply, recreation, fishing, industrial water supply.

Presently 'unproductive' areas may also have a function for recreation or hunting, wildlife/nature reserve. These functions are very difficult to establish consistently and correctly delineated on the basis of aerial photo-interpretation and even in the field at times.

Also, the distinction between subclass 6B (non-improved grazing, unused) and 9 (presently unproductive areas) is often not clear.

It is not practical to 'drag along' all those non-applicable categories in legends for local/regional use and particularly not to reserve colours for them in mapping, which could be better used to clearly differentiate the locally important categories and their subdivisions.

Integration of the agricultural land-use survey results with those of a vegetation survey also gave problems particularly on the point of class 6 (unimproved grazing areas) and also where fallowing areas under class 4 (cropland) were concerned.

For these reasons a classification scheme has been devised as published on Appendix A to Chapter 29.

On the left the land cover forms are shown such as can be identified on aerial photography and other remote sensing images, without any inferences as to their function. After all, the same cover type can have different functions (if any) and the other way round.

I.   *Buildings and artificacts:*
  — buildings;
  — artifacts (such as roads, railway yards, platforms, harbour quays, canals, dams, dykes, etc.);
II. *fields / plantations:*
  — fields: recognizable as such by their geometric forms; term reserved for areas under herbaceous crops, cultivated grasses/legumes, temporary fallow;

— plantations: areas planted to treecrops, grape-vines, shrub crops;

III. *open (semi-)natural vegetation* (according to physiognomy):
— woodland/savanna;
— shrubland;
— dwarf shrubland;
— grassland;
— herb vegetation;

IV. *forest:*
— forest plantation;
— (semi) natural forest;

V. *waterbodies, snow/ice cover;*

VI. *bare areas:*
— bare 'rock';
— 'badlands';
— beaches;
— sanddunes;
— mudflats.

The fifth column gives a classification of (functional) land-uses:

1. *settlement and infrastructure:*
   a) residential
   b) industrial, quarries, mines (above ground part)
   c) transport and communications
   d) sport and recreation
   e) other (watersupply, military, cemetry, educational, historical (ruins)

2. *agricultural land-uses:*
   a) field crops, annual incl. horticultural: cereals, annual rootcrops, cotton, etc.
   b) field crops, perennial: sugarcane, cassava, agave
   c) perennial forage crops, e.g. grassland, grass/legume pastures, alfalfa
   d) tree crops: e.g. oilpalm, rubber, coconut; also banana
   e) shrub crops and vines: coffee, tea, cashew, grapes
   Each may be further differentiated into individual croptypes

3. *grazing:*
   a) intensive
   b) ranching
   c) pastoralism

4. *forestry for:*

a) *timber production*
b) *pulpwood production*
c) *firewood, charcoal, polewood a.o. domestic uses*
d) *other, incl. bark, cork, turpentine, resin*

5. *conservation:*
   a) nature reserve
   b) wildlife reserve
   c) watershed management
   d) dune stabilization
   e) other

6. *hunting, fishing:*
   a) hunting
   b) fishing
   c) aquaculture
   d) other

7. not in use.

Column 3 shows which (functional) land uses in column 5 may correspond with each of the cover forms of column 2. Column 6 shows which covertypes of column 2 may correspond with each of the land-uses of column 5.

From this 'checklist' a legend can be composed, incorporating 'use' (sub)categories inasfar as required and consistently identifiable with the help of an acceptable input of subsidiary information.

A recent North American land-use classification using remote sensing

In 1972, Anderson, Hardy and Roach, in the US 'Geological Survey Circular 671' published *'A Land-use Classification System for Use with Remote Sensor Data'.* This classification system was to provide the framework of a legend to be used in land-use survey throughout the US, based on uniformity of data, scale and categorization at the more generalized first and second level. The classification system has been developed on the assumption that different remote sensors will provide information for different levels of classification. In general the following relations were anticipated (see Table 1).

At level I, based on imagery in scales between 1:1,000,000 and 1:250,000, classification will

Table 1. Classification levels in relation to type of remote sensors.

Classification level	Source of information
I	Satellite imagery, with very little supplemental information
II	High-altitude and satellite imagery combined with topographic maps
III	Medium altitude remote sensing (1:20,000) combined with detailed topographic maps and substantial amounts of supplemental information
IV	Low-altitude imagery, with most of the information derived from supplemental sources

be general and based on major differences in land *cover.*

At level II, based on imagery of around 1:100,000 scale, information transfer to even 1:24,000 US Geological Survey topographic maps is regarded feasible, by which a substantial amount of supplemented input is obtained. This would allow classification on the basis of more specific land-uses than only nine major cover-types. These categories, however, cannot all be interpreted with equal reliability and some may be extremely difficult to interpret from high-altitude imagery alone. Rather than mutilate the categorization or distort it and so reduce the number of useful applications, it seemed preferable to suggest that additional steps be taken to obtain a satisfactory interpretation. Conventional aerial photography and other, non-remote sensor data may be needed for interpretation of difficult areas.

At level III, substantial amounts of supplemental information, in addition to remotely sensed information at scales of 1:40,000 to 1:15,000 would seem necessary. At 1:20,000 scales, transferring immediately onto 1:24,000 topographic maps, surprisingly detailed inventories may be undertaken in which most land-uses, except those of very complex urban areas or very heterogeneous mixtures can be adequately located, measured and coded.

Level IV of the projected classification would call for much more supplemental information besides much large-scale aerial photography.

The classification presented in this publication, for the Inter-Agency Steering Committee on Land-use Information and Classification, however, is limited to the levels I and II. The following is largely quoted from the publication. The paper clearly states that there is no ideal classification of land-use and it is unlikely that one will ever be developed. There is further no logical reason to expect that one detailed inventory will be adequate for more than a relatively short period of time, since land-use patterns change and so do demands for the natural resources which affect the development of land-use patterns.

In attempting to develop a classification system for use with remote sensing techniques that will satisfy the needs of the majority of users, certain guidelines of criteria for evaluation should first be established. To begin with, there is considerable diversity of opinion about what constitutes 'land-use', although *present land-use* is one of the characteristics that is widely recognized as significant for planning and management purposes. One concept that has much merit is that land-use refers to *'man's activities on land which are directly related to the land'.*

Land cover on the other hand, would describe 'the vegetational and artificial constructions covering the land surface'. But: some activities of man can be directly related to the type of land cover, others, like recreational activities (incl. hunting) can be related to land cover *by use of remote sensing techniques* only with difficulty, if at all.

Land cover is therefore the basis for categorization at the first and second levels and the *activity* dimension of land-use will appear at third and fourth levels of categorization.

Even in cover, these are problems: where to draw the line between land and water in the case of seasonally wet areas, tidal flats, or marshes with various kinds of plant cover. How large an area should be in any particular use before we recognize the use? How do we resolve the problem of heterogeneous mixtures of equally significant land-use? Or of multiple uses of an area? Here apparently arbitrary decisions must be taken if descriptions of categories are complete and criteria explained, the inventory process is repeatable. The paper then goes on by giving ten criteria which should be met by a land-use classification system for use with orbital imagery:

1. minimum overall level of accuracy in the image interpretation around 90 %;
2. roughly equal accuracy of interpretation for the individual categories;
3. repeatability of results between different interpreters and times;
4. usable/adaptable for use over extensive areas;
5. permit vegetation and other types of land cover to be used as surrogates for activity;
6. suitable for use with imagery taken at different times of the year;
7. subcategories from ground surveys or larger scale or enhanced imagery should find effective use;
8. collapse of categories (into higher level categories) should be possible;
9. comparison with land-use information compiled in the past or to be collected in the future should be possible;
10. multiple use aspects of land should be recognized when possible.

Greater accuracy than 85–90 % of the time is generally not required at this generalized first and second level and could be attained only at much higher cost. Special attention was given during the development stage of the classification to the hierarchical classification problem of barren lands and natural vegetation (rangeland). It was further realized that when (nearly) complete reliance is placed on remote sensing as an inventory technique, complete compatibility with categorizations used in enumeration and observation techniques used in more traditional approaches would not be possible; the definitions of land-use categories in these approaches may be quite at variance.

The classification system proposed in the paper was only a first approximation, requiring testing in practice.

The classification
— Level I
— Level II

Definitions. The subject content of each class in each category at levels I and II is then given extensively. The reader is referred to the original publication for this important aspect of the classification.

Some noteworthy points in these definitions are the following:
— The 'urban' and 'built-up' land category takes precedence over others when he criteria for more than one category are met. Thus residental areas that have sufficient tree cover to meet 'forest land' criteria will be placed in the 'residential' category. Areas of sparse residential land-use will be included under another category.
— Residential sections may also be included in other use categories where they are integrative parts of the other use.
— Recreational areas are not segregated as such at level II, but may present some problems in identification. Recreational facilities that form an integral part of an institution should be included in the 'institutional' category. A self-contained sports area on the other hand, such as a stadium for professional events is 'commercial'. The intensively developed recreational areas near the resort, would be included in the 'commercial' category. Public and private golf courses, ski and toboggan areas and other recreational facilities, just as riding areas, are classed as 'open land'.

01. On 'extractive land': unused pits or quarries that have been flooded are placed in the 'water' category if the water body is larger than 40 acres (16 ha).
On 'transportation, communications and utilities': overland railway tracks are not in-

cluded in 'rail facilities', unless six or more tracks are joined to give sufficient width for delineation at a scale of 1 : 250,000.

Education, religious, health, correctional and military facilities are the main components of the 'institutional' category.

'Open land' consists of golf courses, some parks, ski areas, cemetries (!) and undeveloped land within urban setting. It may be in very intensive use but a use that does not require structures. 'Open land' includes the small blocks of less intensive or non-conforming uses that become isolated.

02. 'Agricultural land': on high altitude imagery, the chief indications of agricultural activity will be symmetrical patterns made on the landscape by use of mechanized equipment... However, pasture and other lands where such equipment is used infrequently may not show as well-defined shapes. Some urban land-uses, such as parks and large cemetries, may be mistaken for 'agricultural land', especially when they occur on the perifery of the urban areas... The interface of 'Agricultural land' with other categories of land-use may sometimes be a transition zone in which there is an intermixture of land-uses at first and second levels of categorization. Where farming activities are limited by wetness, the exact boundary may also be difficult to locate and 'agricultural land' may grade into swamp 'forest land', 'non-forested wetland' or 'water'.

From imagery alone, it is generally not possible to make a distinction between 'cropland' and 'pasture' with a high degree of accuracy and uniformity, let alone a distinction among the various components of 'cropland', also depending on the timing within the growing season.

Horticultural areas at these levels (I and II) will often have to be included in an other category, generally 'cropland' and 'pasture'. Combination with knowledge on soils and climatical factors needed for these operations, however, may help: waterbodies in close proximity to moderate temperature fluctuations, site selection for 'airdrainage', etc.

Typically American are 'feeding operations': specialized large livestock production enterprises, chiefly beefcattle feedlots and large poultry farms, hog- and fur-bearing animal farms. Their waste disposal problems justify a separate subcategory for these relatively small areas (!).

0.3 'Rangeland' may be defined as 'land where the potential natural vegetation is predominantly grasses, grass-like plants, forbs or shrubs, where natural herbivory was an important influence in its precivilization state, and that is more suitable for management by ecological rather than agronomic principles'.

The level II categories of 'rangeland' in the USA are: 'grass', 'savannas' (Palmetto prairies), 'chapparal' and 'desert shrub'.

04. 'Forest lands' are 'lands that are at least 10% stocked by trees capable of producing timber or other wood products that exert an influence on the climate or water regime'.

NB: in this category we find no distinction between 'natural' and 'planted' forest, but at level II, distinction is made between 'deciduous', 'evergreen' and 'mixed'. In the 'evergreen' subclass, both coniferous and tropical broadleafed evergreens (e.g. the hardwoods as mahogany and ebony) are included.

05. 'Water' includes all areas within the landmass of the US that are predominantly or persistently water covered and at least 200 m wide and covering at least 16 ha.

Waterbodies that are vegetated are placed in the wetland category or under 'forest land' if swamp forests exist.

06. 'Non-forested wetland' consists of seasonally flooded basins and flats, meadows, marshes and bogs. Uniform identification is difficult because the wetland areas change due to long-term drought, high rainfall, seasonal fluctuations in precipitation and diurnal tides. Correlations with tide and weather information should be done for consistent results.

The subclass 'vegetated' non-forested wetland embraces areas with less than 10% forest crown-cover or a non-woody vegetation.

07. 'Barren land': land of limited ability to support life and little or no vegetation: only soil, sand, rocks. Land temporatily barren owing to man's activities is usually included in another land-use category: fallow land, land being tilled, cleared sites for urban development, waste areas and tailings dumps and exhausted areas of extractive and industrial land-use.

Level II categories of 'barren land': salt flats, beaches, sand other than beaches, bare exposed rock, and other.

08. 'Tundra': cold treeless lands (in Alaska) with a vegetative cover of moss and lichen, grasses and herbs.

Distinguishing between 'tundra' and 'non-forested wetland' is difficult where there is a hummocky landscape with intervening areas of standing water in seasonally changing condition of freezing up.

Final observation

It may be concluded that this last classification has again several typically American overtones. It appears, however, quite important in that it objectively and consequently bases itself on space and high altitude imagery interpretation without substantial input from other sources, at the first and second level. What this means with respect to the amount of detail that can be consistently and accurately mapped, will, however, considerably change with:
— the advent of much higher resolution systems (thematic mapper);
— the increasing capability for integrating multi-temporal data;
— the increasing capability of integrating the data from the different remote sensing techniques such as MS scanning (space) photography and radar;
— the increasing capability of integrating the remote sensing data with available spatial information on climatical, terrain and socio-economic data, which will greatly support identification/classification, their accuracy and functionality in response to data requirements.

Mapping land-use using aerial photographs and other remote sensing images

Introduction: the aerial photo-interpretation approach (remote sensing)

This approach requires a high 'specialist reference level' on the part of the interpreter with regard to the subject under study, i.e. land-use, of which agricultural land use is usually the most important aspect.

This specialist knowledge of agriculture should cover the entire complex of agricultural production involving crops, animals, their ways of cultivation and the techniques used therein.

Successful photo-interpretation also requires a sufficiently high 'local reference level' with respect to the specific types of land-use/crops/animals that should be expected in the particular area under survey and the particular levels and techniques of their management, the scale on which and the time-sequence in which operations/crop developments take place. This local reference knowledge has to be built up in advance by studying relevant literature on the area, existing maps, statistical information and/or by interviewing people who know the area well. In some cases there may be a possibility of making an orientation visit the area.

Knowledge of types and properties of aerial photography and other remote sensing techniques is also required in order to judge which techniques will be applied and later to interpret well. The following paragraphs deal with special aspects of interpretation concerned with land use surveys.

Timing of the photography

As said before, land-use is a dynamic process.

357

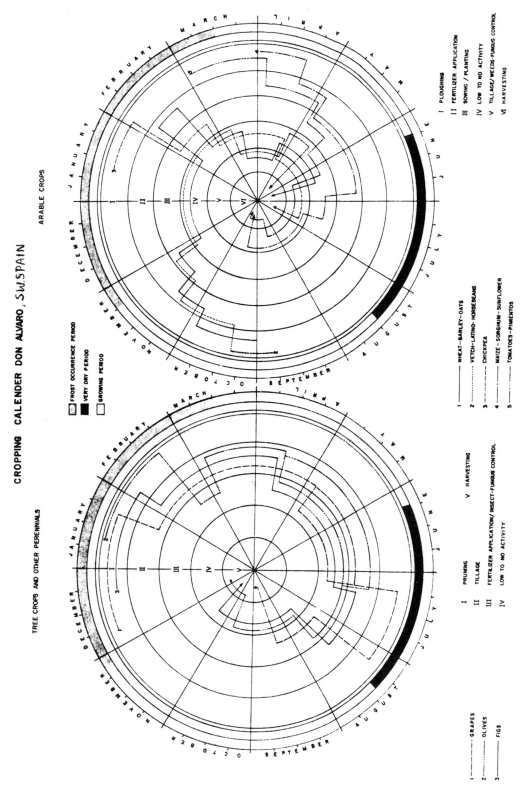

Fig. 1. Crop calender.

In agricultural areas, the aspect of the land surface changes continually: soil preparation, sowing, gradually closing up of the crop, flowering, maturing and harvesting, follow each other in different time-sequence depending on croptype, land elevation/exposition, etc.

A very important tool for determining the most appropriate time for executing the aerial photography or imaging by other remote sensing techniques, is the *agricultural or crop-calender* for the area to be surveyed/mapped. This most appropriate time is that period in the growing season when the crops that we want to map, are in a development stage or have undergone a particular management operation, which makes them well distinguishable or identifiable individually by the characteristics of their image.

The agricultural calender should show for all the important crops in the area when soil preparation takes place, when sowing, etc. Preferably it should also depict symbolically any management operation which is specific for a particular crop/land-use and the result of which should be visible in the photo-image, such as ridging (potatoes, groundnuts) hilling up (yams), spraying, fertilizer application, interrow tillage, pruning, harvesting. Also an indication of planting distance in the case of row crops/trees is useful. It may further be useful to indicate in the crop calender information with regard to wet and dry season to assist in the determination of suitable periods for executing the photography (no cloud cover). If such an agricultural calender is not already available, we shall have to gather all the necessary information to make one ourselves. An example of a crop calender is given in Fig 1.

It is usual that provision is made in a survey for development for photography to be flown for the express purpose of a land use survey. More often than not, the land use surveyor has to make do with already existing photography that has been flown for the purpose of e.g. photogrammetry or soil survey. In that case the timing of the photography within the season may not be ideal for the interpretation of land-use or crops. On the contrary, for photogrammetry as for a soil survey photography is preferably flown at a time when there is as little as possible vegetation cover hiding the terrain features and soil surface. The photogrammetrist moreover wants a 'thin' negative, he is only interested in relief and physical infrastructure and not in the fine 'nuances' in greytone which are so important to the land-use surveyor. The photographs may furthermore be far from recent; this is no problem in the case of relatively static phenomena, but far from ideal for a land-use survey on which to base development planning, particularly for areas of dynamic land-use. In that case, without an undue amount of supplementary fieldwork, we can only make a land-use map showing the *land-use at the time of the photography*.

When the photography is any time of the year, the agricultural calender for the area has again an important function: it will show us which crops we have to expect in the field at that time of the photography, and in what stage of developent they will be. With our specialist (and local) reference knowledge with regard to those crops and their (local) management and with experience in the art of photo-interpretation, we can then figure out what those crops should look like in the photograph at the given time (and scale!) of photography.

Scale of the photography

The degree of detail in land-use classification depends on the ultimate mapping scale required, which is largely determined by the objectives of the survey (and economic considerations); these in turn dictate the scale of the photography required. We can distinguish three different types of surveys:

Reconnaissance level surveys. Here one can see the mapping of broad categories of (present) land-use as determined by predominant land cover such as agricultural land, grassland, forest, waterbodies. These surveys may serve to delinate zones where semi-detailed study is unwarranted and others where future work should concentrate. In other words, a reconnaissance level land-use survey 'by cover' is only adequate for assisting in the identification and justification of follow-up surveys. They are valuable for providing broad level information for equally broad level, national or regional, policy decisions. See also Chapter 34 and 35.

The photo scale is determined by the scale at which the cover-categories to be mapped can be consistently (!) identified with the degree of accuracy required. The 'Manual of Photographic Interperation of the American Society of Photogrammetry' mentions in this respect that scales of 1 : 30,000 or smaller are adequate for this level of land-use survey. Others mention as scale limits 1 : 30,000 to 1 : 100,000.

The mapping may be at a yet smaller scale, depending on the simplicity of the land-use picture. Normally, however, it is best to present the survey data at the same scale as the aerial photographs for optimal presentation for the effort spent. Financial considerations will anyway dictate that we should order photography of no larger scale than required for the consistent interpretation of land-use at the level of detail required at the degree of accuracy as stipulated: doubling of photoscale means quadrupling the number of photographs which hampers the overall view during the photo-interpretation and considerably increases the effort and time to be spent on interpretation and transfer of the information to final mapscale (problems of generalization).

Semi-detailed level surveys. Here the mapping is more detailed: croptypes, classes of grazing land, forest types, details of plantation crops, functional categories are more prominently mapped. Photo- and mapscale is usually around 1 : 20,000. The data from semi-detailed level (pre-feasilibity) surveys serve a variety of planning decisions and provide quantitative (area) information in relation to geographic location on land-use/cover, as a basis for planning changes therein. Conclusions from these data form the basis for national regional policy deci-

sions and decisions on programmes/projects which will then require more detailed (pre-investment) surveys to establish project feasibility and cost/benefit ratios.

Detailed level surveys. These supply data on what crop-species/varieties and other types of vegetation (grass types, tree species, etc.) are in the individual fields. These surveys may need aerial photography at scales of 1 : 10,000 or larger.

Since the cost of photography and interpretation mapping per unit area is relatively high, these surveys are usually confined to representative sample areas. Such detailed land-use data are only required for land reform, reallocation and for the planning of agricultural operations at the individual farm level where investments/loans to farmers are at issue for productivity improvement, specific land-use planning and soil/water conservation. Such sample area information (pure class areas) is also required as an input to 'supervised computer-aided classification' of satellite scanner data.

In general we can say that the success of the photo-interpretation in terms of ease and accuracy, is determined by the combination of proper timing and scale (and quality) of the photography in relation to the character of the land use (scale and complexity of farming and cropping pattern) in the study area, the interpreter's specialist and local reference level, his experience and skill in this work.

Photo-image characteristics employed in land-use interpretation

In the photo-interpretation for mapping land-use, the following characteristics or features observed in the photo-image, are employed for identification and classification:
a) greytone (panchromatic and black and white infrared photography) or colourtone (full colour and colour infrared or 'false colour' photography)
b) shape and size; shadow
c) pattern or texture

d) stereoheight impression
e) associated features

Greytone or colourtone. Photographic grey-tone/colourtone of the main categories of 'objects' at the earth's surface (vegetation, soil and water) can be explained on the basis of the reflectance characteristics of these objects in combination with the characteristics of the film/filter combination used in photography. Photographic greytone is only of relative value since it is, apart from the reflection intensity of the object, also influenced by the photographic process which can even upset greytone comparability within the photograph ('artefacts'). The human eye can, moreover, only distinguish some eight to ten (roughly 2^3) grey levels. Fine differentiation or consistently correct classification of objects on the basis of photographic greytone alone is therefore not possible.

Colourtone as in colour/false colour photography forms a better basis for interpretation; in the first place the human eye can distinguish many more shades of colour, and secondly the 'natural colours' in the full colour photography are more easily associated with the objects. Only if the relationship between radiance from the 'object' and the resulting photographic greytone/colourtone is exactly known (exposure time, characteristics of film and filter used, developing and printing precisely controlled), the absolute reflection intensity (radiance) values can be derived from (negative) film densities measured by micro-densitometry. Modern scanning techniques, however, allow us to register the absolute reflection intensities, and that up to 2^8 levels, and use these for computer-aided classification maps of land-use.

In chapters 18 and 19 the relationship between reflectance of earth's surface objects and the way these are registered by various sensors, is discussed further.

Shape and size; shadow. Shape: many objects representing a land-cover-category can be identified by their characteristic two-dimensional shape seen in the photo-image. Particularly in wide angle photography and away from the center of the photograph we may also observe something of their side-view.

Size: the dimensions of an object in the photo-image depend on the scale of the photography. It is therefore good to compare those dimensions with those of (a) known object(s) seen in the same photograph; in case of doubt, it is recommended to measure the dimensions in the photographic image and multiply these by the scale factor of the photography, to get an idea of the real dimensions in the field, on which to base our identification. Two objects (e.g. a house and a doghouse) may have the same shape in the photograph but to know properly their identity we need their absolute dimensions. What goes for the objects also goes for linear features: road, railway, path, canal or ditch, etc. Also, what are the linear features: plow furrows, crop ridges, mowing swaths or plantrows. It may be important to know the exact distance in the field between plantrows/plants, since these are often characteristic for the crop type or variety.

Shadows can also be of considerable help in our interpretation by revealing shape (in side view) and (relative) height of objects (trees, building, structures). The presence of a belt of shadow along the off-sun side of a field will so give information on the height of the crop grown. On the other hand, shadow can disturb our interpretation by obscuring areas (cloud-shadow) and affecting the greytone of the photographic image. The internal shadows inside a canopy of coniferous forest causes it to be darker than the canopy of a forest of deciduous trees. Also it causes the image of the crown of oaktrees (looser structure) to be darker than that of the compact crowns of beechtrees.

A negative shadow-effect is the so-called 'hotspot'. The 'hotspot' occurs on photographs around the shadow of the airplane where the sun radiation and reflection are absolutely parallel. The hotspot is the result of lack of shadow due to the parall of radiation and reflection. Another phenomenon is reflection on a smooth water surface situated between the sun and the camera. Most surfaces reflect diffusely but on smooth water mirror reflection occurs which can be a help in identifying open water, e.g. irrigated rice fields, swamps, irrigation canals (see also chapter 18 mirror or reflex reflection). In such a hotspot, details of structure disappear (Bunnik, 1978, developed a special method of radiance measurement in specially created hotspots).

Pattern or texture. Pattern is one of the most important photo-image characteristics in the interpretation of cultural land-use. By 'pattern' we mean the spatial arrangement of macro-features in the photo-image. These macro-features are formed by objects, individually visible in the image as dots, squares, lines, etc. These may be arranged in grid patterns, offset patterns, radial patterns, concentric patterns, in parallel line patterns, etc. Patterns can be natural (e.g. wave patterns, cloud patterns, drainage patterns), or man-made (cultural) such as settlement patterns, parcellation or 'field' patterns, tillage patterns, planting patterns, spraying and harvesting patterns.

There are typical spatial arrangements of fields which reveal a particular land-use/crop, such as the large block patterns of tropical estates (sugarcane, oilpalm) in level areas; the often very narrow paddy-fields layed out according to the terrain-contours in hilly/mountainous terrain; the typical subdivision of irrigated rice-fields by small dikes (levees) to maintain water-depth within acceptable limits in areas of gentle slope. Particularly treecrops and vine-crops (grapes) may in specific areas have specific planting patterns: in a grid, in offset or a parallel line patterns, straight in flat terrain and following the contour in hilly terrain.

The planting distance in such a pattern can also be specific (measure in the photo) for the crop category. In the case of fruittrees, the dot (tree crown) size within the same variety may vary with age of the trees but the planting distance is determined by the ultimate size of mature trees and – in semi-arid areas – their water requirements.

We speak of '*texture*' when we cannot detect the individual 'objects/elements' in the photo-image, but can see that they are there by the impression of surface roughness in the photo-image. This can be due to the smallness and close arrangement of the objects or the small scale of the photography. We can describe texture in terms of coarse, medium or fine and granular, linear, etc. Often only subjective terms as 'cloudy', 'fleecy', 'canvas-like', 'fibrous' will properly characterize texture. Absence of texture is 'smooth': in large-scale photography, a still watersurface will be smooth, in small-scale photography, however, a grass-surface will also look smooth. See also Chapter 18, 19 and 20.

Stereoheight impression. Stereoheight can be seen by studying two 'subsequent overlapping' photographs in the run in their proper position under the stereoscope. Thus we get a (slightly exaggerated) height impression of objects/terrain features. This height element is very important to distinguish tall crops from low crops, types of buildings, but also the position of a particular tract of land in the terrain can be of great importance. This touches on the relationship between land-use and landform, which will be discussed later.

The threshold value for stereoscopic detectibility of height for an object (minimum difference in height between the object and its surroundings) can be calculated with the empirical formula

$$\frac{1}{3}C \times 10^{-6} \times N.$$

For the commonly used super wide angle lens this is

$$\Delta h_{min} = 0.50 \times 10^{-4} \times N \text{ meters}$$

This factor 0.50 is for a camera with focal length of $c = 150$ mm; for any other lens this factor becomes $\frac{c}{150} 0.50$.

$C = $ focus in mm. N is the scale-factor of the photography.

For photography with a c = 300 mm lens and scale 1:20,000, the threshold value Δh_{min} will thus be $\frac{300}{150} \times 0.50 \times 20,000 = 2$ meters.

In this case, an object has to be at least 2 m higher than its surroundings, if we are to begin observing it to be higher than its surroundings. This formula is very helpful to reason out what it means in terms of height, whether an object shows stereoheight in the stereomodel or not.

Associated features. These are objects or situations observed in the photographs which themselves do not form an integral part of the land cover, but which by their presence indicate a certain land-use.

Thus, the presence of a combine harvester in a field indicates a grain crop, a milking machine or its tale-telling marks indicate dairy farming, etc. A typical desert environment (sand dunes) suggests the occurrence of date-palms; tea-cultivation is generally associated with a whole network of pathways along the contour for reaching all parts of the area for picking the tea, also by the presence of a large white factory building but it is usually also associated with hilly/mountaineous areas, coconut cultivation with beaches, paddy cultivation with valleys. This use of 'association' is a typical illustration of the value of specialist (and local) reference knowledge on the part of the interpreter.

It is on the basis of varying combinations of above photo-image characteristics as 'converging evidence', that an identification/classification is made. If an identification of land-use/cover is not possible, at least areas or objects can be classified on the basis of the (combinations of) photo-image characteristics.

Photo-interpretation keys

As has been mentioned in Chapter 16 the photo-key interpretation method is appropriate for land-use surveys. For large area surveys and particularly to guide more interpreters, with less experience/knowledge of the local land-use, photo-interpretation keys can be a great help to ensure accuracy and consistency in the interpretation. One type of photo-interpretation key is the 'selection' key, the other the 'elimination' key.

The selection key is built up by cutting from the overlapping portions of stereopair photographs the corresponding parts of areas representative for each of the categories to be interpreted and mapped. These mini-stereopairs are then fixed on a card in such a manner that they can be viewed in stereo with the aid of a pocket-stereoscope. The name and mapping symbol for this category are written above these photos whereas underneath them a systematic description of the most category-specific photo-image characteristics are written.

If one crop type is grown in the area under different management systems, e.g. in the case of rice (transplanted and broadcast), or in the case of fruit trees (as highstem individual trees in a grid system, or as lowstem, pruned) or in lines, trained on wire supports, then, a key-card for each of these is made. If, e.g. due to relief, crops are in various stages of development at the time of the photography, separate cards for these should also be made.

The interpreter employs this key in the following way: each time he comes upon a unit in the photo-cover, of which he has doubts about the land-use/crop, he looks among the cards in his key until he finds one with a photograph in which the image characteristics seem to conform with those of the unit. To make quite sure, he then verifies against the text on the card if the photo-image characteristics in key and photo-cover really agree and then decides on the identity of the land-use/cover.

By occasionally checking, throughout the interpretation, a number of interpretations against the key, consistency and accuracy will be maintained.

An *elimination* key consists of a sequence of paired questions with regard to what can be seen in the photo-image in question, progressing from the general to the specific, narrowing down to the correct interpretation by each time eliminating that possibility which does not apply (agree with the image characteristics seen) in the unit to be classified. This type of key is also called a 'dichotomic' key because of the branching out via successive bifurcations, of the alternatives.

It should be noted that a particular key is only valid for the general area for which it has been prepared (character of land-use and crops), the general time in the season and scale at which the photography was made (the aspect of land-use/crops keeps changing with time, and visibility of characteristics as pattern and stereo impression decrease with decrease in scale).

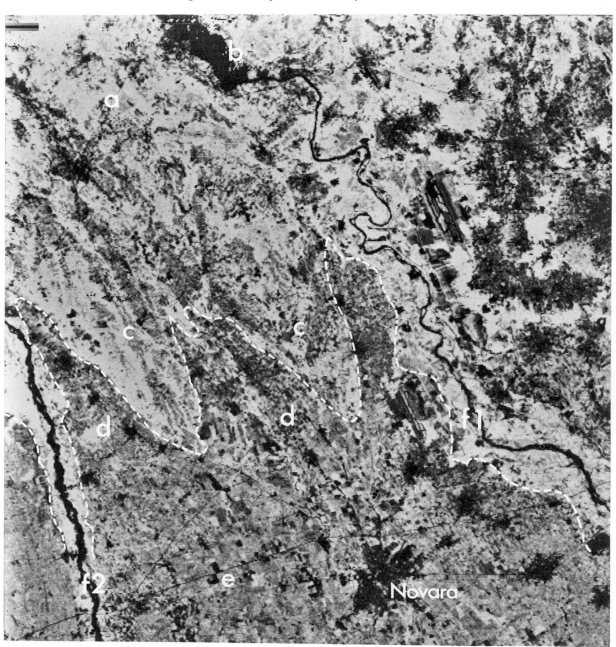

Example 1: N. Italy, Novara area.

Relation land-use/landform

The height perception obtained by viewing stereopairs of photographs under the stereoscope, affords us the opportunity of studying the land-use in its – often very close – relation with the terrain. This point can be well illustrated by two examples.

Example 1. In the Novara area, in Northern Italy, we can distinguish from north to south:
1. mountains and foothills (a) with as land-use/cover: forest on the steeper slopes and meadows in the valleys and on the gentle slopes; rock outcrops;
2. lakes, (b) scoured out by south-moving glaciers, which left elevated tongues of moraine deposits (c) pointing south. The latter, composed of coarse sand and gravel, carry small-scale cultivation of grapes; the flanks of the drainage-ways cutting through this formation, are under forest;

3. infilled valleys and an apron of gently sloping sandy outwash material (d) from (2) are under small-scale dryland farming of small grains, maize, meadows and small box-like plantations of poplar trees;
4. south of this we enter the vast alluvial plain of the Po river (e) and its tributaries (Ticino, Sesia, etc.). This low lying, nearly level area of loamy soils is in use for the large-scale cultivation of mainly rice, but also wheat, maize and some meadows, all completely irrigated and fully mechanized. Within this zone, however, we encounter some (slightly elevated, sandy) remnants of moraine formation; these stand out by the dryland cultivation of maize and wheat mainly, and settlement;
5. the south flowing Ticino and Sesia rivers (f1 and f2) are flanked by belts of lowlying flood-prone pastures, poplar and willow plantations; their beds consist of very coarse material, from pebbles to boulders (e).

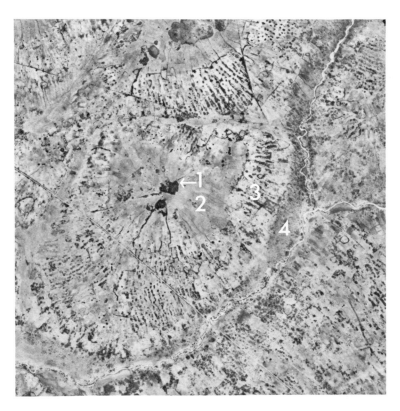

Example 2: Tanzania, Mwanza area.

There is thus a very close relationship between landform, terrain condition, soil type, drainage condition, irrigation possibilities, with scale and technique of farming, field size, shapes and arrangement, and choice of crops.

Example 2. Another example, for the tropical-zone, is in shown for Tanzania, the Mwanza region, southwest of Lake Victoria. The landscape consists of Inselbergs, granite rock outcrops, footslopes with a seepage zone and hardpan soil, sloping down into narrow and broad valleys (mbugas). The toposequence or 'catena' strongly influences land-use possibilities:
— the rock outcrop areas (1) form relatively important catchment areas and have, with their immediate fringe of coarse quartz--sandy soil, to be kept under bush/forest, i.e. no cutting or grazing;
— the footslope zone (2) consists in its upper part of finer sand/loamy sand which has been washed downslope: during the rainy season (March–June), maize, sorghum millet, cassava, groundnut, and some cotton are grown here, usually in very broad-based, high and widely spaced ridges; during the dry season the stubble is grazed;
— below this zone follows a more gently sloping belt of hardpan – underlain shallow soil with sandy soil from above fingering out downwards over the outcropping hardpan. In the contact zone (3) between these two, a seepage zone is found where the water from above is forced to the surface, offering possibilities for small-scale irrigation of rice, sweet potatoes and vegetables. Some drainage channels running down from the rock outcrop are well stabilized by natural vegetation of bush, shrub and grass; their water is also utilized for small-scale irrigation. The hardpan soil is subject to strong surface wash and sheet erosion. It is hardly suitable for any form of crop cultivation and is mainly used for extensive grazing;
— again below this, we find the narrow to very wide valley bottoms (4) with heavy to very heavy, grey to black vertisols (shrinking and cracking when dry, swelling when moist). During the wet season these valleys are in use for rice cultivation there where the flooding is not too deep, and also for maize and sorghum; in the dry season, chickpeas and sweet potatoes are grown on residual moisture. These 'mbugas' also offer very valuable grazing.

In this case forced resettlement into a roadside settlement seems to have disrupted the traditional way of scattered homestead settlement, shifting cultivation and extensive grazing over the entire area, so well adapted to the spatial and temporal variation in soil and moisture conditions. This has resulted in exhaustion by overcultivation and overgrazing around the nucleated settlement areas and abandoning of cotton cultivation in favour of food-crop production.

It goes without saying that the use of correlation with terrain is a help in the preliminary photo-interpretation. Field check, at least samplewise is an absolute must in order to see if the inferred land-use is present (see also Chapter 20 and 29).

Because of the importance of landform/land-use relationships for the delineation and understanding of land-use, due emphasis is put in the Rural and Land Ecology Survey Course at the ITC* on considering and mapping land-use in its relation to landforms. Students here are trained in the analysis and mapping of both these aspects of the land, if possible in one map with one compound legend showing the relations between the two and including information on relevant characteristics of soil, hydrologic conditions, erosion features, etc. Already before the photo-interpretation, therefore also, the photographs are layed out in their proper relative position and a quick stratification of the area is made on a transparent overlay, delineating strata on the basis of landform, drainage and land-use pattern; now and then a representative photopair is studied under the

* ITC: International Institute for Aerial Survey and Eearth Sciences at Enschede, the Netherlands.

stereoscope to verify landform content and boundaries (see Chapter 20 and Appendix 29).

NB: Satellite imagery is also regularly used as a basis for stratification, when available.

Reflectance characteristics of land-use features and their expression in terms of fotographic greytone or colourtone

The use of various film/filter combinations on their own and particularly in conjunction, can considerably assist in the interpretation of land-use/cover types. It may help us to see more clearly particular photo-image characteristics, better differentiate land-use/cover types.

In Chapter 18, the reflectance of vegetation is treated in more detail. The advantage of infrared radiation in combination with light has been mentioned and explained. The common use of false colour photography (colour infrared) for land-use survey is therefore clear (see Table 2).

Table 2. Colour of healthy green vegetation on false colour photographs.

	Reflected	Resulting colour
Healthy green vegetation	(very little) blue little green very little red much infrared	black little blue very little green much red

Not only the fact that the reflectance in the IR is high, but also that insolation is high and the bandwith is broader (0.7–0.9 µm) causes the amount of IR radiation registered from vegetation to be high and thus its image in the false colour photograph to be very red (with only a little blue tinge).

Coniferous forest internal shadows appear in darker red than a broadleafed forest canopy; soilsurface, when moist, appears in bluish black; a dry soil surface in pale blue. Clear water becomes black, like shadow-areas; muddy water appears in blue tones in the false colour photograph. In all these cases the actual colour depends very much on proper exposure, processing and printing.

Due to the absence of a blue-sensitive emulsion layer and the use of a yellow filter in the photography, false colour photography is not affected by atmospheric haze so that the chances for successful photography in false colour are higher than in the case of true colour photography!

The above knowledge on reflectance properties and film characteristics is a great help in interpretation of greytones and colours in photographic images for the mapping of land-use/cover.

Crop condition

Aerial photography can also play an important role in the survey for crop condition. What happens, when plants loose their vigor due to the incidence of diseases, pests, shortage or excess of water or minerals?

In all these cases the activity of the chlorophyl (photo-synthesis) is reduced, resulting in a decrease in the absorbtion of visible radiation and thus an increase of the reflectance in the visible part of the spectrum, particularly the red. The chlorophyll itself may be affected, unmasking the more yellow (= green + red) pigments.

If the above disturbances also affect the mesophyl causing it to collapse or the airspaces to fill up with foreign substances (fungus), then also the reflectance in the infrared within the 0.7–1.35 µ wavelength range will decrease. It should be easy to reason out what the effects on photographic greytone/colourtone will be:
— a lighter greytone in the panchromatic photo-image;
— a darker greytone in the B & W infrared photo-image;
— a more yellow colour in the full colour photo-image;
— a less red colour in the false colour photo-image.
Then, we should not forget that in the field we always have to do with a complex of vegetation and underlying soil or water (irrigated rice). When there is a loss in crop.vigor by above causes or due to the crop entering maturity, the

canopy starts to thin out and where the leafcover disappears, soil (or water) starts getting exposed to the 'sensor', and its reflectance (generally higher in the visible and lower than healthy vegetation in the infrared) starts influencing/dominating the total reflectance from the crop/soil complex. This happens sooner in the infrared (six leaflayers needed for optimal reflectance) than in the visible (only two leaflayers give already optimal absorptance).

The effects of exposure of bare soil (and equally of dead crop litter) enhance this effect on the photographic greytone/colourtone of decrease in crop-vigor because most soils reflect highly in the visible part of the spectrum and less than vegetation in the infrared.

The photo-image of mature small grains like barley or wheat may become very dark in the B & W IR photograph, which is most likely due to strong internal shadowing within the standing crop. The relevant areas can only be distinguished in this photograph from areas of water, on the basis of fieldshape or harvesting pattern. The mature crop is rather light grey in pan, yellow in the full colour and hard (bluish) green in the false colour photo-image.

It is logical that any deposit on the foliage of a crop, such as dust, chemicals, fungus mould, etc., also affects the crop's grey-/colourtone in the photographic image.

It should be realized, though, that any disease or pest attack which does not affect the reflectance pattern of the upper strata of the crop-canopy does not usually find expression in the photo-image! Thus damage by leaf-eating insects will generally not be noticed in the photo-image as long as there is still enough foliage left to cover the soil.

In the same way, disease-attack which starts in the lower leaf-strata remains unnoticed until the damage has reached the upper strata after severe damage may have been caused already. Damage to the roots and water conducts, or certain mineral deficiencies/toxicities which affect the reflection of the upper leaves, does find expression in the photographic greytone/color-tone of the canopy, however.

Flowering can also noticeably change photographic grey-/colourtone. By careful timing, this phenomenon may be utilized for e.g. differentiating between different types of fruittrees (almonds versus citrus), on the basis of differences in their time of flowering.

Land use change

Change in land-use can be established by sequential photography. Repeated observation to watch the character and speed of change is called 'monitoring'. In this way we can 'follow' processes like urbanization, deforestation, erosion, land reform, irrigation, etc. Crops are often better characterized by the way their aspect changes with time in accordance with the crop-calender, than by their aspect at a certain moment. It is usually only for research purposes that repeated photography within one season is done; it would be far too costly otherwise.

Repeated photography, usually multiband photography (e.g. the combination of panchromatic and B & W IR or of panchromatic and colour infrared film) is, however, done on a routine basis in the UK for the detection and monitoring of disease attack in crops. Considerable success has in the past been achieved in this way in establishing the cause for the spread of disease in potatoes, by carefully mapping the infection nuclei and their mode of spreading by way of aerial photography. In general, however, the use of common photography for this purpose is rather expensive. High altitude photography but especially satellite images are cheaper in use after initial cost, because they remain long in space and can make infinite numbers of images for only recording and processing cost.

High altitude photography

High altitude photography, producing photographs at scales from 1:60,000–1:135,000 is extensively used in the USA for resources inventory and pollution detection for agricultural monitoring. Use is made of one camera with panchromatic and one with false colour (not haze-sensitive!) film with a 150 mm lens a photoscale of 1 : 80,000, and with a 210 mm lens a photoscale of 1 : 58,000 is obtained. The quality of the photograph allows blowing up to larger scales, due to the high spatial resolution afforded by photography. The overlapping photography provides stereo-capability! For many areas of the world, however, this type of photography will be too costly to produce compared with Landsat, which can produce 1 : 2,000,000 images. The purchase of e.g. B & W MSS imagery to produce (diazo) false colour composites is a much cheaper possibility within anyone's reach.

The Russian manned spacecraft Salyut 6, in a 260 km orbit, carried an MKF-6 multispectral camera system. Exposed film is recovered and new supplies are brought by the Soyuz and Progress spacecraft. The system consists of six separate cameras of 125 mm focal length and 55 × 81 mm format, with their appropriate filters; ground resolution is 20 m. The configuration is roughly comparable to the S-190A system carried in the American Skylab, which gave a lower resolution because of its higher orbit. Excellent imagery in B & W, full colour and colour infrared can be produced of this multiband photo-material. The same camera system is also being used in aeroplanes.

The American space-shuttle carried a Zeiss metric aerial camera with 30 cm focal length and 23 × 23 cm format as part of the European Space Agency (ESA) and NASA joint venture. From the 250 km altitude, the photographs cover an area of 190 × 190 km. The ground resolution is 20 m; acquisition scale 1 : 820,000 in B & W and Color IR; for some areas, there is 80 % overlap between successive photographs. At a later stage, a 60 cm focal length camera will be used, and still later, the camera-system is planned to operate from free-flying satellites. There are thoughts now of developing a world-wide data base of high resolution stereo-photography in this way.

Multispectral scanning for land-use mapping

We have seen that the photographic film/photograph is not a very good source of information on the intensity of radiation reflected by earth-surface objects: only if the entire process of exposure, processing (and printing) is meticulously controlled, we may obtain absolute values for radiance intensities from photographic greytone (film-density) by the tedious work of micro-densitometry (on the negative). The other photo-image characteristics such as shape, (relative size), pattern texture, stereoheight and associated features are usually much more important in photo-interpretation for land-use mapping.

Shape, pattern and texture can be observed in great detail in photography of the appropriate scale (and timing), because of the high spatial resolution of photography. High quality photographs can even be several times enlarged to show more detail before the 'grain' affects image sharpness. Also the possibility afforded by (overlapping) photography for viewing an area in stereo, thus judging the height of objects and also seeing relief to study land-use in its relation to landform, is a great asset.

The modern technique of multispectral scanning, however, provides a much better opportunity than photography, for utilizing the reflectance characteristics of earth surface objects for their classification. In Fig. 18.4–19.1 we saw that the main classes of objects at the earth surface have quite different reflectance patterns. In the early development stages of multispectral scanning (MSS) it was thought that each category of objects (e.g. each crop) had its own specific pattern of reflection values in the various wavelength bands. The only thing to be done for automatic crop recognition by computerized (reflection) pattern recognition, was to construct a catalogue of the typical reflection patterns of crops. It has by now been realized that crop-types occur in so may (often local) varieties, grow under so many environmental conditions and management systems, and change their aspect so much over the growing period, that 'spectral signatures' (reflection patterns) only have very local and temporally restricted application value.

In several parts in the world, e.g. South

America, strong relief and differences of a socio-economic nature cause great differences within crops and their management even over small distances. In the wet tropics a crop-like rice can, at one point in time within one area, be encountered in any of its development stages; crops are, moreover, seldom grown in pure culture but in a great variety of mixtures with other crops, whereas fields are often so small and their spatial patterns so complex that the sensor cannot pick them out individually. Even in less complex situations, the classification of land-use on the basis of the digital reflection values, with the help of the computer, generally requires intensive interactive work in which the interpreter with his specialist and particularly also local knowledge plays a decisive role. In Chapter 21 the most feasible techniques are treated. Possession of a good agricultural calender for the area is here also a necessity.

The MSS technique can measure and register reflected radiation in narrow portions of the electro-magnetic spectrum which can be carefully defined to correlate with important parameters of crop quality and quantity; this precision in band selection is not possible in photography, whereas the MSS scanner can also measure outside the photographic λ range, further into the reflective and even in the emissive (heat-radiation) infrared. In research work, airborn scanners are used which can measure in 12 and even more separate wavelength bands. It is now generally accepted that the combination of three well located spectral bands will give the optimal amount of information needed; adding more bands usually adds to cost and time required but little to the classification result.

Fig. 19.1 shows in the lower part which bandwidths are employed in the satellite-borne Landsat and SPOT scanners. For Landsat-D only four out of the in total seven bands of the Thematic Mapper have been plotted; these are the most important of the seven, for application in vegetation/crop mapping. Images produced for visual interpretation can be B & W images for the reflection intensities in individual bands. They can be false colour composite images for two or three bands, each coded in a different color, usually blue, green and red for bands 4 (green band), 5 (red band) and band 7 (near IR band), respectively. Other band combinations (e.g. band ratios, principal component values, biomass index, etc.) can also be used as a basis to make (colour) images for interpretation/classification; they enhance differences between, or specific aspects of, land-use categories/crops or their condition, e.g. the intensity of photosynthesis, the leaf area index, green biomass and even crop-temperature which is a measure for transpiration / moisture-stress, which may be closely correlated with plant-production.

In automatic computer classification (see Chapter 21), the computer compares the patterns formed by combinations of (radiometrically corrected) in-band reflection values for all ground-elements (pixels). By applying statistic decision rules it classifies all these pixels on the basis of degree of comparability of their reflection patterns for two or three bands.

In supervised classification (see Chapter 21) for each land-use class to be mapped, a number of representative areas in the field are indicated to the computer by their geographic coordinates. The computer takes from the data file the corresponding reflection values for these areas in the spectral bands considered for the classification. The degree of clustering for each class sample and of separability between classes can be judged by plotting the *sample* pixels per class in for instance the band-7/band-5 'feature space' on the computer screen. The computer operator/interpreter can then try to improve segregation between clusters by applying certain mathematical algorithms and then decide on the statistical decision rules for apportioning all pixels for the study area to one of the categories on the basis of the sample data statistics. It is also possible to define 'thresholds', i.e. only pixels which lie within certain limits around the sample 'central' values will be apportioned to the respective classes, the ones with reflection values beyond these thresholds will be classed as 'unknown'. In the map outprint each class will then be allotted its own colour whereas the 'unknown' pixels will be left blank.

The one thing which should never be forgotten while marveling at the beautiful colourmap so obtained is, to objectively fieldcheck the result for accuracy of the classification.

The system of computer-aided classification can be taken one step further by bringing into the decision procedure non-spectral geographically defined information regarding terrain forms, soil types, etc., which determine the probability of certain land-use/crop types occurring in certain parts of the survey area. This approach has been termed an 'expert system' approach to classification; it will be discussed in greater detail in Chapter 21.

Landsat for land-use/crop surveying

The Landsat multispectral scanning (MSS) data and images as treated in Chapter 19 have advantages and disadvantages for the survey of land-use/crops, in comparison with aerial photography.

As experienced in Chapter 19, advantages of satellite remote sensing for land-use survey also are:
— undisturbed image of large area in synoptic view;
— repetitive coverage giving possibility to detect land-use change;
— the facility of visual presentation and communication, e.g. to demonstrate to the authorities responsible for land-use policy where measures are urgently required.

The MSS images are very useful for a first stratification into main land form/land cover regions or strata: large areas under one land-use (forest, estate plantations) can be easily delineated. It may also be possible to delineate areas of complex, small scale land use which are homogeneous in their composition. Those latter areas are classified as complex land-use categories. Areas with too many 'mixed pixels' because each pixel on the ground contains several land-use types (crops), defy further analysis. In

this case the MSS imagery serves to determine where aerial photography with (the always necessary) fieldwork will have to unravel the land-use picture in more detail if so required. The spatial resolution of MSS data is generally too coarse to detect any of the elements employed so prominently in aerial photograph interpretation for land-use mapping: pattern, texture, shape, shadow and associated features. Future generations however, may have finer resolutions and so come nearer to photographs with synoptic view as the present satellites but with the great advantage over photographs of the possibility of unlimited repetition. Stereovision, the great advantage of photographs, is lacking for vegetation and land-use purposes. The future satellites may provide stereovision to distinguish land forms, but probably not for differences in vegetation height: SPOT!

Integration of satellite data,
selective aerial photography and fieldwork

Integration of satellite data with selective aerial photography and fieldwork is a normal procedure. The aerial photography can take the form of 'flying' only parts of the area, selected on the basis of their representativeness for the land form/land cover types in the study area. With the help of this photography and limited fieldwork, the land-use/crop identities of a number of (large and well locatable) 'fields' per land-use category are established, to serve as training samples for supervised computer classification of the land-use for the entire study area.

Also integration of aerial photography, satellite MS data and radar imagery, both spatially and temporally, is possible, each contributing part of the information required.

Multi-temporal biomass image
for landcover classification

Apart from spectral and spatial (including stereo-height) characteristics of the landcover, as presented in the digital data and/or images, also the way in which the landcover changes with time, may be an important classification criterion.

In an experiment on crop/landuse identification in N. Italy, the Landsat MSS data of three critical points in time during the growing season were employed to test the possibilities of using this factor "change".

By using the "Normalized Vegetation Index" (NVI), it is possible to calculate for each of the three dates for each image-pixel one value which is representative for "Green Biomass" on the corresponding ground-area.

NB: The NVI $= \dfrac{17-15}{17+15}$, in which 17 and 15 stand for the reflection intensity in band 7 and band 5, respectively.

On the basis of the crop-calender (see Fig. 1) for this area three critical periods were determined in which to obtain Landsat MSS data:

— in May all ricefields are under water; winterwheat (and other small wintergrains) form a complete green leafcover; fields for grain-maize are ready for sowing/just-sown, i.e. bare soil.

— in the first half of July the winterwheat would be fully mature, ready for harvesting; the grain-maize, sown in May, would constitute a full green leaf canopy;

— in October, the grapes would still have a leafcanopy, all other crops would have been harvested or are ready for harvesting (rice).

Grassland and forest remain green throughout (with the exception of the odd parcel of grassland/forage crop mown shortly before the Landsat overpass).

By colour-coding each pixel on the basis of its Green Biomass Index value, three biomass pictures made.

By colour-coding biomass for the May coverage in red, that for July in green and that for October in blue, and then carefully electronically registering (super-imposing) the images, a colour-composite image was produced.

This image shows in its compound colour, which areas had a green biomass cover only in May, only in July or only in October; further, which areas had a green biomass cover in May and July or in May and October or in July and October.

Comparison with the crop calender made it possible to decide what landcover type(s) were indicated by the respective colours.

class \ month	May	July	October	Compound colour
water	o	o	o	black
grassland	+	+	+	white
forest	+	+	+	white
rice	−	o/+	o	black/green
wheat	+	o	o	red
grain maize	o	+	o	green
fodder maize	(wheat)	+	o	red
grapes	o	o	+	blue
fodder oats	+	+	o	yellow
Colour coding	red — green — blue			

+ = high biomass
o = no

By overlaying the image with a transparency at the same scale of the main infrastructure (roads, rivers, settlements), the location of a number of large fields per land-use category can be established via the aerial photography and their colour-based interpretation in the multi-temporal Landsat image checked against the real land-use as established by aerial photograph interpretation, supported by fieldwork. For the large fields this Landsat image interpretation turned out to be highly accurate. The area, however, contains a considerable proportion of land under small-scale farming with complex land-use and cropping patterns, resulting in a large proportion of mixed pixels. The areas concerned would have to be tackled by aerial photograph interpretation to get an idea of their class composition or have to be delineated as complexes of a known variety of crops. The much higher spatial resolution of SPOT data might well go far in solving this problem. A considerable problem constituted the rice areas: in July, these show a considerable variation in colour between black (open water, no biomass) and green (full leafcover above the water), overlapping with the colours of other land-uses. Here the use of the band-7 (near IR) image for May proved of great value. In this image all water surfaces appear in black; by subtracting all non-rice waterbodies (rivers, lakes, reservoirs shown in the topographic map), a 'mask' can be made covering all the rice areas. Thus these areas can be 'set aside' and only the non-rice areas need to be considered for further analysis.

This approach can be further refined by inserting more 'expert knowledge' on the probabilities in spatial distribution occurrence of certain land-use/crop categories, obtained from any other source.

Thus, the knowledge on landform/land use relation can be exploited:
— there can be no grape cultivation in the alluvial valley;
— there can be no rice cultivation in the mountains, hills, moraine areas, infilled valleys and outwash apron zone, etc.

The spatial information on the occurrence of settlements, open water, elevation, slope, slope-aspect, supplied by topographic maps can be incorporated in the computer-performed decision process. By inputting all these items of spatially defined information, also those on climate, geology, geomorphology, soil, drainage, irrigation, etc., a complete geo-data base is built up, the data of which can be employed for the above 'expert system' approach in land-use/crop classification and mapping. Information obtained from any other form of remote sensing (and fieldwork) such as aerial photography interpretation and radar can also be utilized to further enhance/refine the classification.

Thermal scanner data for land-use/ crop survey

Thermal band scanners can be used to map areas of moisture stress in vegetation, areas threatened by nightfrost, areas of irrigation – water leakage.

Irrigation water needs of crops could be determined on the basis of crop temperature which can be correlated with transpiration rates. Particularly under conditions of limited water supply as in semi-arid areas, crop transpiration is a main measure for crop growth. Cereals and fruitcrops in general are particularly sensitive to drought during heading and flowering.

When crop transpiration rate is affected by lack of water supply to the leaves, which may be due to water shortage in the soil, salination (high osmotic value of the soil water), rootdamage or interruption of the watertransport in the plant by whatever cause, leaftemperature may rise sufficiently for the sensor to register this (even $\frac{1}{2}$-°C temperature resolution sensors are used). The heat radiation flow in the thermal range is, however, so low that the spatial resolution element in thermal scanning with fast-moving scanner systems as in Landsat 3 and D must be very large; temperatures/temperature differences can only be determined for relatively large areas.

Also soil moisture status may be judged by thermal scanning. In this context, possible locust breeding sites in desert areas may be detected from satellites well before they are betrayed by vegetation development. For the interpretation, classification and mapping of land use, thermal sensing is, however, of very restricted value.

In thermal scanning from aeroplanes, the images have a lot of geometric distortion. The influence of wind on the temperature pattern is great; well known is the influence which obstacles that are warmer than the environment have on the temperature image to their leeward side. Also the passage of clouds remains visible in the thermal image some time afterwards.

For some phenomena which are invisible to the naked eye, however, thermal images are of detective value, such as covered over (illegal) dumps of refuse; these are, however, rather examples of land-misuse than of land-use.

Amazingly, the places where many years ago the newly reclaimed soils in the 'Zuiderzee' (now IJsellake) polders had been 'inoculated' with earth-worms to stimulate their 'ripening', could clearly be traced back on thermal image-ry; the close contact between the gras-sod and the underlying soil caused a different tempera-ture-regime in the worm-sites which had appar-ently spread laterally very little. Such informa-tion is more of research value than for land-use survey.

Radar for land-use/crop survey

Particularly for rapid, large-area surveys of ar-eas which are under cloud cover nearly contin-uously (e.g. the Amazon region) or during most part of the growing season, radar can be a po-tentially very useful main or additional tool in data gathering on land-use/crops.

Radar systems working with larger than 2 cm wavelength provide good penetration through cloud and even rain.

Rapid imaging of vast areas in a continuous, wide swath is provided by this remote sensing system and when the successive swaths are cov-ered in overlapping fashion, they can be viewed stereoscopically. See also Chapter 20.

Land-use parameters which determine the in-tensity of the reflection (greytone in the image) are:
— surface roughness relative to the wavelength used;
— directional orientation (aspect) of the target area;
— moisture status of the objects.
Radar use for detecting land-use is still in an early stage.

For success in crop differentiation, marked growing seasons and pure crop parcels will be a first condition. For calibration, some crops (like sugarbeets) which have a nearly angle-indepen-dent and high backscatter during their entire full ground cover period, or else forests, appear to be quite suitable. The first large radar opera-tion was the RADAM scheme in Amazonia, Brazil. Land-use differences appeared to be mappable also.

Another noteworthy application of radar in land-use studies was that of flood-monitoring for the production of a flooding – susceptibility classification map for one of the large river – basins in Colombia.

The interpretation of the radar images in con-junction with Landsat imagery for the huge floodplain areas was supported by 1 : 10,000 scale aerial photography of small selective areas from which extrapolation via the radar and Landsat imagery was possible for classification of land-use and flooding probability and dura-tion.

Very much research on the ground, as well as from airborne and space platforms is still to be done on the interactions between radar-radia-tion, vegetation and soil, regarding the optimal combinations of wavelengths, depression an-gles, flight direction, polarizations, timing, (pre)processing of the masses of data involved and geometric correction of the data for their combination with data from other sensors, be-fore optimal use of radar for land-use/crop sur-veys can be achieved.

Remote sensing in land-use surveys
in the developing world

Particularly for the developing world, remote sensing can offer great benefits. Availability of trained personnel, means of transportation and accessibility are often seriously hampering ground surveys, whereas the lack of information over vast areas is serious.

With growing population pressure and rapid changes in land-use (urbanization, deforesta-tion) there is a great need for (up-to-date) infor-mation on the status of land-use and the nature and rate of change therein, for management and planning purposes. Also timely information on acreages under important crops, their condition and development for timely and accurate crop production estimates is important for safe-guarding continuity in food supplies.

For restricted area and more detailed surveys, say once in 5 years or more, the use of aerial photography and its interpretation is still pre-ferred. For reconnaissance and semi-detailed surveys involving large areas, however, the combination of space-born sensor data with selective aerial photography and fieldwork, is to be preferred. Particularly also for repeated coverage, for the monitoring of change, or for a multi-temporal approach, space-borne sensor data offer great possibilities. Here, relatively low cost and simple forms of using these data can readily provide much useful information.

Standard B & W satellite images can be diazo-processed into false colour composite transparencies for visual interpretation over a light-table. With simple photographic laboratory facilities, selected parts of the originals can be enlarged to a scale of 1:250,000 and the transparent 'blow-ups' used for making such diazo false colour composites.

Overlaying a negative and a positive image of the same time and area will result in a picture of completely neutral tone. Overlaying the negative of one date with the positive of another date, however, will clearly show up areas where changes have taken place. The same can be done by using different colourcodes for the two periods.

Using digital data for computer-aided classification and mapping will often involve a disproportional financial and technical burden in having to purchase and run very costly hard- and software and face the problem of keeping the equipment going in the face of powercuts (for running and airconditioning) and poor servicing/maintenance facilities.

Non-availability of the data (due to lack of coverage for the area or problems in delivery from a far away receiving station) may be another bottleneck. Digital data processing for the production of images for visual interpretation or for computer-aided classification further requires highly trained personnel with a high specialist and local knowledge of land-use as well.

A solution to the problem of high cost of obtaining and running full-fledged digital processing facilities in each country would appear to be the installation of regional service centers.

The bitterest disputes are, however, between nations in the same region and particularly between those having common borders!

None of the present regional receiving and processing stations, each with its 28 million km^2 reach, together covering nearly the entire world surface, are in their set-up and operation really regional in character.

There is further no unity in reception capability for different spaceborn sensor systems.

Even in the case of obtaining one's processed data/images from outside, it will be necessary to have staff well versed in the techniques and possibilities of digital image processing, who know what products to request for.

To spread the rapidly mounting cost of such products their widest possible (multi-disciplinary) productive use should be assured. This goes even more for national data processing facilities!

The nature of the problem in land-use survey with the aid of remote sensing is often a special one: land-use in the developing world is often characterized by small-scale farming in spatially and temporally complex patterns of land-use and crop production. There are often no distinct cropping seasons, trees and (mixed) crops occur on the same parcel and they are under various forms and stages of crop/fallow rotations. Completely different forms of management, scales of farming and crop types and varieties can be found within a relatively small area, due to differences in relief, elevation and land tenure conditions.

On top of this, large parts are for long periods or continuously hidden by clouds, through which only radar can penetrate. In semi-arid areas, successful imagery can only be obtained during the dry season when boundaries between cultivated, fallow areas and natural grassland are often extremely vague. Under these conditions there is a limit to what can be achieved by remote sensing in land-use/crop surveying.

Spaceborn sensor systems with a high revisit frequency (temporal resolution), selective side-view capability and high spatial resolution may go far to meet some of the above problems but the volume of the data to be transmitted to earth will have to be kept manageable by some form of on-board preprocessing.

26. Mapping the Potential Natural Vegetation

J.T.R. KALKHOVEN and S. VAN DER WERF

A map of the existing vegetation represents the plant cover at the moment of investigation. This can be the very purpose of the map, but it can also be a disadvantage in view of the overall observed phytodynamics. Among various solutions to deal with this problem (see Chapter 23) an often used technique is mapping the climax or the (potential) natural vegetation. Let us first briefly discuss the background and development of these concepts and then describe the principles of construction of the potential natural vegetation. Finally some examples of maps will be given.

Succession and climax

Change in vegetation with time is a universally observed phenomenon. The long-term developments in which plant populations and communities gradually are replaced by others are termed succession. Succession can be progressive, which means generally that structural complexity and species richness as well as biomass increase. A regressive succession means a partial breakdown of the built up complexity (Whittaker and Woodwell, 1972), which is caused by natural or manmade disturbances. The progressive development after such disturbances is called secondary succession. The undisturbed vegetation development in a site that has never supported any vegetation before is termed primary succession. It is the purely natural evolution of the plant communities, in-volving, and caused by, evolution of species and changes of soil and climate. The succession of a particular vegetation can be stopped and maintained in a certain stage by regular natural events or human impact. For example, the flooding of a river may prevent the development of a forest on the borders, and the yearly mowing of grassland will prevent further succession.

Without these disturbances the development is thought to end in a stable vegetation, called climax. Since Cowles' (1899) and Clements' (1916) elaboration of the concepts of succession and climax, an amount of literature has appeared in which the concepts are amplified, criticized and modified (see e.g. Tansley, 1935; Tüxen and Diemont, 1937; Whittaker, 1953). It has resulted in a really confusing terminology. For reviews and references see Knapp (1974), Mueller-Dombois and Ellenberg (1974), Golley (1977) and Miles (1979). Although the different versions of the climax concept are severely questioned today (Miles 1979), most vegetation ecologists accept the idea of a relatively stable, terminal stage in vegetation development, provided that the macroclimate does not change and man does not intervene. This terminal stage is often termed the 'natural vegetation' and is considered to be in a dynamic equilibrium with the actual set of environmental conditions of the site. Although fluctuations often occur the terminal stages are relatively stable communities with self-perpetuating populations which may form the final stages of secondary

A.W. Küchler & I.S. Zonneveld (eds.), Vegetation mapping. ISBN 90-6193-191-6.

succession, steady-states for several decades up to some hundreds of years (Mueller-Dombois and Ellenberg, 1974). These natural communities are recognized by a relative constancy of species composition and structure and one may describe them as climaces, determined predominantly by climatic and edaphic conditions.

In fact there is not absolute stability, due to continuing evolution of species and long-term changes of soil, landforms and macroclimate. These are, however, very slow, taking thousands to millions of years and are therefore negligible in our context. For example, in peat vegetation in Kalimantan (Borneo), Indonesia the species composition remained nearly unchanged since some million years (Anderson and Muller, 1975). For the temperate and boreal zones of the earth this is a less expectable situation, since the climate has not been constant over such a long period, but still long enough to reckon with a relatively stable situation.

Potential natural vegetation: origin and meaning of the concept

In Europe where since neolithic times human cultivation has profoundly affected the natural vegetation, the natural forest vegetation has been replaced by vegetation types which differ in degree of artificiality (Long 1974). Some examples are heathlands, garrigue and maquis, meadows and pastures, planted pine and spruce forests. These communities can exist by means of a regular intervention of man and his domestic animals. After the elimination of this human influence the vegetation shows a succession to a terminal stage which is called the natural vegetation as discussed above. The vegetation types with a partly artificial structure or species composition are termed replacement or substitute communities.

This natural vegetation is not equal to the 'original' vegetation. Then what is original? To take North-America as an example, is it the state of vegetation before the coming of the white men? The Indian people, however, had

an obvious impact on the plant cover much earlier. Is it the state of the vegetation in the time before any human being was living in the area? Since these very early times the climate has changed considerably, as did the soils and therefore the natural vegetation of these old times is not comparable with the present-day natural vegetation (Küchler, 1964). It neither is the vegetation that would have grown in a site if no human impact had occurred. That would be completely theoretical definition with no practical use.

Every period of relative stability has its own natural vegetation in which the plant cover of the different sites are in equilibrium with the climatic, edaphic and hydrological conditions of the sites in these various periods. The present-day terminal stage of vegetation succession which would develop under the actual environmental conditions is called, after Tüxen (1956), the potential natural vegetation of today (die heutige potentielle natürliche Vegetation), because it is a potential and not a real existing situation. To eliminate the effects of future climatic changes and speculations about the time needed to reach this climax stage, Tüxen added to this definition that this terminal stage would be present at once ('schlagartig'). Hence the potential natural vegetation is a hypothetical vegetation, that can be inferred from the existing real vegetation, its developmental tendencies and its site relationships (Mueller-Dombois and Ellenberg, 1974). The potential natural vegetation of a special habitat is at the same time the symbol of all the existing and possible artificial vegetation types of that habitat. In other words it is an indication of the production-potential of the habitat (Tüxen, 1956, 1957; Trautmann, 1966). The habitat is described as the sum of all environmental conditions of a site, being the ecological basis of the plant cover of that site; the biota are only in so far part of the habitat as they determine the production-potential of the site. In this sense habitat is considered equal to the German 'Standort' as described by Trautmann (1966). In the following we use 'climax' and 'terminal stage' as synonyms of 'potential natural vegetation' sensu Tüxen.

The question arises to what extent the human influence is thought to be removed. The impact that has resulted in irreversible changes of the habitat characteristics (erosion and landslides after deforestation, extreme podsolization after forest deformation into heathland, desertification after overexploitation) of course cannot be undone. Man has created new substrates as is the case in the plaggen soils of the Northwest European lowland, which are built up of sods and dung of domestic animals. This resulted in a fertile soil with a thick humus layer that will be different from the original podsol soil for an extremely long time.

In a lot of places, e.g. the prairies of Northern America, the savannas of Africa and the thicket in the Mediterranean areas, the vegetation is adapted to a regular burning by men. Should the influence of fire be excluded from the potential natural vegetation? There is much evidence that in nearly all vegetation zones except the humid deciduous and evergreen forest belts and the arctic and alpine zones, fire caused by lightning is a natural phenomenon, which does not occur as regularly as human-caused burning, but has probably a more widespread influence (Küchler, 1964; Walter, 1968). The Pinus ponderosa forest of Arizona and probably many other natural pine-forests depend fully on natural fires. Even boreal forests and tropical rain forests do not seem to be totally free from (irregular) burning (Whitmore, 1975, Zackrisson, 1977). So absence of man does not eliminate the effect of fire on the fluctuations and species composition of the natural vegetation in areas where human burning is now regular. Depending on vegetation type and locality, fire often should be included as a natural quality of the environment.

Herds of browsers of the primeval prairie, steppe and woodland were a natural part of the climax in their time. It is realistic to think browsers would play a role in the potential natural vegetation of today but they will never be so important as they were.

By building dikes and regulating the course of rivers in Western Europe man has given rise to formation of fixed back swamp soils, that did not exist in the same manner before, and of polders in the coastlands. These are lying below sealevel, protected from inundation by dikes and pumping engines. The present day qualities of the habitats of these fluvial and marine clay soils will support an Elm-Ash or Ash-Alder forest as natural vegetation and not water plant communities of pondweed and seagrass (Tüxen, 1956; Stumpel and Kalkhoven, 1978). With the aim of mapping the potential natural vegetation and its application in planning in mind it is not realistic to see these dikes and large drainage systems as removable parts of the local environment.

Another effect of human impact which cannot be eliminated anymore is the intentional and unintentional introduction of alien plants and animals. Poa pratensis in many parts of North-America (Küchler 1964), Prunus serotina in forests of Northwestern Europe (Baillieux et al., 1977, Trautmann, 1976), Poa annua on the Falkland Islands (Holdgate, 1967) and Agrostis stolonifera on Marion Island (Gremmen, 1981) are only a few examples of organisms that have conquered a permanent and, often, dominant position in the regional communities.

The influence of man to set aside when defining the potential natural vegetation is the direct influence on the plant cover of the local site which has a minor and reversible effect on the production potential of the site (Trautmann, 1966). It concerns activities of man and his domestic animals like: logging, grazing, burning, mowing, hunting, fertilizing, ploughing, digging as well as pollution of soil, air and water.

Remarks and modifications

Since the introduction of the concept of potential natural vegetation and the mapping practice, especially in the Federal Republic of Germany and surrounding countries, several remarks, objections and modifications have been made. The most important may be treated briefly here.

In Northwestern Europe, where forest is considered to be the climax vegetation of nearly all habitats, there are soils that never have supported a natural forest community (Doing, 1974). The polders for example are reclaimed from the sea or drained marshland and are sown immediately with Phragmites australis, other grasses or herbs; raised bogs are totally leveled down, showing a soil that is a mixture of rest peat and sand or boulder clay, which soils are directly used for crops; plaggen soils (see earlier) were during their formation used as fields with cereals and other crops. On all these soils, however, there is a natural growth of herbs, shrubs and trees on small, not cultivated pieces of land. Woody vegetation here is hardly older than 100 years. Is this enough for extrapolating to a climax in the sense of potential natural vegetation?. The extrapolation can only be based on observations on comparable soils in the regio which support (near-)natural vegetation or on observations at habitats that support a comparable set of replacement communities. Kalkhoven et al. (1976) solved this problem by using the concept of plesioclimax (Gaussen, 1955) which is the natural vegetation 100 years after abandonment. This is no real climax for the soil types mentioned above, because the soils will not reach maturity within this period. It is the 'near-by climax'.

Gaussen advised the use of this plesioclimax especially for sites where the soil development is not yet ended and for cases in which the expected succession to a climax in reality should take at least several hundreds of years. The longer the period the more uncertain is the prediction about the terminal stage, especially for habitats where no natural vegetation is left or ever has been. The whole set of vegetation types of a habitat type, with the 100-year-old natural stage as 'final' community, is called a vegetation series ('série de végétation'). The uncertainty about the exact floristic and structural composition of the potential natural vege-tation ought not to be a serious problem if, prevailingly, one will see the potential natural vegetation as a symbol of the direction of the expected succession and of the whole series of well-known successional stages or as a symbol of the qualities of the site. For the planning practice of nature and landscape management this approach will satisfy (Gaussen, 1955; Stumpel and Kalkhoven, 1978).

The ecologists of the CEPE Louis Emberger in Montpellier, have chosen for a different solution in mapping the 'phytodynamics' of a landscape (Long, 1974). They state that in the Mediterranean ratio it is infeasible to predict the vegetation succession on the long and heavily cultivated land over more than a few decades let alone over a century. So they propose to map no more than the actually foreseeable succession stages. Godron and Poisonnet (1973) give some examples of this approach. The basic cartographical unit is the 'sequence of vegetation' ('séquence de végétation'), a concept comprising the developmental stages of a special habitat as far as they are existing in successions on comparable habitats, named after the last known stage. For recognition of this last stage the presence of self-rejuvenating trees and shrubs is important. If there are no trees or shrubs known from a habitat type, then the least artificial herb/grass community is the top of the sequence.

Another problem concerns the size of the area for which the potential natural vegetation should be predicted. The influence of evaporation of the plant cover on the micro- and mesoclimate and the hydrology of the site is size-dependent (Doing, 1974). The definition of Tüxen includes an unchanged habitat, which means that the mesoclimate and hydrological situation of today should be taken into account inalterably. In relation to the aim and use of the map of the potential natural vegetation it is not realistic to expect considerable influence on the climate. As Mueller-Dombois and Ellenberg (1974) state: 'the potential natural vegetation never will exist in the form, the extension projected on the maps'. Actualization of the potentional natural vegetation by an appropriate management in densely populated countries will not extend over a large area, but in most cases only a few square km's or less are available. The size of these patches is comparable with the patches of (near-)natural vegetation from which the potential natural vegetation often is derived. In areas where very large patches of near-natural vegetation are conservated for a development into a natural climax vegetation (e.g. prairies, northern boreal forest) no considerable changes of mesoclimate are to be expected.

Mapping the potential natural vegetation can also stimulate the research on natural communities (Küchler, 1964). Of great importance is a development in several countries of Northwest- and Central-Europe where permanent quadrats are established in (near-)natural forests in order to follow the succession and to observe the steady-state (Glavač, 1972, with references).

Some exotic and disturbance species will be part of the potential natural community for a very long time, especially long living and rejuvenating trees. They do not alter the type of potential natural vegetation but only form varieties of the main type (Van der Werf and Londo, in prep.).

Construction and mapping of the potential natural vegetation

The principles and methods of determination and delineation of the potential natural vegetation units are exhaustively described by Tüxen (1956, 1957) and Trautmann (1966), with an abundance of examples. Reading their papers is highly recommended. The treatment in this chapter can only be a brief reproduction of their work.

1. Starting point is the recognition and description of the natural and near-natural communities. In most of the vegetation zones of the earth vegetation is heavily influenced by man. In the tropical, boreal, arctic and alpine areas some scattered patches are pre-

sumably largely unaffected. The area of near-natural vegetation is in these zones much more extensive. In the highly cultivated temperate and Mediterranean regions several patches of near-natural vegetation and a few remnants of the natural plant cover are still to be found in most countries. These near-natural communities are considered to have been changed only in minor aspects as compared to the real natural vegetation. It is this near-natural vegetation of the forests in Europe that is thought to be rather similar to the potential natural vegetation because of their relative stability in time. Pollen-analytical research is very helpful in investigating the change and stability of the vegetation (e.g. Firbas, 1949/52; Iversen, 1964; Janssen, 1960; Trautmann, 1969). Criteria for recognition of climax communities are discussed by Whittaker (1974).

2. In zones with forest as potential natural vegetation shrubs and trees, natural edges of planted forest and spontaneous species in hedgerows show the local assortment of tree and shrub species that indicate at least the woody elements of the terminal stage. The indicator value of combinations of woody species depends on the region: in the Northwestern European lowland Corylus avellana, Prunus spinosa, Rosa canina and Crataegus monogyna discriminate the moist Oak-Hornbeam forest as potential natural vegetation from the moist Oak-Birch forest area. In the mountainous area of Middle Germany the same species differentiate between some types of beech forest (Trautmann 1966), whereas in the fluvial plains these shrubs discriminate Elm-Ash forests as potential natural vegetation from the flooded willow forest (Kalkhoven et al., 1976). Other examples are given by Neuhäusl (1963). One should be cautious in using the presence of very few individuals of a tree or shrub species or very small patches of woodland as indicator for the potential natural vegetation of a wider area. First of all a check on the

DEUTSCHLAND 1:200.000 — BLATT KÖLN

Natürliche Vegetation

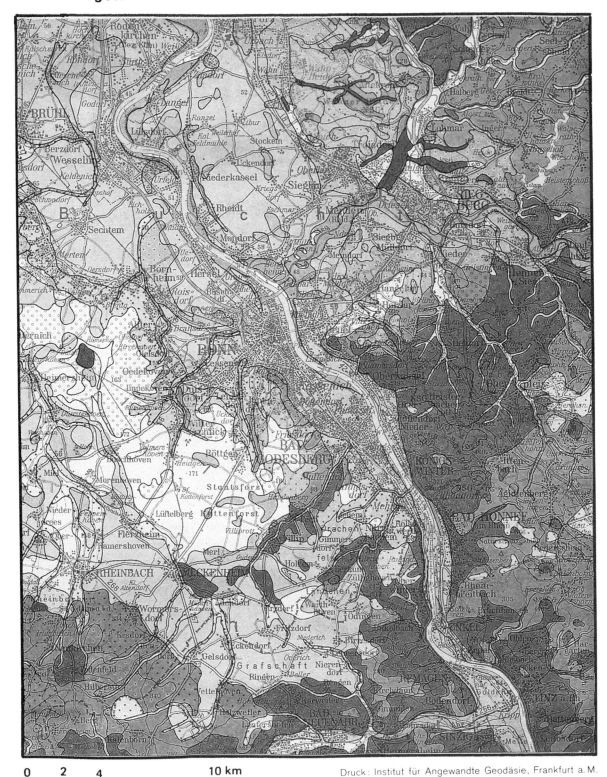

0   2   4   10 km

Druck: Institut für Angewandte Geodäsie, Frankfurt a. M.

Potentielle

Birkenbruchwald des Flachlandes, örtlich Heidemoor, Gagelgebüsch, Eichen - Birken - wald und Eichen - Buchenwald

Feuchter Eichen - Buchenwald des Flachlandes, selten Übergänge zum Eichen - Birkenwald

Feuchter Eichen - Buchenwald des Berglandes

Trockener Eichen - Buchenwald des Flachlandes, selten Übergänge zum Eichen - Birkenwald

Erlenbruchwald des Flachlandes, selten waldfreies Niedermoor (N)

Erlenbruchwald und feuchter Eichen - Buchenwald im Wechsel

Weidenwald und Mandelweidengebüsch

Traubenkirschen - Erlen - Eschenwald, stellenweise mit Erlenbruchwald und Eichen - Hainbuchenwald

Eichen - Ulmenwald (rote Punkte : auf stark entwässerten Standorten im Erfttal)

Artenreicher Sternmieren - Stieleichen - Hainbuchenwald

Stieleichen - Hainbuchen - Auenwald der Berglandtäler einschließlich der bach - und flußbegleitenden Erlenwälder

Maiglöckchen - Stieleichen - Hainbuchenwald der Niederrheinischen Bucht

Maiglöckchen - Stieleichen - Hainbuchenwald und feuchter Eichen - Buchenwald im Wechsel

Trockener Flattergras - Traubeneichen - Buchenwald mit Übergang zum Eichen - Buchenwald

Flattergras - Traubeneichen - Buchenwald

Typischer Hainsimsen - Buchenwald

Hainsimsen - Buchenwald mit Rasenschmiele

Flattergras - Hainsimsen - Buchenwald

Typischer und Flattergras - Hainsimsen - Buchenwald im Wechsel

Maiglöckchen - Perlgras - Buchenwald der Niederrheinischen Bucht, meist auf kiesigen Böden

Maiglöckchen - Perlgras - Buchenwald der Niederrheinischen Bucht, stellenweise Flattergras - Traubeneichen - Buchenwald, auf lehmigen Böden

Maiglöckchen - Buchenwälder des Villeosthanges

Maiglöckchen - Perlgras - Buchenwald und Maiglöckchen - Stieleichen - Hain - buchenwald im Wechsel

Hainsimsen - Perlgras - Buchenwald sowie Perlgras - Buchenwald und Hainsimsen - Buchenwald im Wechsel

Hainsimsen - Perlgras - Buchenwald mit Rasenschmiele

Typischer Perlgras - Buchenwald

Feldaufnahme : H. Wedeck, G. Wolf, K. Meisel u. a. (1965 - 1969)

LEGEND

Vegetatiekaart van Nederland
Vegetation map of the Netherlands
Vegetation series, named after the natural forest types (according to Westhoff & den Held, 1969)

a	**Querco roboris-Betuletum**
ab	**complex** of Querco roboris-Betuletum and Fago-Quercetum
b	**Fago-Quercetum**
c	**Fago-Quercetum**, type poor in species
f	**Fraxino-Ulmetum**
g	**Anthrisco-Fraxinetum**
h	**Circaeo-Alnion**
hk	**complex** of Circaeo-Alnion and Alnion glutinosae
k	**Alnion glutinosae**
l	**Alno-Padion**
lm	**complex** of Alno-Padion and Salicetum albo-fragilis
bl	**complex** of Fago-Quercetum and Alno-Padion
bh	**complex** of Fago-Quercetum and Circaeo-Alnion
cn	**complex** of Fago-Quercetum and Macrophorbio-Alnetum
abhk	**complex** of Querdo roboris-Betuletum, Fago-Quercetum, Circaeo-Alnion and Alnion glutinosae
bhk	**complex** of Fago-Quercetum, Circaeo-Alnion and Alnion glutinosae

Additions

	predominantly dry vegetation types
	predominantly moist and wet vegetation types
	development to peat bog within the Alder forest series
w	**open water**
	urban area

Code for the actual vegetation

The code counts five positions for forest, shrub, semi-natural vegetation, cropland and marsh/water vegetation respectively. The presence of each category is expressed in % of the area of the map unit concerned: 5 = >75%, 4 = 50-75%, 3 = 25-50%, 2 = <25% or many line and point shaped elements, 1 = few line and point shaped elements..

example: code 31142

3.... 25-50% forest
.1... few hedgerows
..1.. few small patches of semi-natural vegetation
...4. 50-75% cultivated land (grassland and arable land)
....2 a mumber of patches and lines with water and marsh vegetation

It depends on the colour, that is the vegetation series, what types of shrub, semi-natural and marsh/water vegetation are present. In the original study this can be checked in a table of vegetation series.

VEGETATIEKAART VAN NEDERLAND

OORSPRONKELIJK SCHAAL 1:200.000

1975

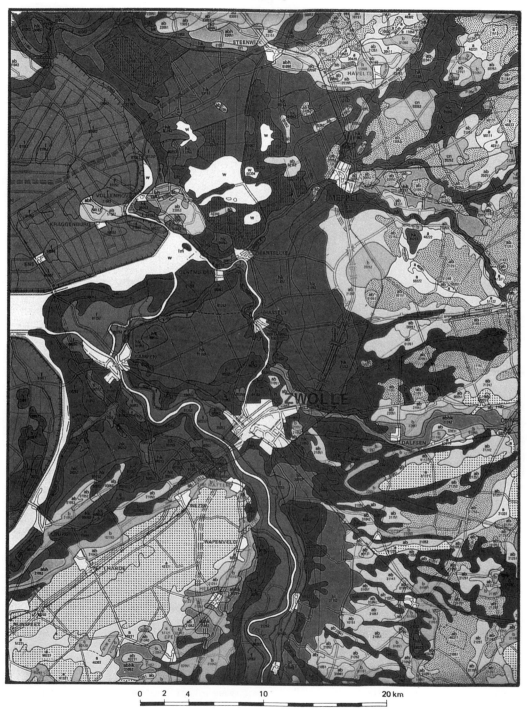

0 2 4   10   20 km

habitat in which these woodlots occur is necessary. Especially in long cultivated areas these little forest elements may grow on sites, that were not worth to be cultivated because of the unsuitable environment and therefore may not be representative for a larger area.

3. According to the degree of human impact on the vegetation of a habitat a set of artificial or anthropogeneous plant communities can be distinguished. They are termed replacement (or substitute) and contact communities ('Ersatzgesellschaften, Kontaktgesellschaften') or secondary communities. Tüxen (1956) presents a distinction in degrees: the replacement communities of the first two degrees are most characteristic for one type of potential natural vegetation. The slightly cultivated areas as heathlands and unmanured grasslands, often summarized under the name of semi-natural vegetation types, belong to these categories. Most weed communities and other vegetation types of the heavily cultivated fields and grasslands are replacement communities of the third and fourth degree. These are least indicative and can also occur on the habitats of related potential natural vegetation types. In fact several series of secondary types, both natural and artificial will be found in a mosaic: for example, cyclic succession stages within a mature system (young, mature and decaying forest and their sylvicultural equivalents), natural succession stages following development of young soils (or rejuvenated by digging), degradation stages after overexploitation, recovery, tilling and other disturbances. From the presence and frequency of indicative replacement communities the type of potential natural vegetation can be concluded and boundaries of the habitat be traced. Helpful in searching these boundaries is also the flowering of populations of species in spring and the distribution of several species in general (Dierschke, 1974; Bohn, 1981).

4. Among the site-factors first of all the characteristics of the soil profile, which are very well observable in the field, are important. Soil maps are undoubtedly very useful in determining type and boundaries of potential natural vegetation units. For medium- and small-scale maps a soil map can be used as basis (Seibert, 1968; Kalkhoven et al., 1976). The validity of the use of soil maps widely depends on the nature of the soil units and the vegetation scientist has always to check the coincidence of soil type and vegetation boundaries (Trautmann, 1966). Geomorphological as well as topographical maps are helpful in locating the potential natural vegetation boundaries. From these maps important landform characteristics as exposition, elevation and inclination, and land-use can be derived.

Examples of potential natural vegetation maps and some characteristics

For several decades in many countries and regions (potential) natural vegetation maps have been made on different scales. Some of these maps will be mentioned here briefly, with a short note on several characteristics. It cannot be an exhaustive review. Tüxen (1956) gives a bibliography of earlier maps on natural vegetation. For excellent reviews of vegetation maps of the various continents see Küchler (1965, 1966, 1968, 1980). A recent review of vegetation maps of this kind in the Federal Republic of Germany is given by Schröder (1984).

First of all the examplary maps of Western Germany are to be mentioned: Trautmann (1966), Trautmann (1973), Krause and Schröder (1979), Bohn (1981), all on scale 1:200,000. The accompanying texts contain full information on the mapped area: extensive descriptions of species composition and structure of the legend units, the replacement communities of every type as well as the distribution of several, indicative plant species. Lists of areas of great botanical value are included. The 1:200,000 maps are accompanied by small maps (1:1,000,000) containing abiotic information on the area (climate, landforms, soil). A section of the map of Trautmann (1973) is shown in Fig. 1 (see colour section).

Based on the same principles and the same scale is the potential natural vegetation map of the Tokyo area (Miyawaki and Okuda, 1974). Nice examples of large-scale mapping are the 1:10,000 actual and potential maps of some Japanese cities (Miyawaki et al., 1973; Miyawaki and Fujiwara, 1973), where much natural and near-natural vegetation can be found in temple forests and parks. The maps are accompanied by volumes richly illustrated with vegetation profiles and tables. In large-scale mapping it may be usefull to include the actual vegetation as a subdivision of the potential vegetation as is described by Van der Werf (Van der Werf and Londo).

Kalkhoven et al. (1976) made a potential natural vegetation map of the Netherlands, scale 1:200,000, based on the plesioclimax concept admitting, with van der Werf (o.c.), a development period of 100 to 200 years. For every mapping unit the existing vegetation was characterized by a five-figure code, which indicates the cover of five groups of replacement communities of the vegetation series belonging to the different legend units (Fig. 2, see colour section). At the same scale Rivas-Martinez (1982) presented a map of vegetation series of the surrounding landscape of Madrid, Spain, whereas Neuhäusl (1963) describes the mapping practice in Czechoslovakia where a series of 1:200,000 scale maps are prepared based on several large-scale maps. The aim, however, was not the potential natural vegetation sensu Tüxen but reconstruction of the natural vegetation, before the profound impact of man (subatlanticum). Faliński (1971) presented guiding principles and methods for mapping the potential natural vegetation maps of Poland, on several scales, starting with 1:200,000 and 1:300,000 scale maps and ending with a generalized map on the scale 1:1,000,000.

On a scale 1:500,000 (Seibert, 1968; Trautmann, 1972) a further generalization is needed.

Potential natural vegetation units are combined in complexes, indicated on the map by 'character' or 'prevailing' communities ('Charaktergesellschaft, Leitgesellschaft'). Used in this way the potential natural vegetation is helpful in characterizing the units of the vegetation-geographical hierarchy of vegetation districts, provinces, etc. (Schmithüsen, 1968).

The units on maps of scale 1:1,000,000 and smaller will come close to the zonal and azonal, extra- and intrazonal vegetation as elaborated by Walter (1954, 1973) after the concepts on zonality in soils. Examples of such maps are those published by Scamoni et al. (1958) of East Germany (1:1,000,000), Tomaselli (1970) of Italy (1:1,000,000), Horvat et al. (1974) of Southeast Europe (1:3,000,000), Küchler (1964) of USA (1:3,168,000) and the vegetation map of the Council of Europe Member States (Council of Europe, 1987) on scale 1:3,000,000.

Summary

As a kind of recipe the different steps in the mapping of the potential natural vegetation will be summarized:

1. classification of the natural and near-natural vegetation of the region. A widely accepted system of classification units should be available;
2. assessment of the relationship between this near-natural vegetation and landform, soil and groundwater regime;
3. assessment of the potential natural vegetation types for the different habitat types of the mapping area;
4. survey of the series of replacement communities belonging to the different types of natural vegetation;
5. establishment of the indicator value of specites in relation to the natural conditions of the site versus anthropogeneous disturbances;
6. mapping of the near-natural vegetation stands and the most indicative replacement

communities, as well as indicator species and species groups;

7. delineation of the map units with help of topographical maps, geomorphological and soil maps, distribution of replacement communities, by preference of the first degree, distribution of various species. In extremely deforested landscapes a survey of tree toponyms can be very helpful.

27. Compiling Small Scale Vegetation Maps from Source Maps

A.W. KÜCHLER

Many vegetation maps at small scales have this in common that they have been compiled from source maps; they are not based directly on field work. For instance, on the vegetation map of the world by Brockmann-Jerosch (1935), the equatorial scale is 1:20,000,000; therefore 1 cm represents 200 km. Many vegetation units on the map are 3 or 4 cm wide and several times as long, hence they correspond to areas 600 to 800 km (370–500 miles) in width and thousands of kilometers (miles) in length. It would therefore be a futile enterprise to base such a map on fieldwork. Any sample plot, even an entire country, would be snuffed out by the reduction in scale.

Small-scale maps are therefore based on compilation. The basic idea is simple enough: the author collects vegetation maps of the component parts of his area and combines them into one new map on a smaller scale.

Source maps

In carrying out such a project, the first step is to obtain complete map coverage for the area. This enables the author to make a detailed study of the vegetation. The more complete the collection of vegetation maps, the better, and an author should not discontinue his search for vegetation maps of a given area because he already has one. Several different vegetation maps are more valuable than just one because they support one another, or else they throw a

new light on the vegetation if they are not in harmony. Where no satisfactory source material is available, the compiler can go one of two ways. He can speculate on what the vegetation is likely to be, and sometimes his reasoning and his meager evidence may permit him to hit the mark within tolerable limits. However, there always remains the very real danger that his speculation was wrong, and his map suffers accordingly. The other way open to the compiler is then to leave the doubtful areas blank and not to show the vegetation there at all. This makes the map more realistic and strengthens the reader's confidence in the remainder of the map. The latter method was successfully employed by Trapnell (1948) in Northern Rhodesia and by Schweinfurth (1957) in the Himalaya.

Usually, it is not difficult to find vegetation maps that show one or several parts of the region to be mapped, but there will be other parts for which good vegetation maps seem to be lacking. A most thorough search is imperative and many of the elusive maps can be discovered by correspondence, by travel to libraries and mapping centers of government agencies, etc., and by a close study of the literature. The search for source maps should continue throughout the duration of the project, and if valuable additional material becomes available when the manuscript is approaching completion, such material should then be incorporated into the body of the map even though this spells new problems and delays. Many maps are ac-

A.W. Küchler & I.S. Zonneveld (eds.), Vegetation mapping. ISBN 90-6193-191-6.

companied by an explanatory text which often contains valuable supplementary information. Wherever such texts exist they should be collected just as diligently as the maps themselves. Acutally, the vegetation of the world has now been mapped, and the mapper should have no problem in finding information and useful material on any region of any continent. Substantial areas have been mapped at the scale of 1:1,000,000 and others at smaller scales. There are smaller but often remarkably detailed maps. Among these, the ones in the Soviet World Atlas (Atlas Mira) are particularly good. All these maps will offer the mapper enough guidance to assure a vegetation map of reasonable quality.

No vegetation map at small scales can be very accurate because of the great degree of generalization. This means that the types of vegetation shown on the map are in fact composites of many different types of communities. They are definitely not so uniform as they appear on the map. The deviation from the indicated character of the vegetation is not shown; it cannot be shown. The reader of such maps must therefore understand that these maps are general guides, broad surveys, and as such very useful. They permit the reader to view large areas simultaneously, they reveal the broad relationships between the vegetation of one country with that of the surrounding ones and allow a basic understanding of the broader features of the vegetation of large regions.

Method of classification

The second step in compiling a vegetation map is to select a classification of vegetation. This, of course, depends largely on the purpose of the map, i.e. on just what features of the vegetation the author wishes to show and in what detail. Obviously, the available classifications are not equally well adapted to given needs and it may become necessary to modify an existing classification or even compose a new one (cf. Sochava, 1954). It is very important that the au-

thor ascertains whether the selected classification can actually be employed throughout the area of his map. Any uncertainty calls for more detailed studies in order to determine the feasibility of the selected classification; if the compiler remains uncertain, he should discard it in favor of another and perhaps more versatile one. When the author is completely satisfied that the selected classification will express the idea he wishes to convey to the reader, and that it can do this in every part of the map, then and only then can he proceed to the next step.

Translation and generalization

The third step consists in changing all legend items of all the maps the author has collected into the terminology of the classification of the new map. This is referred to as the 'translation'. In many instances, this will present no particular problem, but there are times when such a translation is difficult. For example, an author may wish to present the vegetation on a purely floristic basis in an area which includes high elevations. On all source maps at his disposal he finds the vegetation above the timberline described as alpine meadow. Obviously, this term reveals nothing about the species which compose these meadows. The author may then find it useful to study the literature for more information on the vegetation type in question. Correspondence with authorities on the area with the problematical vegetation types is also enlightening.

In the course of translating a map legend from one classification (the first one) into another (the second one), it is obvious that the second classification must be no more complex than the first one; usually it is simpler. If the second classification is simpler than the first one, the translation will result in the need for combining two or more types of the first classification into one broader unit of the second classification. Such combinations must be done with care, and the author should be thoroughly acquainted with the vegetation and its phytocenological character. It is not acceptable to combine any two vegetation types that happen to be contiguous in some places. The types to be combined should be related to each other and so justify the new complex. For instance, the periodically flooded forests on the delta plains of the Mississippi River show differences according to the variations in relief, even the slightest, because the length of the flooding period is dependent thereon. Yet such different types of forests can well be combined, because the inundation gives them certain common characteristics. Elsewhere, different types of pastures with the same soil and water supply are due to differences in management but the common environment nevertheless gives them common features that justify combining them. This procedure leads therefore to the formation of new complexes which are genuine vegetation entities in spite of their composite character.

The author soon learns that a compilation map is not simply a mosaic of source maps. Making translations may present little difficulty in sections for which there is only one source map that completely satisfies the author. This, however, is not often the case. It is usual to find several overlapping source maps and that the various authors may disagree in their interpretation of what they observed. Overlapping maps may therefore be quite unlike in their common sections. If the source maps are contiguous, vegetational boundaries on one map may not be continued on the next map in the appropriate places, if at all.

Various reasons may account for such discrepancies. All maps except those at the very largest scales are generalized more or less, and every author generalizes according to his own point of view and purpose. Therefore differences between authors arise easily. If the various source maps are on different scales then the degree of generalization is also different; hence the maps as a whole must differ as well. Different methods and goals in preparing the source maps nearly always imply different results. Where such source maps cover large areas it is often impossible for one author to study all parts of the area in the same detail. If, however, different parts of the area are investigated by different authors, differences of interpretation are nearly inevitable.

Authors should not be reproached because their results are not in harmony with one another. It is entirely possible that every one of the authors was as careful and painstaking in the preparation of his map as can be expected. The differences result from the nature of the work, and especially from the varied purposes of the maps. Whatever the character of the source maps may be, the translation must be reasonable and must correspond to the facts as closely as the scale permits. The art of compiling a small-scale vegetation map consists above all in finding a way that leads from contrast and contradiction to unity and harmony.

Where detailed source maps are used it is advisable to generalize them before they are translated. This generalization must be done in any case sooner or later, and doing it before the translation means that there will be less to translate, and hence fewer errors and misinterpretations are likely to be made. In so doing, it is useful to have the vegetation maps of the surrounding areas within reach whether or not their legends have been translated. This helps to interpret the vegetation more accurately over larger areas, to assure continuity from one map or one region to the next, and thus to make certain that the quality of the translation remains on the highest level.

If a type covers only a very small area, it is, in the course of generalizing, usually suppressed altogether. But there are cases when the opposite is as much or more justifiable, and instead of reducing a small area to zero, it should be enlarged enough that it can be shown on the map. This applies to all instances where special types of vegetation help to contribute to a better understanding of the map as a whole, as for example in the case of galeria forests in grasslands, alpine meadows, oasis vegetation, and others.

The translation of the source material into the terminology of the new map will be strengthened considerably by consulting soil maps and geologic maps. Where such maps are well done, they can be fine guides in cases of doubt, especially with regard to vegetation types that can form vegetation complexes and also with regard to the location of vegetation boundaries. Ordinary climatic maps are less valuable but phenological maps such as Ellenberg's (1955) map of southwestern Germany can prove to be very useful indeed.

It is not possible to translate the vegetation units of every classification into entities of every other classification. For instance, maps based on Braun-Blanquet's classification can be translated into Rübel's classification, but the reverse is not possible. Where the maps are not comparable, translations may be very difficult or altogether impossible. However, a compiled map is almost invariably a map of small scale which is so generalized that only the broader units of vegetation can be considered, or only the higher ranks in a hierarchical classification.

Küchler (1964) based his map of the United States on what he called a 'classless classification', i.e. one without superior or inferior ranks. He described every vegetation unit by growth forms and taxa, and this method can be applied at any scale. Furthermore, most maps describe the vegetation in such detail that the compiler can glean valuable information from them even though this may be limited to either growth forms or species.

This approach was already foreshadowed by an earlier map of the same area (Küchler, 1953a) and has the great advantage of being highly flexible and adaptable. It permits the translation of almost any vegetation map either wholly or at least in part. The compiler is thus in the fortunate position of being able to use practically all source maps that may come to his attention.

It may be mentioned parenthetically that such a 'classless' method has its close parallels at large scales. The maps on which a given environmental feature is calibrated by the vegetation, do not usually show units of an a priori

classification such as that by Braun-Blanquet. The publications by Walther (1957), Long (1974), and others reveal that unequivocally.

Preparing the new map

The fourth step in the compilation consists of preparing the new manuscript. This is done by changing the material on the many different translated source maps to a uniform scale. It is most readily achieved by transferring the source material to a new outline map. If the area is very large, it is desirable to use a series of base maps. A good example of such base maps are the Aeronautical Charts published by the United States government, or the sheets of the International Map of the World at the scale of 1:1,000,000.

At first, the translated source map should be transferred just as it is, whether it has been generalized or not, and regardless of how the amount of detail compares with other sections of the manuscript map. Discrepancies of this kind can only be dealt with after the first draft of the manuscript has been completed. If any vegetation types were combined as a result of translating them from one classification to another, then only the new combined types (complexes) should be transferred to the map sheets of the new scale.

Where several overlapping source maps are available on which the authors contradict one another, the compiler must resolve the contradiction before he transfers the material to the manuscript sheets of uniform scale. The solution of such a problem may require a considerable amount of research and correspondence, possibly even fieldwork, before a final decision can be made. No effort should be spared to produce the very best possible solution because this will directly affect the quality of the compiled map, and hence its usefulness.

Problems will arise when, for part of the map, the only available source material is done at appreciably smaller scales than that of the new map. Of course, there may be a considerable difference between the scales of the new manuscript map and its eventual form when it is published. If the scale of the source maps is only slightly smaller than that planned for the printed map, the difficulties are not serious, and an enlargement of the source map is feasible. On the other hand, if the scale of the source material is much smaller, then the author must proceed with the greatest caution. It becomes imperative that he utilize every means at his command to raise the scanty information to a more useful level by a most careful and detailed study of the literature and of physical maps of all kinds and by consultation with authorities on the area in question. Personal inspection of the area may contribute much, too, and the author has to allow time and funds in his budget for such emergencies.

The author will find it useful to study carefully the relations between the vegetation types and the major physical features of the landscape in the areas surrounding the problem section. This may enable him to project some of the known information into the involved area. As it is quite possible that serious mistakes are made in the process of such extrapolation, a separate map of the problem area and its surrounding regions should be prepared and submitted for criticism to all authorities familiar with the area. Major mistakes can thus be eliminated and the quality of the map can be maintained.

The result of the fourth step is a unified manuscript, one map or a set of maps, all on the same scale and projection and all employing the one selected classification.

The legend must next be organized and arranged in accordance with the classification, and all items on the map must be checked against the legend. There must be no item on the map that is not also represented in the legend, and every item in the legend must occur somewhere on the map.

The final map must not only have acceptable scientific standards; it must also satisfy esthetic standards and be well balanced. It is, for instance, quite likely that in one section of the map a great deal of detailed information is given whereas in another section very general data have to suffice. One need only compare plains with horizontal geological strata, and hence, a relatively uniform vegetation, with a mountainous terrain where the vegetation is likely to change considerably within short distances due to differences in altitude, slope, exposure, and many other features. It is then not only justifiable but desirable to somewhat generalize the detailed information so as to bring it into a better balance with the rest of the map. These and other changes and corrections necessitate a redrafting of the manuscript, possibly even more than once. The author must prepare for such eventualities, especially when planning and budgeting his project. Remote sensed small-scale imagery may sometimes help in generalizing and updating problematical areas.

The completed manuscript gains a great deal if the author submits it to colleagues and experts for a critical discussion. Of course, this may well imply that sections of the map must be modified, and therefore redrawn, but this work is very much worthwhile. Not only does it strengthen the quality of the map considerably, but it also eliminates a good deal of adverse criticism that might possibly have developed later on.

Continual proofreading

The author cannot be urged enough to spend the utmost care on proofreading at every step and stage, and indeed more than once at each step. This may be a tedious activity but the need for it cannot be overemphasized. It is quite essential that all proofreading is done at least twice because it is not feasible to simply read through a map as one reads a book or an article. The continual shifting from the source map to the manuscript map and back to the source map is very tiring, especially for the eyes, so that in spite of all care many a mistake is overlooked during the first proofreading.

Records

Another feature of significance is the record the author keeps as his work proceeds. He should carefully note in detail every source he ever consults, whether this be a map, a book, an article in a journal, letters from colleagues, conversations, or personal inspections of the vegetation. In certain instances, it is desirable to publish a bibliography, but this need not contain all the different types and bits of information that accumulated in the course of the project. However, while the compilation is still in progress, all these data may be of great value because one does constantly refer back to sources already used or because it is desirable to ascertain on what sources certain items on the map are based.

A special file for problems is also very useful. Whenever a problem arises, it is carefully described and the description deposited in the problem file. The author can then proceed with his work and solve the problem at a more appropriate time. A problem may imply research, correspondance, and even travel. The problem must be solved before its description is removed from the file. Eventually the solution of the problem is added to the description and then transferred to a new file of solved problems. Such a second file may turn out to be very useful indeed at some later stage of the project, especially when similar problems arise or whenever it is desirable to check on how a particular solution was obtained.

Problems of field investigation

There have been repeated references to the possibilities of travel in connection with field investigations of the vegetation types for which information is inadequate or on which the source maps disagree to an extent that the author cannot bridge the gaps and resolve the contradictions. When an author decides to compile a map he must therefore be prepared to include the possibility of such travel in his plans. If he lives in the area to be mapped, and if the various parts of the area are reasonably accessible, travel can be arranged more or less whenever it is convenient. If, on the other hand, the author plans to compile a map of, say, a distant continent, then he must approach the matter differently and with greater caution. First of all, he should travel only after the first draft of the manuscript has been completed because it is only then that he will know definitely where his map is particularly weak and in what sections field studies will be most beneficial. He can then plan his trip according to his needs, thus rendering it most profitable. He must keep in mind while planning his trip that he may possibly be unable to return to any of the places he wants to visit. It is therefore important that he knows prior to going into the field just exactly what information is needed, and while in the field, only the most detailed, comprehensive, and complete notes will assure the author that a second visit is unnecessary. The most careful planning of the trip prior to the departure is imperative.

As a matter of record, a small inset map may be planned to give an indication of the reliability or the degree of detail available for the various sections of the newly compiled map.

These, then are the individual steps that must be followed in the course of compiling a vegetation map. The author of small-scale vegetation maps must face many difficulties. But his efforts will have their reward, for the fruit of his labor will be a visuable document that can serve and benefit a large number of people.

28. Küchler's Comprehensive Method

A.W. KÜCHLER

This method is based on the idea of mapping vegetation in such a comprehensive manner that the results of the fieldwork can be applied to a maximum number of vegetation classifications. In the laboratory, therefore, the field notes can presumably be manipulated so as to fit any given classification that is based exclusively on features of the vegetation (Küchler, 1966, 1967a). Accordingly, comprehensive field mapping is done only once but the result of the field work can be employed as a basis for a variety of vegetation maps, each of which is to serve a different purpose. This method is a special variation of the photo-guided field survey.

First laboratory activity

Before actual mapping in the field can begin, some work must be done in the laboratory. The first step is a careful study of the vegetation of the area to be mapped, based on the literature and all available maps showing vegetation and features of the landscape such as topography, geology and lithology, soils, drainage, macro- and microclimate and land-use.

Simultaneously, the second step consists of the acquisition and careful study of a complete set of aerial photographs of the area. The photographs should be as recent as possible. The scale of the photographs is often 1:20,000, but it is much better to have them done at 1:10,000 and printed on glossy paper.

A good aerial photograph is the very mini-mum that a mapper must have. More useful and elaborate is a set of photographs. If the photographs are specially taken for the preparation of the vegetation map, and the aeroplane is properly equipped, then the mapper should obtain a set of three types of photographs: (1) black-and-white; (2) natural color; (3) false color-infrared. These sets can be taken simultaneously and should be repeated: first photographing period: spring; 2nd period: summer; 3rd period: fall. This is the most elaborate approach to the use of aerial photography. It can be justified by the fact that the combination of these three different types of photographs permits an interpretation of high accuracy which cannot be obtained from a single set. Black-and-white photographs are most commonly obtained, but if the mapper can obtain only one set, and has a choice, then he should select the false color-infrared type. The photographs should be taken in such a manner that there is a 60% overlap. This is important for stereoscopic viewing and proper terrain analysis.

Investigation of the photographs reveals different types of vegetation. The contrasts between the various types range from strong to very subtle but every contrast should be noted. A good stereoscope, preferably of the mirror type, is a great help because the most detailed analysis of the vegetation is essential. Every area which is at all different from neighboring areas is bounded by a line of ink drawn directly on the photograph (Fig. 1). Omissions on the assumption that it will be easier or better to

A.W. Küchler & I.S. Zonneveld (eds.), Vegetation mapping. ISBN 90-6193-191-6.
© 1988, Kluwer Academic Publishers, Dordrecht. Printed in the Netherlands.

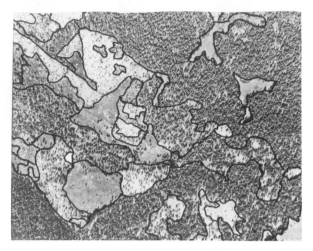

Fig. 1. Black line boundaries separating different vegetation types as distinguished on an aerial photograph.

establish the line when in the field should not be tolerated.

All boundaries of every vegetation unit should be on the photograph before going into the field. This means that there is a definite line on the photograph which can be checked in the field. If the line is acceptable it need not be touched. If the location of the line is unsatisfactory, it can be shifted but such a shift can be made only for specific reasons. Thoughtful observation is then required before a line can be moved on the photograph to a more accurate position.

If, however, the line had not been drawn at all, in the hope of establishing its location while in the field, the difficulties may be serious and the results unsatisfactory. For instance, it is difficult to walk about in a forest of mixed composition and determine where to draw a line. And if a place has been selected for a line, it may be difficult to locate it on the photograph.

The decision where to draw a line on a photograph, separating one type from the next, is not always easy to make. Certain considerations made prior to drafting these boundary lines will save much confusion. One such consideration concerns the problem of transitions, which has been discussed in Chapter 9.

Another problem relates to the smallest size of an area to be shown. When is an area too small to be considered? At times, a given type of vegetation contains an 'island' of a different type. Should this be ignored? That depends on the scale of the map which is to be based on this field investigation. If the scale is not known, it may be assumed to be approximately half of that of the photograph. The problem is easy to solve if the scale is known at least approximately. One need only enlarge the minimal area of the final map to the scale of the photograph. If the 'island' is smaller than this minimal area, then it should be ignored; if it is as large or larger, then it should be shown. The minimal area on the final map, if in color, should be at least 1 mm in diameter if round and not much less than 1 mm wide if the shape is long and narrow. On black-and-white maps the minimal area is usually much larger, depending on the type of patterns to be used.

It is very useful to calculate the actual field size of the minimal area. For instance, the map is to be published at the scale of 2 inches to the mile (1 : 31,680), and the minimal area is to be 2 mm in diameter. Then a distance of 2 mm on the final map corresponds to $2 \times 31,680 = 63,360$ mm and to 63.36 m in the field. This is roughly 200 feet. In this particular instance, the minimal area to be considered in the field should therefore have a diameter of not less than 200 feet, or 63.36 m. This corresponds to 6.3 mm on a photograph with a scale of 1 : 10,000.

When all boundaries are drawn on the photographs, it is important to make sure that all marginal boundaries are continued on the adjacent photographs.

Further preparations for fieldwork include the acquisition of appropriate equipment and a well-trained staff. Equipment includes pencils, fountain pens and ink, paper, clip boards, notebooks, binoculars (the more powerful, the better), an Abney level or its equivalent for measuring the height of trees, measuring tape for measuring the diameter of trees, topographic maps, and whatever is necessary for the personal comfort of each member of the party. An altimeter may prove useful in mountainous areas. Cameras are very desirable. If more than one camera is available, one should be used for color film and one for black-and-white film. For photographs in forests the camera with black-and-white film, placed on a tripod assures the best results. A good exposure meter prevents serious failures. Photographs are useful when analyzing the field notes in the laboratory. In the notebook a special section should be set aside to record the photographs, i.e. the number of the plant community on the Phytocenological Record with numbers of film and individual exposures. It is also useful to take along a topographic map on which the exact place where a photograph is taken can be shown by an angle; the vertex of the angle indicates the photographer's location and the two sides of the angle embrace the width of the photograph. The number of the photograph is placed as closely to the vertex as convenient.

Information gathered in the field should be placed on forms of the Phytocenological Record for standardized recording. A master copy Chapter 4, Table 2, and Chapter 5, Table 1 can be made in the laboratory and copies should be available in plentiful supply. It is best to have an adequate supply of printed forms where such vegetation mapping is done. The number on a given form corresponds to the number of the described phytocenose as recorded on the aerial photograph and later on the base map (see below).

A major consideration for vegetation mapping is the careful selection of an appropriate staff. Under certain circumstances, a party of one is adequate. However, this means that this one person must know how to observe physiognomy, landforms, and many other details. He must have a very thorough knowledge of the flora; he must carry notebook, aerial photographs, clip board, camera, binoculars, and often a raincoat, lunch, and other items. It is much better to have more than one person in the party. A party leader with two assistants approaches the ideal. If the party consists of more

than one person, one should be a well-trained taxonomist. Ideally, such a taxonomist has specialized in the flora of the region to be mapped. His only equipment is the latest edition of a standard flora covering the region, such as Gray's Manual of Botany (Fernald, 1950) for the northeastern United States or A California Flora (Munz and Keck, 1973). It is best to have the names of all species based on the same manual, as there are variations from one manual to the next, and the use of different sources leads to confusion.

The entire party should be thoroughly familiarized with the aims and methods of the fieldwork. All members of the group should be well acquainted with the physiognomic and floristic methods of analyzing plant communities. There are many cases calling for relatively arbitrary decisions in the field, and an intelligent discussion among the staff members helps in making the decisions as reasonable as possible.

Fieldwork

With staff and equipment assembled, the party can proceed to the field. Here it becomes necessary to visit every individual area outlined on the aerial photographs. It must be remembered that the information is to be as complete as possible but that aerial photographs show only the surface layer of the vegetation. What grows under the canopy of a forest can only be discovered by direct observation. In deserts and at the high altitudes of alpine terrain, the paucity of vegetation types and the simplicity of their structure and composition permits one, after much practice, to recognize types accurately from a distance, e.g. across a valley, if strong binoculars are available. In all other cases it is imperative that each vegetational area outlined on the photograph be inspected.

Upon arrival in a given area to be inspected, the first task is to walk about in it, preferably from end to end and across, observing the vegetation critically. Thereupon all relevant data are entered on the Phytocenological Record which is carried on a clip board.

It is necessary to know in advance just what to look for, and then, in the field, to look for it with unwavering consistency. Once it has been established which aspects of the vegetation to consider, it is absolutely necessary to consider them uniformly throughout the region mapped, with no exceptions. For instance, it is not permissible to observe and record in the field notes the height of the vegetation in one area and then ignore it in another area. If the height of the vegetation is to be recorded, it must be recorded in every area, i.e. for each individual synusia of every phytocenose. This insistence on being consistent at all times pays big dividends later on, when the fieldwork has been completed and the maps are being prepared in the laboratory.

Recording good notes is at times difficult in the field. To make sure that the notes remain legible for some time, the script should be reasonably large. It is very desirable to type the field notes as soon as possible. This is important because pencil notes become increasingly difficult to read and are blurred after some time. On the other hand, it is not good to rely on ink because in wet weather, writing with ink is difficult, the paper absorbs moisture, the ink runs, and the notes may soon be illegible.

It is, of course, necessary that every area outlined on the aerial photographs can be correlated with the field notes. For this reason, every areal unit which appears on a photograph must be numbered, and the same number must appear in the Phytocenological Record where the vegetation of this area is described.

When the vegetation of a given area has been inspected critically, its salient features are recorded on the prepared forms. The first step is to fill in the first section of the Phytocenological Record and to give the area its number (Chapter 4, Table 2). No number may be written on the photographs until a given area has been inspected. Then its number is written down on the form and on the photograph simultaneously. The next number is not set down until the next area has been inspected and analyzed. In this manner, area follows area, each with its own number. If an area is inspected in which the vegetation is the same as in another area already inspected, it nevertheless receives its own number, and the character of the vegetation is recorded anew. This is not necessary for such types as barren, urban, etc.

Upon inspection of the phytocenose, the mapper proceeds to describe it both physiognomically and floristically. The physiognomic analysis (cf. Chapter 4) reveals the structure of the plant communities, i.e. height and coverage of every synusia, and, in addition, such special features as may be present. If physiognomic formulas are used they should always be recorded as accurately and as completely as possible.

Often the physiognomic formula reveals characteristics of the vegetation that are not evident from the floristic description. For instance, if the flora is described as *Betula papyrifera* 5 and *Abies balsamea* 5, we can only see that the vegetation consists of two dominant species each of which covers more than three-quarters of the area. This may seem confusing. But the physiognomic formula 'D6E4' gives the explanation. 'D6' stands for a broadleaf deciduous forest 10–20 m tall, and 'E4' means small needleleaf evergreen trees (2–5 m). The formula reveals therefore that the vegetation consists of two layers, an upper one consisting of broadleaf deciduous trees (paper birches) and a lower one of needleleaf evergreen trees (balsam firs).

The coverage of the vegetation is estimated as outlined in Chapter 4. Such estimates rarely present difficulties. The height of the various layers of vegetation can be measured, and after some practice it can be estimated with a considerable degree of accuracy. It is very useful to know the height of 10 cm, 1/2 m, and 2 m with regard to one's body(e.g. one's knee may be 1/2 m above the ground), as these are critical heights 'within reach' of the mapper.

For measuring the height of trees, an Abney level provides the simple solution. Where such an instrument is not available, a measuring tape and a plastic draftsman's triangle (right angle, isosceles) suffice. The height of the tree equals the observer's distance from the tree when he looks along the hypotenuse of the triangle and its extension toward the tree top (Fig. 2). The height of the observer's eye above the ground should be added to the distance from the tree for greater accuracy. Care must be taken that the lower side of the triangle is horizontal. To assure the best results most easily, the height of the observer's eyes above the ground is marked on the bark of the tree (point G in Fig. 2). Then the observer moves away from the trees until the tree top coincides with the upper tip of the triangle. To make sure that the lower side of the triangle is horizontal, it may be placed on the tripod as high as the observer's eyes. The triangle's lower edge must then point to the mark on the tree which indicated the height of the eyes above the ground. If the tree stands on a slope, observations are made along the contour.

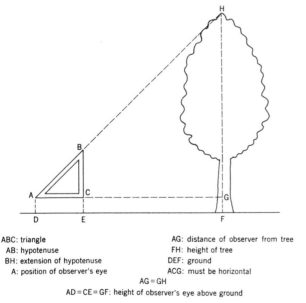

ABC: triangle
AB: hypotenuse
BH: extension of hypotenuse
A: position of observer's eye

AG: distance of observer from tree
FH: height of tree
DEF: ground
ACG: must be horizontal

AG = GH

AD = CE = GF: height of observer's eye above ground

Fig. 2. Establishing the height of a tree with the help of an isosceles triangle.

It takes very little time to become acquainted with the physiognomic classification, and its application in the field is simple and quick. The structural analysis on the Phytocenological Re-

cord permits an accurate recording of all necessary details.

The description of the physiognomy and the structure of the vegetation is adequate if all observational data are recorded on the forms of the Phytocenological Record. When the mapper has no such forms at his disposal, he will use formulas, as shown in Chapter 4. Where the Phytocenological Record is used, the collected information can easily be transcribed into formulas whenever this is desired.

The following points should be kept in mind in analyzing the flora. Each species must be identified by its scientific name, and its coverage must be determined, i.e. estimated. The distribution within the area of herbaceous species may also be recorded.

The values of the coverage and distribution of species and the method of recording them are given in Chapter 5. The mapper is, of course, free to record the basal area instead of coverage. Both are equally acceptable but one method may be preferable to the other under certain circumstances.

The taxonomist is usually able to identify all species on sight. When he is in doubt, he can neverthelass give the species a name that can be entered on the record, followed by a question mark. He can then collect a specimen and identify it accurately upon returning to the camp and to his botanical manual. The correct name should then be entered on the record at once. Doubt may remain because of a lack of diagnostic characteristics due to immaturity or old age. It may also remain because of the taxonomic complexity of the genus, making the specific or subspecific identification the concern of the specialist of the genus. If some doubt about the accuracy of the name persists, the question mark is retained on the record. Only scientific botanical names should be used to identify the plant species, or else the value of the entire work may be seriously impaired.

The floristic composition of stand samples is recorded on the back of the Phytocenological Record (Chapter 5, Table 1). It is always best to break down the phytocenoses into their constituent synusias, each of which is described physiognomically and floristically. Hence, for a given sample, the mapper records first the tree layer. He writes 'tree layer' or simply 'trees' in the first column. Then, under the heading of 'tree layer', all species are listed that have attained tree size. Each species is recorded by its botanical name only and its coverage is given in the column of stand sample No. 1. As other sample plots may contain tree species that do not occur in plot No. 1, the mapper should leave some space for additional names before proceeding to the next-lower synusia. Climbers are listed here if they reach into the tree layer.

The shrub synusia is handled just like the tree synusia. All woody plant species that are not of tree size are here recorded just as the tree species were recorded before. Again each species is listed by botanical name and coverage. If there is a synusia of dwarf shrubs, e.g. 'D2i', then the mapper must not hesitate to enter such a separate synusia.

The procedure is the same for whatever synusia there may be. However, many phytocenologists prefer to list herbaceous species with both their coverage and their distribution even though the latter is not recorded in the other synusias. The moss layer is the place in which mosses and lichens are recorded provided they are not epiphytes.

If a form is inadequate to record all species, the work is continued on a second form. This second form receives the same Phytocenological Record number as the first one, with 'page 2' written after the number. On the first page, under Notes, the mapper should write 'continued on page 2' to indicate that page 1 is incomplete.

Finally, under Notes (Chapter 4, Table 2), all sorts of remarks may be made that help to throw light on the recorded phytocenose. Comments such as 'old burn' or 'recently logged' will prove helpful when the character of the vegetation is to be interpreted, especially if it can be established what forest type was burned or logged. Comments on topography, geology, soil, water conditions (bog, etc.), or any other site quality should always be made in the greatest possible detail. There may be no particular need for any remarks but the mapper is urged not to ignore anything that is at all relevant.

The result of the fieldwork is a set of Phytocenological Records and a series of aerial photographs on which the vegetation types are outlined. Each individual outlined area on the photographs must have received a number which corresponds to the number in the records where the particular type is described. The critical observations, especially with regard to the lower layers of the vegetation, may have necessitated the addition of further boundaries on the aerial photographs. It may also be that boundaries first drawn in the laboratory had to be shifted or removed upon field inspections. If any boundaries have been added or changed on the photographs during the fieldwork, such a change should be made on all photographs on which the boundary appears. This change may have to be repeated on two or even three other photographs, especially if a change on the photograph was made near a corner; this should be done in the field.

Second laboratory activity

The fieldwork concluded, the party returns to the laboratory. The work of the taxonomist comes to an end. If no typewriter was available during the fieldwork, the first step is to type all records exactly as they appear on the field note forms. They are the basis of all information. If it seems desirable to rearrange the various items on the lists, then this should be done afterwards. It is always possible to make any number of changes. But it is of fundamental importance to have the original field data typed just as they were first set down. The contents of the notebook should be typewritten also. As each phytocenose is described and typed on its own form, it is best to type the qualities of its site and all relevant information recorded in the notebook in the space for notes on the form.

The forms are of the same kind as those used for collecting data in the field.

The next step is the preparation of the base vegetation map. This is a map which shows the exact outline of each individual vegetation type, i.e. each area shown separately on the photographs, and the numbers of the areas. In order to make this map it is best to transfer the lines (boundaries of phytocenoses) on the aerial photographs to a topographic map. It is not feasible to trace lines directly from the photographs on white paper because of the distortion of the photographs. The topographic quadrangles with streams, roads, lakes, coastlines, houses, and other features help overcome this difficulty. Once all boundary lines have been transferred to the topographic map, they can be traced on white paper or preferably on a stable base to maintain the constancy of the scale. On this drawing it is best to add rivers, and lake and ocean shores, if any, to assist in orientation. Then the numbers of all vegetation types, i.e. of all areas shown on the photographs, are entered on this map. Scrupulous checking (proofreading) upon completion is quite essential.

The result of all this labor is a base map of the vegetation of the selected area (Fig. 3). It shows the exact outline of every vegetational unit and the number of each unit, which refers to the corresponding number in the records with its detailed description of the vegetation type. All the basic work has now been completed, and the outline map, with its numbers, and the records form the foundation for the vegetation map (or maps) to be drawn.

The preparation of the final vegetation map depends on what particular classification is to be selected. This may be physiognomic or floristic, or a combination of the two. Perhaps forest trees are the only objects on interest, in which case all undergrowth and herbaceous vegetation can be ignored. It all depends on the purpose of the map, and according to this purpose, the information on the records is manipulated so as to produce the most useful categories. When these categories have been established, one need only go through the records, assign each unit to the appropriate category and then trace the new map off the base vegetation map, merging all units which are adjacent to each other and of the same category. Obviously, the comprehensive information on the records lends itself to a great variety of possibilities, and herein lies the chief value of this method.

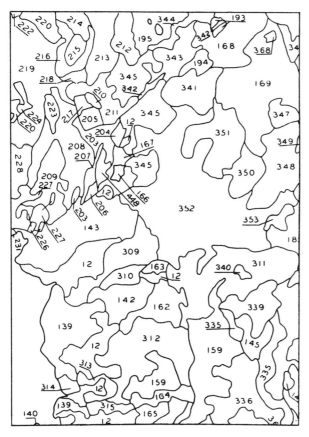

Fig. 3. Section of a base map showing the outlines of every vegetation type and its number referring to the mapper's field notes.

29. The ITC* Method of Mapping Natural and Seminatural Vegetation

I.S. ZONNEVELD

Introduction

The characteristic elements of the ITC method for mapping natural and seminatural vegetation are:
— effective use of the landscape approach applying systematic interpretation of aerial photographs and other remote sensing means;
— stratified random sampling as a means to collect objectively basic information for classification;
— during the sampling a well administered 'quick herbarium' is made, an activity being a compromise between a professional floristic plant collection (which would take too much time) and an utmost minimum necessary for documentation and identification ('no more than 31 seconds/specimen');
— classification of vegetation using the complete species, composition, and not only 'characteristic' or only indicator species. Only in exceptional cases where the flora is too difficult or largely unknown, structural aspects (lifeforms) may be used exclusively;

— the vegetation classification is made ad hoc, based on statistical treatment of samples taken stratified according land units, delineated by means of interpretation of aerial photographs or other remote sensing means, and only compared with a posteriori or translated into existing classification systems;
— in addition to the floristic classification a structural classification is applied. Depending on the character of vegetation and landscape and also on the mapping aim and scale, the structure is being described either for the floristic classification units or independent and as such used in the map legend. Structural classes are defined in such a way that the highest categories can be recognized as much as possible on aerial photographs (see Appendix A);
— the more detailed the survey is, the more 'photo-guided field survey' (see Chapter 16) will be applied;
— legend design is done by use of (1) an annotated map combining the chronological information and classification results in map form; and (2) a cross table showing the content of the preliminary map units in terms of classification units in table form (see Chapter 11 and 11A and Fig. 1);
— results of digital processed teledetection images being the results of 'semi-supervised' or 'unsupervised classifications' (see Chapter 21) are normally used for stratified sampling. It is seldom possible to supervise

* ITC is the acronym for the present "Institute for Aerospace Survey and earth sciences" (formerly "International Training Centre for Aerial Survey) Delft/Enschede, P.O.B. 6, 7500 AA, The Netherlands.

A.W. Küchler & I.S. Zonneveld (eds.), Vegetation mapping. ISBN 90-6193-191-6.

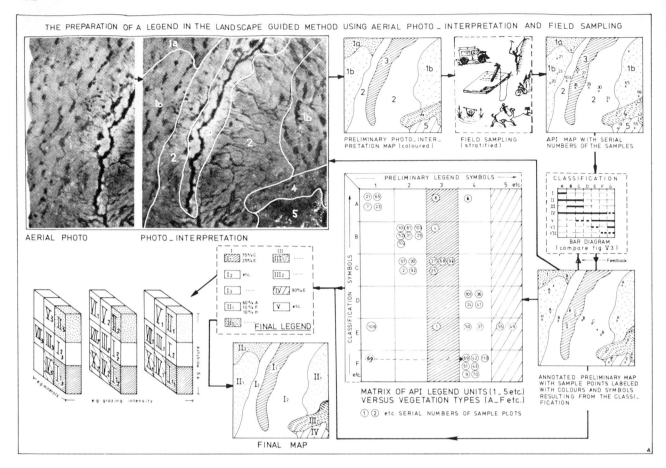

Fig. 1. Scheme of the preparation of a legend in the landscape-guided method using aerial photo-interpretation and field sampling.

a classification so that automated vegetation mapping is possible. Only in special cases (usually in land-use surveys, or pure bio-mass surveys, etc.) the automated processed image can be used as such (see Chapters 20 and 21).

The landscape approach is described in Chapter 16. Below the ITC method will be described somewhat more in practical details via the various steps through which the survey proceeds.

Comparison of these steps with the description of land(scape) unit surveys as in Zonneveld (1979), shows that essentially the same methodology is applied. This follows from the basic philosophy of land units as a starting point of delineation.

The steps of landscape-guided vegetation surveying and mapping using aerial photographs and other remote sensing interpretation*

The preceding chapters contain discussions and basic principles about surveying, sampling and mapping of vegetation. This paragraph deals with some details about the sequences of the activities and some additional remarks about the vegetation survey using aerial photographs as developed at ITC. We shall follow the procedure in the natural sequence as given in a

* See also Fig. 1 in Chapter 16.

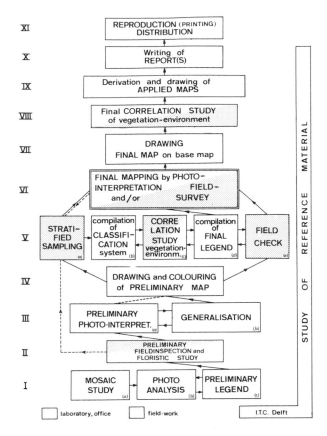

Figure labels (left column roman numerals and blocks):

XI — REPRODUCTION (PRINTING) DISTRIBUTION

X — Writing of REPORT(S)

IX — Derivation and drawing of APPLIED MAPS

VIII — Final CORRELATION STUDY of vegetation-environment

VII — DRAWING FINAL MAP on base map

VI — FINAL MAPPING by PHOTO-INTERPRETATION FIELD-and/or SURVEY

V — STRATI-FIED SAMPLING (a) | compilation of CLASSIFI-CATION system (b) | CORRE-LATION STUDY vegetation-environm. (c) | compilation of FINAL LEGEND (d) | FIELD CHECK (e)

IV — DRAWING and COLOURING of PRELIMINARY MAP

III — PRELIMINARY PHOTO-INTERPRET. | GENERALISATION (b)

II — PRELIMINARY FIELDINSPECTION and FLORISTIC STUDY

I — MOSAIC STUDY (a) | PHOTO ANALYSIS (b) | PRELIMINARY LEGEND (c)

(right vertical label) STUDY OF REFERENCE MATERIAL

□ laboratory, office   □ field-work   I.T.C. Delft

Fig. 2. Scheme of surveying and mapping vegetation using aerial photo interpretation.

'Scheme of Surveying and Mapping Vegetation Using Aerial Photographs'. This scheme is presented in this paper in two ways: in Figure 2 with blocks containing only some key words and in a similar figure but with a little more elaborate text in each block (Fig. 3). This scheme is made for a situation, where the area is assumed to be unknown and ideal survey conditions exist. In practice there will usually be not a fully ideal situation, moreover fully unknown areas no longer exist. The system can and should therefore be adapted, to the real conditions of terrain and working organization. As we have stated before the general interpretation procedure is the same for any other remote sensing image both for satellite and for small-scale photography for radar images and for scanning MSS. Even thermal images can be used in the same way. Also if no photographic images, but computer outprints are available, the same steps can be distinguished. Survey of vegetation dynamics (monitoring) starts in the same way. The success depends on the speed stratified sampling can be done or in how far the need for repeated sampling can be reduced.

We distinguish eleven steps. Each step may consist of various separate activities which, however, belong together and moreover influence each other.

Step I

Step I consists of, first, study of reference material that may be continued in later steps, 'mosaic study' (a), photo-analysis (b) and design of a preliminary legend (c); the aim of this step is to provide as objective a start as possible. Before going to the field the area is on the photo-mosaic roughly subdivided into units (mosaic study) and representative areas (photo-pairs) are chosen for a photo-analysis which can serve as a basis for a photo-interpretation legend.

If the area is rather small (covered by a small number of photographs) all the photographs could be used for such an analysis. 'Mosaic' study does not always require a specially prepared photo-mosaic. A rough laying out of the photographs may serve the purpose. A quick perusal of the photographs by stereoscope is highly advisable (a few minutes per photograph). In mountainous areas mosaics are difficult to make. In flat areas it is advisable to request always an uncontrolled mosaic from the company supplying the photographs. These mosaics can serve as preliminary maps on to which the data from the photographs can be very easily transferred.

If reasonable base maps are available (which should be checked in this phase of the survey), an uncontrolled mosaic is sufficient. For the making of topographical base maps refer to Chapter 8. *In case the vegetation survey is done parallel to general geodetic surveys, controlled mosaics or ortho-photography may be applied. For the mere purpose of a vegetation survey such*

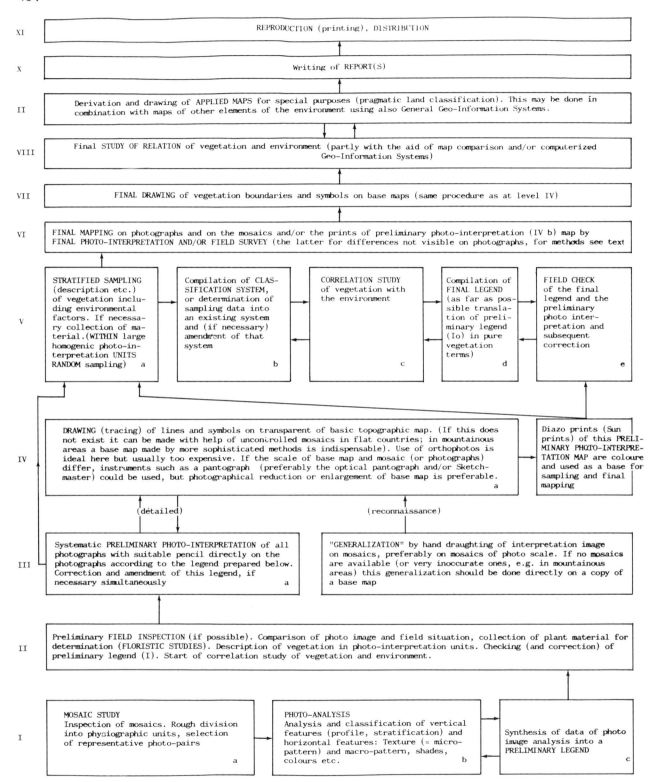

XI — REPRODUCTION (printing), DISTRIBUTION

X — Writing of REPORT(S)

II — Derivation and drawing of APPLIED MAPS for special purposes (pragmatic land classification). This may be done in combination with maps of other elements of the environment using also General Geo-Information Systems.

VIII — Final STUDY OF RELATION of vegetation and environment (partly with the aid of map comparison and/or computerized Geo-Information Systems)

VII — FINAL DRAWING of vegetation boundaries and symbols on base maps (same procedure as at level IV)

VI — FINAL MAPPING on photographs and on the mosaics and/or the prints of preliminary photo-interpretation (IV b) map by FINAL PHOTO-INTERPRETATION AND/OR FIELD SURVEY (the latter for differences not visible on photographs, for methods see text

V —
a. STRATIFIED SAMPLING (description etc.) of vegetation including environmental factors. If necessary collection of material.(WITHIN large homogenic photo-interpretation UNITS RANDOM sampling)
b. Compilation of CLASSIFICATION SYSTEM, or determination of sampling data into an existing system and (if necessary) amendment of that system
c. CORRELATION STUDY of vegetation with the environment
d. Compilation of FINAL LEGEND (as far as possible translation of preliminary legend (Io) in pure vegetation terms)
e. FIELD CHECK of the final legend and the preliminary photo interpretation and subsequent correction

IV —
a. DRAWING (tracing) of lines and symbols on transparent of basic topographic map. (If this does not exist it can be made with help of uncontrolled mosaics in flat countries; in mountainous areas a base map made by more sophisticated methods is indispensable). Use of orthophotos is ideal here but usually too expensive. If the scale of base map and mosaic (or photographs) differ, instruments such as a pantograph (preferably the optical pantograph and/or Sketchmaster) could be used, but photographical reduction or enlargement of base map is preferable.

Diazo prints (Sun prints) of this PRELIMINARY PHOTO-INTERPRETATION MAP are coloured and used as a base for sampling and final mapping

(detailed)   (reconnaissance)

III —
a. Systematic PRELIMINARY PHOTO-INTERPRETATION of all photographs with suitable pencil directly on the photographs according to the legend prepared below. Correction and amendment of this legend, if necessary simultaneously

"GENERALIZATION" by hand draughting of interpretation image on mosaics, preferably on mosaics of photo scale. If no mosaics are available (or very inaccurate ones, e.g. in mountainous areas) this generalization should be done directly on a copy of a base map

II — Preliminary FIELD INSPECTION (if possible). Comparison of photo image and field situation, collection of plant material for determination (FLORISTIC STUDIES). Description of vegetation in photo-interpretation units. Checking (and correction) of preliminary legend (I). Start of correlation study of vegetation and environment.

I —
a. MOSAIC STUDY Inspection of mosaics. Rough division into physiographic units, selection of representative photo-pairs
b. PHOTO-ANALYSIS Analysis and classification of vertical features (profile, stratification) and horizontal features: Texture (= micro-pattern) and macro-pattern, shades, colours etc.
c. Synthesis of data of photo image analysis into a PRELIMINARY LEGEND

Fig. 3. Scheme of surveying and mapping of vegetation using aerial photo-interpretation I.T.C.

means are usually too expensive. The preparation of photographs with centre points, flight lines, etc., for elementary photogrammetric orientation is treated in any handbook on photo-interpretation. Although it seems superfluous to state here, due attention should be paid to proper storage of photographs so that each photograph is readily available at any time for office and field activities.

As in any step, but especially in the first one it is very important to study references, the aims of the survey and all aspects of the area, topographical maps, landscape, geomorphology, soils, vegetation and last but not least: the flora. Study of books about the area and, if available, herbarium material, will enable the surveyor to judge whether it will be possible to do a vegetation survey on a floristic basis. Such a basis will not be chosen if the flora is still largely unknown (not yet sufficiently studied at all) or very complicated. In most cases, however, assistance of a local botanist during the sampling stage is sufficient to overcome difficulties. But even if floristics will have a minor part in the survey, one needs to know the dominant species.

Finally one will decide in which seasons the period of field survey has to be planned. The accessibility of the terrain and the aspect of the vegetation both play a role (often controversial in tropical semi-arid zones) in this respect. The final result of step I is a legend that can be used for the preliminary photo-interpretation. In Appendix B a guideline is given for a systematic handling of a block of photographs to be interpretated.

Step II

Step II will only be possible if the survey area is near the headquarters where photo-interpretation, etc., is done. If this is not the case one should proceed immediately to step III. Also in a very simple area with only a few photographs a very experienced surveyor who is already acquainted with the kind of vegetation in the area may proceed immediately to step III.

In all other cases step II helps the surveyor to become acquainted with the main vegetation and the other landscape features in the field. Moreover, one can start to study the flora and select some herbarium material for determination. One may check the decision made before (step I) about the role of floristics. Especially he should try to get some idea about the dominant covering species in this step in order to be able to use the data during photo-interpretation. The surveyor should, however, avoid making intensive descriptions of sites and vegetation during this step. This can only be done well after the next steps, because one needs a preliminary photo-interpretation as a basis for 'objective stratified sampling' (see former chapters). Never should this step merge into extensive 'air-photo-tourism' = elaborate photo-guided field survey over large areas without sufficient systematic interpretation before (in practice too much time has been wasted in this way!).

Step II involves only a short visit to the terrain, criss-crossing it in some main directions as far as this can be done in relation to accessibility cost and time available. If step II has to be omitted (see above) the activities belonging to this step will be done during stage V. Some of the activities described here may also be done before the survey starts during an orientation visit, if any, during the stage of negotiation with the local authorities. For small-scale remote sensing (satellites or radar interpretation) normal aerial photographs have to be used in this step (see last section of this chapter).

Step III

Step III is fully considered as home (office) work. It is the core of the vegetation survey using aerial photographs. It should be done before the fieldwork. Too often photographs are used just as an additional tool during field surveys. In those cases maximum profit from the aerial survey is not obtained. For this a systematical study of each photograph is required. For the concepts of recognition, analysis and interpretation refer to Chapter 20. For the systematic handling of a block of photographs to be interpreted see Appendix B.

The interpreter should constantly be aware of the final mapping scale and draw his delineations so that they are still cartographically acceptable after reduction. For inexperienced surveyors this is a major problem. The result is photographs or other remote sensing images with lines drawn upon them with grease pencil. It is advisable to avoid using overlays. Overlays may cause loss of detail (texture) and the data on them are less easy to transfer to mosaics. But if only unique copies or only transparent films of colour photographs are available, then this will be necessary. In that case completely transparent overlays should be used. If no mosaics are available copying of transparencies or Xerox copies thereof may be easier than of photographs directly. In all cases where black-and-white paper copies of photographs are used, they should be directly written on, this is after all what they were made for, not to be kept spotless! Moreover if a grease pencil is used and one draws on a hard underlay, photographs can hardly be damaged and can easily be cleaned again. It is advisable to use coloured pencils (at least on black-and-white photographs, never black!): blue for drainage lines, yellow for roads (= transparent and leaves details visible), red for main land unit lines and green for those vegetation features that do not coincide with terrain forms, etc., so are probably man induced vegetation structures. For vegetation survey one should always use originals (direct prints from the original negatives). Copies of copies give a loss in texture that, for vegetation interpretation, reduces the quality too much. The best are diapositives. For fieldwork those transparencies are, however, less easy to handle. A portable field light table as constructed at ITC may reduce the difficulties in this respect. To this stage belongs also the transfer of the data of the photographs to the base map or, preferably, to a mosaic. This basemap or mosaic may have the same scale, but preferably half the scale of photographs. In that case some generalizations may be done during the interpretation, which may prevent unnecessarily going astray into details.

It should be stated that the transfer from the photographs to the basemap or mosaic belongs to the interpretation stage! This is because of the above mentioned generalization process. Moreover after having transferred all data to one map, the whole area as a totality is seen for the first time. The map picture shows immediately errors or some mis-interpretations as far as lines on adjacent photographs appear not to coincide. But also more important matters may appear (e.g. scale problems). Such a check on the map image as a whole can only be done if the preliminary map is coloured. From this it follows that one should never finalize one map sheet before all others of the same survey area have been compared with each other in the prefinal form. If data are directly transferred to a basemap (not via a mosaic) a whole series of instruments are available ranging from simple sketchmasters to more sophisticated optical instruments.

Step IV

Tracing the map from the mosaic (if any) on to the base map is done in step IV. Another essential activity for this step is the colouring of the maps. If no photo-mosaics are used and the interpreter has brought the boundaries himself to the basemap directly from the photographs, only colouring or some special redrafting work is done in this step. Both tracing and colouring can be done by assistants (draftsmen). The surveyor, however, should decide on the colours for himself. In this stage he is free to select the colours solely so that he and his party can understand and read the map (see also step VII).

It should be noted that *tracing* can be done by *helpers*. It should be repeated that the surveyor *himself* will have to *transfer the lines* from the photographs on to the mosaic or, if no mosaic exists, on to a base map, because this is very responsible work requiring more knowledge of vegetation than drawing technique. In flat terrain and using mosaics the most suitable way to do the transfer is by free hand drawing. Small photographic distortions can then be easily corrected especially on mosaics. On topo maps with much detail (too much for a base-map) sometimes free hand sketching may be also possible. Usually, however, assistance is required from low-grade photogrammetric instruments as sketchmaster, stereosketch or, optic pantograph. Photographs with much distortion, as usually is the case in mountainous terrain, require more sophisticated instruments. Orthophotographs are ideal here although usually too expensive. Good satellite imagery can sometimes replace photo-mosaics!

The preliminary map should always be made on a transparant paper from which copies can be made, then the copies can be used for various purposes, e.g. for special notes in the field (step V) and possibly also for later corrections (see steps VI and VII). *A reserve for emergency to be kept in the 'safe' is an absolute necessity* because already quite some investments in time and money have been made. Only one copy will be coloured, so the legend should be noted in symbols in each individual map unit on the transparancy! It is advisable not to use hatchings at least not too heavy ones. This disturbs the possibilities of making black pencil notes on the field map. The main purpose of the coloured copy is to plan the stratified sampling of step V,

Step V

The main fieldwork of the landscape-guided approach is done in this step, especially in the first block (a) which is devoted to sampling. Before starting sampling, the field trips are planned so that at the end a sufficient number of tests will have been made in each important mapping unit. The sample spots are selected on the basis of the preliminary map and, before the fieldwork starts, are indicated in detail on the photographs. In large surveys preferably a series of 'scenario books' are made in which day programmes and fields routes are noted. Careful

planning of routes and sample spots is very important in surveys that require expensive transportation means, e.g. helicopters in tropical rain forest areas.

But also in any other types of survey a rather detailed fieldwork plan is very useful. If the party is a mixed group with other scientists (soil surveyors, geomorphologists, etc.) such planning is also very important in order to avoid frustration, arising from differences in the time required by each team member for observation on the spot. The programmes should not be too rigid and not fixed on calendar data. There should always be more points than strictly necessary because usually certain spots will appear to be inaccessible either according to weather conditions or terrain constraints. It should be possible to arrange the days of fieldwork in a certain area within the total time available for fieldwork, in accordance with terrain and weather conditions.

Points that are representative of essential mapping units should be selected. Also points that are obscure during the interpretation for some reason can be visited. *One should, however, avoid sampling mainly rarities!* It should be possible to derive from the recorded data the 'normal' character of the mapping units! It requires self-discipline to do this properly. In principle it requires random sampling stratified within a map unit. Sample points should, besides being stratified in photo-interpretation units, be clustered or transected as much as possible to avoid traveling time and still sufficiently distributed over all different parts of the map in order to avoid local (climatical) bias (see Chapter 17). Vegetation relevés should always include a collection of herbarium material of unknown plant specimens too, but *only* in the form of *'quick herbarium books'*. Never in the form of classical herbarium collection. In proof cases this leads to either shortage of time or insufficient collections. As a rule '31 seconds per specimen' works reasonably well in practide. The most efficient way is to use files with exchangable leaves, which are, however, fixed during the work; these cannot be blown away and enable to turn over the leaves quickly in search for similar specimen for comparison, etc. In the evening leaves can be examined and regrouped according to results of classification, etc.

As is described in Chapter 17 besides vegetation data one should try to obtain as many data about the environment as is possible from each spot within the available time. At the end of each day the data collected should be properly arranged and systematically stored. For this reason preprinted forms and also punch charts are very useful. Even for notes in the field it is advisable to use preprinted forms, check lists, etc. Use of electronic computer compatable recording instruments in the field may increase in the future; however, plant identification often does not allow direct storage of final data.

In detailed surveys during the field survey one should also do some checking on the delineations made on the photographs in relation to the boundaries in the field. In reconnaissance surveys this is only incidentally possible. The sequence of the blocs (a), (b), (c), (d) and (e) is not a strict sequence in time. In case of a remote survey area that only can be visited once and then only for a short period (compare CSIRO helicopter survey in New Guinea) blocks (a) and (e) and some (c) will be done simultaneously in the field.

The observation data will be taken home and processed as a basis for a classification system (b), in order to do correlation studies with environmental factors (c) and finally in order to make a legend (d). If one lives in or near the survey area the compilation of data for classification and correlation study can be started after the sampling is done and then the area can be revisited for field checking and additional data collection, if any.

If one works in an area where a reasonable classification already exists, sampling time can be reduced by spending less time at each sample spot. A quick determination of the vegetation type in the field with the help of the textbook containing the classification may be sufficient. Even in that case, however, it is advisable to make notes about particularities and in any way

about environmental conditions. If a large area is surveyed in reconnaissance it is advisable to make some detailed field surveys of small representative areas that have, in relation to the survey as a whole, more the character of sampling than mapping. Especially if using small-scale radar or satellite imagery this is most preferable with the help of large-scale photographs of small parts of the total areas. The final result of stage V is a general classification system (in the case that it did not exist before) and a legend derived from a combination of (1) this classification system, (2) the map picture of the preliminary map and (3) data on additional field inspection. Important intermediate documents to arrive to the legend are (i) the annotated map = a photo-interpretation map on which at each sample point the final classification symbol is indicated; (ii) a 'cross table' between photo-interpretation units and classification units (see Fig. 1).

Step VI

During the former step most of the available data and information have been collected. In order to make best use of this information it is in most cases necessary to re-interpret all the photographs or imagery once more, following the same procedure as described for step III. It is not always necessary to draw a totally new map. If the area is of a type that is well-known to the surveyor his preliminary interpretation may be already sufficiently good and needs only corrections on limited points. He will have marked on his maps during the planning of the field survey points where he is not sure about a delineation or about the classification of certain features (see step V) and during the final interpretation he will check mainly this and other areas, where during the field survey discrepancies between interpretation and the field situation appeared. Less experienced interpreters should always do a full new interpretation and compare it with the preliminary one.

Not all important vegetation features can be interpreted from the photographs. Especially in (semi-) detailed surveys in cultivated areas there may be a need to map vegetation differences in arable land that do not correlate with any terrain features sufficiently to be mapped indirectly. In that case an additional classical field survey for which the photograph is used, mainly as a field map, should be done. Therefore in Fig. 2 fieldwork is also indicated. In remote areas this part of the work has to be done in combination with stage V but it is clear that this is difficult in all those cases where a reasonable classification does not yet exist. In those cases only a description in the legend of the occurring vegetation types taking it for granted that no delineation can be made, should be sufficient. In reconnaissance surveys it will usually never be possible to do more delineation work on the map than can be seen on the photograph.

Step VII

After the surveyor has made his final map design and his eventual generalization, the draftsman can take over and draw the final map. The surveyor has still a task in helping to select those cartographic means that give the best expression of what he wants to show. He should see to it that each patch of the map receives a symbol, letter or figure (number), that can be referred to in the text (vegetation names are often long and confusing for non-specialists) and that makes the map readable for colour-blind people as well (inclusive for not colour-blind people under bad light conditions). Especially in the selection of colours the surveyor should cooperate with the cartographer. Colours should be correct both in aesthetical and logical senses. The aesthetical impression may differ from person to person. Therefore if differences of opinion arise the logical consideration should prevail. The best logical criteria are the physical laws of colour. To make a map readable the colours (or black-and-white shades) should express the various gradients in the vegetation classification (see for gradients Chapter 9). Colour physics gives three clear gradients: hue, value and chroma (compare the Munsell Colour chart and similar colour schemes). The hue is determined by the rainbow (the symbol of overcoming the deluge, also the deluge of confusingly coloured maps!). It is nonsense to show a gradient with interchanges the colours from the sequence given by nature. Hue can be used for the most important gradient to be shown, chroma and value for others; other marks like dot and line patterns are used for additional distinctions. Also neutral colours not occurring in the rainbow hue gradient (brown in various values, chromas and also grays) can be used also for certain aspects.

It is impossible to give one fixed colour (hue) system for all units on all maps. In general it seems illogical to use red for moisture-indicative and blue for dryness-indicative vegetations, but here personal appreciations and traditions enter. The physical sequence within hue, chroma and value, however, can be accepted by every intelligent person, because it is logical. To use black-and-white and grey shades instead of colour is still more difficult. The main rule should be that not only the pattern should differ but at the same time also the shade impression. However, ten differences of shades are a maximum. Too often one sees maps showing a number of units each with a different pattern of lines or dots with the same gray impression. Such maps contain 'hidden images' but they cannot be used for the main purpose for which they are made: to allow a visual picture of vegetation (terrain) variations and the relation between the vegetation (terrain) types (see Chapter 10).

The legend should be clear and can be made as a list of unit descriptions. If it is, however, possible to draw the legend in the form of a diagram this should be done. This enhances the readability of the map considerably (see Fig. 1 and Chapter 11A). As will be clear from what will be said about the following step, it is very important to speed up the work in map drawing by all available means. Computerized processes of map drawing, various newly developing geo-information systems will become more and more important. This allows one to make various maps and documents each showing a certain type or combination of information (see Chapter 14). But usually first one basic vegetation map made by hand should be available.

Step VIII

When the map is ready it can serve as document for further study. This study has started already long before, but now there is a complete coverage of the whole area. Moreover the map is now, although not printed, in a form that it can easily be consulted by and demonstrated to other experts. This is very necessary because the preparation for the following step (IX) requires more knowledge than the vegetation surveyor has. He needs especially closer contact with the future users such as agriculturists, engineers, planners, climatologists, biologists, pedologist, etc. During step VIII data have to be collected and correlation studies have to be done to make evaluation of the map possible. If the vegetation survey is part of a more integrated or multidisciplinary survey this stage is the most important one for many other experts. The map produced in step VII is the basic document for discussions with those experts. The remarks about speed of the final map drawing were made because of this essential need when discussing the previous step. Too often consultation with other experts in a multidisciplinary team suffers from the fact that copies of the final maps are not produced in time. The preliminary map produced and used in steps III, IV, and V may fulfill already a purpose for integration if also used in the field by other disciplines. If the survey is carried out as a (holistic) landscape survey, the 'integration' between all land disciplines takes place during stages V, VI, and VII. Vegetation then occupies only one column of the legend table.

Step IX

Maps made for a purely scientific aim as well as maps for a more practical aim are made for a purpose. The recorded vegetation data have to be translated (evaluated) into pragmatic terms.

The previous step is just a start for this evaluation. A clear separation between both steps does not really exist. In Chapter 37 a description of the evaluation procedure is given.

Evaluation can be shown in several ways:
1. in description in the text;
2. in table form supplementary to the map legend; and
3. in special maps with their own legend in terms of that evaluation.

Often for matters of economy in printing cost only the first solution is used. The best solution, however, is the third one. In such maps the practical user is not biased by the, for him, confusing amount of scientific terms and details. Many units in the map may be analogues for a special evaluation aim and will have the same symbol and colour on the evaluation map for that purpose. Step IX is devoted to the compilation of this type of map.

Steps X and XI

Steps X and XI are as essential as all the other steps in vegetation survey. It is therefore regrettable that in planning the budget and time schedule of the survey they are in many cases underestimated. We shall not discuss these steps in detail here. For this refer to textbooks and lecturenotes on reporting and cartography. We only want to state that the author should always bear in mind that he writes (and prints maps) for readers. Not all the readers are the same. Usually he has to write for two very different groups of people:
1. Scientific colleagues, who are interested in the methodology of his works and the scientific results. They want a clear documentation and a scientific, sound account of the conclusions. They are prepared to criticize (even unjustly) if this is not done properly;
2. practical users, or scientific users from other sciences, who are not interested in the vegetation aspects as such but want a clear and concise statement of the results and the possibilities of application for their purpose. They are easily annoyed if too much 'science' is incorporated in the text.

In many cases two different publications are required for these two purposes. At least a subdivision into clearly separated parts will be necessary, especially for the practical user. Among these a separation has to be made between the broad policymakers and the planners of more details. The former has only time to read a few pages of headlines, the latter needs more elaborate data and maps.

In reconnaissance surveys using very small-scale imagery or photographs the stratified sampling meets with difficulties because not sufficient detail can be seen to locate sample plots. This may be the case using satellite imagery, very high altitude photography and also radar imagery. In these cases it is very profitable to use a 'cascade' of scales. So satellite imagery may be used in combination with high altitude photography as second, low altitude photography as the third step. The use of two scales is already commonly applied. The ideal case is to start with the smallest scale. These photographs or images are used applying stages I, II, III and IV. Instead of apllying steps V areas are selected on the base of the map resulting of step IV where larger scale photography will be flown. The selection is done in such a way that on an as limited area as possible, all the relevant properties of the area are covered with large-scale photography. So the large-scale photographs and the interpretation of these comes in the place of sampling. With the large-scale photographs interpretation is done and also the stratified field sampling. So with the large-scale photography the steps I, II, III, IV and also V and VI are carried out for the area covered by the new photographs. Then, with the help of the knowledge gained in the small-scale photography the interpretation of the small-scale images or photographs is redone in step VI after in step V of the large-scale and V (d) of small-scale photography a new legend for the small-scale

survey is designed. So the following scheme is used:

Small Scale   small scale
I II III IV   Vd VI VII VIII
          large scale
          I II III IV V VI

If a cascade of three scales is used a similar deviation is made from step IV of what is called above the large scale (then is the middle scale), etc. The ideal case is that the second flight can be determined by the results of the first interpretation (step IV). However, due to logistics, weather, financial and administrative conditions the risk to postpone the more detailed flight (or flights) to a later stage in the project is often too great. One often sees that the more detailed flights that have to assist in interpreting the small-scale ones are flown simultaneously (in the same period at least) with the small-scale photography. In that case a rather random distribution is to be advised unless the area is already so well-known that one can easily select the areas to be flown in detail before. But also then preferably first the small-scale and later the more detailed photographs should be studied. It is clear that if one applies a pure key type of survey (as in land-use surveys see Chapters 16 and 25), the large-scale (detailed) photography has to be available before the small-scale.

Appropriate scales for the stratified checks with large-scale photographs for vegetation survey are 1:5,000 to 1:10,000. If the whole area is flown in black-and-white pan, which is in many cases for reconnaissance survey still the best advice, the most detailed test strip should preferably be made on false colour or true colour film. In many cases rather good plant details can be seen on coloured diapositives which can be studied under magnification. The step from satellite imagery to 1:5,000 scale is a bit too large. An intermediate step of ca 1:50,000 scale will be useful. In that case the scheme is:

Satellite : I (II) III IV   Vd VI VIII etc

1:50,000:   I II III IV   Vd VI
1:5,000 :   I II III IV V VI

Finally, close observation with one's eyes, direct observation from small airplanes as practiced in systematic reconnaissance flights, may be combined as sampling methods in multistage systems (see Chapter 16A; see also Hutchinson C.S., ITC Journal, 1983). In many cases rangeland management requires scales not smaller than 50–100,000. Scale 1:50,000 is then sufficient, as largest, a two-step approach is then enough.

Legend design in survey by landscape approach using photo-interpretation

The process of designing a final legend follows the preliminary photo-interpretation and the classification (see Fig. 1). It is done with the help of two intermediate documents: (i) the annotated map and (ii) the cross table.

The annotated map is a clear specimen of the preliminary photo-interpretation map, showing also the exact position of the sample plots. This map is coloured and at each sample point the final classification symbol is indicated with, on a distance clearly readable, symbols, preferably in colour.

The use of coloured thick 'felt pens' is advocated because the map is to be used in an iterative quick thinking combination, and interpretation process that requires recognition in a glance (it means that part of science that never can be replaced by computer work). If done well, this map illustrates in a spatial way the amount of correlation between classification units and preliminary photo-interpretation units.

The second document, the 'cross table' is a relatively large piece of paper with a grid to be used as a matrix. In one-dimension e.g. horizontally, each block represents a preliminary photo-interpretation unit, then vertically are given the classification units, resulting from the classification process of sample data. In this cross table, all samples (relevé's) are entered. This table gives a kind of graphical-statistical picture of correlation between photo-interpreta-tion and classification. One should use a large piece of paper because one should indicate each relevé by its administrative number. Then it is possible immediately to look after the position of that relevé in the classification and on the map, and get it from the file to study it more precisely in all those cases where its position in the matrix deviated from the idea one had about it.

One should not immediately be panic stricken if a cross table does not show nice diagonal forms or shows large overlaps, or irregular distribution of classification units over the preliminary legend, and vice versa. This is normal for the first cross table exercise. In fact the making of a cross table and annotated map is still part of the classification process. The doubtful relevé's can be re-examined and be found to be transitional and be better placed in a neighbouring category. But also the preliminary photo-interpretation may appear to be liable to alternation in a way that better would fit the classification. So both maps and classification can be corrected by this process. Aerial photographs (and the other remote sensing means if applied) cannot be missed in this iterative process. The human mind has to be quick and concentrated on spatial (chorological) as well as abstract (topological) aspects and therefore should not be disturbed by unclear presentations (see above).

Both the annotated map indicating all relevé spots, with clear classification symbols in clearly intelligently coloured map units, and the cross table with a good presentation of all relevé's in a matrix of preliminary legend and classification and indicated by their administrative number is indispensable. After cleaning up the cross table and the annotated map (including extra photo-interpretation) the cross table shows clearly which legend units exist from complexes of classification units. Also it may appear that different legend units have the same classification. One should decide if the apparently different photo-image that lead to differentiation is of importance or if it was just caused by less relevant (temporary) features and can be disregarded, which means that the two units are combined to one. In Fig. 1 the process beginning with the photo-interpretation and ending in final mapping for the method of landscape-guided survey is given. The figure can be subdivided in four quadrants.

Top left gives the chorological data collection, top right shows the collection of semantic information and classification, bottom right represents the real legend formation by combination semantics and chrology.

In key method and photo-guided field survey (see Chapter 16) the procedure deviates from the one sketched above, but is so evident that no discussion about it is necessary.

APPENDIX A

ITC WORLD LAND COVER and LAND-USE CLASSIFICATION

Code	LAND COVER FORMS Main entrance for aerial photo and other remote sensing immage interpretation	Land use related to the cover forms (examples)	Code	LAND USE As distinguishable in aerial photo-interpretation or interpretation of other remote sensing images and checked in field survey	Cover forms related to the land uses
I	*Building and artifact* – buildings – artifacts – canals – dams, dykes	1	1	*Settlement and infrastructure* (a) residential (b) industrial, quarrying, mining (c) transport and communications (d) recreational (e) others	I, V, VI
II	*Fields/plantations* – fields: arable crops, fallow, cultivated forage crops plantations: tree crops, shrub crops and grapes vines	2a, 2b 2c, (6)	2	*Agriculture* (a) annual herbaceous crops; irrigated or dry-land (b) perennial herbaceous crops; irrigated or dryland e.g. alfalfa, grass/legumes mixtures, sugar cane, sisal, bananas (c) tree and shrub crops; irrigated or dryland; incl. cacao, rubber	II
III	*Open natural vegetation* – woodland/savanna – shrubland – dwarf shrubland – grassland + herb vegetation	3, 4, 5, 6 or 7	3	*Extensive grazing* (a) ranching (b) pastoralism	III, (IV)
IV	*Forest* – forest plantation – (semi) natural forest	2c, 4, 3, 5, 6 5, 6	4	*Forestry for* (a) timber (b) pulp-wood (c) wood; fire wood, pole wood, charcoal and other domestic uses (d) others: e.g. bark, terpentine, tannin, cork	(II), IV, (III)
	Water body, snow/ice cover	5, 6, 7 (1b), (1d) (1c) (2) (3)	5	*Conservation* (a) nature reserve, National Park, etc. (b) game reserve (c) watershed management (d) dune stabilisation (e) others	III, IV, V, (VI)
V	*Bare land* – barren rock – beaches – 'badlands' – mudflats	7 (5)	6	*Hunting and fishing* (a) hunting (b) fishing (c) aquaculture (d) other uses	III, IV, V, (II)
			7	*Not used*	II, III, IV, V, VI

COMPLEX COVER FORM:
Two or more cover categories on different land for cartographical reasons put in a single land mapping unit, (e.g. indicated on map legend description as complex of 'open natural vegetation and bare land' (III + VI).

The used codes could be used on final maps as legend symbols. In case of complexes *it might be better to use a seperate code* (see Ch. 11).

From I.T.C.: Landecology Group, 1982, modified

Appendix B

Guide for systematic interpretation
of a block of aerial photographs [1]

The following sequence of manipulating a block of photographs is recommended:

A. *Preliminary inspection* (step Ia) [2]
A1 Checking the position of the photographs with the help of an indexmap or photo-index.
A2 Preparation of a lay-out of alternate photographs in their approximately correct position.
A3 Preliminary study of main land(scape) units on the lay-out of all photographs or photo-mosaics; design of a sketch land unit or vegetation map; preliminary inspection of photo-pairs of sample areas under the mirror stereoscope.

B. *Study of sample photo-pairs and control runs* (step Ib + c)
B1 Initial study of purpose, required information and of all relevant data available from maps (topographical, geological, vegetation, soils), documents and experts.
B2 Drawing of matchlines.

COMPLEX LAND USE:
Two or more land use categories on different fields but for cartographical reasons put in a single land mapping unit: indicate on map legend description as, for example: 2c + 2a

MIXED LAND USE:
Two or more land use sub-categories are carried out on the same field; for example:
intercropping: woody crops (olives) with annual crop (small grains); indicate on map legend: 2c/2a

MULTIPLE LAND USE:
Two or more main land use categories simultaneously or periodically on the same land; for example:
dehesa: Extensive grazing/Forestry/Agriculture (3/4/2)
shifting cultivation: Agriculture/Forestry (2/4) or Agriculture/Extensive grazing (2/3).

B3 Detailed analyses of photo-pairs of sample areas in each main land unit and in their transitional zones (use Appendices C and D).
B4 Production of a preliminary legend for the imagery interpretation map.
B5 Detailed analyses of sample photo-runs to check preliminary legend.

C. *Study of all photographs* (step III)
C1 Systematic interpretation of all photos together with comparison of data from other sources (see B1).
C2 Editing of the legend of the photo-interpretation map.

D. *Transfer of data, map and covering report* (step III + IV)
D1 Transfer of the interpretation data from the photographs to photo-mosaics or base maps (procedure given in separate instruction) and final check of the interpretation.

Instructions:
A. *Preliminary inspections of AP's*
A1 The index is checked to see the general system of flying and the approximate relative position of the area. Overlapping runs and photographs are marked and where necessary eliminated.
A2.1 Alternate photographs of the area are laid out in their approximately correct overlapping positions. It should be possible to

[1] From ITC Land Ecology Group and ITC Soils Survey Group.
[2] Steps as described in Fig. 2.

do this from the index. This lay-out remains the starting point for the whole subsequent interpretation.

A2.2 The lay-out is covered with glass plates.

A3.1 On the lay-out (covered with glass plates) those parts which seem to represent rather homogeneous areas (distinct land units) are roughly delineated with grease pencil. These land units should be given preliminary symbols, and descriptive names e.g. using terrain (Zuidam, 1979) and land cover form: Terminology as applied in the ITC World land cover and Landuse classification (Appendix C).

A3.2 Make a description of the different land units in terms of terrain, vegetation and/or land-use. For this purpose some sample photo-pairs are roughly scanned under the mirror stereoscope.

A3.3 The above mentioned analysis is roughly copied on a very much smaller scale on a piece of paper. The sketchmap is mainly used as a reference map during the procedure of systematic analyses. Therefore, also the approximate position of the various AP's (simply represented by the line of flight through their pp's) and the photonumbers should be indicated on the sketchmap.

The copying of the sketch by application of a grid system can be carried out as follows: strings are extended over the layout in a certain grid, which, in the desired reduction, is drawn on a piece of paper. Now the analysis on the glass plates is copied by hand, square after square. Some main topographic features are copied in this way as well.

B. *Study of sample pairs and control runs*

B1.1 The purpose of the map is clearly defined and a list is made of the items of information that can and should be derived:
  i) direct from the photograph (compare Appendix A);
  ii) by stratified field sampling on the base of the photo-interpretation map.

B1.2 All relevant data available from maps, literature and from other sources with respect to the area are studied in advance before carrying out the interpretation.

B2 The end match line is drawn on the right-hand part of each AP. The side match lines are drawn on all AP's, simply by connecting the wing points.

B3.1 From each preliminary land unit (see sketchmap), a representative area covering approximately one photograph is chosen.

B3.2 This area is analyzed stereoscopically in detail, in order to find typical characteristics of each land unit and to classify as much detail as possible and necessary according to the scale of the survey. Especially in large-scale surveys one will find that the land-use, the land cover form and the terrain, will not always coincide with each other.

If these differences are too large to be eliminated by generalization one should use two types of boundaries with a different colour (if only black is used and also on the final interpretation map: a different line width or a different type of line interruptions). One colour then is used for airphoto-visible vegetation form boundaries and another for terrain (physiographic) boundaries that seem to be of land ecological interest and may, or may not, appear to be boundaries later on between floristic vegetation units invisible directly on the aerial photograph but that can be observed in the field. It depends on scale and purpose of the map whether the final generalization will take more in account the vegetation or more the terrain (physiographic) boundaries. For drawing technique, etc., see C1.1

B4 A classification of land mapping units is made. This *preliminary legend* is to be used in the following stage of the systematic interpretation.

B5.1 The draft of the preliminary legend is tested out by means of steroscopic examination of a few runs or parts of runs through the main distinguished land(scape) units including their transitional zones.

B5.2 Afterwards, the legend is modified and added where necessary. With reference to the analyzed and classified sample areas and control runs a semi-final legend is set up for the whole area. A description based on vegetation form, terrain form and land-use features in as much relevant detail as possible is given in each unit.

The obtained semi-final legend should serve as a basis for the systematic block analysis to be made next. Depending on the final scale of the photo-interpretation map, the detailed legend of the analyses of sample pairs and control runs, should be simplified by a *logical* combination of units. The units are described in terms of important elements and in terms of various other aspects important for the interpretation. Only wherever well-known vegetation form and terrain form systems and their units have been *clearly* recognized, these should be used in the description. If there is any doubt use fancy names.

B5.3 A list is made of the mapping units symbols, indicating the most logical and convenient order in which the systematic analysis can proceed. First, the clearly recognizable patterns and finally the more complicated or doubtful ones. In view of the same purpose, a list of topographic symbols is made.

C. *Study of all photographs*

C1.1 It is preferred to draw directly on the photograph surface with special grease pencil instead of using an overlay.

Use always a very hard *table* or *glass plate,* no carton, to support the photograph, in order not to disturb the photographs with the pencil. If diapositives, unique or expensive material is used, overlays are necessary. Use always the full transparent ones (never kodatrace, etc., but acetate foli or other types) that are as transparent as glass: otherwise too much information is blurred. The analysis should cover the area between the end matchlines as well as between the side matchlines of each photo. Although only one (the right-hand) matchline is drawn on a photograph, the other one will be seen as soon as the photo-pair is observed stereoscopically.

Besided the necessity to do systematic analysis on basis of a classification as discussed previously, the analyses should have the proper detail with regard to the final scale desired. The interpreter should therefore keep in mind, that the mapping units on the map of final scale should have a diameter no less than 5 mm, or in case the units are extremely elongated, no less than 2 mm. To check these measures during the photo-analyses, a small piece of transparent paper, with the proportionally enlarged minimum unit sizes drawn on it, is very helpful.

Important details of smaller size (e.g. isolated important trees, certain landmarks for orientation during fieldcheck, etc.) can be indicated with special symbols (dot, cross, line, etc.).

C1.2 When the analysis of the top is finished, the detail along the bottom matchlines of that run is transferred stereoscopically to the corresponding matchlines of the next run (second run).

C1.3 A systematic analysis is made now of the second run and the detail along the matchline in common with the next run is transferred. This is repeated until all runs have been analyzed.

During the stereoscopical examination of all photographs, at the same time, every unit of the analysis is classified according to the semi-final legend. Where necessary small modifications or additions to the semi-final legend may be made. Where necessary the previously analyzed photographs are once more scanned for introducing the new modifications. One will always come across areas that are too difficult to interprete for the time being. Experience teaches us that after having seen the total context, problems are easily being solved. Therefore finish first the clear parts on every photograph. After this is done go back to the difficult areas on the same or even earlier interpreted photographs. The analysis per run is sometimes difficult because of the repeated occurrence, often in small strips of the transition zones.

C1.4 When the analysis of all runs is completed, a lay-out is made again and the following points are carefully checked:
  i) whether details on adjacent photographs and on adjacent runs are matching and continuous;
  ii) whether all units are completely bordered by a unit boundary;
  iii) whether all units are marked by a legend symbol.

C2 The semi-final legend classification together with the additions and modifications obtained during the block analyses are reviewed to complete the editing of the final interpretation legend which will serve for the photo-interpretation map. These brief descriptions need to be 'to the point', leaving no doubt in the reader's mind.

D. *Transfer of data, map and covering report*

D1 *Transfer of interpretation data from photographs to the draft map (mosaics or topo-maps)*
The data obtained from the systematic photo-analysis and topographic data are now transferred from the photographs to a (preliminary) base map by means of optical pantograph, Radial Line Plotter, Zeiss Aero sketchmaster, or stereosketch. This base map may be a not simplified topographical map, or alreay a final base map (see D2). The best preliminary base map is a photomosaic or orthophoto map. Transfer to such a photomosaic can best be done by free hand drawing without using the above mentioned instruments. Even if the photomosaic or orthophoto map has a smaller scale than the AP used. This allows also direct generalization. This is a great advantage of photomosaics and orthophoto maps. [4]

It is very useful, during field check, to have the approximate situation of the principle point of each AP on the map indicated with AP number.

D2 *Drawing and colouring of the photo-interpretation map*

D2.1 (First prepare a final basemap).
Put tracing paper over an existing topographical map or a mosaic [5] (the preliminary base map) and fix with tape. In this case semi-map paper is useful (kodatrace if high accuracy is needed; in most cases common tracing paper is sufficient at this stage). Trace the essential topographic features with ink.

Internationally accepted standard symbols for the topographic features should be used (see cartographic textbooks).

Care should be taken that enough topographic features are mapped, so that a good relative orientation of the interpretation lines to those is possible. The top of the map preferably should be directed towards the north. Prepare transparent prints (Radex-red) of this base map. Reserve several copies for later derived maps and put one on top of the draft map (on mosaic or topo map).

---

[4] Unfortunately the latter are usually not available (too expensive) see Chapter 29, Step III.
[5] If no topomap or mosaic is available one should make a mosaic oneself or use a template method to prepare a base.

D2.2 Trace on the Radex-red the results of the photo-interpretation. At this stage distinguish between clearly observed photo-interpretation boundaries and less certain photo details.

D2.3 Add the photo-interpretation symbols to the map, using the established legend.

D2.4 Add the following information to the map:

Title:

If no field survey is going to be made, the map is simply called: 'photo-interpretation map for vegetation or land survey purposes of an area in .....'. The title is drawn at the extreme top of the map.

Date:

i) Of photographs used (in hour, day, month, year);

ii) if field observations will be made: (exact date to be filled in later, but already now) a space should be reserved for season and year.

On a vegetation map, the date is a very important item and should be mentioned in large figures.

Scale:

If the base map was compiled without adequate ground control, the word 'approximate' is added, e.g. 'approximate scale 1:20,000'.

Photo lay-out:

Scale, number and location of the photo runs and number of first and last AP's in each run are indicated in a small diagram. Preferably the principle point of each photo is given on the map itself with a dot or cross.

Location of the area:

A small map of the country or region is drawn in which the investigated area is indicated by means of a black dot or rectangle.

North arrow:

A simple arrow is drawn in the geographic or true north direction.

Legend of the mapping units:

In the legend short descriptive names of the vegetation and/or land(scape) units are written. A lengthy description of the single units is not necessary since they are fully described in the covering report. On the map simply a symbol appears besides a *short* descriptive name of the unit.

Reference legend:

In the reference legend the symbol for various types of boundaries, roads, rivers, towns, etc., are indicated.

Note on the method of base map compilation:

The following can be noted on the map: 'base map compilation by (slotted) template method with (without) ground control. Plotting by means of Zeiss vertical sketchmaster (and/or Radial Line Plotter), mosaic', etc. If no ground control was available during the triangulation the following should be added: 'Because of absence of ground control, the scale should be regarded as very approximate'.

Note of map compilation:

In the lower right hand corner information about the author(s) is added to the map: e.g. 'Map compiled and drawn by J.Johnson, group of land ecology, ITC, Enschede, The Netherlands, 19..'.

Frame:

A frame of a single or double line is drawn around the whole map.

D2.5 Three prints are made on double weight paper.

D2.6 One of the prints is coloured with colour pencil or with Ecoline fluid paint in accordance with the grouping of the map legend. The colouring also of the preliminary map is a great help in finding possible mistakes made during the drawing of the map! Most of all the preliminary map image should start to be a 'reality' in the surveyor's head. It is not a series of lines and symbols, but a holistic pattern that has to feed the intuition when deciding what to do and where to go in the field; *without colours, this simply does not work.*

The colour scheme for an interpretation of vegetation or land(scape) units is not bound by standarized rules. Designing a colourscheme will always be a kind of art: trying to transfer information via form and colour to somebody else.

Generally, related mapping units should have a related colour in terms of colour theory. This allows for a three-dimensional system using: hue, chroma and value. A three-dimensional figure (see also Fig. 1 and Chapter 11A, Legend formation) can be helpful in designing the colourscheme.

D2 *Writing of interim-report on photo-interpretation stage*

A short covering report may be written after the completion of the photo-interpretation and the map construction. Either as a student exercise, or in practice, as a basic document for the survey team. An outline mentioning the main items which have to be discussed in the covering report, is as follows:

Introduction:

A short general description of the area, scale of map and purpose of the study is given and the main purpose for which the map has been made.

Characteristics of the AP's given:

Number, size, scale, image, quality, data of exposure, quality of printing, kind of photo-paper and photo surface, camera data, flying height, etc., are briefly discussed.

Working methods in this particular study:

The methods (incl. instruments and their advantages or disadvantages) used in this particular study are briefly discussed, emphasizing the deviations from the general outline of working methods.

Description and classification of mapping units:

A description of the various elements used during the analysis and their relative importance in the 'build up' of the mapping units is given, followed by a detailed description and classification of the mapping units of the legend.

Results of the study and the final map:

The results of the study are discussed in terms of character of mapping units, expected ecological gradients, vegetation properties and their significance for a study in the field.

Field sampling design:

A field sampling design is made on the base of the coloured preliminary photo-interpretation map. Sample locations are indicated with a circle, alternative locations are indicated in case some locations are inaccessible. An activity scheme for fieldwork is made.

Note:

The next stage of the mapping procedure is a stratified sampling in the field, random within the indicated circles, and a field check of the systematic photo-interpretation. This will eventually result in the production of a vegetation or land(scape) unit map. Date of this report will later be incorporated in the final report of the completed survey.

Appendix C

The ITC method of analyzing and classifying the structure of vegetation in reconnaissance and semi-detailed vegetation surveys. HEIN A.M.J. VAN GILS and WILLEM VAN WIJNGAARDEN

Introduction

Vegetation structure is a major source of information on aerial photographs and even more so in Landsat multispectral scanner (MSS) imagery in all land resource surveys, especially in rangeland surveys. So (even in no specific vegetation surveys) systematic description and typification of vegetation structure is necessary for the interpretation of photographs and images, as well as ground data. In this chapter, two widely used vegetation typifications are reviewed and the

420

ITC approach to vegetation typification according to three dimensions is described. The principle of this approach is to plot vegetation cover values per vegetation layer in a two- or three-dimensional feature space and then classify the vegetation structure. Examples are given from ITC surveys.

Modern vegetation mapping is carried out with the systematic use of aerial photographs and/or satellite imagery. Stereo aerial photographs, e.g. directly supply various types of data such as landform, drainage plattern and land cover (= vegetation, buildings and artifacts, water (open or underneath vegetation), bare land) and field patterns.

Landsat MSS indicates mainly land cover and drainage and field macro patterns, but lacks relief.

Within land cover, seven main categories have been distinguished in the ITC world land cover land-use classification (see Appendix A of this chapter).

1. buildings and artifacts;
2. fields/plantations;
3. open natural vegetation;
4. forest;
5. water bodies, snow/ice cover;
6. bare land;
7. burned land.

These main land cover categories can usually be recognized in coloured Landsat MSS products by interpretations of colour, as well as shape, pattern, location, etc.

Categories 2, 3 and 4 represent vegetation proper. The others are, however, also important indirectly in vegetation mapping.

In this chapter, we propose a systematic but local or regional approach to vegetation classification according to structure.

Vegetation structure as opposed to floristic composition of vegetation

Vegetation structure and composition do not usually correlate with each other in an absolute sense in rangeland areas. A structural type can include various floristic sub-types, and floristic types can include various structural sub-types (see Fig. 1 and Tables 1 and 2). This has important consequences for photo-interpreters because the structural types of vegetation they distinguished on the aerial photographs may not necessarily indicate the floristic vegetation type(s). Vegetations with different floristic compositions may show identical structures, and vice versa (CARAP, 1980; Wijngaarden, 1984).

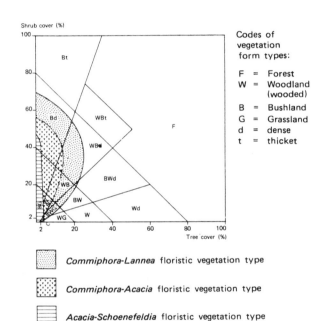

Fig. 1. Relationship of vegetation structure types and floristic types in a semi-arid savanna vegetation in southeast Kenya. Total woody cover values should be read along diagonals (Wijngaarden 1984).

Table 1. Matrix of vegetation structural types versus floristic types from sample area in Libya.

Structural type	Floristic type							
	1	2	3	4	5	6	7	8
Forest	7	1	1					
Woodland	3	3	3					
Shrubland			2	10	8	6	1	
Dwarf shrubland		1			1	1	2	13
Grassland								20

Table 2. Matrix of vegetation structural types versus floristic types from sample area in Botswana.

Structural type	Floristic type					
	1	2	3	4	5	6
Woodland					1	
Shrub savanna and shrubland	10					
Dwarf shrubland		6	4			
Grassland		1	7	7	3	3

The fact that vegetation structure and composition do not correlate in rangeland areas also has implications for the hierarchical vegetation classification system as applied in the 'Braun-Blanquet approach' (for some inconsistencies which such systems have caused in Europe, see van Gils, 1977 and 1978).

Structure and physiognomic vegetation typifications

There are many vegetation typifications according to structure and physiognomy (Unesco, 1973; Eiten, 1968; CCAS, 1956). In trying to apply them during many years of vegetation mapping on different continents, we found that they are not sufficiently systematic for application in any specific survey area. Their most significant disadvantage for our purposes is that they are obviously designed for use on exploratory scales, i.e. for national, continental or global mapping. The surveys for which ITC develops methods are mainly on reconnaissance or semi-detailed scale for a specific region within a country, and where only a limited number of structural vegetation types are found. These types have to be subdivided further than the global typifications of Eiten and Unesco to arrive at useful regional maps.

Tabulated and graphic reconstructions of the Unesco and the Eiten vegetation typifications are shown in Tables 3 and 4 and Fig. 2 and 3. A first general observation to be made is that the Eiten typification uses a single woody component (trees and shrubs). This has two disadvantages: first, it is often difficult from aerial photographs to differentiate shrubs and trees. Further, in many vegetation types (e.g. in the Mediterranean scrub woodland, 'maquis', and the African savanna) the distinction between shrubs and trees is unclear. Small trees and shrubs are

Table 3. Reconstruction of the Unesco physiognomic vegetation typification in tabular format

Structura. type		Percent cover per layer			
		Trees (height > 5 m)	Shrubs (height 0.5–5 m)	Dwarf shrubs (height < 0.5 m)	Herbs
I	Forest	interlocking	n.a.	n.a.	n.a.
II	Woodland	> 40 but not interlocking	n.a.	n.a.	n.a.
III	Shrub				
	a thicket	n.a.	interlocking	n.a.	n.a.
	b shrubland	n.a.	not-interlocking	n.a.	n.a.
IV	Dwarf shrub				
	a thicket	n.a.	n.a.	interlocking	n.a.
	b shrubland	n.a.	n.a.	individual clumps	n.a.
V	Herbaceous vegetation				
	a woody grassland*	10–40	n.a.	n.a.	dominant
	b grassland	< 10	n.a.	n.a.	dominant

* 'savanna'

Table 4. Reconstruction of the Eiten structural vegetation typification in tabular format.

Structural type	Cover (percentage)		
	Woody species (trees + shrubs)	Tree proportion of woody species	Herbs and dwarf shrubs
Forest	> 30	10–60	n.a.
Woodland			
woodland	10–60	> 60	n.a.
sparse woodland	< 10	n.a.	< 10
Scrub woodland*	10–60	< 60	n.a.
Savanna	< 10	n.a.	> 10
Scrub	< 60	0	< 10
	> 60	< 10	n.a.
Plant field	absent	n.a.	n.a.

* includes Eiten's category 14 (open scrub).

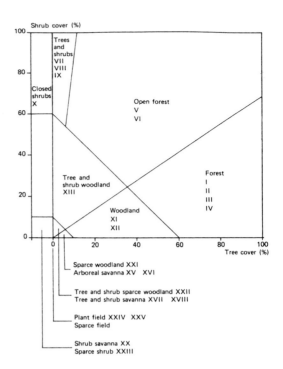

Fig. 3. Eiten vegetation typification in graphic format. The Roman numerals correspond with those of Eiten. Total woody cover values should be read along diagonals.

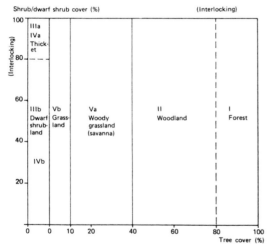

Fig. 2. UNESCO (1973) vegetation typification in graphic format.

within the woody component – trees, shrubs and dwarf shrubs by height.

Too much emphasis is usually placed on the high (tall) woody component (trees, high shrubs) and not enough differentiation is given to the low woody component and herbs for the purposes of livestock grazing and/or erosion control. This bias may be the result of the impact of foresters in this field and/or the greater visibility of the high woody component on aerial photographs.

Structural typification in aerial photo-interpretation

The most striking aspect of vegetation in aerial photographs is the cover of the woody component, regardless of the height of the woody plants, and it is relatively easy to distinguish classes of cover of 100 to 80%, 80 to 40%, 40 to 20%, 20 to 2% and 2 to 0% (see Table 5).

often multi-stemmed, and non-multistemmed forms can be so low that tree-like forms may be classified as shrubs.

Several authors (Unesco, 1973; Eiten, 1968) use 'cover' of trees, shrubs, etc., to define their units, but trees, shrubs, etc., are seldom defined themselves. We therefore propose defining –

Table 5. Cover class boundaries of woody vegetation component in aerial photographs.

	Aerial photographs	Physiognomy on the ground
Cover woody vegetation component (%)	Crown distance / crown radius ratio	
80	0, crowns interlocking	closed canopy
40	1	semi-open canopy, difficult to pass if at chest height
20	2	open canopy
2	10, very widely spaced	woody component not important

If the cover class ranging from 2 to 20% occupies large tracts of land and is of importance for the purpose of the survey, tree densities (number of trees/ha) can be used to subdivide this class (11).

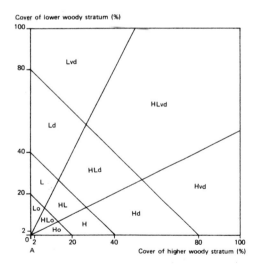

Fig. 4. Structural vegetation typification for aerial photo-interpretation of area where two horizontal layers (strata) can be observed. Total woody cover values should be read along diagonals.

H = Higher woody stratum dominant
HL = Higher and lower woody stratum dominant
L = Lower woody stratum dominant
A = Absence of woody stratum
o = Open
d = Dense
vd = Very dense

The second vegetation feature to be observed on aerial photographs is horizontal layering (strata), but more than two horizontal layers can rarely be detected. The distinction of two horizontal layers combined with their relative cover may lead to the following classes:
1. higher stratum dominant;
2. higher and lower stratum of approximately equal cover;
3. lower stratum dominant.

The combination of total cover and relative cover per stratum can result in a structural vegetation typification for aerial photo-interpretation as presented in Fig. 4.

The grass cover within woody vegetation cannot be directly assessed from aerial photographs. Even in pure grasslands, the grass cover is usually difficult to assess from semi-detailed photographs (1:25,000 to 1:75,000). Depending on the contrast in photo features, it is sometimes possible to distinguish the soil surface and grass, but only on large-scale aerial photographs.

Structural typification based on field sampling

In many cases, it is preferable to start a field survey without a preconceived structural typification. We are accustomed to collecting data on cover of the various layers in the field and reducing those data pragmatically (see examples in Tables 6 and 7, and Fig. 5 and 6). Such vegetation types can better represent the local varieties in a typification for exploratory scales. For example, the structural types in Fig. 5 cannot easily be placed and named in the Unesco or Eiten typifications (Fig. 2 and 3, respectively). These typifications also do not readily accommodate a third dimension – for example water or grass cover.

Considerable information on variations in the low woody and grass components (and sometimes forbs or lichen) can be obtained by processing field samples. It is then necessary to relate the vegetation typifications based on field samples to land features visible in aerial photo-

Table 6. Matrix of vegetation form types (rows) versus aerial cover in percent per plant growth for a specific Mediterranean area (Extramadura, Spain).

Structural type	Percent cover per layer		
	Trees	Bushes	Grasses
Forest plantation	> 60	n.a.	n.a.
Woodland (cork oak)	> 25	n.a.	n.a.
Dehesa woodland (stone oak)	15–25	< 10	n.a.
Bushed woodland	5–25	> 10	n.a.
Bushland	< 5	> 20	< 20
Wooded grassland	5–15	< 10	> 40
Grassland	< 5	< 20	> 50
Sparsely vegetated land	< 5	< 10	< 10

The class delineations are established after completion of data collection in the field on cover per layer; this typification does not cover all theoretical possibilities but contains actually sampled forms.

Table 7. Matrix of structural vegetation types (rows) versus estimated aerial cover in percent per plant growth structure for a specific Mediterranean area; the classes are distinguished in such a way that they can be interpreted on aerial photographs scale 1:50 000 [9].

Structural type	Trees	Percent cover		
		Shrubs	Dwarf shrubs	Herbs (gresses and forbs)
Forest	⩾ 50	n.a.	n.a.	n.a.
Woodland	10–50	n.a.	n.a.	n.a.
Shrubland	< 10	⩾ 10	n.a.	n.a.
Dwarf shrubland	< 10	< 10	< 30	b.a.
Grassland	< 10	< 10	< 30	⩾ 20
Sparsely vegetated land (not sampled)	< 10	< 10	< 30	< 20

graphs to incorporate both low wood and grass cover in the legend of the resource map.

Introduction of grass as a third characteristic does not necessarily require a three-dimensional graph, however. It seems reasonable to hypothesize that grass cover in a woody vegetation will in some cases be negatively correlated with the shrub cover, since they compete for light water and nutrients. We tested this hypothesis using a data set of Dehesa vegetation in Spain (Musa et al., 1983) and found that the grass cover indeed showed a strong negative correla-

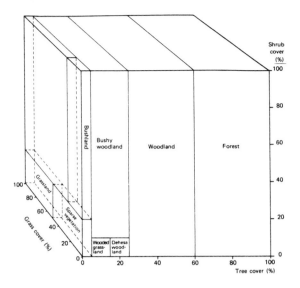

Fig. 5. Three dimensional graphic representation of structural vegetation types of Table 6.

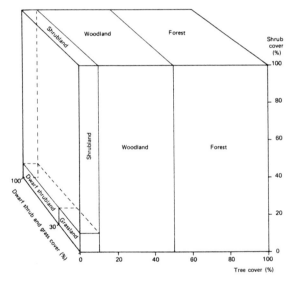

Fig. 6. Three dimensional graphic representation of Table 7.

tion (correlation coefficient of 0.80) with shrub cover. Such a correlation would exist only in very stable areas, but this relationship implies that the grass cover could be included in the graph with the shrub cover (x-axis) against the tree cover (y-axis) of Fig. 7, without a three-dimensional presentation. Various environmental factors (e.g. fire, grazing and human influences), however, may affect the grass to such an extent that the woody cover only sets a maximum for the grass cover.

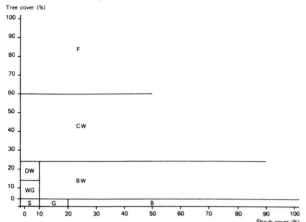

Fig. 7. Two dimensional graphic representation of Table 7. The boundary between S and G depends on the third variable, grass cover. The grass cover shows a negative correlation with shrub cover (grass), if the tree cover is below 25 percent.

F = Forest
CW = Cork oak woodland
DW = Dehesa woodland
BW = Shrubby woodland
WG = Wooded grassland
B = Bushland
G = Grassland
S = Sparse vegetation

The presentation of structural vegetation typifications can be given in tabular and/or graphic forms. The graphic forms of Fig. 2, 3 and 4 show only two axes, tree cover against shrub cover. If there is a third variable, i.e. the cover of another vegetation component such as grass or dwarf shrubs, a three-dimensional graph becomes necessary. If the grass cover is directly related to shrub or shrub and tree cover, the structural types identified in aerial photographs can be specified in terms of grass cover.

Conclusion

Structural vegetation types are not directly related to compositional vegetation types. The implication for mapping land resources, including vegetation, is that vegetation structure can-not represent the vegetation composition and the latter should be studied in the field.

The structural vegetation typifications designed for exploratory scales are unsuitable for surveys on reconnaissance and semi-detailed scales, but typifications appropriate for reconnaissance and semi-detailed surveys cannot be pre-conceived for the whole world; they should be based instead on local or regional field and photo data.

Some older structural vegetation typifications use only two dimensions (shrub and tree cover) in ordering and classifying vegetation structure. A three-dimensional classification including grass cover (or, for example, tree height) provides other essential information, especially for rangeland resource and erosion hazard inventories.

Essential information on variations in vegetation structures (e.g. grass cover, grass height) often cannot be seen on aerial photographs. A structural vegetation typification in such cases should be based on field samples.

Appendix D

Twenty-one things to remember in photo-interpretation for vegetation survey

General

1. Photo-interpreter and field-surveyor should be *the same person* (and a vegetation scientist).
2. The *photo-image* expresses as the *character of a landscape* more or less like a photograph of somebody that can reveal something of his (her) character, if combined with groundtruth.
3. Photo-interpretation for *vegetation survey* requires full knowledge of *landscape* (especially geomorphology).
4. The most *universal vegetation survey tool* for reconnaissance survey is *black-and-white superwide angle pan airphotography*, scale 1 : 20,000 – 1 : 50,000 with yellow filter.

426

Interpretation

5. First *aim of photo-interpretation* is *stratified sampling*. Second aim: delineation of map units.

6. In executing photo-interpretation for vegetation survey *five different types of information* are indicated on the photographs (preferably also with different colours):
   a) human infrastructure (roads, villages, fences, etc.): (yellow);
   b) physiotope (landform, relief, terrain: (red);
   c) drainage ways (valley bottoms, rivers): (blue);
   d) divide (crestlines, etc.): (brown or dotted red);
   e) vegetation structure, as far as deviating from b (mainly man-made vegetation patterns land-use): (green).

7. *Start with the easy things.* The more difficult aspects will be more easily solved at the end.

8. In mountainous areas aspect (exposure) is an important preliminary photo-interpretation element. However, do not indicate this by symbols, but by indicating divide and drainage way (valley bottom) with lines.

9. Use preferably *no transparent* for photo-interpretation. If transparent is necessary (expensive or unique photo-material) use only *full transparent* (no kodatrace, etc.).

10. For reconnaissance survey of *flat terrain,* the most economic transfer system from photograph to map is an *uncontrolled mosaic* to which the photo-features should be transferred by hand without instruments.

Fieldwork

11. The field-surveyor should have a *coloured* preliminary photo-interpretation map in the field.

12. Field-sampling for *vegetation survey* requires necessariliy also observation of *physiotope* (rock, slope, landform, exposure, altitude, at least simple soil description by auger).

13. Fieldwork is *description of content* of map units, only a minor part is 'field-check'.

14. *Detail of sampling* is *not* necessarily correlated with detail of *mapping.*

15. In general *mass* (production) measurements should be made of a *limited selection* of the observation points, preferably *after* classification is done.

16. In general a *'quick herbarium'* should be made (at the spot in ring-files) of all unknown plants which get a code (relevé number and sequence letter) and a fancy name. Never try to make a real sophisticated herbarium during relevé execution.

Classification and legend

17. Classification procedures should *only* be done by *computer* if sufficient facilities are available, so that the surveyor is able to do it him/herself interactively. The result should always manually be corrected using common sense. Sometimes the table method (pref. using mechanical table device) may be still cheaper and safer.

18. Classification should result in *sociological groups,* preferably illustrated in a bar-scheme. Hierarchical grouping of communities can be done much later (by comparing existing systems).

19. The crucial tools for legend making are the:
    a) *'cross table'* (between classification units and preliminary interpretation);
    b) *'annotated map'* (coloured photo-interpretation map with clear from distance readable indication of final classification symbol at each sample point).

20. Preliminary as well as final map colours are the most important expression means. *Map colours* should be *logical.* Guidelines may be ecological gradients, in some case also structural.

21. *Hierarchical levels* of legend and general typology do *not* necessarily correlate.

30. Vegetation Mapping in Japan

A. MIYAWAKI and K. FUJIWARA

Introduction

The oldest vegetation map of Japan is a 'Vegetation map of Mt. Fuji' by Hayata (1911), based on physiognomy. Vegetation mapping using phytosociological units was first conducted by T. Suzuki (1954) on the Ozegahara Moor. Since the 1960s, the field of vegetation science has been steadily developed, especially phytosociological studies including vegetation mapping based on the method of Braun-Blanquet and Tüxen.

On the other hand, serious social problems in Japan such as the destruction of nature and environmental pollution, have been appeared at the beginning of the 1970s. Accordingly, because of such environmental catastrophies, the results of vegetation investigation were greatly needed. Especially the mapping of regional vegetation is important for the diagnosis and prescription for conservation, reasonable use and new creation with vegetation of biological environmental conditions and natural resources.

According to several objectives for the conservation and preservation of nature and biological environments, national and regional land-use projects, and similar practical purposes of both applied and pure science, vegetation mapping has been conducted at various scales in Japan. In the early stages of vegetation mapping, it was mostly carried out with legends inferred from the physiognomy and the dominant species. But during the 1970s in Japan, vegetation mapping genuinely established the hierarchical system of plant communities (association, alliance, order and class as well as their lower units like subassociation and variants), based on the species combinations of relevés (Miyawaki, Okuda and Mochizuki, 1978, 1983).

From the end of the 1960s to the present, the number of maps of the actual and potential natural vegetation already published in Japan has exceeded 900. Furthermore, maps of vegetation naturalness and of suitable plantations introduced on the basis of the maps of actual and potential natural vegetation were also produced, as a basis for the construction of environmental forests and greenbelts for environmental protection (see Table 2 and Fig. 6).

Significance of vegetation mapping

Scientific significance

Today in Japan, land development programs such as nature conservation, landscape management and regional planning are practically based primarily on vegetation mapping. Vegetation maps are very important for various land uses related to plant environments as well as scientific objectives.

There are many ecophysiologically determined results for intrinsic properties (such as productivity, sensibility, resistance ability and so on) of plant communities in a special or certain vegetation. These results can be widely ap-

A.W. Küchler & I.S. Zonneveld (eds.), Vegetation mapping. ISBN 90-6193-191-6.

Fig. 1. Map of the actual vegetation of Japan. Legend:

A. Natural vegetation

    I. *Vaccinio-Piceetea* region (Alpine, subalpine and subarctic conifer forest zone)

        1. *Vaccinio-Pinion pumilae,* dwarf-scrub heath, windward grassland, etc.

        2. *Piceion jezoensis* (Hokkaido)

        3. *Abieti-Piceion* (Honshu, Shikoku)

    II. *Fagetea crenatae* (Summergreen broad-leaved forest zone)

        4. *Saso-Fagion crenatae* in Hokkaido (*Tilia maximowicziana-Quercus mongolica* var. *grosseserrata* community, etc.)

        5. *Saso-Fagion crenatae* under Japan Sea climate in Honshu

        6. *Sasamorpho-Fagion crenatae* under Pacific ocean climate in Honshu, Shikoku and Kyushu

        7. *Tsugion sieboldii*

    III. *Camellietea japonicae* (Evergreen broad-leaved forest zone, including coppice)

        8. *Quercion acuto-myrsinaefoliae*

        9. *Maeso japonicae-Castanopsion sieboldii*

      10. *Psychotrio-Castanopsion sieboldii*

B. Substitute vegetation

      11. *Quercetum acutissimo-serratae, Castaneo-Quercetum serratae, Castaneo-Quercetum grosseserratae,* etc. (firewood or charcoal forest)

      12. *Rhododendro reticulati-Pinetum densiflorae,* etc.

      13. Afforestation (*Pinus thunbergii, Cryptomeria japonica, Chamaecyparis obtusa, Larix kaempferi, P. lutchuensis,* etc.)

      14. *Miscanthetea sinensis* grassland (including *Sasa* spp., *Arundinaria* spp)

      15. *Chenopodietea* (field weed community)

      16. *Oryzetea sativae* (paddy field community)

C. Others

      17. Urban area

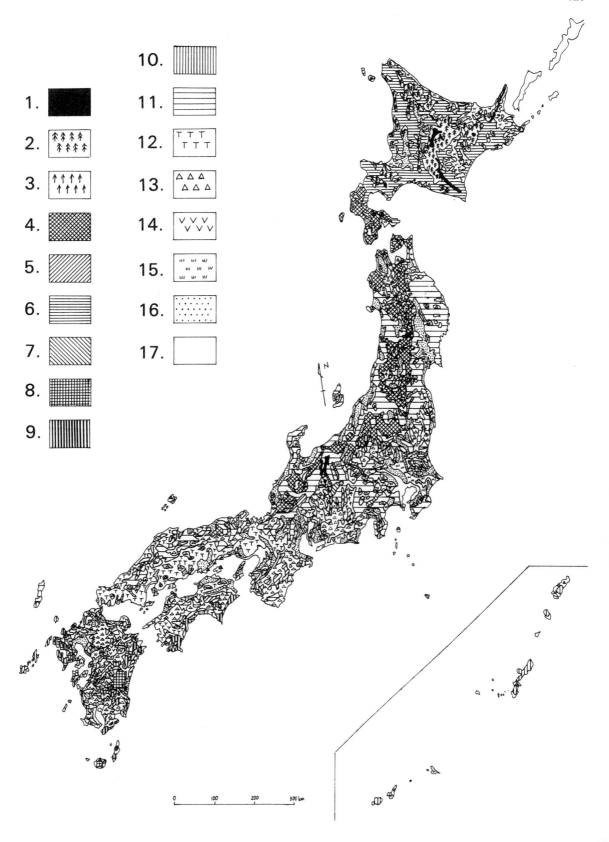

430

Fig. 2. Map of the potential natural vegetation of Japan. Legend:

I. *Vaccinio-Piceetea* and *Dicentro-Stellarietea nipponicae* region, etc. (alpine, subalpine and subarctic conifer forest zone)
    1. *Vaccinio-Pinion pumilae* (including dwarf-scrub heath, windward grassland, etc.)
    2. *Piceion jezoensis* (Hokkaido) and *Abieti-Piceion* (Honshu, Shikoku)

II. *Fagetea crenatae* region (deciduous broad-leaved forest zone)
    3. *Saso-Fagion crenatae,* etc. in Hokkaido (*Tilia maximowicziana-Quercus mongolica* var. *grosseserrata* community, etc.)
    4. *Saso-Fagion crenatae* under Japan Sea climate (Honshu)
    5. *Sasomorpho-Fagion crenatae* under Pacific Ocean climate (Honshu, Shikoku & Kyushu)

III. *Camellietea japonicae* region (evergreen broad-leaved forest zone)
    6. *Quercion acuto-myrsinaefoliae*
    7. *Maeso japonicae-Castanopsion sieboldii*
    8. *Psychotrio-Castanopsion sieboldii*
    9. *Psychotrio manillensis-Acerion oblongii*

IV. Common to various vegetation classes
    10. *Alnetea japonicae* or *Salicetea sachalinensis*

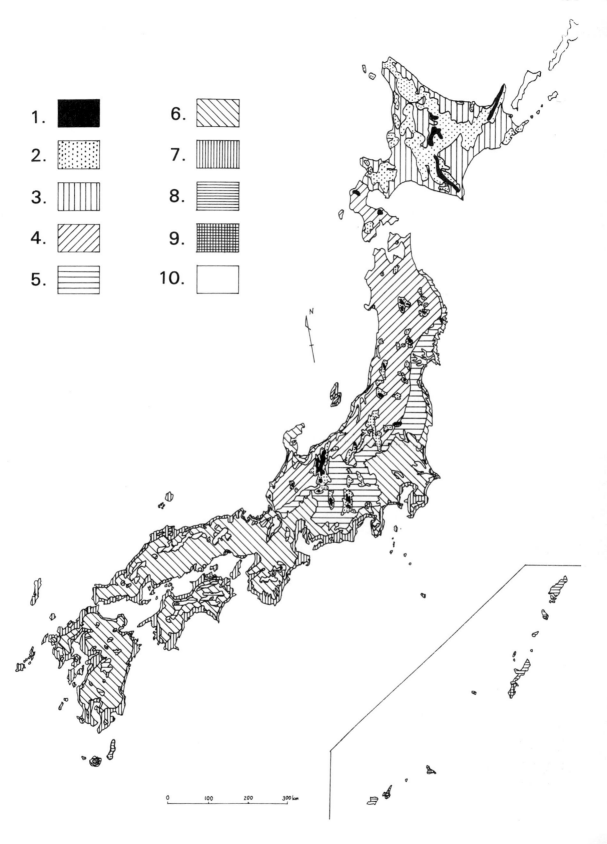

plied over regions with the same communities recognized on the vegetation map. The vegetation maps prepared earlier in Germany may also greatly contribute to related branches of science, such as plant geography, soil science, field zoology, forestry, civil and agricultural engineering, etc.

Vegetation maps are one of the basic research materials for all further development of biological studies in the field of science.

Applied significance

In Japan vegetation mapping has made rapid progress since the 1970s, as this country was touched by the environmental crisis caused by natural destruction through industrialization and urbanization.

In the past, human activities such as land cultivation and settlement were marked by conservatism, at a small scale and over a limited area. However, in recent years these jobs are quickly performed by heavy equipment at every place from the lowland of evergreen broad-leaved forest regions to mountain regions. This led inevitably to the large-scale disturbance of natural and semi-natural vegetation. Accordingly, now the last native and semi-natural forests are under serious threat distributed as they are only on many small islands located around the four major Japanese islands (Honshu, Shikoku, Kyushu and Hokkaido). It is recognized that the vegetation maps of their remaining forest areas are very important and effective for ecological diagnosis and prescription about the disturbed biological environments. As mentioned above, vegetation mapping in Japan should be carried out as soon as possible.

Until the 1950s the vegetation science including vegetation mapping was regretfully very poor and nobody was interested in producing the needed vegetation maps in Japan.

In 1964, vegetation mapping of the natural forest vegetation according to phytosociological principles, at the 1:5,000 scale, was accomplished for the first time for designation of a quasi-national park on the Dodaira district of the Tanzawa Mountains, located at 40 km NW from Yokohama (Miyawaki, Ohba and Murase, 1964). Since 1966, many maps of actual vegetation have been conducted for the purpose of natural conservation and management at most nature and semi-natural areas (Miyawaki and Okuda 1966a,b,; Miyawaki et al., 1967, Miyawaki and Fujiwara 1969a,b,c, 1970a,b,c, 1971a,b,c,d, 1972a,b,c,d, etc.).

Since 1967 the mapping of potential natural vegetation has been conducted based on the corresponding map of the actual vegetation and with the same scale, for reasonable land-use and urbanization planning (Yokoyama, Ide and Miyawaki, 1967; Miyawaki, Ide and Okuda, 1968; Miyawaki and Fujiwara, 1968; Miyawaki, 1969; Miyawaki and Fujiwara, 1969; Miyawaki, Fujiwara et al., 1971; Miyawaki and Ohno, 1972; Miyawaki, Harada et al., 1973; Miyawaki and Fujiwara, 1974; Miyawaki and Okuda, 1974; etc.).

Since the 1970s the Tokyo bay area, Osaka, Nagoya, N Kyushu and other areas have been rapidly industrialized and urbanized, which led to the decrease of vegetated areas corresponding to the increasingly dense population. The remnant actual vegetation must be protected and native environmental forests created. The Japanese people have lived with native forests situated traditionally at Shinto shrines, Buddist temples, and near settlements and villages. These traditional native forests in such places are considered potential natural vegetation of today, which should be rehabilitated around them in the cities, industrial sites and their surroundings. The map of potential natural vegetation, drawn on the basis of native or nearly original forests, is one of the important diagnostic maps. According to above purposes, many separate maps of potential natural and actual vegetation, at the same scale have already been produced (Miyawaki and Harada, 1974; Miyawaki, 1974; Miyawaki and Fujiwara, 1975; Miyawaki, Okuka and Suzuki, 1975a,b; Miyawaki and Okuda, 1975; Miyawaki and Suzuki, 1975; Miyawaki and Tohma, 1975; etc.).

Until now, there are many successful exam-

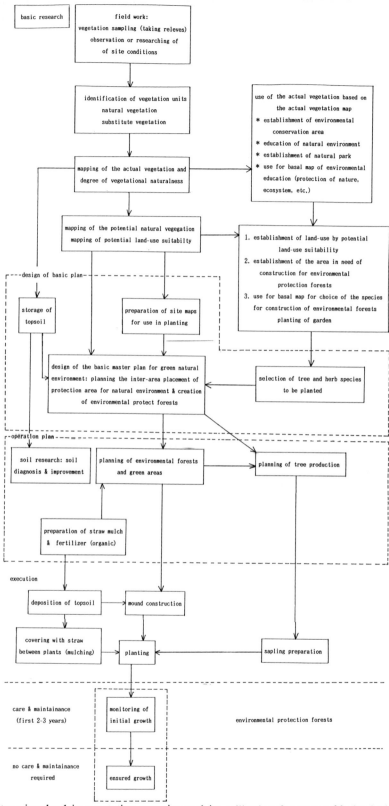

Fig. 3. Flow-chart of steps involved in vegetation mapping and its utilization for reasonable land planning and creation of environmental protection forests and other green areas.

ples of the creation of native forests based on the maps of potential natural and actual vegetation. These were created as environmental protection forests at city construction and industrial sites such as iron works, several hydric, thermal and atomic power stations, car and textile companies, etc. Fig. 3 presents a flow-chart of the steps involved, from vegetation mapping to research planning and creation of environmental protection forests. The most significant application of the maps of potential natural and actual vegetation is their use for natural and environmental protection and the creation of green shelters for the improvement of human life in the cities.

Japanese vegetation mapping

Mapping of actual vegetation

Most of the plant communities occurring today in Japan have been arranged as phytosociological units (associations, alliances, orders and clases, based on the species composition of relevés and the hand-sorting table work using the raw materials studied in the field. Thereafter, the vegetation maps were drawn, with legends derived from these community units.

Many maps of actual vegetation at various scales have been produced by Miyawaki et al. (1964–1987). Vegetation mapping using a scale of 1:200,000, covering the whole Japan, was conducted by the Cultural Agency of the Japanese Ministry of Education. Science and Culture (1969–1976) and the Japanese Environment Agency (JEA, 1975–1977). More detailed vegetation mapping, on a 1:50,000 scale, has been carried out by JEA since 1985, supported by all active phytosociologists and many plant ecologists.

Not only the national government but also many local governments, such as prefecture and city, now apply for the mapping of actual vegetation, which is indispensable for diagnosing the plant environment and their planning of

environmental protection and regional development (Fig. 3) (Miyawaki et al., 1986).

Mapping of potential natural vegetation

Since the development of the concept of potential natural vegetation by Tüxen (1956), the mapping of potential natural vegetation in Japan (since the 1970s) has been generally recognized as very significant (Miyawaki and Fujiwara, 1968, 1969, 1970, and many others).

The potential natural vegetation represents the theoretical natural vegetation in a landscape under human impact. The map of the potential natural vegetation of Japan is the integrated vegetation-ecological expression of the current site conditions, assuming all human influences on the vegetation in the landscape could be excluded (Miyawaki, 1984, p. 96) (see Table 1).

The mapping of potential natural vegetation is not as easy as the mapping of actual vegetation, especially in those parts of Japan which have a long history of human cultivating activities and are now developing into industrial and urbanized areas. One example is the Tokyo Bay district, the location of the Japanese capital city Tokyo, and many other metropolitan areas with dense population, such as Yokohama and Kawasaki.

In large urban areas with low-grade natural conditions, natural vegetation has been replaced by substitute vegetation, which is also under heavy human impact. In Japan, from old times up to about 1940, not all the natural environment and vegetation was exterminated, when settlements were established, maybe due to taboos. Because of religious beliefs new villages and cities kept natural forests or created them anew around Shinto Shrines and Buddist temples and also around residences (Fig. 4).

Therefore, we can recognize the potential natural vegetation even in big cities such as Tokyo, Yokohama, Osaka, Kyoto, etc. (Miyawaki; vegetation maps 1984, 1985, 1986), because of the remaining small native forests which can be found scattered throughout the cities even up

Fig. 4. The natural forests have remained as the Shinto shrine forests (Miyazu city, Pref. Kyoto).

until the present day. That is to say, the potential natural vegetation can be evaluated everywhere on the Japanese Archipelago by recognizing the differences and relations between the natural and corresponding substitute vegetation (Table 1).

Until now, in most parts of the country, the potential natural vegetation has been mapped at various scales. The potential natural vegetation will now be mapped throughout Japan, however at the small scales of Fig. 2. The map of Japanese potential natural vegetation, from the phytosociological point of view, distinguishes the following major forest vegetation regions.

Camellietea japonicae region (evergreen broad-leaved forest zone). Central and SW-Honshu, Shikoku and Kyushu down to the Okinawa Is-

lands, from the coast to about 700 m (in the north, e.g. Kyushu, even higher). The potential natural vegetation is evergreen broad-leaved forest, *Camellietea japonicae* forests. At the base of alluvial slopes and on other mesic sites, usually with deeper soils, the *Polysticho-Perseetum thunbergii,* and in parts of southern Japan (Kyushu and Shikoku) the *Arisaemato ringentis-Perseetum thunbergii* and other Persea thunbergii forests occur as the potential natural vegetation. On ridges and other somewhat drier sites near the coast (to 15 km inland), especially on thin soils, the potential natural vegetation is made up by *Ardisio-Castanopsietum sieboldii,* including the *Psychotrio-Castanopsietum sieboldii* with several *Castanopsis sieboldii* associations in the Ryukyu Islands.

Evergreen Quercus (Cyclobalanopsis) forests cover extensive inland areas in the *Camellietea japonicae* region, including the *Quercetum myrsinaefoliae* in the Kanto Plains near Tokyo and Yokohama, and the *Lasiantho-Quercetum gilvae, Photinio-Castanopsietum cuspidatae, Distylio-Cyclobalanopsietum, Nandino-Quercetum glaucae,* and *Aucubo-Quercetum salicinae* in western Honshu, Shikoku, and Kyushu.

On ridge crests, steep slopes, and other, similarly extreme sites near the upper limit of the *Camellietea japonicae* region, where pure evergreen broad-leaved forests can no longer dominate, conifer-dominated forests occur in some places. These include *Illicio-Abietetum firmae, Ilici-Tsugetum sieboldii,* and *Lindero-Cryptomerietum.*

A schematic representation of the vertical distribution of the most important alliances and classes of the actual and potential natural vegetation in central Honshu is shown by Miyawaki et al. (1977, Fig. 5).

More than 2,000 years ago, the Japanese people began to colonize the Japanese islands, settling initially in the evergreen broad-leaved forest region, that is the *Camellietea japonicae* region (Miyawaki and Itow, 1966). The Japanese intensive agriculture and overall land-use practices, mainly with limited agricultural impact, were excluded in the areas of unstable topogra-

Table 1. Guide for Reconstructing the potential natural vegetation of Kanagawa Precture.

	Potential natural vegetation	Character species (including planting & remaining trees) & differential species		Substitute vegetation
		Characteristic	General	
1	Ardisio-Castanopsietum sieboldii, typical subassociation	Damnacanthus major, Asarum kooyanum, Elaeagnus glabra	Castanopsis cuspidata var. sieboldii Persea thunbergii Cinnamomum japonicum	Daphno-pseudo-mezereum-Quercetum serratae
2	Ardisio-Castanopsietum sieboldii, subassociation with Quercus acuta	Quercus acuta, Osmanthus heterophyllus, Cymbidium goeringii, Quercus salicina	Osmanthus heterophyllus Ilex integra Camellia japonica Ligustrum japonicum Aucuba japonica	
3	Ardisio-Castanopsietum sieboldii (supply and recover the topsoil)		Trachelospermum asiaticum	no vegetation (bare land)
4	Arachniodo-Castanopsietum sieboldii	Arachniodes aristata Arachniodes standishii Maesa japonica some other ferns		Cornus controversa community
5	Polysticho-Perseetum thunbergii, typical subassociation	Persea thunbergii Polystichum polyblepharum Cornus controversa		Cornus controversa community Neolitsea sericea-Aucuba japonica community
6	Polysticho-Perseetum thunbergii, subassociation with Zelkova serrata	Zelkova serrata Celtis sinensis var. japonica, Acer palmatum Aphananthe aspera		
7	Euonymo-Pittosporetum tobira	Euonymus japonicus Pittosporum tobira Rhapiolepis umbellata var. integerrima, Elaeagnus macrophylla, Pinus thunbergii		Imperata cylindrica var. koenigii-Miscanthus sinensis community
8	Quercetum myrsinae-foliae, subassociation with Zelkova serrata	Quercus myrsinaefolia Nandina domestica Dryopteris uniformis Houttuynia cordata Zelkova serrata		Cornus controversa community Neolitsea sericea-Aucuba japonica community
9	Quercetum myrsinae-foliae, typical subassociation	Thea sinensis Trachycarpus fortunei		Quercetum acutissimo-serratae, typical subassociation
10	Quercetum myrsinae-foliae, subassociation with Abies firma	Abies firma, Vaccinium oldhamii, Fraxinus sieboldiana, Rhododendron kaempferi, Pinus densiflora		Quercetum acutissimo-serratae, subassociation of Pinus densiflora Castanea-Quercetum serratae
11	Illicio-Abietum firmae typical subassociation	Abies firma, Quercus salicina, Illicium religiosum, Callicarpa mollis, Carex reinii	Quercus salicina Camellia japonica Osmanthus heterophyllus Ardisia japonica	
12	Illicio-Abietum firmae subassociation with Tsuga sieboldii	Tsuga sieboldii	Trachelospermum asiaticum	

General	Types of land use	Soil	Geomorphology, climate, etc.
Pasania edulis afforestation, Pinus thunbergii aff., Cryptomeria japonica aff., Prunus lannesiana var. speciosa aff., Cinnamomum camphora aff., Phyllostachys heterocycla var. pubescens stand	secondary coppice forest afforestation orchard (Cornus unshiu, etc.) field	yellow brown forest soil	ridges (with little Kanto loam) of Tertiary rocky ridges & steep slopes adjacent to Polysticho-Perseetum thunbergii

stable ridges with thin soil layers, better in the accumulation of soils or litter between Illicio-Abietetum firmae and Quercetum myrsinaefoliae than the typical subassociation |
| No vegetation (bare land) | residence industrial estate | no or little topsoil | topsoil lost by extensive housing construction with heavy machines |
| Puerario lobatae-Humuletum scandens, Erigeron canadensis-Erigeron sumatrensis community, Pinellio ternatae-Euphorbietum pseudochamaesyceae | native forest residence industrial estate grassland fallow field | yellow forest soil | valleys or gentle slopes, like terraces talus areas accumulated with Tertiary stones

alluvial or lower parts of slopes on plateaux, with soil accumulation, mesic or hydric habitat, mainly alluvial or reclaimed land piling up soil

alluvial or gentle slopes along ravines with deep soil accumulation, similar to the habitat of Quercetum myrsinaefoliae and recovering to original Persea thunbergii forest after the first cutting |
| Arundinario chino-Miscanthetum sinensis | grassland coastal park remnant forest for protecting the hinterland from seashore | | coastal cliffs or steep slopes facing the sea winds, with thin soil layers exposed to the winds, old coastal dunes |
| | orchard Thea sinensis garden | Kuroboku soil (black colored soil) | plateaus and hills with loam accumulation

steep slopes, talus, areas, natural levees well draining mesic sites

mesic gentle slopes on the plateaus or hills |
| Pinus densiflora afforestation | forests for charcoal or firewood | Kuroboku soil | ridges with thin loam accumulation dry steep slopes slightly oligotrophic sites |
| Arundinario chino-Miscanthetum sinensis | | | |
| | forests for charcoal or firewood mowed grassland afforestation | | 600–700 m in elevation steep slopes of 22~45° thin but mesic and stable soil layers parent rocks of Tertiary age

above 700 m borderline with Fagetea crenatae |

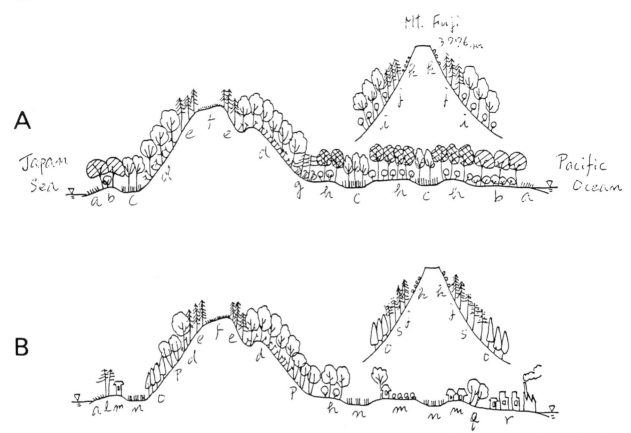

Fig. 5. Schematic diagram of the vertical distribution of the most important vegetation of central Japan (Miyawaki, Suzuki and Fujiwara, 1977, a little changed). (A) Potential natural vegetation; (B) Actual vegetation. a = dune vegetation; b = *Polysticho-Perseetum thunbergii*; c = *Alnion japonicae*; d = *Saso-Fagion*; e = *Abietetum mariesii*; f = *Sphagnum* bog; g = *Illicio-Abietetum firmae*; h = *Quercetum myrsinaefoliae*; i = *Sasamorpho-Fagion*; j = *Abietetum veitchio mariesii*; k = alpine meadow; l = *Pinus thunbergii* forest; m = residental district; n = *Oryzetea sativae*; o = *Cryptomeria japonica* forest; p = *Pruno pilosae-Quercetum serratae*; q = *Quercetum acutissimo-serratae*; r = urban district.

phy such as steep slopes, landslide areas, ravines, etc. Other areas and shorelines with water bodies are potentially subject to the greatest impact of human activity. Traditionally, cultivating the more stable and more productive land areas has been common: 16% of the Japanese countryside is farmland, and in addition about 15% or more is under other forms of extensive use.

Vegetation like secondary forests, grasslands, plantations, remnants of almost virgin natural vegetation or only slightly modified natural vegetation, are to be found in old villages or town sites in the so-called 'Shinto shrine' forests, in estate forests, and in sanctuary forest reserves (Fig. 4).

Within the last 30 years a new industrial era has brought the development or construction of new cities, of new major traffic facilities and of other large-scale industrial complexes. These recent developments have had a severe impact upon the land and have injured the natural vegetation of Japan. The native diversity of natural environments has been lost in this industrialization process. The traditional landscape of farmlands has disappeared and has been substituted by urbanized deserts and industrial complexes. This situation is indicated when comparing the maps of potential natural vegetation with that of actual vegetation at the same scale (Fig. 5).

Fagetea crenatae region (summergreen broad-leaved forest zone). The *Fagetea crenatae* region in central Honshu, reaches from 700 to about 1,600 m above sea level. The horizontal width of this belt increases steadily to the north, toward northern Honshu and Hokkaido. To the south, in western Honshu and Kyushu, the potential *Fagetea crenatae* region rises steadily to higher elevations.

An important summergreen forest grouping in the *Fagetea crenatae* region is that of the *Saso-Fagetalia crenatae* communities, where the ground is almost always thickly covered with characteristic small *Sasa* species. Due to the climatic conditions, especially the abundant snow associated with the winter monsoon, *Saso kurilensis-Fagion crenatae*, with *Aucubo-Fagetum crenatae* and other Fagus forests, are widespread on the windward, Japanese Sea side of Honshu. On the Pacific side, with less snow grow *Sasamorpho-Fagion crenatae* communities, with *Corno-Fagetum crenatae*, *Sapio japonicae-Fagetum crenatae*, *Fagetum crenato-japonicae*, and other associations.

In the *Fagetea crenatae* region north of Kuromatsunai in southern Hokkaido, Fagus species can no longer occur. In place of the *Fagetea crenatae* such forests as *Quercus mongolica* var. *grossesserata-Quercus dentata* forests and the *Tilia maximowicziana-Quercus mongolica* var. *grosseserrata* community appear. In wet places along streams or mires (moor), *Fraxino-Ulmetalia* with *Pterocaryon rhoifoliae* and *Ulmion davidianae* are widespread.

In the *Fagetea crenatae* region, the coastal plains of north Japan or the hilly, mountain areas of central Japan are occupied by many conifer forests composed mainly of *Cryptomeria japonica*, *Chamaecyparis obtusa*, and *Larix leptolepis* planted widely as substitute vegetation. A new intensive forest management program has been promoted by the Japanese National Forest Service.

Such clear results can be found only through both maps of actual vegetation and potential natural vegetation, which are used in Japan for several theoretical and applied purposes (see references, Miyawaki et al., 1969–1987).

Vaccinio-Piceetea region (subalpine and subarctic conifer-forest region). In Honshu, the upper summergreen broad-leaved forest belt borders directly on the subalpine conifer-forest belt. On Honshu, Shikoku and Hokkaido this is the *Abieti-Piceion*. In central Honshu (Japanese Alps), the *Abietetum mariesii* and *Abietetum veitchii* occur above 1,600 m. On Shikoku the *Abietetum sikokianae* can be found above 1,800 m on the mountains Ishizuchi (1,981 m) and Tsurugi-sa (1,893 m).

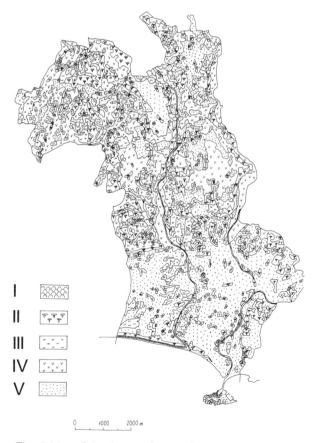

I
II
III
IV
V

0   1000   2000 m

Fig. 6. Map of the degree of vegetation naturalness (Fujisawa city, after Miyawaki and Fujiwara, 1975).

440

Table 2. Legend for the degree of vegetation naturalness (Fujisawa city, after Miyawaki and Fujiwara, 1975).

Vegetation naturalness		Vegetation type	Concrete exemple	
V	10	Natural Vegetation	Natural grasslands (mires, alpine meadow, wind-exposed grassland, etc.)	Natural vegetation with single layer in community structure: alpine heathland, wind-exposed grassland, dune vegetation, etc.
	9		Natural forest (climax forest or naturally regenerated forest composed of natural species)	Natural multilayer vegetation: Picea jezoensis-Abies sacchaliensis, Aucubo-Fagetum crenatae, Polysticho-Perseetum thunbergii, etc.
IV	8	Secondary forest	Secondary forest (similar to natural forest on species composition.)	Substitute vegetation closely similar to natural forests such as young natural forest of Fagus, Castanopsis, evergreen Quercus, etc.
	7		Substitute secondary forest	Firewood & Charcoal forest such as Castaneo-Quercetum serratae, Quercetum acutissimo-serratae, Castaneo-Quercetum grosseserratae, etc.
	6		Afforestation	Afforestation of conifers (Cryptomeria japonica, Pinus thunbergii, etc.), evergreen broad-leaved trees (Cinnamomum camphora, Pasania edulis, etc.), etc.
	5	Secondary grassland (perennial)	Tall grassland	Sasa or Arundinaria thickets, Miscanthus communities, Artemisia communities, etc.
II	4		Short grassland	Zoysia communities (lawn) or Plantago asiatica communities under treading, etc.
	3	Cultivated land-annual weed area)	Orchard, seedling field for garden, residence with plantation	Fruit orchards, mulberry plantations, tea gardens, and other horticultural areas.
II	2		Field & paddy field	Cultivated area with annual plant communities such as Pinellio ternatae-Euphorbietum pseudochamaesycis (field weeds) Sagittario-Monochorietum (paddy weeds), etc.
I	1		Urban area, developed tracts	Urban bare land, developed tracts, or urban weed area, etc.

On Hokkaido the *Dryopterido-Abietetum mayrianae, Saso-Piceetum jejoensis,* and other boreal or subarctic needle-leaved forests grow on the mountain Daisetsu (2,290 m), in central Hokkaido) and on other mountains. In nor-thern and eastern Hokkaido, however, the conifer stands sometimes descend almost to the seacoast or to the nearby bogs.

The character of the subalpine and boreal conifer forests is physiognomically rather monotonous, due to the small number of dominant conifer species and the only occasional admixture of *Betula ermanii.*

The natural vegetation of the Japanese subalpine and boreal conifer belt has been greatly altered recently through clear-cutting, road construction, and other human influences. This conifer vegetation is generally in better condition, for the most part, than the vegetation of the *Camellietea japonicae* and the *Fagetea crenatae* regions. Thus, the potential natural vegetation in this region is easier to ascertain in the field than in the other, lower-lying regions.

Alpine vegetation belt. The central Japanese mountain ranges in Honshu are called the 'Japanese Alps' and include more than ten peaks above 3,000 m in elevation. On this highest of the Japanese mountains, above 2,500 m, and on the mountains of Hokkaido above 1,500 m, there occur various alpine heath and meadow communities mixed in among the shrub-like *Pinion pumilae* krummholz communities, which still belong to the subalpine *Vaccinio-Piceetea.*

Strictly speaking, the areas of alpine vegetation in Japan are simply not large enough to be represented on the map. Instead, on the map of the potential natural vegetation of the Japanese Islands, the areas of the *Pinion pumilae* communities and the micro-mosaic of alpine meadow and heath communities are combined as alpine vegetation.

31. The International Vegetation Map (Toulouse, France)

F. BLASCO

Three main centers for vegetation mapping were developed in France, immediately after the end of the second world war, i.e. Toulouse, Montpellier and Grenoble.

In Montpellier the aim is to produce very accurate maps at 1:50,000 or 1:20,000 scale, based on Braun-Blanquet's principles using the quadrat method. This is one of the most famous floristic classifications based on the concept of plant associations (see Chapters 5, 6, 6A).

Mapping activities in Grenoble are not very different from those recommended by Gaussen for his Toulouse school (Ozenda, 1963; Dobremez, 1973).*

Toulouse has become a sort of holy place for French-speaking vegetation mappers, mainly because many maps of arid, tropical, equatorial countries have been prepared in this town of Southern France, where Gaussen founded two establishments for mapping vegetation; their activities are different:

— one is in charge since 1947 of the vegetation map of France at 1:200,000 scale;
— the other one created in 1961, is mainly involved in tropical vegetation mapping at samller scales (1:1,000,000).

For the vegetation map of France the same method has been applied for more than 30 years with conspicuous success. However, in extremely diversified tropical countries we had to forge a high level scientific compromise involving a great amount of data and, since a few years new sophisticated means of processing.

This chapter illustrates, with selected examples, the applicability of our cartographic methods and that the most formidable obstacle for vegetation mappers is to produce a reachly documented, complex and precise, but nevertheless clear, map.

Basic data concerning Gaussen's method

Gaussen's school has swarmed all over the continents. Following its fundamentals, tens of vegetation maps have been prepared. Everywhere this type of cartography appears mainly as a technical tool for a graphic representation of complex biological phenomena such as ecology, plant succession, physiognomy, etc.

Though P. Legris (1959), A.W. Küchler (1965), V.M. Meher-Homji (1968, 1973) gave excellent accounts of Gaussen's method it is essential to recall its main principles.

Since the publication of 'The causes of vegetation cycles' by Cowles, in 1911, followed by Clements' classical 'Plant succession' (1916) tens of thousands of pages have been written concerning the continuous evolution of vegetation. Though it is one of the most speculative topics related to vegetation science *the idea to map dynamic features* can be considered as one

---

* H. Gaussen born on 14th July 1891 at Cabrière d'Aygues (Vaucluse-France) deceased at Toulouse 27th July 1981. His biography has been written by de Ferre (1982).

A.W. Küchler & I.S. Zonneveld (eds.), Vegetation mapping. ISBN 90-6193-191-6.
© 1988, Kluwer Academic Publishers, Dordrecht. Printed in the Netherlands.

of the most fruitful. 'As vegetation develops, the same area becomes successively occupied by different plant communities. This process is termed plant succession. Within a region, the same final or climax stage results *from this series* of successive stages whether they start in open water, or on solid rock or denuded land' (Weaver and Clements, 1929, p. 60). Inspired by this concept of series of vegetation, Gaussen has proposed to map each phase of each series with their dynamic relations in order to avoid the static character of the majority of vegetation maps. He gave the following example (1955):

On a given barren soil (a°), the abandoned vegetation evolves through progressive phases, herbaceous (a_1), under-shrubby (a_2), followed by shrubby stages upto the tallest one, which can be a forest (A); for instance a forest of Oak (*Quercus sessiliflora*). These successive stages constitute the 'series of oak'.

Another barren soil (b°) located at a different elevation, will lead to another forest type (B), a Beech forest (*Fagus silvatica*); this is the 'series of Beech'.

"Apparently 'a°' and 'b°' are analogous but their dynamic capability, which is an indicator of land potentialities, being different, it is necessary to separate them carefully" (Rey and Gaussen, 1955). The dynamic of each phase is not only closely related to the principles of dispersal in higher plants (Pijl, 1972), but also to all the complex phenomena interfering in the processes of ecological successions (Golley, 1977).

That is why climaxes are hypothetical. They imply an unrealistic condition for tropical countries, which is a total protection against human activities including cessation of fire.

The meaning of the colour is essential in Gaussen's method.

Each series of vegetation has its own colour. Colours are selected in such a way that they give an immediate indication of the average value of *essential ecological factors* such as temperature, dryness (rainfall, length of the dry season, etc.), nature of soils. Theoretically similar colours applied to two distant geographic areas imply that essential ecological factors are analogous in both areas.

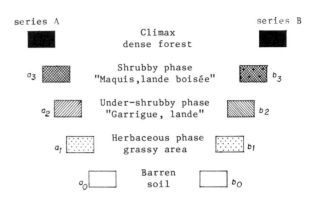

Fig. 1. The series of vegetation. According to Rey and Gaussen (1955), two identical herbaceous phases a_1 and b_1 belonging to two distinct vegetation series must be separated according to their evolutionary tendencies. However, their floristic components can be analogous.

As shown in Fig. 1, the *selection of patterns* and the way in which the colour is applied depend on the *physiognomy and density* of each vegetation type. Stippled designs (a_1 and b_1) are used for grassy types, but the colour will vary from one series to another. Ruled designs represent woody open types, whereas plain colour is used for a dense forest (A and B). Cultivated areas are left in white, bearing some agricultural symbols having statistical values.

Criteria for Gaussen's classification of vegetation are basically, ecological, dynamic, physiognomic and floristic. This system is well adapted to 1:200,000 to 1:1,000,000 scales. It is rather different from those suggested by Ellenberg (1956), Dansereau (1958), Holdridge (1959), etc., but it comes near those adopted by the UNESCO (1973) and later on, by the FAO (1983).

The presentation of each map with six inset maps of the same region, at a very small scale, give a good amount of thematic data such as topography and administrative divisions, geology and lithology, soils, bioclimates, agriculture and climaxes or potentialities of the region.

The explanatory booklet which accompanies each vegetation map (or sheet) gives all the essential data which cannot be mapped either on the main map or on the inset maps. This booklet gives useful information on the techniques employed for the preparation of the sketch maps (or of the final version) as well as floristic lists or chemical properties of soils or comments on the future scientific utilization of the map (biogeography, ecology, climatology, pedology, agronomy, forestry, etc.).

The world-wide applicability of the method has been proved.

Gaussen's method was primarily applied to temperate and Mediterranean countries, where climaxes are well defined by a single or a few dominant species, closely related to characteristic ecological conditions. In tropical areas where the floristic composition of plant communities and environmental conditions are much more complex, some alterations of the basic principles had to be introduced.

The example of the vegetation map of France

Obviously, it is easier to produce a vegetation map for a well-known western country than for a tropical region for which scientific data are scarce. Since the publication of the first 'Carte Botanique de la France', by Lamarck and de Candolle (1815), an enormous amount of scientific material has been gathered almost for each square kilometer of the country. Nevertheless it took no less than 35 years to complete the cartographic work which includes 63 sheets at 1:200,000 scale (see Fig. 2). This scale was selected because it offers, among other advantages, the capability to express clearly, even in complex montainous areas like the Alps (five sheets) or the Pyrenees (five sheets too), sufficiently accurate facts, in a synoptic way. The synthetic cartographic expression concerns several topics such as botany, forestry, agronomy, ecology, etc.

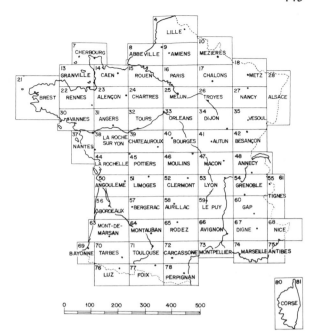

Fig. 2. The vegetation map of France. The concept of 'series of vegetation' is the corner-store of Gaussen's method applied on the 63 sheets, at 1:200,000 scale.

The C.E.R.R. (Centre d'Ecologie des Ressources Renouvelables)* has now almost completed the coverage of main land of France, using various kinds of aerial photographs, and a very accurate ground survey completed by an intensive analysis of ecological data.

The 63 sheets constitute a rather unique piece of work, gathering on the same cartographic document, all over the country, the extent and floristic composition of every natural vegetation type, from forests to heaths and grasslands, as well as all kinds of crops and manmade secondary, more or less degraded, shrubby or grassy types. The result is extremely rich in scientific data, beautiful and I doubt if there is any other example of more overloaded vegetation map in the world (see also Fig. 3, colour section).

Since a few years researches are being carried out in order to apply Gaussen's method to large-scale vegetation studies and mapping.

* 29, rue Jeanne Marvig, 31055 Toulouse Cedex – France; formerly S.C.V., Service de la carte de la végétation.

446

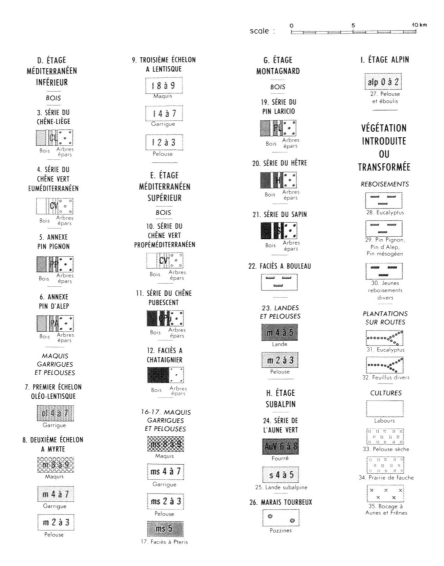

Fig. 3. A fragment of the map "CORSE" (DUPIAS et al. 1965). A specimen of its simplified legend.

The preparation of each sheet of the vegetation map of France, at 1:200,000 requires:
— a good classification system remaining homogeneous all over the country; in the adopted system, ecological data and dynamic features play an essential role;
— a good knowledge of the floristic components in each phase of vegetation series. *The name of each series is given by the species which dominates the others in the climax;*
— an appropriate remote sensing technique based on the exploitation of several kinds of aerial photographs; the commonest being black-and-white stereoscopic assemblages at 1:30,000 scale.

Adaptations of the method to tropical countries

How far the genuine Gaussen's method, initially meant for temperate countries, can be applied to warm countries, tropical or mediterranean? *The following comments concern the collection of maps known as 'International Map of the Vegetation and of Environmental Conditions'* which is published at 1:1,000,000 scale for many non-temperate countries such as Mexico, Algeria, Cameroun, Burkina Faso, Madagascar, Sri Lanka, India, Cambodia, Sumatra (Indonesia), etc.

The common denominator of all these *small-scale vegetation maps* is to give a synoptic view of the global distribution of important vegetation types and their relationships with determining ecological factors. Another peculiarity of this collection of maps is their technical preparation which includes almost invariably the following sequential operations:
— elaboration of bioclimatic and edaphic maps covering the concerned area;
— compilation of agricultural and forest maps generally provided by the Agriculture Departments of States and by the Forest Departments (Forest Working plans);
— a completion of a sketch map using the available aerial photographies and satellite or radar images;

— routine ground truth control and collection of new field data concerning the floras, forest structural diversity, shifting cultivation practices, etc.;
— preparation of a *basic map* which will be used for the production and drawing of the final map.

The 'vegetation map of India'

Regarding the 'vegetation map of India', there is not much to say on its methodology because Gaussen's principles have been applied, right from the beginning, since the publication of the first sheet 'Cape Comorin', in 1961. "In a given ecological region, degraded stages if protected from destructive activities of man and his cattle, tend to evolve towards a floristically and physiognomically definite maximum type named 'plesioclimax'. The successive physiognomic stages ranging from barren soil to the 'plesioclimax' (usually a forest), go to form a 'series of vegetation'" (Meher-Homji, 1973, p. 72).

About 80% of the national territory have already been mapped, in 12 sheets (Fig. 4). Some difficulties have been encountered, particularly for an objective delineation and characterization of some series. For instance, it has been difficult to give a satisfactory ecological explanation for the distinction of twins series like '*Albizzia amara* + *Acacia* and *Acacia* + *Capparis decidua*. According to Legris and Meher-Homji (1973) soils and geology intervene to distinguish them. Moreover, the fieldwork based on lists of species and on traditional profiles of vegetation has become insufficient since the concept of stratification is disputed and battled for breaching by the majority of modern phytogeographers (see page 715 aldo Chapter 16, 29 and 34).

Nevertheless the immediate interest of such a type of vegetation map, in overpopulated countries like India or Sri Lanka, where the natural vegetation has been very much depleted is to show the evolutionary trends of the vegetation provided it has been put under an effective pro-

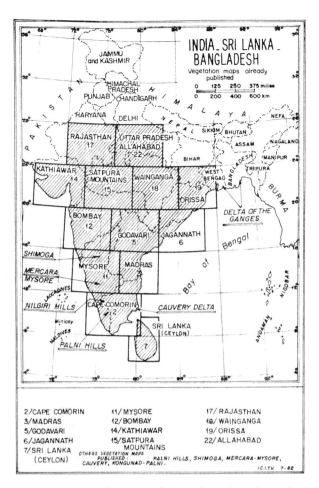

Fig. 4. The vegetation map of the Indian subcontinent. One of the most impressive accomplishments under the direction of Dr. Pierre Legris. It includes 13 published sheets at 1:1,000,000 scale. In spite of unavoidable difficulties the concept of 'series of vegetation' has been used.

tection. 'It shows more than the purely physiognomic maps of vegetation because of the correlations they suggest between the ecological conditions (climatic, edaphic, biotic) and the present state of natural vegetation. The floristic complements are contained in the explanatory booklet' (Legris and Meher-Homji, 1968, p. 39).

All the 12 published sheets of the Indian subcontinent have been prepared at the French Institute of Pondicherry (605001 – India) in collaboration with the Indian Council of Agricultural Research.

The vegetation map 'Madagascar'

The vegetation map 'Madagascar' has also been prepared and published by the French Institute, at Pondicherry. With this map an important turning has been taken for the application of the series of vegetation though the scale of the map has not changed (1:1,000,000). Instead of using *conspicuous species of trees* for the floristic characterization of each series we have been compelled to select endemic families or genera or species as well as poorly representative but highly represented groups in the climax.

Examples:
— series with Didiereaceae endemic family and *Euphorbia* cosmopolitan genus;
— series with *Uapaca bojeri* endemic species and Claenaceae endemic family;
— series with *Tambourissa* endemic genus and *Weinmannia* almost pantropical genus.

This unorthodox concept of 'series' denotes two essential facts:
— in some tropical areas it is very difficult to apply the concept of 'series' because it is almost impossible to find objectively a few species which are both conspicuous and characteristic in the climax of a given phytogeographical territory. The failure is mainly due to the very complex and heterogeneous floristic composition of tropical vegetation, particularly in Madagascar where more than 7,400 species of higher plants are recorded, among which 66% are endemic.

The preparation of the vegetation map 'Madagascar' (three sheets) was considered as a natural follow-up of the work carried out by botanists for the completion of the flora of Madagascar and Comoro Islands (Humbert, 1936–1965);
— some groups of *endemic plants* (families, genera or species) are much more representative of a series than a few conspicuous species which very often have a very wide biogeographical distribution.

The preceding remarks announce a wave of

450

unavoidable local adaptations which began in 1972 with the publication of the map 'Cambodia' (P. Legris and F. Blasco) in which the term 'series' has disappeared. Such a voluntary modification has also been introduced on the maps 'Guadalajara Tampico; Mexico' (Puig, 1979), 'Sumatra' (1983) and Cameroun (Letouzey, 1984).

For almost all these maps *the first level of subdivision in the classification is still of ecological order*. The principal factors retained for this distinction being climatic, physiographic or edaphic. *The bioclimatic limits* are determined by the criteria of rainfall, its regime (season of occurrence or rains), the length of the dry period, the relative humidity and the temperature. Drawing of these limits therefore necessitates, in first place, a complete and detailed exploitation of all the available climatic data. At this stage, our zoning comes close to the notion of 'life-zones' of Holdridge (1959).

The subsequent levels are physiognomic, phenologic and dynamic; for instance in the map 'Cambodia'.

The *sub-humid types* include the following gradation:
1. deciduous dry forest (forêt sèche décidue);
2. degraded woodland (forêt claire dégradée);
3. dense thicket (fourré dense);
4. discontinuous thicket (fourré discontinu).

To sum up we can say that the *floristic definition* of the vegetation series has been voluntarily withdrawn from the latest vegetation maps *without influencing the fundamental dynamic character of our mapping system.*

The map 'Sumatra'

The map 'Sumatra' is the last born of our collection: 'International map of vegetation'. The sheet 'South Sumatra' has been published in November 1983 and two additional sheets are now being finalized at Toulouse and at the BIOTROP (P.B. 17, Bogor, Indonesia). These maps deserve specific comments because the employed field methods are entirely new to our school.

Considering this vegetation map as a first phytogeographical survey of Sumatra covering large little known areas, it was decided to adopt a certain level of generalization compatible with the scale of 1:1,000,000 and to distinguish some forest communities according to their architectural properties following Oldeman's method (1974). This method is also applied in the Western Ghats of India (Pascal, 1983).

General principles applied to Sumatra. Practically all biological observations were obtained from a field survey based mainly on detailed local investigations (Y. Laumonier, 1983), in selected sample plots distributed all over the mapped area. For the sheet 'South Sumatra' alone, in addition to frequent scientific fieldwork, the architecture of 14 'biostatic' (Oldeman) plots has been studied in the various ecological zones of the island.

There is a universal agreement on the fact that, roughly, the plant cover in Sumatra can be subdivided into three main groups, i.e.:
— plant cover almost unmodified by anthropic factors;
— secondary forests and other derived types (Belukar);
— cultivated types or, more generally, agricultural areas (*Ladangs* and plantations).

We do not know as yet the relative territorial importance of each of these large groups; it will be known only when their mapping is completed and this is one of the aims of our mapping programme.

The first group of *almost unmodified* types mainly involves primary lowland Dipterocarp rain forests (*Shorea, Hopea, Anisoptera, Vatica, Dipterocarpus*). Practically, we considered that the majority of these rainforests, corresponds to the climax; the colour used on the maps to depict these forests, depends on the local rainfall, temperatures and drainage conditions. Naturally the derived secondary types are shown in the same colour as the climax but they bear different patterns according to their density and floristic composition.

Lowland Malaysian rainforests are not uniform, neither floristically nor structurally. This is a very well-known fact illustrated by the great variations of their standing stock, between 40 and 80 m³/ha in Sumatra, between 50 and 160 m³/ha in Kalimantan. The classification has to account for this diversity. The reasons for these changes, at a regional scale are not clearly explained. However, large edaphic variants have been studies and described, in relation to local edaphic properties determining such types as 'evergreen forests with semi-evergreen tendency', 'fresh water swamps forests', 'peat swamp forest', etc. (Whitmore, 1975; Brunig, 1970).

Swamp and peat forests as well as mangroves can be called 'edaphic types' because their very distinct floristic composition is due first of all, to an almost permanent flooding by fresh water or, in the case of mangrove stands, by brackish waters. These large groups of edaphic types include diverse floristic communities sometimes called 'facies' which have not been distinguished on the map owing to their very limited size.

The largest areas covered by edaphic types are located on the eastern alluvial plains, although some much smaller patches exist also on the western coastal plains (Indrapura and Bengkulu).

Naturally, undisturbed phytocenoses are also found at various elevations, on the Barisan range. 'Moist evergreen mountain forests' (1800–2700 m) or 'evergreen low forests and thickets of the sub-alpine zone' (above 2700 m) belong to this group. In such cases the altitude and temperature play determining roles.

Our present knowledge on the taxonomy and floristic composition, in the various mapped areas is too insufficient. We were not in a position to base our classification on floristic considerations only, though a complete list of taxa found in sample plots is given *in the explanatory notice of the map* to be published in 1988. It is highly probable that the different geomorphological or geographical units of the island, harbour an undetermined number of endemic taxa (endemic species and, supposedly also,

some endemic genera), in relation to the degree of physical insulation of each unit or 'block' (Pijl, 1969; Carlquist, 1965). The orogenic activity in Sumatra, particularly in the Barisan Range, is a complex phenomenon which began in the Palaeogene and which still continues. At present 31 volcanoes are active or in semi-activity. One of the essential tectonic phases for biogeographers is probably the orogenic activity and block-faultage wich occurred during the Middle-Miocene, although subsequent orogenic periods have been reported in the Barisan Range, until the Plio-Pleistocene. It is obvious that these tectonic movements accompanied by the separation of individual blocks are an essential factor for speciation. Further taxonomic and phytogeographical research will fix the identity and number of taxa having a limited geographical distribution and which can be considered as characteristic of each geomorphological 'block' in which they are confined (endemic taxa).

We wanted our classification to be not only physiognomically and structurally oriented but also floristically oriented. Though botanical inventories have given good results it is too early to give a list of characteristic or endemic plants of each geomorphological unit. Hence, we have introduced in our classification and legend of the map a *geomorphological aspect too*. Our first concern was to find, among the rare geomorphological works on Indonesia, particularly on Sumatra, comprehensive data on lithology, stratigraphy, tectonics, land forms, as well as a sketch on the main geomorphological units of the island. Verstappen's contribution 'A geomorphological reconnaissance of Sumatra and adjacent islands' (1973) provides a sufficient background for the regional distribution of the main biogeographical units and has been used for the vegetation map.

We should add four more remarks of scientific and practical interest:
— The hierarchy of classes adopted in the classification is brief. We have considered that at the scale of 1:1,000,000, it was not essential to subdivide the complex mosaic of sec-

452

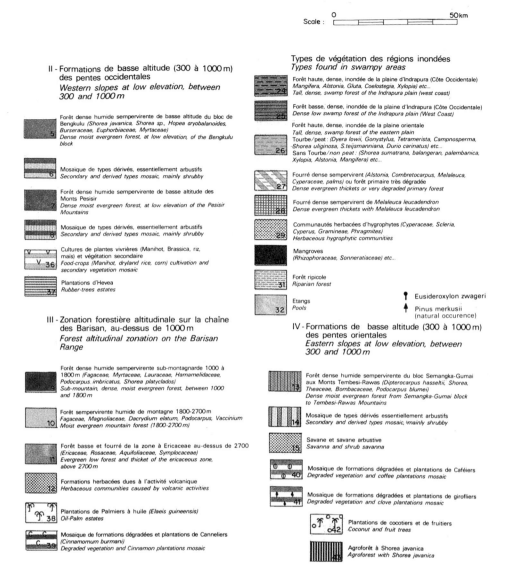

Scale : 0 ——— 50 km

II - Formations de basse altitude (300 à 1000 m) des pentes occidentales
Western slopes at low elevation, between 300 and 1000 m

5 Forêt dense humide sempervirente de basse altitude du bloc de Bengkulu (Shorea javanica, Shorea sp., Hopea aryobalanoides, Burseraceae, Euphorbiaceae, Myrtaceae)
Dense moist evergreen forest, at low elevation, of the Bengkulu block

6 Mosaïque de types dérivés, essentiellement arbustifs
Secondary and derived types mosaic, mainly shrubby

7 Forêt dense humide sempervirente de basse altitude des Monts Pesisir
Dense moist evergreen forest, at low elevation of the Pesisir Mountains

8 Mosaïque de types dérivés, essentiellement arbustifs
Secondary and derived types mosaic, mainly shrubby

36 Cultures de plantes vivrières (Manihot, Brassica, riz, maïs) et végétation secondaire
Food-crops (Manihot, dryland rice, corn) cultivation and secondary vegetation mosaic

37 Plantations d'Hévéa
Rubber-trees estates

III - Zonation forestière altitudinale sur la chaîne des Barisan, au-dessus de 1000 m
Forest altitudinal zonation on the Barisan Range

9 Forêt dense humide sempervirente sub-montagnarde 1000 à 1800 m (Fagaceae, Myrtaceae, Lauraceae, Hamamelidaceae, Podocarpus imbricatus, Shorea platyclados)
Sub-mountain, dense, moist evergreen forest, between 1000 and 1800 m

10 Forêt sempervirente humide de montagne 1800-2700 m Fagaceae, Magnoliaceae, Dacrydium elatum, Podocarpus, Vaccinium
Moist evergreen mountain forest (1800-2700 m)

11 Forêt basse et fourré de la zone à Ericaceae au-dessus de 2700 (Ericaceae, Rosaceae, Aquifoliaceae, Symplocaceae)
Evergreen low forest and thicket of the ericaceous zone, above 2700 m

12 Formations herbacées dues à l'activité volcanique
Herbaceous communities caused by volcanic activities

38 Plantations de Palmiers à huile (Elaeis guineensis)
Oil-Palm estates

39 Mosaïque de formations dégradées et plantations de Canneliers (Cinnamomum burmani)
Degraded vegetation and Cinnamon plantations mosaic

Types de végétation des régions inondées
Types found in swampy areas

24 Forêt haute, dense, inondée de la plaine d'Indrapura (Côte Occidentale) Mangifera, Alstonia, Gluta, Coelostegia, Xylopia) etc...
Tall, dense, swamp forest of the Indrapura plain (west coast)

25 Forêt basse, dense, inondée de la plaine d'Indrapura (Côte Occidentale)
Dense low swamp forest of the Indrapura plain (West Coast)

26 Forêt haute, dense, inondée de la plaine orientale
Tall, dense, swamp forest of the eastern plain
Tourbe/peat : (Dyera lowii, Gonystylus, Tetramerista, Campnosperma, Shorea uliginosa, S.teijsmanniana, Durio carinatus) etc...
Sans Tourbe/non peat : (Shorea sumatrana, balangeran, palembanica, Xylopia, Alstonia, Mangifera) etc...

27 Fourré dense sempervirent (Alstonia, Combretocarpus, Melaleuca, Cyperaceae, palms) ou forêt primaire très dégradée
Dense evergreen thickets or very degraded primary forest

28 Fourré dense sempervirent de Melaleuca leucadendron
Dense evergreen thickets with Melaleuca leucadendron

29 Communautés herbacées d'hygrophytes (Cyperaceae, Scleria, Cyperus, Gramineae, Phragmites)
Herbaceous hygrophytic communities

Mangroves
(Rhizophoraceae, Sonneratiaceae) etc...

31 Forêt ripicole
Riparian forest

32 Etangs
Pools

⚑ Eusideroxylon zwageri

⚑ Pinus merkusii (natural occurence)

IV - Formations de basse altitude (300 à 1000 m) des pentes orientales
Eastern slopes at low elevation, between 300 and 1000 m

13 Forêt dense humide sempervirente du bloc Semangka-Gumai aux Monts Tembesi-Rawas (Dipterocarpus hasseltii, Shorea, Theaceae, Bombacaceae, Podocarpus blumei)
Dense moist evergreen forest from Semangka-Gumai block to Tembesi-Rawas Mountains

14 Mosaïque de types dérivés essentiellement arbustifs
Secondary and derived types mosaic, mainly shrubby

15 Savane et savane arbustive
Savanna and shrub savanna

40 Mosaïque de formations dégradées et plantations de Caféiers
Degraded vegetation and coffee plantations mosaic

41 Mosaïque de formations dégradées et plantations de girofliers
Degraded vegetation and clove plantations mosaic

42 Plantations de cocotiers et de fruitiers
Coconut and fruit trees

Agroforêt à Shorea javanica
Agroforest with Shorea javanica

Fig. 5. A fragment of the map " South Sumatra ". The venerable concept of " series of vegetation " has desappeared. A specimen of its legend (Laumonier 1983).

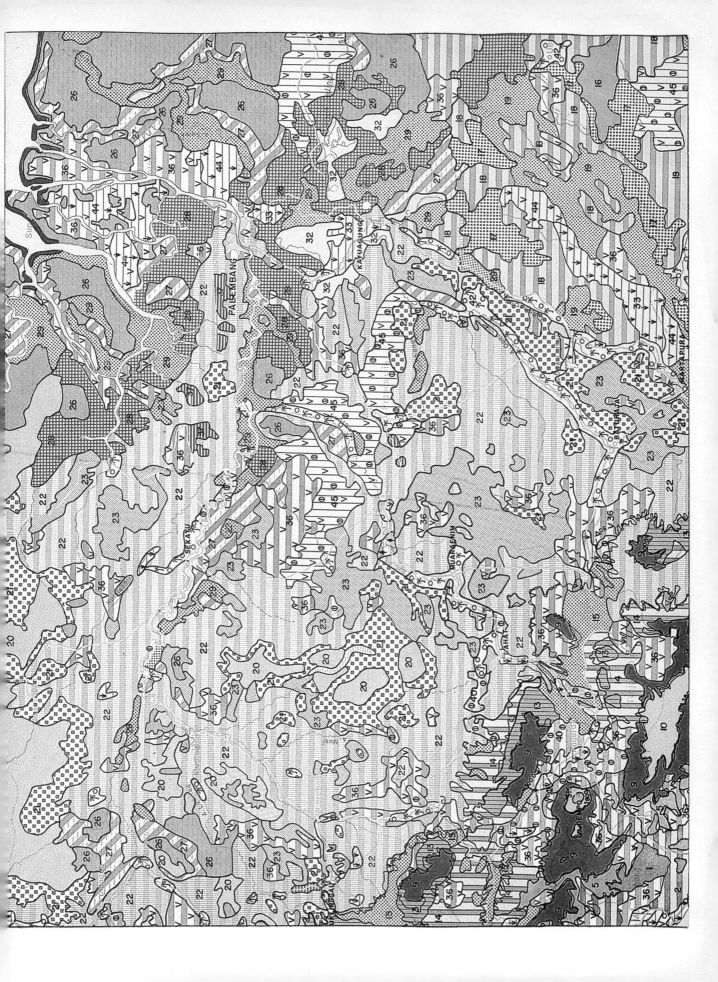

ondary phytocenoses into an undetermined number of subgroups, derived by degradation (and regeneration) processes, from various climaxes. However, we have tried to show the links between the secondary and the primary types. *In the legend, every secondary stage is related by ecological, floristic and dynamic links to one primary rainforest from which it has been derived;*

— on the sheet 'South Sumatra', two more noteworthy peculiarities are found at low elevation. First, contrasting combinations of species are unknown, as far as we know, because contrasting ecological conditions also are practically unknown. In the second place, all primary forests are evergreen; deciduous types are unknown. In a way the classification was simplified since we did not have to deal with the delicate problem of semi-deciduous (or semi-evergreen) forest types which create unavoidable dilemma to vegetation mappers (Blasco, 1983). This means, however, that the gradient between the various natural forest types is extremely gradual. Consequently, some boundaries that we had to draw, were problematic;

— the classification should be applicable to all Sumatran vegetation types. For practical purposes we had to limit the number of selected criteria to 6 or 7 (physiognomy, architecture, phenology, flora, ecology, dynamism, local geomorphology). Each of these criteria is by itself exceedingly complex;

— we did not try to produce a classification applicable to all Indonesian vegetation types which include more than 100 million hectares of dense forests among which 89 million are rain forest types. Steenis (1935, 1958) and Kartawinata (1980) have published interesting broad classifications, useful for very small-scale mapping purposes at the national level (191 million hectares).

Fig. 5 (see colour section) is an illustration of the classification adopted for Sumatra, by Toulouse school and BIOTROP.

Fieldwork related to the vegetation map 'Sumatra'. Prior to the fieldwork we had an idea of the existing vegetation types by outlining the largest vegetation types either on aerial photographs or on the MSS 5 of the Landsat satellite imagery. This broad knowledge of the possible 'ground truth' was an essential step to map out the route for our field surveys. It was practically the only way to select properly our itineraries in a region where the few existing roads are in an extremely poor state, often soaked by floods. Moreover, good topographic maps are also lacking.

At this stage we were able to distinguish only the main primary evergreen forests from secondary types (Belukar) though, in many cases, the texture of the aerial photo, or even of the satellite imagery, gave interesting distinctions inside a given broad vegetation unit. This is typically the case of the two swamp forest types distinguished in the Indrapura plain (West coast).

In this way we established the location and extension of the main vegetation units, including crop land (Ladang) and plantations. This is generally the level of knowledge of every rather inexact sketch-map, which needs to be completed and enriched by the introduction of greater precision obtained from the field.

Our field trips were not concentrated in a given season. In Southern Sumatra, there is nowhere a pronounced dry season even if in some years a short dry season is recorded in Lampung areas. August and September (sometimes July) which are the less rainy months record at least 80 mm.

The aims of the fieldwork are:
— to characterize properly, structurally, floristically and ecologically each mapped unit;
— to compare their actual boundaries with those drawn on the provisional sketch map;
— to incorporate properly each distinguished vegetation type in the legend of the map.

We can subdivide our fieldwork into two main broad stages.

The general physiognomic and floristic survey does not include any systematic relevee. This is

455

only a general description of each surveyed forest or vegetation type, giving an average size and density of each 'strata' (a passive expression of the stratification) and the identification of most conspicuous trees, herbs, climbers, epiphytes, etc. This general description of the vegetation is sufficiently accurate to permit some comparisons with the other forest types of Asia described by other foresters, botanists or biogeographers. After this general survey, we proceeded with a methodological study of a carefully selected sample plot.

The methodological study of the structure and the architecture. The method adopted for the characterization of Sumatran forest types is that of Oldeman (1974). It is mainly a graphic expression of the architecture of each phytocenose, in which the different individuals constituting the forest are grouped into three *sets* (assemblage or *ensemble*): the set of future, the set of present and the set of past. Each of these sets has a precise floristic composition and a dynamic meaning (Laumonier, 1980):

— *the set of future* (see Fig. 6A–D), naturally includes young trees, with linear axis and tapering upwards. They practically do not bear epiphytes, and the development of their buttresses has not yet begun. In plain, an interesting relationship was established between their height (H) and their diameter (d). Practically we always found as Oldeman in French Guyana: H > 100 d.

'As long as the homeostatic state of the forest plot is maintained the set of the future does not move and just survives' (Halle et al., 1978);
— *the set of present* includes only trees which have reached their maximum growth. They occupy their maximum space and volume in the stand. Often, their crowns are flattened. Large lianas and numerous epiphytes are found on their boles and crowns. In their case the relationship between their height and diameter is H < 100 d;
— *the set of past* is composed of trees being eliminated either because they have reached their maximum size and growth since a long

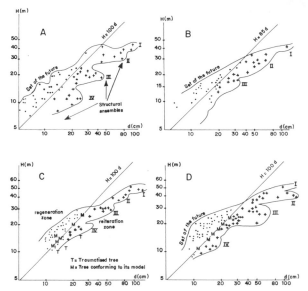

Fig. 6. H/d relation in four forest plots (South Sumatra). Log tree height against log trunk diameter above breast height (after Laumonier, 1980). (A) Sub-mountain zone at 1,400 m elevation; (B) swamp forest; (C) lowland, west coast at 250 elevation; (D) lowland, Jambi region 90 m elevation. M = tree conforming to its model; T = traumatised tree.

time, or simply because they suffered lethal wounds or important traumatisms (fall of another tree or of a big branch, etc.). Their chances for survival are seriously impaired.

The graphic representation of the architecture of each forest plot is easy (Fig. 6). It allows a comparative study of the various evergreen forest types and their classification.

Moreover, each forest or vegetation type can be analyzed by this method in its geological and pedological context.

Each forest plot covers a strip of 100 m × 20 m encompassing an area of 0.2 ha. This size is quite large for a complete description, measurement and taxonomic identification and drawing of each individual.

This method has the enormous advantage to quantify the structure and to give an indication of the dynamic position of each mapped forest type. It is a real progress when compared with the simple profile-diagram used by many scientists in tropical countries (Blasco, 1971; Brunig,

1970; Richards, 1952; Robbins and Wyatt-Smith, 1964; etc.). It is now being applied in almost all tropical countries.

Other maps and monitoring of vegetation types

The vegetation *map of South America* has been prepared in Toulouse following three distinct but complementary types of studies: preliminary preparation of a bioclimatic map; analysis of about 600 satellite images; classification and mapping of vegetation formations. The basis mapping was carried out at 1:1,000,000 scale on the Overseas Navigation Chart. Then, every basic sheet was reduced at 1:5,000,000. The final map has been printed by UNESCO (1981) in two sheets, which cover all the countries South of Panama.

Hence, the vegetation map of South America, which is very detailed in places, in spite of its very small scale, bears a classification based on ecology, physiognomy, phenology and dynamics. The majority of cartographic types are covered by the UNESCO classification (1973) from which the terminology has been extracted. At this stage the floristic composition of types is passed over in silence, on the maps, mainly because the available data concerning characteristic plants of each cartographic type or ecological zone, were far from being sufficient. Nevertheless, this map has proved very useful for an evaluation of forest resources in South America (Lanly, 1984).

An almost similar methodology is now being adopted by the FAO (Division of Forest Resources) for a classification, mapping and monitoring of vegetation types in South East Asia and the Far East (Blasco and Legris, 1983). However, sufficiently detailed botanical investigations are now available for almost all existing vegetation types of that part of the world. Hence, when each ecological zone can be characterized by a group of plants which are characteristic or endemic (unknown elsewhere) it becomes an *eco-floristic zone* which can be considered as a fundamental cartographic unit. When each eco-floristic zone has been properly

delineated on a map, it can be subsequently subdivided into several vegetation types on a physiognomic, phenologic and dynamic basis.

The size of an eco-floristic zone can be very small. The mountainous zone of South Indian Hills constitute an excellent example, with specific humid cold climatic conditions, bearing a large number of endemic plants. In the Nilgiri Hills, this eco-floristic zone covers less than 5,000 km^2 bearing no less than 80 endemic plants. On the contrary, some eco-floristic zones are very wide. The territory of Kalimantan (Indonesian Borneo) covering thousands of square kilometers has a great ecological and floristic uniformity. In both cases, the satellite provides good information concerning the *actual limits* of main vegetation types i.e.: dense almost untouched evergreen forests, logged over forests, savanas, etc. *The physiognomy* is mainly extracted from large-scale (about 1:20,000) aerial photographs, allowing the distinction between 'tall dense forests and heath dense forests' or between 'shrub and tree savannas'. Finally the structure and floristic composition are mainly provided by appropriate field surveys, sometimes also by the available literature.

The era of satellite

Since 1972, several earth observation satellites have been launched. Today, their general technical and practical properties are well known (Manual of Remote Sensing, 1975). With the exception of Landsat 5 (Landsat D') launched on March 1, 1984, from which we are still awaiting the products, the ground resolution or size of each picture element (pixel) is not very accurate (79 × 56 m). For vegetation mappers the basic principle is that each ground unit, theoretically, has its own spectral signature. This means that Landsat sensors are sensitive to various wavelengths reradiated by each vegetation type or ground feature. Naturally, each ground unit receiving the sunlight reradiates only a part of the received energy and the four bands give different responses to the various wavelengths.

Thousands of publications are now available concerning the study of satellite imagery. European laboratories as well as many centers around the tropical world are being equipped with computers and gadgetry. A new era has come. We have to adapt our methods to these new technologies and 'commodities'.

Vegetation mapping through visual analysis of Landsat imageries

Vegetation mapping through visual analysis of Landsat images has been carried out for the *'Vegetation Map of South America'* edited by the UNESCO (1981) at 1:5,000,000 scale, in two sheets. The method followed takes into account the size of the area (8,000 km from north to south and about 5,000 km east to west, at the widest part of the map) and the requirements of three essential tools:
— the preparation of a bioclimatic map at the same scale and with the same projection. Bioclimatic data have been completed by essential edaphic results given by the soil map on the same scale. These data were essential for the understanding of the present vegetation;
— the study of satellite imagery has involved no less than 600 frames at 1:1,000,000 scale, mainly MSS 5 (multispectral scanner, spectral band 5, 0.6 to 0.7 µm). The 'visual *or* manual' interpretation of these satellite

data was drawn on toposheets at a scale of 1:1,000,000 belonging to the Overseas Navigation Chart leading to a basic mapping at 1:1,000,000, later reduced to 1:5,000,000;
— the specific classification has a major objective which is to group 'in any climatic zone types which are physiognomically very different, but which are very often derived from each other through the degradation of the strongest formations compatible with the bioclimate and soils of the zone in question' (UNESCO, 1976).
On the other hand M.F. Bellan (1980) and J.P. Pascal (1983) have published interesting vegetation maps of India at larger scale (1:250,000), using a skillful manual interpretation of Landsat data. Hence, according to the scale and the aims of mapping (floristics, potentialities, vulnerability, dominant ecological factors, land-use units, etc.) and according to the amount of available field data it is possible to produce reliable vegetation maps of many kinds using low resolution Landsat data.

Vegetation mapping through digital analysis of satellite products

We shall not repeat here the fundamentals which have been commented in Chapter 20 and 21. It is, however, essential to insist on the fact that spectral criteria can be used for the classifications of land features. Until now, however,

Table 1. Synoptic view of some characteristics of first and second generation remote sensing satellites.

Instrument	1st Generation multispectral scanner (Landsat 1–2–3)	2nd Generation	
		Thematic mapper (Landsat 4)	High resolution visible (SPOT)
The spectral range is expressed in µm	band 4 0.5–0.6 band 5 0.6–0.7 band 6 0.7–0.8 band 7 0.8–1.1	band 1 0.45–0.52 band 2 0.52–0.60 band 3 0.63–0.69 band 4 0.76–0.90 band 5 1.55–1.75 band 6 10.4–12.5 band 7 2.08–2.35	band S_1 0.50–0.59 (green) band S_2 0.62–0.68 (red) band S_3 0.79–0.89 (infrared) band P 0.51–0.73
Ground resolution	80 m	band 6: 120 m others: 30–40 m	band S: 20 m band P: 10 m P = panchromatic

advanced remote sensing techniques have not yet been directly applied to our mapping system and all our work carried out is at an experimental and research stage.

Among the most interesting results those obtained from SPOT* simulations in Bangladesh are of special interest, particularly in the densely wooded deltaic complex of the Ganges called the 'Sunderbans'. The satellite itself was launched in February 1986 from Kourou launching site.

At this stage the aim of the experiment was to have a realistic assessment of SPOT capabilities for the land-use surveys, in tropical deltaic areas.

The design of SPOT specifies a ground resolution of 10 m for panchromatic mode and 20 m for multispectral mode. Table 1 gives a synoptic view of some characteristics of first and second generation remote sensing satellites.

According to CNES (Centre National d'Etudes Spatiales), the satellite will operate in a circular sun-synchronous near polar orbit at an altitude of 832 km (14 orbits per day). Like Landsat, SPOT will overfly a given area at the same solar time. The swath of each high resolution visible (HRV) unit is 60 km. However, thanks to the use of a mirror, steerable by ground control, the system can be pointed within a range of 950 km. Hence, as the viewing axis can be oriented, SPOT can produce a steroscopic coverage. This capability added to its high resolution are two of major characteristics of this French satellite. Moreover, during the 26 days period separating two successive satellite passes over a given point on the earth's surface and taking into account the steering capability of the instruments, *the point in question could be observed on seven different passes if it were on equator and on 11 if at a latitude of 45°.*

* SPOT = Système Polyvalent d'Observation de la Terre.

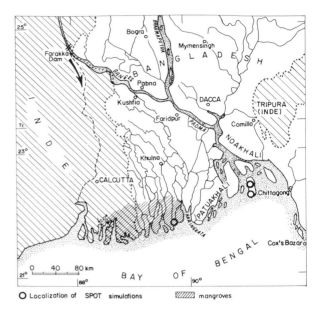

Fig. 7. Delta of Ganges and Brahmaputra.

O Localization of SPOT simulations   ///// mangroves

SPOT simulations

They have been carried out in several temperate and tropical areas. For South East Asia, however, only Bangladesh has got such simulated data which have been acquired using a Daedalus Scanner, on board a Mystere 20 Aircraft flying at an altitude of 7,000 m. From these Daedalus data, a lot of images have been computed simulating the three characteristic SPOT spectral bands.

Four different test sites have been worked out, covering mangrove forest areas near the mouth of the Ganges (Fig. 7). For the sake of clarity one simulated SPOT product in the form of a colour composite (Fig. 8, see colour section) illustrates the following comments. The processing of SPOT simulated CCT, involves two main steps, i.e.:

— *a monospectral thresholding,* aiming at removing the water from the three channels. Being mainly interested in vegetation units, we tried to eliminate all information dealing with water. In other words the lowest radiance values in channel S_3 have been used in order to generate a mask on the three channels. After this thresholding, the stretching was applied (using all grey levels) outside

1 - Natural dense mangrove (mainly Sonneratia apetala)
2 - Plantations (about 5 years old)
3 - Plantations (2 to 4 years old)
4 - Orchards with villages
5 - Dry harvested fields
6 - Dry fields with some dry crops including pulses, tomatoes, chillies
7 - Very wet fields

8 - Almost barren wet muds (intertidal zone)
9 - Recent deposits, scattered grasses, new cropland

A, A' Brackish water (sea water) deep (A), turbid and shallow (A')
⌒ Embankments consolidated by trees
- - - Footpath and embankment under construction
⌒ Drainage network

Fig. 8. SPOT simulations in the Sunderbans (Halya island-Ganges) and their interpretation.

the mask; this means only for pixels concerning mangrove forests and permanently emerged lands;

— *the multidimensional decorrelation,* is a methodological approach which has been proposed by Pousse, Saint and Podaire (1982). It has been applied to channels S_1, S_2 and S_3. Its aim is to dissociate as far as possible the information borne by original channels which are considered to be too much correlated, particularly channels S_1 and S_2 ($p > 0.9$). It consists in generating new pseudo-channels S'_1, S'_2, S'_3 entirely decorrelated.

The present test using only the method of multidimensional decorrelations leading to the production of colour composite films has shown a very broad range of discriminations, at scales of about 1:80,000. The following classes are only a partial enumeration of what can be distinguished and mapped from SPOT simulations in Bengal estuarine environment:

— water: with several classes of turbidity;
— probable tidal amplitude;
— natural and man-made drainage network;
— at least four classes of spontaneous mangrove forests in which *Heritiera fomes, Sonneratia alba, Nypa fruticans* and *Excoecaria agallocha* are dominant;
— new dry sandy alluvium;
— tanks or fish ponds (?);
— forest nursery;
— harvested paddy fields;
— places invaded by aquatic weeds (hygrophytes);
— embankments consolidated by 'jacket' trees or new barren embankments under construction;
— clusters of planted trees along channels (with houses under the trees);
— reafforestation, on new deposits (*Sonneratia alba* mainly).

It is interesting to compare these rather futuristic results with those obtained for the same area with a digital analysis of a Landsat 2, CCT (March 17, 1977). As expected, the digital processing of Landsat data allows the extraction of only five main, reliable, but heterogeneous classes, i.e.:

— water;
— croplands, almost barren during the dry season;
— dense high mangrove types on soils of lower salinity in Bangladesh with *Heritiera* and *Nypa;*
— less dense low mangrove types on over-saline soils, on the Indian side, almost without *Heritiera* and *Nypa;*
— new alluvial deposits.

The obtained map, at 1:250,000 scale (Deshayes, 1981) could not have been much more accurate, mainly because of the low resolution of Landsat 1, 2 and 3, belonging to the first generation.

Conclusion

The 'Institut de la Carte Internationale du Tapis Végétal' of Toulouse along with its associated institutions of the tropical world (French Institute, Pondicherry; BIOTROP at Bogor, etc.) have produced an impressive amount of vegetation maps which take into account our scientific knowledge and local social needs. During many years the methodology has rather faithfully followed Gaussen's principles with unavoidable modifications.

Today, however, we are compelled to evolve faster than ever because remote sensing technologies and computers are providing vegetation mappers with an unceasing flow of potential facilities which play an increasing role at every level of almost all societies. Theoretically, a better knowledge of renewable resources thanks to new technical advantages, can lead to a better management and may help to reach some development objectives sooner.

Until now aerial photographs coupled with a very intensive ground survey have constituted the corner-stone of our vegetation mapping activities. But the introduction of data provided by the second generation of satellites puts our institutions on the eve of a jump towards a yet undefined technological future.

32. Other Contributions to Methodology

A.W. KÜCHLER

The belt method

Schmid (1940, 1948) introduced what he called chorological units or vegetation belts on his vegetation map of Switzerland. These belts are composed of species that occupy similar areas (Schmid, 1955). This approach is closely related to the Arealkunde by Meusel (1943). It is based on a careful investigation of the area occupied by every species of the country. By comparing these areas, it becomes evident that many species occupy the same area whereas other species are grouped together in another area. They seem to be mutually exclusive and form 'belts' which can extend over very long distances. For instance, Schmid sees certain belts reach throughout the Alpides from the Alps to the Himalaya Mountains and cites the larch-pine forests at the alpine timberline as an example, containing in its various parts such genera as *Rhododendron, Vaccinium, Delphinium, Prenanthes, Aconitum, Anemone, Ranunculus,* sometimes even the same species.

Some species occur throughout a belt, others in several individual parts of the belt, and still others only in one part (Schmid, 1940). All these make up the species combination of the belt. Of course, there are many species which are not limited to one belt: they are termed 'bizonal' if limited to two belts, and 'plurizonal' if spread over more than two belts. Obviously, the plurizonal species do not lend themselves to the establishment of belts.

Belts are concrete units and the floristic difference between two adjacent belts is much greater than the difference between any two sections of the same belt, even if these sections should be hundreds and even thousands of miles apart. A belt contains a large number of heterotypical phytocenoses, conditioned by climatic and other local circumstances. Every phytocenose is assigned to the belt with which it shares the largest number of species. A vegetation belt is a chorological, geographical and historical (geological, geomorphological, climatic, etc.) unit. In contrast, the biocenose or phytocenose is a synecological entity which greatly depends on the adjustment of its constituent species to one another.

The floristic and semistatistical method of Braun-Blanquet reveals associations, alliances, orders, and classes. But it never leads to chorological entities like the vegetation belts. For these are heterotypical by definition, containing various biocenoses that result from climatic and local conditions (Schmid, 1940).

The method of mapping the vegetation of, say, Switzerland, consists therefore in comparing the areas of distribution of all species involved and in grouping them according to the common occurrence. The resulting belts can be checked in the field and, in the case of Switzerland, correspond to reality to a remarkable degree.

Schmid's method has been applied at similar and larger scales elsewhere (Sappa and Charrier, 1949; Schwarz in Moor and Schwarz, 1957; Galiano, 1960). But the larger the scale and

A.W. Küchler & I.S. Zonneveld (eds.), Vegetation mapping. ISBN 90-6193-191-6.

hence the smaller the area to be mapped, the more difficult is it to distinguish between Schmid's belts and phytocenoses. Eventually, the belts vanish and only the component phytocenoses remain. Indeed, on the map of the Creux du Van area by Schwarz at 1:10,000, the vegetation belts are not mentioned at all. The method is therefore best adapted to medium and small scales (Schmid, 1948, 1949).

Schmid's method of vegetation mapping is certainly most interesting, particularly because it is based ultimately on historical and evolutionary considerations, on routes of travel and migration during climatic fluctuations, phytopaleontological observations as well as synecological relations including both the relations of species to the physical environment and to competing species and communities. This broad foundation makes Schmid's method very stimulating; it should be the object of further experimentation.

The place-name method

Leo Waibel, one of the most eminent German geographers during the first half of the twentieth century, combined a deep insight into the geography of vegetation with broad experiences in four continents and many fields of learning. His highly original imagination was stimulated by the least observation; and when, by chance, he came across a quotation by Carl Troll (1936, p. 275) from the Enciclopedia universal ilustrada (1929) to the effect that the term 'savanna' as applied to various tropical grassland types is not of Spanish but West Indian aboriginal origin, his fertile mind at once swung into action and the ensuing research led to his famous vegetation map of Cuba (Waibel, 1943).

The method of employing place names in the study of the natural vegetation was not new to Waibel, as it had been developed in Germany to a high degree of perfection. But no part of the Western Hemisphere had been investigated in such a manner that its place names could be successfully employed for the preparation of a vegetation map and, at least on this side of the Atlantic Ocean, Waibel's contribution was a real piece of pioneering.

In contrast to the usual ways of studying and mapping vegetation by analyzing the various plant communities, Waibel consulted a large variety of sources which included historical accounts, old and new maps, reports of botanists, military surveys, soil maps, geological maps, geographical descriptions, analyses of the location and site of settlements, especially with regard to their central or marginal location on certain vegetation types, archeological data, and accounts of travelers. Another most important part was the careful study of the Cuban place names and their etymological interpretation. This latter part is, of course, the core of Waibel's method and certainly proved to be more enlightening than might have been anticipated. His investigation of the Carib term 'savanna' was particularly useful, as were all Spanish and Indian terms of vegetation types related to savanna, such as semi-savanna, mixed savanna, parkland, havanna, sabana nueva, sabanita, sabanilla, sabanazo, sabaneton, quemado, ceja, ciego, sao, and others.

The result of his endeavour was a vegetation map of Cuba at the scale of 1:2,000,000 on which he distinguished the following details:

1. hardwood forests, 1906–1907;
2. pine forests;
3. parklands, 1906–1907;
4. sabanas;
5. cacti-thornshrub formation;
6. swamps;
7. former hardwood forests;
8. scattered pine trees;
9. former parklands;
10. cultivated lands or original vegetation unclassified;
11. strand or littoral vegetation (on map only, not in legend).

The map also shows the consecutive numbers 1–169 and their location, one number each for the names of the places that were used in the reconstruction of the original vegetation. The date '1906–1907' in legend items no. 1 and 3 is the date of the American military map of Cuba at 1:62,500 which shows the distribution of hardwood forests and parklands as of that date.

Waibel's work was a library investigation and did not include any fieldwork. Circumstances did not permit Waibel to check his results by studying the Cuban landscape, but a vegetation mapper who desires to emulate Waibel and test his method in some other region should definitely plan on numerous field checks, carefully prepared and executed. Waibel's method implies research in history, geography, phytocenology, and Indian linguistics. This multiple approach is most stimulating to the imagination and of particular value to regional geographers. Indeed, if a vegetation map is to be related to historical or regional studies, it is fair to say that Waibel's method holds out a special promise for satisfying results.

Parcelle mapping

From France comes a method of mapping vegetation that has proved valuable, especially for agronomists. It evolved at Montpellier in the Mediterranean region of France. This is an area which has been inhabited and cultivated for a very long time, and the historical aspect strongly enters into this method of mapping. The natural vegetation, of course, has vanished long ago, but land use varies with time, and the history of the land can be read by interpreting its vegetation. This history, indeed, may not be ignored in the evaluation and the utilization of the land.

The minor cadastral divisions of France are sections. They must not be confused with an American section of 640 acres (1 square mile). The French sections are divided into units called 'parcelles'. These are small parcels of land of irregular outline and comprising only a small number of acres. In the cadastral records, every parcelle has its number and description with regard to land-use, e.g. meadow, field, forest, etc., with various degrees of elaboration.

This land classification is periodically revised because of changes in land-use and because taxes, governmental farm supports, etc., are based on it. To map the vegetation of parcelles implies, therefore, mapping at very large scales.

The method of mapping the vegetation of parcelles was introduced by Kuhnholtz-Lordat (1949) who, as director of the Agricultural College at Montpellier, was interested in aiding local farmers to obtain better yields. He saw the need for vegetation maps at 1:20,000 as produced at Montpellier (Carte des groupements végétaux de la France) but found that this scale was too small(!) for many agronomic purposes. He, therefore, uses scales of 1:1,250, 1:2,500, and 1:5,000. A scale of 1:10,000 is considered to be so small as to form the extreme limit within which his observations can be shown, but this scale is too marginal and is usually discarded in favor of a larger scale.

Kuhnholtz-Lordat felt that the usual variety of vegetation map may fulfill its usual tasks. But there are others. For instance, the plant communities are usually shown in a scientific manner which is as it should be. However, when the mapper makes a vegetation map of pastures, should he show the vegetation from the point of view of the scientist or perhaps from the point of view of the sheep that have to find their food there? Obviously, a sheepherder would find it more desirable to have a map which shows the distribution of grasses and forbs arranged in the order of their desirability for sheep. Furthermore, the actual value and the agricultural and fiscal potential of any parcel of land is strongly affected by the manner in which it was managed during the past; for while a farmer can improve his land considerably, he can also ruin it.

The cover of the landscape is divided into fields, pastures, and forests, which are called by their Latin equivalents. Thus, the cultivated field is termed ager, applying to orchards, vineyards, olive and chestnut groves and the like. Pasture is referred to as saltus; this comprises all grass and brushland, as well as forests used

464

for grazing. Finally there is the forest, called silva. This term implies only those forests which are protected from fire and are not open to grazing.

Under pressure of circumstances, many once-flourishing vineyards have been abandoned in southern France. They turn into pastures and eventually into forests wherever they are provided protection against fire and overgrazing. And these abandoned lands have received particular attention. Well-managed and cultivated farms may well be left to continue as they are but the abandoned, impoverished, and rather unproductive old vineyards could bear a more thorough study so as to permit their reintegration into a more harmonious and fruitful landscape. The new type of vegetation map records their various stages of evolution toward the climax forest of oaks, including the degrees of erosion and the species which are characteristic for each individual evolutionary phase of the vegetation. As, in different parts of the region, different species may imply the same phase of vegetational evolution, they are considered homologous species and are given the same color. The colors therefore have a dynamic quality.

Kuhnholtz-Lordat's method of mapping the vegetation of the parcelles is here presented as it was developed in and as it applies to the Mediterranean part of southern France.

Mapping is done on copies of cadastral sheets. At first, the mapper makes a general survey of the area and examines which species may be used as homologous indicators of dynamic phases of the vegetation. Each group of homologous species is then assigned a color and both are recorded on the margin of the map. Then each individual parcelle is examined and properly described. A systematic procedure is advisable so as to avoid any oversights. Kuhnholtz-Lordat recommends the following sequence:

1. note the topography. If there are no contours on the map, indicate slopes by conventional symbols or arrows;

2. map the homologous species in the following order: herbaceous species – dwarf shrubs (chamaephytes) – leguminous shrubs – other shrubs – species of forest trees including seedlings – water features;

3. prepare a clean copy oriented with north at the top and showing in ink the outlines of every color and every symbol. Whenever the map tends to become crowded, it is better to present a part of the information on transparent overlays.

The ager's value is shown by what it produces. There is no difficulty, and a symbol suffices: a little circle for an olive tree, a grey dot pattern for vineyards, etc. These are plants that remain for long periods. Orchards may be less permanent but, as the cadastral scale is large, it is best to mark the orchards as such and in addition mention the type of fruit. From a phytocenological point of view the information is more interesting and complete if the weeds are listed with the crops.

The saltus consists of a series of postcultural phases evolving toward the climax forest. The sequence begins usually with herbaceous communities of considerable nutritive value for sheep, followed by others that are much less valuable. These latter communities are burned periodically because the young shoots are more palatable. Then follow shrubby communities, in spite of the fire, or indeed, because of it; they are usually preforest communities (preclimax) maintained as saltus (pasture) by their management. A common feature of the saltus is the occurrence of chamaephytic communities of thyme, lavender, etc., on stony ground. The many stones may have been brought to the surface by the plough, or else they accumulated as the finer soil particles eroded away. This floristic type of vegetation mapping is based on the value of the plant communities and their place in the sere. The various phases have been given colors as follows: the first rich postcultural communities are tan (light and dark); the following stage, periodically burned, is shown in purple; the chamaephytic communities are yellow on the map, and orange, red, and yellowish green are used for the preforest stages.

The silva can be shown in its floristic detail

and according to its value. There are some well-managed closed forests but many more forests consist of short, widely spaced trees. This permits much light to penetrate between the crowns, resulting in an undergrowth of heliophytic shrubs and sometimes even of an herbaceous ground synusia. In mapping the silva, the forbs and grasses are shown first, next the shrubs, and finnaly the trees. Deciduous tree species have their own color, i.e. some shade of green; pines are shown in black symbols.

To illustrate the many-sided usefulness of the parcelle vegetation maps, the value of different vegetation types to beekeepers may serve as an example. Apiculture thrives most where nectar-producing plant species occur in masses, especially if each plant bears many flowers and the flowering season is long. In the region of Montpellier, *Rosmarinus officinalis* belongs in this category, with *Thymus vulgaris, Lavandula stoechas,* and *Dorycnium suffruticosum* nearly as important. Renaud (1949) has prepared a vegetation map of some parcelles near Montpellier (Domaine de Fontcaude) with overprinted bee symbols. No such symbol indicates that the vegetation discourages apiculture, one symbol encourages it, and two symbols promise optimal conditions. An analysis of the map reveals that the postcultural stages include few nectar-producing species, at least not in sufficient quantity. There are no forests on this map, but it is well-known that beekeepers do not frequent forests except perhaps when the chestnuts and the acacias are in bloom. On the other hand, the preforest stages of the saltus turn out to be rich in species that attract bees, especially two such stages: one with much *Rosmarinus,* and one with much *Thymus, Lavandula,* and *Dorycnium.* The map reveals at a glance the location and extent of the most favorable communities as well as their composition and their place in the sequence of the succession. This helps beekeepers decide where to place their apiaries; it also suggests to them what type of vegetation to look for in other parts of their region.

The use of Kuhnholtz-Lordat's method of mapping the vegetation of parcelles is now spreading through France. It is said to be easy to learn; it is also a fast and inexpensive method. Its strongly historical character makes it interesting to many people other than agronomists. Especially in the field of phytogeography and, indeed, in human geography, this method should receive a warm welcome.

Mapping marine biocenoses

One of the least explored aspects of vegetation mapping is that of marine vegetation. Until recently, marine biological research was strongly focused on taxonomy, physiology, autecology, and similar features, but the idea of mapping marine biocenoses signifies a rather new development. When modern scientific methods were employed to explore marine biocenoses, the logical result of marine vegetation maps became inevitable. Roger Molinier prepared such maps for Grand Ribaud (1954) and the Cape Corse area (1961). Peres and Picard (1955) published a map of the Gulf of Marseille. Molinier and Picard (1959) published their techniques for mapping the vegetation on the floor of the ocean, thereby simulating similar mapping projects elsewhere. Molinier (1961) then summarized all relevant material and brought it up to date at the Toulouse Colloquium. The following paragraphs are based directly on the observations and conclusions of these authors.

One of the basic differences between terrestrial and marine vegetation is that on land the maps are usually limited to show phytocenoses whereas maps of the ocean floor present biocenoses. This is only natural, considering the great importance of animal species which are sessile or nearly so. Indeed, when Möbius first introduced the term 'biocenose', he was dealing with marine communities. Of course, plants are particularly important in the upper marine zones if the substratum is relatively solid. But where the latter is mobile and in greater depth, the animal species are often of greater importance than the plants.

466

Another important difference between terrestrial and marine vegetation is the stratification above and below sea level. The rapid decline in the intensity of sunlight with growing depth results in a series of more or less clearly observable strata, all of which are quite thin when compared with the altitudinal belts of terrestrial vegetation. In fact, the lowest altitudinal vegetation belt on land can be of a vertical extent several times that of the entire photic zone in the ocean, which rarely exceeds 200 m in depth. This is particularly true of many parts of the humid tropics.

It has been found that methods of studying the terrestrial vegetation can also be used on the ocean floor for the purpose of establishing biocenoses. Marine biocenoses are distinct not only in terms of their physiognomy and floristic and faunistic composition but also with regard to ecology and distribution on the ocean floor. Transitions, of course, occur in the sea as they do on land. As in the case of most terrestrial vegetation, the marine biocenoses are recognized on the basis of qualitative criteria rather than absolute counts or weight determinations. However, the use of the quadrat method is limited to biocenoses on a solid substratum. On a mobile substratum, dredging (see below) seems to give the only satisfactory results.

In using the quadrat method, the minimal area should not be less than 25 cm^2, but it rarely exceeds 1 m^2. It is also convenient to indicate the vertical distribution within the community with the help of some abbreviations. Molinier and Picard (1959) distinguish two strata and epiphytes, thus:
— SE: strate élevée (upper stratum);
— SS: sous-strate (lower stratum);
— E: épiphyte.
Another item requiring special attention is the use of sociability classes for animal colonies. The latter may be massive, ramified, and spread over the ground, or branched in arboreal fashion. An entire colony is considered a unit and the sociability figures apply to the distribution of colonies rather than individual organisms. But the following distinction is also important:
1. separate colonies occupying a given area together. Their sociability is recorded in the usual manner, i.e. in five classes (1–5);
2. a single colony may occupy an area quite as large. It is often necessary to examine such a colony very carefully in order to establish whether there really is only one colony to begin with, or perhaps several colonies. Where a single colony occupies such a large area, its sociability is shown by a cross surrounded by a circle: ⊕. The cross indicates a sociability close to zero as there is only a single specimen; the surrounding circle indicates that the area involved is covered by a single colony only. Of course, in the tables that always go with the quadrat method, the ⊕ is preceded, as usual, by the coverage coefficient.
Where biocenoses have been established by actual inspection on the ocean floor and by the sample plot method, the extent of such biocenoses can be determined by additional sampling without need for further descents. When repeated sampling reveals a change in the composition of the biocenose, the mapper must dive again and take new records.

The dredge used in obtaining biological specimens is not an instrument which scoops up part of the ocean floor. Rather, it is a device, called a drague, consisting of four heavy metal bars joined to form a rectangle and equipped with two knives which correspond to the two longest dimensions. At one end, a net is attached which is lined with cloth so as to retain samples of small organisms. The samples are taken at regular intervals along parallel lines; then the process is repeated along parallel lines that cross the first ones at right angles, so as to form a grid. Only a dense grid will give much detail and this is necessary where the biocenoses are of small extent and change frequently. Where the biocenoses are uniform over large areas, the grid may be wide. Even though the outline of areas occupied by given biocenoses can thus be established satisfactorily, it is wise to take additional samples within the individual grid squares.

Molinier and Picard propose some ideas concerning the cartographic representation of the marine biocenoses with the help of colours. Starting with Gaussen's ideas on the use of colors for relating vegetation to the characteristics of the environment, Molinier and Picard found it advisable to limit themselves to a single sequence. The practice of superposing colors to indicate the combined effects of several environmental factors did not seem feasible to them. But they also found that sea level is an unsatisfactory boundary and that it is desirable to show terrestrial phytocenoses and marine biocenoses on the same map. They related terrestrial phytocenoses to xericity, but as crucial as this factor is on land, it is obviously meaningless in the ocean. The authors therefore change at sea level to another climatic factor of equal if not more decisive significance: insola-tion. Their color scheme is therefore arranged as follows: for terrestrial phytocenoses red to green, with red at the most xeric end of the sequence; for marine biocenoses from green to purple, with green for the lightest parts of the ocean, i.e. closest to the shore and to the surface of the sea. From here the colors change to blue and purple with growing depth and distance from the shore. Within such light zones, expressed by appropriate colors, different biocenoses are shown by different shades of the zonal color. Transitions and dynamic aspects are shown by alternating color bars.

Molinier's (1954, 1961) maps are done on the basis of direct inspection and observation. This, however, is feasible only down to a depth of approximately 40 m. As the depth increases appreciably beyond this contour, it becomes ever more necessary to rely on mechanical means of obtaining the data required for a map of biocenoses.

33. Ecological Vegetation Maps and their Interpretation

A.W. KÜCHLER

The recognition of phytocenoses and their distribution is fundamental for an understanding of vegetation. Indeed, establishing a natural order of the biogeocenoses is a goal of primary significance in basic science as well as many of its applications. Size and extent of the plant communities, structure and composition, and relations to each other and to the landscape make up the very basis of phytocenology. Phytocenoses are an expression of site qualities; this aspect recently developed into a crucial and revealing field of research because of its many practical applications. In particular, vegetation maps clarify our knowledge of the living conditions and the ecology of phytocenoses because the close relations of plant communities to environmental features (soil, exposure, altitude, etc.) can often be observed directly on them.

In discussing the significance of vegetation maps in research Ozenda (1964) said: 'In the study of vegetation and its relations to the environment, the vegetation map is a tool of exceptional value. It may even be considered indispensable. This will be increasingly so because the evolution of modern forms of scientific expression tends to progressively play down texts and to replace them with more synthetic and more readily exploited forms of representation. In fact, the vegetation map expresses more facts than a text can, it expresses them more clearly, i.e., it can be used easier and faster, and it expresses them more objectively.'

A problem of great importance is to establish the degree of consistency with which given phytocenoses are tied to certain biotopes and to discover the causes of this consistency. Progress in such research can be materially aided by vegetation maps and indeed, simplified and expedited better than by any other means. The results of such investigations must always be verified in the field and on the vegetation map. This strengthens both the research method and the usefulness of the map. The recent developments in mapping aquatic vegetation open up an additional and vast field of research that is just beginning to be appreciated.

One of the significant features of a vegetation map is that it forces the author to include every part of the landscape. As the map covers the entire area, communities and biotopes are observed which might have been missed by spotty observations. A phytocenologically sound vegetation map therefore prevents a subjective emphasis on favored communities. In addition, the mapper discovers most effectively to what extent the units of his classification of vegetation actually correspond to the communities he observes.

For instance, Zonneveld (1963) and his collaborators mapped the vegetation of the Biesbosch, a freshwater tidal area in the Rhine delta region in the Netherlands. The map was made for purely practical purposes: to serve as a basis for melioration. Nevertheless, it not only revealed important scientific problems but contributed directly to their solution. The most significant results of mapping the vegetation and of the research based on it are that they:

A.W. Küchler & I.S. Zonneveld (eds.), Vegetation mapping. ISBN 90-6193-191-6.

470

1. increased our knowledge of the phytocenoses;
2. revealed new combinations of species to form phytocenoses with ecological indicator value;
3. extended our knowledge of the autecology and synecology of several species and the phytocenoses they form;
4. contributed materially to our knowledge of the geomorphology and evaluation of a region and the features of sedimentation;
5. showed the stages of deposition and maturation of the soil, used in delimiting the maturation phases on the soil map and potentially important in classifying the area for land-use;
6. clearly revealed the suitability of the area for planting reeds, rushes, willows, and poplars;
7. greatly facilitated the evaluation of cultural measures and their results;
8. permitted an insight into the original natural landscape in spite of strong human influences;
9. revealed the effect of aeration of the soil and of the flooding intensity on the morphology of growth forms;
10. showed the relation between the aeration of the substratum and the ability of phytocenoses to tolerate flooding;
11. gave a new insight into the nature of succession in a highly dynamic environment.

Krause (1950) has concerned himself at length with vegetation maps as ecological research tools, especially in analyzing the causes responsible for the geographic pattern of vegetation as we observe it today. He gives the following example.

He asks: 'Why does a species occur at a particular site?'. Using heather (*Calluna vulgaris*) as an example, he observes on his vegetation maps that it occurs as a member of the 'Calluneto-Genistetum' on the sandy plains of northwest Germany, on the sandstone mountains of central Germany in the *Calluna-Antennaria* association, and as a *Calluna* facies in drained peat bogs. He demonstrates that *Calluna vul-*

garis finds on all three sites essentially the same relations between its own specific constitution and certain site qualities, which are repeated throughout, although each time in a different guise. Such generalization of many individual cases is one of the goals of scientific research. In an instance like the above, some of the critical requirements of the species are revealed, as well as certain important site qualities leading to the plant communities of which *Calluna vulgaris* is a prominent member.

Mason and Langenheim's (1957) penetrating analysis of the environment resulted in their conclusion that the environment consists of everything that affects a plant community. The environment is therefore enormously complex. Many ecologists therefore limit their consideration of the environment to climate and soil, but even then the complexity remains staggering. Walter (1977) includes the following information on his 'climatic diagrams' for a given station:

Table 1. Walter's climatic criteria.

1. Elevation above sea level
2. Mean montly temperature for every month
3. Mean annual temperature
4. Mean daily temperature for the coldest month
5. Absolute temperature minimum
6. Mean daily temperature maximum of the warmest month
7. Absolute temperature maximum
8. Mean daily temperature variation
9. Cold season: months with a mean daily temperature minimum below 0 °C
10. Months with an absolute minimum below 0 °C
11. Mean duration of the frostfree season
12. Mean monthly precipitation for every month
13. Mean annual precipitation
14. Relative period of drought
15. Relative humid season
16. Reduced supplementary precipitation curve and dry period

At first sight, this climatic list should please every ecologist. It is quite comprehensive. But from the point of view of plants, average temperature values are not too meaningful. The real question is whether a taxon can tolerate the extremes, and Walter lists these as well. In addition, climatic values are interesting when comparing conditions at one place with those at another. In order to be comparable, such values must be obtained under standard conditions, i.e. from instruments placed in instruments shelters. But plants to do not grow in instrument shelters and must tolerate quite different conditions if they are to survive.

Heat may serve to illustrate the case. The matter seems simple because temperature can be measured readily with the help of a thermometer. But what temperature? There is the temperature of the air in sun and shade, the temperature of the leaves which may fluctuate greatly with every passing of a cloud, the temperature of the inside of a trunk which fluctuates rather on a daily basis, and the temperature at the tips of the roots deep below the surface of the soil which fluctuates only with the seasons. Yet all must be tolerated as well as the temperature gradients within the plant.

Soils, of course, are just as complex as the climate.

Mapping site characteristics

A number of scientists have endeavored to express the physical environment with the help of vegetation maps. At small scales, such maps can imply only a few broad features, as is perhaps best illustrated by Pina (1954) on his map of Portugal. Such a map may be considered an introduction to a country. Krause (1952) published an excellent study in which he presents his *Gross-Standort,* a term which extends certain site characteristics over large areas, retaining nevertheless a considerable degree of unity in the basic features; it may be translated as '[vegetational] site region'. Troll (1941, 1943, 1955, 1956) in a series of imaginative papers presents world-wide relations between vegetation and sites; his basic thought of appreciating the quality of a site or of site regions is fundamentally the same as that of the other authors even though his approach is different. Small-

scale maps usually cannot show individual plant communities but are limited to more or less abstract generalization. However, they do reveal the regional correlations of which the mapper loses sight all too readily.

Close acquaintance with the features of the landscape is achieved by focusing attention on a restricted region. Gaussen's (1948) maps are particularly original; they are veritable regional geographies and among the most comprehensive pictures of an area that can be produced cartographically at the scale of 1:200,000. Troll (1939) shows great detail on his map of the Nanga Parbat. The purpose of this map is to show the three-dimensional distribution of vegetation types, and in addition, he succeeded in organizing his material in such a fashion that the more outstanding site qualities are always apparent and indeed an integral part of the individual legend items. Troll's map deserves therefore much more appreciation than it has received in the past.

More recently, a number of efforts have been made to bring site qualities into a sharp focus. Although the aims of the various authors overlap, their methods and results vary considerably. At this point, vegetation maps become particularly valuable because they permit the recognition and clear formation of phytocenological problems. They sharpen the phytocenologist's power of observation appreciably and permit him, first, to take note of what seems obvious, and then to advance from here to observe the finer, more delicate, and much less obvious relations between the plant communities and their sites. The basic advantage of a vegetation map, according to Krause (1950), is that it reveals not only the mosaic of plant communities in the landscape but also the orderly fashion of the spatial distribution of the mosaic's tesseras.

The vegetation maps of Schlitz (Seibert, 1954) and of Leonberg (Ellenberg and Zeller, 1951) attempt to give a very intimate insight into the landscape. It is not necessary here to discuss the usefulness of these maps; both are examples of the highest and most advanced levels in analyzing and mapping site qualities. However, it is useful to compare the maps.

Seibert uses Braun-Blanquet's phytosociological method. The map, at a scale of 1 : 16,000, is remarkably detailed and therefore can point out very fine variations in the vegetation. It shows the site qualities by various vegetation types, i.e. by implication. This is, of course, the usual approach; it is quite justified because vegetation reacts to neatly to environmental variations and change.

In contrast, however, Ellenberg and Zeller (1951) present a vegetation map which shows the character of the sites directly rather than by implication only and thereby departs from the more traditional practice. They even do this at a smaller scale (1 : 50,000) than the map of Schlitz. Their professed goal is a description of the natural conditions controlling agricultural production, and they emphasize that this description must be based on the requirements of the plants. In this, Ellenberg's thinking follows Walter (1951), who rejects the conventional anthropocentric approach to ecology, and who sees the plant as the center from which to view the environment.

Seibert's map of Schlitz shows phytosociological associations and their subdivisions; Ellenberg and Zeller show site types. The latter's way is therefore more immediately revealing of what the former only implies. Both maps require elaborate reports and techniques to utilize them to the greatest benefit. Somewhat later, Tüxen (1956a) goes beyond implications and reveals site qualities to a marked degree on his vegetation map of Baltrum.

A vegetation map should show site qualities as comprehensively as our present state of knowledge and the map scale permit. Therefore, a site classification as the one used by many foresters is sometimes insufficient; it is based exclusively on the annual increment of one or very few tree species. Site class or site quality for pine is indicated by the average height of dominant or uncrowded trees at 50 years of age. The great advantages of this system are its simplicity and the cheapness of its execution. Its

crudeness is justified by the extensive nature of forestry as practiced in the United States and many other nations. Dunning (1942) attempted a refinement of this method while retaining its basic ideas, and McArdle (1949) adjusted it to the conditions of western Oregon and Washington (Table 2). It is used in California as well but in most regions of the world even the tallest trees do not generally exceed 100 feet appreciably.

Table 2. McArdle's site classification.

Site class symbol	Site index (height in feet of dominant and codominant trees at 100 years)
I	200
II	170
III	140
IV	110
V	80

This table therefore demonstrates clearly how limited the areal applicability of such system must be. In essence, this type of site classification attempts to evaluate site qualities by the annual growth of one or a few given tree species, assuming that the latter expresses the former; and within varying limites, this is so, but the system is not reliable. Hills (1952) studies the problem of site characteristics and mapping in Canada, whereas Westveld (1951, 1952, 1954) sought to apply some of the Finnish ideas to site mapping in New England. Ultimately, they all map vegetation and describe the sites through it.

At times, there have been doubts concerning the efficiency of vegetation maps in studying site qualities. Such doubts almost invariably arise from insufficient information or inadequate exploitation of mapping techniques. In all such cases, additional checking is necessary before any conclusions can be finalized.

A detailed vegetation map will reveal the site qualities to a remarkable degree. But a detailed map is possible only at a large scale; necessarily, therefore, it portrays only a small area. The question then arises as to how far the information on the vegetation map is applicable beyond the borders of the map. Theoretically, the answer to this question is that the vegetation map is applicable where the same ecological conditions prevail. However, these conditions may change imperceptibly yet steadily, and the user of vegetation maps must exercise discretion. The safest method is perhaps to relate the site to the vegetational site regions (sensu Krause's *Gross-Standort*). Within such a region, ecological conditions are related and sufficiently alike to permit their interpretation with reasonable accuracy. But the information on the vegetation map should not be applied to other areas of the same vegetation site region without considering the variations of the ecological conditions, and it should not be applied at all to sites in a different region. Troll (1971) discusses the geographical methods of landscape-analysis and concludes that the landscape is a complex of biotopes. Geographical research and biological research therefore meet at this important junction, as 'site' and 'biotope' are largely synonymous. Site portrayal and site analysis are among the most important tasks and contributions of the field of vegetation mapping, and probably will be for a long time to come for, as Gaussen (1957) put it, 'the vegetation map is a means of integrating all environmental factors'.

Ecological vegetation maps

One of the reasons for making vegetation maps has always been that vegetation faithfully portrays the character of the environment. Indeed, mapping the vegetation is the only effective method to present the ecological order of our living space. This is so because only on a map can the geographical location, extent and distribution of plant communities be shown objectively.

One has come to distinguish vegetation maps sensu stricto, i.e. vegetation maps which present vegetation exclusively, from 'ecological vegetation maps' which may be defined as vegetation maps which show the geographical distribution of vegetation types and in addition, attempt to relate this distribution to one or more features of the environment.

Environmental information was introduced on vegetation maps almost from the beginning, but it was not always enlightening. Vague ecological terms are used on many vegetation maps, e.g. swamp, dry, moist, subtropical, subalpine, premontane and many others. Frequently, these terms have a different meaning in different regions. Usually, they are not defined, and if defined, the definition applies only to the mapped area.

On many maps, ecological information appears only in a few legend items. Such maps are therefore only partially ecological. The practice continues today. Thus, the vegetation map of California (Küchler, 1977) includes such items as subalpine forest, riparian forest, tule marsh, desert salt bush. Such names are long established, locally employed terms, and their use on a map is often more a matter of convenience, aiding local readers rather than an effort to introduce precise ecological information. Similarly, Richard (1978), on his map of Chamonix-Thonon les Bains, brings such soil features as mesophile, neutrophile, externe, intern, montane, fresh, etc.

In all these cases, the authors arbitrarily and quite unsystematically select a single 'controlling' feature of the environment for a given phytocenose. The advantage of this system lies in the complete freedom of the author to select the environmental feature that should help the reader understand the geographical distribution of phytocenoses. He is also free to choose when and when not to make such elaborations. The map thereby becomes subjective. The system is descriptive and need not imply a weakness of the map as long as the reader is aware of this subjectivity. Yet, it does not allow any critical comparison of one legend item with another and contributes relatively little to the systematic interpretation and exploitation of the vegetation map.

The situation changes dramatically when environmental information on vegetation maps is introduced in a systematic manner, i.e. when every type of vegetation shown on the map is related to every environmental feature mentioned in the legend. Two types of such systematically prepared ecological vegetation maps can be distinguished: (1) those on which the vegetation is related to only one environmental feature or quality, and (2) those with more than one environmental quality.

Single-quality ecological vegetation maps

It happens frequently that an environmental feature imparts to the biotope a quality of significance for a given form of management. The relationship of various phytocenoses to various degrees of intensity of this environmental quality can be established with relative ease and hence also mapped. Ozenda and Pautou (1980) state categorically: 'Whenever a correlation between a given vegetation unit and a given environmental phenomenon can be established and verified statistically, the spatial distribution of the phenomenon and often its seasonal development can be deduced from that of the vegetation and hence from the latter's representation on maps: this is the principle of ecological cartography.'

The correlation is so close that the distribution of plant communities permits conclusions about the particular environmental quality with astonishing accuracy. Thus, Walther (1957) related the plant communities of the Weser flood plain near Damnatz in northern Germany to the seasonal fluctuations of the water table. Although the range of the fluctuations was relatively uniform in different places, the depth of occurence varied considerably and was clearly revealed by the phytocenoses. Zonneveld (1960, 1963) also related the vegetation to water fluctuations and Emberger, Gaussen and Rey (1955) showed that the various heath types of the Landes in southwestern France indicated different drainage conditions. The correlation of

plant communities with a single feature of the environment permits an accurate calibration which is particularly useful in various management considerations.

However, such calibration is not possible when the character of the environmental feature is not described consistently, as for instance, if soils are shown as wet soils, acid soils, sandy soils, eutrophic soils, etc., because the various categories are not comparable. On the other hand, single-quality ecological vegetation maps can be very useful indeed, especially in land-use planning, whenever the relationship between vegetation and a given feature of the biotope can be calibrated.

Multi-quality ecological vegetation maps

Many authors have sought ways and means to portray the ecological conditions of an area. The idea of mapping ecosystems, while not new, became popular during the second half of this century. The results, however, varied little: basically the maps portrayed the vegetation which implied the character of the biotopes. One or a few features of the environment were often supplemented as fitted the purpose of he authors. For good examples, one need only peruse the various numbers of the excellent 'Documents de cartographie écologique' (Ozenda, 1963ff.).

By far the best of the more comprehensive presentations of vegetation-climate relations come to us from Walter (1976, 1977, 1979). In fine detail, he shows the enormous complexity of the biotope and its relations with the vegetation. Walter is able to explain many features of many vegetation types ecologically. But his work also reveals the great variety of environmental features exerting different influences and producing different effects in different plant communities. Walter's work offers the great advantage of being applicable on a worldwide scale. Box (1981), inspired by Walter, attempts a similar though much more limited procedure in an effort to apply his results to modeling.

Bailey (1976), on the other hand, regionalized ecosystems by correlating them directly with the vegetation on the vegetation map of the United States by Küchler (1967b).

The ecosystem, and indeed its components, are so complex that it is not possible to map them all individually. Authors therefore become selective, and the number of possibilities and combinations is almost infinite. As a result, there is a great variety of multi-quality ecological vegetation maps, and this is not astonishing considering the variety of authors, techniques and needs. The more modest maps relate the vegetation only to two or three features of the biotope whereas the more ambitious ones try to include as many features as possible. However, even the most comprehensive maps legends cannot portray an ecosystem in all its intricacies.

The vegetation-site map

An important map was presented by Ellenberg and Zeller (1951), showing the close interrelation of vegetation, climate, soil, and geological features. This 'vegetation-site map' is intended to show site conditions in their relation to the cultivation and production of crops produced in fields and meadows, in orchards and vineyards. It is, therefore, an excellent basis for agricultural planning in all its ramifications. It is equally useful for teaching purposes at agricultural colleges and in courses on phytocenology and ecology.

To supplement their vegetation map with data on soil properties, they observed, as Pallmann (1948) had done before them, that minor characteristics of the soil can be more significant for high crop yields than those important in soil systematics. They decided that only those features of the soil are important in this kind of work that affect growth and yield of crops. Due to the considerable local relief, the geological origin of the soil becomes a major feature of consideration.

Where soils are characterized by two layers, one above the other, the two layers are shown in alternating bars. This is done only where the surface layer is so thin that the lower layer appreciably affects the crop plants. For instance, a layer of clay may rest on a layer of sand.

Climatic data are deemed most important but meteorological stations cannot supply the needed information because they are too widely spaced and too few in number. Adequate climatic information can only come from the plants themselves. Hence, phenological observations are indispensable. By singling out several key species and observing key ontogenetic phases while cruising through the landscape in a car, the phenological zonation of the region becomes evident. Of course, such a record is valid only for the tested species and for the season and year of observation. But by repeating such phenological observations the following one or more years, and by adjusting the phenological boundaries to the sites, it is possible to show the resulting climatic zones within the local landscape with sufficient accuracy.

The relation of yields to slopes and exposures, to passages of cold air by air drainage, etc., is well established. Gentle and steep slopes are therefore distinguished on the map, as well as the direction of their exposure. This shows the farmer and the planner where the most and the least amount of insolation may be expected and over how large an area. If a danger like a late spring frost is common to the entire area, then there is no need to show it on the map. But if there are individual sections where steady strong winds from one direction prevail during the main vegetative period, i.e. while the shoots are growing, such areas must be indicated because the wind's effect on the crop yield is so great.

The vegetation-site map is therefore a map of phytogeocenoses showing not only the potential natural vegetation and its substitute communities but also the climatic, edaphic, and geological features of the various biotopes.

The legend of the map comes in tabular form of four columns, relating the plant communities to the properties of the soils and the geological base of their respective sites with the help of colors and symbols.

The colors are the means of presenting data, of which the reader is conscious first and foremost. They must therefore show here the major qualities of the biotopes. Ellenberg and Zeller established a scale from one end of the spectrum to the other and related it to the water conditions of the sites. Dry and relatively warm sites are shown in red hues; sites with alternating moisture conditions are purple; medium-moist and relatively favorable sites are shown in yellow or brown; moist areas, especially those endangered by flows of cold air, are green; finally, wet and usually cool sites are given in blue. The degree of lightness or darkness of a given color relates the site to its utility for agricultural purposes. The darker a color, the less adapted is the site to cultivation (e.g. dark red, dark green, dark brown, etc.) and the lighter a color, the better adapted is the site. A variety of letter, number, and other symbols is used to indicate local features of importance to land-use and planning.

The vegetation-site map permits a deep insight into the prevailing environmental conditions. For the preparation of the vegetation-site map and its associated table, the phytocenologist requires the cooperation of a wide variety of collaborators, especially experts on the ecological requirements of crop varieties, agronomists, and others. At present, the needed and very detailed information is not available outside some very small portions of our world, but as more data accumulate the contribution made by Ellenberg and Zeller will be emulated in ever-widening circles.

It is not necessary to present many examples of multi-quality ecological vegetation maps. What makes such examples interesting is the variety of environmental features which authors use to explain the geographical distribution of vegetation types. Table 3 gives some examples. On eight multi-quality ecological vegetation maps, selected at random from the International Collection of Vegetation Maps, there appear 20 different features of the biotope. Of these,

eleven occur only once, five occur twice, and only four occur more than twice. Table 3 illustrates the great variability of multi-quality ecological vegetation maps.

Table 3. multi-quality ecological vegetation maps.

Davis (1943),	
Southern Florida:	vegetation and geology
Gaussen (1948),	
Perpignan:	vegetation and climate
Hueck (1960),	
Venezuela:	vegetation only (physiognomic)
Küchler (1964),	vegetation only
USA:	(physiognomic and floristic)
Trapnell (1948),	
Northern Rhodesia:	vegetation and soils
Tüxen (1956a),	
Baltrum:	vegetation only (floristic)
Wieslander (1934),	vegetation, fire,
Ramona, CA:	and economics

The Nootka-Nanaimo map

The most elaborate of all multi-quality ecological vegetation maps is undoubtedly the map of Nootka-Nanaimo by Klinka (1977). On this map of the central part of Vancouver Island, British Columbia, the vegetation is shown in four biogeoclimatic zones: (1) alpine tundra; (2) mountain hemlock; (3) coastal Douglas fir; (4) coastal western hemlock. These are the biogeoclimatic zones of Krajina (1973) which he introduced on his map of British Columbia. These four biogeoclimatic zones are divided into six biogeoclimatic subzones and subdivided in 14 biogeoclimatic variants. Each of these has its descriptive title and color. In the color boxes there is also a brief formula of letters and numbers which reappear on the map. This greatly helps to identify the vegetation of any given area.

In a separate table, every zone, subzone and variant of the vegetation is floristically described in some detail. In a second table, the various vegetation units are related to climatic conditions. These are shown in Table 4, which reveals that Klinka shows no climatic extremes but does include information on evapotranspiration, etc. Every unit of vegetation (except alpine tundra) is related to every feature of the climate which is presented by mean values as well as standard deviations of differentiating climatic characteristics for all subzones and variants. The values of climatic characteristics differentiating between subzones are printed on a blue background, while those differentiating between variants are shown on a yellow background. This makes the table intelligible and revealing.

Table 4. Klinka's climatic criteria.

1. The type of climate according to the Köppen classification
2. Index of continentality
3. Mean radiation during the growing season
4. Accumulated degree days over 5.6 °C
5. Frost-free period
6. Mean temperature of the coldest month
7. Mean temperature of the warmest month
8. Mean annual temperature
9. Number of months with temperatures over 10 °C
10. Number of months with less than 10 °C
11. Number of months with a water surplus
12. Number of months with a water deficit
13. Mean annual precipitation
14. Mean precipitation for each month from April through September
15. Mean precipitation of the driest month
16. Mean precipitation of the wettest month
17. Number of months with snow
18. Maximum snow depth
19. Actual evapotranspiration for each month from February through October
20. Potential evapotranspiration
21. Actual/potential evapotranspiration ratio

All major tree species are needle-leaved evergreen and are represented by a name and a tree-shaped symbol. The symbol's size expresses their productivity. There are four classes of forest productivity: good, medium, poor, and low. A cross-section (profile) relates the species and their productivity to elevation above sea level as well as to exposure.

Finally, every unit of vegetation is related to the soil. The latter is shown in all horizons, their relative thickness and their major physical and chemical features. Every vegetation type has its own soil profile thus assisting the reader in his efforts to correlate the phytocenoses with their respective substrates.

In this manner, Klinka shows in considerable detail the individual plant communities, the climatic conditions, the prevailing soil types and the topography. His map may therefore be considered a long step forward toward the preparation of scientifically sound ecological vegetation maps. It is about equivalent with a text on the ecology of the mapped area.

What makes all this information so valuable is that the tables and figures do not simply present some factual data but, most importantly, succeed in establishing the correlations between the various phytocenoses and their climatic and edaphic environment. This explains to the reader not only what plant communities occur in the mapped area but also why given phytocenoses occur on given biotopes.

The interpretation of ecological vegetation maps

Ecological vegetation maps are valuable to the extent that they are interpreted correctly. The environment of the vegetation is usually expressed in terms of climate, substrate and topography, any one of these alone or in combination with one or both the others. The climate is less often described and then often only by implication, as for instance in altitudinal zones. Water is often recognized as of prime importance and is expressed in one or several of its many aspects. While it is relatively easy to express some climatic values quantitatively, features of the substrate are described nearly always qualitatively, which limits comparisons and calibrations. Land forms can be calibrated as slopes and altitudes can be expressed quantitatively, but this is rarely done. Seibert (1980) describes the aim of interpreting ecological vegetation maps as follows: 'The goal of ecological interpretation is to determine in what manner ecosystems can serve to maintain and to improve environmental qualities for the physical and spiritual well-being of man, his domesticated plants and animals, and his institutions'. This anthropocentric approach should not be surprising: it is after all for the benefit of man that vegetation is mapped.

The problem of interpreting ecological vegetation maps is not that it cannot be done but that, in order to do it correctly, is must be done in such a manner that the possibility of overlooking the true cause and effect relations between vegetation and environment is reduced to a minimum. The more comprehensively both phytocenoses and biotopes are described, the better are the chances of approaching an optimum interpretation.

The interpreter of ecological vegetation maps must know his facts well, and even then must exercise great caution. For instance, on many maps of the Alps, a calcareous substrate is distinguished from a silicious one, each bearing a variety of plant communities. Such maps are generalized to the extent of suppressing the details needed to explain the distribution of phytocenoses within the calcareous or silicious regions. The incomplete presentation of the environment is simply a matter of scale. Some of these maps, however, limit the calcareous-silicious contrast to higher elevations. Limestone can be just as common at low elevations but is ignored and replaced by some edaphic terms. The authors give the impression that limestone is suddenly no longer significant, or else that soils are of no importance at higher altitudes. Both implications are misleading, and the comparability of the legend items is lost.

If a vegetation map is to be interpreted for a given purpose then clearly defined phytocenoses and equally clearly defined biotopes facilitate the process. The best definition of a phytocenose is based on the most detailed description of the structure (the spatial distribution pattern of its growth forms) and the floristic composition (the aggregate of taxa). Both are needed because only one may change in response to a change of the biotope. In addition, it remains important to remember that the ho-

locenotic character of the ecosystem and the very large number of its features must necessarily lead to selections by the author.

This selectivity is further forced upon the mapper by the scale. Generalization affects not only the vegetation but also the various environmental features. It results in the sacrifice of details. But which features of the biotope are so unimportant that they may be ignored? Davis (1943), on his vegetation map of southern Florida, disregarded all environmental features but one for each vegetation type. To what extent can that be justified? The impressive accuracy of Walther's (1957) map at 1:5,000 is quickly lost when the scale shrinks, and the usefulness of the ecological information may diminish even faster.

In Fig. 1, individual plant communities on sites at right can be distinguished on large-scale maps. The scale shrinks toward the left, and broad vegetation types on wet sites (swamp forest, bog meadow, etc.) are all that remains when the scale has shrunk substantially. The mapper's task then consists in selecting the features he considers most responsible for the geographical distribution of the phytocenoses on his map. It means that he must know in advance what vegetation units he plans to show and which environmental features seem most responsible for their distribution. This is important if he wishes to avoid the situation where a plant community is shown on the map but the environmental feature most responsible for its distribution has been suppressed.

The mapper's selectivity makes the results subjective, and no matter how many and which features are selected, it ignores all others. Is that acceptable? Considering the reaction of plant communities to the entire environment, i.e. to

Fig. 1. Ecological scale relations: degrees of generalization; the map scale grows from left to right.

all its features, is a conclusion based on an ecological vegetation map scientifically valid when, admittedly, it ignores major parts of the environment? The reader's spontaneous answer may be negative but a closer look is useful.

It is often difficult to establish what feature or what combination of features is responsible for the presence of a given phytocenose. The interpreter may be aided in his quest by studying the boundaries of occurance and ask himself, why does the phytocenose end here? Answers to such questions can help explain the distribution of many plant communities. A good example may be found on the vegetation map of Isaac-Comet in Queensland (Story, 1967) which was published by the CSIRO in conjunction with three other maps: geology and geomorphology, soils, and pasture lands. There are substantial differences between the vegetation map and all others although area and scale are the same on all. A closer look, however, reveals that the distribution of vegetation types does agree here with some soils, there with some geological aspect, etc. It means that the vegetation types do end for definite environmental reasons in harmony with the changes of the biotopes.

Phytocenoses can indicate the gradations of a single environmental feature with extraordinary precision. In the case of single-quality ecological vegetation maps, the interpreter need only ascertain the accuracy of the definitions of the phytocenoses and of the measurements of the environmental quality. There are no obstacles to a good interpretation of single-quality ecological vegetation maps.

The matter is different on multi-quality ecological vegetation maps. The gradients of different environmental features do not run in the same direction and so produce conditions of extreme complexity. Mappers should therefore choose the very largest scale they can justify and describe the environment as comprehensively as possible. Listing two or three qualities of the environment is valuable but listing seven or eight qualities is more valuable. The purpose of the map significantly affects the number and kind of features of both vegetation and environment needed to adequately interpret the map.

The strictly scientific, systematic exploration of the ecological relations of the mapped vegetation, independent of any utilitarian motives, remains the most difficult task because it requires the greatest comprehensiveness. For example, Walter (1977) prepared a list of what he considered the most important climatic features responsible for the geographical distribution of phytocenoses. Klinka (1977) proposed a different list. If the two lists were combined, most ecologists would agree that the result is relatively complete. Yet, it ignores the features mentioned above when discussing the problem of recording heat meaningfully from the point of view of plants. It may not be possible to obtain such information, and the mapper must do without it. Still, the new list remains incomplete, and Seibert (1980) pointed out that a 'comprehensive analysis of ecological systems with regard to their function and dynamism will not be possible in the foreseeable future'.

On the other hand, even a few environmental qualities can help so much in explaining the distribution of phytocenoses that any increase in their number will be beneficial. The ultimate solution comes when all environmental features are known and their effects understood. Increasing information means that the solution is being approached asymptotically, although it can never be reached. But then, it is not necessary to reach it. Long before reaching it, the interpretation of multi-quality ecological vegetation maps can be so accurate as to make further investigations unprofitable, not unlike the species-area curve used in determining the most efficient size of sample plots: the curve continues to rise long after the size of the plots has been fixed.

It may be difficult and expensive to prepare a scientifically sound multi-quality ecological vegetation map. But once it is done, its comprehensiveness permits one to derive many special purpose maps directly from it, saving much expense, time and energy.

34. Landscape (Ecosystem) and Vegetation Maps their Relation and Purpose

I.S. ZONNEVELD

Introduction

Beside pure vegetation maps also other maps may show vegetation information that also can be interpreted ecologically. The aim of vegetation mapping as such is to serve phytocenology (vegetation science) and/or to be applied to planning and managing the use of natural resources and to many other purposes. For the first purpose, a map giving vegetation data in the form of a description of the characteristics of floristics and/or structure and physiognomy may be wanted. Such a map may serve the vegetation scientist but also the geographer as a tool of study. Such a study may involve the comparison of such an as objective as possible recorded map with e.g. a soil map, a geomorphological map or with climatic data, etc., and so learn about possible causal relationships. In addition, the distribution of vegetation units may teach the vegetation scientist about genesis of vegetation and inter-relations between units.

For application to management of resources, such maps have value as far as they concern the aspects of occurrence of valuable vegetation functions and materials (e.g. shelter, mass, fodder, fuel, fibre, erosion control, etc.). For application (as for part of the pure scientific use in other sciences), however, besides the value of the vegetation itself, it often has the indicator value the users ask for. It may be useful, therefore, to edit the legend of the map in such a way that not only the properties of the vegetation in terms of floristic composition, structure and physiognomy is described, but also the environmental factors indicated by the vegetation. This can, of course, be done only if the mapper has sufficient knowledge of the relations between the vegetation units and their environment.

The image of a map with such a legend may be identical to that of a map of the same area made with the same methodology of delineating units purely according to the field characteristics of vegetation (floristics and structure). Only the map legend gives also the interpretation of the units in their indication value. An example of such a map surveyed in the field using exclusively floristic composition and structure mapping and classification characteristics, is the map of 'The Brabantse Biesbosch' by Zonneveld (1957, 1960) (treated already in chapter 33). There some features of the environment are indicated by giving the legend the form of an ecological diagram which helps the reader in interpreting the ecological meaning of the vegetation units.

In spite of this addition, such a map remains a floristic or a floristic-structural vegetation map. Nevertheless, such maps are sometimes called ecological vegetation maps. One should notice here that the word 'ecological' is meant as 'mesological ecological' in the sense of Pavillard (1936). That means study of the relation seen from the environment (in the case of vegetation, that is soil, climate, water and animals, etc.). This is the pendant of the other aspect of

A.W. Küchler & I.S. Zonneveld (eds.), Vegetation mapping. ISBN 90-6193-191-6.
© 1988, Kluwer Academic Publishers, Dordrecht. Printed in the Netherlands.

ecology; 'ethological ecology', that is the study of the reaction of the organism(s) to the environment. In the case of vegetation, the study of the life form composition (as a reaction to mesological factors) is such an ethological ecological approach.

A well-known example of pure floristic structural maps with mesological information is the vegetation map of France 1:200,000. The mesological (ecological) data are there, however, put on the map by adding inset maps at 1:1,000,000 on the same sheet indicating geology, soils, climate, relief, drainage patterns, etc., as well as land-use. The map of Leemans and Verspaendonk (1980, see also Figs. 1 and 2) is a beautiful example of a purely floristic map, containing an ecological diagram of legend units indicating the environment and moreover having inset maps indicating soil and grazing intensity (the latter derived from the map itself) and in addition dynamic aspects by adding insets made by photo-interpretation of earlier stages (10 to 400 years ago) showing the development.

Part of the inset maps on the two above-mentioned maps contain information that was not gathered by the vegetation mapper but derived from other sources (soil map, geological map, etc.); other inset maps are derived from the main map (grazing type and pressure). The main map in both cases remains, however, in spite of the added inset map and the ecological diagram shape of (part of) the legend, a purely floristic vegetation map.

Maps do exist where mesological data of the environment are listed in table form after vegetation classification names in the legend, see Spiers (1978). The purely floristic or structurally defined vegetation units are also often grouped under the main (mesological) geomorphological headings: mountain vegetation, floodplains, marshes, dune complex, etc. Still, such maps may be generated purely by field survey using flora and structure as the only criteria of delineation deserving fully the name *vegetation map*. It becomes different, however, if a map is the result of combined field data collection on land from (geomorphology) soils and vegetation, and the map units are delineated and classified according to all three of these land attributes. Well-known examples are the series of 'Land system maps' of Australia and New Guinea made by CSIRO. Most of these maps are made for applied purposes.

Land unit (landscape or landsystem) maps

There is a great variety of maps of land unit maps either purely dominated by geomorphologist or soil or vegetation scientist, or more purely holistic with a more balanced emphasis on the land ecology. Suitability for practical or scientific purposes vary accordingly. Zonneveld, (1979) gives an outline of the methodology of this survey methodology. These maps became and still are very popular because for most applied aims not only the structure and composition of vegetation is important, but especially the indication value in combination with other land data like soils, water, climate and relief. The survey and mapping methodology of soil and vegetation survey are very similar, especially when aerial photographs or other remote sensing means are used (see Zonneveld, 1979).

For soils as for vegetation, aerial photograph interpretation of geomorphological data is necessary. So if one needs the data of the three attributes for the final aim and at the same time one can assist each other by mutual exchange of ideas and data during the mapping stage and in the field of logistics (same car or helicopter or camp in remote reconnaissance surveys where transport cost are the bulk of the expenditure), it is a small step to do *land*(scape) *surveys* instead of separate vegetation, soil and geomorphological surveys. There are many more maps of this type made of large parts of the world, as there are published. Many development schemes in the world use maps of this kind. Students of applied vegetation mapping are encouraged to produce maps of this type as a base for land evaluation (ITC). (See also chapter 37).

It is interesting to note that it often appears useful to produce separate thematic maps on vegetation, soils and geomorphology all based on the same combined survey in spite of the purely holistically done survey. The CSIRO maps used to have these separate maps sometimes on a smaller scale added to the larger landsystem maps, but there are sometimes also separate maps printed at the same scale. Here we see the reverse of the ecological vegetation map described above. The survey is based on the landscape as a whole (especially geomorphology, vegetation and soils), and the map legend on such maps is purely expressed in one attribute, so-called physiographic soil survey (see Edelman, 1980). So the soil map of the Netherlands, scale 1:200,000 is mapped using land-use, vegetation and geomorphology but purely expressed in soil classification terms.

Such derived vegetation maps made by CSIRO (see also Sombroek and Zonneveld, 1971 about physographic soil survey in Nigeria) and also ITC land surveys e.g. Kaarta, Mali (see Zonneveld, in I.T.C. 1977) may have an identical pattern as the 'mother land unit map' that has served to make a soil map with a soil legend as well as a vegetation map. A derived soil and vegetation map made in such a way, has not only a different legend but also some differences in map image in the final shape. Those land-use differences which have little influence on the soil type (at least in the used soil classification) will be deleted from the land unit map in the process of deriving a soil map. On the vegetation map, however, it should be maintained, if it is to be a recent map of the present vegetation. On the other hand, if during the field survey and subsequent classification stage, vegetation does not show a reaction on certain terrain forms or soil differences, these could be deleted from the land unit map during the process of interpreting it into a vegetation map.

From this reasoning it is clear that the difference between pure vegetation maps and pure land unit maps becomes vague as soon as pho-to-interpretation or other remote sensing imagery is used. In Chapter 16 and 29 the delineation of boundaries especially in reconnaissance surveys is explained – how it is done on the base of terrain forms where geomorphology is as important a criterium as vegetation structure. But even in the past, before the time of aero-space image products, the surveyor in the field, the soil surveyor as well as the vegetation surveyor had to look at the land (the ecosystem) as a whole to orient himself and to put lines on those places where he could not access near enough. So the question whether a map is a pure vegetation map, a (holistic) land(scape) map or a vegetation-landscape map has not always a clear answer. This is the case especially if one has only the map at his disposal and no text about the history of it. One may state that in all cases where a vegetation scientist contributes substantially to the whole or even does the whole survey him(her)self, we have to do with vegetation survey or even vegetation mapping and not even one of the least effective types of it.

The ITC method (see Chapter 29) includes surveys executed purely by vegetation scientists to collect (simple) soil, geological and geomorphological data at most sample plots in order to have at least material to interpret the indication value Ch. 17, (Fig. 1a + b). Depending on the skill of the surveyor in these other sciences, his (her) map may acquire more the character of a full land unit map, be it still with the main emphasis on vegetation. For pure holistic surveys, it is advisable, however, to work with a team, even with a certain hierarchy (see Zonneveld, 1979).

Land qualities on vegetation and ecological maps

The ecological information incorporated in the legend is to be used in most cases of application in one or another way to help in assessing the suitability of the site for a certain use. An intermediate concept between pure vegetation (or ecological) data and the suitability assessment is the 'quality' (a concept defined among the basic principles of land evaluation, see Chap. 37). 'Qualities' stands in the context of land evaluation for a (combination of) properties that determine the usefulness of the site for a special defined use.

So is 'grazing capacity' a complex quality of rangeland for range management. A simpler quality is the amount of nitrogen available in the vegetation. Together with the mass and percentage of palatability. This determines the even more complex grazing capacity to a certain extend. It is clear that these qualities are related to the vegetation properties as used in the vegetation classification, such as floristic composition and structure combined with some specific assessments. Also the indication value plays a role here because nitrogen content of vegetation does not depend on species only but also on the fertility status of the soil, which in turn can be indicated by the vegetation.

In forestry, site-quality is often expressed in a simple height figure related to age. A relation can be found between vegetation unit and such a site-quality ('Bonitaet'). Such qualities could and are in certain cases also indicated on vegetation maps. Thus average data of the range of production and quality of fodder per vegetation unit are incorporated in various ITC student- and consulting vegetation maps of Tanzania, Mauritania and Mali (see DHV, Ilaco, 1976 and Beck, i.p.).

As is stated in Chapter 37 (land evaluation) vegetation maps should remain basic documents and not the evaluation map itself. The evaluation in terms of suitability or definition of management actions is always a subjective judgement influenced by various economic and social criteria that are debatable and may change rapidly. Evaluation data should be, if put on the map legend itself, be published as a separate table after the legend, clearly as an interpretation of the objective data contained in the map. Nevertheless the objective data depicted in the map and its legend can be influenced by the ultimate aim.

Not all maps have to serve a general aim. Maps made for a specific purpose can gain accuracy and detail of desired information if that special aim is taken into account during the survey and mapping. So Bannink, Leys and Zonneveld (1973, 1974) designed a special vegetation classification of purely floristic character for the Dutch arable weed vegetations and the planted coniferous forest, together taking more than half of the Dutch area (whereas the more (semi) natural vegetation in the Netherlands only covers a few per cent). The classification was made to serve vegetation mapping of these areas in such a way that optimal environmental factors like chemical fertility and moisture regime (both very difficult to grasp in detail by soil survey) could be indicated. The existing purely scientific vegetation classification for general purposes did not satisfy the needs of optimal recognition possibilities in the field and sharp indication. So the authors designed a system on the base of several thousands of relevées and comparative studies of soils, water, climate and vegetation which purposely has a hierarchic structure in which fertility determined the first main subdivision, moisture a second level, and further subdivisions were made according to climate and structural constraints (in forestry). (Sociological groups which could also be interpreted ecologically act as diagnostic characteristics, see Chapter 6A and 29).

So maps made with these classifications can immediately be translated into certain ecological factors. It appeared further that the forest classification could be translated into site-quality expressed as height/age figures using a combination of fertility figures and moisture figures derived from the vegetation classification. Soil characteristics may be indicative for moisture regionally. The relation purpose and mapping may also lead to a 'non-consistent' content of the legend. For instance, a map may be made for grassland management especially. The area may contain, apart from grasslands, arable land and forest, etc. For matters of economy one may classify and map the grasslands on such a map in a much more detailed way (as far as scale and data allow), than the forest or arable land data.

For comparable reasons, all steep slopes and mountainous areas (recognizable under the stereoscope) can be indicated as such, with or without vegetation information. For similar reasons one group of vegetations may be treated only physiognomically, whereas a different group with more interest for the user will be classified in more floristic detail. This does not mean that such deviations of the rule to be consistent in classification, is always allowed. One should only deviate if the final result suits the users and at the same time does not add to confusion. Any map made for practical application, be it a pure map or a map with other information concerning the environment remains a scientific document which should be healthy scientifically and thus be able to serve pure scientific purposes, if that does not cost too much extra.

Applied science is no less a science than pure science and both have to fertilize each other. Mapping costs relatively a lot of money, and if without much extra cost an applied map can contribute to development of science in general, one should never waste such an opportunity.

The holistic dune maps of Doing

A special type of landscape-vegetation survey has been developed by Doing 1963 and 1974. The landscape of coastal dunes along the European coast is so intricate that even at map scales larger than 1:10,000, vegetation map units are still complexes of vegetation types that have to be generalized. In fact they often occupy no more than a few square meters. The generalization in this landscape presents special problems, as it does in other areas with irregular intricate relief differences short distances.

Doing developed holistic landscape mapping with emphasis on vegetation, applied on scales from 1 to a few thousand to more than 1:250,000 but so far mainly applied at scale 1:25,000. The mapping units distinguished on

these maps form a hierarchic system, with a limited number of 'main types' divided into 'subtypes' of various orders, permitting details as far as desirable according to the local situation. These subdivisions in scale units depend very much on the scale of mapping. The units are defined and delineated according to vegetation community complexes, land form, soil associations, geomorphological, geological and vegetation listing and geographical and climatological position.

The landscapes are named after one or more plant species. So the main landscapes are:
A. Ammophyla landscape;
C. Corynephorus landscape;
E. Hydrocolyle landscape;
H. Hippophad landscape;
K. Koeleria landscape.
These lands are subdivided and then have names as:
Ar. Ammophyla-Rubus caeseus landscape;
At. Ammophyla-Cirsium arvense landscape, etc.
The method is based on the following principles:
1. integration of disciplines (of land attributes) should not be postponed until more thematic maps are produced. Instead 'landscapes' (geosystems, ecotypes) should be described directly in an early stage;
2. single geosystems (ecotopes) cannot be used as mapping units (as is the case in such intricate landscapes as the coastal dunes of Europe). Landscape mapping units on map scales should be complexes of geographically correlated ecotopes (geosystems). Thus landscape units may be described in the form of lists of dominant, differential and characteristic geosystems. Spatial patterns of ecotopes (geosystems) may be mapped in key areas. Such landscape maps moreover do not lose their validity as quickly as traditional vegetation maps (e.g. because of successional processes);
3. delineation of landscape units is also determined by boundaries in geomorphology, land-use and local climates (e.g. altitudinal

zones). No fixed order of criteria should be established beforehand;

4. landscape units should have a functional background based on a study of their origin, land-use, hydrology, erosional processes, etc. The main lines of succession must be derived from historical data, old maps and photographs, interviews, etc. These maps appear useful for regional planning and management.

The method is to be recommended as a special form of holistic land unit maps with emphasis on vegetation especially for regions with intricate morphology.

Conclusion

So ecological landscape and vegetation maps made for all kinds of special aims have in the past and shall in the future contribute much to the cause of vegetation science as to their intended aims.

35. Introduction to the Application of Vegetation Maps

I.S. ZONNEVELD

The character of vegetation in relation to application

Vegetation (natural as well as man-induced) is the green mantle of the earth and by that a part and building stone of the landscape. It is a land attribute also valuable in itself in its contribution to the human ecosystem by supplying food, construction materials and shelter, fuel, conservation of soil, and aesthetics in the scenery. It is also an integrated expression of all other land attributes and of the environment = ecosystem = landscape in its entiry. Therefore, the 'green mantle' (vegetation) can be used as an indicator of the environment in its spatial variations (patterns) and its qualities. Vegetation reacts more distinctly and quickly to temporary variation (process) than the other land attributes, in spite of the mutual interactions between these attributes. This is the reason why inventory of vegetation is useful for both scientific and practical purposes. The strong impact of man on the environment has by the same reasons an overwhelming influence on vegetation. It is at present even so that hardly untouched vegetation remains. In many cases one can speak about devastation and at least there is a strong demand for management and/or protection of man-induced as well as (almost) natural vegetation. Inventory of 'natural' but also cultural (man-induced) vegetation resulting in maps is an important activity that should serve as a basis for these activities.

The aim of vegetation mapping

Vegetation, 'the green blanket of the earth' is an attribute of land and consists of plant individuals either spontaneously sown and growing or sown or planted and managed by man. Those vegetation types existing without strong influence by man are called 'natural vegetation' those for which existence depends mainly on conscious human activity (pastures, orchards, gardens, tree plantations, agricultural crops) can be called cultural vegetation. In between there is a wide scale of trasitional semi-natural vegetation types, such as not purposely sown but managed rangelands, forest, shrublands, etc. Even pure cultural vegetation like cereals on arable land contain quite a number of species that are spontaneously sown (weeds). In general therefore natural and cultural vegetation consists of more than one species. The individuals belonging to these species are the building stones of the three-dimensional body of the vegetation. This three-dimensional body has form and shape (structure) and can be recognized, classified and mapped by properties derived from this and from the building stones (plant taxa) itself. This holds in principle for natural as well as cultural vegetation.

Vegetation is of interest for man because:
— it is the original *source* of food, shelter, raw material for industry products, fuel, energy in general for animals and man;
— it is a main *constituent* of the landscape *influencing* also the other abiotic land attri-

A.W. Küchler & I.S. Zonneveld (eds.), Vegetation mapping. ISBN 90-6193-191-6.

488

butes such as stability of land and its soil and relief (preventing soil erosion) influencing the hydrological regime, sedimentation processes, etc., influencing micro topo even macro climate;

— it is, as dependent on other land factors, an *indicator* for environmental conditions that cannot always be readily observed directly;

— it can also be a constraint for men and animals, e.g. weeds in crops, habitat diseases and pests, and vegetation structure influencing the accessibility of the terrain. Vegetation maps and other geo-information systems containing vegetation data are meant to serve as inventory and information bases in order to apply this knowledge in practice as well as science.

In Fig. 1 (circle) it is shown that each land attribute is at the same time a building stone, acting factor and dependent in the whole complex of the land or 'ecosystem'. Important to realize, is that a strict relation exists between all attributes only in four-dimensional space including time as the fourth dimension. In other words the adaptation of one factor to another needs time. So at one moment certain relations may seem less strict than others. Change (adaptation to new conditions) of one factor (depended) may be more rapid than of another. This is the main reason for the fact that for survey of the land it is necessary to map and classify not only the soil or only the vegetation or only other attributes of the complex ecosystem of the land, but to pay attention to all of them. So it is often

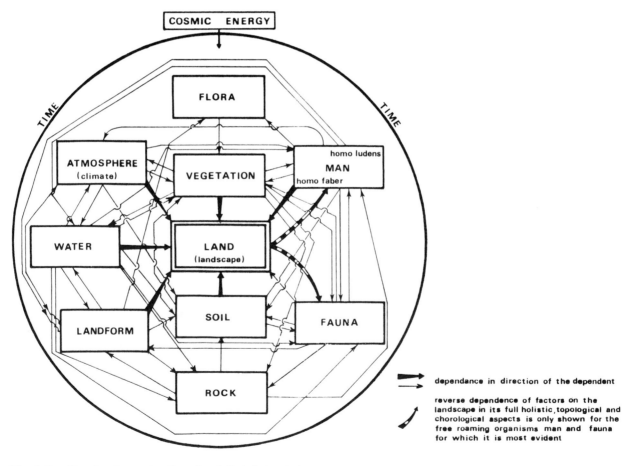

Fig. 1. Landforming factors (attributes) and their interrelation.

useful not to make separate vegetation maps but to incorporate vegetation data (from photo-interpretation and field sampling) into wider landunit concepts (see Chapter 34, also Zonneveld, 1979). The pattern in the time dimension (process) makes it also necessary and possible to make prognoses (see further Chapters 36–38).

Before vegetation is described as the 'green mantle of the earth' be it natural or planted. This means that this book deals with purely natural spontaneous vegetation as well as with land-use, as long as this is 'green'. For mapping of crop pattern; distribution of forest, grassland, arable land cover types also belongs to the subject of this volume. A special treatment for this aspect is given in Chapter 25.

An important step between inventory survey and application is land evaluation (land appraisal). Crucial qualities can be directly or indirectly derived from vegetation data (see Chapter 37).

So vegetation and land-use survey maps as well as vegetation data in other documents can provide basic information for land evaluation as part of land-use planning (see Chapter 37, e.g.):

— evaluating land for domestic livestock and/or wildlife grazing;
— forest and watershed management (as part of an inventory for these purposes);
— various types of land improvement for agricultural purposes (reclamation, drainage, irrigation, conservation, re-allocation of land);
— conservation of 'natural' areas;
— recreation;
— technical design (e.g. road construction, open cast mining, military applications);
— scientific (usually including educational) purposes.

Planning can therefore make use of land unit maps including vegetation, in its various stages, from global master planning down to detailed planning for direct implementation. In Chapter 2 we see how during the history of vegetation mapping all these applications developed.

An important aspect of vegetation mapping is often not realized. A popular belief is that the result of a vegetation survey is a map, that then can be used for the above-mentioned purposes. However, the whole series of activities of hand and head that preceed the map are a chain of building up of knowledge. As much of that knowledge should be expressed in the map, but by far not all can be stored there. Vegetation survey for vegetation mapping is first of all a *way of scientific research* that is most necessary even if no final map would result from it! That knowledge is indispensable for the science itself (landscape ecology, vegetation science, see Zonneveld, 1979) as well as for the practical aim of the very survey itself. This knowledge is about classification of vegetation units and about its relation to environment in a topological sense. This means the relation with the underlying rock, soil, relief the locally acting adding animals, man, water and climate. But especially the mapping activity forces to a 'chorological' attitude too, that means to see the surface of earth in its horizontal relations (see Neef, 1967; also Zonneveld, 1963, 1979; Naveh and Liebermann, 1983; and other landscape ecological literature). Hundreds of exixting vegetation maps and the reports accompanying them give evidence of this statement, too many to be referred to individually. This knowledge should be also transmitted to users by other means, such as the map or other used geo-information systems, by writing or (in practical development projects) also orally. Maps are just products, not the main purpose of vegetation mapping surveys. They belong to the most important intermediate but not final aims of that survey. Mapping as such, is one of the most important tools of vegetation science and landscape ecology in general, in a purely scientific sense as well as for application.

The special purpose of vegetation maps may be different for each individual vegetation. It is most important to consider that purpose before starting the work. The field of application can be divided into:

490

Maps as means for further study of vegetation and its environmental relations

For thorough study of vegetation and its relation to the environment, presentation on a map is very useful and can even be considered as necessary. Mapping of data is, in a way, making an inventory of what exists. Moreover, certain aspects can only be represented and studied in a map image such as pattern study, relation with other factors as climate, soil, water and other data that often are available only in maps.

Maps as means for transfer of information

The knowledge, and results of research by the vegetation scientist can be transferred to other scientists and to the users of ecological data in the practical field. Just as soil maps, geological maps, settlement maps, town plans, climate maps, overall land unit maps and so on, the vegetation map may serve as one of the bases of land evaluation or land appraisal (the classification of the land in terms of man's use of it and measures that can or should be taken to improve it for more efficient use).

It should be kept in mind that no vegetation map can serve all possible needs. For different uses, different basic vegetation maps using different mapping criteria may be needed (see also Chapters 16, 19, 25–34).

When do we need a vegetation map?

It is not always necessary to include vegetation maps in a report on natural resources. The information provided by vegetation (as is stated already) can also be used and incorporated in documents (e.g. maps) of soils, climate, general land units. hydrological units, etc. It goes without saying that in these reports it should always be mentioned how and according to what principles, vegetation has been used as an indicator. If vegetation is not used, the reason for neglecting this source of valuable information would have to be mentioned. In many cases, however, it will be useful to compile and publish vegetation maps either as basic documents or as directly applicable information. The latter method is the obvious choice when the vegetation itself has to be used as the main natural resource.

So in forestry the items to be used (trees) are part of the vegetation and special attention should be paid to them. Grazing areas can best be presented in the form of a vegetation map e.g. with emphasis on the agrostological aspects. General land-use maps are special types of vegetation maps if the legend is to be expressed in vegetation and crop cover classification systems even if as guiding principle the type of management is used (see Chapter 25).

36. Environmental Indication

I.S. ZONNEVELD

Introduction

The need of man to adapt to or remodel his environment has induced not only the study of the vegetation as resource, but also as an indicator of that environment. He has learned to use observable reactions of living things to inform him about quality of climate, soil and water. To use vegetation mapping for this, is one of the means of biological indication, in this case on landscape scale. Other ways are using organelles, cells, tissues, single plants and animals, genotypical as well as pheno-typical (life-) forms, see for this Best and Haeck (eds.) 1982 and Environmental Monitoring and Assessment 3 (1983), Reidel Publ. Comp. and Steubing and H.J. Jäger (ed.) Junk Publ., The Hague 1982. The vegetation map contains information about environment in two ways. One is the information described in the legend on vegetation types about we know the ecological indication value. The other way is the form of the patterns of units that directly have to say something to the experienced reader.

This use of life as indicator replaces more or less direct measurement of these environmental factors themselves. Biological indication has been applied for a very long time. Within living memory hunters, shepherds and farmers have recognized the quality of land by the growth of plants and by the behaviour of animals, when they choose the place of their settlements, their hunting grounds, their fields and pastures. Pellerwoinen, the Finnish mythological figure,

sowed Alder on the wet spots, Fir on the dry land and the Finns knew so. The pre-technological people of Western Africa still recognize good soil by the Gau-tree (*Acacia albida*), the Gaya-grass (*Andropogon gayanus*) and also the Roan-antelope (*Hippotragus equinus*). The Dutch name 'Zorggras' (sorrow grass), farmers-name for *Holcus mollis*, points to the sorrowful situation in agricultural sense, if this grass appears in the arable fields: a sign of depletion of fertility. The old Dutch farmer who says that he in pitch-darkness can tell where the good soils are by walking on his socks, knows that thistles (*Cirsium arvense*) only occur on the best soils, be it that it reveals a somewhat shoddy farmer.

Biological indication is possible because there are 'correlative complexes' (relation-systems), in which the behaviour, the form, the existence of biological features of various kinds are connected causally, direct and indirect, with actions of the environment (see former chapter and Fig. 1). These actions can be abiotic in nature (heat and cold, dryness and wetness, the presence of minerals which are necessary or poisonous) but also of biological nature such as grazing, trampling, manuring by animals. In all cases the main law in ecology holds good: for each action one can distinguish a minimum required and a maximum tolerated. Between these is the optimum level for the influence of each environmental factor. An excess of the maximum tolerable amount of minerals is called: poisoning. Being insufficient in mineral

A.W. Küchler & I.S. Zonneveld (eds.), Vegetation mapping. ISBN 90-6193-191-6.

requirement is called: deficiency. There are in general poisonous doses, rather than real poisons. Notably the plant-nutrients such as nitrogen, potash, phosphor can be deficient, but also poisonous, dependent on the doses.

Why biological indication

Why is it that one uses biological indicators even for in principle physically and chemically measurable factors? There are six reasons:

(i) — Often it concerns cumulative processes of strongly fluctuating factors which cannot be measured by one single observation using a chemical or physical method. The classical examples are groundwater, presence of nitrogen in soil and climate properties.

(ii) — Physical and chemical methods may be too time-consuming and/or too costly to repeat them often in space and time. For instance, gradients and processes in the vegetation or fauna can help to extrapolate a limited number of physical/chemical measurements in space and time. Important examples are soil qualities, climate zones.

(iii) — Sometimes the quantity and/or intensity of the working agent is thus (low) that chemical and physical assessments are very complex and at any rate not accurate enough. With biological indication especially with vegetation maps often gradients can be indicated. By chemical/physical measurements of the extremes of such a gradient the bio-indication can then be relatively quantified.

(iv) — Sometimes the combination of effects is more important than the separate factors. For instance, the indication of soil moisture is always a reaction on both the total availability of water in the ground not on the easy measurable phreatic level only. Fertility indication is in most cases a combined reaction on Potash, MN, N, PH, etc. The total effect can be different from the mere sum of all separate actions (synergy).

(v) — It is also important to realize that various factors are very difficult to measure with respect to their proper direct (operational) action. Van Wirdum (1981) distinguishes between 'operational' and 'conditional' factors. The latter are complex circumstances to which the concerning real operational factors are connected directly or indirectly, but which are not the real agents themselves.*

In various cases it may be stated that the proper operational actions are still unknown. By measuring the effect one gets a more realistic image than by measuring some 'presented' agents themselves. Moreover the so-called direct measurements are in essence extremely rough. For instance, what is the relation between the subtle process of the actual ion-exchange between soil and plantroot, compared to the coarse chemical methods to determine these ions after demolition by grinding fine with fierce force of soil samples, then devoid of structure and life?

Examples are the relation between texture and fertility and humidity of soil as indicated by the vegetation. Fertility depends on availability of diverse nutrients of which the quantity and availability (absorption complex) is related to the texture. Moisture-holding capacity is especially determined by the structure which is again connected to certain extent by the texture. Thus texture is a simple conditional factor. Structure and making of the absorption complex inclusive the reaction on this by the plants is a complicated conditional situation which determine diverse operational effects. By Bannink et al. (1973) an other example is given of the difficult measurableness of operational factors such as the phosphate-supply in forest grounds and the solution of the problem by vegetation mapping (see also Chapter 34).

It appears that within a definite type of soil, a clear correlation exists between the 2N-HCl solvable phosphate both with growth of planted trees and the spontaneous forest floor-vegetation. The relation appears to exist in three groups of different soil types: 'plaggenboden'

---

* See for his 'positional' factors page 497.

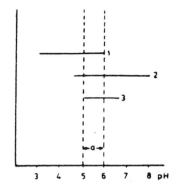

Fig. 1. Use of a few species for indication of an environmental factor. 1, 2, and 3: actual ecological amplitudes of three different species with respect to pH. (a) pH section covering the habitat of the three species.

vegetation. An absolute correlation between P and vegetation type (neither tree production) does not exist however. Evidently the measured quantity of phosphate is not identical with the 'operational' quantity. Other factors, specific for each type of soil, under which probably the form in which humus and iron are present in the soil, define the operationality of the phosphate, which is defined only coarsely via a rough method of destruction and subsequent dissolution in relatively strong acids (2N-HCl).*
See Fig. 2.

* The various ways to determine soluble phosphate, P-citron, P.Al, P.water appear to be all just approximations. They never are really revealing the quantity available for plants. Even if they approach the ideal, the content of P in natural soils may be so low that the determination figures do lie in the range of determination errors. Only in arable soils relatively high P contents are available even in poor soils. For these the existing methods give acceptable results.

(old arable land), 'brown forest' soils and a group of '(veld)podzol' soils, with the understanding that in each of the three (groups) of soil types a correlation exist between P and

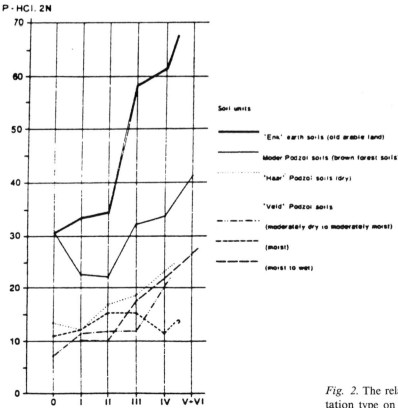

Fig. 2. The relation between P-HCL2N-value and the vegetation type on various soil units in Douglas forests in The Netherlands.

It is also generally known that a total N percentage of the soil can hardly reflect anything of the operational availability of this nutrient, although here also, high values tend to indicate to probably better availability than low values. This follows also from the observation that the decomposition of organic matter with higher percentage of nitrogen proceeds also often more rapid than that of organic matter with a high C/N ratio, experiments are required to find out to what extent the differences in vegetation indicate phosphate and/or nitrogen or something else. From fertilizer tests it appears that phosphate application can cause the differences in vegetation of the kind mentioned before. Agricultural fertilizing advice is usually given empirically. The chemical analysis of soil can only be used if one takes into account all kinds of conditional factors.

(vi) — Finally there is a sixth important argument for biological indication. This can be formulated, one should 'ask the patient herself how she is feeling'. This often is done whenever the quality of a certain environment for the means of indication itself or related processes of life is concerned, or for the total ecosystem of which the means of indication is a main attribute. Such is the case e.g. if spontaneous vegetation in forests, arable fields and pastures are concerned. For deciding the actual situation this is the ideal way of doing.

However, there are also disadvantages: a patient may not know 'why' he feels good or bad and 'what there is to be done'. In order to state the latter further investigation is necessary, in order to access what factor is too much or too little (and how much the excess of shortage is). Indeed, biological indication is often lacking quantitative data. In general, a combination of biological and chemical/physical methods is the most ideal way.

Principles of bio-indication

Limitation of the use of vegetation as an indicator

The use of organisms and mutatis mutandis of vegetation as indicator is restricted by limitations arising from the four following main laws:
— the law of Baas Becking-Beyerinck: 'Everything (diaspores) is everywhere but the environment selects' confirms indicatory value on presence or absence of organisms but neglects the restrictions of accessibility;
— the law of the physiological (potential) and ecological (actual) amplitude. Competition and tolerance, priority and primarity determine also the occurrence of species (taxa);
— the law of relative site constancy (Walter). Species behave more critically towards a site factor the further they are removed of their optium (centre of their plant-geographical distribution area);
— the complex character of factors. Correlation with measured factors differs depending on interacting of other factors. The difficulty in measuring operational factors compared with conditional factors plays a role here, as well as synergy.

The first-mentioned law of Baas Becking and Beyerinck only applies to organisms with very light and numerous diaspores. Even here the current windsystems cause differences in accessibility. For most other organisms various barriers like mountain chains and oceans hamper the transport. This means that certain organisms do not occur on many places where suitable niches are available.

Competition is at least as important for the absence of a species at a certain site. Its place is occupied by another species because the latter is stronger or because it was by chance just a bit earlier and strong enough to resist others, or because at that site an individual of another species as relict from a former succession stage is growing, which may be less suitable for the site, but for the time being holds itself (primarily). Competition, priority and primarity determine together the above-mentioned law which states that the ecological (actual) and physiological (potential) site amplitude differ. This means that results of laboratory experiments with single or small group of plants may not be extrapolated to the field situation. Nevertheless, one is used to distinguish e.g. nitrogen-indicators, moisture-indicators, etc. Even handbooks do exist with indication values for taxa, irrespective the competition circumstances (Ellenberg, 1974; Londo, 1975, and others). These lists are reasonably valid within the local area for which they are developed based on field experience (mainly estimation).

The law of relative site constance is postulated by Walter (1973) especially for the phenomenon that many plants, originating from relative humid climates 'withdraw' themselves in the dryer climates to local humid topoclimates and/or more humid soils. An example is the gallery forest along rivers in savannas with clear relation to the rain forest on the upland in the more humid tropical climates. Species growing at the fringes of their distribution areas, become in general more accurate indicators of certain environmental factors than they are in the center (Hengeveld and Haeck, 1982). Arable weeds are also good examples.

Finally the complex character of environmental factors with their interrelations, makes it very difficult to determine precise relations between species and the real operational factors. So certain species like *Eupatorium cannabium, Epilobium hirsutum, Sambucus nigra* and *Fraxinus excelsior* react on nutrients liberated from rapid decomposing organic matter (N and P especial). This situation occurs in areas with strongly fluctuating eutrophic waters but also in soils rich in lime, far away from any groundwater influence. As a consequence the same species are found among the 'phreatophytes' indicating (fluctuating) groundwater and as lime-indicators as well. They do in reality 'simply' react on the operational factors phosphorus and nitrogen, which are both conditioned by rapidly decomposing organic matter.

The answer to the restriction mentioned in this chapter is the use of a combination of species instead of single ones, the use of vegetation types preferably in combination with other observable land attributes like relief and soils. This is treated further on in this volume.

The advantage of using more than one species as indicator is:
a) by using more species, indication will be sharper, even if competition and synergy do not exist (Fig. 1);
b) by using more species local deviation in behaviour of one of the species will be less relevant;
c) the vegetation unit as a whole points to a complex environmental situation usually correlated with certain conditional factors. Certain operational factors may in turn depend on these conditional factors.

A combination of the use of vegetation units (possibly also land units) together with single species is the so-called 'çoincidence or calibration' method introduced by Tüxen (1958) ('Koinzidenzmethode, Eichungsmethode'). By using the phytocenological table (or an other matrix-) method, the coincidence has been assessed between the occurrence of a species and a factor considered as operational, which is physically measured (groundwater N, P, K, etc.) within a certain syntaxonomic vegetation unit (possibly also in combination with a soil unit or land-use unit or landform). By doing this it is guaranteed that influences of competition and the law of relative site constant will be considerably reduced.

We (Bannink et al. 1973, 1974) elaborated examples of these methods for groundwater indication and general chemical fertility by arable weeds and also forestfloor vegetations of coniferous forest plantations in the Netherlands. In this way indicator species could be assessed, which could be allocated together in so-called ecological groups, composed of species (taxa) with similar ecological amplitude for certain environmental factors. By having a number of indicator plants for a certain factor one has a reasonably reliable means, because although on each concrete site certain species may be absent for unknown reasons, still the others care for the indication. The value of indication by single species and vegetation units is also discussed by De Boer (1983) and Oosterveld (1983) (see Environmental monitoring and assessment 3, 1983). There exists a wealth of literature on the use of vegetation communities for site-indication in forestry and grazing culture (e.g. Ellerberg, 1974 and 1978, Speranza and Testoni (ed.), 1982, and see also Chapter 33 and 34).

Structure, life form and growth form as indicators

Structure, growth- and lifeform of vegetation are of high value for the purpose of indication, Lifeforms are heridatery properties that can be interpreted as genetically fixed adaptations to the environment. Various lifeform systems do exist and more could be thought of. The hydrotype systems of Iversen (1936) describes adaptations to the factor water. The spectrum of various types of adaptations can be used as a measure of the hydrological regime (compare Zonneveld, 1959, 1960, and Zonneveld and Bannink, 1960). The system of Raunkiear (1937) depends on the adaptation to the most unfavourable season (hibernation strategy). This characteristic makes it very suitable for climate indication, especially of biomes, but also of plant associations.

Only a few examples may be mentioned. Zonneveld (1959, 1960) gives examples of indication of hydrological factors such as duration, frequency of flooding and mechanical influence of currents on lifeforms that can be expressed as frequency diagrams per vegetation type. Not published studies by the same author (Zonneveld and Bannink, 1960) show clear correlation of Iversens and Raunkiear lifeforms with factors as (ground)water fluctuation, humus formation and sand deposition in inland dune and heathland areas in the Southwest Netherlands and Belgium. Unfortunately various authors have tried to develop the Rauniear system into a general one and made it as indication system by this less workable. It is much better to use a simple system for climate indication (e.g. Raunkiear in original form) and another one for the water factor (Iversen), etc.

On aerial photographs horizontal structural (pattern) image are usually informative. So dot-patterns, be it small vegetation elements in a homogeneous matrix, but also 'patchiness' (bare areas in a dense vegetation) indicate extreme conditions, like salinity or other extremely 'dynamic' factors. The vertical structure in forests is strongly related to the humidity regime via the climate. From the tropical rain forest towards the steppe via the savannas, one observes a gradual simplification of the structure. Half way forest vegetation composed by two or three strata predominate. To the humid side these grades via more strata into a complete, the space filling, profile of the ideal rain forest where only the lowest strata are rather open. Towards the dry side the tree (and shrub) layers become more and more open until only a grass/herb layer remains in steppe-like vegetations often also with annuals and xerophytic chamaephytes, along the fringes of the almost pure bare deserts. However, the use of structure as an indicator in detail is in many places hampered by the fact that human influence in the past and in an increasing way recently, has changed the structure so much that reliable observation becomes difficult. Floristic properties change also, but much slower and they still give better possibilities for general soil and climate indication.

The (growth) form of individual plants can be genotypical (as in lifeforms) but also phenotypical. Also the latter can be used. One and the same species may indicate certain environmental conditions by its phenotypic growth form. This may be due to seasonal differences in climate at present or in the past. In subarctic and subalpine areas the form of trees may indicate thickness of the snow cover. The same species looks quite different depending on the influence of snow, the wind and the browse intensity. Schreiber (1977) used the deformation of branches caused by frost damage to buds to indicate climatic fluctuations in the past. Growth rings in wood are used for the same purpose, but than for a much greater period.

It is clear that such phenotypical reaction of plants and vegetation structures are especially important for monitoring immediate action of men and animals, more than the lifeform and general structure which react in a slower way on long-term effects.

Use of the land concept and positional factors

The use of the 'land' (scape) concept is a far reaching applicating of integrated ecological indication. Beside vegetation data also abiotic land attributes as relief, soil, rock, groundwater, etc., are utilized, together composing 'land units' at, certain scale also called 'land systems'. These land units then are indicative for a whole series of properties that as 'qualities' are important for land evaluation (Zonneveld, 1979; see also Chapter 37).

Another example is the 'lifezones' concept by Holdrige (1959), where by means of a combination of climate stations – with various kinds of data about temperature and precipitation are being measured – and via vegetation classification the quantitative data over large areas are extrapolated. The same principle is applied with the UNESCO's bioclimatic maps. Still the reliability of such methods depends strongly on how well the vegetation classification and field survey of vegetation data is done. Doing developed a landscape mapping method using, terrain data, plant communities and single species to characterize landscape units in Dutch dune areas (see chapter 34).

The use of 'potential actions' (see Van Leeuwen, 1981b; Van Wirdum, 1981) as indication of the environment, depends also at the integration of abiotic land factors. Here we deal with spatial circumstances pointing to the existence of certain operational factors. The simplest example is the relief. The lowest places receive material from above (water and/or nutrients) due to gravimetrical powers. In this re-spect Van Leeuwen (1966, 1981a, b, c) proved that the potential value for natural values (diversity, occurrence of rare organisms) coincides with such gradient situations where an oligotrophic environment 'rules over' (is situated above) an eutrophic environment. In the opposite situation, the eutrophic environment will spoil quickly the lower situated oligotrophic vegetation. The observation can be done by abiotic means (e.g. assessment that peat occurs above limestone) or biotic means (one maps the vegetation and by indication, one observes that an oligotrophic (peat) vegetation lies over any calciphylous vegetation. Then one can predict that the transitional zone will have a high valuable character or (it may be disturbed by present land-use) at least the potential to develop in that direction.

Quantitative versus qualitative

In the foregoing the possibilities and restrictions of the use of vegetation as indicator have been discussed. A warning has still to be given against efforts to use these in a too quantitative way. It may be clear from what is said, that real quantitative data, even those obtained with the most delicate chemical and physical methods, are difficult to assess, because of the fact that so many real operational factors do not respond to such measurements. This is contrary to some conditional factors of which the measurements, however, only supply indirect data.

Although the biological indicator may react directly to the operational factor, real quantitative results, however, cannot be expected due to the complex nature of the cybernetic system of the green blanket and its communities. In most cases one should satisfy oneself with a diagnosis of what is happening. If real quantitative data are required in order to interfere into an ecosystem, a combination of bio-indication and physio-chemical assessment methods will be necessary in most cases. Still empirical work and experiments will be also unavoidable (see also Zonneveld, 1982, 1983).

Map image as indication

Besides the indication value of the floristic composition and the structure of the vegetation, the macropattern – the arrangement of the mapping units and their particular form – have an own indication value.

The arrangement of certain vegetation units in a zonal pattern indicate a certain ecotone. The form of the pattern elements may reveal something about the operational agent causing it. Rectangular and other clear simple geometric forms use to point to human origin of at least one important environmental factor (land-use patterns). Fire patterns have (various) special particular forms that may indicate, their origin after having been mapped. All kinds of geomorphological forms depicted in the vegetation map image may reveal the dependance of vegetation on soil, hydrological or (topo-)climatological factors that are correlated with such geomorphological forms. The map elements as such may show this, but also smaller image properties on aerial photographs within the units may do so. Various examples are given by Zonneveld, de Leeuw and Sombroek (1971) about aeolic erosion and termite-induced patterns.

37. Basic Principles of Land Evaluation using Vegetation and other Land Attributes

I.S. ZONNEVELD

Introduction

Reclamation, land reform, land-use management is identical with changing, and/or manipulating or replacing vegetation.

The basic discipline at the base of such activities is determination of the 'value', the suitability for a certain aim, of vegetation that has to be changed, manipulated or replaced. This 'appraisal' should be done for the actual situation existing at present, but also for a potential situation.

For that latter case one should also evaluate the vegetation (land-use) determining factors of the environment (climate, water, soil, etc.). So land evaluation is in principle done for *land* on the base of land *units*. Vegetation is only one attribute of this land (see Fig. 1 Chapt. 35). Vegetation data are, however, not only important for the evaluation of the vegetation as attribute, the indication value (see Chapter 36) gives also information about other land attributes. So even in cases where (semi-)natural vegetation will be fully changed to be replaced by anthropogene plantations, vegetation data are wanted to serve as a base for land evaluation.

In this chapter the principles of modern land evaluation will be outlined. Special emphasize will be given to the role of vegetation, the main subject of this handbook.

For the same reason land evaluation for rangeland land utilization types, where semi-natural and natural vegetation are aim of maintenance and/or management, will be treated in somewhat more detail, than land evaluation for other land-use where natural or semi-natural vegetation only may play a role as indicator.

On the other hand this chapter shows that although vegetation mapping may play an important role, more than just knowledge of vegetation is needed for proper land evaluation of range lands.

Outline of land evaluation

Land evaluation in its broadest sense includes the whole process of data gathering (inventory), systematizing (classification) and further elaboration to judge and plan optimal land-use. In its narrow sense it is the process of determining the value of a certain stretch of land for a certain land-use. On the basis of the results of land evaluation, land-use planners can make decisions with respect to desired land-use changes.

In various countries regional systems of land classification or land evaluation have been developed. Summaries of a number of these systems are presented in 'Approaches to Land Classification' (FAO, 1974) and in 'Land Classifications' (Olsen, 1974). One of the widely propagated and adapted systems is the United States Department of Agriculture (USDA) Land Capability Classification (Klingebiel and Montgomery, 1961).

This is an interpretative classification based on the combined effect of climate and permanent soil characteristics on the productive ca-

A.W. Küchler & I.S. Zonneveld (eds.), Vegetation mapping. ISBN 90-6193-191-6.

500

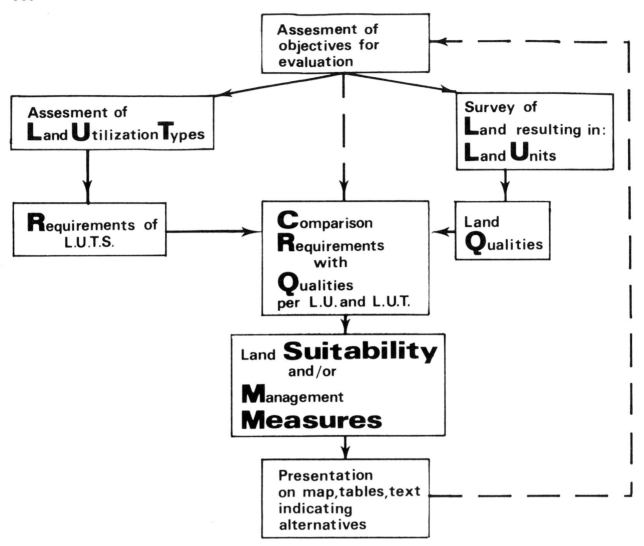

Fig. 1. Steps and Feed bach in land evaluation.

pacity, risk of erosion and soil management requirements. The system emphasizes soil erosion and conservation aspects. Local versions of this classification system have been developed in several countries to suit their specific environmental conditions. A disadvantage of this system is that the uses for which the land is classified are defined in general terms only.

Moreover an objective comparison of alternative land-use possibilities for the same stretch of land is not possible with the USDA system. Grazing moreover is just considered as a rest activity. No proper judgment for land-use plan-

ning neither management can be made on base of its classes.

Bennema and Beek (1972) developed a land evaluation methodology that is flexible and fulfills much better planning requirements (see also Beek, 1978). It has been presented as an international useful tool at the FAO-Wageningen Workshop (1972; see Brinkman and Smyth, 1973).

FAO had adopted the main lines of this methodology for application in all its projects and published the framework for land evaluation (Soils Bulletin 32), based on the publication mentioned earlier (see FAO, 1978). In the meantime at the initiative of the International Society for Soil Sciences (ISSS) a series of workshops were and are organized in order to facilitate the introduction of the methodology for the various types of land-use such as engineering agriculture, irrigated agriculture (see FAO, 1973, 1974, 1977, 1979) and forestry (see Laban (ed.), 1981). The workshop in Addis Abeba (October–November, 1983; see Siderius (ed.), 1984) deals with extensive rangeland evaluation. Here vegetation as an attribute with value in itself is most pronounced, compared with all former ones, even forestry where the 'crop' may be produced by plantations, which is never the case in extensive rangeland.

Land evaluation in its broaded sense is an answer to four questions. What is there, where is it, when, and how does it work. After that we want to know, what do we have to do where, when and how. For a proper answer to the first four questions we need first:

— data from the relevant land properties in terms of 'what', 'where', 'when' and 'how'; and subsequently

— evaluation of these data in terms of "what', 'where', 'when' and 'how' in relation to the alternative land-uses considered or the management measures to be taken.

The first mentioned set of needs are fulfilled by *land inventory*. Vegetation survey is one of the contributions to this inventory.

The second one is called *land evaluation* sensu stricto.*

If we call the activity concerning proper management within one individual holding (ranch, farm, etc.) internal land evaluation then we could name the activity for (integrated) 'land-use planning' of a region 'external land evalua-

tion' or 'regional land evaluation'. The latter is the main tool for land-use planning.

In Fig. 2 it is shown that evaluation has land ecological, sociological, economical and technological aspects. 'Man' can express his desires (certain types of land utilization). Land ecology determines what is ecologically possible (by 'physical land evaluation'). Generally or locally available technology narrows down what is possible. Finally economic factors determine what part of the wishes can be fulfilled (economic land evaluation). Land evaluation normally starts with the land ecological evaluation, be it within a certain general socio-economic framework.

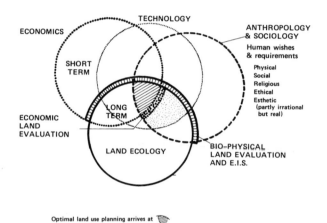

Optimal land use planning arrives at 🦅

Fig. 2. The circles of human requirements, land ecological possibilities and economic and technical boundaries. (Short-term economic reasoning and too heavy technological impact usually violate land ecological and human values).

One distinguishes between actual value and potential value. Technological requirements are usually higher in the latter case. The details of economic evaluation are usually done after the ecological evaluation has stratified the general possibilities. Also more detailed sociological evaluation may have to follow.

Internal land evaluation is done for proper design of management measures in the relevant land utilization type. For extensive grazing, e.g. nomadism and transhumance, such a distinc-

---

* Sometimes inventory and land evaluation are both treated under the same name 'land evaluation'. Also the name 'land classification' is utilized for the whole or for the evaluation part proper. In this handbook, we shall try to use the word land evaluation as defined above: evaluation of inventory data for a certain land-use; in order to be able to oversee problems with other, competing land-use.

tion in external and internal evaluation need hardly to be made. For managing ranches and nature reserves, however, it may be important to make that distinction. The result will be for the range manager measures to be taken. A general administration that has to keep an eye on land-use policy in which ranches fit, will, however, need also an external land evaluation.

The essential character of the land evaluation methodology presented here is, in contrast to the mentioned existing ones, that it presents only a set of principles and concepts that can be used in any evaluation. It does not contain preconceived judgments about qualities of land in relation to specific land-use types and also no proposed hierarchy in those types of land-use. The results are essentially expressed in terms of suitabilities of a certain (single or composed) tract of land for a series of alternative land-uses (measures) in such a way that planners and managers have a clear survey of possibilities from which they can make their choice.

Since the international introduction of the method, quite some experience has been gained by it as a tool for selection of land for reclamation, re-allotment, land consolidation and land reform projects ('regional' or 'external' land evaluation as defined in the previous chapter).

The main steps in such evaluations are:
1. formulation of objectives;
2. formulation of wanted land utilization types and their requirements on land;
3. inventory (survey of soil, vegetation, current land-use, land form (relief), climate and other land attributes);
4. interpretation of these data in 'qualities' including also limitations of land in relation to actual and potential use;
5. matching of existing qualities with requirements of land utilization types resulting in suitabilities and/or internal land evaluation measures to be taken;
6. a possible next step is formulation of recommended use.

In Fig. 1 the steps and the feedbacks in between are schematically pictured.

The final product in external evaluation is a map and list of land units with an indication of suitability for an alternative set of land utilization types as a selection base for recommended use. For internal land evaluation the final result may be a map and list of land units with an indication of measures to be taken (type and numbers of animals to be allotted, measures of burning, mowing, etc.).

Monitoring can be considered as a repeated land evaluation activity. It will be most effective if certain parameters can be automated and simplified in such a way that relatively high frequence can compensate the lack of detailed observation (see also Chapter 24).

In designing a land evaluation, methodology to be applied also for extensive grazing, some specific aspects have to be considered so far not yet considered in land evaluation for other land-use types. (1) The multistage aspect of grazing industry: primary production (the vegetation, forage source) secondary production (animals) and sometimes even tertiary production (tourism). For each stage, land evaluation procedures may have to be applied. (2) Another aspect is the mobility of the produce medium in relation to the strong dynamics of the climate in areas where extensive grazing is common. Season aspects in land utilization have to be incorporated in the land utilization types either as subdivision or as a specific characteristic.

Land

In terms of land evaluation, land has a comprehensive, a holistic meaning. It is synonymous with the holistic use of the term landscape in 'landscape ecology' or 'landscape science' (which means much more than just scenery or utterly appearance). A short original description goes back to Von Humboldt: 'The total character of an area on the earth's surface'. The FAO Framework description is longer: 'An area of the earth's surface, the characteristics of which embrace all reasonable stable, or predictably cyclic, attributes of the biosphere, the soil and underlying rock, the water,* the plant and animal populations, and the results of past and present human activity, to the extent that these attributes exert a significant influence on present and future uses of the land by man'.

— Land (mapping) units are spatial classification units. They represent parts of a study area which are more or less homogeneous with respect of certain land characteristics or which exhibit a well defined pattern of variation in this respect. The best way to define land units is done by mapping. This can be done by separate execution of soil, geomorphological, vegetation or other land attribute surveys that eventually will be combined or by directly compiling holostic land mapping units;

— Land characteristics are properties of land used to indicate intrinsic properties, determining (causing) the character of the land. They are also used to distinguish classification units from each other (diagnostic characteristics). Preferably, they should be properties which can be measured or estimated. Examples are degree and length of slope, rainfall, soil texture, soil depth, vegetation form and vegetation composition.

For land unit classification various selection principles are possible, e.g. land genesis and spatial occurrence are, for all mapping units, important selection principles.

For land evaluation, however, properties that are directly related to the typical value have to be used. This means the selection principle is of a socio-economic and technological character ('which properties determine what use can be made of the land?'). These selected properties are called 'qualities';

— Land qualities.* It is usually difficult or impossible to employ land characteristics, as used to distinguish the land mapping units, directly for a land evaluation. Many land properties interact in their influence on the suitability of land for a certain use. Moreover, the operational acting factors are often not directly used in characterizing the land units. Because of this and the problem of interaction, the concept of 'land qualities' has been introduced.

A 'Land quality' is a (usually compound) property of land which is selected because of its specific influence on the suitability of land for a specific kind of land-use. Qualities are also used as diagnostic values to classify land in classes of usefulness for a certain purpose. Qualities are hence the 'diagnostic characteristic' in the pragmatic classification of land. Just as selecting land characteristics the selection of qualities is an abstraction from the properties of the

---

* The terms 'characteristic' and 'quality' are derived from the FAO Framework (1976). The meaning differs from colloquial English language. 'Characteristic' is used here in the sense of a 'property' that determines or at least contributes to the character of an item. A property can also be used for recognition of an item even if that very property does not contribute something essential to a certain thing. Also such properties are often used as 'characteristics' in classification systems. In that case the term 'diagnostic characteristics' is appropriate. In soil classification e.g. colour is not essential for the character of a soil, however, it is an often used diagnostic characteristic, even occurring in the name of soil types.

The term 'quality' in land evaluation jargon points to those properties that are especially relevant for the evaluation process such as required or limiting factors for plant growth. Colour is in this case irrelevant, the same holds even for phreatic level or texture. The available water and oxygen supply, however, are clear qualities. They are usually strongly depending on the characteristics, phreatic level and structure, which in turn depend partly on texture. So qualities determine the character of land as relevant to evaluation criteria. At the same time qualities are used as diagnostic properties (e.a. subdivided in classes) to delineate suitability classes (the classification aspect of land evaluation).

There are languages in which different words are used for the two meanings of characteristics (a.o. in Dutch 'kenmerk' for diagnostic characteristics and 'karakteristieke eigenschap' for the intrinsic characteristic).

---

* In the FAO Framework the terms 'geology' and 'hydrology' are used instead of 'rock' and 'water'.

land. But the guiding principles for land qualities are direct answers to the physical, biological and technological requirements of the land-user. There are a large number of land qualities but only those relevant to the land-use alternatives considered, need to be determined.*

They may act in a positive, marginal or negative way for a certain land-use. In the latter case qualities are called limitations. A property can be a (positive or negative) quality in respect to one land utilization type and neutral (hence no quality at all) for another.

Examples of land qualities are: water availability, oxygen availability (drainage condition), nutrient availability, workability or ease of cultivation of the land, resistance to erosion, flooding hazard, temperature regime, climatic hazards affecting the growth plants or animals, availability of drinking water for animals, fire hazard, forest production, forage availability of rangeland, and crop yield in agriculture.

Carrying capacity of rangeland is a very complex quality of which the determination deserves much attention to internal as well as to regional land evaluation for grazing. In the section 'Land qualities (for rangeland evaluation)' of this chapter a list of rangeland qualities is given. Other examples of land qualities are shown in Chapter 2.4 of the FAO Framework FAO 1976 (see also Zonneveld, 1979). The interpretation of land properties (intrinsic characteristics) into qualities' is the first step of land evaluation proper, following the inventory stage.

Kinds of land-use

Land evaluation involves relating the land qualities of the mapping units with specified kinds of land-use. The types of land-use selected for evaluation are generally limited to those which are the most relevant under the general physical, economic and social conditions in the study area. They may be defined for

* See page 503.

either the current situation or for envisaged future conditions, i.e. after the implementation of land improvements (irrigation, drainage, conservation, etc.) and/or infrastructural and institutional improvements. Relevant types of land-use should be selected early during the land evaluation study. The FAO Framework recognizes:
— major kinds of land-use; these are broad subdivisions of rural land-use. Examples of major kinds of land-use are rainfed agriculture, irrigated agriculture, rangeland grazing, forestry, recreation, nature reserve, national park, etc. Major kinds of land-use are mainly used in land evaluation studies at broad reconnaissance levels;
— land utilization types (LUTs); these are kinds of land-use defined in greater detail than the major kinds of land-use (see above).

According to FAO (1976), see also Beek (1976), a land utilization type is 'a kind of land-use in a given physical, economic and social setting (current or future) described or defined in a degree of detail greater than that of a major kind of land-use'. It is characterized by special properties so-called 'key attributes' (Beek, 1976).

Although it is sometimes presupposed that a LUT is almost identical with a farming system or an other management unit, those are often composed within space and/or time out of several LUTs. For example the main key attribute of the LUT extensive grazing are:
1. produce: medium (animal species, species mixture, herd/flock structure, etc.);
2. produce: functions and products of media (transport, traction, status, milk, meat, wool, etc.);
3. mobility (spatial and temporal arrangements of grazing orbits, permanency of domicile);
4. land-use rights and land tenure;
5. rights to animals and their produce;
6. size of holding (stock, land, water, etc.);
7. labour:
   — source (family, hired, age, sex, etc.);

— tasks (kind, permanent or seasonal, etc.);

— intensity (hours/holding, hours/output, etc.);

8. market orientation (trade, subsistence, sales, exchange, etc.);

9. income (cash, susbsistence);

10. management:
    — attitudes (production, objectives, etc.);
    — knowledge (skills, education, etc.);
    — technology (kind, levels, source, etc.);

11. capital investments (internal and external investments);

12. infrastructural and institutional facilities (credit, markets, input delivery systems).

At the base of such key attributes one could distinguish land utilization types for extensive grazing.

The main categories of rangeland utilization types (range-LUTs) are:

i) hunting

ii) pastoralism

iii) ranching

These categories may be differentiated as in Table 1 (Ingold, 1982):

Table 1. Main categories of rangeland utilization types (range-LUTs).

Characteristics rangeland utilization types	Access to animals	Access to land	Associated characteristics
Hunting	common	common	
Pastoralism	divided	common	herdsman
Ranching	divided	divided	fences, predator elimination, control of stocking density, meat oriented

The differentiation between two categories of herbivores, 'wildlife' and 'lifestock' is not very fundamental in this respect. Cattle is hunted in some cases; beef cattle on ranches may be pretty 'wild', reindeer may be hunted, herded as well as ranched.

i) Hunting is often a multiple land utilization type on agricultural, forestry or livestock rangeland and may be subdivided in:
   — subsistence hunting;
   — commercial hunting;
   — recreational hunting;

ii) pastoralism can be subdivided according to various 'key attributes', such as produce (a), stock type (b), objectives (c) and mobility (d):

   a) produce:
      — meat pastoralism (small amounts of mik may be used by herdsman);
      — milk pastoralism (including self-evidently meat produced from male calves);
      — wool may be a by-product from one or the other (rarely the dominant produce);

   b) stock type:
      — small stock pastoralism (shoat = sheep + goat);
      — cattle pastoralism;
      — reindeer pastoralism;
      — llama and alpaca pastoralism;

   c) objectives:
      — subsistence pastoralism;
      — commercial pastoralism;
      — amenity pastoralism (preservation of open landscape for tourism e.g. in The Netherlands, also preserve settlements in remote rural areas depending on pastoralism e.g. Scotland, Switzerland);

   d) mobility:
      — village communal grazing;
      — transhumance;
      — nomadism;
      — ranching;*

iii) ranching: here domestic but also wild animals may be kept and managed in various levels of intensity. Ranching was generally introduced in the last century. Ranches can only exist in areas where variation in space

* Ranching is a subdivision according to the main (management) system as well as to the mobility over the territory.

(within the boundaries of the ranch area) is large enough to overcome the variation in seasons. That means the ranch should contain dry season grazing areas, natural or at least (semi-)artificial. In arid and semi-arid areas it can only exist if large areas of many thousands of hectares with suitable land units are included. Ranches can be open or closed (the latter is often too expensive). Compound management systems, cluster of LUTs and multi-purpose LUTs are possible as well as subdivision in seasonal LUTs. See further Zonneveld, 1984 (Addis Abeba) and Weda, 1984 and Siderius, 1984.

It should also be clear that the suitability of a tract of land for a certain land utilization type may depend on the use of the adjoining tract of land and its use. Compare the arable land of the primitive agricultural system in N.W. Europe that demanded the fertilizing by sods and dung from the sheep range system in combination.* One should not call such a compound land-use system a land utilization type (LUT) in the sense of land evaluation, but a management system composed of several (mutual influencing) LUTs. Extensive grazing may be part of such a cluster. Modern agricultural systems tend to have a more monoland utilization type character with all advantages and disadvantages of it. For land evaluation the advantage of such systems is that a farming system, especially if one evaluates it on a not too detailed scale, may be synonymous with one land utilization type and not a cluster as mentioned above.

The same tract of land may be used for more than one purpose. In that case a land utilization type is to be defined as a *multi-purpose LUT*. In forestry it is often the case (timber production, conservation, recreation, sometimes even grazing), see Van Andel et al. in Laban (ed.), 1981. Extensive grazing can be part of such multi-purpose land utilisation types, but then not an independent one. Especially in these cases of complex and multi-purpose land utilization types one needs a good vegetation and land-use map (or land unit maps with proper indication of those land attributes) as a base for evaluation, as well as designing land utilisation types.

The main types of grazing that can be distinguished in extensive grazing according to mobility over the territory differ from each other, because their different adaptions to seasonality in relation to the land factors, especially vegetation quality, in combination with water availability. They can be grouped in a matrix with as one variable scale and predictability (irregularity) of variations in time (dynamics) and at the other side, scale of spatial variation in area (see Fig. 3). The main groups of land utilization types are mentioned below (see also Fig. 4):

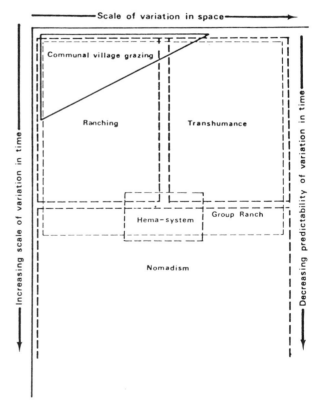

Fig. 3. Territoriality in pastoralism as dependent on variation of the environnement in space and time.

---

* All primitive permanent agriculture depends on minerals brought to it by grazing animals.

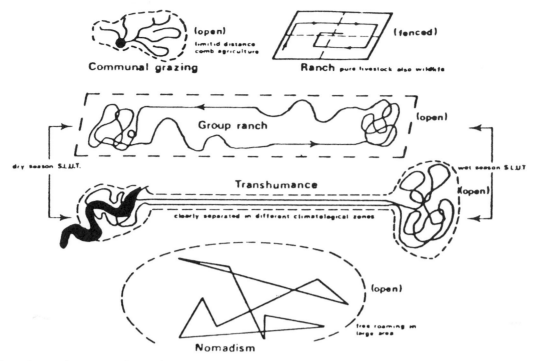

Fig. 4. Extensive grazing systems (land utilization types) mainly determined by spatial and temporal variation in vegetation, and also drinking availability.

— village communal grazing: situated around a village, walking distance usually nor further than 10 to 20 km;

— transhumance: a yearly trek between dry season and wet season grazing areas, or summer and winter areas (in the mountains). Distance may well exceed 100's of km. In one part of the area people may have a permanent dwelling, while they live in the other region in tents or temporary huts;

— nomadism: people constantly roam through the area in search for the scarce vegetation. They may have rather fixed areas, sometimes over many hundreds of kilometers, sometimes also over shorter distances, but their movement is as irregular as the climate to which the system is adapted. They live permanently in tents. They may apply also a certain agriculture, e.g. wheat and barley in the deserts of the Near East and North Africa. In relative moist years they use the grain for human consumption ('blé dur'), in relative dry years they use the crop as fodder for the animals.

Quality and requirements determination

Requirements

Before one can start to translate land properties (intrinsic land characteristics) into 'qualities', one should know the requirements of the selected LUTs. Let us take as an example rangeland evaluation. Here, in contrast to common arable agriculture we deal with a multi-production stage system: primary production (vegetation), secondary production (animals) and sometimes even tertiary production in case of 'tourist industry' based on animal viewing (hunting) and other attractions in rangeland (see Van Wijngaarden in Siderius, 1984).

The requirements for primary production refer to the basic factors of plant growth such as mechanical support by a substratum, not too little, not too much, just enough energy by radiation, kinetic energy (temperature) moisture, minerals, oxygen in soil, genes (available flora to build up a vegetation). Also resistance against too unfavourable quantities of heat (fire), minerals (salinity), mechanical stress (wind, desiccation, trampling, grazing). The land attributes: atmosphere (macro and micro climate), landform, rock, water, soil, flora, fauna; all have to contribute to fulfil these requirements.

The requirements for the secondary production (animals) are related to the type of digestion system of the animal, ('bulk feeders', 'intermediate feeders', 'concentrate selectors'), amount of the forage needed, quality of the forage wanted, frequency needed for watering, quality of the water tolerated, need for shelter, vulnerability to diseases, resistance against climatic hardship (need for shelter, etc.), ability to overcome rough terrain, ability to walk long distances, social animal behaviour (to same species and to other species), tolerance and/or preference to various other environmental aspects.

For a tertiary level, besides requirements for presence and absence of certain factors for animals, there may be also other produce media that should be considered. Rangelands used as national parks or nature reserves may need on top of available animals and plants also qualities such as
— scenery (natural beauty);
— recreational qualities (as e.g. swimming possibilities, etc.);
— conservation needs about rare or otherwise important plants and animals and other natural features;
— natural diversity might be a special requirement.

Before and during the inventory stage one should study these requirements in order to collect sufficient data on land properties related to them.

Land qualities (for rangeland evaluation)

The answer to the requirements are the land qualities. These are listed below separately for the primary and secondary (and tertiary) production level. They are derived from land properties (intrinsic characteristics). *The ones that are directly derived from vegetation survey are marked with**. The ones derived from land attributes partly indicated by vegetation or by integrated land characteristics including vegetation are marked with*.*

A. Primary production level:
   — related to plant growth:
   1. moisture regime*;
   2. temperature regime*;
   3. radiation regime*;
   4. nutrient conditions** and*;
   5. oxygen availability to roots*;
   6. rooting conditions*;
   7. surface sealing (as affecting natural re-seeding);
   8. flood hazard*;
   9. soil toxicities*;
   10. excess of salts*;
   11. genetic potential of vegetation**;
   — related to vegetation:
   12. fire susceptibility**;
   13. ease of control of undesirable species (bush clearance)**;
   14. conditions for hay and silage*;
   15. soil workability;
   16. potential for mechanization*;
   — related to soil conservation:
   17. erosion hazard under grazing conditions*;
   18. susceptibility to trampling*;
B. secondary production level:
   1. forage availability:
   — quantity/unit** (both herbage and browse);
   — quality: digestibility**;
   — palatability**;
   — acceptability**;
   — toxicity*;
   — nutrient value:
     – crude and digestible protein**;
     – minerals (macro- and micro-nutrients);
   — yearly and seasonal variation*: all data for an average rainfall year and for one or two consecutive drought years (defined according to local climatic data);

2. water:
   — quantity;
   — quality (conductively + mineral composition);
   — distance;
   — yearly and seasonal variation (as mentioned above for forage);
3. biological hazards:
   — poisonous plants**;
   — endemic and epidemic diseases;
   — sleeping disease (tsetse);
   — fasciolasis (liver fluke);
   — tickbound diseases;
   — predation;
   — yearly and seasonal variation;
4. climate hardship:
   — temperature/wind/rain, as inducing need for shelter;
   — yearly and seasonal variation;
5. accessibility (inclusive surface conditions):
   — slope;
   — stonyness;
   — hindrance by vegetation**;
   — other land-use*;
   — flooding*;
6. ease of fencing or hedging:
   — availability of natural materials**;
   — ease of establishment;
   — conditions for maintenance;
7. shelter and protection for mating areas (related to 4 and 5);
8. location (mainly in relation to market):
   — kind;
   — distance*;
   — yearly and seasonal variation*;
C. tertiary production level:
   1. scenic aspects*;
   2. recreational aspects (swimming possibility, etc.)*;
   3. easy view places (animal observations) and others*.

Most of these qualities in the above-mentioned three groups are of bio-physical nature: forage quantity and quality, water in all its aspects, pests and diseases, edaphic conditions, erosion hazard, climatic hardship and accessibility. Availability of vegetation data is a must for the assessment of many of these. The presence of political borders, unfavourable legislation in relation to certain practices and land tenure aspects, are purely social of character, but should be taken into account in any realistic land evaluation, even if it is meant to be mainly physical of nature. Especially also present land-use that may interfere with the wanted land utilization type can be considered as a quality or limitation. Vegetation survey may show the extend and interfering patterns thereof.

A very important quality for the secondary production level of very complex character, that has been the concern of all people involved in range management, is carrying capacity. This is the base of the most important rangeland management tool: the stocking rate. The carrying capacity is in fact an integration of nearly all the above-mentioned qualities for the secondary production level. See for detailed discussion Siderius (ed.), 1984.

In most of the land qualities the aspect of variation in time, over the seasons, as well as the different years plays an important role. This is a very specific character of land qualities in relation to extensive grazing. In agriculture and more technologically sophisticated systems one can overcome dynamic aspects of the environment by technological means. In case of extensive grazing one should adapt. Before we have seen already that the main character and differences of the grazing system types are caused by differences in dynamic (variation) in time. The variation in space is used to counteract this and gives rise to such different solutions as transhumance versus closed ranch systems. It may be clear to any surveyor and range manager that the following three scales play an essential role here:
— scale of management unit;
— scale of temporal changes;
— scale of land unit pattern.
So the concept of scale should be clearly presented in quality presentation and also later in the matching with requirements in order to ar-

rive at suitabilities and/or management measures.

Rating of qualities and requirements

The next chapter will describe how requirements and qualities have to be compared. A first condition is a rating and weighing of qualities. About weighing a general rule cannot be given. Rating is often done by estimating a quality in a scale of 5. For grazing purposes in some cases working with the actual dimensions or figures may have to be advocated (e.g. width of zones (distances) to watering points, etc.; see section 'Comparison of the requirements of a LUT with the qualities of a land unit' in this chapter).

Land suitability classification

Definition

Land suitability is the fitness of a land mapping unit for a defined use. The rating of classification of the suitability of a particular land mapping unit depends on the extent to which its land qualities satisfy the land-use requirements. The suitability of each land mapping unit is classified for each land-use that is considered to be relevant for the study area.

Suitability is assessed by comparison of the land use requirements with the land qualities. This can be achieved with actual land qualities, but it is also possible to estimate quantities after a proposed improvement, e.g. the establishment of a borehole, the (re)seeding of a feedlod, etc. This comparison leads to two kinds of classifications for a certain land mapping unit: (i) the classification of the current suitability (without improvements) for a defined land use in its present condition, and (ii) the classification of the potential suitability (with land improvements) for a specified use under future conditions. In types of land utilization for agriculture, intensive grazing, forestry, etc., one distinguishes between major and minor improvements which are well defined (see FAO, 1976). This distinction is irrelevant for evaluation for extensive grazing.

Major improvements would change the land utilization type in such a way it is not extensive grazing anymore. Moreover, an important aspect of the grazing system evaluation is the internal evaluation in order to take measures which are a kind of ad hoc 'improvements' of the situation. The judgment of the improving character of the measure is inherent to the measure and its choice (see section 'Internal land evaluation (for management)' in this chapter).

Land suitability is determined by physical and socio-economic factors (qualities). One can try to match all this factors at once and so arrive at the suitability of a certain LUT in the total physical, social and economic context of a region. For regional planning (in which the grazing systems should fit) this would be ideal. Also for the internal land evaluation, the integrated dealing with physical and economic (and social) repercussions and aims of the measure is to be strived for. However, especially in the regional planning case, social and economic aspects are so complicated that one usually chooses a two-step procedure, in which in the first place mainly physical qualities are considered (the capability of the bio-physical land) and in a following, second step details of economy and social aspects are taken into account. (Compare also Fig. 2 where the different positions of bio-physical and socio-economical land evaluation is shown.)

In range management a special problem is the general land-use policy of a whole nation or even a combination of nations (extensive grazers – often cross political boundaries) has to be considered in such a procedure. All this does not mean that economic and social considerations are absent in the first mainly physical step of the evaluation.

The overall social aspect of extensive grazing in many parts of the world is evident. The classical clash between 'Abel' and 'Cain' is neither purely economic, nor a purely physical one. In judgement of importance of bio-physical qualities and limitation, common sense in relation to economic aspects is incorporated. Theoretically it may be possible to provide water anywhere by car or by pipeline, or even transported from the polar ice, if necessary. In rich countries like Saudi Arabia this supposedly is feasible for not too remote parts of the desert. In Mali, however, it is out of the question to suppose such a possibility. The recognition of LUTs implies already socio-economic considerations. Within that context in practice most land evaluations start with a bio-physical appraisal. Economic evaluation is done later including social consideration when overall output, in alternative LUTs or total systems, can be compared with input using econometric methods. Presuppositions may differ according to the political and social contexts for which they are made.

The multi-production stage character caused by the sequence of primary, secondary and tertiary production, as mentioned in section 'Quality and requirements determination' – 'Requirements', has also an important influence on the suitability classification.

The FAO-framework method, as described here, can be applied most efficiently in the secondary and tertiary production stages. This means that requirements of the LUTs will be compared with the qualities from group B of section 'Land qualities (for rangeland evaluation'. For the primary production one could use the same procedure, however, the data of the primary production can be derived in a different (sometimes parametric) way and may lead directly to qualities (like grazing capacity on fodder basis, etc.) to be used in the second stage.

Comparison of the requirements of a LUT with the qualities of a land unit

The bio-physical requirements of land-uses for which land is evaluated are determined on a preliminary basis at an early stage of the land evaluation process (see section 'Kinds of land-use'). If these data are not sufficient, then during the process of evaluation, more precise information is collected on land qualities, relationships between land qualities and land-use requirements and on possibilities for land improvement.

A 'review' of the available data is generally made during the final stage of the study. This review includes the checking of the relevance and the refinement of the descriptions of the originally broadly defined types of land-use, their requirements and management properties.

Observed relationships between land properties and the input/output of a particular land-use in- or outside the study (surveyed) area determine the selection of the qualities and their values. Exact information on these relations is scarce. Procedures for estimating inputs and outputs are therefore an essential part of a land evaluation study. As to the land evaluation of rangeland this has to be done separately for the primary and secondary production.

The procedures generally include the construction of 'conversion tables' in which critical values of the land qualities are related to different degrees of suitability for the land-use. Conversion tables may be based on (i) quantitative data from the study area, or (ii) empirical assessment of assumed relationships between benefits and land qualities. Quantitative information on inputs and outputs of the land-use may be obtained for different types of land, for instance through direct measurements on trial sites, interviews with herdsmen and local officials or from (local) experiment stations.

Subsequently for each land utilization type the requirements are expressed in terms of the quality figures or classes. Also all the land units are expressed as a list (spectrum) of quality figures of classes. It is now possible to 'match' the 'quality spectrum' of a land unit with the quality requirement spectrum of a land utilization type. This is illustrated in Table 2, in which the qualities are expressed in five classes: 1 is high, 5 is low value. We can consider this table as an example of the evaluation for secondary production. This means that qualities belong to the B group of section 'Land qualities (for range-land evaluation)' see page 508. The meaning of the suitability as indicated on the righthand side is explained in section 'Land suitability classification' page 515.

Table 2. Matching of requirements and qualities to arrive at suitability.

LUT	Required qualities					Suitability of LUT	
	a	b	c	d	e	A	B
A	5	5	4	4	1		
B	1	1	4	4	5		

Land unit	Available qualities					Suitability of LUT	
	a	b	c	d	e	A	B
I	5	5	5	5	5	N	N
II	2	1	4	4	1	S1	S2a
III	3	3	3	4	3	N or S2e	N or S3 (a.b.)

One can use the principle that one limiting factor determines fully the suitability. This may be the case for quality 'e' of land unit III for LUT A. So III is unsuitable (N) for LUT A in spite of the fact that the requirements a, b, c, d are met. One could also make an expert's subjective judgment. If one knows that 'e' is not so important, III is reasonably suitable with some restrictions. (Se2) etc. (see for suitability, section 'Kinds of land-use'). Land unit I is unsuitable for A as well as for B. Land unit III is possible also unsuitable for B (two deficient qualities of the five (a and b)) or at the most marginally suitable, S3. Land unit II is suitable for A and has only deficiency (a) for B, so it is reasonably suitable with some restrictions.

An example of a dominating quality is the availability of drinking water. A high potential carrying capacity based on primary production and other land qualities is useless, if the source of drinking water is too far away. The same holds for absence of cover (bush) for those animals who will never go further than a certain distance in the open field (e.g. the bushbok). This introduces also a specific element in land evaluation for extensive grazing: zones determined by points or lines. So many types of animals have a critical distance to the water or shelter source beyond which they will not move. Therefore around point-shaped sources (boreholes, tanks, small ponds) a circle is created with a radius equal to that critical distance. Outside the circle the suitability of the environment for that animal is zero, irrespective of any other value. For line elements (like rivers) there is a zone and a limit parallel to that river (see Fig. 5).

Obstacles influencing accessibility and any other limitation occurring in the lands units may modify this pattern. The already mentioned demand for nearby shelter may cause similar zones at a critical distance from a dense bush. These zones can be considered as diagnostic characteristics of land units. That means that a certain combination of vegetation, soil, etc., within such a zone will be considered as a different land unit, compared with the same soil-vegetation combination outside that zone. In that case the zones should be clearly indicated on the maps and suitability calculation will be done per zone, then considered as a land unit.

On very small-scale maps the zones may be small compared with mapping units. In that case one may calculate per mapping unit which percentage of the unit is affected by the (e.g. negative) influence of that zone and use those data as a quality (limitation) for the mapping unit as a whole, 100% being the worst, 0% being the most favourable (see Baig, 1977).

For qualities such as biological hazards, accessibility and climatic hardship experience teaches how to rate these for each type of animal, respectively LUT (or seasonal LUT). The simplest rating is: 'yes or not present', the more advanced rating may give certain classes of severeness, the minimum and maximum values of tolerance, and requirements.

The crucial quality is forage availability and its nutritive quality. For this quality certain maximum limits of toxic components or minimum nutrient content (N%) can be used in the quality rating. If from the (S)LUT and possible

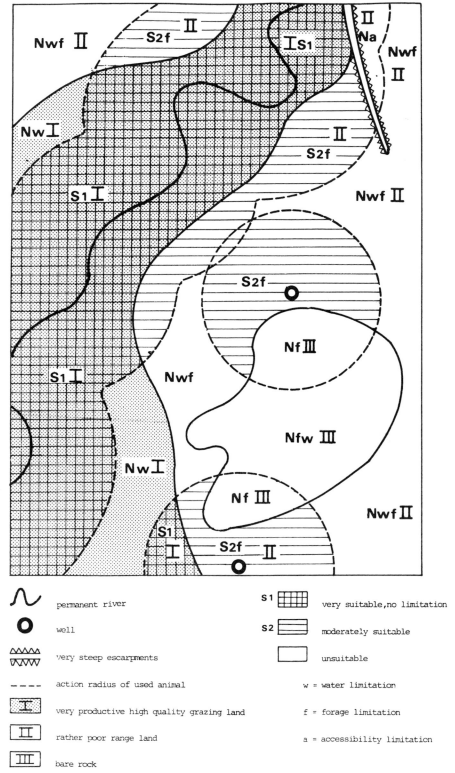

Fig. 5. Land suitability for grazing, depending on: vegetation, productivity and quality, water availability, accessibility, and action radius of the used animal.

other considerations (economic, social?) a minimum production (or number of animals) can be calculated, any amount of forage that is too low to support such a secondary production, can be considered as below the minimum level hence rated as zero. This means also that the quality 'forage availability' should be converted in terms of 'animals/ha' – provided all other factors are favourable. Such a quality is often called: 'grazing capacity'. An explanation of this concept often mentioned as a main practical aim of vegetation survey in its various meanings in relation to suitability classification, is given in the next section.

Grazing (carrying) capacity as land quality

Grazing capacity can be defined as the number of animal units that can graze on a certain area without doing damage, based on the amount of forage available, supposing all other factors are favourable.

'Carrying capacity' is sometimes used in a wider concept. The amount of animal units, (numbers or total production) that can be maintained per area unit as is determined by all site factors (not only forage, but also water, accessibility, etc.). The term grazing capacity is sometimes also used in this way.

In the former section we mentioned already that the forage-based grazing capacity is an important quality in the suitability determination. Such a quality can be derived from actual forage and animal requirements, but can also be a result of primary production evaluation. In that case it may also be a potential value. A good example is the estimation of production of nitrogen (and also phyto-mass) based on various parameters by the P.P.S.-project (Penning--de Vries and Djiteye, 1982; see also Breman, Ketelaars and Van Keulen, Part III and V, Working Group D and others in Siderius (ed.), 1984).

The wider concept of carrying capacity determination is more a complex quality, it can be a full bio-physical land evaluation in itself. An example of such a classification is the one of Thalen, 1979 (see also Baig, 1977).

$$G_a = \frac{(P_h \times p_h \times n_n) + (P_b \times p_b \times n_b)}{R_a} \times$$
$$\times f_1 \times f_2 \times ... f_n$$

G_a = unknown grazing capacity for animal type (a) for a land mapping unit expressed in animal units per unit area;

P_h = production of forage in the herbaceous layer in the land mapping unit;

p_h = proper use factor for the herbaceous layer;

n_n = correction factor for nutritive value in the herbaceous layer;

P_b = production of forage in the form of browse;

p_b = proper use factor for the browse;

n_b = correction factor for nutritive value of the browse;

R_a = forage requirement of animal type (a);

f_1, f_2, f_n = multipliers for relevant land quantities.

The accessibility of the zones due to the distance of the watering points or shelter as a percentage of a land unit, mentioned in the former section may in this formula be found among the multipliers. It should be emphasized, however, that if such a formula is used, this is done instead of a quality comparison according to Table 2. Moreover, socio-economic considerations are difficult to take into account in such a formula. For a (semi)-quantitative approach it is useful to calculate the numbers of animal units in a (possible) area in order to manage the stocking rate. In fact it is a suitability classification based on one very complex land quality (a broadly based carrying capacity). The suitability classification is in this case nothing else than a rating of quality into the arbitrary classes: high, medium or low, with boundaries adapted to the special aim. Especially for an evaluation of internal aspects of a rangeland management system, this may be an efficient approach (see section 'Internal land evaluation (for management)').

Land suitability classification

The FAO Framework does not offer a recipe for land classification for general application. It restricts itself to providing a hierarchical structure for classification systems with possibilities for adaptation to local conditions and objectives. Usually three hierarchical levels of land suitability with decreasing generalization are recognized: orders, classes and subclasses.

There are two basic land suitability orders: S = suitable and N = not suitable. Land classified in the 'suitable' order is expected to yield benefits which justify the inputs without unacceptable risk of damage to the and. Land suitability classes reflect the degree of suitability within orders. The number of classes to be recognizes depends on the purpose and the scale of the land evaluation study. In qualitative studies, for instance, three classes are often distinguished in the 'suitable' order: S1 = highly suitable, S2 = moderately suitable and S3 = marginally suitable. The minimum requirements of the land which determine its classifications in a certain order or class for a particular use are generally shown in a conversion table. Land suitability subclasses indicate the kinds of limitations of land which are classified in classes other than S1. For instance, subclasses S2w and S2d indicate land moderately suitable (s2) because of the not optimally available water (w), and land moderately suitable (S2) because of drainage deficiencies (d), respectively (see Table 3). Land can be classified for its current or potential suitability for a certain use (see above, section 'Definition' p. 510).

These classifications may be qualitative or quantitative. In a quantitative land suitability classification the ratings of the performance of the uses are usually expressed in economic terms.

Presentation of the results

The results of a land evaluation are presented in the form of a report and maps. One of the maps included in the report usually shows the land mapping units together with a legend or table which indicates their suitability for each of the land-uses considered (Table 3).

Responsibilities

External land evaluation is an activity between more scientific and technical work on the one hand and political desiderata on the other. Care should be taken to divide responsibilities properly. Analysis and synthesis of natural resources and the technical methods to exploit them is the task of scientists and technicians, trained in the various land(scape) ecological methods.

Table 3. Example of a generalized classification of land units for kinds of land-use, to be compared for internal and external land evaluation.

Unit symbol	Short land unit description	Kinds of LUT			
		Irrigated foddercrop	Summer grazing	Winter grazing	Production forest
1.	flood plain	S1	N	S1	N
2.1.	South Sahel open grassland sandy soil	S1	N	S2	N
2.2.	South Sahel woodland	S2	N	S2	S2
3.	North Sahel sandy soils	N	S1	N	N
4.	rocky terrain, semi-humid climate	N	N	S2	S1

A special case is the 'appreciating' of land for other than technical or economic or ecological reasons, usually the 'tertiary production stage'. This holds for esthetics, emotional value, historical ties, etc. Here the judgement of those who care for these values should be asked through associatons or action groups dealing with these entities.

The society (consisting of: administrators, politicians and the common citizen) should not interfere. That means surveys should be done by surveyors, the assessment of requirements for land utilization types by experts. However, the choice between land utilization types, and the decision which alternative land utilization types should be considered, is not a privilege for the technicians, scientists and esthetists alone. In case of regional planning it is the government, as representative of the society, that has to decide. It depends on the type of government in which way the 'simple man in the street' (or on the 'land') can influence

this decision-making progress. Surveyors belong in this state to the latter category. In Fig. 6 the role of society, government, technicians and scientists is given schematically.

Internal land evaluation (for management)

For management purposes the classification in degree of suitability is not sufficient. Within a certain land utilization type measures have to be taken for management. The evaluation here deals with the judgement of what kind of measures are necessary (minimum required activity) and the opposite, what should not be done (maximum tolerated activity).

This type of evaluation is done especially for extensive grazing and for conservation purposes, and also for the management of recreation areas, national parks and nature reserves. It deals with the period and intensity of herding or culling, or any other means of influencing the

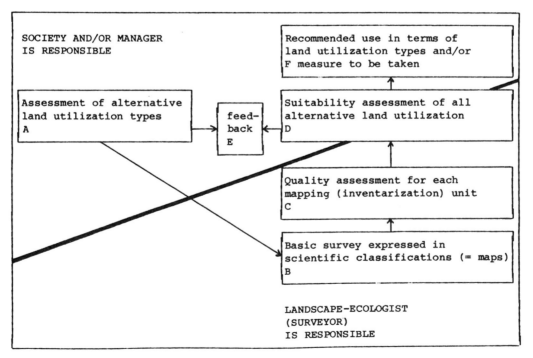

Fig. 6. Steps and responsibilities in land evaluation. In case of internal range management within e.g. a ranch, it is the range manager who has the final decision.

stocking rate in a certain period in a certain area; also prescribed burning or other measures may be planned in such a way.

For each (seasonal) land utilization type evaluation leads to assessment of measures (if minimum requirements are not fulfilled but may be supplied) or assessment of limitations that may lead to the non application of certain utilization types. For these types of management, especially also for conservation management, the overall land utilization type is a given fact. In this case evaluation does not lead to a choice between LUTs, but to measures to be taken to keep favourable factors above the minimum required level and the unfavourable ones below the maximum tolerated ones. Vulnerability and required action is to be evaluated before action can be taken. Pollution, lowering ground-water, uncontrolled fires, excess of people or overgrazing are common factors exceeding the tolerance limits in nature reserves in addition to oxygen, minerals and water in soil, etc. Adaption of the stocking rate is a common tool in management. Calculation of the acceptable stocking rate as given in examples in section 'Grazing (carrying) capacity as land quality' is a useful evaluation tool (for more details see Siderius (ed.), 1984).

38. Some Examples of Application

A.W. KÜCHLER

Vegetation maps show the location, extent and geographical distribution of units of vegetation and indicate directly or by implication something about the nature of the sites on which the vegetation units occur. The significance of such maps lies in the fact that they portray areas of equal biological potential or what Rey (1962c) called 'isopotential zones'. Such zones consist therefore of ecologically equivalent biotopes and form the natural divisions of the landscape. These, in turn, can be read directly on a map of the natural vegetation. The important point here is that those in charge of land-use should have a map that reveals the areas of biological potential. Whether these areas happen to be occupied by one or another kind of substitute communities at a given time is of secondary importance because even the cultural vegetation fits neatly into the framework of the natural phytocenoses, and the laws of phytocenology apply just as much to cultural vegetation as they do to natural plant communities. In the words of Molinier (1951): 'Phytocenology does not perform miracles, it does better than that. A miracle, by definition, is both incomprehensible and exceptional. But phytocenology is in the front rank of all considerations concerning land use because of the possibilities it makes available to man.'

Agriculture

The farmer is becoming ever more intricately enmeshed in a technological economy over which he has no control. Revolutionary changes exert their increasing pressures on him from all sides, and more and more he comes to rely on experts in specialized fields, scientists in agricultural experiment stations, chemists and geneticists, technicians and government agents. In addition, he is harrassed by hordes of representatives from commercial concerns who explain to him why he can no longer afford to do without their goods, and these goods come in a bewildering variety of machinery, fertilizers, herbicides, pesticides, crop insurance and feeds with chlortetracycline, sulfamethazine, and a host of equally baffling admixtures.

The one stabilizing factor in the farmer's life is his land, the biotopes on which he produces his crops. The land is there, season after season, awaiting his treatment and his care. It is the most critical factor in this whole business of farming. Vegetation mappers have become deeply involved in agricultural problems as their maps reveal the quality and condition of the various biotopes. These maps apply to all agricultural pursuits, be they field crops and their respective weed communities, orchards, vineyards or market gardens, etc.

In spite of the use of herbicides, weeds remain common. Like other plants, weed species do not grow just anywhere, and they, too, occur in definite combinations. Weed communities may be better indicators of the conditions of the cultivated soil than the crop plants, revealing areas of equal potential yield, whether

A.W. Küchler & I.S. Zonneveld (eds.), Vegetation mapping. ISBN 90-6193-191-6.

of crops already established or yet to be planted. Weed communities must be observed quite comprehensively and completely because many weed species are pioneers with wide ranges of tolerance for environmental conditions (Ellenberg, 1950).

The more intensive the farming, the finer is the coordination between the farmer and his land. The vegetation mapper cannot tell the farmer how to farm but he can tell him which sites may be considered alike and hence likely to succeed (or fail) uniformly throughout the combined areas of these sites. The farmer can then decide how to proceed under the given circumstances.

The usefulness of vegetation maps in agriculture becomes more obvious where perennial crops are produced, as in orchards etc. Tropical tree and shrub plantations of coconut palms, oil palms, coffee, rubber, cacao, bananas and others parallel the production of apples, cherries, apricots, grapes, etc., of the middle latitudes. Such pursuits are usually considered an intensive form of agriculture but the areas involved in individual establishments may be large in all latitudes. The giant fig and peach orchards of up to 16,000 ha in the Central Valley of California are a case in point.

The masterly contribution by Ellenberg and Zeller (1951) on Kreis Leonberg represents one of the most successful techniques in relating vegetation mapping to agricultural pursuits. One of the stated goals of this work was to assure the farmer the greatest possible sustained yields, and hence income, on the basis of a rational ecological use of all available biotopes. It has remained unsurpassed among the scientific publications of its kind.

There are many examples showing the intimate link between agriculture and vegetation maps. Just one of these may suffice to illustrate the usefulness of such maps. *Suaeda fruticosa* dominates the vegetation of parts of southern France. Light cultivation is carried on profitably although this perennial indicates a slightly saline soil. It was found that this same species dominates the vegetation in parts of Algeria and this was taken to mean that there, to, the land is suitable for light cultivation. As a result, elaborate plans for large scale settlement were prepared and pushed forward to the point of laying out streets, building houses for settlers and constructing public utilities. The failure of repeated attempts at making this land productive cost the government as well as private individuals a fortune. Afterwards vegetation maps of both areas were prepared and they revealed that the vegetation of one area was quite different from that of the other, and that the dominant *Suaeda fruticosa* was one of the few species that occurred in both. It so happened that this species has wide ecological tolerances and is therefore unsuited to serve as an indicator. It was a powerful proof that environmental features should not be evaluated by the presence or absence of one particular species but always and exclusively by the character of the entire phytocenose. It also emphasized the need for preparing vegetation maps in the earliest planning stages rather than much later or not at all.

Grasslands

Grasslands represent a herbaceous type of vegetation with its own characteristics. Their distribution in the landscape and their yields reflect a close relationship to their respective biotopes more readily than cultivated fields do. Quite in general, the floristic composition of grasslands is the result of natural selection in which only those species survive that can compete successfully on the given biotopes and at the same time withstand the harsh treatment by man and beast. Grazing and trampling, burning, mowing, and sometimes fertilizing may all be practiced individually or in combination, but in the course of years, the ranges become adjusted to such conditions and can be remarkably stable. For instance, the tall grass prairie of eastern Kansas can maintain its character and high productivity more or less indefinitely provided the owner maintains his herds well within the carrying capacity of the range in rainy periods as well as in periods of prolonged drought.

Vegetation mappers often find that they must map range lands at large scales because the herbaceous vegetation responds readily to minor variations in site qualities (cf. Chapter 22). Until fairly recently, the vast ranges of Argentina, Australia, the Soviet Union and the western United States, to name but a few, were relatively natural, having evolved through very long periods. Intensified use, especially during the last hundred years, has profoundly affected these grasslands, and all too often, the change has been for the worse. The pastures were overgrazed, undesirable species immigrated and spread at the expense of desirable ones, some of the top soil was carried away by wind and water, and altogether the value of the great ranges declined.

It is now possible to establish within reasonable limits the potential natural vegetation of an area. The requirements and ranges of tolerance of species can be determined, as well as their nutritive value. It is therefore possible to observe quite accurately the degree to which a range has been damaged. A vegetation map can now reveal the carrying capacity of different parts of the range according to the conditions and the quality of the component species. It can also help a great deal in rehabilitating the run-down ranges and in maintaining a high level of production by showing the character of the sites, what species to employ in reseeding the range, what plant communities to promote in order to control undesirable species, and to assure a maximum water supply. This last point is particularly important in most of the great grasslands of the world, and the water relations of herbaceous plant communities stimulate some of the most significant and rewarding studies. These include such topics as the calibration of phytocenoses with regard to precipitation and run-off, the water-holding capacity of the soil, and the quality, fluctuations and movements of the ground water.

Vegetation maps therefore can reveal directly the distribution of herbaceous plant communities characterized by a given quality and productivity.

Molinier (1951) reported on the use of vegetation maps in the Crau region of southern France where hay of a particular type and quality is highly esteemed and which therefore commands a high price. Gradually, hay from other regions and of inferior quality was sold under the name of the better hay. The government was asked to regulate the matter, and the result was a vegetation map which indicated the floristic composition of the meadows. One could see at a glance where and only where the superior hay could be produced. The name of the better hay is now restricted to the proper areas.

Forestry

Modern silviculture is fast becoming a form of applied ecology. Only a comprehensive recording and mapping of all synusias by growth forms and taxa will permit an adequate recognition of phytocenoses. This implies specifically that foresters must map not only trees but also what is underneath: the shrubs and forbs, including even the mosses, etc. Of course, foresters are primarily interested in trees, but tree species often have broad ecological tolerances and will occur on various biotopes. It happens therefore quite regularly that the finer distinctions between site qualities are shown more effectively by the herbaceous synusia. Such complete mapping applies not only to more or less natural forests but also to semi-natural ones and even to plantation forests.

In contrast to agricultural crops, forest trees mature slowly, taking 30 or many more years. Foresters must therefore know how to manage the forests, as a mistake may not be recognized until many years after planting time and the resulting losses can be severe. No quick correction is possible for any change of direction again takes long periods before producing the desired results. The forester is therefore extremely interested in the character of his area and he must know climate and soil in detail and as comprehensively as possible. As the environment cannot be grasped in all its facets, it is both simpler and more useful to map the vegetation. Plant communities portray faithfully whatever benefits an environment may have to offer, as well as its limitations. Large-scale vegetation maps showing trees, the undergrowth and even the ground cover are the surest device for understanding the nature of the local environmental conditions. With such vegetation maps at hand, the forester can avoid many mistakes and at the same time manage the forest for a sustained maximum yield.

Westveld (1954) observed that the ground cover under the trees is of particular value to foresters. He says: 'A ground-vegetation type map is the practical equivalent of a forest type management map. Therein lies its real value, for ground-vegetation types can be translated directly into site potentialities. Not only do vegetation types provide the index to timber yield potential, but they also provide a sound basis for proper orientation of silvicultural goals. Knowing where the most favorable site types occur for the growth of specific tree species is one of the principle requisites of profitable timber production.'

For example, the ground-vegetation type equivalent to the red spruce-yellow birch type of the northeastern United States is characterized by *Oxalis-Cornus.* Even should the present tree cover, owing to past cutting practices, consist mainly of hardwoods, the presence of *Oxalis acetosella* and *Cornus canadensis,* key indicators of softwood sites when they occur in combination, warns the forester against managing primarily for hardwoods. Here silviculture should be aimed at reestablishing the normal softwood representation.

A leading forester of Oregon considers vegetation maps one of the best investments foresters can make. He found that a vegetation map is to the forester what a house plan is for a builder, permitting a detailed analysis of every part of the forest. He also observed that vegetation maps eliminate guess work and save large expenditures by showing location, extent and outline of areas involving pest control, volume estimates, fire control, thinning, etc., and discovered that a vegetation map serves many purposes unsuspected before it was available. Mraz and Samek (1963) confirm that vegetation maps are economical even when their preparation is expensive.

In many parts of Europe, vegetation maps have become essential tools in forestry, and in California a state forester concluded that silviculture without vegetation maps has become unthinkable.

Examples of additional applications

The significance of vegetation maps for such fields as crop production, pastures and hay meadows, orchards, plantations and forests is obvious as they all relate to some form of natural or cultural vegetation. But, as mentioned in the preface, the circle of users of vegetation maps has widened greatly in the course of the last half century.

Climatology, geology and pedology have been and will continue to be important in the study of the landscape, and phytocenology is simply a later addition to this group of sciences that permits us to appreciate the intricacies of our environment. But phytocenological methods are superior in analyzing the complexities of sites and site types because they make vegetation the basis of investigation, and it is, after all, the cultural vegetation and its welfare on which man's interest is now so eminently focused. Plants are rooted in the soil and exposed to the daily weather conditions of all seasons and can therefore report the nature of the environment much more comprehensively than any instruments ever can. The presence or absence of given species, their appearance, the pattern of their distribution, their association with other species, and many other features tell a story which is of fundamental value to all who deal with the land and what grows on it. Indeed, nothing permits one to appreciate the character of the land more readily and completely than vegetation itself.

To illustrate: it is difficult, even for trained pedologists, to establish the areal extent of wet soils of similar nutrient value. An endless number of chemical analyses is inevitable. But the vegetation reveals at a glance the entire environmental complex, including the soil type, the physical and chemical characteristics of the soil water as well as the climatic and biotic features of the habitat.

While vegetation maps reflect the qualities of the environment, they can also serve quantitatively with regard to individual features of a site. Such a site feature is then calibrated, i.e. related to different plant communities according to the degree of its intensity. Such calibration is done by carefully recording the plant communities in the field and by measuring the site quality in every sample plot where the vegetation is being recorded. The floristic tables of these communities are placed in order of increasing or decreasing values of the site quality. It then becomes evident which species combinations correspond to which values of the site quality. As a result, given species combinations portray a given feature of the environment quantitatively. This opens new vistas for vegetation mapping with calibrated units of vegetation which reveal clearly outlined areas of given environmental values (Tüxen, 1958; Walther, 1957, 1960).

As vegetation is so closely related to its environment, it is not astonishing that authors should attempt to focus on such relationships. For instance, Küchler (1963b) distinguishes between aclimatic, quasi-climatic and climatic vegetation maps. The difference between these is the degree to which vegetation types are related to climate: not at all in the first case, occasionally in the second case, and systematically in the third case. The last case is, of course, the most interesting, and the maps by Ellenberg and Zeller (1951), by Holdridge (1959), and by Gaussen (1948ff.) are good examples.

The intimate relationship between vegetation and the substrate has been demonstrated many times (Küchler, 1967a, 1975b). Vegetation and soils therefore show rather precise parallels with regard to their distribution. However, such a close correlation depends on the features of the vegetation and of the soil that are mentioned in the legend of the map.

An excellent example comes to us from Story (1963) who worked with the staff of the Commonwealth Scientific and Industrial Research Organization (CSIRO), and who mapped the vegetation of the Hunter Valley region in New South Wales. On the Land Systems maps of the CSIRO, vegetation is combined with soil, relief, geology and certain aspects of the climate. Every unit on the map describes all these features. It is therefore an easy task to produce a vegetation map of this area by simply omitting any reference to everything but the vegetation. The same can be done with the soils. The result would be two maps, one showing the vegetation and one the soils, and they would agree in every respect. It so happens, however, that there are two large inset maps, one of the vegetation and one of the soil, and these two maps disagree considerably! They reveal that a change in a particular feature of the vegetation corresponds to a change of a particular feature of the soil. If such a feature is not mentioned, then maps of the vegetation and the soil cannot be expected to agree. Thus two different phytocenoses may indicate a difference not reflected on the soils map. The maps cannot explain such a difference if, for instance, it is caused by a high water table in one case and a low one in the other. If the water table is not included in the environmental features shown on the map, then the harmony between vegetation and soil can not be demonstrated.

There is also a considerable correlation between vegetation maps and geological maps. However, disagreement is common because the latter show the age of the rocks rather than their quality. A lithological map is much more revealing to phytocenologists for it makes a great deal of difference to plant communities whether the substrate is limestone or shale. Whether such materials are of triassic or cretaceous age is largely irrelevant. A good and detailed lithological map greatly promotes the collaboration between phytocenologists and geologists.

The map is the geographer's most important tool, and he more than anyone else will appreciate Schmithüsen's (1942) observation on the particular advantage vegetation maps have over other maps. This great advantage consists in the fact that the mapper can actually walk along the boundaries of every plant community, and outline and establish the actual area occupied by a given vegetation type. This areal feature is rather unique because climatic maps, for instance, are based on data obtained at a limited number of points of observation. Similarly, geological and soil maps rest on inspections of occasional outcrops, drillings or borings. Only the vegetation units are open to and based on areal observation. This is a significant advantage in any regional or landscape study.

A biotope is the smallest component of the landscape. It has comparatively uniform features throughout its area and hence a relatively uniform type of natural vegetation. Every biotope implies a certain potential for plant life in general and for land-use (economic development) in particular. For each type of natural vegetation there are one or several crop types for optimal land-use. Man devotes his land to a variety of crops but whichever he chooses, each of these will have certain ecological features that correspond to the type of natural vegetation on the site of which they occur. A map of the cultural vegetation can therefore reveal site qualities nearly as well as a map of the natural vegetation.

Schmithüsen (1942) showed that at times a map of the cultural vegetation can reveal the contrasts between biotopes even better than the natural vegetation. For instance, the forest of an alder bog can merge rather imperceptibly with forests on the adjacent drier sites. But in the cultural landscape, a meadow replaces the alder bog and then the moist site will contrast sharply with the surrounding forests or fields with their lower water table. As land-use becomes more highly developed, more complex, and more carefully attuned to the physical and biological features of the landscape, the natural biotopes are reflected in the cultural landscape with increasing clarity and in ever finer detail.

The insight into the landscape achieved through maps of natural and cultural vegetation can be deepened considerably by making comparative studies of two or more regions. For instance, in many respects, the wormwood steppes of the Aralo-Caspian plain are startlingly like the Great Basin sagebrush; the relations of the California chaparral to the Corsican macchia have been pointed out many times, the paramos of the northern Andes reappear on the East African volcanoes, and the Dakota prairie has its counterpart in the Ukraine. Such comparisons can be refined by observing on the vegetation map how the phytocenoses are related to soils, microclimate, relief and land-use. Such comparative studies result in a greater appreciation of the characteristic features of the geographical region and reveal new problems, stimulating research. Whatever its scale and purpose may be, an accurate vegetation map is essentially a significant geographical achievement because of the sensitive relations between vegetation, climate, soil and land-use. Indeed, Krause (1955) emphasized that the greatest success in the interpretation and use of vegetation maps may be expected from one who possesses a thorough knowledge of the nature of vegetation and who, simultaneously, is a comprehensively trained geographer.

Another expanding application of vegetation maps occurs in many so-called 'underdeveloped' nations. Vegetation mapping is nothing new in such countries as the former colonial powers have often done a remarkably fine job in mapping the vegetation of their possessions. Nevertheless, the need for such maps continues, and the comments by Devred (1960) and Lebrun (1961) on mapping the vegetation of Zaire are enlightening. A number of small-scale vegetation maps have been published with scales of less than 1:100,000. Their purpose is to serve research, to define and to delimit the areas where the vegetation can guide land-use planning, to reveal the general relations between vegetation and the natural and anthropic features of the environment, and to formulate projects for the exploitation of the plant resources. Such maps also serve as a basis for the introduction of crop species from other parts of the world as they show the ecoclimatic analogies between the countries of origin and the various regions of Zaire. Finally, these maps are the best means to show and, indeed, to define the isopotential vegetation zones of the country.

The large-scale maps of Zaire are prepared in areas selected on the small-scale maps; they serve as a basis for detailed planning. This is concerned with the production of industrial and food crops, tree and shrub plantations of oil palms, rubber, coffee, etc., the management of natural, improved or artificial pastures, forestry, the location of villages, social, commercial and industrial centers, agricultural experiment stations, the study of water resources, the stabilization and rationalization of agriculture and erosion control. Even around urban centers, large-scale vegetation maps are useful in selecting sites for parks, a suburban belt of market gardens and intensive livestock production, mixed agricultural and residential areas for the families of industrial or retired laborers, for forests devoted to the production of firewood, charcoal, etc.

Public and private agencies everywhere, concerned with the construction of dams, canals, reservoirs, and irrigation or drainage projects frequently pay large indemnities to land owners on whose land the level of the water table has changed as a result of these activities. In such instances, careful planning calls for vegetation maps, to be prepared before the projects are started and again several years after their completion. The vegetation maps will reveal the character and exact extent of the change. Sometimes the supposed damage does not exist, or it is limited to small areas. In some cases, there may even be an improvement with greater yields than before. Ellenberg (1952) illustrates this problem: the Aller, a tributary of the Weser River, was regulated by making one cut across a loop and by erecting four barrages. Vegetation maps prepared both before and after the regulation of the river revealed the changes in the depth of the water table and the subsequent deterioration or improvement of the pastures and hay meadows. Then, in the same valley, another barrage was built but, surprisingly, the resulting changes in the depth of the water table could not be traced on the vegetation maps. Investigations revealed that, both before and after construction, the water table was so deep that it always remained out of reach of the shallow-rooted grassland communities. All indemnity claims could therefore be dismissed! In a similar case near Braunschweig, the use of vegetation maps saved the government indemnity payments amounting to approximately £ 240,000 (Buchwald, 1951). Vegetation maps are therefore valuable tools in determining a fair compensation for possible damage to fields and meadows resulting from public works, and more than pay for themselves.

Finally, a word about the application of vegetation maps to military considerations is appropriate. Military men must be able to act promptly under emergency conditions and must plan for these. They are concerned with the movement of men and equipment, and such movement must proceed as rapidly as possible even though regular communications may not be available. Furthermore, such movements may have to be concealed. Cross-country movements and the aspects of camouflage and concealment create a vital interest in many aspects of vegetation, especially its physiognomy and structure. Obviously, a forested mountainous terrain presents aspects totally different from those in a vast plain of grasslands. One need only compare the immense distances covered by the campaigns on the Russian steppes or the Libyan deserts during World War II with the bloody and obstinate battles for every yard in the jungles of Guadalcanal or in the Hürtgen Forest on Germany's western frontier.

From a military point of view, the vegetation map offers the great advantage of allowing an officer to study the vegetational features of the terrain. He can therefore plan his operations with much greater confidence of moving efficiently while avoiding dangerous obstacles and surprises. For instance, lateral visibility is often significant and can frequently be recognized on vegetation maps with a surprising degree of accuracy because such maps, especially when they show the physiognomy and structure of vegetation, indicate the height and density of the plant communities, the density of a forest, the amount of undergrowth, how 'open' shrub formations are, and the like. An officer also wants to be aware of all kinds of vegetational obstacles in the form of closely growing trees, lianas and vines, swamp vegetation or jungle-like thickets when light or heavy equipment is to be moved. Such obstacles to traffic can also be gleaned from a vegetation map.

The complexity of vegetation must be related to the complexities of military activities, and a system for describing the vegetation must fulfill all requirements of these activities. The requirements make it obvious that a floristic approach to vegetation is not practical. From a soldier's point of view, it makes no difference whether a forest is composed of red oaks (*Quercus rubra*) or of black oaks (*Quercus veluntina*). Only the physical character of life forms can be considered as it alone affects military activities. Thus it may be vital to know whether the trees are so close that the equipment cannot be moved through the forest, or whether the trees are so young and small that tanks can knock them down.

The following features of vegetation, called 'prime parameters', are considered to have potential military significance:

1. growth form;
2. height;
3. coverage;
4. stem characteristics;
5. branching habit;
6. root habit;
7. foliage characteristics;
8. armature;
9. distribution;
10. stem spacing.

These prime parameters can be broken down wherever more detailed divisions and subdivisions are useful. Other problems concern the relation of vegetation to food supply and health aspects. For instance, some potential food plants have edible leaves and poisonous roots, others have poisonous leaves but edible roots, still others are poisonous when raw. The problems are numerous. The description of vegetation based on the above system is designed to serve military interests. In some respects, this leads to some unconventional interpretations. For instance, a large snag (dead tree) is considered equivalent to a deciduous tree during the period of defoliation; a deciduous tree is said to have a leafy crown even though the tree has shed its leaves if it plays host to a large number of evergreen epiphytes; slash is considered a transition between vegetation and microrelief!

39. The Outlook: Future Needs and Possibilities

I.S. ZONNEVELD

Needs

Since a few thousands of years the natural occurring changes in the vegetation cover of the earth were relatively strongly accelerated and deviated by human influence, especially in the (semi-) arid zones and the limited occupied areas elswhere; but never in history the alterations were so intensive as the last half of the century. Now especially the humid tropics and also the temperate zones are the theater of changes. If the tropical rain forests are dwindling away and elswhere the natural forests and shrublands and grasslands are increasingly influenced by man, what is the need for vegetation survey at this era of human dominance over the world ecosystems? This need is higher than ever before.

According to the recently most common estimations, the world population will increase the next century to the maximum amount that the world can support by the combination of ecological sociological, technological, political, economical factors, being about 12 billion which is three times as much as the present (1985) population.

Also that people will have to live from the food and fibre, produced by the vegetation cover and to rely on the stability of the environment caused by the vegetation as consolidating landfactor.

Still at present a large part of the green cover is subject to destruction, it needs repair. In order to support three times as many people as at present, one needs to 'improve' it (make it more suitable than it is now for that purpose) and most of all it needs to be managed to keep it in the best condition, and prevent any future degeneration or destruction.

The stabilization of the world population density is a proces that will be effected by a system of forces of which only one will be the increasing effort or even difficulty to produce more food and fiber. Other forces are the problems to find sufficient means and space to dump sewage and other disposal, to find sufficient space and environmental quality for other human activities than those induced by the basic food and reproduction instincts, for 'recreation in the cultural sense'. Also to preserve areas were human activity has to be excluded in order to let other aspects of the world ecosystem (the 'creation') survive.

It is certain that this play of forces will affect the vegetations. This influence will depend also on the mechanism that will directly cause the decrease of biological reproduction: either birth control or catastrophies. At present both are taking place side by side. Catastrophies are partly due to destruction of the vegetation cover by a combination of normal climatological factors and extreme human overuse (sahel, floods). What, however, may be the case or procedure, one thing is sure, we should monitor (that is to watch in order to warn about) the precious blanket we all depend on, the vegetation.

The only means to monitor effectively the vegetation is repeated vegetation mapping of at

A.W. Küchler & I.S. Zonneveld (eds.), Vegetation mapping. ISBN 90-6193-191-6.

least those vegetation features that are crucial to base the warning on, in terms of what should happen.

Secondly we need to repair destroyed parts of that precious cover, and if possible we need to improve it.

Mending a tissue or improving it, requires good knowledge about that cover and again watching how the process of repairing or improvements proceeds. Sequential vegetation mapping before, during and after that process is a logical aspect of this.

All these activities about repair and improvement are already common work for vegetation surveyors. A third task is mapping as a base for management, to keep the optimum situation constant.

Whatever the character and status of the vegetation will be, how strong the pressure of man on the green blanket is, knowledge of that vegetation, especially in its spatial pattern and geographical distribution and made available in geo-information systems like maps and other devices compatable for planning purposes, is a conditio sine qua non for good management. Vegetation mapping of any kind including formal land-use (besides functional aspects of utilization of land) will be an essential tool to increase that knowledge.

The various chapters of this handbook may give together a review how this tool could be used for the final aim.

Beside this aspect of need of vegetation mapping as base for management of vegetation one should also not forget the more purely scientific aim 'to know for the satisfaction of to know'.

There is no strict limit between the various ways of usefulness of knowledge, between pleasure and necessity. But any cultural human has a need to know the character and the background of the main element of his environment, the vegetation. Beside the actual aspect especially also the relicts of the past have his interest. It is very important that especially that relicts are studied by the methodology of vegetation survey. For the pleasure of to know, but at the same time as a means to learn from the past about the possibilities of the future. Any knowledge of (also pure) natural ecosystems with the vegetation as the core of the biosphere on which animals and man depend can also be of use for the material aspect of human life.

Possibilities

We have discussed in this book that since a few thousands of years man has recorded his environments in the form of maps. The past century classical cartography came to florish lush and bright and also vegetation maps appeared. This book is an evidence about the possibilities.

Geographers took the lead in producing maps with vegetation data and especially the last half a century vegetation science developed and vegetation scientist became active in producing an abundance of vegetation maps.

Never in history of the world the possibilities for vegetation mapping has been so large as at present.

There is a large number of vegetation scientists and students who want to become so. Managers and planners are becoming aware of the practical usefulness of vegetation and land-use maps.

Modern means of transport, but especially observation platforms varying from lightweight aircraft to satellites, take the place of the mountain tops from where earlier surveyors (comp. Mozes from the mount Nebo about 1500 A.D.) tried to do some remote sensing.

The hand drawn and -painted final map later replaced by sophisticated multicolour printed products is gradually making place for electronic computer stored data banks, that can produce any map information in an integrated but also in an analytical form. The earlier maps were most costly products, in price comparable with the 'Old Masters' paintings as Rembrandt and Michael Angelo, and also almost as beautiful, but because of price and procedure of achievement not easy to reproduce and to be used at the spot where they are needed.

More reproduction means have bridged already the gap between cartographer and user and brought maps already directly in user's hand as a routine product. The future developments will make the manipulation of geobased data even easier via the land information systems of which vegetation and formal land-use data are dominating ones. Those planners and managers who up till now may not have seen the importance to take recent vegetation data for their work, may now start to use them because of the relative limited effort needed.

In our enthusiasm about modern developments in techniques applicable for vegetation survey, we should, however, be aware of science fiction. The last decades we have seen enough of this around the modern remote sensing techniques.

Technicians, with sophisticated knowledge of electronics but not hampered by too much view on ecology and land cover, propose already now so-called 'integrated systems' where remote sensing data fully automatically can be lead into geo-information systems (GIS) where all knowledge of the world ecosystem, needed for solving problems, would be integrated and stored. One should not forget that computerized geo-data banks in essence never will be able to do much more than adding and subtracting with some cybernetic feedback multiplications here and there. The 'noise' generated by the non fully 'registerable' images of different thematic maps of the same area that should but do not coincide, give due to orientation, observation and classification errors already large technical problems to be solved. In this book (Chapter 14) is stated that, contrary to probably other GIS literature, the benefit of electronic geo-information systems is not *only* even not mainly that separate gathered information can be 'integrated' (read 'added'). An important application will be that from a large pile of integrated land information, such as holistic land-unit maps, but also vegetation maps and soil maps with holistic (that is in a systematic classification expressed) information, via electronic geo-information systems separate parameters (needed for the solving of special problems) quickly can be derived in map form.

Real integration of land problems can only be done by the human brain because there are to many factors involved that cannot be programmed before hand into a computer 'brain', for the simple reason that the land ecosystems consisting of so many interwoven subsystems, are so very complicated, the factors so interrelated and virtualy so little known. So in many cases integration of land data should be done during the survey before storing the data in the electronic information system databank.

This means that remote sensing and geo-information systems are most valuable tools, especially if they can be used interactively. They facilitate work, they may save time or at least increase production quality and possibilities per time unit. But they can never, in a routine survey and nowhere replace the brain of the scientist.

This does not mean that a certain automation will not be possible. This, however, is only true in case simple single parameters are of interest. So satellite images could be automatically transformed, even already in the future satellites before radiated back to receiving stations, into vegetation cover maps or into so-called 'biomass' maps. Seasonal changes can be recorded in this way as well as gradual changes over the years.

It is even not to sophisticated science fiction to suppose that it will be possible to have in the not too far future your weekly vegetation over image in colour on your TV screen, say each Saturday, in the same way as we now, since a decade, are used to our daily weather satellite image. For areas where rangelands are of common interest it would be at least an issue for television companies to think about.

Such automated maps will never be complete vegetation maps. The floristic data will have to be collected in the field as is done already in many places for a routine on benchmark plots, once a year or with a shorter or longer interval as wanted.

All these activities, for the interpretation of the single parameter maps as well, are vegetation scientist jobs, in the same way as a technician cannot replace the meteorologist in inter-

530

pretation of the weather satellite images. For all kinds of mapping of vegetation, be it classical floristic structural maps or just simple single parameter maps, the vegetation scientist has to be there and not only in the beginning making the programs for the machines. He (she) has also to guide and watch and manipulate the tools used, and moreover has to be in the field in the middle of the vegetation cover itself regularly and intensively.

This brings us to another constraint in the development of vegetation mapping in the future, in the field of science policy. Vegetation mapping has to be done by vegetation scientists (assisted by technicians). Vegetation science is a science on a high integration level. Especially vegetation surveyors are (should be) scientists of a generalizing kind.

To be a 'generalist' or a specialist depends probably on original character as well as from education and training. In all science centres in the world the antagonism between specialist-analyst on one hand and 'generalist'-integration minded people on the other is evident through the centuries. Both ways of scientific activities and attitudes are necessary. The specialist way sometimes seems (superficially) more exact, leads easier to concrete results, is often easier in approach and therefore probably, anyhow, has the largest number of applicants.

In a democratically ruled system the danger exists that these people will rule out the 'generalists' (there are many examples in past and present about this).

Vegetation scientist and all other representatives of sciences on a higher integration level, should therefore continuously be aware of this danger and defend their science against unsound attacks. It is especially the applications where high integration level sciences have an important task, so also for the direct applicability of science the development of high integration level scientists is a must.

A relative new development that may induce a certain optimism in this context is the increasing interest in general system theory, the more mechanistic part as well as the more holistic philosophical side of it, but especially also the land(scape) ecology.

At this place we shall only mention the various survey and mapping activities related with this concept. Landscape ecological maps, land unit maps, 'land system maps' produced by sometimes so-called 'integrated surveys', having a legend compiled with data as well on geomorphology, soils as vegetation and land-use are in the field of land planning and management development taking the place of monothematic landform, soil and vegetation maps. By certain vegetation (as well as soil- and landform) mappers this can be seen as a competition. They would prefer to make vegetation maps.

In foregoing chapters, however, we have already seen how related the methods and techniques are of making soil, landform or vegetation maps if remote sensing means are used. We have also seen how in the modern geo-based information systems all land data can and should be stored in an integrated way and how the systems can be used to abstract from the total that what should be used at that moment. That may be a vegetation map or only a map of one vegetation parameter like cover or a complete land unit map.

We see that the technical means, both of recording as well as interpretation as well of map and information system making, lead us to multidisciplinary (to integrated) approaches forcing the specialists to work more together. Therefore even in a book on vegetation mapping there is no need to try to make a plea for keeping vegetation mapping pure. Pure vegetation maps will indeed remain necessary for various purposes at least educative ones. But the essence of the vegetation mapping technology is as much a part of more integrated multidisciplinary activities that may be named land(scape) ecological surveys. Vegetation surveys will have an even more fruitful task if working in an integrated team together with soil scientists, geomorphologists, agronomists and others.

With this we give this book into the hands of students and also of the leaders of universities, as well as of those that may forget from time to time that the higher integration levels of science should get sufficient attention.

40. The UNESCO Classification of Vegetation

This classification is designed to serve primarily vegetation maps at the scale of 1:1,000,000
In the Unesco classification, units of unequal rank are distinguished from one another by letters and
numbers as follows:

I, II, etc. = FORMATION CLASS
A, B, etc. = *FORMATION SUBCLASS*
1, 2, etc. = FORMATION GROUP
a, b, etc. = *Formation*
(1), (2), etc. = Subformation
(a), (b), etc. = Further subdivisions.

	I	CLOSED FOREST

I CLOSED FOREST
Formed by trees at least 5 m tall with their crowns interlocking. [1]

I.A. *MAINLY EVERGREEN FOREST,* i.e. the canopy is never without green foliage. However, individual trees may shed their leaves.

I.A.1 TROPICAL OMBROPHILOUS FOREST. (Conventionally called tropical rain forest.) Consisting mainly of broad-leaved evergreen trees, neither cold nor drought resistant. Truly evergreen, i.e. the forest canopy remains green all year though a few individual trees may be leafless for a few weeks. Leaves of many species with 'drip tips'.

1 I.A.1a *Tropical ombrophilous lowland forest.* Composed usually of numerous species of fastgrowing trees, many of them exceeding 50 m in height, [2] generally with smooth, often thin bark, some with buttresses. Emergent trees or at least a very uneven canopy often present. Very sparse undergrowth, and this composed mainly of tree reproduction. Palms and other tuft trees usually rare, lianas nearly absent except pseudo-lianas (i.e. plants originating on tree branches, subsequently rooting in the ground, or vice versa). Crustose lichens and green algae are the only constantly present epiphytic life forms; vascular epiphytes are usually not abundant; abundant only in extremely humid situations, e.g. Sumatra, Atrato Valley (Colombia) etc.

2 I.A.1b *Tropical ombrophilous submontane forest.* Emergent trees largely absent and canopy relatively even. In the undergrowth, forbs common. Vascular epiphytes and pseudo-lianas abundant, e.g. Atlantic slopes of Costa Rica.

[1] In reproductive stage or as immature secondary growth temporarily less than 5 m tall, but individuals of scapose life form, i.e. real trees, not shrubs. In subpolar conditions, the limit may be only 3 m, in tropical ones 8 or 10 m.
[2] Height limits are only a generalized guide, not an absolute criterion.

	I.A.1c	*Tropical ombrophilous montane forest.* Abundant vascular and other epiphytes. Tree sizes usually less than 50 m; crowns extending relatively far down the stem. Bark often more or less rough. Undergrowth abundant, often represented by rosulate nano- and microphancrophytes (e.g. tree ferns or small palms); the ground layer rich in hygromorphous herbs and cryptogams, e.g. Sierra de Talamanca, Costa Rica.
3	I.A.1c(1)	Broad-leaved.
4	I.A.1c(2)	Needle-leaved.
5	I.A.1c(3)	Microphyllous.
6	I.A.1c(4)	Bamboo, rich in tree-grasses replacing largely the tuft micro- or nanophanerophytes.
7	I.A.1d	*Tropical ombrophilous 'subalpine' forest.* (Excluding cloud forest or woodland. Considered unique by some investigators, but probably not important. Definition required.)
	I.A.1e	*Tropical ombrophilous cloud forest.* Tree crowns, branches and trunks as well as lianas burdened with epiphytes, mainly chamaephytic bryophytes. Also the ground covered with hygromorphic chamaephytes (e.g. *Selaginella* and herbaceous ferns). Trees often gnarled, with rough bark and rarely exceeding 20 m in height, e.g. Blue Mountains, Jamaica.
8	I.A.1e(1)	Broad-leaved, most common form.
9	I.A.1e(2)	Needle-leaved.
10	I.A.1e(3)	Microphyllous.
	I.A.1f	*Tropical ombrophilous alluvial forest.* Similar to submontane forest (I.A.1b), but richer in palms and in undergrowth life forms, particularly tall forbs (e.g. Musaceae); buttresses frequent, e.g. Amazon Basin.
11	I.A.1f(1)	Riparian forest on low, frequently flooded, river banks, mostly dominated by fastgrowing trees; herbaceous undergrowth nearly absent; epiphytes rare, poor in species, e.g. the Amazonian igapó.
12	I.A.1f(2)	Occasionally flooded on relatively dry terraces accompanying active rivers. More epiphytes than in (1) and (3), many lianas.
13	I.A.1f(3)	Seasonally water-logged along the lower river courses, where the water accumulates for several months on large flats, often behind low natural levees; trees frequently with stilt roots; canopy density not uniform; as a rule poor in undergrowth, except in more open places, e.g. the Amazonian várzea.
	I.A.1g	*Tropical ombrophilous swamp forest.* Not along rivers, but on edaphically wet habitats, which may be supplied with either fresh or brackish water. Similar to alluvial forests (I.A.1f), but as a rule poorer in tree species. Many trees with buttresses or pneumatophores; mostly taller than 20 m, e.g. in eastern Sumatra.
14	I.A.1g(1)	Broad-leaved, dominated by diocyledonous plants.
15	I.A.1g(2)	Dominated by palms, but broad-leaved trees in the undergrowth, e.g. the *Raphia taedigera* swamps of Costa Rica.
16	I.A.1h	*Tropical evergreen bog forest* (with organic surface deposits). Poor in tree species, with canopy often forming a pattern of tall trees at the bog fringe to shorter trees near the centre. Trees often have thin diameters and are commonly equipped with pneumatophores or stilt roots.

| | I.A.2 | TROPICAL AND SUBTROPICAL EVERGREEN SEASONAL FOREST. Consisting mainly of broad-leaved evergreen trees. Foliage reduction during the dry season is noticeable, often as partial shedding. Transitional between I.A.1 and I.A.3. Subdivisions a-c largely similar to those under I.A.1, the tropical ombrophilous forest. |

17 I.A.2a *Tropical or subtropical evergreen seasonal lowland forest.*

I.A.2b *Tropical or subtropical evergreen seasonal submontane forest.*
18 I.A.2b(1) Broad-leaved.
19 I.A.2b(2) Needle-leaved.

20 I.A.2c *Tropical or subtropical evergreen seasonal montane forest.* In contrast to I.A.1c no tree ferns; instead, evergreen shrubs are most frequent.

21 I.A.2d *Tropical or subtropical evergreen dry 'subalpine' forest.* Physiognomically resembling the winter-rain evergreen sclerophyllous dry forest (I.A.8a), usually occurring above the cloud forest (I.A.1e). Mostly evergreen sclerophyllous trees, smaller than 20 m with little or no undergrowth. (If not opened by human activity). Poor in lianas and epiphytes, except lichens.

I.A.3 TROPICAL AND SUBTROPICAL SEMI-DECIDUOUS FOREST. Most of the upper canopy trees drought-deciduous; many of the understorey trees and shrubs evergreen and more or less sclerophyllous. However, evergreen and deciduous woody plants are not always separeted by layers. They may occur mixed within the same layer, or shrubs may be primarily deciduous and trees evergreen. Nearly all trees with bud protection; leaves without 'drip tips'. Trees show rough bark, except some bottle trees, which may be present.

22 I.A.3a *Tropical or subtropical semi-deciduous lowland forest.* The taller trees may be bottle trees (e.g. *Ceiba*). Practically no epiphytes present. Undergrowth composed of tree reproduction and true shrubs. Succulents may be present (e.g. in form of thin-stemmed caespitose cacti). Both therophytic and hemicryptophytic lianas occur occasionally. A sparse herb layer may be present, mainly consisting of graminoid hemicryptophytes and forbs.

23 I.A.3b *Tropical or subtropical semi-deciduous montane or cloud forest. Similar to a, the semi-deciduous lowland forest, but canopy lower and covered with xerophytic epiphytes (e.g. Tillandsia usneoides).* Within the semi-deciduous group I.A.3, a submontane formation cannot be clearly distinguished.

I.A.4 SUBTROPICAL OMBROPHILOUS FOREST. Present only locally and in small fragmentary stands, because the subtropical climate is typically a climate with a dry season. Where the subtropical ombrophilous forest occurs, e.g. Queensland (Australia) and Taiwan, it usually grades rather inconspicuously into the tropical ombrophilous forest. Some shrubs may grow in the understorey. The subtropical ombrophilous forest should however not be confused with the tropical ombrophilous montane forest, which occurs in a climate with a similar mean annual temperature, but with less pronounced temperature differences between summer and winter. Consequently, seasonal rhythms are more evident in all subtropical forests, even in the ombrophilous ones.

The subtropical ombrophilous forest is physiognomically more closely related to the tropical than to the temperate one. Therefore the subdivisions conform more or less to the tropical ombrophilous forest, item I.A.1a to h.

24	I.A.4a
25	I.A.4b
26	I.A.4c(1)
27	I.A.4c(2)
28	I.A.4c(3)
29	I.A.4c(4)
30	I.A.4d
31	I.A.4e(1)
32	I.A.4e(2)
33	I.A.4e(3)
34	I.A.4f(1)
35	I.A.f(2)
36	I.A.4f(3)
37	I.A.4g(1)
38	I.A.4g(2)
39	I.A.4h

40   I.A.5   MANGROVE FOREST. (Occurs only in the tidal range of the tropical and subtropical zones.) Composed almost entirely of evergreen sclerophyllous broad-leaved trees and/or shrubs with either stilt roots or pneumatophores. Epiphytes in general rare, except lichens on the branches and adnate algae on the lower parts of the trees. (Subdivisions possible; transitions to 1g, the tropical ombrophilous swamp forest, exist). E.g. coasts of Borneo, New Guinea, etc.

       I.A.6   TEMPERATE AND SUBPOLAR EVERGREEN OMBROPHILOUS FOREST. Occurs only in extremely oceanic, nearly frostfree climates of the southern hemisphere, mainly in Chile. Consisting mostly of truly evergreen hemisclerophyllous trees and shrubs. Rich in thalloepiphytes and in ground-rooted herbaceous ferns.

       I.A.6a   *Temperate evergreen ombrophilous forest.* Vascular epiphytes and lianas may be present; height generally exceeds 10 m.

41   I.A.6a(1)   *Mainly broad-leaved trees, e.g. Nothofagus forests of New Zealand.*

42   I.A.6a(2)   With needle-leaved trees admixed.

43   I.A.6a(3)   Mainly needle-leaved or scale-leaved trees, e.g. *Podocarpus* forests of New Zealand.

44   I.A.6b   *Temperate evergreen ombrophilous alluvial forest.* Rich in forbs, e.g. western New Zealand.

       I.A.6c   *Temperate evergreen ombrophilous swamp forest.*

45   I.A.6c(1)   Needle- or scale-leaved. Dense, tall (up to 50 m and more) scale-leaved lowland forest. Buttresses common. Closed to open ground cover of graminoids (mainly sedges) and forbs (mostly ferns). Rich in vascular and bryoid epiphytes; some lianas, e.g. *Podocarpus dacrydioides* communities of New Zealand.

46	I.A.6c(2)	Broad-leaved. Tall (up to 50 m and more) broad-leaved lowland forest. Trees densely spaced so that crowns touch, but canopy permits much light to pass. Fairly open shrub synusia. Epiphytes lacking in canopy, e.g. *Eucalyptus ovata* forests of Victoria.
47	I.A.6d	*Subpolar evergreen ombrophilous forest.* In contrast with I.A.6a, the temperate broad-leaved forest, vascular epiphytes lacking and canopy height much reduced (in general less than 10 m). Also leaf-size generally reduced, e.g. beech forests of New Zealand.
	I.A.7	TEMPERATE EVERGREEN SEASONAL BROAD-LEAVED FOREST (with adequate summer rainfall). Consisting mainly of hemi-sclerophyllous evergreen trees and shrubs. Rich in herbaceous chamaephytic and hemicryptophytic undergrowth. Very few or no vascular epiphytes and lianas. Grades into subtropical (I.A.4) or temperate ombrophilous forests (I.A.6) or into winter-rain evergreen broad-leaved sclerophyllous forests (I.A.8). Probably includes subpolar types. (Subdivisions possible as under tropical and subtropical seasonal foests, I.A.2a to d.)
48	I.A.7a	
49	I.A.7b(1)	
50	I.A.7b(2)	
51	I.A.7c	
52	I.A.7d	
	I.A.8	WINTER-RAIN EVERGREEN BROAD-LEAVED SCLEROPHYLLOUS FOREST. (Often understood as Mediterranean, but present also in south-western Australia. Chile, etc. Climate with pronounced summer drought.) Consisting mainly of slcerophyllous evergreen trees and shrubs, most of which show rough bark. Herbaceous undergrowth almost lacking. No vascular and only few cryptogamic epiphytes, but evergreen woody lianas present.
53	I.A.8a	*Winter-rain evergreen sclerophyllous lowland and submontane forest.* Giant eucalypts, e.g. *Eucalyptus regnans* in Victoria and *E. diversicolor* in Western Australia. Dominated by trees over 50 m tall.
54	I.A.8b	Largely as described in I.A.8 but less than 50 m tall, e.g. Californian live-oak forests.
55	I.A.8c	Alluvial and swamp forest of this type perhaps existing, but not sufficiently known.
	I.A.9	TROPICAL AND SUBTROPICAL EVERGREEN NEEDLE-LEAVED FOREST. Consisting mainly of needle-leaved or scale-leaved evergreen trees. Broad-leaved trees may be present. Vascular epiphytes and lianas are practically lacking.
56	I.A.9a	*Tropical and subtropical lowland and submontane evergreen needle-leaved forest,* e.g. the pine forests of Honduras and Nicaragua.
57	I.A.9b	*Tropical and subtropical montane and subalpine evergreen needle-leaved forest,* e.g. the pine forests of the Philippines and southern Mexico.
	I.A.10	TEMPERATE AND SUBPOLAR EVERGREEN NEEDLE-LEAVED FOREST. Consisting mainly of needle-leaved or scale-leaved evergreen trees, but broad-leaved trees may be admixed. Vascular epiphytes and lianas practically lacking.

58	I.A.10a	*Evergreen giant forest.* Dominated by trees higher than 50 m, e.g. *Sequoia* and *Pseudotusaga* forest in the Pacific West of North America.

I.A.10b   *Evergreen forest with rounded crowns.* Dominated by trees 5–50 m high, with more or less broad, irregularly rounded crowns, e.g. *Pinus* spp.

59   I.A.10b(1)   With evergreen sclerophyllous understorey.

60   I.A.10b(2)   Without evergreen slcerophyllous understorey.

61   I.A.10c   *Evergreen needle-leaved forest with conical crowns.* Dominated by trees 5–50 m high (only exceptionally higher), with more or less conical crowns (like most *Picea* and *Abies*), e.g. California red fir forests.

62   I.A.10d   *Evergreen forest with cylindrical crowns* (boreal). Similar to I.A.1c but crowns with very short branches and therefore very narrow, cylindro-conical.

I.B   *MAINLY DECIDUOUS FOREST.* Majority of trees shed their foliage simultaneously in connexion with the unfavourable season.

I.B.1   TROPICAL AND SUBTROPICAL DROUGHT-DECIDUOUS FOREST. Unfavourable season mainly characterized by drought, in most cases winter-drought. Foliage is shed regularly every year. Most trees with relatively thick, fissured bark.

63   I.B.1a   *Drought-deciduous broad-leaved lowland and submontane forest.* Practically no evergreen plants in any stratum, except some succulents. Woody and herbaceous lianas present occasionally, also deciduous bottle-trees. Ground vegetation mainly herbaceous (hemicryptophytes, particularly grasses, geophytes and some therophytes), but sparse, e.g. the broad-leaved deciduous forests of northwestern Costa Rica.

64   I.B.1b   *Drought-deciduous montane (and cloud) forest.* Some evergreen species in the understorey. Drought-resistant epiphytes present or abundant, often of the bearded form (e.g. *Usnea* or *Tillandsia usneoides*). This formation is not frequent, but well developed, e.g. in northern Peru.

I.B.2   COLD-DECIDUOUS FORESTS WITH EVERGREEN TREES (OR SHRUBS) ADMIXED. Unfavourable season mainly characterized by winter frost. Deciduous broad-leaved trees dominant, but evergreen species present as part of the main canopy or as understorey. Climbers and vascular epiphytes scarce or absent.

65   I.B.2a   *Cold-deciduous forest with evergreen broad-leaved trees and climbers* (e.g. Ilex aquifolium and *Hedera helix* in western Europe and Magnolia spp. in North America). Rich in cryptogamic epiphytes, including mosses. Even vascular epiphytes may be present at the base of tree stems. Lianas may be common on flood plains.

66   I.B.2b   *Cold-deciduous broad-leaved forest with evergreen needle-leaved trees,* e.g. the maple-hemlock forests of New York.

I.B.3   COLD-DECIDUOUS FORESTS WITHOUT EVERGREEN TREES. Deciduous trees absolutely dominant. Evergreen chamaephytes and some evergreen manophanerophytes may be present. Climbers insignificant but may be common on flood plains, vascular epiphytes absent (except occasionally at the lower base of the tree); thalloepiphytes always present, particularly lichens.

67	I.B.3a	*Temperate lowland and submontane broad-leaved cold-deciduous forest.* Trees up to 50 m tall, e.g. the Mixed Mesophytic Forest (U.S.A.). Primarily algae and crustose lichens as epiphytes.
	I.B.3b	*Montane or boreal cold-deciduous forest* (including lowland or submontane in topographic positions with high atmospheric humidity). Foliose and fruticose lichens, and bryophytes as epiphytes. Trees up to 50 m tall, but in montane or boreal forest normally not taller than 30 m.
68	I.B.3b(1)	Mainly broad-leaved deciduous.
69	I.B.3b(2)	Mainly needle-leaved deciduous (e.g. Siberian *Larix*) forests.
70	I.B.3b(3)	Mixed broad-leaved and needle-leaved deciduous.
	I.B.3c	*Subalpine or subpolar cold-deciduous forest.* In contrast to I.B.3a and b, the lowland and montane cold-deciduous forest, the canopy height significantly reduced (not taller than 20 m). Tree trunks frequently gnarled. Epiphytes similar as in b, but in general more abundant. Often grading into woodland (see Formation Class II).
71	I.B.3c(1)	With primarily hemicryptophytic undergrowth.
72	I.B.3c(2)	With primarily chamaephytic undergrowth. May merge with forests admixed with conifers.
	I.B.3d	*Cold-deciduous alluvial forest.* Flooded by rivers, therefore moister and richer in nutrients than cold-deciduous lowland forest, I.B.3a. Trees and shrubs with high growth rates and vigorous herbaceous undergrowth.
73	I.B.3d(1)	Occasionally flooded; physiognomically similar to I.B.3a, with tall trees and abundant macrophyllous shrubby undergrowth.
74	I.B.3d(2)	Regularly flooded; trees not as tall and dense as in I.B.3a, but herbaceous undergrowth abundant and tall. (In Eurasia *Salix*- or *Alnus*-species frequently dominating.)
	I.B.3e	*Cold-deciduous swamp or peat forest.* Flooded until late spring or early summer, surface soil organic. Relatively poor in tree species. Ground cover of varied growth forms mostly continuous. (Subdivisions as under I.B.3b, the boreal cold-deciduous forest).
75	I.B.3e(1)	
76	I.B.3e(2)	
77	I.B.3e(3)	
	I.C	*EXTREMELY XEROMORPHIC FOREST.* Dense stands of xeromorphic phanerophytes, such as bottle trees, tuft trees with succulent leaves and stem succulents. Undergrowth with shrubs of similar xeromorphic adaptations, succulent chamaephytes and herbaceous hemicryptophytes, geophytes and therophytes. Often grading into woodlands. (See Formation Class II.)
78	I.C.1	SCLEROPHYLLOUS-DOMINATED EXTREMELY XEROMORPHIC FOREST. Life form combination as above, except for predominance of sclerophyllous trees, many of which have bulbose stem bases largely embedded in the soil (xylopods).
	I.C.2	THORN-FOREST. Species with thorny appendices predominate.
79	I.C.2a	*Mixed deciduous-evergreen thorn forest.*
80	I.C.2b	*Purely deciduous thorn forest.*
81	I.C.3	MAINLY SUCCULENT FOREST. Tree-formed (scapose) and shrub-formed (caespitose) succulents very frequent, but other xero-phanerophytes are usually present as well.

| | II | WOODLAND (open stands of trees) |

Composed of trees at least 5 m tall with crowns not usually touching but with a coverage of at least 40 per cent. A herbaceous synusia may be present. See Formation Group V.A.1 if the coverage of the trees is less than 40 per cent and there is a herbaceous synusia. The boundary of 40 per cent coverage is convenient because it can be estimated with ease during the field-work: when the coverage of the trees is 40 per cent the distance between two tree crowns equals the mean radius of a tree crown.

	II.A	*MAINLY EVERGREEN WOODLAND,* i.e. evergreen as defined in I.A.
82	II.A.1	EVERGREEN BROAD-LEAVED WOODLAND. Mainly sclerophyllous trees and shrubs, no epiphytes.
	II.A.2	EVERGREEN NEEDLE-LEAVED WOODLAND. Mainly needle- or scale-leaved. Crowns of many trees extending to the base of the stem or at least very branchy.
	II.A.2a	*Evergreen needle-leaved woodland with rounded crowns (e.g. Pinus).*
83	II.A.2a(1)	With evergreen sclerophyllous understorey (Mediterranean).
84	II.A.2a(2)	Without evergreen sclerophyllous understorey.
85	II.A.2b	*Evergreen needle-leaved woodland with conical crowns prevailing* (mostly subalpine).
86	II.A.2c	*Evergreen needle-leaved woodland with very narrow cylindro-conical crowns* (e.g. *Picea* in the boreal region).
	II.B	*MAINLY DECIDUOUS WOODLAND* (see I-B)
	II.B.1	DROUGHT-DECIDUOUS WOODLAND. (Subdivisions more or less like forests.)
87	II.B.1a	
88	II.B.1b	
	II.B.2	COLD-DECIDUOUS WOODLAND WITH EVERGREEN TREES (see I.B.2).
89	II.B.2a	
90	II.B.2b	
	II.B.3	COLD-DECIDUOUS WOODLAND WITHOUT EVERGREEN TREES (see I.B.3). Most frequent in the subarctic region, elsewhere only on swamps or bogs.
91	II.B.3a	*Broad-leaved deciduous woodland.*
92	II.B.3b	*Needle-leaved deciduous woodland.*
93	II.B.3c	*Mixed deciduous woodland* (broad-leaved and needle-leaved).
	II.C	*EXTREMELY XEROMORPHIC WOODLAND.* Similar to I.C, the only difference being the more sparce stocking of individual trees. (Subdivisions as under I.C.)

94	II.C.1
95	II.C.2a
96	II.C.2b
97	II.C.3

| | III | SCRUB (shrubland or thicket) |

Mainly composed of caespitose woody phanerophytes 0.5 to 5 m tall. [1]
Each of the following subdivisions may be either:
Shrubland: most of the individual shrubs not touching each other; often with a grass stratum; or
Thicket: individual shrubs interlocked.

| | III.A | *MAINLY EVERGREEN SCRUB* (evergreen in the sense of I.A). |

| | III.A.1 | EVERGREEN BROAD-LEAVED SHRUBLAND (or thicket). |

| 98 | III.A.1a | *Low bamboo thicket* (or, less frequently, shrubland). Lignified creeping graminoid nano- or microphanerophytes. |

| 99 | III.A.1b | *Evergreen tuft-tree shrubland* (or thicket). Composed of small trees and woody shrubs (e.g. Mediterranean dwarf palm shrubland or Hawaiian tree fern thicket). |

| 100 | III.A.1c | *Evergreen broad-leaved hemisclerophyllous thicket* (or shrubland). Caespitose, creeping or lodged nano- or microphanerophytes with relatively large and soft leaves (e.g. subalpine *Rhododendron* thickets, or *Hibiscus tiliaeceus* matted thickets of Hawaii). |

| 101 | III.A.1d | *Evergreen broad-leaved sclerophyllous shrubland* (or thicket). Dominated by broad-leaved sclerophyllous shrubs and immature trees (i.e. chaparral or macchia). May often merge with parkland, grassland or heath. |

| 102 | III.A.1e | *Evergreen suffruticose thicket* (or shrubland). Stand of semi-lignified nanophanerophytes that in dry years may shed part of their shoot systems (e.g. *Cistus* heath). [2] |

| | III.A.2 | EVERGREEN NEEDLE-LEAVED AND MICROPHYLLOUS SHRUBLAND (or thicket). |

| 103 | III.A.2a | *Evergreen needle-leaved thicket* (or shrubland). Composed mostly of creeping or lodged needle-leaved phanerophytes (e.g. *Pinus mugo*, 'Krummholz'). |

| 104 | III.A.2b | *Evergreen microphyllous shrubland* (or thicket). Often ericoid shrubs (mostly in tropical subalpine belts). |

| | III.B | *MAINLY DECIDUOUS SCRUB* (deciduous in the sense of I.B). |

| 105 | III.B.1 | DROUGHT-DECIDUOUS SCRUB WITH EVERGREEN WOODY PLANTS ADMIXED. |

---

[1] Not to be confused with developing second growth forests, see footnore relating to I. Sometimes, scrub may reach more than 5 m in height.
[2] Occasionally less than 50 cm tall, thereby grading into IV.A.1a.

| 106 | III.B.2 | DROUGHT-DECIDUOUS SCRUB WITHOUT EVERGREEN WOODY PLANTS ADMIXED. |

III.B.3   COLD-DECIDUOUS SCRUB.

107   III.B.3a   *Temperate deciduous thicket* (or shrubland). More or less dense scrub without or with only little herbaceous undergrowth. Poor in cryptogams.

III.B.3b   *Subalpine or subpolar deciduous thicket* (or shrubland). Upright or lodged caespitose nanophanerophytes with great vegetative regeneration capacity. As a rule completely covered by snow for at least half a year.

108   III.B.3b(1)   With primarily hemicryptophytic undergrowth, mainly forbs (e.g. subalpine *Alnus viridis* thicket).

109   III.B.3b(2)   With primarily chamaephytic undergrowth, mainly dwarf shrubs and fruticose lichens (e.g. *Betula tortuosa* shrubland at the polar tree line).

III.B.3c   *Deciduous alluvial shrubland* (or thicket). Fast-growing shrubs, occuring as pioneers on river banks or islands that are often vigorously flooded, therefore mostly with very sparse undergrowth.

110   III.B.3c(1)   With lanceolate leaves (e.g. *Salix,* mostly in lowland or submontane region).

111   III.B.3c(2)   Microphyllous.

112   III.B.3d   *Deciduous peat shrubland* (or thicket). Upright caespitose nanophanerophytes with *Sphagnum* and (or) other peat mosses.

III.C   *EXTREMELY XEROMORPHIC (SUBDESERT) SHRUBLAND.* Very open stands of shrubs with various xerophytic adaptations, such as extremely scleromorphic or strongly reduced leaves, green branches without leaves, or succulent stems, etc., some of them with thorns.

III.C.1   MAINLY EVERGREEN SUBDESERT SHRUBLAND. In extremely dry years some leaves and shoot portions may be shed.

III.C.1a   *Evergreen subdesert shrubland.*

113   III.C.1a(1)   Broad-leaved, dominated by sclerophyllous nanophanerophytes, including some phyllocladous shrubs (e.g. mulga scrub in Australia).

114   III.C.1a(2)   Microphyllous, or leafless, but with green stems (e.g. *Retama retam*).

115   III.C.1a(3)   Succulent, dominated by variously branched stem and leaf succulents.

III.C.1b   *Semi-deciduous subdesert shrubland.* Either facultatively deciduous shrubs or a combination of evergreen and deciduous shrubs.

116   III.C.1b(1)   Facultatively deciduous (e.g. *Atriplex-Kochia*-saltbush in Australia and North America).

117   III.C.1b(2)   Mixed evergreen and deciduous, transitional to III.C.2.

III.C.2   DECIDUOUS SUBDESERT SHRUBLAND. Mainly deciduous shrubs, often with a few evergreens.

118   III.C.2a   *Deciduous subdesert shrubland without succulents.*

119   III.C.2b   *Deciduous subdesert shrubland with succulents.*

IV		DWARF-SCRUB AND RELATED COMMUNITIES

Rarely exceeding 50 cm in height (sometimes called heaths or heathlike formations). According to the density of the dwarf-shrub cover are distinguiqhed:

Dwarf-shrub thicket: branches interlocked;

Dwarf-shrubland: individual dwarf-shrubs more or less isolated or in clumps;

Cryptogamic formations with dwarf-shrubs: surface densely covered with mosses or lichens (thallochamaephytes); dwarf-shrubs occurring in small clumps or individually. In the case of bogs locally dominating graminoid communities may be included.

	IV.A	*MAINLY EVERGREEN DWARF-SCRUB.* Most dwarf-shrubs evergreen.
	IV.A.1	EVERGREEN DWARF-SHRUB THICKET. Densely closed dwarf-shrub cover, dominating the landscape ('dwarf-shrub heath' in the proper sence).
120	IV.A.1a	*Evergreen caespitose dwarf-shrub thicket.* Most of the branches standing in upright position, often occupied by foliose lichens. On the ground pulvinate mosses, fruticose lichens or herbaceous life forms may play a role (e.g. *Calluna* heath).
121.	IV.A.1b	*Evergreen creeping or matted dwarf-shrub thicket.* Most branches creeping along the ground. Variously combined with thallochamaephytes in which the branches may be embedded (e.g. *Loiseleuria* heath).
	IV.A.2	EVERGREEN DWARF-SCHRUBLAND. Open or more loose cover of dwarf-schrubs.
122	IV.A.2a	*Evergreen cushion shrubland.* More or less isolated clumps of dwarf-shrubs forming dense cushions, often equipped with thorns (e.g. *Astragalus-* and *Acantholimon* 'porcupine'-heath of the East Mediterranean mountains).
	IV.A.3	MIXED EVERGREEN DWARF-SHRUB AND HERBACEOUS FORMATION. More or less open stands of evergreen suffrutescent or herbaceous chamaephytes, various hermicryptophytes, geophytes. etc.
123	IV.A.3a	*Truly evergreen dwarf-shrub and herb mixed formation* (e.g. *Nardus-Calluna-*heath).
124	IV.A.3b	*Partially evergreen dwarf-shrub and herb mixed formation.* Many individuals shed parts of their shoot systems during the dry season (e.g. *Phrygana* in Greece).
	IV.B	*MAINLY DECIDUOUS DWARF-SCRUB.* Similar to IV.A, but mostly consisting of deciduous species.
125	IV.B.1	FACULTATIVELY DROUGHT-DECIDUOUS DWARF-THICKET (or dwarf-shrubland). Foliage is shed only in extreme years.
	IV.B.2	OBLIGATORY, DROUGHT-DECIDUOUS DWARF-THICKET (or dwarf-shrubland). Densely closed dwarf-shrub stands which lose all or at least part of their leaves in the dry season.
126	IV.B.2a	*Drought-deciduous caespitose dwarf-thicket.* Corresponding to IV.A.1a.

127	IV.B.2b	*Drought-deciduous creeping or matted dwarf-thicket.* Corresponding to IV.A.1b.
128	IV.B.2c	*Drought-deciduous cushion dwarf-shrubland.* Corresponding to IV.A.2a.
129	IV.B.2d	*Drought-deciduous mixed dwarf-schrubland.* Deciduous and evergreen dwarf-schrubs, caespitose hemicryptophytes, succulent chamaephytes and other life forms intermixed in various patterns.
	IV.B.3	COLD-DECIDUOUS DWARF-THICKET (or dwarf-schrubland). Physiognomically similar to IV.B.2, but shedding the leaves at the beginning of a cold season. Usually richer in cryptogamic chamaephytes. (Subdivisions similar to IV.B.2. Transition into IV.D and E possible.)
130	IV.B.3a	
131	IV.B.3b	
132	IV.B.2c	
133	IV.B.2d	
	IV.C	*EXTREMELY XEROMORPHIC DWARF-SHRUBLAND.* More or less open formations consisting of dwarf-shrubs, succulents, geophytes, therophytes and other life forms adapted to survive or to avoid a long dry season. Mostly subdesertic. (Subdivisions similar to III-C, extremely xeromorphic shrublands.)
134	IV.C.1a(1)	
135	IV.C.1a(2)	
136	IV.C.1a(3)	
137	IV.C.1b(1)	
138	IV.C.1b(2)	
139	IV.C.2a	
	IV.D	*TUNDRA.* Slowly growing, low formations, consisting mainly of dwarf-shrubs, graminoids, and cryptogams, beyond the subpolar tree line. Often showing plant patterns caused by freezing movements of the soil (cryoturbation). Except in boreal regions, dwarf-shrub formations above the mountain tree line should not be called tundra, because they are as a rule richer in dwarf-shrubs and grasses, and grow taller due to the greater radiation in lower latitudes.
	IV.D.1	MAINLY BRYOPHYTE TUNDRA. Dominated by mats or small cushions of chamaephytic mosses. Groups of dwarf-shrubs are as a rule scattered irregularly and are not very dense. General aspect more or less dark green, olive green or brownish.
141	IV.D.1a	*Caespitose dwarf-shrub–moss tundra.*
142	IV.D.1b	*Creeping or matted dwarf-shrub–moss tundra.*
143	IV.D.2	MAINLY LICHEN TUNDRA. Mats of fruticose lichens dominating, giving the formation a more or less pronounced grey aspect. Dwarf-shrubs mostly evergreen, creeping or pulvinate. Dwarf-shrub–lichen tundra.

	IV.E	*MOSSY BOG FORMATIONS WITH DWARF-SHRUB.* Oligotrophic peat accumulations formed mainly by *Sphagnum* or other mosses, which as a rule cover the surface as well. Dwarf-shrubs are concentrated on the relatively drier parts or are loosely scattered. To a certain extent they resemble dwarf-shrub formations on mineral soil. Graminoid hemicryptophytes, geophytes with rhizomes and other herbaceous life forms may dominate locally. Slowly growing trees and shrubs can grow as isolated individuals, in groups or in woodlands, which are marginal to the bog or may be replaced by open formations in a cyclic succession. The following subdivisions correspond to the classification of bog types adopted in Europe.
	IV.E.1	RAISED BOG. By growth of *Sphagnum* species raised above the general ground-water table and having a ground-water table of their own. Therefore not supplied by 'mineral' water (i.e. water having been in touch with the inorganic soil), but only by rain water (truly ombrotrophic bogs).
144	IV.E.1a	*Typical raised bog* (suboceanic, lowland and submontane). Mosses dominating throughout, except on locally raised dry hummocks, which are dominated by dwarf-shrubs. Trees rare and, if present, concentrated on the marginal slopes of the convex peat accumulation. Mostly surrounded by a very wet, but less oligotrophic swamp.
145	IV.E.1b	*Montane (or 'subalpine') raised bog.* Growing slower than the typical raised bog (or formed in an earlier period with a warmer climate and actually 'dead' or being destroyed by erosion). Often covered with sedges or evergreen dwarf-shrubs. Micro- or nanophanerophytes (e.g. *Pinus mugo*) locally dominating.
146	IV.E.1c	*Subcontinental woodland bog.* Temporarily covered by open wood of low productivity, which in a sequence of wetter years may be replaced by *Spagnum* formations similar to IV.E.1a.
	IV.E.2	NON-RAISED BOG. Not or not very markedly raised above the mineral-water table of the surrounding landscape. Therefore in general wetter and not as oligotrophic as I.V.E.1. Poorer in mosses than the typical raised bog, 1a, to which various forms of transitions are possible.
147	IV.E.2a	*Blanket bog* (oceanic lowland, submontane or montane). The micro-surface of the bog is less undulating and less rich in actively growing mosses than in IV.E.1a. Evergreen dwarf-shrubs are scattered as well as caespitose hemicryptophytes (sedges or grasses) and some rhizomatous geophytes.
148	IV.E.2b	*String bog* (Finnish 'aapa' bog). Flat oligotrophic with string-like hummocks in the boreal lowlands. The Finnish name indicates an open bog without or with only a few trees of very poor vigour, which grow on narrow and low elongated hummocks, the so-called strings. These peat strings are formed by pressure of the ice covering the more or less flooded bog from early fall to late spring. Only these strings are covered by dwarf-shrubs and are rich in *Sphagnum*. The main part of the bog is similar to a wet sedge swamp.

Herbaceous vegetation

The classification of herbaceous vegetation requires special consideration mainly because of (a) the continual seasonal changes in the physiognomy of the communities; (b) the problems of distinguishing many tropical from non-tropical grasslands; (c) the management of grasslands which can seriously affect the structure of the vegetation and which can be changed frequently; (d) the problems of distinguishing between natural and managed grasslands.

There are two major types of herbaceous growth forms: graminoids and forbs. Graminoids include all herbaceous grasses and grass-like plants such as sedges (*Carex*), rushes (*Juncus*), cat's-tails (*Typha*), etc. Forbs are broad-leaved herbaceous plants such as clover (*Trifolium*), sunflowers (*Helianthus*) ferns, milkweeds (*Asclepias*), etc. Usually all non-graminoid herbaceous plants are included in the forbs.

The Unesco classification frequently employs the term grassland to signify a herbaceous type of vegetation dominated by graminoid growth forms. As the graminoids include numerous taxa other than grasses (*Gramineae*), the term grassland must here be understood to mean a physiognomic vegetation type without floristic implications.

As with woody communities, height is a major feature to characterize a plant community dominated by herbaceous growth forms. The great seasonal fluctuations in the height of herbaceous plants require that height measurements (or estimates) be made at the time of flowering, i.e. when the inflorescences are developed. Inflorescenses may not develop where steady or heavy grazing prevails; their height must then be estimated.

Coverage is another important factor in characterizing herbaceous vegetation. However, at the scale of 1:1,000,000 or less, all herbaceous communities are assumed to have a more or less continuous coverage, and no mention is made of this in the map legend. The exeption is very low density. In that case, the community is called open.

The growth forms of graminoids are significant and it is common to distinguish sod forms, bunch (tussock) forms and combinations of these. In the Unesco classification, all graminoid communities are considered to be more or less of the sod form, and this need therefore not be mentioned. However, when bunch grasses dominate, they affect the physiognomy of the vegetation profoundly. The bunch grass form must then be included in the description of the plant communities involved. Such descriptions parallel the descriptions of the crown forms of needle-leaved evergreen trees, as for instance in items I.A.9c and d.

Herbaceous vegetation types often include a synusia of woody plants that gives the type a special character. Frequently, therefore, such synusiae are used to characterize the herbaceous communities. Some of the more important features of a woody synusia include height and density, whether evergreen or deciduous, needle-leaved, broad-leaved or aphyllous (essentially without leaved), etc. Such features prevail in a great variety of combinations and graphically describe the general physiognomy of the vegetation. Other features are admissible if they characterize a community that is extensive enough to be mapped at the scale of 1:1,000,000 or less, e.g. the sclerophyllous nature of the woody synusia of many *Eucalyptus*-grass combinations in Australia.

There are numerous, frequently used vegetation terms such as savannah, steppe, meadow, etc. They have been avoided in the definitions of the Unesco classification because they have too many conflicting interpretations. Occasionally, however, they have been added in parentheses where this helps the reader to identify the category. It is always best to use the Unesco terminology on the vegetation maps. Locally established terms meaningful to the inhabitants of the respective regions (e.g. *campo cerrado*) may be added to but do not replace the Unesco terms in the map legends. In this manner, vegetation maps are meaningful to local users as well as to a world-wide audience. This is especially important for comparative studies.

	V	HERBACEOUS VEGETATION

V   HERBACEOUS VEGETATION

V.A   *TALL GRAMINOID VEGETATION.* In tall grasslands, the dominant graminoid growth forms are over 2 m tall when the inflorescendes are fully developed. Forbs may be present but their coverage is less than 50 per cent.

V.A.1   TALL GRASSLAND WITH A TREE SYNUSIA COVERING 10–40 PER CENT, with or without shrubs. This is somewhat like a very open woodland with a more or less continuous ground cover (over 50 per cent) of tall graminoids. For categories with a tree synusia covering over 40 per cent see Formation Class II.

149   V.A.1a   *Woody synusia broad-leaved evergreen.*

150   V.A.1b   *Woody synusia broad-leaved semi-evergreen,* i.e. composed of at least 25 per cent each of broad-leaved evergreen and broad-leaved deciduous trees.

V.A.1c   *Woody synusia broad-leaved deciduous.*

151   V.A.1c(1)   Similar to the above but seasonally flooded, e.g. in north-east Bolivia.

V.A.2   TALL GRASSLAND WITH A TREE SYNUSIA COVERING LESS THAN 10 PER CENT, with or without shrubs. Subdivisions V.A.2a to c as in V.A.1.

152   V.A.2a

153   V.A.2b

154   V.A.2c

155   V.A.2d   *Tropical or subtropical tall grassland with trees and/or shrubs growing in tufts on termite nests* (termite savannah).

V.A.3   TALL GRASSLAND WITH A SYNUSIA OF SHRUBS (shrub savannah). Subdivisions as in V.A.2.

156   V.A.3a
157   V.A.3b
158   V.A.3c
159   V.A.3d

V.A.4   TALL GRASSLAND WITH A WOODY SYNUSIA CONSISTING MAINLY OF TUFT PLANTS (usually palms).

160   V.A.4a   *Tropical grassland with palms,* e.g. the palm savannahs of *Arocomia totai* and *Attalea princeps* north of Santa Cruz de la Sierra, Bolivia.

161   V.A.4a(1)   Similar to above, seasonally flooded, e.g. the *Mauritia vinifera* savannahs, Llanos de Mojos, Bolivia.

V.A.5   TALL GRASSLAND PRATICALLY WITHOUT WOODY SYNUSIA.

162   V.A.5a   *Tropical grassland* as described in V.A.5 as in various low-latitude regions of Africa.

163   V.A.5a(1)   Similar to above, seasonally flooded, e.g. the Campos de Várzea of the lower Amazon Valley.

164   V.A.5a(2)   Similar to above, wet or flooded most of the year, e.g. the papyrus (*Cyperus papyrus*) swamps of the upper Nile Valley.

V.B.   *MEDIUM TALL GRASSLAND,* i.e. the dominant graminoid growth forms are 50 cm to 2 m tall when their inflorescences are fully developed. Forbs may be present but cover less than 50 per cent. Divided and subdivided as V.A.1 to V.A.3d.

546

165	V.B.1a	
166	V.B.1b	
167	V.B.1c	
168	V.B.2a	
169	V.B.2b	
170	V.B.2c	
171	V.B.2d	
172	V.B.3a	
173	V.B.3b	
174	V.B.3c	
175	V.B.3d	
176	V.B.3e	*Woody synusia consisting mainly of deciduous thorny shrubs,* e.g. the tropical thorn bush savannah of the Sahel region in Africa with *Acacia tortilis, A. senegal* and others.
	V.B.4	MEDIUM TALL GRASSLAND WITH AN OPEN SYNUSIA OF TUFT PLANTS, usually palms.
177	V.B.4a	*Medium tall subtropical grassland with open groves of palms,* e.g. in Corrientes, Argentina.
178	V.B.4a(1)	Similar to the above, seasonally flooded, e.g. *Mauritia* palm groves in the Colombian and Venezuelan llanos.
	V.B.5	MEDIUM TALL GRASSLAND WITHOUT WOODY SYNUSIA.
179	V.B.5a	*Medium tall grassland consisting mainly of sod grasses,* e.g. the tall-grass prairie in eastern Kansas.
180	V.B.5a(1)	Wet or flooded most of the year, e.g. *Typha* swamps.
181	V.B.5a(2)	On sandy soil or dunes, e.g. communities of *Andropogon hallii* in the Nebraska Sand Hills.
182	V.B.5b	*Medium tall grassland consisting mainly of bunch grasses,* e.g. the hard tussock (*Festuca novae-zelandiae*) grasslands in New Zealand.
	V.C	*SHORT GRASSLAND,* i.e. the dominant graminoid growth forms are less than 50 cm tall when their inflorescences are developed. Forbs may be present but they cover less than 50 per cent. Divided and subdivided as V.B.1 to V.B.4.
183	V.C.1a	
184	V.C.1b	
185	V.C.1e	
186	V.C.2a	
187	V.C.2b	
188	V.C.2c	
189	V.C.2d	
190	V.C.3a	
191	V.C.3b	
192	V.C.3c	
193	V.C.3d	

194	V.C.3e
195	V.C.4a
196	V.C.4a(1)

197   V.C.5a   *Tropical alpine open to closed bunch-grass communities with a woody synusia of tuft plants (Espeletia, Lobelia, Senecio),* microphyllous to leptophyllous dwarf-shrubs and cushion plants, often with woolly leaves. Above timberline in low latitudes: Páramo and related vegetation types without snow in the alpine regions of Kenya, Colombia, Venezuela, etc.

198   V.C.5b   Similar to V.C.5a but very open and without tuft plants. Frequent nocturnal snowfall (snow gone by 9 a.m.), the Super-Páramo (i.e. above Páramo) of J. Cuatrecasas. [1]

199   V.C.5c   *Tropical or subtropical alpine bunch-grass vegetation with open stands of evergreen with or without deciduous shrubs and dwarf shrubs,* e.g. Puna.

      V.C.5c(1)   With numerous succulent growth forms.

200   V.C.5d   *Bunch grass vegetation of varying coverage with dwarf shrubs.*

      V.C.5d(1)   With cushion plants that may be locally more important than the dwarf shrubs, e.g. Puna south of Oruro, Bolivia.

      V.C.6   SHORT GRASSLANDS PRACTICALLY WITHOUT WOODY SYNUSIA.

201   V.C.6a   *Short-grass communities.* They may fluctuate in structure and floristic composition due to greatly fluctuating precipitation of semi-arid climate, e.g. short-grass (*Bouteloua gracilis* and *Buchloë dactyloides*) prairie of eastern Colorado.

202   V.C.6b   *Bunch-grass communities,* e.g. blue tussock (*Poa colensoi*) communities of New Zealand, and alpine dry Puna with *Festuca orthophylla* of northern Chile and southern Bolivia.

      V.C.7   SHORT TO MEDIUM TALL MESOPHYTIC GRASSLAND (meadows).

203   V.C.7a   *Sodgrass communities,* often rich in forbs, usually dominated by hemicryptophytes. Occurring chiefly in lower altitudes with a cool, humid climate in North America and Eurasia. Many plants may remain at least partly green during the winter, even below the snow in the higher latitudes.

204   V.C.7b   *Alpine and subalpine meadows of the higher latitudes,* in contrast to the Páramo and Puna types of vegetation in the lower latitudes. Usually moist much of the summer due to melt water.

205   V.C.7b(1)   Rich in forbs, e.g. on the Olympic Peninsula, Washington.

206   V.C.7b(2)   Rich in dwarf shrubs, e.g. on the Rocky Mountains of Colorado.

207   V.C.7b(3)   Snow-bed communities. Open communities, rich in small forbs and/or forblike dwarf shrubs (e.g. *Salix herbacea*). High latitude equivalent of the low latitude super-Páramo (cf. V.C.5b).

208   V.C.7b(4)   Avalanche meadows, occurring as narrow strips of grassland between forests on steep slopes of high mountains where avalanches, descending annually in spring, prevent forest growth. Variable in structure; may have some shrubs and damaged trees.

---

[1] 'Páramo Vegetation and its Life Forms', *Colloquium Geographicum,* Vol. 9, Bonn, F. Dümmlers Verlag, 1968, 223 p.

	V.C.8	GRAMINOID TUNDRA. As in the case of the dwarf-shrub tundra (Formation subclass IV.D), the use of the term graminoid tundra should be limited to high latitudes, i.e. beyond the polar timberline.
209	V.C.8a	*Graminoid bunch-form tundra.* Most graminoids grow in tussock form. Mosses and/or lichens often grow between the tussocks. *Eriphorum* tundra in northern Alaska.
210	V.C.8a(1)	Seasonally flooded.
211	V.C.8b	*Graminoid sod-form tundra.* The graminoids form a more or less dense sod, often with a variety of forbs and dwarf shrubs (both caespitose and creeping). Lichens may be common. Tundra near Lake Iliamna, Alaska.
212	V.C.8b(1)	Seasonally flooded.
	V.D.	*FORB VEGETATION.* The plant communities consist mainly of forbs, i.e. the coverage of forbs exceeds 50 per cent. Graminoids may be present but they cover less than 50 per cent, often much less.
	V.D.1	TALL FORB COMMUNITIES. The dominant forb growth forms are more than 1 m tall when fully developed.
213	V.D.1a	*Mainly perennial flowering forbs, and ferns.*
214	V.D.1a(1)	Saline substrate, and/or wet much of the year, e.g. the *Batis-Salicornia* marshes of Florida and the *Heliconia-Calathea* formations of Central America.
215	V.D.1b	*Fern thickets,* sometimes in nearly pure stands, especially in humid climates, e.g. *Pteridium aquilinum.*
216	V.D.1c	*Mainly annual forbs.*
	V.D.2	LOW FORB COMMUNITIES, dominated by forb growth forms less than 1 m tall when fully developed.
217	V.D.2a	*Mainly perennial flowering forbs, and ferns,* e.g. the *Celmisia* meadows in New Zealand and the Aleutian forb meadows, Alaska.
218	V.D.2b	*Mainly annual forbs.*
219	V.D.2b(1)	Ephemeral forb communities in tropical and subtropical regions with very little precipitation where, from autumn to spring, clouds moisten vegetation and soil, e.g. in the coastal hills of Peru and northern Chile. Dominated by annual forbs which germinate at the beginning of the cloudy season, growing abundantly until the end of it, and giving the landscape a fresh green look. The dry season aspect is desert-like. Geophytes and cryptogamic hemicryptophytes or chamaephytes are always present and may dominate locally. Some phanerophytes may occur as relicts of natural cloud woodland.
220	V.D.2b(2)	Ephemeral or episodical forb communities of arid regions: the 'flowering desert'. Mostly fast growing forbs, sometimes concentrated in depressions where water can accumulate. Sometimes these communities can form ground synusiae in shrub or dwarf shrub formations of arid regions (cf. I.V.C), e.g. in Sonoran Dsert.

221	V.D.2b(3)	Episodical forb communities. Irregular communities of varying structure and composition, developing in the dry parts or river beds during low-water periods of more than two months, or on temporarily dry areas in beds where the river tends to change channels more or less frequently, e.g. in the river beds of the major rivers of the llanos of Colombia and Venezuela.
	V.E.	*HYDROMORPHIC FRESH-WATER VEGETATION* (aquativ vegetation). As in the case of the aquatic mangrove communities, so helophyte communities are not classified in this category. Thus, stands of *Typha* are considered medium tall graminoid communities, wet or flooded most of the year (cf. V.B.5a(1)).
	V.E.1	ROOTED FRESH-WATER COMMUNITIES. Composed of aquatic plants that are structurally supported by the water, i.e. not self-supporting in contrast to helophytes.
222	V.E.1a	*Tropical and subtropical forb formations without appreciable seasonal contrast, e.g. Victoria regia communities of the Amazon.*
223	*V.E.1b*	*Middle and higher latitude forb formations* with major seasonal contrasts, e.g. *Nymphaea* and *Nuphar* communities.
	V.E.2	FREE-FLOATING FRESH-WATER COMMUNITIES.
224	V.E.2a	*Tropical and subtropical free-floating formations, e.g. Pistia, Eichhornia, Azolla pinnata,* etc.
225	V.E.2b	*Free-floating formations of the middle and higher latitudes, disappearing during the winter,* e.g. *Utricularia purpurea, Lemna.*

BIBLIOGRAPHY

Note:

The items of the published literature cited in this volume are listed in the following pages. In addition, other material is presented as well, on the assumption that it will help the reader gain a deeper understanding of vegetation mapping. The material listed here includes therefore publications which are not necessarily the most recent ones, but the value of which remains unquestioned.

The number of vegetation maps cited here may seem small and, of course, vast numbers of such maps could have been listed. This seemed inappropriate. Readers interested in information on vegatation maps are therefore invited to consult the International Bibliography of Vegetation Maps (IBVM). In the bibliography of this volume, reference to the IBVM is made in the item of Küchler, ed., 1965ff. More recent information is available from the author who continues to collect material for future volumes of the IBVM.

In addition, large numbers of vegetation maps prepared for application are never printed for distribution. They exist in limited numbers as appendices of project reports. However they still may contain important information.

Abbate, G., G.C. Avena, C. Blasi and L. Veri. 1981. Studio delle tipologie fitosociologische del Monte Soratte (Lazio) e loro contributo nella definizione fitogeografico dei complessi vegetazionali centro – appenninici. (With colour map 1:10.000). Consiglio Nazionale delle Ricerche. AQ/1/125 Roma.

Abramova, T.C. 1962. The importance of a medium scale geobotanical map for the geobotanical subdivision for agricultural purposes.

Addor, E.E. 1963. Vegetation description for military purposes. Vicksburg, MS. United States Army Engineer Waterways Experiment Station. Miscellaneous Paper No. 3–610, pp. 98–128.

Adriani, M.J. and E van der Maarel. 1968. 'Voorne in de branding'. (Met kaart 1:25.000-kleur) Stickting Wetenschappelijk Duinonderzoek Oostvoorne.

Agency for Cultural Affairs, Ministry of Education 1969–1976: The vegetation map and the map of natural monuments. 1:200,000 with color. 129 leafs. Tokyo.

Aichinger, E. 1954. Statische und dynamische Betrachtung in der pflanzensoziologischen Forschung. Zürich. Geobotanisches Forschungsinstitut Rübel. Veröffentlichungen, vol. 29.

Albertson, F.W., G.W. Tomanek and A. Riegel. 1957. Ecology of drought cycles and grazing intensity on grasslands of the central Great Plains. Ecological Monographs, 27:27–44.

Aldrich, R.C. 1979. Remote Sensing of Wildland Resources: A State-of the-Art Review, General Technical Report RM-71, Fort Collins, CO: U.S. Forest Service Rocky Mountain Forest and Range Experiment Station.

Andere, D.K. 1981. The Kenya Rangeland ecological Monitoring Unit (KREMU) Low-level aerial techniques Rep. Int. Workship 6–11 Nov. 1979, Nairobi. pp. 59–69. I.L.C.A. Addis-Abeba.

Anderson, A. 19xx.Vegetationskartor Siljan (4 maps in colour 1:50.000) Meddelande fran Naturgeografiska institutionen Stockholms universiteit Nr. 160.

Anderson, J.A.R. and J. Muller. 1975. Palynological study of a holocene peat and a miocene coal deposit from NW Borneo. Rev. Palaeobot. Palynol. 19:291–351.

Andersson, L. and T. Rafstedt. 19xx. Fjällens Vegetation Kartpr Del: 1 Jämtlands län (6 maps in colour 1:100.000) Meddelande från Naturgeografiska institutionen Stockholms universiteit Nr. 162.

Andrews, H.J. and R.W. Cowlin. 1936. Forest type maps of Oregon and Washington. Portland, OR. Pacific Northwest Forest & Range Experiment Station.

Anonymous. 1934. Areas characterized by major forest types: Alabama, Florida, Georgia, Louisiana, Mississippi. New Orleans, LA. Southern Forest Experiment Station.

Anonymous. 1940. Areas characterized by major forest types: North and South Carolina. Asheville, NC. Appalachian Forest Experiment Station.

A.W. Küchler & I.S. Zonneveld (eds.), Vegetation mapping. ISBN 90-6193-191-6.
© 1988, Kluwer Academic Publishers, Dordrecht. Printed in the Netherlands.

Anonymous. 1941a. Areas characterized by major forest types: Virgina. Asheville, NC. Appalachian Forest Experiment Station.

Anonymous. 1941b. Areas characterized by major forest types: northern Michigan. St. Paul, MN. Lake States Forest Experiment Station.

Anonymous. 1948a. Lac Letondal. Ottawa. Department of Mines & Resources, Canadian Forest Services.

Anonymous. 1948b. Wildlife and cover map of Arkansas. Little Rock, AR. Arkansas State Game and Fish Commission.

Anonymous. 1955. Field manual: grassland sampling. Berkeley, CA. California Forest & Range Experiment Station.

Anonymous. 1973. International classification and mapping of vegetation. Paris. United Nations Educational, Scientific and Cultural Organization. Ecology and Conservation. No. 6.

Anonymous. 1976. Montana: climax vegetation. Helena, MT. United States Department of Agriculture, Soil Conservation Service.

Apinis, A.E. 1963. Der Wert der Vegetationskarte für die grundlegenden bodenmikrobiologischen Untersuchungen. In: Reinhold Tüxen, 1963a, pp. 185–194.

Arrigoni P.V. and Di Tommaso. 1981. Carta della vegetazione dell'Isola di Giannutri. (Provincia di Grosseto) (With colour map 1:5.000) Consiglio Nazionale delle Richerche AQ/1/130. Roma.

Aubert de la Rüe, E., F. Bourlière and J-P. Harroy. 1957. The Tropics. Alfred A. Knopf, Inc. New York, NY.

Aubréville, A. 1961. De la nécessité de fixer une nomenclature synthétique des formations végétales tropicales avant d'entreprendre la cartographie de la végétation tropicale. In: Henri Gaussen, 1961a, pp. 37–47.

Augarde, J. 1957. Contribution à l'étude des problèmes de l'homogénéité en phytosociologie. Montpellier. Service de la carte phytogéographique. Bulletin, Serie B, 2:11–23.

Avena, G.C. and C. Blasi. 1980. Carta della vegetazione del Massiccio del Monte Velino. Appennino Abruzzese. (With colour map 1:25000) Consiglio Nazionale delle Richerche. AQ/1/35 Roma.

Ayasse, L. and R. Molinier. 1955. Carte des groupements végétaux des environs de la Motte-du-Caire au 1/10,000. Revue Forestière Française, 9–10:697–707.

Baig, M.S. 1977. Inventory and evaluation of rangeland. An example for rangelands in arid zones worked out for Quetta Pishin areas, Baluchistan, Pakistan. M.Sc. thesis ITC, Enschede, The Netherlands, 132 pp + app.

Bailey, R.G. 1980. Description of the ecoregions of the United States. Washington, DC. U.S. Dept. of Agriculture, Forest Service. Miscellaneous Publication No. 1391.

Bailey, R.G. 1976. Ecoregions of the United States. Fort Collins, CO. U.S. Dept. of Agriculture, Forest Service. Rocky Mountains Forest & Range Experiment Station.

Bakker, T.W.M., J.A. Klijn and F.J. van Zadelhoff. 1979. 'Duinen en duinvalleien' (With colour maps 1:100.000) Een landschapsecologische studie van het Nederlandse duingebied. Centrum voor Landbouwpublicaties en landbouwdocumentatie Wageningen. pp. 201.

Baillieux, P., P. André and C. Lheureux. 1977. Prunus serotina Ehrh. intérêt, exigences et traitement. Bull. Soc. Roy. For de Belgique 84(4):197–208.

Bannink, J.F., H.N. Leys and I.S. Zonneveld. 1973. Vegetation, habitat and site class in Dutch conifer forst. Bodemk. Stud. 9 Stichting v. Bodem Kartering, Wageningen. Versl. Landbouwk. Onderzoek 800, 188 pp.

Bannink, J., H.N. Leys and I.S. Zonneveld. 1974. Weeds as Environmental Indicators Especially for Soil Conditions. Versl. Landbouwk. Onderzoek 807, Bodemk. Studies 11, Stiboka, Wageningen, 88 pp.

Barbero, M. and P. Ozenda. 1979. Carte de la végétation potentielle des Alpes piémontaises. Grenoble. Documents de cartographie écologique, 21: 139–162.

Barker, G.R. 1982. 'FRIS: St. Regis' Forest Management Approach," Chapter 40 in Remote Sensing for Resource Mangement, C.J. Johannson and J.L. Sanders, eds., Ankeny, IA: Soil Conservation of America, pp. 454–469.

Barnhart, Dr. 1930. Cambridge. 5th International Botanical Congress. Proceedings, p. 559.

Baumgardner, M.F. 1974. Remote sensing information in the service of ecology. Proceeding first International Congress of Ecology. 274–278. Den Haag.

Beadle, N.C.W. and A.B. Costin. 1952. Ecological classification and nomenclature. Proceedings of the Linnean Society of New South Wales, 67:61–82.

Beard, J.S. 1944. Climax vegetation in tropical America. Ecology, 25:127–158.

Beard, J.S. 1972. 'The Vegetation of the Newdegate and Bremer Bay areas, Western Australia.' Maps and explanatory Memoir 1:250.000 Series (With 2 monocolor maps.) Vegmap Publications, Sydney.

Beard, J.S. 1974–1981. The vegetation survey of Western Australia. Nedlands, W.A. University of Western Australia Press.

Beek, K.J. 1978. Land evaluation for agricultural development. Diss. Wageningen, ILRI Publ. 23, Wageningen.

Beek, K.J. and J. Bennema. 1972. Land evaluation for agricultural land use planning. An ecological methodology. Dept. Soil. Sci. & Geol., Agric. Univ. Wageningen, Netherlands, 70 p.

Bellan, M.F. 1979. Cartographie de la végétation à l'aide de l'image-satellite. Exemple en Inde: Carte de la végétation des Palni à 1/250 000. Thèse de Doctorat de 3e cycle, Univ. Paul Sabatier, Toulouse; 84 p., 2 images satellites, 1 carte.

Belov, A.V. 1983. Some aspects and perspectives of geobotanical mapping and prognostication in Siberia. Leningrad. Academy of Sciences of the U.S.S.R., Komarov Institute of Botany. Geobotanical Cartography, pp. 18–24.

Bennet, C.F. 1963. Notes on the use of light aircraft in mapping the vegetation in the American tropics. Professional Geographer, 15:6:21–24.

Benson, D.H. 1981. Vegetation of the Agnes Banks sand deposit, Richmond, N.S.W. Cunninghamia, 1:35–57.

Berezin, A.M. 1962. On the choice of a scale of aerophotos and of kinds of aerorecords for the deciphering and mapping of forests. In: Victor B. Sochava, 1962a, pp. 223–231.

Berg, A. van den et al. 1984. MAP: a set of computer programs that provide for input, output and transformation of cartographic data. in: IALE, the International Association for Landscape Ecology. Proceedings of the first international seminar on Methodology in Landschape Ecological Research and Planning. Theme III Methodology of data analysis. Roskilde University Centre, Denmark.

Berg, J. van der. 1985. MONA 1984 Rapport over de vorderingen van het Mona-project, deelproject van het Moneragala District. Survey Project. ITC Enschede.

Bertrovic, S. 1973. Pflanzensoziologische Kartierungen in Kroatien und andern Teilen Jugoslawiens. In: Reinhold Tüxen, 1963a, pp. 231–243.

Bharucha, F.R. 1952. Vegetation cartography. Indian Forester, 78:6.

Bharucha, F.R. 1955. Les problèmes cartographiques du sud-est asiatique. In: Henri Gaussen, pp. 189–193.

Bhimaya. 1967. Studies on detailed survey methods for management of range lands in North Western arid zone of India. Int. Seminar on Integrated Survey of Natural Grazing areas.

Bickmore, D.P. 1982. Mapping the vegetation of Britain. Cartographica, 19:2:100–107.

Bie, S.W. 1983. Project Criteria/Soft- and hardware for Global Land/soil Monitoring System Consultats rep. UNEP. Nairobi, Kenya.

Bie, S. and J.J. Kessler. 1981. An aerial resource inventory of the National Park 'Boucle du Baoule' (Mali, West-Africa). Nature Conservation Department, Agricultural University, Wageningen.

Billings, W.D. 1952. The environmental complex in relation to plant growth and distribution. Quarterly Review of Biology, 27:251–265.

Blaeu, J. 1640-1654. Theatrum Orbis Terrarum. Amsterdam. Novus Atlas, vols. 1–6.

Blasco, F., F. Lavenu and G. Saint. 1983. Résultats des simulations de SPOT au Bangladesh. Colloque 'Géographie, Aménagement et Télédétection spatiale' CNES-CNRS, Chantilly 18–19.

Blasco, F., Y. Laumonier and Purnajaya. 1983. Tropical vegetation mapping. Sumatera. Biotrop Bulletin No. 18. Bogor. Indonesia. 71 p.

Blasco, F. 1982. The transition from open forest to savanna in continental South East Asia. in: F. Bourlière – tropical savannas, Ecosystems of the world 13:167–181, Elsevier. Amsterdam.

Blasco, F. 1971. Montagnes du Sud de l'Inde; forêts, savanes, écologie. Thèse de Doct. Etat, Univ. Paul Sabatier, Toulouse et Inst. français de Pondichéry, Trav. Sect. scient. Techn., T. X, 436 p., 3 cartes.

Bleeker, P. 1975. Explanatory Notes to the Land Limitation and Agricultural Land Use Potential Map of Papua New Guinea. (With colour maps, Scale 1:1.000.000) Land Research Series N. 36, pp. 80. Commonwealth Scientific and Industrial Research Organization, Autralia.

Bleydenstein, J. 1969. Grassland and Vegetation Surveys for Economic Development. Physical Resource Investigations for Economic Development. O.A.S. Case Book OAS Washington.

Boer, T.A. de. 1954. Die Grünlandvegetationskartierung in den Niederlanden. Festschrift Aichinger, pp. 1232–1234. Springer Verlag. Wien.

Boer, T.A. de 1963. Die Grünlandvegetationskartierung als Grundlage für die landwirtschaftliche Verbesserung ganzer Gebiete in den Niederlanden. In: Reinhold Tüxen, 1963a, pp. 441–455.

Boer, Th. A. de. 1983. Vegetation as an Indicator of Environmental Changes. Environm. Monitoring and Assessm. 3.

Boerboom, J.H.A. 1960. De plantengemeenschappen van de Wassenaarse Duinen. Mededelingen van de Landbouwhogeschool te Wageningen. Nederland 60 (10) 1–135 pp. 135. (With colour maps, Scale 1:5.000).

Boere, G.C. and W.K.R.E. van Wingerden. 1983. Integratie en Perspectiven van inventariseren en monitoring van Flora en Fauna in Nederland. Med. W.L.O. 9 (1982) N. 1. 5–10.

Bohn, U. 1981. Vegetationskarte der Bundesrepublik Deutschland 1:200 000 – Potentielle natürliche Vegetation – Blatt CC 5518 Fulda. Schr. Reihe Vegetationskunde 15.

Borza, A. 1963. Bibliographie der Vegetationskarten Rumäniens. Excerpta Botanica, sectio B: Sociologica, 5:103–107.

Boughey, A.S. 1957. The physionomic delimitation of West-African vegetation types. Journal of the W.-African Sc. Association 3, 2, (pp. 148–165).

Box, E.O. 1981. Macroclimate and plant forms. Dr. W. Junk Publishers. The Hague, NL.

Brakebusch, L. 1983. Über die Bodenverhältnisse des nordwestlichen Teiles der argentinischen Republik mit Bezugnahme auf die Vegetation. Petermanns Geographische Mitteilungen, vol. 39, No. 7, Tafel 11.

Braun, E.L. 1950. Deciduous forests of eastern North America. Blakiston Co. Philidalphia, PA.

Braun-Blanquet, J. 1932. Pflanzensoziologie. Jul. Springer, Berlin. Plant Sociology (transl.) by Fuller and Conard, 1932. McGraw-Hill-Book Co., New York-London, (see also 1966).

Braun-Blanquet, J. 1944. Sur l'importance pratique d'une carte détaillée des associations végétales de la France. Station internationale de géobotanique méditerranéenne et alpine. Montpellier. Communication 86.

Braun-Blanquet, J., L. Emberger and R. Molinier. 1947. Instructions pour l'établissement de la carte des groupements végétaux. Centre National de la Recherche Scientifique. Paris.

Braun-Blanquet, J. and Y.T. Chou. 1947. Carte des groupe-

554

ments végétaux de la France, région nordouest de Montpellier. Station internationale de géobotanique méditerranéenne et alpine. Montpellier.

Braun-Blanquet, J. 1964. Pflanzensoziologie, 3rd ed. Springer Verlag. Wien.

Brinkman, J.W.E. 1972. Studiekartering van het landgoed Schovenhorst te Putten. Scriptie voor de afdelingen Bodemkunde, Houtteelt en Natuurbeheer in het kader van de ingenieursstudie aan de Landbouwhogeschool te Wageningen.

Brinkman, R. and A.J. Smyth. 1973. Land evaluation for rural purposes. *ILRI Publ.* 17, Wageningen, Netherlands, 115 p.

Brockmann-Jerosch, H. 1935. Vegetation der Erde. In: Hermann Haack, Physikalischer Wand-Atlas. Perthes Verlag. Gotha.

Broek van den, J.M.M. en W.H. Diemont. 1966. Het Savelsbos. Bosgezelschappen en Bodem. (With colour maps Scale 1:5.000). Verslagen van Landbouwkundige Onderzoekingen 682 RIVON Verhandeling Nr. 3. deel XXIII pp. 120.

Bromme, T. 1851. Atlas zu Alexander von Humboldts Kosmos. Krais & Hoffmann Verlag. Stuttgart.

Brown, D. 1954. Methods of Surveying and Measuring Vegetation. Bulletin 6. C.A.B. Books, England.

Brown, D.E. and C.H. Lowe. 1977. Biotic communities of the South-west. Fort Collins, CO. U.S. Dept, of Agriculture, Forest Service. General Technical Report RM-41.

Brullo, S., F. Fagotto and G. Cicero Lo. 1980. Esempi di cartografia della vegetazione di alcune aree della Sicilia. (With colour maps Scales: 1:10.000; 1:12.500; 1:25.000). Consiglio Nazionale delle Ricerche AQ/1/37–40. pp. 66. Roma.

Brünig, E.F. 1970. Stand structure, physiognomy and environmental factors in some lowland forests in Sarawak. Trop. Ecol. 11:26–43.

Brünig, E.F. 1973. Bioecology of tropical forests. *In:* E.F. Brünig, Silvics and silviculture management in humid tropical forests. 68 p.

Brush, G.S., C. Lenk and J. Smith. 1976. Vegetation map of Maryland. Baltimore, MD. Johns Hopkins University, Dept. of Geography.

Bruyn, C.A. de. 1984. USEMAP. Operations Manual Volume I, general description. ITC Department of Urban Survey. ITC/US/CAB/840119.

Buchwald, K. 1951. Vegetationskarten als Grundlage für die Landschaftsplanung. Pflanze und Garten, 1:11–13.

Buchwald, K. 1953. Generalwasserpanung auf Grund natürlicher Standortkartierung am Beispiel des Argengebietes. In: Wirksame Landschaftspflege durch wissenschaftliche Forschung. Pp. 57–62. Dorn Verlag. Bremen-Horn.

Buchwald, K. 1954. Grünlandkartierung im Rahmen des ERP Grünlandförderungsprogramms. Landwirtschaft, angewandte Wissenschaft, No. 21.

Buell, P.F. and P. Dansereau. 1966. Studies on the vegetation of Puerto Rico. Mayaguez. University of Puerto Rico, Institute of Caribbean Science.

Buks, I.I. 1975. The principles for compiling a correlational ecologo-phytocoenological map of Siberia and the Far East. XII. International Botanical Congress. Institute of Geography of Siberia and the Far East. (Irkutsk, U.S.S.R.) Leningrad, pp. 5.

Bunnik, N.J.J. 1978. 'The multispectral reflectance of shortwave radiation by agricultural crops in relation with their morphological and optical properties'. H. Veenman & zonen b.v. Wageningen.

Burke, T.F. and H. Shelton. 1953. The United States of America. Denver, CO. Jeppeson & CO.

Burrough, P.A. 1984a. Geografische Informatie Verwerking. Diktaat bij het kollege F302: Geografische informatie systemen, cursus 1984/85. Vakgroep Fysische Geografie. R.U. Utrecht. (english text).

Burrough, P.A. 1984b. The use of geographical information systems for cartographic modelling in Landscape Ecology. in: IALE, the International Association for Landscape Ecology. Proceedings of the first international seminar on Methodology in Landscape Ecological Research and Planning. Theme III Methodology of data analysis. Roskilde University Centre, Denmark.

Burtt-Davy, J. 1938. The classification of tropical woody vegetation types. Oxford. Imperial Forestry Institute Paper No. 13.

Butorina, T.N. 1962. Principles of compilation of forest type maps. In: Victor B. Sochava, 1962a, pp. 103–109.

Cain, S.A. and G.M. de Oliveira Castro. 1959. Manual of vegetation analysis. Harper & Brothers. New York, NY.

Camarasa, J.M., R. Folch i Guillen, R.M. Masalles and E. Velasco. 1976. Paisatge vegetal del delta de l'Ebre. Institució Catalana d'Historia Natural. Barcelona.

Caneva, G., G. De Marco and L. Mossa. 1981. Analisi fitosociologica e cartografia della vegetazione (1:25.000) dell'isola di S. Atioco. (Sardegna sud-occidentale) (With colour map 1:25.000) Consiglio Nazionale delle Ricerche. AQ/1/124. Roma.

Caniglia, G., F. Chiesure, L. Curti, G.G. Lorenzoni, S. Marchiori, S. Razzara and N.T. Marchiori. 1978. Carta della vegetazione di Torre Colimena Salento. Puglio Meridionale. (With colour map 1:25.000). Consiglio Nazionale delle Ricerche. AQ/1/8. Roma.

Cano, E. and R. Gómez Cadret. 1968. La vegetación de la República Argentina (Bibliography of vegetation maps). Buenos Aires. Instituto Nacional de Technología Agropecuaria. Centro Nacional de Investigaciones Agropecuarias, Instituto de Botánica Agrícola.

CARAP. 1980. Countrywide Animal and Range Assessment Project, Botswana. DHV/ITC, Amersfoort/Enschede, Vol. III.

Carnahan, J.A. 1975. Problems of vegetation mapping in Australia. 12th International Botanical Congress, Leningrad. Section 8: Ecological Botany.

Carnahan, J.A. 1976. 'Natural Vegetation' (With colour map 1:6.000.000). Atlas of Australian Resources. 2a. Series Geographic Section Division of National Mapping.

Department of National Resources. Camberra, Australia. pp. 26.

Carnegie, D.M., S.D. de Gloria and R.N. Colwell. 1975. Usefulness of Landsat data for monitoring plant development and range conditions in California's annual grassland. Houston, TX. NASA Johnson Space Center. Proceedings of the NASA Earth Resources Survey Symposium, 1A: 19–41.

Cate, R.B., J.A. Artley and D.E. Phinney. 1980. Quantitative Estimation of Plant Characteristics Using Spectra Measurement: A Survey of the Literature, Tecnical Report N. JSC-16298; SR-LO-0048, Houston, TX: NASA/AgRISTARS/Lockheed Engineering and Management Services, Inc.

CCAS. 1956. CSA specialist meeting on phytogeography. Yangambi 28 July – 8 August 1956. CCTA Publ 22, pp. 1–40.

Centre National de la Recherche Scientifique. 1961. Méthodes de la Cartographie de la Vegetation. Toulouse 16–21 Mai 1960. pp. 322.

Chabrol, P. 1961. Méthodes cartographiques et plans forestiers dans l'œuvre de la réformation de la Grande Maitrise des Eaux et Forêts de Toulouse au 17e siècle. In: Henri Gaussen, 1961a, pp. 219–226.

Chamberlin, T.C. 1983. Vegetation map of Wisconsin. Madison, WI. Atlas of the Geological Survey, plate 3.

Champion, H.G. 1936. A preliminary survey of the forest types of India and Burma. Indian Forest Records, N.S.

Christian, C.S. and G.A. Stewart. 1953. General report on the survey of the Katherine-Darwin region, 1946. Melbourne. Commonwealth Scientific and Industrial Research Organization. Land Research Series No. 1.

Cibula, W.G. 1981. Computer implemented land classification using Landsat MSS digital data, II: Vegetation and other land cover analysis of the Olympic National Park. NSTL Station, NS. NASA Earth Resources Laboratory, Report No. 193.

Clausman, J. Zorg voor het Milieu. 'Vegetatiekartering' pp. 7.

Clausman, P.H.M.A. and A.J. den Held. 1982. Vergelijking in de tijd in het kader van het vegetatie-onderzoek van de provincie Zuid Holland. Med. W.L.O. 9 n. 1: 11–18.

Cleef, A.M. 1981. The vegetation of the páramos of the Colombian Cordillera Oriental. Diss. Utrecht. Statement 1.

Clements, F.E. 1916. Plant succession: an analysis of the development of vegetation. Carnegie Institution of Washington, Publ. No. 242, 512 p.

Clements, F.E. 1928. Plant succession and indicators. A definitive edition of plant succession and plant indicators. Hafner Pub. Co Ltd. 453 p.

Clements, F.E. 1936. Nature and structure of the climax. Journ. Ecol. 24: 252–284.

Cobb, S. 1976. The distribution and abundance of the large herbivore community of Tsavo National Park. Ph.D. thesis, Oxford University.

Cochrane, G.R. 1963. A Physiognomic Vegetation Map of Australia. University of Australia. J. Ecol. 51, Nov./63 Blackwell Scientific Publications Oxford. pp. 639–655.

Cochrane, G.R. 1967. 'A Vegetation Map of Australia'. Some explanatory notes. F.W. Cheshire Melbourne, Canberra, Sydney. pp. 34.

Cochrane, G.R. 1967. The Description and Mapping of Vegetation in Australia: a Contribution to the International Biological Programme. pp. 299–316.

Colwell, J. 1974. Vegetation canopy reflectance. Remote Sensing of Environment, 3: 175–183.

Colwell, R.N. 'Manual of remote sensing'. American society of photogrammetry, Sheridan Press, 1983, applications (2111–2384).

Cook, G.H. and J.C. Smock. 1978. Vegetation map of New Jersey. Trenton, NJ. Geological Survey of New Jersey.

Corbetta, F. and G. Pirone. 1981. Carta della vegetazione di Monte Alpi e zone contermini. (Tavoletta 'Latronico' della carta d'Italia Consiglio Nazionale delle Ricerche. AQ/1/122. Roma. (With colour map Scale 1:25.000).

Costin, A.B. 1954. Study of the ecosystems of the Monaro region of New South Wales. Government Printer. Sydney, N.S.W.

Cottam, G. and J.T. Curtis. 1956. The use of distance measures in phytosociological sampling. Ecology, 37: 451–460.

Council of Europe 1979. Vegetation map of the Council of Europe member states (ed. P. Ozenda). Nature and Environment Series no. 16.

Coupland, R.T. 1959. Effects of changes in weather conditions upon grasslands in the northern Great Plains. In: Grasslands. American Association for the Advancement of Science.

Couteaux, M. 1974. Essai de cartographie écologique du bas Vivarais. Grenoble. Documents de cartographie écologique, 13: 49–68.

Cowles, H.C. 1899. The ecological relations of the vegetation on the sand dunes of Lake Michigan. Bot. Gaz. 27: 95–391.

Cowles, H.C. 1911. The causes of vegetation cycle. Bot. Gaz. 51: 161–183.

Cox, J.A., R. Higler and B. van den Hoff. 1983. Land Ecological Survey of South-Monaragala, Sri Lanka. Unpubl MSc thesis, ITC, Enschede.

Credaro, V., C. Ferrari, A. Pirola, M. Speranza and D. Ubaldi. 1980. Carta della Vegatazione del Crinale Appenninico dal Monte Giovo al Corno alle Scale. (Appennino Tosco-Emiliano). (With colour map 1:25.000). Consiglio nazionale delle Richerche. AQ/1/81. Roma.

Croze, H. and M.S. Gwynne. 1978. Rangeland monitoring: function, form and results. Discussion paper prepared for the West African Rangeland. Coordination Meeting, Regional Centre for Remote Sensing, Ouagadougou.

Croze, H., M. Norton-Griffiths and M.D. Gwynne. 1978. Ecological monitoring in East Africa. New Scientist, Vol. 77 (1088), 283–285.

Cunningham, R.N. and H.C. Moser. 1939. The distribution of forests in the upper peninsula of Michigan. St. Paul,

556

MN. Lake States Forest Experiment Station.

Cunningham, R.N. and H.C. Moser. 1940. The distribution of forests in northern Minnesota. St. Paul, MN. Lake States Forest Experiment Station.

Curran, P. 1980. Multispectral remote sensing of vegetation amount. In: B.W. Adkinson, ed., Progress in physical geography. Pp. 315–341. Edward Arnold. London.

Curtis, J.T. and R.P. McIntosh. 1951. An upland forest continuum in the prairie-forest border region of Wisconsin. Ecology, 32: 476–496.

Curtis, J.T. 1959. The vegetation of Wisconsin. Madison, WI. University of Wisconsin Press.

Curtis, J.T. 1959. The vegetation of Wisconsin. An ordination of plant communities. The University of Wisconsin Press. Madison.

Curtis, M.A. 1860. Vegetation map of North Carolina. Raleigh, NC. Geological and Natural History Survey of North Carolina, part 3.

Cushman, M. and M. Luck. 1949. Vegetation-soil and timber stands and vegetation elements, Mendocino Country. Berkeley, CA. University of California. California Forest & Range Experiment Station.

Dangermond, J. 1982. Software Components commonly used in Geographic Information Systems. Environmental Systems Research Institute, California.

Dansereau, P. 1957. Biogeography. Ronald Press. New York, NY.

Dansereau, P. 1958. An universal system for recording vegetation. Contribution de l'Institut de Botanique de l'Université de Montréal N° 72, 52 p., 1 carte.

Danserau, P. 1961. Essai de représentation cartographique des éléments structuraux de la végétation. In: Henri Gaussen, 1961a, pp. 233–255.

Davis, J.H. 1943. Southern Florida. Florida Geological Survey. Bulletin No. 25.

Davis, J.H. 1967. The natural vegetation of Florida. Gainesville, FL. University of Florida, Agricultural Experiment Station.

Deering, D.W. and R.H. Hass. 1980. Using Landsat Digital Data for Estimating Green Biomass, Nasa Technical Memorandum 80727, Greenbelt, MD: NASA Goddard Space Flight Centre.

De Marco, G. and L. Mossa. 1980. Analisi fitosociologica e cartografia della vegetazione (1:25.000) dell'Isola di S. Pietro (Sardegna sud-occidentale) (colour map.) Consiglio Nazionale delle Ricerche. AQ/1/80. Roma.

Demchenko, L.A. 1962. An essay on the compilation of a medium scale map for a district atlas. In: Victor B. Sochava, 1962a, pp. 211–214.

Deshayes, M. 1981. Traitement numérique des données LANDSAT. Applications à la cartographie automatique de la végétation tropicale. Thèse Docteur-Ingénieur, Univ. Paul Sabatier, Toulouse, 121 p.

Devred, R. 1955. Kwango. Brussels, Institut National pour l'Etude Agronomique du Congo belge. No. 10.

Devred, R. 1960. La cartographie de la végétation au Congo belge. Bulletin agricol du Congo belge et du Ruanda-Urundi, 51: 529-542.

DHV/ILAO. 1976. Shinyanga regional integrated development plan – phase II.

Dierschke, H. 1974. Zur Abrenzung von Einheiten der heutigen potentiell natürlichen Vegetation in Waldarmen Gebieten Nordwestdeutslands. In: Tatsachen und Probleme der Grenzen in der Vegetation. Ber. Intern. Symp. Rinteln 1968 305–325.

Dijkema, K.S. and W.J. Wolff. 1983. Flora and vegetation of the Wadden Sea Islands and coastal areas. Rotterdam. A.A. Balkema, Publishers. Final Report of the section 'Flora and vegetation of the islands.' Wadden Sea Working Group, Report No. 9.

Dijkema, K.S. 1980. Toward a vegetation and landscape map of the Danish, German and Dutch Wadden Sea islands and ainland coastal areas. Acta Botanica Neerlandica, 29: 523–531.

Dobbar, H.F. 1982. Monitoring van Flora en Vegetatie. Med. W.L.O. n. 1: 5–10.

Dobremez, J.F. 1972. Mise au point d'une méthode cartographique d'étude des montagnes tropicales. Le Népal, écologie et phytogéographie. Thèse doct. Etat, Univ. Grenoble, 373 p., 3 cartes.

Dobremez, J.F., J. Girel, J.P. Guichard, M.C. Vartanian and F. Vigny. 1976. Bourg-en-Bresse. Documents de cartographie écologique, 18: 11–42.

Dobremez, J.-F. 1971ff. Carte écologique du Nepal. Documents pour la carte de la végétation des Alpes, 9ff: 147–190.

Dombremez, J.-F. and F. Vigny. 1982. Matériaux pour une carte écologique régionale. Grenoble. Documents de cartographie écologique, 25: 1–25.

Doing, H. 1961. Systematische Ordnung und floristische Zusammensetzung Niederländischer Wald und Gebüschgesellschaften. Wentia 8, 1–85.

Doing-Kraft, H. 1963. Eine Landschaftskartierung auf vegetationskundlicher Grundlage in den Dünen bei Haarlem. In: Reinhold Tüxen, 1963a, pp. 297–312.

Doing, H. 1964. Recreatie en Natuurbescherming in het Noordhollands Duinreservaat. Supplement 2: Vegetatie. Mededeling Nr. 69c/1964. pp. 52. (With colour maps Scale 1:25.000).

Doing, H. 1974. Landschapsecologie van de duinstreek tussen Wassenaar en IJmuiden. Meded. Landbouwhogeschool Wageningen 74–12.

Doing H. and S. van der Werf. 1962. Overzicht der Nederlandse Vegetatiekaarten. Laboratorium voor Plantensystematiek en -geografie, Landbouwhogeschool, Wageningen, Nederland. H. Veenman & Zonen N.V. Wageningen.

Doing, H. en S. van der Werf. 1962. Overzicht der Nederlandse Vegetatiekaarten. Medede. Landbouwhogeschool, Wageningen 62(6), 1–30 (62) pp. 30.

Donita, N., V. Leandru and E. Puscaru-Soroceanu. 1960. Harta geobotanica of Romania. Bucharest. Academia Republicii Populare Romini.

Donselaar, J. van. 1965. An ecological and phytogeographical study of Northern Surinam Savannas. Wentia 14:1–169.

Drude, O. 1907. Die kartographische Darstellung mitteldeutscher Vegetaionsformationen. Dresden. Bericht des freien Vereins für systematische Botanik und Planzengeographie.

Drude, O. 1905. Die Methoden der speziellen pflanzengeographischen Kartographie. Wien. 2nd International Botanical Congress. Wissenschaftliche Eergebnisse.

Dunning, D. 1942. A site classification for the mixed conifer selection forests of the Sierra Nevada. Berkeley, CA. California Forest & Range Experiment Station, Research Note 28.

Dupias, G., H. Gaussen, M. Izard and P. Rey. 1965. Carte de la végétation de la France à 1/200 000e; feuille 80–81: Corse.

Du Rietz, G. 1921. Methodologische Grundlage der modernen Pflanzensoziologie. Akademische Abhandlung, Upsala.

Du Rietz, G.E. 1931. Life forms of terrestrial flowering plants. Acta phytogeographica suecica, vol. 3.

Du Rietz, G.E. 1932. Vegetationsforschung mit Soziationsanalytischer Grundlage. Abderhaldens Handbuch der Biologischen Arbeitsmethoden 11 (5), pp. 293–480.

Duvigneaud, P. 1960. Le Laboratoire de Botanique Systématique et de la Phytogéographie de l'Institut Botanique Léo Errera. Comm. Léo Errera Univ. Libre Bruxelles 1958: 135–184, Bruxelles.

Duvigneaud, P. 1961. Application de la méthode des groupes écologiques à la cartographie au 1/50,000 des forêts de la Lorraine belge. In: Henri Gaussen, 1961a, pp. 83–86.

E. Afr. Agric. For. J. 1969. Special Issue. E. Afr. Agric. For. J., 34.

Edelman, C.H. 1950. Inleiding tot de Bodemkunde van Nederland (Introduction to the Soil science of the Netherlands). Noord-Hollandsche Uitgevers mij. Amsterdam.

Egler, F.E. 1951. A commentary on American plant ecology. Ecology, 32:673–694.

Egler, F.E. 1954. Vegetation science concepts I. Vegetatio, 4:412–417.

Ehwald, E. 1950. Über das Zusammenwirken von Standortskunde und Pflanzensoziologie bei der forstlichen Standortskartierung. Allgemeine Forstzeitschrift, 5:416–418.

Eiten, G. 1968. Vegetation forms. São Paulo, SP. Boletim do Instituto de Botânica, No. 4.

Eiten, G. 1968. Vegetation Forms. A classification of vegetation based on structure, growth form of the components, and vegetative periodicity. Bol. Inst. Bot. 4, Sao Paulo, Bazil, 33 pp.

Eklundh, J.D. and A. Rosenfeld. 1981. 'Image smoothing based on neighbour linking'. IEEE vol. PAMI 3, No. 6, (679–683).

Ellenberg. H. 1950. Unkrautgemeinschaften als Zeiger für Klima und Boden. Stuttgart. Eugen Ulmer Verlag.

Ellenberg, H. 1952. Auswirkungen der Grundwassersenkung auf die Wiesengesellschaften am Seitenkanal westlich Braunschweig. Stolzenau. Angewandte Pflanzensoziologie No. 6.

Ellenberg, H. 1953. Physiologisches und ökologisches Verhalten derselben Pflanzenarten. Berichte der Deutschen Botanischen Gesellschaft, vol. 65.

Ellenberg, H. 1955. Südwestdeutschland: Wuchsklimakarte. Stuttgart. Reise- und Verkehrsverlag.

Ellenberg, H. 1956. Aufgaben und Methoden der Vegetationskunde. Stuttgart. Ulmer Verlag.

Ellenberg, H. 1974. Zeigerwerte der Gefäszpflanzen Mitteleuropas. Scripta Geobotanica 9, Göttingen, Verlag Erich Goltze. Z. Auflage 122 pp.

Ellenberg, H. 1978. Vegetation Mitteleuropas mit den Alpen in ökologischer Sicht, Stuttgart, Ulmer Verlag, 981 pp.

Ellenberg, H. and O. Zeller. 1951. Die Pflanzenstandortkarte. – am Beispiel des Kreises Leonberg – (With colour maps Scale 1:50.000). Institut der Landwirtschaftlichen Hochschule Stuttgart-Hohenheim. pp. 11–49.

Ellenberg, H. and D. Mueller-Dombois. 1966. Tentative physiognomic-ecological classification of plantformations of the earth. Ber. geob. Inst. ETH. Stiftg. Rübel Zürich, 37, 1965/66.

Ellenberg, H. and F. Klötzli. 1966. Vegetation und Bewirtschaftung des Vogelreservates Neerachter Riet. (With colour map, scale 1:2.500) Inst. ETH. Stiftung Rübel. Zürich. 37 (1965/66) pp. 88–103.

Emberger, L. 1939. Carte phytogéographique du Maroc. Zürich. Geobotanisches Forschungsinstitut Rübel. Veröffentlichungen, vol. 14.

Emberger, L. 1961. Evolution, principes actuels et problèmes de la technique cartographique pratiques au service de la carte des groupements végétaux. In: Henri Gaussen, 1961a, pp. 211–217.

Emberger, L. Principes de la Méthode de Travail du Service de la Carte des Groupements Végétaux du C.N.R.S. Série B. tome III, fascicule 2. pp. 91–99.

Emberger, L., H. Gaussen and P. Rey. 1955. Service de la carte phytogéographique. Paris. Centre National de la Recherche Scientifique.

Emberger, L. and G. Long. 1959. Orientation actuelle au service de la Carte des Groupements Végétaux de la cartographie phytosociologique appliquée. Bulletin du Service de la carte phytogéographique, serie B, 4:119–146.

Emberger, L. and R. Molinier. 1955. Les couleurs. Paris. Centre National de la Recherche Scientifique, Service de la carte phytogéographique.

Environmental Agency 1975. 1977. Report of preservation for natural environments (basic investigation), with maps of actual vegetation and natural grades after vegetation with colour 129 leafs and while & black natural grades maps after vegetation 129 leafs. Tokyo.

Environmental Agency 1981, 1982. Actual vegetation map 1:50,000. (from Hokkaido to Okinawa). The 2nd national survey on the natural environment (vegetation). With colour vegetation map. 1249 leafs. (Japanese). Tokyo.

Environmental Agency 1982. The natural environment of

558

Japan. With 8 colour vegetation maps, 248 pp. (Japanese/English). Tokyo.

Esselink, P. and H.A.M.J. van Gils. 1985. Ground-based reflectance measurements for standing crop estimates. ITC Journal, 1985-1:47–52.

Etter, H. 1947. Vegetationskarte des Sihlwaldes der Stadt Zürich. Zeitschrift der Schweizer Forstverwaltung, Beiheft 24.

Falinski, J.B. 1971. Methodical basis for Map of Potential Natural Vegetation of Poland. Acta Soc. Bot. Poloniae. Vol. XL (1):209–222.

FAO, 1974. Approaches to land classification. *Soils Bulletin* 22, FAO, Rome 120 p.

FAO, 1976. A framework for land evaluation. *Soils Bulletin* 32, FAO, Rome. 72 pp.

FAO, 1976. A Framework for land evaluation. Soils bulletin no. 32, F.A.O., Rome 72 pp.

FAO, 1977. Land evaluation standards for rainfed agriculture. Report of an expert consultation. *World Resources Reports* 49, Rome.

FAO, 1979. Report of an expert consultation on land evaluation criteria for irrigation: Feb.-March 1979. *World Soil Resources Reports* 50, Rome.

FAO, 1980. Land evaluation guidelines for rainfed agriculture. M-59. Report of an expert consultation held in Rome, December 1979.

FAO, 1982. Project on classification of tropical vegetation types in Asia. First draft. Methodology and applications. Project F.A.O. No. 30, 113 p.

Fenton, E.W. 1947. The transitory character of vegetation maps. Scottish Geographical Magazine, 63:129–130.

Feoli, E., M. Langonegro and L. Orloci. 1984. Information analysis of vegetation data. Dr. W. Junk Publishers, The Hague.

Fernald, M.L. 1950. Gray's manual of botany, 8th ed. American Book Co. New York, NY.

Ferré, Y. de 1982. Eloge de Monsieur le Professeur Henri Gaussen. Mém. Acad. Sci. Insc. Belles-Lettres Toulouse, vol. 144:27–34.

Fiori, A. 1940. Formazioni vegetali d'Italia. In: Giotto Daninelli, Atlante fisico-economico d'Italia, p. 21. Milano.

Firbas, F. 1949/52. Spät- und nacheiszeitliche Waldgeschichte Mitteleuropas nördlich der Alpen. 1. Bd.: Allgemeine Waldgeschichte. 2. Bd.: Waldgeschichte der einzelnen Landschaften. Fischer Verlag, Jena.

Fisher, J.R. and K.E. Bradshaw. 1957. Uses of soil-vegetation survey information in road construction. Soil Science Society of America. Proceedings, 21:115–117.

Flahault, C. 1894. Projet de carte botanique, forestière et agricole de la France. Bulletin de la Société Botanique de la France, vol. 41.

Flahault, C. and C. Schröter. 1910. Referate aund Vorschläge betreffend die pflanzengeographische Nomenklatur. Bruxelles. 3rd International Botanical Congress.

Fleming, M.D. and R.M. Hoffer. 1979. 'Machine Processing of Landsat MSS Data and DMA Topographic Data for Forest Cover Type Mapping,' Proceedings of Machine Processing of Remotely Sensed Data Symposium, West Lafayette, In: Purdue University, LARS, pp. 377–390.

Floret, M.C. 1967. Conception et Réalisation Cartographiques, a Moyenne Échelle du Thème Phyto-Écologique en Tunisie Septentrionale. (Exposé prononcé lors de la 29e réunion commune des Sections A et B, le 22 février 1967). pp. 3.

Floret, C. 1968. La carte Phyto-Ecologique à l'échelle de 1:200.000 de la Tunisie Septentrionale, son Utilisation pour le Developpment du Territoire. Extrait des Annales de l'Institut National de la Recherche Agronomique de Tunisie. Vol. 41. Fasc. 1 – 1968. pp. 41–48.

Floret, Ch. and D. Schwaar. 1966. Cartographie Phyto-Écologique a petite échelle et photo-interprétation en Tunisie du Nord. IIe Symposium International de Photo-Interprétation Paris-1966 pp. 10.

Floret, C. and D. Schwaar. 1968. La carte d'utilisation du sol à l'échelle de 1:50.000 de la Tunisie Septentrionale. Extrait des Annales de l'Institut National de la Recherche Agronomique de Tunisie. Vol. 41. Fsc. 1 – 1968; pp. 85–100.

Floret, Ch. and D. Schwaar. Cartografie Phyto-Ecologique à petite échelle et Photo-Interpretation en Tunisie du Nord. pp. 17.

Foncin, M. 1961. Representation de la végétation sur les cartes anciennes. In: Henri Gaussen, 1961a, pp. 147–155.

Fontès, J. 1984. Carte Internationale du Tapis Végétal et des Conditions Ecologiques à 1/1 000 000e. La Haute Volta en préparation à l'I.C.I.T.V.

Fosberg, F.R. 1958. Mapping of vegetation types. Paris. UNESCO. Study of tropical vegetation, pp. 219–220.

Fosberg, F.R. 1961a. What should we map? In: Henri Gaussen, 1961 a, pp. 23–35.

Fosberg, F.R. 1961b. The study of vegetation in Europe. American Institute of Biological Sciences. Bulletin, June, pp. 17–19.

Fosberg, F.R. 1961c. A classification of vegetation for general purposes. Tropical Ecology, 2:1–28.

Fosberg, F.R. 1967. A pragmatic approach to pratical vegetation mapping. Leningrad. Komarov Institute of the Academy of Sciences of the U.S.S.R. Geobotanical Cartography, pp. 9–17.

Fosberg, F.R. 1967. A classification of vegetation for general purposes. IBP Handbook No. 4. Blackwell Scientific Publications Oxford/Edinburgh.

Fosberg, F.R., B.J. Garnier and A.W. Küchler. 1961. Delimitation of the humid tropics. Geographical Review, 51:333–347.

Fox, L. and K.E. Mayer. 1981. 'Using Ecological Zones to Increase the Detail of Landsat Classifications,' Proceedings ASP Fall Technical Conference, Falls Church, VA: American Society of Photogrammetry, pp. 113–124.

Frank, T.D. 1984. The effect of change in vegetation cover and erosion patterns on albedo and texture of Landsat images in a semiarid environment. Annals of the Association of Americain Geographers, 74:393–407.

Friedel, H. 1934. Boden- und Vegetationsentwicklung am Pasterzenufer. Carinthia, 2:29–41. Klagenfurth.

Friedel, H. 1956. Die alpine Vegetation des obersten Mölltales, Hohe Tauern. Wagner Verlag. Innsbruck.

Fukarek, P. 1963. Die Ausarbeitung einer detaillierten Waldkarte Bosniens und der Herzegowina auf pflanzensoziologischer Grundlage. In: Reinhold Tüxen, 1963a, pp. 363–386.

Gaillard, P.J. and G.P. Baerends. 1968. Voorne in de Branding. Stichting Wetenschappelijk Duinonderzoek, Oostvoorne.

Galiano, E.F. 1960. Mapa de vegetación de la provincia de Jaén (mitad oriental). Instituto de Estudios Gienenses. Jaén.

Galiano, E.F. 1961. Etat actuel de la cartographie botanique en Espagne. In: Henri Gaussen, 1961a, pp. 179-186.

Galkina, E.A. 1961. Special aspects in swamp vegetation mapping. In: Victor B. Sochava, 1961 a, pp. 121–130.

Galloway, R.W., H.M. van de Graaf and R. Story. 1963. Land systems of the Hunter Valley area, New South Wales. Commonwealth Scientific and Industrial Research Organization, Division of Land Research and Regional Survey. Canberra.

Gams, H. 1918. Prinzipienfragen der Vegetationsforschung. Vierteljahresschrift der naturforschenden Gesellschaft in Zürich, 63:293–493.

Gams, H. 1927. Von den Follatères zur Dent des Morcles. Bern. Beiträge zur geobotanischen Landesaufnahme der Schweiz, vol. 15.

Gams, H. 1936. Die Vegetation des Grossglocknergebietes. Abhandlungen der Zoologisch-Botanischen Gesellschaft in Wien, vol. 16, No. 2. Inserted in back.

Gardner, R.A. 1955. Interpretation of soil-vegetation survey data. California Forest & Range Experiment Station. Berkely, CA.

Gardner, R.A. and A.E. Wieslander.. 1957. The soil-vegetation survey in California. Soil Science Society of America. Proceedings, 21:103–105.

Gates, D.M., H.F. Keegan, J.C. Schleter and V.R. Weidner. 1965. Spectral properties of plants. Applied Optics 4:11–20.

Gauch, Jr., H.G. 1985. Multivariate analysis in community ecology. Cambridge Univ. Press. 298 pp.

Gaussen, H. 1936. Signes conventionels pour le travail sur le terrain. Centre National de la Recherche Scientifique. Service de la carte phytogéographique. Paris.

Gaussen, H. 1945. Tapis végétal de la France. Editions géographiques de France. Paris. Atlas de France, pp. 30–33.

Gaussen, H. 1949. Carte de la végétation de la France, feuille Perpignan. Service de la carte de la végétation de la France. Toulouse.

Gaussen, H. 1949. Projets pour diverses cartes du monde a 1:1,000,000: la carte écologique du tapis végétal. Annales Agronomiques, vol. 19.

Gaussen, H., G. Roberty and J. Trochain. 1950. Carte de la végétation de l'Afrique Occidentale Française, feuille Thiès. Office de la Recherche Scientifique Outre-Mer. Paris.

Gaussen, H. (ed.). 1955. Les divisions écoloques du monde. Centre National de la Recherche Scientifique. Paris. 59th International Colloquium, 1954.

Gaussen, H. 1955. Rapport général sur la cartographie écologique. Ann. Biol. 31(5–6):221–231.

Gaussen, H. 1957. Les cartes de végétation Inst. français de Pondichéry. Trav. Sect. Sci. Techn. I(2):51–87.

Gaussen, H. 1958. Integration of data by means of vegetation maps. Proceed. of the Ninth Pacific Congress, vol. 20:67–74, Bangkok.

Gaussen, H. and A. Vernet. 1958. Carte internationale du tapis végétal, feuille Tunis-Sfax. Service de la carte de la végétation de la France. Toulouse.

Gaussen, H. 1959. The vegetation maps. Pondicherry. Institut Français de Pondichery. Travaux de la section scientifique et technique, 1:155–179.

Gaussen, H. 1960. Ombrothermic curves and xerothermic index. Bull. Intern. Soc. for Tropical Ecology, T. I(1):25–26.

Gaussen, H. 1960. L'emploi des couleurs dans la cartographie de la végétation. Colloques Internationaux du C.N.R.S. 'Méthodes de la cartographie de la végétation', Toulouse: 137–145.

Gaussen, H. (ed.). 1961a. Méthodes de la cartographie de la végétation. Centre National de la Recherche Scientifique. Paris. 97th International Colloquium, Toulouse, 1960.

Gaussen, H., P. Legris, M. Viart and V.M. Meher Homji. 1961. Carte Internationale du Tapis Végétal et des Conditions Ecologiques; feuille Cape Comorin à 1/1 000 000e. Notice in Inst. français Pondichéry; Trav. Sect. scient. Technique, Hors Série N° 1, 108 p.

Gaussen, H. 1961b. L'emploi des coleurs dans la cartographie de la végétation. In: Henri Gaussen, 1961a, pp. 137–145.

Geiger, R. 1961. Das Klima der bodennahen Luftschicht. 4th ed. Braunschweig.

GEMS/PAC. 1979. An introduction to ecological monitoring. Information paper prepared for Meeting of UNDP Resident Representatives, Dakar. Hielkema, J.: ITC Journal.

GEMS/PAC. 1979. Ecological monitoring for desertification. Background paper prepared for the Expert Meeting on Methodology for Desertification Assessment and Mapping. Geneva.

Giacomini, V. 1960. La Cartografia della Vegetazione per la Conoscenza della Vegatazione Forestale. From: Accademia Italiana di Scienze Forestali. Vol. IX.-1960. pp. 323–356.

Giacomini, V. 1961. Le problème du choix des échelles en cartographie de la végétation. In: Henri Gaussen, 1961a, pp. 127–135.

Gillman, C. 1949. A vegetation types map of Tanganyika Territory. Geographical Review, vol. 39.

Gils, H.A.M.J. van. 1978. Standengesellschaft mit Geran-

560

ium sanguineum und Trifolium medium in den (sub)montanen Stufen des Walliser Rhônetals (Schweiz). Folia Geobot. Phytotax, Praha, pp. 351–369.

Gils, H.A.M.J. van. 1977. On types of tension zones between deciduous forest (Querco-Fageta) and grassland (Festuco-Brometed). Nat Canada 104, Quebec, pp. 173–176.

Gils, H.A.M.J. van (ed.), H.G.J. Huizing, W. van Wijngaarden and I.S. Zonneveld. 1982. Introduction to Land Ecology. Lecture note, ITC, Enschede, 46 pp.

Gils, H.A.M.J. van and I.S. Zonneveld. 1982. Vegetation and Rangeland Surveys (3rd ed.). Lecture note, ITC, Enschede, 49 pp.

Gils, H.A.M.J. van, A. Kannegieter and D.v.d. Zee. 198x. I.T.C. World landcover and landuse classification for image interpretation.

Ginzberger, A. and J. Stadlmann. 1939. Pflanzengeographisches Hilfsbuch. Springer Verlag. Wien.

Glavač, V. 1972. Zur Planung von geobotanischen Dauerbeobachtungsflächen in Waldschutzgebieten. Natur und Landschaft 47(4): 139–143.

Godron, M. and J. Poissonet. 1973. Quatre thèmes complémentaires pour la cartographie de la végétation et du milieu. Bull. Soc. Lang. de Geographie, Montpellier 6: 329–356.

Godron, M. 1964. Carte phyto-écologique, Argent-sur Sauldre, Sologne. Centre National de la Recherche Scientifique. Centre d'Etudes Phytosociologiques et Ecologiques Louis Emberger. Montpellier.

Godron, M., G. Grandjouan, A. Heaulme, E. Le Floc'h, J. Poissonet and J.P. Wacquant. 1964. Carte phyto-écologique et carte de l'occupation des terres de Sologne. Centre National de la Recherche Scientifique. Centre d'Etudes Phytosociologiques et Ecologiques Louis Emberger. Montpellier.

Goel, N.S. and D.E. Strebel. 1983. 'Inversion of vegetation canopy reflectance models for estimating agronomic variables, I: problem definition and initial results using the Suits model.' Remote Sensing of the environment, 13: 487–507.

Goetz, A.F.H., and B.N. Rock. 1983. 'Remote Sensing for Exploration: An Overview' Economic Geology and the Bulletin of the Society of Economic Geologists. Vol. 78, No. 4. Jet Propulsion Laboratory, California Institute of Technology, Pasadena, California 91109 pp. 573–684.

Goetz, A.F.H., B.N. Rock and L.C. Rowan. 1983. Remote sensing for exploration. Economic Geology, July.

Golley, F. (ed.). 1977. Ecological succession. Benchmark papers in Ecology. Vol. 5, XI + 375 p., Dowden Hutchinson & Ross, Inc. Stroudsburg, Pennsylvania.

Goodall, D.W. 1953. Objective methods for the classification of vegetation, II. Fidelity and indicator value. Austr. Journ. Bot. 5, pp. 434–56.

Goodall, D.W. 1954. Objective methods for the classification of vegetation, III. An essay in the use of factor analysis. Australian Journal of Botany, Volume 2, No. 2. pp. 30–324.

Gooijer, H.H. 1982. 'DE HILVER'. Een ecologische vegetatiekartering. Centrum voor Agrobiologisch Onderzoek Wageningen. Karteringsverslag Nr. 203. April 1983. pp. 31. (Met Bijlagen).

Gorbachev, B.N. and O.S. Gorozhankia. 1962. The compilation of maps of natural (reconstructed) vegetational cover using plant indicators. In: Victor B. Sochava, 1962a, pp. 77–86.

Gordon, M., J. Poissonet and P. Poissonet. 1967. Méthodes d'Etudes des Formations Herbacées Denses. Essais d'Application a d'Etude du Dynamisme de la Végétation. Document N. 35 pp. 28. Centre National de la Recherche Scientifique. Centre d'études Phytosociologiques et Ecologiques Montpellier.

Gossweiler, J. 1939. Carta fitogeográfica de Angola. Edição do Governo Geral de Angola. Lisboa.

Gounot, M. 1956. A propos de l'homogénéité et du choix des surfaces des relevés. Service de la carte phytogéographique. Bulletin, Série B, 1: 7–17.

Gounot, M. 1961. Les méthodes d'inventaire de la végétation. Service de la carte phytogéographique. Bulletin, Serie B., 6: 7–73.

Grabau, W.E. and W.N. Rushing. 1968. A computer comptable system for quantitatively assessment of the physiognomy of vegetation assemblages. Land evaluation page: 263–275. Papers on CSIRO-UNESCO symposium Canberra Australia.

Graetz, R. and M. Gentle. 1982. The relationship between reflectance in the Landsat wavebands and the composition of an Australian semi-arid shrub rangeland. Photogrammetric Engineering and Remote Sensing, 48: 1712–1730.

Graetz, R.D., M.R. Gentle, R.P. Pech and J.F. O'Callaghan. 1982. 'The Development of a Land Image-Based Resource Information System (LIBRIS) and its Application to the Assessment and Monitoring of Australian Arid Rangelands'. Proceedings of the International Symposium on Remote Sensing of Environment – first Thematic Conference on Remote Sensing of Arid and Semi-Arid Lands, Ann Arbor, MI: Environmental Research Institute of Michigan, pp. 257–275.

Grandjouan, G., C. Floret, M. Buisson, Charbert, L. Emberger and G. Long. Écologie Végétale et Développement du territoire. Centre National de la Recherche Scientifique. Montpellier. pp. 20.

Greig-Smith, P. 1964. Quantitative Plant Ecology. 2nd ed. London.

Gremmen, N.J.M. 1981. Alien vascular plants on Marion Island (Subantarctic). Colloque sur les Ecosystèmes Subantarctiques, C.N.F.R.A., no. 51: 315–323.

Grisebach, A. 1872. Die Vegetationsgebiete der Erde. In: Die Vegetation der Erde. Engelmann Verlag. Leipzig.

Grosser, K.H. 1964. Die Wälder am Jagdschloß bei Weißwasser (OL). Waldkundliche Studien in der Muskauer Heide. Abhandlungen und Berichte des Naturkundemuseums Görlitz, Band 39, N. 2. Leipzig, 1964. pp. 101. (colour map 1: 10.000).

Grosser, K.H. 1966. Brandenburgische Naturschutzgebiete.

With colour map Scale 1 : 5.000. Beitrage zur wissenschaftlichen ErschlieBung der Naturschutzgebiete in Berlin und in den Bezirken Potsdam, Frankfurt (Oder) und Cottbus. C-Ww3, pp. 40.

Grubb, P.J., J.R. Lloyd, D.T. Pennington and T.C. Whitmore. 1963. A comparison of montane and lowland rain forests in Ecuador. Journal of Ecology, 51 : 567–599.

Grundsten, T., T. Rafstedt and U. von Sydow. 'FJällens Vegetation Kartor' (5 maps in colur 1 : 100000). Del 3a: Norr ottens län, södra delen. Meddelande från Naturgeografisca institutionen Stockholms universitet Nr. 164. Stockholm.

Guinochet, M. 1955. Logique et dynamique du peuplement végétale. Paris.

Guinet, P. 1953. Carte de la végétation de l'Algérie, feuille Béni-Abbès. Service de la carte de la végétation de la France. Toulouse.

Gupta, Raj K. and C.T. Abichandani. 19xx. Air photo Analysis of Plant Communities in Relation to Edaphic Factors in the Arid Zone of Western Rajasthan. Division of Basic Resource Studies, Central Arid Zone Research Institute, Jodhpur, India, pp. 57–66.

Guricheva, N.P., Z.V. Karamysheva and E.I. Rachkovskaia. 1967. On the compilation of a legend for a large scale vegetation map in the desert-steppe subzone of Kazakhstan. Academy of Sciences of the U.S.S.R., Komarov Institute. Leningrad. Geobotanical Cartography, pp. 57–67.

Gwynne, M.D. and H. Croze. 1975a. East African habitat monitoring practice: a review of methods and application. In the Proceedings of the International Livestock Centre for Africa (ILCA) Seminar on Evaluation and mapping of Tropical African Rangeland. Bamaco, March 1975, 95–142.

Gwynne, M.D. and H. Croze. 1975b. The concept and practice of ecological monitoring over large areas of land: the Systematic reconnaissance Flight (SRF). Paper presented at the Ibadan/Garoua International Symposium on Wildlife Mangement in Savanna Woodland. Ibadan, September 1975.

Haffner, W. 1968. Vegetationskarte des mittleren Nahetals. Geographisches Institut der Universität. Bonn.

Hagihara, J.S. and J. Linn. 1982. A field evaluation report on the use of an ecological land classification for the United States for classifying potential and existing vegetation: Saval Ranch, Elko, Nevada. United States Department of the Interior, Bureau of Land Management. Washington, DC.

Hallé, F., R.A. Oldeman and P.B. Tomlinson. 1978. Tropical trees and forests; an architectural analysis. 441 p. Springer Verlag, Heidelberg, Berlin, New York.

Hardy, N.E. 1976. Examination and analysis of vegetation boundary zones using aerial and orbital data sources. University of Kansas, Department of Geography. Lawrence, KS. Ph. D. Dissertation.

Hare, F.K. 1959. A photo-reconnaissance survey of Labrador-Ungava. Geographical Branch, Mines and Technical Surveys, Memoir No. 6. Ottawa.

Harlan, J.C., D.W. Deering, R.H. Haas and W.E. Boyd. 1979. Determination of Range Biomass Using Landsat. Proceedings of the Thirteenth International Symposium on Remote Sensing of Environment, Ann Arbor, MI: Environmental Research Institute of Michigan, pp. 659–673.

Harlan, L.C., W.E. Boyd, C. Clarke, S. Clarke, and O.C. Jenkins. 1979. Rangeland Resource Evaluation from Landsat, Report RSC 3715-2, Houston, TX: NASA and USDA Nationwide Forestry Application Program.

Harms, W.B. and M. Damen. 1978. Die Anwendung von Vegetationskomplexen in einem Landschaftsplanungsprojekt. Rijksinstituut voor onderzoek in de bos- en landschapsbouw 'De Dorschkamp' Wageningen, Overdruk Nr. 23 pp. 435–453.

Harper, L. 1857. Map of the prairie above Tibley creek. Jackson, MS. Preliminary Report on the Geology and Agriculture of the state of Mississippi, plate 5.

Harper, R.M. 1943. Forests of Alabama. Geological Survey of Alabama, Monograph No. 10.

Hashem, M. and P. Hermelink. 1981. Landscape Ecological Surveys of Kauf National Park, Libya. Unpubl. MSc. thesis. ITC. Enschede. 50 pp.

Hayata, B. 1911. The vegetation of Mt. Fuji. 125 pp. with coloured vegetation map. Tokyo.

Hayden, E.V. 1878. Colorado and parts of adjacent territories. United States Geological and Geographical Survey, Washington, DC. 10th Annual Report.

Heiselmayer, P. 1981. Die Vegetationskarte als Grundlage für ökologische Kartierungen. Angewandte Pflanzensoziologie, 26 : 59–73.

Heiselmayer, P., W. Schneider and H. Plank. 1982. Vegetationskundliche Luftbildauswertung am Beispiel der Umgebung des Glocknerhauses. Carinthia, 225–240.

Hejny, S. 1960. Oekologische Charakteristik der Wasser- und Sumpfpflanzen in den Slowakischen Tiefebenen. Verlag der Slowakischen Akademie der Wissenschaften. Bratislava.

Hejny, S. 1963. Die Wege und Methoden der Vegetationskartierung in Böhmen und Mähren. In: Reinhold Tüxen, 1963a, pp. 261–263.

Heller, R.C., ed. 1975. Evaluation of ERTS-1 Data for Forest and Rangeland Surveys, Research Paper PSW-112, Berkeley, CA: USDA Pacific Southwest Forest and Range Experiment Station.

Hempenius, S.A. 1974. How can ecology prepare itself for remote sensing. Proc. First Int. Congress Ecology, The Hague, 1974. Also I.T.C. Journal 1974 no. 4.

Hengeveld, R. and J. Haecj. 1982. The Distribution of Abundance, I. Measurements. J. Biogeography 9, pp. 303–316.

Heusden, W. van. 1983. Monitoring changes in heathland vegetation using sequential aerial photographs. ITC Journal 1983-2 p: 160–165.

Heyligers, P.C. 1963. Structure Formulae in Vegetation Analysis on Aerial Photographs and in the field. Re-

562

printed from: The Symposium on Ecological Research in
Humid Tropic Vegetation. Kuching, Sarawak. pp. 249–
255.

Heyligers, P.C. 1963. Vegetation and Soil of a white sand
Savanna in Surinam. N.V. Noord-Hollandsche Uitgevers
Mij. Amsterdam Verh. Kon. Ned. Akad v. Wet. afd. Nat.
Tweede reeks LIV. No. 3.

Heyligers, P.C. 1968. Quantification of Vegetation Structure
on Vertical Aerial Photographs Land evaluation. Papers
CSIRO-UNESCO.

Hildebrand, O. 1939. Pflanzensoziologische Reichskartie-
rung. Mitteilungen des Reichsamts für Landesaufnahme,
No. 1.

Hills, G.A. 1952. The classification and evaluation of site
for forestry. Ontario Department of Lands and Forests.
Research Report No. 24.

Hoffer, R.M. 1980. 'Computer-Aided Analysis of Remote
Sensor – Data Magic Mystery or Myth?' Proceedings of
Remote Sensing for Natural Resources: A Symposium,
Moscow, ID: University of Idaho College of Forestry,
Wildlife and Range Sciences, pp. 156–179.

Hoffer, R.M. and C.J. Johannsen. 1969. Ecological Poten-
tials in Spectral Signature Analysis. In: Remote Sensing in
Ecology. University Georgia Press. Athens. pp. 1–16.

Holdgate, M.W. 1967. The influence of introduced species
on the ecosystems of temperate oceanic islands. Proceeed.
and Papers IUCN, 10th technical meeting, I.U.C.N. Publications, New Series 9: 151–176.

Holdridge, L.R. 1947. Determination of world plant formations from simple climatic data. Science, vol. 105, No.
2727: 367–368.

Holdrige, L.R. 1959. Ecological Indication of the Need for a
New Approach to Tropical Land Use. Econ. Bot. 13, pp.
271–280.

Holdridge, L.R. 1959. Mapa ecológico de El Salvador and
Mapa ecológico de Guatemala. Instituto Interamericano
de Ciencias Agrícolas. San José, C.R.

Holdridge, L.R., W.C. Grenke, W.H. Hatheway, T. Liang
and Josef A. Tosi. 1971. The forest environments in tropical life zones. Pergamon Press. New York, NY. 747
pp.

Hommel, P.W.F.M. 1986. Landscape ecology of Udjung
Kulon (West Java). Diss. Wageningen. 206 p. appendices,
coloured map. Publ. privat. 1986.

Hopley, D. 19xx. Aerial photography and other remote sensing techniques. Dep. of Geography, James Cook University, Townsville, Queensland, Australia 4811, pp. 23–43.

Horvat, A.O. 1963. Phytozönologische Waldkartierung im
Mecsek Gebirge bei Pecs (Fünfkirchen) in Südungarn. In:
Reinhold Tüxen, 1963a, pp. 245–259.

Horvat, I. 1958. Carte des groupements végétaux de la Croa-
tie du sud-ouest. Forest Research Institute. Zagreb.

Horvat, I. 1963. Vegetationskarte Europas. In: Reinhold
Tüxen, 1963a, pp. 334–346.

Horvat, I., V. Glavac and H. Ellenberg, 1974. Vegetation
Südosteuropas. Geobotanica Selecta Band IV. Fisher Verlag. Stuttgart.

Hou, H.-Y. 1979. The vegetation of China. Academia Sinica,
Institute of Botany. Beijing.

Houten de Lange, ten S.M. 1977. Rapport van het Veluwe-
onderzoek. Centrum voor Landbouwpublikaties en Landbouwdocumentatie, Wageningen. pp. 263. Met kaartbij-
lagen.

J.A. Howard. 1971. The Reflective Foliaceous Properties of
Tree Species, Application of Remote Sensors on Forestry.
International Union of Forest Research Organizations.
Section 25. p: 127–146.

Howard, A.J. 1970. Aerial photo-ecology. Faber and Faber.
London.

Hsu, S-Y. 1978. Texture-tone analysis for automated land-
use mapping. Photogrammetric Engineering and Remote
Sensing, 44:1393–1404.

Hubbard, J.C.E. and B.H. Grimes. 1972. The analysis of
coastal vegetation through the medium of aerial photography. Reprinted from Medical and Biological Illustration,
Vol. 22, No. 3. July 1972. British Medical Association
Tavistock Square. London, WC1H 9JR. pp. 182–190.

Hubbard, J.C.E. and B.H. Grimes. City University and the
Nature Conservancy 'Coastal Vegetation Survey' pp. 11.

Hueck, K. 1939. Vegetationskarte des Riesengebirges (nörd-
licher Teil). In his: Botanische Wanderungen im Riesengebirge. Gustav Fischer Verlag. Jena.

Hueck, K. 1943. Vegetationskarte des Deutschen Reiches,
Blatt Berlin. Neumann Verlag. Berlin-Neudamm.

Hueck, K. 1950. Natüliche Pflanzendecke von Niedersachs-
en. Atlas von Niedersachsen, p. 19. Dorn Verlag. Bre-
men.

Hueck, K. 1955. Nouvelles cartes de la végétation sud-amé-
ricaine et leur signification pour l'agriculture et la sylvicul-
ture. In: Henri Gaussen, 1955, pp. 181–188.

Hueck, K. 1956. Mapa fitogeográfico do estado de São Paulo. Boletim Paulista de Geografia, 22:19–25.

Hueck, K. 1957. Die Ursprünglichkeit der brasilianischen
campos cerrados und neue Beobachtungen an ihrer Süd-
grenze. Erdkunde. 9:193–203.

Hueck, K. 1960. Mapa de vegetación de la República de
Venezuela. Mérida. Instituto Forestal Latino-Americano
de Investigación y Capacitación.

Huizing, H. 1984. Analysis of Landuse Changes in relation
to land qualities by means of a geoinformation system. (A
case study in the Lemele area, E. Netherlands) Poster N.
II, 4 in: IALE (etc.) pp. 135–136.

Huizing, H.G.J. and I.S. Zonneveld. 1980. (with ass. M.C.
Bronsveld). Monitoring primary production (survey of the
fourth dimension). Environment monitoring for the Arab
world. Proceedings of a Seminar on Methodology and
Application of Remote Sensing from Space, Light Aircraft
and the ground for Monitoring Natural Resources. Am-
man – Sudan 26 – 29 Oct. 1980. Pollution and Climate.
Page 152–191.

Humbert, H. 1936. Flore de Madagascar et des Comores.
Muséum National d'Histoire Naturelle, Laboratoire de
Phanérogamie, Paris.

Humbert, H. 1951. Les territoires phytogéographiques du

Nord de Madagascar. C.R.SOC. Biogéographie, Paris, N° 246: 176–184.

Humbert, H. 1955. Projet de la carte de la végétation de Madagascar. In: Henri Gaussen, 1955. pp. 49–60.

Humbert, H. 1960. Projet de carte de végétation de Madagascar au 1/1 000 000e. In Colloques Internationaux du C.N.R.S. 'Méthodes de la Cartographie de la Végétation', Toulouse: 49–60.

Humbert, H. and G. Cours Darne. 1965. Carte Internationale du Tapis Végétal et des Conditions Ecologiques à 1/1 000 000e: Madagascar. Notice in Inst. français de Pondichéry; Trav. Sect. scient. Techn., Hors Série N° 6, 162 p.

Humboldt, A. von. 1807. Ideen zu einer Physiognomik der Gewächse. Tübingen.

Humbolt, A. von and A. Bonpland. 1895. Géographie des plantes équinoxiales: tableau physique des Andes et pays voisins. In their: Essai sur la géographie des plantes. Levrault, Schoell et Co. Paris.

Hutchinson, C.F. 1982. 'Techniques for Combininb Landsat and Ancillary Data for Digital Classification Improvement,' Photogrammetry Engineering and Remote Sensing, 48: 123–130.

Ignatiev, E.I., and O.V. Shkurlatov. 1962. The importance of geobotanical maps in medico-geographical studies. In: Victor B. Sochava, 1962a, pp. 204–207.

Ihse, M., T. Rafstedt and L. Wastenson. 1981. Air photo-interpretation for vegetation mapping. Handbook i flygbildsteknik och fjärranalys, kap. 9. Stockholm Ed. Nämden för skoglig flygbildsteknik.

Ihse, M. 1978. Aerial photo interpretation of vegetation in south and central Sweden – a methodological study of medium scale mapping. Stockholms universitet. Naturgeografiska institutionen Stockholms, Sweden. SNV PM 1083. NR. A. 93, pp. 165.

Ihse, M. and L. Wastenson. 1975. Aerial photo-interpretation of Swedish mountain vegetation. Swedish Environmental Protection Board. Stockholm. PM 596.

Ihse, M., L. Wastenson. Aerial photo interpretation of Swedish Mountain vegetation – a methodological study of medium scale mapping. (With colour maps.) Stockholms universitet Naturgeografiska institutionen pp. 134 SNV PM 596.

ILCA. 1981. Low-Level Aerial survey techniques. ILCA Monograph 4, ILCA, Addis-Abeba.

Iljina, I.S. 1968. A dynamic principle of compiling a large-scale geobotanical map. Academy of Sciences of the U.S.S.R., Komarov Botanical Institute. Leningrad. Geobotanical Cartography, pp. 21–37.

Ionescu, T. 1956. A propos de la cartographie des groupements végétaux des terres cultivées en zone semi-aride. Service de la carte phytogéographique. Montpellier. Bulletin, Série B, 1: 19–23.

Ionescu, T. 1958. Etude phytosociologique et écologique de la plaine de Doukkala. Ministère d'Agriculture du Maroc, Division de la mise en valeur et du génie rural. Rabat.

Isachenko, A.G. 1962. Some connections of landscape mapping with geobotanical mapping. In: Victor B. Sochava, 1962a, pp. 169–177.

Isachenko, T.I. 1962. Principals and methods of generalization in geobotanical mapping in large, medium and small scale. In: Victor B. Sochava, 1962a, pp. 28–46.

Isachenko, T.I. 1969. The structure of the vegetation cover and mapping. Academy of Sciences of the U.S.S.R., Komarov Botanical Institute. Leningrad. Geobotanical Cartography, pp. 20–33.

I.T.C. 1963. Tree and Stand Height Measurements and Estimations on vertical Photographs. Vol. X, Chapter X.2.1. pp. 35.

I.T.C.-Unesco. Aerial Forest Inventory and Land Use Survey Project. Conservator of Forest Peshawar. pp. 17.

I.T.C. 1977. Project de développement rural integré de la Région Kaärta. Rep. du Mali. Phase de reconnaissance du Volet VIII. Cartographie. I.T.C. Enschede (avec des cartes colorés).

ITC. 1979. An Introduction to land evaluation. RUR. S.7.

Ivanova, E.N. and N.N. Rosov. 1962. Some works on general soil cartography of Siberia and their connection with geobotanical studies. In: Victor B. Sochava, 1962a, pp. 194–199.

Iversen, J. 1936. Biologische Pflanzentypen als Hilfsmittel in der Vegetationsforschung. Levin & Munksgaard. København. 224 pp.

Iversen, J. 1936. Biologische Pflanzentypen als Hilfsmittel in der Vegetationsforschung. Diss. Köbenhavn, Meddeleser fra Skaling-Laboratoriet, 4, Köbenhavn.

Iversen, J. 1964. Retrogressive vegetational succession in the post-glacial. J. Ecol. 52 (Suppl.): 59–70.

Jackson, R.D. 1984. Spectral indices in n-space. Remote Sensing of the environment, 13: 409–421.

Jacquinet, J.C. 1969. Étude Écologique intégrée de l'Unité Régionale de Développement de Soliman (Tunisie). Document N. 54. Centre Nationale de la Recherche Scientifique. Montpellier, pp. 250. (With diazo copy map scale 1: 50.000).

Jaeger, H. 1952. Standortserkundung und Standortskartierung als Grundlagen der Forsteinrichtung. Der Wald, 2: 82–85.

Janssen, C.R. 1960. On the Late-Glacial and Post-Glacial Vegetation of South Limburg (Netherlands). Diss. Utrecht: 112 p. Wentia IV.

Jenks, G.F. and D. Knos. 1961. The use of shading patterns in graded series. Annals of the Association of American Geographers, 51: 316–334.

Jensen, H.A. 1947. A system for classifying vegetation in California. California Fish and Game, 34: 199–266.

Johns, E. 1957. The surveying and mapping of vegetation on some Dartmoor pastures. Geographical Studies, 4: 1: 129–137.

Johnson, G.R. and W.G. Rohde. 1981. 'Landsat Digital Analysis Techniques Required for Wildland Resourses Classification,' Proceeding of the Arid Land Resource Inventories Symposium, General Technical Report WO-28, Washington, D.C.: USDA Forest Service, pp. 204–213.

Johnson, P.L. (ed.). 1969. Remote Sensing in Ecology. Various Chapters University Press. University of Georgia Athens.

Johnston, W.B. 1961. Locating the vegetation of early Canterbury: a map and its sources. Transactions of the Royal Society of New Zealand, 1:2:5–15.

Jones, R.G.B. 1969. Specific Application of Air Photo-Interpretation in Agricultural Development Planning. The Journal and Proceedings of the Institution of Agricultural Engineers. Vol. 24. N. 2–1969 England. pp. 14.

Kalkhoven, J.T.R., A.H.P. Stumpel and S.E. Stumpel-Rienks. Environmental Survey of The Netherlands. A landscape ecological Survey of the Natural Environment in the Netherlands for physical planning on National Level. pp. 141. (With serveral colour maps Scale 1:200.000).

Karamysheva, Z.V. and E.I. Rachkovskaia. 1968. On the compilation of a small scale map of a steppe territory in Kazakhstan. Academy of Science of the U.S.S.R. Leningrad. Geobotanical Cartography, pp. 5–21.

Karamysheva, Z.V. and T.I. Isachenko. 1983. The advance of geobotanical mapping in the Komarov Botanical Institute of the Academy of Sciences of the U.S.S.R. Academy of Sciences of the U.S.S.R. Leningrad. Geobotanical Cartography, pp. 3–18.

Kartawinata, K. 1980. The classification and utilization of forests in Indonesia. Conference on forest land assessment for sustainable uses, East-West Center, Honolulu, Hawaii, June 19–28 1979. BioIndonesia No. 7, 95–106.

Kaul, R.E. 1975. Vegetation of Nebraska. University of Nebraska, Institute of Agriculture and Natural Resources. Lincoln, NE.

Kauth, R.J. and G.S. Thomas. 1976. 'The Tasselled Cap – A Graphic Description of the Spectra-Temporal Development of Agricultural Crops as Seen by Landsat,' Proceeding of the Machine Processing of Remotely Sensed Data Symposium, West Lafayette, In: Purdue University, LARS, pp. 4B–41–4B-51.

Keay, R.W.J. 1959. Vegetation map of Africa south of the tropic of Cancer. Oxford University Press. London.

Keizer, P. de., J. Steenhamer and D. van der Zee. 1984. MONA 1983. Rapport over de volderingen van het Mona-project, deelproject van het Moneragala District. Survey Project. ITC. Enschede.

Kelner, U.G. 1962. Utilization of complementary data when compiling geobotanical maps for comprehensive district atlases of the U.S.S.R. In: Victor B. Sochava, 1962a, pp. 259–264.

Kenya. Kindaruma area Kenya, Soil, Vegetation and Land Evaluation. (With colour and monocolour maps Scales: 1:100.000; 1:250.000).

Kenya. 1974. A vegetation map of the Masai Mara Game Reserve (colour). Scale 1:125.000 Sheets 144/2,4; 145/3,4; 158/1,2,4 Draw by Survey of Kenya from Information supplied by Research Division of the Game Department, 1974.

Kernervon Marilaun, A. 1863. Das Pflanzenleben der Donauländer (Innsbruck). 2nd edition by F. Vierhapper, 1929.

Khan, M.A. 1954. Sliced maps. Indian Forester, 80:103–116.

Kharin, N.G. 1962. New methods of vegetation deciphering on aerophotos. In: Victor B. Sochava, 1962a, pp. 232–236.

Kitchen, L. and A. Rozenfeld. 1984. Scene analyses using region based constraint filtering. Pattern Recognition Vol. 17, No. 2, (189–203).

Kleopov, G.N. and E.M. Lavrenko. 1942. Geobotanical map of the Ukraine. Kiev. Ukrainian Research Institute for Agricultural Botany.

Klingebiel, A.A. and D.H. Montgomery. 1961. Land capability classification. Agricultural Handbook 210. Soil Conservation Service, U.S. Govt. Printing Office, Washington DC.

Klink, H.J. 1973. Die natürliche Vegetation und ihre räumliche Ordnung im Puebla-Tlaxcala Gebiet, Mexico, Erdkunde, 27:213.

Klinka, K. 1977. Biogeoclimatic units, Nootka-Nanaimo. Ministry of Forests, Information Services Branch. Victoria, B.C.

Knapp, R. 1951. Zur Bedeutung pflanzensoziologischer Karten für die Forst- und Landwirtschaft und die Vegetationskartierung in Hessen. Akademie für Raumforschung und Landesplanung. Bremen-Horn. Forschungs- und Sitzungsberichte, 2'63–69.

Knapp. R. 1965. Die Vegetation von Nord- und Mittelamerika und der Hawaii Inseln. Gustav Fischer Verlag. Stuttgart. 373 pp.

Knapp. R. 1973. Die Vegetation von Afrika. Gustav Fischer Verlag. Stuttgart. 626 pp.

Knapp, R. (ed.). 1974. Vegetation Dynamics. Handbook of Vegetation Science, Part VIII. Junk, The Hague.

Knipling, E.B. 1967. Physical and Physiological basis for differences in reflectance of healthy and diseased plants. Proc. Workshop on Infrared Col. Photography in Plant sciences. Florida Dept. Agr. Div. Plant. Industry Winter Haven Florida 1967.

Knipling, E.B. 1969. Lead reflectance and Image Formation on Color Infrared Film. Remote Sensing in Ecology. University Georgia Press. Athens.

Könekamp, A.H. und F. Weise. 1952. Pflanzensoziologie und Grünlandkartierung im Dienste der Landwirtschaft. Braunschweig-Volkenrode. Schriftenreihe der Forschungsanstalt für Landwirtschaft, 5:7–21.

Kop, L.G. 1959. Proefstation voor de akker- en weidebouw. Wageningen. Verslag van het Internationale Symposium over Vegetationkartering te Stolzenau (W.Dld) van 23 t/m 26 maart 1959. Intern Rapport Nr. 37, pp. 6.

Köppen, V. 1931. Grundriss der Klimakunde. Walter de Gruyter Verlag. Berlin.

Koriba, K. 1958. On the periodicity of tree growth in the

tropics, with reference to the mode of branching, the leaf fall and the formation of the resting bud. Singapore. The Gardens Bulletin, Series 3, 17:11–81.

Kosmakova, O.P. 1960. Editorial work in the representation of vegetation on topographic maps. Geodesy and Cartography, 1–2:27–31.

Krajina, W.J. 1959. Classification of ecosystems of forests. University of British Columbia. Vancouver, B.C.

Krajina, V.J. 1973. Biogeoclimatic zones of British Columbia. British Columbia Ecological Reserves Committee. University of British Columbia. Vancouver, B.C.

Krause, W. 1950. Über Vegetationskarten als Hilfsmittel kausalanalytischer Untersuchung der Pflanzendecke. Planta, 38:296–323.

Krause, W. 1952. Das Mosaik der Pflanzengesellschaften und seine Bedeutung für die Vegetationskunde. Planta, 41:240–289.

Krause, W. 1954a. Grünlandkartierung im Rahmen des ERP-Grünlandförderungsprogramms 1951/1953. Landwirtschaft und Angewandte Wissenschaft, 21:1–5.

Krause, W. 1954b. Zur ökologischen und landwirtschaftlichen Auswertung von Vegetationskarten der Allmendweiden im Hochschwarzwald. Springer Verlag. Wien. Festschrift Aichinger, pp. 1076–1100.

Krause, W. 1955. Pflanzensoziologische Luftbildauswertung. Stolzenau/Weser. Angewandte Pflanzensoziologie, vol. 10.

Krause, A. and L. Schröder. 1979. Vegetationskarte der Bundesrepublik Deutschland 1;200 000 – Potentielle natürliche Vegetation – Blatt CC 3118 Hamburg-West. Schr. Reihe Vegetationskunde 14.

Krippelová, T. and R. Neuhäusl. 1963. Bibliographie der Vegetationskarten der Tschechoslovakei. Excerpta Botanica, Sectio B, 5:203–213.

Krylov, G.V. 1962. Forest map and forest site map of western Siberia. In: Victor B. Sochava, 1962a, pp. 110–113.

Küchler, A.W. 1947. Localizing vegetation terms. Annals of the Association of American Geographers, 37:198–208.

Küchler, A.W. 1950. Die physiognomische Kartierung der Vegetation. Petermanns Geographische Mitteilungen, 94:1–6.

Küchler, A.W. 1951. The relation between classifying and mapping vegetation. Ecology, 32:275–283.

Küchler, A.W. 1952. Toward a solution of the problems in mapping the vegetation of the United States at the scale of 1:1,000,000. Washington, DC. International Geographical Union, 17th Congress. Proceedings, pp. 257–260.

Küchler, A.W. 1953a. Natural vegetation of the United States and southern Canada. Chicago, IL. Rand McNally & Co. Goode's World Atlas, pp. 52–53.

Küchler, A.W. 1953b. Vegetation mapping in Europe. Geographical Review, 43:91–97.

Küchler, A.W. 1954a. Some considerations concerning the mapping of herbaceous vegetation. Kansas Academy of Science, Proceedings, 57:449–452.

Küchler, A.W. 1954b. Vegetation maps at the scale from 1:200,000 to 1:1,000,000. Paris. 8th International Botanical Congress. Rapports et communications, section 7, pp. 107–112.

Küchler, A.W. 1955. Projet d'une carte physionomique de la vegetation du monde. In: Henri Gaussen, 1955, pp. 163–168.

Küchler, A.W. 1955. A Comprehansive Method of Mapping Vegetation. Annals of the Association of the American Geographers, Vol. XLV. Nr. 4, pp. 404–415.

Küchler, A.W. 1956a. Notes on the vegetation of southeastern Mount Desert Island, Maine. University of Kansas Science Bulletin, 38:1:335–392.

Küchler, A.W. 1956. Classification and Purpose in Vegetation Maps. (With three separate maps). The Geographical Review, Vol. XLVI, N. 2, 1956. pp. 155–167.

Küchler, A.W. 1957. The New Soviet Vegetation Maps. Ecology, Vol. 38, Nr. 4, October, 1957. p. 671.

Küchler, A.W. 1959. Vegetation Maps in Geographical Research. From: The Professional Geographer, Vol. XI, Nov./59 Nr. 6, pp. 6–9.

Küchler, A.W. 1960. Mapping the Dynamic Aspects of Vegetation. Centre National de la Recherche Scientifique. Méthodes de la Cartographie de la Végétation Toulouse, 16–21 Mai 1960. pp. 187–201.

Küchler, A.W. 1960a. Vergleichende Vegetationskartierung. Vegetatio, 9:208–216.

Küchler, A.W. 1960b. Vegetation Mapping in Africa. Annals of the Association of American Geographers, 50:74–84.

Küchler, A.W. 1961. Mapping the dynamic aspects of vegetation. In: Henri Gaussen, 1961a, pp. 187–201.

Küchler, A.W. 1963a. Die Zusammenstellung von Vegetationskarten kleinen Masstabs. In: Reinhold Tüxen, 1963a, pp. 39–46.

Küchler, A.W. 1963b. Vegetation maps as climatic records. Pau 3rd Biometerorological Congress. Biometeorology, 2:953–964.

Küchler, A.W. 1964. The potential natural vegetation of the conterminous United States. New York, NY. American Geographical Society, Special Publication No. 36.

Küchler, A.W. ed. 1965ff. International bibliography of vegetation maps. Lawrence, KS. University of Kansas Publications, Library Series, Nos. 21. 26, 29, 36 and 45: vol. 1: North America, 1965; vol. 2: Europe, 1966; vol. 3: U.S.S.R., Asia, Australia, 1968; vol. 4: Africa, South America, World, 1970; 2nd ed.: vol. 1: South America, 1980.

Küchler, AW. 1966. Analyzing the physiognomy and structure of vegetation. Annals of the Association of American Geographers, 56:112–127.

Küchler, A.W. 1967a. Vegetation Mapping. New York, NY. Ronald Press. 472 pp.

Küchler, A.W. and J.O. Sawyer. 1967. A study of the vegetation near Chiengmai, Thailand. Kansas Academy of Science. Transactions, 70:281–348.

Küchler, A.W. 1967b. Potential natural vegetation of the United States. Washington, DC. United States Geological Survey. National Atlas, pp. 89–90.

Küchler, A.W. 1969a. The vegetation of Kansas on maps.

Kansas Academy of Science, Transactions, 72:141–166.

Küchler, A.W. 1969b. Natural and cultural vegetation. Professional Geographer, 21:383–385.

Küchler, A.W. 1970. 'A Biogeographical Boundary: the Tatschl Line'. Transactions of the Kansas Academy of Science, Vol. 73, N. 3, pp. 298–301.

Küchler, A.W. 1972. The oscillations of the Mixed Prairie in Kansas. Erdkunde, 26:120–129.

Küchler, A.W. 1973. Problems in classifying and mapping vegetation for ecological regionalization. Ecology, 54:512–523.

Küchler, A.W. 1974. A new vegetation map of Kansas. Ecology, 55:586–604.

Küchler, A.W. 1975a. Boundaries on vegetation maps. In: W.H. Sommer and Reinhold Tüxen, eds., Tatsachen und Probleme der Grenzen in der Vegetation. Lehre. J. Cramer Verlag.

Küchler, A.W. 1975b. The substrate on vegetation maps. In: H. Dierschke, ed., Vegetation und Substrat. Vaduz. J. Cramer Verlag.

Küchler, A.W. 1977. The map of the natural vegetation of California. In: Michael S. Barbour and Jack Major, eds., The terrestrial vegetation of California. New York, NY. John Wiley & Sons.

Küchler, A.W. 1981a. The Argentinian vegetation on maps. Phytocoenologia, 9:465–472.

Küchler, A.W. 1981b. The organization of the content of small scale vegetation maps. Documents de cartographie écologique, 24:35–38.

Küchler, A.W. 1982. Brazilian vegetation on maps. Vegetatio, 49:29–34.

Küchler, A.W. 1984. Ecological vegetation maps. Vegetatio, 55:3–10.

Küchler, A.W. and J.M. Montoya Maquin. 1971. The UNESCO classification of vegetation: some tests in the tropics. Turrialba, 21:98–109.

Kuhnholtz-Lordat, G. 1949. La cartographie parcellaire de la végétation. Paris. Institut National de la Recherche Agronomique.

Kuminova, A.V. 1962. Geobotanical mapping during a comprehensive study of pastures and haylands on state and collective farms. In: Victor B. Sochava, 1962a, pp. 131–138.

Kurtz, F. 1905. Mapa Fitogeográfico de la Provincia de Córdoba. In: Manuel E. Río and L. Achaval. Geografía de la Provencia de Córdoba. Córdoba. Publicación Oficial. Atlas.

Kunznetsov, N.I. 1928. Geobotanical map on the European part of the Soviet Union. Leningrad. Central Botanical Garden, Geobotanical Section.

Laban, P. (ed.). 1981. Proceedings of the workshop on land evaluation for forestry. IRLI Publ. 28, 355 p., Wageningen.

Laclavère, G. and J. Dejeumont. 1961. Sur l'impression des cartes de la végétation. In: Henri Gaussen, 1961a, pp. 265–273.

Lamarck, J.B. and A.P. de Candolle. 1813. Flore française. 3e édition. Tome II, 600 p. Desray, rue Hautefeuille N° 4, Paris.

Langdale-Brown, I. 1959a. The vegetation of Buganda. Uganda Department of Agriculture. Memoirs of the Research Division, Series 2, No. 2.

Langdale-Brown, I. 1959b. The vegetation of Eastern Province, Uganda. Uganda Department of Agriculture. Memoire of the Research Division, Series 2, No. 1.

Langdale-Brown, I. 1960a. The vegetation of the West Nile, Acholi and Lango Districts of the Northern Province of Uganda. Uganda Department of Agriculture. Memoirs of the Research Division, Series 2, No. 2.

Landale-Brown, I. 1960b. The vegetation of Uganda (excluding Karamoja). Uganda Department of Agriculture. Memoirs of the Research Division, Series 2, No. 6.

Langdale-Brown, I. 1964. The vegetation of Uganda and its bearing on land-use. London. Uganda High Commission. Also: Entebbe. The Government Printer.

Lanly, J.P. 1984. Les ressources forestières de l'Amérique du Sud tropicale. Thèse de doct. Etat Univ. Paul Sabatier. Toulouse. 365 p.

Lauer, W. and H.J. Klink. 1973. Vegetationsgebiete am Ostabfall der zentralmexikanischen Meseta. In: H.J. Klink 1973.

Laumonier, Y. 1980. Contribution à l'étude écologique et structurale des forêts de Sumatra. Thèse de 3e cycle Univ. Paul Sabatier. Toulouse. 137 p.

Laumonier, Y. 1983. International map of the vegetation at 1/1 000 000 scale, 'Sumatra'. Biotrop. Bogor. Indonesia.

Lausi, D. and R. Gerdol. 1980. Mappe della vegetazione degli ambienti umidi subalpin delle Alpi Giulie occidentali. Friuli-Venezia Giulia (Provincia di Udine) (With colour maps 1:500). Consiglio Nazionale delle Richerche. AQ/1/78. Roma.

Lautenschlager, L.F. and C.R. Perry, Jr. 1981. An Empirical, Graphical and Analytical Study of the Relationships Between Vegetation Indices. Report n. EW–j1–04150; JSC–17424, Houston, TX: NASA/AgRISTARS EW/CCA, Johnson Space Centre.

Lavenu, F. 1984. Télédétection et végétation tropicale; exemple du Nord-Est de la Côte d'Ivoire et des mangroves du Bangladesh. Thèse de 3e cycle, Univ. Paul Sabatier. Toulouse. 287 p.

Lavergne, D. 1963. Notes et Documents. 8. Les Causses d'Aquitaine (aperçu phytogéographique). Centre National de la Recherche Scientifique. Service de la Carte de la Vegetation. pp. 10.

Lavrenko, E.M., T.I. Isachenko, S.A. Gribova and R. Neuhäusl. 1981. A new project of vegetation map of Europe. Documents de cartographie écologique, 24:7–9.

Lavrenko, E.M. and L.E. Rodin. 1956. Vegetation map of central Asia. Academy of Sciences of the U.S.S.R. Komarov Botanical Institute. Leningrad.

Lavrenko, E.M. and V.B. Sochava. 1956. Vegetation map of the U.S.S.R. Academy of Sciences of the U.S.S.R. Komarov Botanical Institute. Leningrad.

Lebrun, J. 1961. La cartographie de la végétation: une méthode de développement des pays tropicaux. In: Henri Gaussen, 1961a, pp. 111–126.

Lebrun, J. and R. Devred. 1961. La cartographie de la végétation au Congo belge. In: Recent advances in Botany. University of Toronto Press. Toronto.

Leemans, J. and B. Verspaandonk. 1980. Saftinghe. Vegetatiekaart 1:10.000. 1972. Stichting Het Zeeuwse Landschap. Heikenszand. pp. 24 (coloured map).

Leeuwen, Chr. G. v. 1966. A relation theoretical approach to pattern and process in vegetation. Wentia XV, pp. 25–46.

Leeuwen, Chr. van 1981a. College Syllabus. Technical Univ. Delft, 79 pp (in Dutch).

Leeuwen, Chr. van 1981b. From Ecosystem to Ecodevice. Proc. Int. Congr. Neth. Soc. Landscape Ecology Veldhoven. pp. 29–34.

Leeuwen, Chr. van 1981c. Nature Technics. – Ecology and Nature Technics (5) Tijdschrift Kon. Ned. Heidemij. 92 7/8, pp. 297–306 (in Dutch).

Legris, P. and M. Viart. 1959. Documentation and method proposed for vegetation mapping at 1/1,000,000 scale. Institut français de Pondichéry. Pondicherry. Travaux de la section scientifique et technique.

Legris, P. 1961. Botanical and ecological cartography in India. Institut français de Pondichery. Pondicherry. Travaux de la section scientifique et technique.

Legris, P. & V. M. Meher Homji. 1968. Vegetation maps of India. Proceed. Symposium on recent advances in tropical ecology (Ed. R. Misra & G. Gopal) vol. 1: 32–41.

Legris, P. & F. Blasco. 1972. Carte internationale du Tapis végétal à 1/1 000 000; feuille 'Cambridge'. Notice explicative in Inst. français Pondichéry; Trav. Sect. scient. techn., Hors Série N° 11, 240 p.

Legris, P. & V. M. Meher Homji. 1973. The Deccan trap country and its vegetation patterns. Proceed. Symposium on 'Deccan Trap Country'. Bull. Indian Nat. Sci. Academy, No. 45: 108–128.

Le Houérou, H. N. 1955. Contribution à l'étude de la végétation de la région de Gabès. Annales du Service Botanique et Agronomique de Tunisie, 28: 141–180.

Le Houérou, H. N., J. Claudin, M. Haywood et J. Donadieu. 1975. Etude phytoécologique du Honda. (Volume I. Etudes des ressources naturelles et experimentation et démonstration agricoles dans la région du Honda). UNDP/FAO Rome. 154 pp. carte phyto-écologique du *Honda* 2 feuilles colorés.

Leiberg, J. B. 1899. Vegetation map of Montana. Washington, DC. U.S. Geological Survey, Report, part 5, facing p. 256.

Leneuf, B. et Ch. Rosetti. A propos d'un Programme de Stage sur L'application des Methodes Photographiques Aeriennes a l'étude de certains facteurs de la mise en valeur agricole, pastorale et forestière. pp. 7.

Leys, H. N. 1965. Een Vegetatiekartering van het Liesbosch. Overdruk uit De Levende Natuur, maart '65. pp. 63–72.

Leys, H. N. 1978. Handleiding ten Behoeve van Vegetatiekarteringen. Wetenschappelijke mededelingen K.N.N.V. nr. 130 December 1978. Koninklijke Nederlandse Natuurhistorische Vereniging. pp. 52.

Lindeman, J. C. 1953. The Vegetation of the coastal region of Surinam. Dissertation. Utrecht (col. maps).

Linkola, K. 1941. Die Kartierung der Flora und Vegetation Finlands. Helsinki. Sitzungsbericht der finischen Akademie der Wissenschaft, 1938.

Lipatova, V. V. 1962. A contribution towards the bibliography of the problem of vegetation mapping. In: Victor B. Sochava, 1962a, pp. 265–266.

Little, E. L. 1965. Clave Preliminar de las familia de los árboles en Costa Rica. Revista Interamericana de Ciencias Agrícolas. 'Turriaalba' Vol. 15, N. 2 pp. 119–139.

Livingston, B. E. and F. Shreve. 1921. The distribution of vegetation in the United States as related to climatic conditions. Carnegie Institution of Washington. Washington, DC. Publication No. 284.

Lohmayer, W. 1963. Erfahrungen bei der Verwendung von Luftbildern für die Vegetationskartierung. In: Reinhold Tüxen, 1963a, pp. 129–137.

Londo, G. 1975. Dutch list of hydro-, phreato- and aphreato-phytes. Report RIN, Leersum, 52 pp (in Dutch).

Long, G. 1959. Possibilités actuelles d'application pratique de la cartographique et écologique au Service de la Carte des Groupements Végétaux. Service de la Carte des Groupements Végétaux. Montpellier.

Long, G. 1974. Diagnostic phyto-écologique et aménagement du territoire, vol. 1: principes généraux et méthodes. Masson et Cie., Editeurs. Paris.

Long, G. 1959. Possibilites Actuelles d'Application Pratique de la Cartographie Phytosociologique et Ecologique au Service de la Carte des Groupements Végétaux de France. Symposium sur la Cartographie de la Végétation, Stolzenau/Weser, 23–26 mars 1969. pp. 6.

Long, G. 1968. Conception Générales sur la Cartographie Biogéographique Intégrée de la Végétation et de son Écologie. Document N. 46 Centre National de la Recherche Scientifique. Montpellier 1968. pp. 77.

Long, G. 1969. Perspectives Nouvelles de la Cartographie Biogéographique Végétale Intégrée. Separatum Vol. XVIII, 16–V–1969., Vegetatio. Fasc. 1–6 pp. 44–63.

Long, G. 1974. Diagnostic phyto-écologique et aménagement du territoire. I. Principes généraux et méthodes. Collection d'écologie 4. Mason et cie., Ed., Paris.

Lorentz, P. G. 1876. Mapa fitogeográfico de la parte noroeste de la República Argentina. In: Kurt Kueck: Urlandschaft, Raublandschaft und Kulturlandschaft in der Provinz Tucumán. Bonner Geographische Abhandlungen, 10: 1–102.

Lorenzoni, G. G., C. Caniglia, S. Marchiori and S. Razzara. 1980. Carta della vegetazione di Maruggio, S. Pietro e S. Isidoro. (Salento, Publia Meridionale) (colour map 1:25.000). Consiglio Nazionale delle Richerche. AQ/1/123. Instituto di Botanica, Universitá di Padova. Roma.

Louis, H. 1939. Das natürliche Pflanzenkleid Anatoliens.

568

Spemann Verlag. Stuttgart.

Lubimova, E.L. 1961. The use of toponyms for the compilation of botanical maps. In: Victor B. Sochava, 1962a, pp. 64–67.

Lüdi, W. 1921. Genetisch-dynamische Vegetationskarte des Lauterbrunnentals (Sukzessionskarte). Bern. Beiträge zur geobotanischen Landesaufnahme der Schweiz, vol. 9.

Lukicheva, A.N. 1962. Principles of the choice of color designation for small scale geobotanical maps. In: Victor B. Sochava, 1962a, pp. 244–253.

Lukicheva, A.N. and D.N. Saburov. 1969. Revealing the relations between vegetation and landscape in the source of large scale vegetation mapping. Academy of Sciences of the U.S.S.R. Komarov Botanical Institute. Leningrad. Geobotanical Cartography, pp. 33–42.

Lundegårdh, H. 1957. Klima und Boden, 5th ed., Gustav Fischer Verlag. Jena.

Luti, R. 1976. Vegetación de la provincia de Córdoba. Universidad Nacional, Catedras de Ecología, Agricultura y Geobotánica. Córdoba.

Maack, R. 1950. Mapa fitogegráfico do estado do Paraná. Serviço de Geologia e Petrografia do Instituto de Biologia e Pesquisas technológicas. Curitiba.

Maarel, E.v.d. 1966. Over vegetatiestructuren, -relaties en -systemen in het bijzonder in de duingraslanden van Voorne. On vegetational structures, relations and systems, with special reference to the dune grasslands of Voorne (The Netherlands). Diss. Univ. of Groningen.

Maarel, van der, E. and V. Westhoff. 1964. The Vegetation of the Dunes near Oostvoorne (The Netherland) (With a coulour vegetation map) Wentia 12 (1964) 1–61. pp. 61.

MacConnell, W.P. and L.E. Gravin. 1956. Cover mapping a state from aerial photographs. Photogrammetric Engineering, 22:702–707.

Mahler, P.J. (ed.). 1970. Manual of multi-purpose land classification. Publ. 212, *Soil Institute of Iran, Min. of Agric.*, Teheran, 81 p.

Maire, A., J.P. Bourassa and A. Aubin. 1976. Cartographie écologique des milieux à larves de moustiques de la région de Trois-Rivières, Quebec. Grenoble. Documents de cartographie écologique, 17:49–71.

Major, J. 1963. Vegetation mapping in California. In: Reinhold Tüxen, 1963a, pp. 195–218.

Malin, J.C. 1947. Grasslands, 'treeless' and 'subhumid.' Geographical Revue, 37:241–250.

Manual of photointerpretation. 1960. American Society of Photogrammetry. Washington.

Mangenot, G. 1955. Ecologie et représentation cartographique des forêts équatoriales et tropicales humides de l'Afrique occidentale. In: Henri Gaussen, pp. 149–156.

Mangenot, G. 1956. Les recherches sur la végétation dans les régions tropicales humides de l'Afrique occidentale. United Nations Educational, Scientific and Cultural Organization. Paris. Humid Tropics Research: Tropical Vegetation.

Marincek, L., I. Puncer and M. Zupancic. 1980. Die Vegetationskartierung in Slowenien. Documents de cartographie écologique, 23:15–16.

Marrès, P. 1952. La cartographie parcellaire de la végétation. Annales de Géographie, 41:363–366.

Marschall, F. 1963. Vegetationskartierung und Güterzusammenlegung an einem Beispiel aus dem oberen Engadin. In: Reinhold Tüxen, 1963a, pp. 473–480.

Marschner, F.J. 1974. The original vegetation of Minnesota. North Central Forest Experiment Station. St. Paul, MN.

Martius, C.F.P. 1858. Flora brasiliensis. Oldenburg Verlag. Leipzig.

Marvet, A.V. 1968. On the construction of a legend for detailed large scale maps reflecting vegetation dynamics. Academy of Sciences of the U.S.S.R., Komarov Botanical Institute. Leningrad. Geobotanical Cartography, pp. 38–44.

Mason, Herbert L. and J.H. Langenheim. 1957. Language analysis and the concept environment. Ecology, 38:325–340.

Matuszkiewicz, W. The vegetation of the Forest of the environment of Lvov. Anab. Univ. Maria Curie-Sklodowska, Lublin-Polonia, Vol. III. 5. pp. 119–198.

Matuszkiewicz, A. 1961. Bibliographie der Vegetationskarten Polens. Excerpta Botanica, Section B: Sociologica, 3:68–77.

Matuszkiewicz, A. 1963. La cartographie phytosociologique en Pologne. In: Reinhold Tüxen, 1963a, pp. 347–352.

Maxwell, E.L. 1976. A remote rangeland analysis system. Journal of Range Management, 29:66–73.

Maxwell, E.L. 1981. 'Biomass Measurement from Landsat – Drought and Energy Application,' Proceedings of the Second Eastern Regional Remote Sensing Application Conference, NASA Conference Publication 2198, Greenbelt, MD: NASA Goddard Space Flight Centre, pp. 51–72.

Mazade, A.V. 1981. Ten-Ecosystem Study. Final Report. Houston, TX: NASA and USDA Nationwide Forestry Application Program.

Mazing, V.V. 1962. Some problems of large scale vegetation mapping. In: Victor B. Sochava, 1962a, pp. 47–53.

McArdle, R.E. 1949. The yield of Douglas fir in the Pacific North-west. United States Department of Agriculture, Forest Service. Technical Bulletin No. 201.

McCoy, R.M. 1982. Models in remote sensing: an approach to mapping vegetation in arid lands. American Society of Photogrammetry. Falls Church, VA. Pecora VII Symposium, Proceedings, pp. 427–441.

McDaniel, K.C. and R.H. Haas. 1981. Classifying and characterizing natural vegetation on a regional basis with Landsat MSS data. United States Department of Agriculture, Forest Service. Weshington, DC. Technical Report No. WO-28, pp. 197–203.

McDaniel, K.C. and R.H. Haas. 1982. Assessing mesquitegrass vegetation condition from Landsat. Photogrammetric Engineering and Remote Sensing, 48:441–450.

McGee, W.J. 1889. Primeval forests and swamps of northeastern Iowa. Washington, D.C. United States Geological Survey, 11th Annual Report, plate 22.

McIntyre, G.A. 1952. A Method for unbiassed selective sampling using ranked sets. Austr. J. Agric. research 3: p. 385–390.

McNaughton, S.J. 1979. Grassland-Herbivore Dynamics. In: Sinclair, A.R.E. and M. Norton-Griffiths, Serengeti: dynamics of an ecosystem. University of Chicago Press, Chicago.

Meher Homji, V.M. 1973. Phytogeography of the Indian Subcontinent. In: Misra (R.) & al. (ed.) Progress of plant ecology in India vol. I: 10–88. Banaris Hindu Univ., Varanasi, India.

Meisel, K. 1954. Wasserstufenkarte des Emstales zu Dalum und Hespe. (Colour map Scale 1:5.000).

Meisel, K. 1960. Die Auswirkung der Grundwasserabsenkung auf die Pflanzengesellschaften im Gebiete um Moers (Niederrhein) (With colour map Scale 1:25.000) Stolzenau/Weser, pp. 105.

Meisel, K. 1963. Die Vegetationskarte als Grundlage für die Beurteilung von Wasserschaden. In: Reinhold Tüxen, 1963a, pp. 423–430.

Melstser, L.I. 1980. The representation of heterogeneous vegetation of west Siberian tundras on medium scale maps. Academy of Sciences of the U.S.S.R., Komarov Botanical Institute. Leningrad. Geobotanical Cartography, pp. 11–24.

Mennema, J. and E.J. Weeda. Flora and Vegetation of the Wadden Sea islands and coastal areas. Report 9 (With colour maps scale 1:100000) Vegetation Islands. pp. 382.

Mensching, H. 1950. Verbreitungskarten von Pflanzengesellschaften als Hilfsmittel für den Morphologen am Beispiel des Wesertales. Stolzenau/Weser. Mittelungen der floristisch-soziologischen Arbeitsgemeinschaft, Neue Folge, vol. 2.

Merchant, J.W. 1983. Utilizing Landsat MSS Data in forest and range management, a guide to selected literature. Lawrence, KS. University of Kansas, Space Technology Center.

Meulen, F. v.d. 1979. Plant sociology of the Western Transvaal Bushveld. South Africa. A syntaxonomic and synecological study. Dissertations Botanicase Band 49, Nijmegen.

Meusel, H. 1943. Vergleichende Arealkunde. Bornträger Verlag. Berlin-Zehlendorf.

Miles, J. 1979. Vegetation Dynamics. Outline Studies in Ecology. Chapman and Hall. London.

Miller, G.E. 1981. A look at the commonly used Landsat vegetation indices. NASA/Agristars/Lockheed Mangement and Engineering Services. Houston, TX.

Miller, W.A. and M.M. Shasby. 1982. 'Refining Landsat Classification Results Using Digital Terrain Data'. Journal of Applied Photographic Engineering 8:35–40.

Ministerie van CRM. 1979. Natuurwaarden en Cultuurwaarden in het landelijk gebied. (Coloured maps) The Hague. Netherlands.

Miyawaki, A. 1966. Bibliographie der Vegetationskarten Japans. Excerpta Botanica B. 7:54–59. Gustav Fischer Verlag. Stuttgart.

Miyawaki, A. (ed.) 1967. Vegetation of Japan. Encyclopedia Sci. Technology 3: 535 pp. with vegetation map. (Japanese) Tokyo.

Miyawaki, A. 1968. Heutige potentielle naturliche Vegetation und ihre Ersatzgesellschaften im Kanto-Flachland (Mitteljapan). JIBP: 89–95. Sendai.

Miyawaki, A. 1968. Japan. in: Küchler, A.W. (ed.) International bibliography of vegetation maps 3:204–229. Kansas.

Miyawaki, A. 1968. Abschätzung der potentiellen natürlichen Vegetation und ihre Anwendungsmöglichkeiten für die Landnutzung. Berich. Forsch. Studien 22:25–54. (japanisch) Tokyo.

Miyawaki, A. 1968. Typen von Vegetationskarten und ihre Anwendung für die Beurteilung des Standortes. Map 6(2): 1–9. (japanisch mit deutscher Zusammenfassung) Yokohoma.

Miyawaki, A. 1969. Vegetationkundliche Studien im Neustadtbezirk Minamitama westlich Tokyo. Stud. Veget. Grünplanung Neustadtbezirk Minamitana westl. Tokyo. Teil: 1–94. mit 2.frb. Vegetationskarten. (japanisch mit deutscher Zusammenfassung) Tokyo.

Miyawaki, A. 1969. Naturschutz und Vegetationskarte. Berg und Museum: 2–3. (japanisch) Ohmachi.

Miyawaki, A. 1970. Anfertigung der japanischen Vegetationskartierung. Rep. JIBP (CT) 6:64–66. (japanisch) Sendai.

Miyawaki, A. 1970. Vegetationskarte. Die Bedeutung des heutigen Tourismus und seine Zukunft: 95–105. (japanisch) Tokyo.

Miyawaki, A. 1970. Zustand der Vegetationskartierung und ihre Anwendung für die Industrie. 17 pp. (japanisch) Tokyo.

Miyawaki, A. 1971. Über die Karte der realen Vegetation im Hakone-Gebiet. Rep. Map. 9(4):16–17 (japanisch) Tokyo.

Miyawaki, A. 1971. Bibliographie der Vegetationskarten Japans II. Excerpta Botanica Sec. B, Sociologica Band 11(3):238–240. Gustav Fischer Verlag, Stuttgart.

Miyawaki, A. 1972. Voruntersuchung für die Beurteilung und Gestaltung der Umwelt aus ökologischer Sicht. Forsch. Ber. über die rechte Flächennutzung in der Präf. Kanagawa 32 pp. (japanisch) Tokyo.

Miyawaki, A. 1973. Begrünung der Umgebung von Schulen. Rep. of the Ministry of Education 1145: 21–29 (japanisch) Tokyo.

Miyawaki, A. 1974. Phytosociological studies on creation of environmental protection forests around schools-based on field surveys at 158 schools throughout Japan. Rep. Stud. Creation of Environ. Protec. For. around schools. 116 pp. with 2 colored vegetation maps. (with co-workers) (Japanese) Yokohama.

Miyawaki, A. 1974. Vegetation der Präfektur Nagano. Pflanzensociologische Studie – Heft 2, 79 pp. mit 4 farb. Vegetationskarten. (Mitarbeiter) (japanisch) Nagano.

Miyawaki, A. 1975. Entwicklung der Umweltschutz-Pflanzungen und Aussaten in Japan. In: Tüxen, R. (Edit.) Sukzessionsforschung: 273–254. Vaduz.

Miyawaki, A. 1975. Outline of Japanese vegetation. JIBP Syntheses 8:19–27 with 1 colored vegetation map. Tokyo.

Miyawaki, A. 1977. The potential vegetation maps of the Chubu and Kinki regions (central Japan). Environment and Human Survival 4:296–308. (Japanese) Tokyo.

Miyawaki, A. 1977. Vegetation of Japan. Compared with other regions of the world. 535 pp. (with co-workers) (Japanese) Verlag Gakken, Tokyo.

Miyawaki, A. 1977. Vegetation science and environmental protection. 577 pp. (co-edit. Miyawaki, A. & Tüxen, R.) Verlag Maruzen, Tokyo.

Miyawaki, A. 1977. Die Vegetation in der Stadt Sakura (Chiba Präfektur in der Nähe Tokyos). 132 pp. mit 2 farb. Vegetationskarten und Tabellenheft. (Mitarbeiter) (japanisch mit deutscher Zusammenfassung) Sakura.

Miyawaki, A. 1978. Handbook of Japanese Vegetation. 850 pp. (edit.) (with Okuda, S., Mochozuki, R.) (Japanese with scientific plant names) Shibundo Co. Lfd. Tokyo.

Miyawaki, A. 1978. Sigmassoziationen in Mittel und Süd-Japan. In: Tüxen, R. (Edit.) Assoziationkomplexe: 241–265. Vaduz.

Miyawaki, A. 1984. A vegetation-ecological view of the Japanese archipelago. Bull. Inst. Envir. Sci. Tech. Yokohama Nat. Univ. 11:85–101. Yokohama.

Miyawaki, A. 1980. Vegetation of Japan; Vol. 1: Yakushima (edit.). 376 pp. with 4 color vegetation maps and supplement tables. (Japanese with German summary) Shibundo. Tokyo.

Miyawaki, A. 1981. Vegetation of Japan; Vol. 2: Kyushu (edit.). 483 pp. with 3 color vegetation maps and supplement tables. (Japanese with German summary) Shibundo. Tokyo.

Miyawaki, A. 1982. Vegetation of Japan; Vol. 3: Shikoku (edit.). 539 pp. with 3 color vegetation maps and supplement tables. (Japanese with German summary) Shibundo. Tokyo.

Miyawaki, A. 1983. Vegetation of Japan; Vol. 4: Chugoku (edit.). 540 pp. with 3 color vegetation maps and supplement tables. (Japanese with German summary) Shibundo. Tokyo.

Miyawaki, A. 1984. Vegetation of Japan; Vol. 5: Kinki (edit.). 596 pp. with 3 color vegetation maps and supplement tables. (Japanese with German summary) Shibundo. Tokyo.

Miyawaki, A. 1985. Vegetation of Japan; Vol. 6: Chubu (edit.). 604 pp. with 3 color vegetation maps and supplement tables. (Japanese with German summary) Shibundo. Tokyo.

Miyawaki, A. 1986. Vegetation of Japan; Vol. 7: Kanto (edit.). 643 pp. with 3 color vegetation maps and supplement tables. (Japanese with German and Englisch summary) Shibundo. Tokyo.

Miyawaki, A. and K. Fujiwara. 1968. Pflanzensoziologische Studien im westlichen Neubaugebiet der Stadt Fujisawa bei Yokohama. 466 pp. mit 2 frb. Vegetationskarten. (japanisch mit deutscher Zus.) Fujisawa.

Miyawaki, A. and K. Fujiwara. 1969. Pflanzensoziologische Studien im Ise-Shima-Nationalpark. Forsch. Ber. Ise-Shima Nat. Park: 101–143. mit 1 frb. Vegetationskarten. (japanisch mit deutscher Zus.) Tokyo.

Miyawaki, A. and K. Fujiwara. 1969. Vegetation und Vegetationskartierung über das Moor Ozegahara (Mitteljapan). Ann. Rep. JIBP-CT (P) Fisc. Year 1968: 6–11 (japanisch mit deutscher Zus.) Sendai.

Miyawaki, A. and K. Fujiwara. 1969. Ein Begrünungs- und Resaturierungsplan im westlichen Neubaugebiet der Stadt Fujisawa bei Yokohama. mit 1 frb. Vegetationskarte. 38 pp. (japanisch mit deutscher Zus.) Fujisawa.

Miyawaki, A. and K. Fujiwara. 1969. Vegetationskundliche Studien über den Quasi-Nationalpark Meiji no mori bei Mino (Osaka-Präfektur). 58 pp. mit 1 frb. Vegetationskarte. (japanisch mit deutscher Zus.) Tokyo.

Miyawaki, A. and K. Fujiwara. 1974. Vegetation der Stadt Itami (Präf. Hyogo). Vegetationskundliche Studien zu einer Bestandesaufnahme der natürlichen Umwelt und zur Wiederherstellung einer naturgemässen Umwelt. 136 pp. mit 2 farb. Vegetationskarten und Tabellenheft. (japanisch mit deutscher Zus.) Itami.

Miyawaki, A. and K. Fujiwara. 1975. Ein Versuch zur Kartierung des Natürlichkeitsgrades der Vegetation und der Anwendungsmöglichkeit dieser Karte fur den Umwelt- und Natuschutz am Beispiel der Stadt Fujisawa. Phytocoenologia 2 (3/4): 429–436. Stuttgart-Lehre.

Miyawaki, A. and K. Fujiwara. 1976. Vegetation der Umgebung von Wakasa-Ohwi und Mihama, Fukui Präfektur. -- Eine pflanzensoziologische Studie zum Umweltschutz und zur Erhaltung und zur Schaffung von Umweltschutzwäldern. – Bull. Yokohama Phytosoc. Soc. 3: 114 pp. mit 2 farb. Vegetationskarten und Tabellenheft. (japanisch mit deutscher Zus.) Yokohama.

Miyawaki, A. and K. Fujiwara. 1979. Vegetation and vegetation maps. In: Matsunaka, S. (Edit.) Pollution and biotic indicators: 30–35. (Japanese) Verlag Asakura. Tokyo.

Miyawaki, A., K. Fujiwara and H. Harada. 1976. Vegetation des Hama-dori (Bezirk Futaba) in der Präf. Fhkushima. Bull. Yokohama Phytosoc. Soc. 2: 70 pp. mit 3 farb. Vegetationskarten und Tabellenheft. (japanisch mit deutscher Zus.) Yokohama.

Miyawaki, A., K. Fujiwara and H. Harada. 1976. Vegetation der Stadt Kashiwazaki und ihre Umgebung in der Präf. Niigata an der Küste des Japanischen Meeres von Mittel-Honshu. Pflanzensoziologische Studie im nördlichen Grenzgebiet der Castanopsis cuspidata var. sieboldii-Wälder. Bull. Yokohama Phytosoc. Soc. 9: 120 pp. mit farb. 2 Vegetationskarten. (japanisch mit deuscher Zus.) Yokohama.

Miyawaki, A., K. Fujiwara, H. Harada, T. Kusunoki and S. Okuda. 1971. Vegetationskundliche Untersuchungen in der Stadt Zushi bei Yokohama. Besondere Betrachtung der Camellietea japonicae Wälder Japans. Zushi Educat. Comm.: 151 pp. mit 2 farb. Vegetationskarten (japanisch mit deutscher Zus.) Tokyo.

Miyawaki, A., K. Fujiwara and R. Mochizuki. 1977. Vegetation der Ubayashiki in N. Honshu (Iwate Präf.). Bull. Yokohama Phytosoc. Soc. 7: 82 pp. mit 2 farb. Vegetationskarten und Tabellenheft. (japanisch mit deutscher Zus.) Yokohama.

Miyawaki, A., K. Fujiwara and Y. Murakami. 1984. Vegetation der Stadt Fujisawa. 168 pp. mit farbigen Vegetationskarten. (japanisch mit deutscher Zus.) Fujisawa.

Miyawaki, A., K. Fujiwara, S. Okuda, L. Minowa, K. Tsurumaki, T. Koshinaka, Y. Aizawa, Y. Senuma, K. Yamamoto and R. Mochitsuki. 1980. Die Vegetation von Kashiwazaki und Umgebung im Radius von 30 km. Bull. Yokohama Phytosoc. Soc. 24:1–71. mit 2 farb. Vegetationskarten und Tabellenheft. (japanisch mit deutscher Zus.) Yokohama.

Miyawaki, A., K. Fujiwara, S. Suzuki and H. Harada. 1971. Vegetation der Stadt Fujisawa (Kanagawa Präf.) Eine Pflanzensoziologische Studie für den Umweltschutz der Stadt Fujisawa: 117 pp. mit 2 frb. u. 1 Schwarz u. Weiss Vegetationskarten. (japanisch mit deutscher Zus.) Fujisawa.

Miyawaki, A. and H. Harada. 1974. Pflanzensoziologische Studie zur Schaffung einer grünen Umwelt der Stadt Kamakura und ihre Erhaltung. 44 pp. mit 1 Vegetationskarte. (japanisch mit deutscher Zus.) Kamakura.

Miyawaki, A., H. Harada and S. Suzuki. 1971. Pflanzensoziologische Studien für rationelle Nutzung und Erhaltung des Ohbashiroyama-Bezirks (Fujisawa bei Yokohama). Stadt Fujisawa: 43 pp. mit 1 farb. u. 1 Schwarz u. Weiss Vegetationskarten. (japanisch mit deutscher Zus.) Fujisawa.

Miyawaki, A., H. Harada and H. Ude. 1979. Vegetationskundliche Untersuchung zur Schaffung von Umweltschutzwäldern um Industrieanlagen, erläutert am Beispiel der 11 Fabriken der Toray-Industrie-AG. Bull. Yokohama Phytosoc. Soc. 8: 50 pp. mit 11 farb. Vegetationskarten. (japanisch mit deutscher Zus.) Yokohama.

Miyawaki, A., H. Ide and S. Okuda. 1968. Vegetation und Vegetationskundliche Untersuchungen im Kohoku-Bez. (Yokohama). Grundlagenforschung über Natur und Standortbedingungen im Kohoku-Neustadt-Gebiet: 47–86. mit 2 farb. Vegetationskarten. (japanisch mit deutscher Zus.) Yokohama.

Miyawaki, A. and S. Itow. 1966. Phytosociological approach to the conservation of nature and natural resources in Japan. with color vegetation map. Pacific Sci. Congress: 1–5. Tokyo.

Miyawaki, A., S. Itow and S. Okuda. 1967. Pflanzensoziologische Studien über die Vegetation der Umgebung von Aizukomagatake u. Tashiroyama (Präf. Fukushima). Rep. Nat. Conserv. Soc. Japan 29:15–43. mit 1 farb. Vegetationskarte. (japanisch mit deutscher Zus.) Tokyo.

Miyawaki, A. and J.W. Kim. 1985. Phytosociological study of the vegetation of Shonan seashore, Kanagawa Prefecture. with 1 color vegetation map. Bull. Inst. Sci. Technol. Yokohama Natn. Univ. 12: 105–124. Yokohama.

Miyawaki, A., Y. Nakamura and S. Okuda. 1978. Die Potentielle natürliche Vegetation in der Gegend vom Jouetsu (Shibukawa Minakami). ökologischer Forschungsbericht zum Bau des Jonetsu-Shinkansen: 176–226. mit 4 farb. Vegetationskarten. (japanisch mit deutscher Zus.) Tokyo.

Miyawaki, A., Y. Nakamura, S. Okuda and S. Suzuki. 1982. Vegetation der Stadt Handa: 121 pp. mit 2 farb. Vegetationskarten.

Miyawaki, A., Y. Nakamura, K. Fujiwara and Y. Murakami. 1984. Die potentielle natürliche Vegetation der Stadt Fuji. 254 pp. mit farbiger Vegetationskarte. (japanisch mit deutscher Zus/Englisch summary) Fuji.

Miyawaki, A., T. Ohba and S. Okuda. 1969. Pflanzensoziologische Studien über die alpine und subalpine Stufe des Norikura-dake in Mittel Japan. Rep. Nat. Conserv. Soc. Japan 36:50–103. mit 1 farb. Vegetationskarte. (japanisch mit deutscher Zus.) Tokyo.

Miyawaki, A., T. Ohba, S. Okuda, K. Nakayama and K. Fujiwara. 1969. Pflanzensoziologische Studien über die Vegetation der Umgebung von Echigo-Sanzan und Okutadami (Präf. Niigata und Fukushima). Sci. Rep. Echigo-Sanzan Okutadami and its Vicinity, Niigata and Fukushima Prefecture: 57–152. mit 1 farb. Vegetationskarte. (japanisch mit deutscher Zus.) Tokyo.

Miyawaki, A., T. Ohba and N. Murase. 1969. Scientific studies on the Hakone and Peninsula Manauzu, Kanagawa Prefecture. Sci. Rep. Hakone-Manazuru-Peninsula. 59 pp. with 2 color vegetation maps (in Japanese with German summary). Yokohama.

Miyawaki, A. and K. Ohno. 1972. Pflanzensoziologische Studien für Vegetationsgutachten und Grünplanung auf dem Wakabadai in Yokohama. 44 p. mit 2 farb. Vegetationskarten. (japanisch mit deutscher Zus.) Yokohama.

Miyawaki, A., K. Ohno and S. Okuda. 1974. Pflanzensociologische Studie über den Berg Daisen in der Präf. Tottori (W-Honshu), Japan. Bulletin Inst. Sci. Technol. Yokohama Nat. Univ. 1(1): 89–122. mit 1 farb. Vegetationskarte und Tabellenheft. (japanisch mit deutscher Zus.) Yokohama.

Miyawaki, A. and S. Okuda. 1966. Die Vegetation der Umgebung des Katsuoji-Tempels. Osaka Präf.: 3–15. mit 1 farb. Vegetationskarte (japanisch mit deutscher Zus.) Osaka.

Miyawaki, A and S. Okuda. 1966. Karte der realen Vegetation des Staatlichen Naturparks für Naturstudien in Tokyo. Ecol. Studies Biotic Comm. Nat. Study 1:1–14. mit 1 farb. Vegetationskarte (japanisch mit deutscher Zus.). Osaka.

Miyawaki, A. and S. Okuda. 1972. Pflanzensoziologische Untersuchungen über die Auenvegetation des Flusses Tama bei Tokyo, mit einer vergleichenden Betrachtung über die Vegetation des Flusses Tone. Vegetatio 24(4–6):229–311. Denn Haag.

Miyawaki, A. and S. Okuda. 1974. Karte der potentiellen natürlichen Vegetation der Präf. um Tokyo. mit 1 farb. Vegetationskarte (japanisch mit deutscher Zus.). Yokohama.

Miyawaki, A. and S. Okuda. 1975. Vegetation der Umgebung der Wakasa-Bucht, Fukui Präf. Sci. Rep. Natur Conserv. Soc. Japan 47. 25–111. mit 6 farb. Vegetationskarten (japanisch mit deutscher Zus.). Tokyo.

Miyawaki, A. and S. Okuda. 1976. Methode der Vegetationskartierung. In: Numata, M. (Edit.) Handbook of the Nature Conservation: 258–268. Tokyo.

Miyawaki, A. and S. Okuda. 1976. Die potentielle natürliche Vegetation der Präf. um Tokyo. Bull. Inst. Environm. Sci. Yokohama Natn. Univ. 2(1):95–114. mit 1 farb. Vegetationskarte (japanisch mit deutscher Zus.). Yokohama.

Miyawaki, A. and S. Okuda. 1977. Diagnose und Vorschläge auf Grund der Vegetationkunde für künftige Massnahmen für den Umweltschutz für die Umgebungvon Tokyo. In: Miyawaki, A., Tüxen, R. (Edit.) Vegetation Science and Environmental Protection. Tokyo.

Miyawaki, A., S. Okuda and K. Fujiwara. 1970. Pflanzensoziologische Studien über die Vegetation der Tsugaru-Halbinsel, des Berges Iwaki und des Juniko Sees. Sci. Rep. Tsugaru Peninsula Mt. Iwaki National Park: 1–40. mit 1 farb. Vegetationskarte (japanisch mit deutscher Zus.). Tokyo.

Miyawaki, A., S. Okuda and K. Fujiwara. 1971. Pflanzensoziologische Studien über die Vegetation des Numappara-Moores und seiner Umgebung, Tochigi Präf., Mittel Japan. Rep. Nat. Conserv. Soc. Japan. 38:135–182. mit 1 farb. Vegetationskarte (japanisch mit deutscher Zus.). Tokyo.

Miyawaki, A., S. Okuda, K. Fujiwara and K. Inoue. 1977. Vegetation der Sarobetsugenya NO-Hokkaido. 47 pp. The Tourist Resources Conservation Foundation. mit 1 farb. Vegetationskarte (japanisch mit deutscher Zus.). Tokyo.

Miyawaki, A., S. Okuda, H. Harada and Y. Nakamura. 1977. Potentielle natürliche Vegetation des Chubuken des Tokai Gebietes in Mittel-Japan. Bull. Envir. Sci. Techn. Yokohama Nat. Univ. 3(1):77–109. mit 1 farb. Vegetationskarte (japanisch mit deutscher Zus.). Yokohama.

Miyawaki, A., S. Okuda and K. Inoue. 1975. Vegetation des SO-Teils der Präf. Saitama. 86 pp. mit 6 farb. Vegetationskarten (japanisch mit deutscher Zus.). Saitama.

Miyawaki, A., S. Okuda and K. Inoue. 1980. Pflanzensoziologishe Untersuchungen in den Wäldern des Leiji-Schreins in Tokyo. Interdisziplinare Untersuchungsergebnisse über den Meiji-Schrein: 269–333. mit 1 farb. Vegetationskarte (japanisch mit deutscher Zus.). Tokyo.

Miyawaki, A., S. Okuda and K. Suzuki. 1975. Küstenvegetation in der Bucht von Tokyo. 119 pp. mit 2 farb. Vegetationskarten (japanisch mit deutscher Zus.) Tokyo.

Miyawaki, A., S. Okuda and K. Suzuki. 1975. Pflanzensoziologische Studien über die Vegetation im SO-Teil von Chiba und Chiharadai (Präf. Chiba). 93 pp. mit 2 farb. Vegetationskarten (japanisch mit deutscher Zus.) Tokyo.

Miyawaki, A. and Y. Sasaki. 1981. Planung von Umweltschutzwäldern entlang der staatlichen japanischen Eisenbahn (Japanese National Railway) im Hashihara Misana-Bezirk (Mittel-Japan) Heft III. Bull. Yokohama Phytosoc. Soc. 30:1–41. mit 2 farb. Vegetationskarten. Yokohama.

Miyawaki, A. and Y. Sasaki. 1982. Die potentielle natürliche Vegetation des Enrei-Hochlandes. Bull. Yokohama Phytosoc. Soc. 39:1–63. mit 1 farb. Vegetationskarten. Yokohama.

Miyawaki, A., Y. Sasaki and K. Fujiwara. 1971. Bericht über eine Vegetationsaufnahme für den Grünplan und die Landschaftspflege des Waldparkes auf dem Musashi-Hugel nördlich von Tokyo. mit 2 farb. Vegetationskarten (japanisch mit deutscher Zus.) Tokyo.

Miyawaki, A., Y. Sasaki and K. Inoue. 1976. Pflanzensoziologische Untersuchung des Plateaus von Hiki, Präf. Saitama. Bull. Yokohama Phytosoc. Soc. 5: 47 pp. mit 2 farb. Vegetationskarten (japanisch mit deutscher Zus.) Yokohama.

Miyawaki, A., Y. Sasaki and M. Kimura. 1979. Planung von Umweltschutzwäldern entlang der japanischen Staatseisenbahn in Hashibara, Misawa-Bezirk (Mittel-Japan) Bull. Yokohama Phytosoc. Soc. 12: 19 pp. mit 2 farb. Vegetationskarten (japanisch mit deutscher Zus.) Yokohama.

Miyawaki, A., Y. Sasaki and R. Kobayashi. 1982. Vegetation der Stadt Atsugi in der Präf. Kanagawa: 153 pp. mit 2 farb. Vegetationskarten (japanisch mit deutscher Zus.) Atsugi.

Miyawaki, A. and K. Suzuki. 1974. Die potentielle natürliche Vegetation an der Küste Anan (Präf. Tokushima) und ihre Standorte. Forsch. Ber. über die Vegetation der Küsten-Erholungsgebiete. II 49 pp. mit 2 farb. Vegetationskarten (japanisch mit deutscher Zus.) Tolushima.

Miyawaki, A. and K. Suzuki. 1974. Die Vegetation der Stadt Chiba. Eine pflanzensoziologische Studie zur Erhaltung und zur Schaffung einer vegetationsreichen Stadt. 92 pp. mit 2 farb. Vegetationskarten und tabellenheft. (japanisch mit deutscher Zus.) Chiba.

Miyawaki, A. and K. Suzuki. 1975. Vegetation der Halbinsel Uragami, Kumanonada in der Präf. Wakayama. Bull. Yokohama Phytosoc. Soc. 1: 102 pp. mit 2 farb. Vegetationskarten und Tabellenheft. (japanisch mit deutscher Zus.) Yokohama.

Miyawaki, A. and K. Suzuki. 1979. Karte der Sigmassoziationen des Flussgebietes Sagami (Präf. Kanagawa in Mittel-Japan). Bericht über Umweltforschung im Flussgebiet von Sagami in der Präf. Kanagawa. mit 1 farb. Vegetationskarte. Tokyo.

Miyawaki, A. and K. Suzuki. 1980. Process of phytosociological studies and vegetation mapping. Bull. Inst. Envir. Yokohama Nat. Univ. 6:65–76. Yokohama.

Miyawaki, A., K. Suzuki and K. Fujiwara. 1977. Human impact upon forest vegetation in Japan. Naturaliste Canadien 104:97–107. (with French summary) Quebec.

Miyawaki, A., K. Suzuki, Y. Ogawa and M. Kimura. 1979. Vegetation des Bezirks Tsuruga, Fukui Präf. Eine Pflanzensoziologische Studie zum Umweltschutz und zur Erhaltung und Schaffung von Umweltschutzwäldern. Bull. Yokohama Phytosoc. Soc. 15: 74 pp. mit 2 farb. Vegetationskarten und Tabellenheft. (japanisch mit deutscher Zus.) Yokohama.

Miyawaki, A., K. Suzuki, Y. Sasaki, K. Fujiwara and H.

Harada. 1972. Vegetation des Gebietes von Tanoura in Wakasatakahama, Präf. Fukui. Pflanzensoziologische Untersuchung zum Pflanzen von Umweltschutzwäldern. 74 pp. mit 2 farb. Vegetationskarten und 1 Standortkarte. (japanisch mit deutscher Zus.) Osaka.

Miyawaki, A., H. Sugawara and T. Hamada. 1967. Pflanzensoziologische Studien über die Vegetation am Südhang des Fujiyama. Wiss. Ber. Südhang des Fujiyama: 1–40. mit 1 farb. Vegetationskarte (japanisch mit deutscher Zus.) Shizuoka.

Miyawaki, A., H. Sugawara and T. Hamada. 1971. Vegetation of Mt. Fuji. Result of the Co-operative Scientific Survey of Mt. Fuji: 665–721. with 1 color vegetation map. (in Japanese with English summary) Tokyo.

Miyawaki, A., H. Sugawara, T. Hamada and M. Ishizuka. 1969. Pflanzensoziologische Studien über die Vegetation auf dem Nordhang des Berges Fuji (Yamanashi Präf.). Sci. Rep. Mt. Fuji: 1–48. mit 1 farb. Vegetationskarte (japanisch mit deutscher Zus.) Koufu.

Miyawaki, A. and H. Tohma. 1975. Vegetation und Vegetationskarte im W.Teil der neuen Stadt Tama. Ökologosche Studien für Umweltschutz im W-Teil der neuen Stadt Tama: 1–92. mit 2 farb. Vegetationskarten. Yokohama.

Miyawaki, A., H. Tohma and Y. Sasaki. 1973. Vegetationskundliche Untersuchung in der Umgebung von Higashitakane, Stadt Kawasaki. Forsch. Ber. über das Kulturgut in der Präf. Kanagawa 35:1–17. mit 1 farb. Vegetationskarte (japanisch) Yokohama.

Miyawaki, A., H. Tohma and K. Suzuki. 1979. Pflanzensoziologische Untersuchung in der Hainen der Shinto-Schreine und Buddistischen Temple in der Präf. Kanagawa (Hauptstadt: Tokohama) II: 167 pp. (japanisch mit deutscher Zus.) Yokohama.

Miyawaki, A. et al. 1971. Vegetationkarte und pflanzensoziologische Studien im Ise-Shima-Nationalpark (Mie-Präf.). National Park Society of Japan: 15 pp. mit 1 farb. Vegetationskarte (mit Mitarbeiter) (japanisch) Tokyo.

Miyawaki, A. et al. 1971. Pflanzensoziologische Studien über die Vegetation der Izumi-Katsurage Bergkette in den Präf. Wakayama und Osaka. Sci. Rep. the Izumi-Katsuragi Range Natural Park: 37–70. mit 1 farb. Vegetationskarte. (mit Mitarbeiter) (japanisch mit deutscher Zus.) Tokyo.

Miyawaki, A. et al. 1972. Vegetation der Stadt Yokohama. Eine pflanzensoziologische Studie für den Umweltschutz und die Schaffung einer vegetationsreichen Stadt. 141 pp. mit 2 farb. Vegetationskarten und Tabellenheft. (mit Mitarbeiter) (japanisch mit deutscher Zus.) Yokohama.

Miyawaki, A. et al. 1972. Reale Vegetation der Präf. Kanagawa. 788 pp. mit 44 farb. Vegetationskarten und Tabellenheft. (Mitarbeiter) (japanisch mit deutscher Zus.). Yokohama.

Miyawaki, A. et al. 1972. Vegetation der Stadt Kamakura. Eine pflanzensoziologische Studie über die Erhaltung der historischen Landschaft der alten japanischen Hauptstadt Kamakura. 114 pp. mit 2 farb. Vegetationskarten und Tabellenheft. (mit Mitarbeiter) (japanisch mit deutscher Zus.) Yokohama.

Miyawaki, A. et al. 1973. Vegetation der Halbinsel Oga, Präf. Akita in Nord-Honshu, Japan. Rep. Nat. Conserv. Soc. Japan 44:101–145. mit 2 farb. Vegetationskarten und Tabellenheft. (Mitarbeiter) (japanisch mit deutscher Zus.) Tokyo.

Miyawaki, A. et al. 1973. Vegetationskundliche Untersuchung der Küste Anan, Präfektur Tokushima. Forsch. Ber über die Vegetation des Erholungsgebietes Küste von Anan. 28 pp. mit 2 farb. Vegetationskarten und Tabellenheft. (Mitarbeiter) (japanisch mit deutscher Zus.) Tolushima.

Miyawaki, A. et al. 1973. Vegetation des SO-Teils der Präf. Saitama. 77 p. (Mitarbeiter) Urawa.

Miyawaki, A. et al. 1973. Phytosociological studies on creation of environmental protection forests around schools-Based on field surveys at 158 schools throughout Japan. Rep. Stud. Creation of Environ. Protec. For. around Schools. 116 pp. with 2 colored vegetation maps. (with co-workers) (in Japanese). Yokohama.

Miyawaki, A et al. 1974. Forschungsbericht über den natürlichen Umweltschutz im geplanten Mutsu-Ogawara Kultivierungs-Gebiet. 92 pp. mit 6 farb. Vegetationskarten (Mitarbeiter) (japanisch) Sendai.

Miyawaki, A. et al. 1976. Bericht über die Vegetation im West-Teil der Präf. Kumamoto, Kyushu. Forsch. Ber. über die Anlage eines Grüngürtels um den Flughafen Kumamoto. 87 pp. mit 4 farb. Vegetationskarten (Mitarbeiter) (japanisch mit deutscher Zus.) Kumamoto.

Miyawaki, A. et al. 1976. Die potentielle natürliche Vegetation in der Präf. Kanagawa. 407 pp. mit 44 farb. Vegetationskarten (Mitarbeiter) (japanisch mit deutscher Zus.) The Board of Education of the Kanagawa Pref. Yokohama.

Miyawaki et al. 1977. Vegetation der Präf. Yamanashi. 237 pp. mit 10 farb. Vegetationskarten und Tabellenheft. (Mitarbeiter) (japanisch mit deutscher Zus.) Yamanashi.

Miyawaki et al. 1977. Vegetation der Insel Satsuma-Iow (Südteil der Präf. Kagoshima). Bull. Yokohama Phytosoc. Soc. 6: 25 pp. mit 1 farb. Vegetationskarte. (Mitarbeiter) (japanisch mit deutscher Zus.) Yokohama.

Miyawaki et al. 1977. Vegetation der Präf. Tohyama. 289 pp. mit 6 farb. Vegetationskarten und Tabellenheft. (Mitarbeiter) (japanisch mit deutscher Zus). Tohyama.

Miyawaki et al. 1977. Karte der potentiellen natürlichen Vegetation in der Präf. Nagano. Heft 1: 134 pp. mit 4 farb. Vegetationskarten (Mitarbeiter) (japanisch mit deutscher Zus.) Nagano.

Miyawaki et al. 1977. Karte der potentiellen natürlichen Vegetation in der Präf. Nagano. Heft 2: 122 pp. mit 4 farb. Vegetationskarten (Mitarbeiter) (japanisch mit deutscher Zus.) Nagano.

Miyawaki et al. 1977. Vegetation in the surroundings of the Hamaoka-atomic-power plant (Pref. Shizuoka) 60 pp. with 3 colored vegetation maps (with co-workers) (Japanese) Shizuoka.

Miyawaki et al. 1978. Vegetation der Umgebung des Enrei-Tunnels in Mittelhonshu. Forsch. Ber. über die Umwelt

574

zwischen Okaya und Shiojiri: 115–192. mit 2 farb. Vegetationskarten (Mitarbeiter) (japanisch) Yokohama.

Miyawaki et al. 1978. Potentielle natürliche Vegetation des Kinkiken (Umgebung von Kyoto, Osaka, Kobe und Halbinsel Kii). Bull. Inst. Environm. Sci. Techn. Yokohama Nat. Univ. 4(1):113–148. mit 1 farb. Vegetationskarte (Mitarbeiter) (japanisch mit deutscher Zus.) Yokohama.

Miyawaki et al. 1979. Vegetation der Stadt Kashima und ihrer Umgebung in den Präf. Ibaraki und Chiba. Bull. Yokohama Phytosoc. Soc. 129 pp. mit 2 farb. Vegetationskarten (Mitarbeiter) (japanisch) Yokohama.

Miyawaki et al. 1980. Vegetation am Mittel- und Oberlauf des Hijikawa-Flusses und seiner Umgebung in Shikoku. Bull. Yokohama Phytosoc. Soc. 24:1–129. mit 2 farb. Vegetationskarten (Mitarbeiter) (japanisch mit deutscher Zus.) Yokohama.

Miyawaki et al. 1980. Vegetation des Genkais und seiner Umgebung in NW-Kyushu. Bull. Yokohama Phytosoc. Soc. 14:1–189. mit 4 farb. Vegetationskarten (Mitarbeiter) (japanisch mit deutscher Zus.) Yokohama.

Miyawaki et al. 1980. Potentielle natürliche Vegetation des Chugoku-Gebietes (West-Honshu). Bull. Inst. Envir. Yokohama Nat. Univ. 6:77–118. mit 1 farb. Vegetationskarte (Mitarbeiter) Yokohama.

Miyawaki et al. 1981. Vegetation von Futsu und seiner Umgebung an der Bucht von Tokyo in den Präf. Chiba und Kanagawa. Bull. Yokohama Phytosoc. Soc. 18:1–135. mit 2 farb. Vegetationskarten (Mitarbeiter) (japanisch mit deutscher Zus.) Yokohama.

Miyawaki et al. 1981. Vegetation of Tsurugashima-cho in Saitama Pref. Office for Compiling the History of Tsurugashima Town. 78 pp. with 2 colored vegetation maps. (with co-worker) Tsurugashima.

Miyawaki et al. 1981. Vegetationskundliche Forschungen über die Veränderungen der Umwelt in der Hironogegend. 2. Heft. Bull. Yokohama Phytosoc. Soc. 176 pp. mit 4 farb. Vegetationskarten (Mitarbeiter) (japanisch) Yokohama.

Miyawaki et al. 1981. Vegetation der Stadt Kawasaki und ihrer Umgebung. Bull. Yokohama Phytosoc. Soc. 24:1–211. mit 4 farb. Vegetationskarten (Mitarbeiter) (japanisch mit deutscher Zus.) Yokohama.

Miyawaki et al. 1983. Potentielle natürliche Vegetation des Stadt Sakata, Präf. Yamagata an der Küste des Japanischen Meeres in Nord-Honshu: 132 pp. mit farb. Vegetationskarten (Mitarbeiter) (japanisch mit deutscher Zus.) Sakata.

Miyawaki et al. 1983. Vegetation der Umgebung der Stadt Takahata in der Präf. Yamagata, Japan: 116 pp. mit farb. Vegetationskarten (Mitarbeiter) (japanisch mit deutscher Zus.) Takahata.

Miyawaki et al. 1983. Ökologische und vegetationskundliche Untersuchungen zur Schaffung von Umweltschutzwäldern in den Industrie-Gebieten Japans. Heft II. Bull. Yokohama Phytosoc. Soc. 22:1–151. mit farb. Vegetationskarten (Mitarbeiter) (japanisch mit deutscher Zus.) Yokohama.

Miyawaki, A. M. Yokoyama and H. Ide. 1967. Grundlagenforschung für die Erstellung der Karte der potentiellen natürlichen Vegetation und einer pflanzensoziologischen Vegetation und eines pflanzensoziologischen Standortgutachtens. Housing Corporation. 20 pp. mit 1 farb. Vegetationskarte. (japanisch mit deutscher und englischer Zusammenfassung) Tokyo.

Molina, L.C.E. 1973. SLAR En la mapificación de los bosques húmedos tropicales de Colombia. Bogotá. CIAF 1973.

Molinier, R. 1951. La cartographie phytosociologique au service de la prospection agronomique. Association Française pour l'Avancement des Sciences, 70th Congress, Proceedings, Fasc. 4.

Molinier, R., R. Molinier and H. Paliot. 1951. Cartes phytogéographiques à diverses échelles de la forêt domaniale de la Sainte Baume (Var). Association Française pour l'Avancement des Sciences. 70th Congress, Proceedings, No. 4.

Molinier, R. 1952. Carte des groupements végétaux de la France, feuille Aix S.O. Montpellier. Service de la carte des groupements végétaux.

Molinier, R. 1954. Etude des biocénoses marines du Cap Corse. Vegetatio, 9:121–192 and 217–312.

Molinier, R. 1957. L'intérêt Pédagogique de la Carte des Groupements Végétaux. (1:20.000). Centre National de la Recherche Scientifique. pp. 14.

Molinier, R. and J. Picard. 1959. Délimitation et cartographie des peuplements marins benthiques de la mer Méditerranée. Service de la carte phytogéographique. Bulletin, série B, 4:73–84.

Molinier, R. 1961. Carte des associations végétales terrestres et des biocénoses marines dans le sud-est de la France. In: Henri Gaussen, 1961a, pp. 157–170.

Moll, E.J. and L. Bossi. 1983. Vegetation of the Fynbos biome. Capetown. University of Capetown, Botany Department.

Montoya-Maquin. J.M. 1970. Situación de la cartografía de la vegetación en el contexto de estudios de recursos para el desarrollo. Mérida. Universidad de los Andes. Centro de Estudios Forestales.

Moor, M. and U. Schwarz. 1957. Die kartographische Darstellung der Vegetation des Creux-du-Vent Gebietes. Bern. Beiträge zur geobotanischen Landesaufnahme der Schweiz, vol. 37.

Moore, C.W.E. 1953. Vegetation map of the southeastern Riverina, New South Wales. Australian Journal of Botany, vol. 1, No. 3.

Mouat, D.A. and C.F. Hutchinson. 1980. Applied Remote Sensing Program. Office of Arid Lands Studies. University of Arizona, Tucson, Arizona. 'Techniques for Vegetation Mapping in Semiarid Regions Case' Cocoyoc, Morelos, Mexico. October 6, 7 and 8, 1980. pp. 22.

Movia, C.P. 1974. Cuenca del Río de la Plate: alta cuenca del Río Bermejo. Washington, DC. Organization of American States, General Secretariat.

Mraz, K. and V. Samek. 1963. Beiträge zum Problem der Vegetationskartierung mit besonderer Rücksicht auf ihre forstliche Anwendung. In: Reinhold Tüxen, 1963a, pp. 385–393.

Mulder, N.J. 1982. 'Generalised operators for geometric correction resampling and deconvolution' Proc. Symposium ISP comm. III, Helsinki.

Mulder, N.J. 1982. 'Methodology of colour coding MSS and other data' Proc. Symposium ISP comm. III, Helsinki.

Mulder, N.J. 1983. 'Classification and decision making'; Proc. ISP comm. III spec. workshop pattern recognotion, Graz.

Mulder, N.J. 'Spectral correlation and natural colour...' I.T.C. Journal. Enschede 1981-3.

Mulder, N.J. 'Geodata processing' I.T.C. Journal. Enschede. 1983/2.

Mulder, N.J. 1985. 'Decision making and classification' Photogrammetria, 40 (1985) 95–116.

Mulder, N.J. and S.A. Hempenius. 'Data compression and data reduction techniques for the visual interpretation of multispectral images' I.T.C. Journal. Enschede, 1974/3.

Müller-Dombois, D. and H. Ellenberg. 1974. Aims and Methods of Vegetation Ecology. John Wiley & Sons, New York and London, 547 pp.

Munz, P.A. and D.D. Keck. 1973. A California flora. Berkeley, CA. University of California Press.

Musa, A., G.O. Odoh, N. Mengistu, M. Liberman, M. Fadl E and Y.A. Yath. 1983. Land Ecology and Land Evaluation Tajo-Tietar, Province of Caceres, Spain. Land inventory of an area around Serradilla and Torrejon el Rubio. Unpubl. MSc thesis, ITC, Enschede, 74 pp.

Nago, M. and T. Matsuyama. 1980. 'A structural analysis of complex photographs' Plenum Press. New York.

Nasonova, O.N. 1962. Pasture and hayland maps and principles of their compilation. In: Victor B. Sochava, 1962a, pp. 145–151.

Neuhäusl, R. 1963a. Vegetationskarte von Böhmen und Mähren. Zürich. Geobotanisches Institut Rübel. Bericht No. 34, pp. 107-121.

Neuhäusl, R. 1963b. Kartierung der natürlichen Vegetation Mährens. In: Reinhold Tüxen, 1963a, pp. 265–278.

Neuhäusl, R. 1982a. Die Vegetationskarte der Tschechoslowakei 1:200,000 und ihre geographische Interpretation. Berlin. Archiv für Naturschutz und Landschaftsforschung, 22:145–150.

Neuhäusl, R. 1982b. Das 2. Internationale Kolloquium über die Vegetationskarte Europas. Czechoslovak Academy of Sciences. Pruhonice.

Nicklfeld, H. 1980. Vegetationskartierung im Gebirge. Grenoble. Documents de cartographie écologique, 23:1–23.

Nijland, G. 1974. 'Nieuw Overzicht van de Nederlandse Vegetatiekaarten' Mededelingen Landbouwhogeschool Wageningen Nederland. 74-20. H. Veenman & Zonen B.V. – Wageningen.

Noirfalise, A. 1963. Objectifs et problèmes de la cartographie des végétations en Belgique. In: Reinhold Tüxen, 1963a, pp. 95–101.

Noirfalise, A. 1963. Le Centre de Cartographie Phytosocio-logique. Institute Agronomique de Gembloux. Extrait de 'Le Mouvement Scientifique en Belgique'. N. X–1963 p. 8.

Nomokonov, L.I. 1962. The vegetation map in the comprehensive atlas of the Irkutsk region. In: Victor B. Sochava, 1962a, pp. 208–210.

Norton-Griffiths, D.v.M. 1972. Serengeti ecological monitoring program. African Wildlife Leadership Foundation. Washington D.C.

Norton-Griffiths, M. 1978. Counting Animals. Handbook No. 1, African Wildlife Leadership Foundation, Nairobi.

Norton-Griffiths, M., T. Hart and M. Parton. 1983. Sample surveys from light aircraft combining visual observation and very large scale colour photography. ITC Journal, 1983-1: 17–20.

Norwine, J.R. and D.H. Greegor. 1983. Vegetation classification based on advanced very high resolution radiometer satellite imagery. Remote Sensing and Environment, 13:69–87.

Oberdorfer, E. 1957. Eine Vegetationskarte von Freiburg im Breisgau. Freiburg. Berichte der Naturforschenden Gesellschaft, 47:2.

Oefelein, H. 1960. Vegetationskartierung: Helvetia. Excerpta Botanica, Sectio B: Sociologica, 2:215–218.

Okuda, S. und A. Miyawaki. 1966. Reale Vegetationskarte des Staatlichen Naturparks für Naturstudien in Tokyo. (colour map 1:1.000) from: Ecological Studies of Biotic Communities in the National Park for Nature Study. Nr. 1: 1–14 (1966). pp 14.

Oldeman, R.A.A. 1974. L'architecture de la forêt guyanaise. Mémoires ORSTROM N° 73, 204 p., Paris.

Oldeman, R.A.A. 1983. Tropical rain forest, architecture, silvigenesis and diversity. Reprint from Tropical rain forest: ecology and management. Special publication, no. 2 of the British Ecological Society, pp. 139–150. Blackwell Scientific Publication. Oxford.

Olson, G.W. 1974. Land classifications. Search, Agriculture 4(7), 34 p. Cornell University, Ithaca, New York.

Oosterveld, P. 1983. Taraxacum Species as Environmental Indicators for Grassland Management. Environmental Monitoring and Assessment: 3.

Ozenda, P. 1961a. La publication de coupures provisoires: raison d'être, techniques possibles. In: Henri Gaussen, pp. 257–264.

Ozenda, P. 1961b. Le représentation cartographique de la végétation à moyenne échelle à l'aide de trames. Comité français de techniques cartographiques. Bulletin, 11:177–182.

Ozenda, P. 1963. Principes et objectifs d'une cartographie de la végétation des Alpes à moyenne échelle. Documents pour la carte de la végétation des Alpes, 1:5–18.

Ozenda, P., ed. 1963–1972. Documents pour la carte de la végétation des Alpes, vol. 1–10, and Documents de cartographie écologique, 1973ff. Université Scientifique et Médicale de Grenoble. Grenoble. Laboratoire de Biologie Végétale.

576

Ozenda, P. 1964. Biogéographie végétale. Editions Doin, Deren et Cie. Paris.

Ozenda, P et G. Pautou. 1980. Cartographie écologique de l'environnement. Bulletin d'Ecologie, 11:1.

Ozenda, P. 1981. Colloque international sur la cartographie de la végétation à petite échelle. Grenoble. Documents de cartographie écologique, 24:1–134.

Paijmans, K. 1965. Typing of Tropical Vegetation by Aerial Photographs and Field Sampling in Northern Papua. Division of Land Research, C.S.I.R.O., Camberra A.C.T. (Australia). Photogrammetria Vol. 21, N. 1, Febr. 1966. pp. 1–25.

Paijmans, K. 1969. Land Evaluation by Air Photo Interpretation and Field Sampling in Australian New Guinea. Photogrammetria, 26 (1970) 77–100 pp. 8. Elsevier Publishing Company, Amsterdam.

Paijmans, K. 1975. Explanatory Notes to the Vegetation Map of Papua New Guinea (With colour maps.). Land Research Series N. 35. Commonwealth Scientific and Industrial Research Organization, Australia.

Pannekoek, G. 1981. 'GOIRIE' Een ecologische vegetatiekartering. Centrum voor Agrobiologisch Onderzoek Wageningen. Karteringsverlag Nr. 195, Maart 1981. pp. 39.

Pallmann, H. 1948. Über die Zusammenarbeit von Bodenkunde und Pflanzensoziologie. St. Gallen. Verhandlungen der schweizer naturforschenden Gesellschaft.

Pascal, J.P. 1982. Forest map of South India; scale 1/250 000. 'Shimoga' and 'Mercara-Mysore'. Published by the Karnataka and the Kerala Forest Departments and the French Institute of Pondicherry.

Pautou, G., J. Girel and G. Ain. 1979. Recherches écologiques dans la vallée du haut Rhone français. Documents de cartographie écologique, 22:5–63.

Pavillard, J. 1935. Eléments de sociologie vegetale (phytosociologie) I. Paris 1935.

Pavlidis, T. 1977. 'Structural pattern recognition' Springer Verlag. Berlin.

Peden, D. 1985. Estimating maize yield in Kenya using airborn digital photometer. ITC Journal 1985. Praet van: Senegal.

Pedrotti, F. 1967. Carta fitosociologica della vegetazione dei piani de Montelago. Instituto di Botanica, Università di Camerino. Camerino.

Pedrotti, Franco. 1975. Carta fitosociologica della vegetazione della palude di Colfiorito. Società Geografica. Firenze.

Pedrotti, F. 1976. Esperienze di cartografia della vegetazione (1965–1975). Estratto da Studi Trentini di Scienze Naturali Rivista del 'Museo Tridentino di Scienze Naturali'. Vol. 53, N. 6 B, 1976. pp. 205–214.

Pedrotti, F. 1982. Überblick über die Vegetationskartierung in den Alpen und Apenninen. Grenoble. Documents de cartographie écologique, 25:89–95.

Peer, T. 1981. Die aktuellen Vegetationsverhältnisse Südtirols. Wien. Angewandte Pflanzensoziologie, 26:151–168.

Pereda, N.P. 1975. A physiognomic vegetation map of Sri Lanka (Ceylon) Journal of Biogeography 2:185–203.

Pérès, J.M. and J. Picard. 1955. Biotopes et biocénoses de la Méditerranée occidentale comparés à ceux de la Manche et de l'Atlantique nordoriental. Archives de zoologie experimentale, vol. 92, No.1.

Pettinger, L.R. 1982. Digital classification of Landsat data for vegetation and land cover mapping in the Blackfoot River watershed, southeastern Idaho. United States Geological Survey. Washington, DC. Professional Paper No. 1219.

Pickard, J. 1983. Vegetation of Lord Howe Island. Cunningshamia, 1:133–266.

Pignatti, E., P. Nimis and A. Avanzini. 1980. La vegetazione ad arbusti spinosi emisferici: Contributo alla interpretazione delle fasce di vegetazione delle alte montagne dell'Italia mediterranea. Consiglio Nazionale delle Richerche. AQ/1/79. Roma.

Pignatti, S. 1980. Über Vegetationskomplexe und ihre kartographische Darstellung. Documents de cartographie écologique, 23:17–18.

Pina Manique e Albuquerque, J. de 1954. Carta ecologica de Portugal. Direcção Geral dos serviços agrícolas; serviço editorial da repartição de estudos, informação e propaganda. Lisboa.

Pirola, A., C. Montanari and V. Credaro. 1980. Valutazione speditiva del grado di protezione del mantello vegetale contro l'azione delle acque cadenti e dilavanti. Esempio condotto sul piccolo bacino del Rio Grande (Valle del Sillaro, Appennino Bolognese). (With colour map Scale 1:10.000). Consiglio Nazionale delle Ricerche AQ/1/75. Roma. pp. 20.

Pitschmann, H., H. Reisigl, H.M. Schiechtl and R. Stern. 1970ff. Karte der aktuellen Vegetation von Tirol. Documents pour la carte de la végétation des Alpes, 8:7–34.

Poissonet, P. 1967. Place de la Photo-Interprétation dans un Programme d'Etude détaillée de la Flore, de la Végétation et du Milieu. IIᵉ Symposium International de Photo-Interprétation. Paris, 1966. pp. 52–55.

Poldini, L. 1980. Carta della vegetazione del Carso Triestino (zona dell'accordo di Osimo) (With colour, 1:15.000) Consiglio Nazionale delle Ricerche AQ/1/82. Roma.

Poore, M.E.D. 1963. Problems in the classification of tropical rain forests. Journal of Tropical Ecology, 17:12–19.

Poore, M.E.D. 1982. Vegetation Mapping by Computer. Cartographica, 19:2:71–107.

Portecop, J. 1979. Phytogéographie, cartographie écologique et aménagement dans une île tropicale: le cas de la Martinique. Grenoble. Documents de cartographie écologique, 21:1–78.

PRORADAM, 1979. La Amazonia Colombiana y sus Recursos. Proyecto Radargrammetrico del Amazonas, Bogotá.

Puig, H. 1976. Végétation de la Huasteca, Mexique. Mission Archéologique et Ethnographique Française au Mexique.

Puig, H. 1979. Carte Internationale du Tapis Végétal et des conditions écologiques. Feuille 'Guadalajara-Tampico (Mexique)' Echelle: 1/1 000 000. Notice explicative in

Inst. fr. Pondichéry; trav. Sect. scient. techn., Hors Série N° 16, 142 p.

Puppi, G., M. Speranza and A. Pirola. 1980. Carta della vegetazione dei dintorni del lago Brasimone – Emilia Romagna. (With colour map 1:25000) Consiglio Nazionale delle Ricerche. AQ/1/74. Roma.

Quintanilla, V.G. 1974. Les formations végétales du Chili tempéré. Grenoble. Documents de cartographie écologique, 14:33–80.

Radam. 1974. Ministerio das Minas e Energia. Departamento Nacional da Produção Mineral. Prieto Radam. Rio de Janeiro. Levantamento de Recursos Naturais Vol. n. 4.

Rafstedt, T. and L. Anderson. 1981. Air photo interpretation of mires. SNV (Swedish Environmental Protection Board), PM 1433.

Rafstedt, T. and L. Andersson. ■. 'Flygbildstolkning av myrvegetation' En metodstudie for oversiktlig kartering. Rapport snv pm 1433. Meddelanden fran Naturgeografisca. Institutionen Vid Stockholms Universitet. Nr. A 117. Stockholm. Sweden. pp. 106.

Raimondo, F.M. 1980. Carta delle vegetazione di Piano della Battaglia e del territorio circostante. (Madonie, Sicilia) Consiglio Nazionale delle Ricerche. AQ/1/89. Roma, 1980. (With colour map Scale 1:4.000).

Raman, K.G. 1962. Classification of geographical complexes in Latvia and the possible use of these principles in geobotanical mapping. In: Victor B. Sochava, 1962a, pp. 178–185.

Raunkiaer, C. 1934. The life forms of plants and statistical plant geography. Clarendon Press. Oxford.

Raunkiaer, C. 1937. Plant Life Forms. Clarendon Press. Oxford. 104 pp.

Reeves, R.G., A. Anson and D. Landen. 1975. Manual of Remote Sensing. 2 vol., 2144 p. American Soc. of Photogrammetry. Falls Church, Virginia, U.S.A.

Renaud, P. 1949. Cartographie parcellaire appliquée à l'apiculture. In: G. Kuhnholtz-Lordat, 1949.

Retzer, J.L. 1953. Soil-vegetation survey of wild-lands. Journal of Forestry, 51:615–619.

Rey, P. and H. Gaussen. 1955. Service de la carte de la Végétation de la France au 200 000ᵉ. Centre National de la Recherche Scientifique, Service de la carte phytogéografique pp. 11–34. 29 rue Jeanne Marvig, 31400 Tououse.

Rey, P. 1955. Recensement cartographique des milieux et analyse écologique des cartes de la végétation. In: Henri Gaussen, 1955, pp. 169–180.

Rey, P. 1957. Initiation à l'utilisation scientifique et pédagogique des cartes de la végétation. Service de la carte phytogéographique, Bulletin, Serie A, 2:73–86.

Rey, P. 1958. La cartographie botanique en couleurs. Service de la carte phytogéographique, Bulletin, Serie A, 3:11–19.

Rey, P. 1961. De la clarté en toute chose, même en cartographie de la végétation. In: Henri Gaussen, 1961a, pp. 283–288.

Rey, P. 1962a. Ecologie et agronomie: la cartographie de la végétation à l'épreuve de l'agronomie. Bulletin technique d'information des services agricoles, 172:1–3.

Rey, P. 1962b. Les perspectives fondamentales de la cartographie de la végétation. Comité français de techniques cartographiques, 14:69–73.

Rey, P. 1962c. Généralisation cartographique de la végétation. Toulouse. Service de la carte de la végétation. Notes et Documents, No. 5.

Rey, P. 1962d. Recherche biogéographique, carte de la végétation et aménagement de l'espace rural. Toulouse. Service de la carte de la végétation. Notes et documents, No. 6.

Rey, P. 1966. La place de la photographie aérienne dans les Méthodes d'Etude du Milieu Vivant. IIᵉ Symposium International de Photo-Interprétation Paris. pp. 67–62.

Rey, P. 1967. La carte de la Végétation, base d'un Recensement des Ressources Biologiques. Réflexions sur la Régionalisation de l'Aménagement des Ressources Naturelles. Congrès d'Ottawa. pp. 5.

Richard, L. 1978. Carte écologique des Alpes: feuille Chamonix – Thonon les Bains. Documents de cartographie écologique, 20:1–39.

Richard, L. 1981. Quelques propositions sur la cartographie des groupements végétaux indicateurs. Grenoble. Documents de cartographie écologique, 24:112–116.

Richard, L. 1983. Nouvelles données pour la zonation écologique des Alpes nord-occidentales et contribution à la notice de la carte écologique à 1:50,000 'Saint Gervais.' Grenoble. Documents de cartographie écologique, 26:83–116.

Richards, J.A., O.A. Langrebe, Swain. 1981. 'Pixel labeling by supervised probabilistic relaxation' IEEE vol. PAMI-3, No. 2 March 1981, (188–191).

Richards, P.W., A.G. Tansley and A.S. Watt. 1940. The recording of structure, life form and flora of tropical forest communities as a basis of their classification. Journal of Ecology, 28:224–239.

Richards, P.W. 1952. The tropical rain forest; an ecological study. 450 p. University Press. Cambridge.

Richardson, A.J. and C.L. Wiegand. 1977. 'Distinguishing vegetation from soil background information'; Photogr. Eng. and Rem. Sens., vol. 42, no. 5, pp. 679–684.

Rivas-Martínez, S. 1982. Mapa de las Series de Vegetación de Madrid. Escala 1:200 000. Diputación de Madrid.

Robbins, R.G. 1958. The montane vegetation of New Guinea. Tuatara 8(3):121–133.

Robertson, J.S. 1982. Vegetation surveying and mapping in Scotland. Cartographica, 19:2:74–82.

Roberty, G. 1961. La végétation des régions dépourvues de tradition agricole précisément définie et sa représentation cartographique. In: Henri Gaussen, 1961a, pp. 103–110.

Robus, M.A. 1983. Geobotanical mapping. In: Woodward-Clyde Consultants, The Lisburne Development Area (Prudhoe Bay). Anchorage, AK. Environmental Studies Report prepared for ARCO Alaska, Inc. 3:1–16 and 11:1–2.

578

Romariz, Dora de A. 1981. Le projet Radambrasil. Grenoble. Documents de cartographie écologique, 24:117–120.

Rosenberg, V.A. 1962. Principles of the compilation of forest maps. In: Victor B. Sochava, 1962a, pp. 98–102.

Rosetti, C. 1962. Un dispositif de prises de vues aériennes à basse altitude et ses applications pour l'étude de la physiognomie de végétations ouvertes. Service de la carte phytogéographique, Bulletin, Série B, 7:211–238.

Rosetti, C. 1963. Etude de communautés végétales ouvertes à l'aide de photographies aériennes à grande échelle. Centre d'études phytosociologiques et écologiques, Montpellier, France. pp. 495–499.

Rossetti, C. 1964. Réflexions sur l'utilisation des photographies aériennes pour l'étude du couvert végétal. pp. 6.

Rossetti, C. 1965. Réalisation par Photo-Interprétation d'un Inventaire Expédié des Peuplents de Pin d'Alep de Tunisie et Exament Critique de ses Résultats. pp. 15.

Rossetti, C. 1966. Réflexions sur l'utilisation des Photographies Aériennes pour l'Etude du Couvert Végétal. pp. 6.

Rossetti, C., P. Kowaliski and N. Havé. 1967. Relations entre les caractéristiques de réflexion spectrale de quelques espèces végétales et leurs images sur des photographies en couleur, terrestres et aériennes. Acte IIe Symp. int. de photo-interpret. Paris, 1966.

Rouse, J.W., R.H. Haas, J.A. Schell and D.W. Deering. 1973. Monitoring vegetation systems in the Great Plains with ERTS. Third NASA ERTS Symposium, NASA SP–351, 1:309–317.

Roussine, N. and C. Sauvage. 1961. Afrique du Nord. Excerpta Botanica, Sectio B: Sociologica, 3:48–50.

Rouvillois-Brigol, M., C. Nesson and J. Vallet. Oasis du Sahara algérien. Études de photo-interprétation N. 6. Institut Géographique National. pp. 4.

Rowe, J.S. 1959. Forest regions of Canada. Department of Northern Affairs and National Resources, Forestry Branch. Ottawa.

Rowe, J.S. 1961. The level-of-integration concept and ecology. Ecology, 42:420–427.

Rübel, E. 1916. Vorschläge zur geobotanischen Kartographie. Bern. Beiträge zur geobotanischen Landesaufnahme der Schweiz, vol. 1.

Rübel, E. 1930. Die Pflanzengesellschaften der Erde. Huber Verlag. Bern.

Ruthsatz, B. and C.P. Movia. 1975. Relevamiento de las estepas andinas del noreste de la provincia de Jujuy. Fundación para la Educación, la Ciencia y la Cultura. Buenos Aires, (República Argentina) With colour maps, Scales 1:50.000 and 1:200.000. Cátedra de Botánica, Dep. de Ecología, Facult. de Agron. de la Universidad de Buenos Aires. Argentina.

Rützler, K. Photogrammetry of reef environments by helium balloon. Dep. of Invertebrate Zoology, National Museum of Natural History, Smithsonian Institution, Washington, D.C. 20560, U.S.A. pp. 45–52.

Sappa, F. and G. Charrier. 1949. Carta della vegetazione della val Sangone. Firenze. Nuovo Giornale botanico italiano, N.S., vol. 56.

Sargent, C.S. 1884. Forets, prairie and treeless regions of North America exclusive of Mexico. Washington, DC. 10th Census of the United States, vol. 9. Maps of individual states (with the pages they face given in parentheses): Alabama (524), Arkansas (544), Florida (522), Georgia (519), Louisiana (536), Maine (496), Michigan (550), Minnesota (558), North Carolina (514), South Carolina (519), Texas (541), West Virgina (512), Wisconsin (544).

Satyanarayan, Y. and V.V. Dhruvanarayan. Use of Aerial Photographs in Surveying Ground-water And Vegetation Resources in the arid zone of India. pp. 505–507.

Saxton, W.T. 1924. Phases of vegetation under monsoon conditions. Journal of Ecology, 12:1–38.

Sayago, M. 1957. La cartografía botánica en colores. Córdoba, Argentina. Revista de la facultad de ciencias exactas físicas y naturales, 19:1–16.

Scamoni, A. 1950. Kriterien bei der Standortskartierung im Bereich des Dilluviums von Mecklenburg, Brandenburg und Sachsen-Anhalt. Allgemeine Forstzeitschrift, 5:435–438.

Scamoni, A. et al. 1958. Karte des natürlichen Vegetation der Deutschen Demokratischen Republik. – 1. Ergänzungsband zum Klima-atlas der Deutschen Demokratischen Republik. Berlin.

Scamoni, A. 1958. Zur Karte der natürlichen Vegetation der Deutschen Demokratischen Republik. Remagen. Berichte zur deutschen Landeskunde, 21:53–74.

Scamoni, A. 1963. Prinzipien der Karte der natürlichen Vegetation der Deutschen Demokratischen Republik. In: Reinhold Tüxen, 1963a, pp. 47–59.

Scharfetter, P. 1932. Die kartographische Darstellung der Pflanzengeseelschaften. In: Emil Abderhalden, Handbuch der biologischen Arbeitsmethoden, Abteilung XI, Teil 5, 1. Hälfte, pp. 77–164. Urban und Schwarzenberg. Berlin.

Schimper, A.F.W. 1898. Pflanzengeographie auf physiologischer Grundlage. Gustav Fischer Verlag. Jena.

Schimper, A.F.W. and F.C. von Faber. 1935. Pflanzengeographie auf physiologischer Grundlage, 3rd ed. Gustav Fischer Verlag. Jena.

Schatterer, E.F. 1983. Proceedings of the workshop on southwestern habitat types. Rocky Mountains Forest & Range Experiment Station. Fort Collins, CO.

Schlüter, H. 1976. Geobotany and vegetation mapping as a basis of the analysis of ecosystems and landscapes. Academy of Sciences of the U.S.S.R., KomarovBotanical Institute. Leningrad. Geobotanical Cartography, pp. 22–31.

Schmid, E. 1940. Die Vegetationskartierung der Schweiz im Masstab 1:200,000. Zürich. Geobotanisches Forschungsinstitut Rübel. Bericht für das Jahr 1939, pp. 76–85.

Schmid, E. 1948. Vegetationskarte der Schweiz. Huber Verlag. Bern.

Schmid, E. 1949. Vegetation des Mediterrangebiets. Orell-Füssli. Zürich.

Schmid, E. 1955. Principes de cartographie mondiale aux échelles du 200,000^e et du 1,000,000^e basés sur les unités biocénologiques. In: Henri Gaussen, 1955. pp. 157–161.

Schmithüsen, J. 1942. Vegetationsforschung; ökologische Standortslehre in ihrer Bedeutung für die Geographie der Kulturlandschaft. Zeitschrift der Gesellschaft für Erdkunde zu Berlin, pp. 113–157.

Schmithüsen, J. 1957. Anfänge und Ziele der Vegetationsgeographie. Mitteilungen der Floristich-soziologischen Arbeitsgemeinschaft. N.F. Heft 6/7, pp. 52–78.

Schmithüsen, J. 1957. Probleme der Vegetationsgeographie. (With colour maps). Deutscher Geographentag. Würzburg. 29, Juli bis 5. August 1957. pp. 72–84.

Schmithüsen, J. 1959. Allgemeine Vegetationsgeographie. De Gruyter & Co. Berlin.

Schmithüsen, J. 1963. Der wissenschaftliche Inhalt von Vegetationskarten verschiedener Maßstäbe. International Society for Plant Geography and Ecology. Vegetationskartierung vom 23.–26.3.1959 in Stolzenau/Weser. pp. 321–329.

Schmithüsen, J. 1968. Allgemeine Vegetationsgeographie. Lehrbuch der Allgemeinen Geographie IV. Gruyter. Berlin.

Schlüter, H. 1966. Archiv. für Naturschutz und Landschaftsforschung. (With colour map Scale 1:10.000) Band 6, Heft 1/2, 1966. Deutsche Demokratische Republik. Deutsche Akademie der Landwirtschaftswissenschaften zu Berlin, pp. 44.

Schouw, J.F. 1823. Grundzüge einer allgemeinen Pflanzengeographie (mit Atlas). Berlin.

Schouw, J.F. 1838. Pflanzengeographische Karte der Erde. Berghaus' physikalischer Atlas.

Schreiber, K.F. 1977. Landscape Planning and Protection of the Environment. Appl. Sci. Dev. 5, pp. 128–135.

Schröder, L. 1984. Kartenübersicht zur potentiellen natürlichen Vegetation und realen Waldvegetation in der Bundesrepublik Deutschland. Natur und Landschaft 59(7/8):280–283.

Schröter, C. 1985. Die Thalschaft St. Antönien im Prättigau in ihren wirtschaftlichen und pflanzengeographischen Verhältnissen. Landwirtschaftliches Jahrbuch der Schweiz, vol. 9.

Schröter, C. 1910. Über pflanzengeographische Karten. Bruxelles. 3rd International Botanical Congress. Proceedings, 2:97–154.

Schulz, G.E. 1962. Phenological maps and their use in geobotany. In: Victor B. Sochava, 1962a, pp. 87–91.

Schweinfurth, U. 1957. Die horizontale und vertikale Verbreitung der Vegetation im Himalaya. Bonn. Ferdinand Dümmler Verlag. Bonner Geographische Abhandlungen, vol. 20.

Schweinfurth, U. 1966. Neuseeland. Ferdinand Dümmler Verlag. Bonn. Bonner Geographische Abhandlungen, vol. 36.

Schwickerath, M. 1954. Die Landschaft und ihre Wandlung. Georgi Verlag. Aachen.

Schwickerath, M. 1963. Assoziationsdiagramme und ihre Bedeutung für die Vegetationskartierung. In: Reinhold Tüxen, 1963a, pp. 11–38.

Seibert, P. 1954. Vegetationskarte des Graf Görtzischen Forstbezirks Schlitz. Stolzenau/Weser. Angewandte Pflanzensoziologie, vol. 9.

Seibert, P. 1958. Die Pflanzengesellschaften im Naturschutzgebiet 'Pupplinger Au' (With Colour and monocolour maps Scales 1:16.000 and 1:8.000). Landschaftspflege und Vegetationskunde, Heft 1. München. pp. 79.

Seibert, P. 'Vegetation und Landschaft in Bayern' (With colour map Scale 1:500.000). Erläuterungen zur Übersichtskarte der natürlichen Vegetationsgebiete von Bayern. Band XXII, Lfg. 4, 1968, Bonn.

Seibert, P. 1968. Uebersichtskarte der natürlichen Vegetationsgebiete von Bayern 1:500 000 mit Erläuterungen. Schr. Reihe Vegetationskunde 3.

Seibert, P. 1980. Ökologische Bewertung von homogenen Landschaftsteilen, Ökosystemen und Pflanzengesellschaften. Forstwissenschaftliche Fakultät der Universität. München.

Sendtner, O. 1854. Die Vegetationsverhältnisse Südbayerns nach den Grundsätzen der Pflanzengeographie und mit Bezugnahme auf die Landescultur geschildert. München.

Shantz, H.L. 1911. Natural vegetation as an indicator of the capabilities of land for crop production in the Great Plains area. United States Department of Agriculture, Bureau of Plant Industry, Bulletin No. 201.

Shantz, H.L. 1923. The natural vegetation of the Great Plains region. Annals of the Association of American Geographers, 13:81–107.

Shantz, H.L. and R.L. Piemeisel. 1924. Indicator significance of the natural vegetation of the southwestern desert region. Journal of Agricultural Research, vol. 28, No. 8.

Shantz, H.L. and R. Zon. 1923. Natural vegetation of the United States. United States Department of Agriculture. Washington, DC. Atlas of American Agriculture.

Shasby, M.B., R.R. Burgan and G.R. Johnson. 1981. 'Broad Area Forest Fuels and Topography Mapping Using Digital Landsat and Terrain Data,' Machine Processing of Remotely Sensed Data Symposium, West Lafayette. In: Purdue University, LARS, pp. 529–538.

Shchelkunova, R.P. 1962. The use of aerophotography in the compilation of geobotanical pasture maps in the Far North. In: Victor B. Sochava, 1962a, pp. 164–168.

Shimwell, D.W. 1971. Description and classification of Vegetation. Sidgwick and Jackson. London.

Shreve, F. 1915. The vegetation of a desert mountain range as conditioned by climatic factors. Carnegie Institution of Washington. Publication No. 217.

Shreve, F. 1917. Vegetation areas of the United States. Geographical Review, 3: facing p. 124.

Shreve, F. 1942. The desert vegetation of North America. Botanical Review, 8:199–246.

Shumilova, L.V. 1962. Vegetation mapping as a basis of phytogeographical subdivision (rayonization). In: Victor B. Sochava, 1962a, pp. 68–71.

Sicco Smit, G. 1979. SLAR Mosaic Interpretation for Forestry Purposes. A case study of the interpretation of a SLAR mosaic of Nigeria, without additional information. United Nations/FAO Regional Training Seminar on Re-

580

mote Sensing Applications. Ibadan, Nigeria, Nov. 1979.

Sicco Smit, G. 1975. Will the road to the green hell be paved with SLAR. ITC-Journal 1975–2.

Siede, E. 1960. Untersuchungen über die Pflanzengesellschaften im Flyschgebiet Oberbayerns. (With colour map 1:8.000) Landschaftspflege und Vegetationskunde, Heft 2. Munchen, 1960. pp. 59.

Sinclair, T.R., M.M. Schreiber and R.M. Hoffer. Pathway of solar radiation through leaves.

Sissingh, G. and P. Tideman. 1960. De planten Gemeenschappen uit de Omgeving van Didam en Zevenaar. Mededeling van de Landbouwhogeschool te Wageningen Nederland. 60(13), 1960. pp. 30. (With colour map Scale 1:25.000).

Sobolev, L.N. 1962. Principles in the compilation of a pasture and hayland map. In: Victor B. Sochava, 1962a, pp. 139–144.

Sotchava, V. 1967. Geobotanical Mapping 1967. pp. 96.

Sochava, V. 1968. Geobotanical Mapping. pp. 87.

Sochava, V. 1975. The Content of Vegetation Maps and how to enrich it. XII International Botanical Congress. Section 8. Ecological Botany Leningrad, pp. 7. (U.S.S.R.).

Sochava, V.B. 1954. Les principes et les problèmes de la cartographie géobotanique. Academy of Science of the U.S.S.R. Leningrad. In: Essais de Botanique, pp. 273–288.

Sochava, V.B. 1958. The most significant achievements in mapping the vegetation of the U.S.S.R. in the past 40 years. Izvestia of the All-Union Geographical Society, 90:109–117.

Sochava, V.B. 1961. Quelques conclusions méthodiques d'après les travaux sur la cartographie de la végétation de l'U.R.S.S. In: Henri Gaussen, 1961a, pp. 203–210.

Sochava, V.B. 1962a. Principles and methods of vegetation mapping. Academy of Sciences of the U.S.S.R. Leningrad.

Sochava, V.B. 1962b. Mapping problems in geobotany. In: Victor B. Sochava, 1962a, pp. 5–27.

Sochava, V.B. 1967. Achievements in thematic cartography and vegetation maps. Academy of Sciences of the U.S.S.R. Komarov Botanical Institute. Leningrad. Geobotanical Cartography, pp. 3–9.

Sochava, V.B. 1968. A complex study of the natural environment and the geobotanical map. Academy of Sciences of the U.S.S.R. Komarov Botanical Institute. Leningrad. Geobotanical Cartography, pp. 3–4.

Sochava, V.B. 1976. Logical principles of construction and improvement of the information content of vegetation maps. Academy of Sciences of the U.S.S.R., Geobotanical Cartography, pp. 12–18.

Sombroek, W.G. and I.S. Zonneveld. 1971. Ancient dune fields and fluvratile deposits in the Rima-Sokoto river basin (N.W. Nigeria) Soil Survey Papers no. 5. Soil Survey Institute Wageningen. The Netherlands. 109 pp. maps.

Solovei, I.N. 1962. Vegetation mapping during large scale soil mapping in Belo-Russia. In: Victor B. Sochava, 1962a, pp. 200–203.

Soo, R. von. 1954. Angewandte Pflanzensoziologie und Kartierung in Ungarn. Festschrift Aichinger. Springer Verlag. Wien. Pp. 337–345.

Soo, R. von. 1960. Bibliographia Phytosociologica: Hungaria. Vegetationskartierung. Excerpta Botanica, Sectio B: Sociologica, 2:120–121.

Sougnez, N. and R. Tournay. 1963. Cartographie de la végétation, Belgium. Excerpta Botanica, Sectio B: Sociologica, 5:250–257.

Specht, R.L. 1980. South Australia. Woods & Forest Department. Adelaide.

Speidel, B. 1963. Vegetationskartierung als Grundlage zur Melioration salzgeschädigter Wiesen an der Werra. In: Reinhold Tüxen, 1963a, pp. 457–468.

Speranza, M. and P. Testoni (ed.). 1982. La Comunità Vegetali come Indicatore Ambiental. Inst. ed Orto Botanico dell Università di Bologna. Bologna.

Spiers, B. 1978. Vegetation survey of semi-natural grazing lands (Dehesas) near Merida, SW Spain as a basis for land use planning. ITC Journal 4, pp. 649–679. Map.

Steenis, C.G.G.J. van. 1958. Vegetation map of Malaysia. United Nations Educational, Scientific and Cultural Organization. Paris.

Stellingwerf, D.A. 1964. Textbook of Photo-Interpretation. Compilation of Forest and Vegetation Maps of Vertical Photographs. Vol. X, Chapter X.4, pp. 60.

Stellingwerf, D.A. 1966. Practical Application of Aerial Photographs in Forestry and Other Vegetation Studies. I.T.C. Publications B 37 Summer 1966, pp. 23. I.T.C. Publications B 38 Summer 1966. pp. 27.

Stellingwerf, D.A. 1966. Practical Applications of Aerial Photographs in Forestry and Other Vegetation Studies. I.T.C. Publications. Serie B. Number 36 pp. 60.

Stellingwerf, D.A. 1966/68. Applications of aerial photographs in forestry. ITC publications, 36, 37, 38, 46, 47, 48.

Stellingwerf, D.A. 1968. Vegetation Mapping From Aerial Photographs. 'Invited paper of the Conference on the Use of Light Aircraft in East-African Wildlife Management. – Kenya – December 1968'. pp. 14.

Stellingwerf, D.A. 1969. Vegetation Mapping from Aerial Photographs. Reprinted from the East African Agricultural and Forestry Journal, Vol. XXXIV – Special Issue. pp. 80–86.

Stellingwerf, D.A. 1969. Kodak ektachrome infrared aero film for forestry purposes. I.T.C. publication B. 54.

Stellingwerf, D.A., G. Sicco Smith and J.M. Remeyn. 1984. Applications of aerial photographs and other remote sensing imagery in Forestry (temperate regions). ITC publ. Number 3. Part. 1.

Steubing, L. 1963. Ergebnisse einer Vegetationskartierung in Windschutzgebieten. In: Reinhold Tüxen, 1963a, pp. 469–472.

Steubing, L. and H.J. Jäger (eds.). 1982. Monitoring of Air Pollutants by Plants: Methods and Problems. Dr. W. Junk Publishers. The Hague. 172 pp.

Stewart, R.E. and J.W. Brainerd. 1945. Patuxen Research

Refuge. United States Fish and Wildlife Service. Washington, DC.

Stichting voor Bodemkartering. 1965. De Bodem van Nederland (Soils of the Netherlands) 292 text + 11 maps 1 : 200.000 + several seperate text books. Stichting voor Bodemkartering. Wageningen.

Stoddart D.R. 19xx. Mapping reefs and islands. Department of Geography, Cambridge University, Cambridge CB2 3EN, United Kingdom. pp. 17–21.

Stokkom van, H.T.C. 1980. Toelichting op de vegetatiekaart van het Verdronken Land van Saeftinge van 1979, Schaal 1 : 10.000. Rijkswaterstaat Meetkundige Dienst Afd. Fotointerpretatie. With a diazo copy map. Delf.

Story, R. 1963. The Hunter Valley area, New South Wales: Vegetation. C.S.I.R.O. Melbourne. Land Research Series No. 8.

Story, R. 1967. Vegetation. In: Land of Isaac-Comet area, Queensland. Commonwealth Scientific and Industrial Research Organization, Series 19.

Strahler, A.H. 1981. 'Stratification of Natural Vegetation for Forest and Rangeland Inventory Using Landsat Digital Imagery and Collateral Data,'. *International Journal of Remote Sensing* 2 : 15–41.

Strahler, A.H., T.L. Logan and N.A. Bryant. 1978. 'Improving Forest Cover Classification Accuracy from Landsat by Incorporating Topographic Information', Proceedings of the Twelfth International Symposium on Remote Sensing of Environment, Ann Arbor, MI Environmental Research Institute of Michigan, pp. 927–942.

Stumpel, A.H.P. and J.T.R. Kalkhoven. 1978. A Vegetation Map of the Netherlands, Based on the Relationship Between Ecotypes and Types of Potencial Natural Vegetation. Research Institute for Nature Management, Leersum, The Netherlands. Vegetatio, 37, 3: pp. 163–173.

Sukachev, V.N. 1947. The basis of forest biogeocenology. Jubilee Sbornik of the Academy of Sciences of the U.S.S.R. Moscow.

Sukachev, V.N. 1954. Quelques problèmes théoriques de la phytocénologie. In: Essais de botanique. Academy of Sciences of the U.S.S.R., Komarov Botanical Institute. Leningrad. Pp. 310–330.

Sukachev, V.N. 1960. The correlation of the concept 'forest ecosystem' and 'forest biogeocenose' and their importance for the classification of forests. Silva Fennica, 105 : 94–57.

Sweet, E.T. 1880. Geology of the western Lake Superior District. *In:* Geology of Wisconsin, Survey of 1873–1879. Madison, WI. Commissioners of Public Printing, vol. 3, facing p. 325.

Suits, G.H. 1972. 'The calculation of the directional reflectance of a vegetation canopy.' Remote Sensing of the environment, 2 : 117–125.

Tansley, A.G. 1935. The use and abuse of vegetational concepts and terms. Ecology, 16 : 284–307.

Tansley, A.G. and T.F. Chipp. 1926. Aims and methods in the study of vegetation. British Empire Vegetation Com-

mittee. London.

Terechov, N.M. 1962. Coordination of geobotanical maps with physiographic and other maps. In: Victor B. Sochava, 1962a, pp. 254–258.

Thalen, D.C.P. 1981. The integration of data collection techniques in small scale vegetation survey for land-use planning. Documents de cartographie écologique, 24 : 128.

Thalen, D.C.P. 1978. Complex mapping units, geosyntaxa and the evaluation of grazing areas. Internationale Vereinigung für Vegetationskunde. Berichte, 1977 : 491–514.

Thalen, D.C.P. 1979. On photographic Techniques in Permanent Plot Studies. Vegetatio, Vol. 39, 3 : pp. 185–190. I.T.C., Enschede, The Netherlands.

Thalen, D.C.P. 1977. Complex Mapping Units, Geosyntaxa and the Evaluation of Grazing Areas. (Rinteln 4.– 7.4.1977). pp. 491–514. FL-9490 Vaduz.

Thalen, D.C.P. 1982. Ecologische monitoring systemen international gezien: in het bijzonder met toepassing van remote sensing technieken. Med. W.L.O. 9 n. 1.

Thalen, D.C.P. 1981. An approach to evaluating techniques for vegetation surveys on rangelands. Low-level aerial survey techniques. Report Int. Workshop 6–11 Nov. 1979 Nairobi ILCA, Kenya min. envorment as Nat. resources UNEP : 85–99 ILCA, Addis Abeba 1981.

Thalen, D.C.P. 1979. Ecology and utilization of desert shrub rangelands in Iraq, Dr. W. Junk b.v., The Hague, 448 pp.

Thalen, D.C.P. 1978. Complex mapping units, geosyntaxa and the evaluation of grazing areas, pp. 491–514 in: Assoziationskomplexe (Sigmeten) und ihre praktische Anwendung (int. Symp. Rinteln April '77) Reindhold Tüxen (ed.), J. Cramer Verlag, Vaduz.

Thill, A. 1961. La cartographie des végétations et son application forestière. Bulletin de la Société Royale Forestière de Belgique. Juillet.

Tideman, P. 1963. Vegetationskartierung als Grundlage für Verwaltungspläne in Naturschutzgebieten in den Niederlanden. In: Reinhold Tüxen, 1963a, pp. 481–490.

Tivy, J. 1954. Reconnaissance vegetation survey of certain hill grazings in the Southern Uplands. Scottish Geographical Magazine, 70 : 21–33.

Tom, C.H. and L.D. Miller. 1980. 'Forest Site Index Mapping and Modelling,' *Photogrammetric Engineering and Remote Sensing,* 46 : 1585–1596.

Tomaselli, R. 1958. Plant communities of the western half of the University of Kansas Natural History Reservation. Casa Editrice Renzo Cortina. Pavia.

Tomaselli, R. 1970. Note illustrative della carta della vegetazione naturale potenziale d'Italia (prima approssimazione). Collana verde 27. Minist. Agric. e Foreste. Roma.

Tomaselli, R. 1970. Vegetazione naturale potenziale d'Italia. Pavia. Istituto Botanico dell'Università.

Tosi, J. jr. 1966. Economics and the natural environment. Seminar on Integrated Survey, 1966. Publications of the ITC-UNESCO Centre for Integrated Survey, S. 13.

582

Tosi, J. A. 1969. República de Costa Rica incl. Isla del Coco: mapa ecológico. Centro Científico Tropical. San José, C.R.

Tongeren, O. van. 1968. FLEXCLUS an interactive program for classification and tabulation of ecological data. Acta Neerlandia 35(3) August 1986: 137–143.

Townshed, J. and C. Justice. 1981. 'Information Extraction from Remotely Sensed Data,' *International Journal of Remote Sensing* 2: 313–329.

Traets, J. 1961. Verklarende Tekst bij het Kaartblad. Lillo 14 E. Vegetatiekaart van België. (With colour map 1:20.000) pp. 79.

Traets, J. 1960. Verklarende Tekst bij het Kaartblad. Kalmthoutse Hoek 6 W. Vegetatiekaart van België. (With colour map 1:20.000) pp. 61.

Traets, J. 1959. Verklarende Tekst bij het Kaartblad. Kalmthout 6 E. Vegetatiekaart van België. pp. 76.

Trapnell, C. G. 1948. Vegetation-soil map of Northern Rhodesia. Department of Agriculture, Forestry Branch. Lusaka.

Trapnell, C. G. 1969. Vegetation: Kenya. Overseas Development Administration, Land Resources Development Center. Surbiton.

Trautmann, W. 1963. Methoden und Erfahrungen bei der Vegetationskartierung der Wälder und Forsten. In: Reinhold Tüxen, 1963a, pp. 119–127.

Trautmann, W. 1972. 'Vegetation'. (Potentielle natürliche Vegetation). Deutscher Planungsatlas Band I: Nordrhein-Westfalen. Lieferung 3. Gebrüder Janecke Verlag. Hannover.

Trautmann, W. 1969. Zur Geschichte des Eichen-Hainbuchenwaldes im Münster-land auf Grund pollenanlytischer Untersuchungen. Schr. Reihe Vegetationskunde 4: 109–129.

Trautmann, W. 1966. Erläuterungen zur Karte der potentiellen natürlichen Vegetation der Bundesrepublik Deutschland 1:200 000, Blatt 85 Minden. Schr. Reihe Vegetationskunde 1.

Trautmann, W. 1973. Vegetationskarte der Bundesrepublik Deutschland 1:200 000 – Potentielle natürliche Vegetation – Blatt CC 5502 Köln. Schr. Reihe Vegetationskunde 6.

Trautmann, W. 1976. Veränderungen der Gehölzflora und Waldvegetation in jüngerer Zeit. Changes in the flora of woods and in the woodland vegetation of the Federal Republic of Germany in recent decades. Schr. Reihe Vegationskunde 10: 91–108.

Trautmann, W. 1972. Vegetation (Potentielle natürliche Vegetation). Deutscher Planungsatlas Bd. 1: Nordrhein-Westfalen, Lieferung 3. Veröff. Akad. Raumforsch. Landesplan., Hannover.

Trochain, J. L. 1961. Représentation cartographique des types de végétation intertropicaux africains. In: Henri Gaussen, 1961a, pp. 87–102.

Troll, C. 1936. Termiten-Savannen. Stuttgart. In: Länderkundliche Forschung: Festschrift Norbert Krebs, pp. 275–312.

Troll. C. 1939. Vegetationskarte der Nanga Parbat Gruppe. Leipzig. Deutsches Museum für Länderkunde. Wissenschaftliche Veröffentlichungen, N.S., No. 7.

Troll, C. 1941. Studien zur vergleichenden Geographie der Hochgebirge. Bonn. Friedrich Wilhelm Universität, Bericht No. 23, pp. 49–96.

Troll, C. 1943. Die Frostwechselhäufigkeit in den Luft- und Bodenklimaten der Erde. Meteorologische Zeitschrift, 60: 161–171.

Troll, C. 1948. Der asymmetrische Aufbau der Vegetationszonen und Vegetationsstufen auf der Nord- und Südhalbkugel. Geobotanisches Forschungsinstitut Rübel. Zürich. Bericht für das Jahr 1947, pp. 46–83.

Troll, C. 1951. Das Pflanzenkleid der Tropen in seiner Abhängigkeit von Klima, Boden und Mensch. Verlag des Amtes für Landeskunde. Remagen.

Troll, C. 1955. Der jahreszeitliche Ablauf des Naturgeschehens in den verschiedenen Klimagürteln der Erde. Studium Generale, 8: 713–733.

Troll, C. 1956. Der Klima- und Vegetationsaufbau der Erde im Lichte neuer Forschungen. Akademie der Wissenschaften und Literatur. Mainz. Abhandlungen, pp. 216–229.

Troll, C. 1971. Landscape ecology and biogeocenology. Geoforum, 8: 43–46.

Troy, J. P., H. Gaussen, P. Legris, V. M. Meher Homji and F. Blasco. 1968. Carte internationale du Tapis Végétal et des conditions écologiques à 1/1 000 000e feuille 'Kathiawar'. Notice explicative in Inst. français de Pondichéry; trav. Sect. scient. techn., Hors Série N° 9, 96 p.

Tucker, C. J. 1979. 'Red and Photographic Infrared Linear Combinations for Monitoring Vegetation,' *Remote Sensing of Environment* 8: 127–150.

Tucker, C. 1979. Red and photographic infrared linear combinations for monitoring vegetation. Photogrammetric Engineering and Remote Sensing, 5: 781–784.

Tueller, P. T. 1982. 'Remote Sensing for Range Management,' Chapter 12 in *Remote Sensing for Resourse Management*. C. J. Johannsen and J. L. Sanders, eds., Ankeny, IA: Soil Conservation Society of America, pp. 125–140.

Tüxen, R. and H. Diemont. 1937. Klimaxgruppe und Klimaxschwarm, ein Beitrag zur Klimaxtheorie. Jahresber. naturhist. Ges. Hannover 88–89: 73–87.

Tüxen, R. 1950. Neue Methoden der Wald- und Forstkartierung. Stolzenau/Weser. Mitteilungen der Floristisch-soziologischen Arbeitsgemeinschaft, N.F., vol. 2.

Tüxen, R. and E. Preising. 1951. Erfahrungsgrundlagen für die pflanzensoziologische Kartierung des norddeutschen Grünlandes. Stolzenau/Weser. Angewandte Pflanzensoziologie, vol. 4.

Tüxen, R. 1951. 'Angewandte Pflanzensoziologie'. Arbeiten aus der Zentralstelle, für Vegetationskartierung. (With colour map Scale 1:10.000). Stolzenau/Weser 1951. pp. 72.

Tüxen, R. 1951. Angewandte Pflanzensoziologie. N. 2. Bruchwaldgesellschaften im Großen und Kleinen Moor Forstamt Danndorf (Dromling). Wit a colour map Scale

1:10.000). Stolzenau/Weser 1951. pp. 46.

Tüxen, R. 1952. Ein einfacher Weg zur nachträglichen Feststellung von Entwässerungsschäden. Stolzenau/Weser. Mitteilungen der Floristisch-soziologischen Arbeitsgemeinschaft, N.F. vol. 3.

Tüxen, R. 1952. Angewandte Pflanzensoziologie. N. 5 (Colour map Scale 1:10.000). Stolzenau/Weser 1952, pp. 77.

Tüxen, R. and G. Hentschel. 1954. Bibliographie der Vegetationskarten Deutschlands. Stolzenau/Weser. Mitteilungen der floristisch-soziologischen Arbeitsgemeinschaft, N.F., vol. 5.

Tüxen, R. 1954a. Pflanzengesellschaften und Grundwasserganglinien. Stolzenau/Weser. Angewandte Pflanzensoziologie, vol. 8.

Tüxen, R. 1954b. Über die räumliche, durch Relief und Gestein bedingte Ordnung der natürlichen Waldgesellschaften am nördlichen Rande des Harzes. Vegetatio, 5–6:454–478.

Tüxen, R. 1954. Die Wald- und Forstgesellschaften im Graf Görtzischen Forstbezirk Schlitz (With colour maps 1:15.000). Stolzenau/Weser 1954. pp. 63.

Tüxen, R. 1956a. Baltrum. Stolzenau/Weser. Bundesantalt für Vegetationskartierung.

Tüxen, R. 1956b. Die heutige potentielle natürliche Vegetation als Gegenstand der Vegetationskartierung. Stolzenau/Weser. Angewandte Pflanzensoziologie, 13:5–42.

Tüxen, R. 1956. Die heutige potentielle natürliche Vegetation als Gegenstand der Vegetationskartierung. 8 mit 3 Abb. und 10 Tabellen/mit 1 fbg. Karte 1:25.000). pp. 55.

Tüxen, R. 1957. Die heutige natürliche potentielle Vegetation als Gegenstand der Vegetationskartierung. Remagen. Berichte zur deutschen Landeskunde, 19:200–246.

Tüxen, R. 1958. Die Eichung von Pflanzengesellschaften auf Torfprofiltypen. Stolzenau/Weser. Angewandte Pflanzensoziologie, 15:131–141.

Tüxen, R. 1958. Die heutige potentielle natürliche Vegetation als Gegenstand der Vegetationskartierung. Pflanzensoziologie 13: pp. 5–42. Stolzenau/Weser 56. (See also Ber. Dtsch. Landsh. Remagen, 1958).

Tüxen, R. 1958. Die koinzidenzmethod-Eichung von Pflanzengesellschaften auf edaphischen Faktoren. Angew. Pflanzensoziologie 15, pp. 1–10.

Tüxen, R. and H. Meissner. 1959. Vegetationskartierung. Excerpta Botanica, Sectio B: Sociologica, 1:50–51.

Tüxen, R. 1959. Bibliographie der Verbreitungs- und Arealkarten von Pflanzengesellschaften. Excerpta Botanica, Sectio B: Sociologica, 1:227–261.

Tüxen, R. 1959. Typen von Vegetationskarten und ihre Erarbeitung. Bericht über das Internationale Symposion für Vegetationskartierung, vom 23.–26.3.1959 in Stolzenau/Weser. pp. 139–154. 500 pp.

Tüxen, R. 1961. Bemerkungen zu Einer Vegetationskarte Europas. In: Méthodes de la Cartographie de la Végétation Toulouse, 16–21 Mai, 1960. pp. 61–72.

Tüxen, R. Methodisches Handbuch für Heimatforschung in Niedersachen. Vegetationskartierung. pp. 153–168.

Tüxen, R. 1963a. Bericht über das Internationale Symposium für Vegetationskartierung vom 23. bis 26.3.1959 in Stolzenau an der Weser. Weinheim. Cramer Verlag.

Tüxen, R. 1963. Bericht über das Internationale Symposion für 'Vegetationskartierung' vom 23.–26.3.1959 in Stolzenau/Weser. International Society for Plant Geography and Ecology. pp. 481–490.

Tüxen, J. 1963. Vegetationskarten als Hilfsmittel der Altlandschaftsforschung am Beispiel des Messtischblattes Stolzenau an der Weser. In: Reinhold Tüxen, 1963a, pp. 313–319.

Tüxen, R. 1974. Tatsachen und Probleme der Grenzen in der Vegetation. J. Cramer Verlag. Lehre.

Ubaldi, D. 1978. Carta della vegetazione di Vergato Bologna. Emilia-Romagna (With colour map 1:25.000). Consiglio Nazionale delle Ricerche. AQ/1/4. Roma.

UNESCO. 1973. Classification internationale et cartographie de la végétation. Ecologie et Conservation N° 6, 96 p. Also in English + Spanish.

UNESCO. 1981. Carte de la Végétation d'Amérique du Sud. 1 carte à 1:5 000 000 en 2 feuilles + notice explicative, 183 p. Recherches sur les Ressourses Naturelles N. 17.

U.S. Bureau of Reclamation. 1953. *Bureau of Reclamation Manual* Vol. V. Irrigated land use Part 2, land classification, 132 p. Dept. of Interior, Washington DC.

Van der Pijl, L. 1972. Principles of dispersal in higher plants. 2nd. Edition. Springer Verlag. Berlin. 162 p.

Van Dijk et al. 1984. 'AVHRR training set' vol. 1 NDAA/NESDIS/AISC June 1984.

Van Steenis, C.G.G.J. 1935. Maleische Vegetatieschetsen. Tijdschr. Ned. Aardrijksk. Genoots 52 25–67; 171–203; 363–398.

Van Steenis, C.G.G.J. 1957. Outline of vegetation types in Indonesia and some adjacent regions. Proceed. of the 8th Pacific Science Congress, 4:61–97.

Van Steenis, C.G.G.J. 1958. The vegetation map of Malesia. Scale 1:1 000 000. UNESCO, Paris.

Vane, G., A.F.H Goetz and J.B. Wellman. Airborne Imaging Spectrometer: A New tool for Remote Sensing. Jet Propulsion Laboratory, California Institute of Technology Pasadena, California, U.S. pp. 5.

Van der Werf, S. and G. Londo (in Prep.) Natuurtechnisch Bosbeheer. Vol. 4. Rijksinstituut voor Natuurbeheer, Natuurbeheer in Nederland. PUDOC, Wageningen.

Veatch, J.O. 1953. Soils and Land of Michigan. Michigan State College Press. East Lansing, MI.

Veri, L. and F. Tammaro. 1980. Aspetti vegetazionali del Monte Sirente. (Appennino Abruzzese) (With colour map 1:25.000). Consiglio Nazionale delle Ricerche AQ/1/83. Roma.

Veri, L., V. La Valva and G. Caputo. 1980. Carta delle vegetazione delle isole Ponziane (Golfo di Gaeta) (With colour map. Scale 1:14.000). Cinsiglio Nazionale delle Ricerche AQ/1/41. Roma, 1980. pp. 25.

584

Verstappen, H. Th. 1973. A geomorphological reconnaissance of Sumatra and adjacent islands. Verhandelingen of Royal Dutch Geographical Society. 182 p.

Vervoorst, F. B. 1967. Las comunidades vegetales de la depresión del Salado (Provincia de Buenos Aires). INTA. Buenos Aires. La vegetación de la República Argentina, Serie Fitogeografía, No. 7.

Victorov, S. V. 1962. Geobotanical indication maps and methods of their compilation. In: Victor B. Sochava, 1962a, pp. 72–76.

Victorov, S. V., A. Vostokova and D. D. Vishivkin. 1964. A short guide to geobotanical mapping. The Macmillan Co. New York, NY.

Victorov, S. V., Y. A. Vostokova and D. D. Vyshivkin. 1964. Short guide to Geo-Botanical Surveying. Int. Series of Monographs on Pure and Applied Biology, Volume 8. Botany Division. Pergamon Press.

Vinogradov, B. V. 1962. The major problems of desert vegetation mapping using aerophotos. In: Victor B. Sochava, 1962a, pp. 215–222.

Vinogradov, B. V. 1966. Aeromethods in studies of vegetation of arid zones (only Russian text). Academy of Sciences of the USSR. Laboratories of aeromethods of the Ministry of Geology of the USSR.

Vinogradov, B. V. 1978. Geobotanical Mapping. pp. 22–34 Leningrad. U.S.S.R.

Volkova, V. G. 1969. Detailed plans of vegetation and the method of complex ordination. Academy of Sciences of the U.S.S.R. Komarov Botanical Institute. Leningrad. Geobotanical Cartography, pp. 42–51.

Voronov, A. G. and A. M. Cheltsov-Bebutov. 1962. On the methods of biogeographical mapping of open landscapes. In: Victor B. Sochava, 1962a, pp. 186–193.

Vries A. A. de. 197x. Paper on false colour and dutch-elm disease detection. Amsterdam.

Vries, D. M. de. 1949. Objective combination of Species. Acta bot. Neerlandica, I. 4.

Wagner, H. 1948. Die Bedeutung der Vegetationskartierung für Forschung und Praxis. Wien. Jahrbuch der Hochschule für Bodenkultur, 2: 23–26.

Wagner, H. 1961a. Die Fassung der Gesellschaftseinheiten auf Grund der grossmasstäbigen Vegetationskartierung. In: Henri Gaussen, 1961a, pp. 171–178.

Wagner, H. 1961b. Bibliographie der Vegetationskarten Österreichs. Excerpta Botanica, Sectio B: Sociologica, 3: 305–315.

Wagner, H. 1980. Aufgaben der Vegetationskartierung in Forschung und Praxis. Documents de cartographie ecologique, 23: 21–22.

Wagner, H. 1981. Zur Farbenwahl in der Vegetationskartierung. Wien. Angewandte Pflanzensoziologie, 26: 277–281.

Wagner, P. L. 1957. A contribution to structural vegetation mapping. Annals of the Association of American Geographers, 47: 363–369.

Waibel, L. 1943. Place names as an aid in the reconstruction of the original vegetation of Cuba. Geographical Review, 33: 376–396.

Walker, D. A., K. R. Everett, P. J. Webber and J. Brown. 1980. Geobotanical atlas of the Prudhoe Bay region. United States Army Corps of Engineers. Hanover, NH. Cold Regions Research and Engineering Laboratory. Report No. 80–14.

Walter, H. 1951. Grundlagen der Pflanzenverbreitung. Ulmer Verlag. Stuttgart.

Walter, H. 1958. Klimax und zonale Vegetation. Angew. Pflanzensoz. Festschr. E. Aichinger I: 144–150. Wien.

Walter, H. 1968–1973. Die Vegetation der Erde. Gustav Fischer Verlag. Stuttgart.

Walter, H. 1968. Die Vegetation der Erde in öko-physiologischer Betrachtung, Bd. II. De gemässigten und arktischen Zonen. Fischer Verlag. Jena.

Walter, H. (transl. by Joy Wieser). 1973. Vegetation of the Earth in relation to Climate and the Eco-Physiological Conditions. Heidelberg Science Library.

Walter, H. 1976. Die ökologischen Systeme der Kontinente. Gustav Fischer Verlag. Stuttgart.

Walter, H. 1977. Vegetationszonen und Klima, 3rd ed. Eugen Ulmer Verlag. Stuttgart.

Walter, H. 1979. Vegetation of the earth and ecological systems of the geobiosphere. Springer Verlag. Heidelberg.

Walther, K. 1957. Vegationskarte der deutschen Flusstäler: mittlere Elbe oberhalb Damnatz. Bundesstelle für Vegetationskartierung. Stolzenau/Weser.

Walther, K. 1960. Pflanzensoziologie und Kulturtechnik. Zeitschrift für Kulturtechnik, 1: 65–76.

Walther, K. 1963. Die Vegetationskartierung in den einführenden pflanzensoziologischen Lehrgängen der Bundesstelle für Vegetationskartierung. In: Reinhold Tüxen, 1963a, pp. 155–165.

Wangerin, W. 1915. Vorläufige Beiträge zur kartographischen Darstellung der Vegetationsformationen im norddeutschen Flachland unter besonderer Berücksichtigung der Moore. Berichte der deutschen botanischen Gesellschaft, 33: 168–199.

Warren, P., K. Reichardt and D. Mouat. 1981. Remote Sensing Newsletter. *University of Arizona. 81–2 October 1981.* 'Vegetation Mapping of the Grand Canyon National Park.' pp. 11.

Watson, R. M. and J. J. Beckett. 1976. A pilot aerial inventory of the woody vegetation in the Loka region of E. Equatoria province, Sudan. Resource Mangement & Research, Nairobi.

Weaver, J. E. and F. E. Clements. 1929. Plant ecology. Second edition, 601 p., Mac Graw Hill Co. Ltd. Bombay.

Wells, P. 1962. Vegetation in relation to geological substratum and fire in the San Luis Obispo quadrangle, California. Ecological Monographs, 32: 79–103.

Wenzel, A. 1963. Technische Erfahrung in der Vegetationskartographie. In: Reinhold Tüxen, 1963a, pp. 167–172.

Werf S. van der. 1957. 'Vegetatiekartering'. 'Kruipnieuws' Orgaan van N.J.N. Sociologengroep. 19e Jaargang – N. 2 pp. 25–43.

Western, D. and J.J.R. Grimsdeel. 1979. Measuring the distribution of animals in relation to the environment. Handbook No. 2, African Wildlife Leadership Foundation, Nairobi.

Westhoff, V. 1954. Die Vegetationskartierung in den Niederlanden. Sonderdruck aus 'Angewandte Pflanzensoziologie', Veröffentlichungen des Kärntner Landesinstituts für angewandte Pflanzensoziologie in Klagenfurt. Festschrift Aichinger, II. Band. pp. 1.223.1.234.

Westhoff, V. 1955. Vegetatie-Kartering. Overdruk uit T.N.O. Nieuw Nr. 107. pp. 61–67.

Westhoff, V. 1957. Een Gedetailleerde Vegetatiekartering van een deel van het Bosgebied van Middachten. Laboratorium voor Plantensystematiek en geografie der Landbouwhogeschool). pp. 57. (With colour map, scale 1:6493).

Westhoff, V. 1954. Die Vegetationskartierung in den Niederlanden. Wien. Festschrift Aichinger. Springer Verlag. Pp. 1223–1231.

Westhoff, V. 1958. Boden- und Vegetationskartierung von Wald- und Forstgesellschaften im Quercion robori-petraeae-Gebiet der Veluwe, Stolzenau/Weser. Angewandte Pflanzensoziologie. Vol. 15.

Westhoff, V. 1961. Bibliographia phytosociologica. Neerlandia: Vegetationskartierung. Excerpta Botanica, Sectio B: Sociologica, 3:140–141.

Westhoff, V. & A.J. den Held. 1975. Plantengemeenschappen van Nederland. Thieme, Zutphen. 324 pp.

Westveld, M. 1951. Vegetation mapping as a guide to better silviculture. Ecology, 32:508–517.

Westveld, M. 1952. A method of evaluating forest site quality. Northeastern Forest Experiment Station, Upper Darby, PA. Station Paper No. 48.

Westveld, M. 1954. Use of plant indicators as an index to site quality. Society of American Foresters, New England Section. Boston.

White, F. 1981. The vegetation of Africa. Paris. United Nations Educational, Scientific and Cultural Organization. Natural Resources Research, vol. 20.

Whitmore, T.C. 1975. The Tropical Rain Forest of the Far East. Oxford University Press. Oxford.

Whitmore, T.C. 1975. Tropical rain forest of the far East. 282 p. Clarendon Press. Oxford.

Whittaker, R.H. 1953. A consideration of climax theory: the climax as a population and pattern. Ecol. Mon. 23:41–78.

Whittaker, R.H. 1974. Climax concepts and recognition. In: Knapp, R. (ed.) Vegetation dynamics. Handbook of Vegetation Science, part VIII. Junk, The Hague.

Whittaker, R.H. 1962. Classification of natural communities. Botanical Review, 28:1–239.

Whittaker, R.H. 1970. Communities and ecosystems, 2nd ed. Macmillan. New York, NY.

Whittaker, R.H. 1978. Classification of plant communities. Dr. W. Junk Publishers. The Hague.

Whittaker, R.H. and G.M. Woodwell. 1972. Evolution of Natural Communities. In: Ecosystem Structure and Function (ed.) J.A. Wiens. Proceed. 31th Annual Biology Colloquium. Oregon State University Press.

Wieslander, A.E. 1934. Vegetation types of California, Ramona quadrangle. California Forest & Range Experiment Station. Berkely, CA.

Wieslander, A.E. 1937. Vegetation types of California, Tujunga quadrangle. California Forest & Range Experiment Station. Berkely, CA.

Wieslander, A.E. 1944. Forest survey type maps. California Forest & Range Experiment Station. Berkely, CA.

Wieslander, A.E. 1949. The timber stand and vegetation-soil maps of California. California Forest & Range Experiment Station. Berkely, CA.

Wieslander, A.E. and R.E. Storie. 1952. The vegetation-soil survey in California and its use in the management of wild lands for yield of timber, forage and water. Journal of Forestry, 50:521–526.

Wieslander, A.P. and R.E. Storie. 1953. Vegetational approach to soil surveys in wild land areas. Soil Science Society of America. Proceedings, 17:143–147.

Wijngaarden W. van. 1984. Soils and vegetation of the Tsavo area. Reconnaissance Soil Survey Rep R-7. Kenya Soil Survey, Nairobi.

W.v. Wijngaarden. 1985. Elephants – Trees – grass – grazer. Relationships between climate, soil, vegetation and large herbivores in a semi-arid savanna ecosystem (Tsaro, Kenya). Diss. Agricultural University Wageninge.

Wikstrom, J.H. and R.G. Bailey. 1983. Land systems classification. Renewable Resources Journal, 1:4.

Williams, D.L. and L.D. Miller. 1979. Monitoring Forest Canopy Alteration Around the World With Digital Analysis of Landsat Imagery, Greenbelt, MD: NASA Goddazrd Space Flight Centre.

Williams, Don L., C. King. N.E. Hardy and J. Cihlar. 1974. Vegetation of the University of Kansas Natural History Reservation and the Nelson Environmental Study Area according to the UNESCO classification. University of Kansas. Lawrence, KS. Field Facilities Committee.

Williams, R.J. 1955. Vegetation regions (of Australia). Canberra. Department of National Development. Atlas of Australian Resources.

Wilson, J.G. 1962. The vegetation of the Karamoja District, Northern Province of Uganda. Uganda Department of Agriculture, Memoirs of the Research Division, Series 2, No. 5. In pocket.

Wirdum, G. van. 1981. Design for a Land Ecological Survey of Nature Protection. Proc. Int. Congr. Neth. Soc. Landscape Ecology. Veldhoven, pp. 245–251.

Wittich, W. 1950. Die Auswertung der Standortskartierung durch die Forsteinrichtung. Allgemeine Forstzeitschrift, 5:413–416.

Wolff, G. 1967. Aerial photography in the assessment of fertilizing results in forestry and in detecting diseases of trees. Agricultural Aviation Intern. Agr. Aviat. Centr. The Haque 9(4):114–122.

Wooster, L.C. 1876. Vegetation map of the lower St. Croix district, WI. In: T.C. Chamberlin: Geology of Wisconsin.

586

1882. Madison, WI. P. 146.

Wraber, Maks. 1963. Allgemeine Orientierungskarte der potentiellen natürlichen Vegetation im slowenischen Küstenland als Grundlage für die Wiederbewaldung der degradierten Karst- und Flyschgebiete. In: Reinhold Tüxen, 1963a, pp. 369–384.

Wyatt-Smith, J. 1962. Some vegetation types and a preliminary vegetation map of Malaya. Kuala Lumpur. Regional Conference of Southeast Asian Geographers.

Wyatt-Smith, J. 1964. A premilinary vegetation map of Malaya with description of the vegetation types. J. of tropical Geography, 18:200–213.

Yurkovskaya, T.K. 1968. Some principles of compilation of a bog vegetation map. Academy of Sciences of the U.S.S.R. Komarov Botanical Institute. Leningrad. Geobotanical Cartography, pp. 44–51.

Zackrisson, O. 1977. Influence of forest fires on the North Swedish boreal forest. Oikos 29:22–32.

Zanotti, A.L., F. Corbetta and L. Aita. 1980. Carta della vegetazione della Tavoletta 'Trivigno' (Basilicata). (With colour map 1:25.000). Consiglio Nazionale delle Ricerche. AQ/1/84. Roma.

Zee, D. van der. 1984. Monitoring Mneragala. A landscape under Pressure in: IALE, the International Association for Landscape Ecology. Proceedings of the first international seminar on methodology in Landscape Ecological Research and Planning. Theme II: Methodology and Techniques of Inventory And Survey, pp. 85–96 Roskilde University Centre, Denmark.

Zee, D. van der. 1985. Monitoring Moneragala. Een landschap in verandering. Public lecture given on the Sri Lanka Seminar of the KNAG. (Royal Netherlands Geographical Society) departm. of geography of developing countries. January 11th. 1985 in Amsterdam.

Ziani, P. 1963. Die Vegetationskarte als Grundlage der Verbesserung der degradierten Bodenflächen im Ost-Mittelmeergebiet. In: Reinhold Tüxen, 1963a, pp. 395–407.

Zohary, M. 1947. A vegetation map of western Palestine. Journal of Ecology, 34:1–18.

Zohary, M. 1981. Südliche Levante: Vegetation. Ludwig Reicher Verlag. Wiesbaden. Tübinger Atlas des Vorderen Orients.

Zohary, M. and G. Orshan. 1966. Vegetation of Crete. Israel Journal of Botany, vol. 14, Supplement.

Zonneveld, I.S. 1959. The Relation between Soil and Vegetation Research. Boor en Spade, pp. 35–58 (in Dutch).

Zonneveld, I.S. 1959. Rapport over het 'Internationale Symposium voor Vegetatiekartering' gehouden van 23 tot 26 Maart 1959 te Stolzenau/Weser West-Duitsland. pp. 16.

Zonneveld, I.S. 1960. A study of soil and vegetation of a freshwater tidal delta. The Brabantse Biesbosch, Vol. I + II. pp. 209 + 396; 3 coloured maps appendices (Dr-Thesis). Wageningen, Med. Landb. Onderz. No. 65, 20.

Zonneveld, I.S. 1963. Vegetationskartierung eines Süsswassergezeitendeltas (Biesbosch). In: Reinhold Tüxen, 1963a, pp. 279–296.

Zonneveld, I.S. 1966. Zusammenhänge Forstgesellschaft – Boden – Hydrologie und Baumwuchs in einigen Niederländischen Pinus-Forsten auf Flugsand und auf Podsolen (with summary). Anthropogene Vegetation: 312–335. Junk. Den Haag, 1966.

Zonneveld, I.S. 1966. The role of plant ecology in Integrated surveys of the natural environment. Sem. on Integrated Surveys, P. 2. I.T.C. Delft. The Netherlands. 1966.

Zonneveld, I.S. 1970. The contribution of vegetation science in the exploration of natural resources. ITC-UNESCO publication P. 3. I.T.C. Delft. The Netherlands. 1970.

Zonneveld, I.S. 1973/75. A photo report on 1/4 century vegetation succession in a fresh water tidal area (Biesbosch) Netherlands. Sukzessionforschung. Berichte Int. Symp. Int. Ver. Veg.

Zonneveld, I.S. 1974. Aerial photography, remote sensing and ecology. Proceed. First Int. Congress on Ecology, The Hague, 1974. Also ITC Journal 1974-4.

Zonneveld, I.S. 1974. 25 years of sequential photographic monitoring of a tidal environment. I.T.C. Journal 1974.

Zonneveld, I.S. 1977. Classificeren en evalueren van bos mede met behulp van de spontane vegetatie. Ned. Bosbouwtijdschrift. Jrg. 1977. pp 44–65.

Zonneveld, I.S. 1979. Landscape Science and Land Evaluation. ITC-textbook VII-4, 2nd ed., 134 pp.

Zonneveld, I.S. 1980. Some aspects of the mutual relation between climate and vegetation in the Sahel and Sudan (zones) and the consequences for research and development. Symposium Epharmony of Vegetation. Rinteln. 1979: 409–454. ITC Journal, 1980(2): 255–296.

Zonneveld, I.S. 1981. The role of single land attributes in forest evaluation. In: Laban, P. (ed.). Proceedings of the workshop on land evaluation for forestry. *ILRI Publ.* 28, 76–94, Wageningen.

Zonneveld, I.S. 1982. Principles of Indication of Environment through Vegetation. L. Steubing and H.J. Jäger (eds) Monitoring of Air Pollutants by Plants: Methods and Problems, Junk Publishers, The Hague.

Zonneveld, I.S. 1984. Conclutions and Outlook I.A.L.E. Congres Landscape Ecology. Roskilde. Oct. 1984. 18 pp.

Zonneveld, I.S. (1986) i.p. Scope and Concepts of Landscape Ecology, an Emerging Science. Inv. Pap. IVth International Congress of Ecology INTECOL. Syracuse U.S.A. aug. 1986. In Forman & Zonneveld (ed.). Trends in Landscape Ecology. Chapt. 1. Springer Verlag (i.p.).

Zonneveld, I.S. (1987) i.p. Landscape Survey and evaluation. Invited paper Binational Workshop on Desert Ecology. "What is special about desert ecology". Sede Boger. Proceedings i.p.

Zonneveld, I.S. and J.F. Bannink. 1960. Studies van Bodem en Vegetatie op het Nederlandse deel van de Kalmthoutse Heide. Stichting voor Bodemkartering Wageningen, pp. 115.

Zonneveld, I.S., H.A.M.J. van Gils and D.C.P. Thalen. 1979. Aspects of the ITC approach to vegetation survey.

Doc. Phytosoc. IV. Lille pp: 1029–1063.

Zonneveld, I.S., P.N. de Leeuw, and W.G. Sombroek. 1970. Ecological Interpretation of Aerial. Photos in a Savanna Region in N. Nigeria. ITC Publication B63.

Zonneveld, I.S. and D.C.P. Thalen. 1976. Methods in Vegetation survey for Development. – The Present State of the Art –. International Symposium to mark the Twenty-fifth Anniversary of the I.T.C. (9–15 Dec. 1976). pp. 18.

Zonneveld, I.S. and D.C.P. Thalen. 1980. Interpretation and application of the vegetation and landscape map. In: K.S. Dijkema and W.J. Wolff. 1983. Flora and vegetation of the Wadden Sea Islands and coastal areas. A.A. Balkema Publishers. Rotterdam. Wadden Sea Working Group, Report No. 9.

Zorn, H.C. 1965. An Instrument for Testing Stereoscopic Acuity. Photogrammetria I.T.C. pp. 230–238. Oct. 1965.

COLOUR SECTION

The following 32 pages renders the black and white photographs, where indicated through-out the book, in colour.

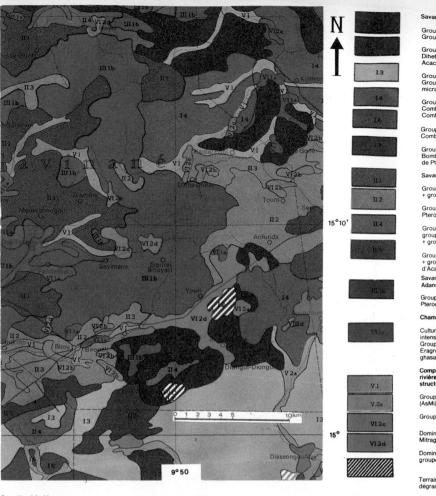

		Pourcentage par unité de légende
Savane arborées dominées par Combretum ghasalense.		
Groupement de Combretum ghasalense & Eragrostis tremula (GE) + Groupement de Combretum ghasalense & Acacia senegal (GAe)	GE GAe	70% 25%
Groupement de Combretum ghasalense, Eragrostis tremula & Diheteropogon (GED) + Groupement de Combretum ghasalense, Acacia senegal & Diheteropogon (GAeD)	GED GAeD	80% 20%
Groupement de Bombax costatum & Eragrostis tremula (BE) + Groupement de Bombax costatum, Eragrostis tremula & Combretum micranthum (BEM) + groupement de Anogeissus (BAo + AoAy)	BE BEM BAo AoAy	80% 15% 2% 2%
Groupement de Combretum ghasalense & Acacia sénégal (GAe) + Combretum ghasalense & Acacia nilotica (GN) + groupement de Combretum ghasalense, Acacia senegal & Diheteropogon (GAeD)	GAe GN GAeD	35% 25% 30%
Groupement de Bombax & Eragrostis (BE) + groupement de Combretum ghasalense, Acacia sénégal & Diheteropogon (GAeD)	BE GAeD	66% 33%
Groupement de Bombax & Eragrostis (BE) + groupement de Bombax, Eragrostis & Combretum micranthum (BEM) + groupement de Pterocarpus lucens & Eragrostis tremula (PE)	BE BEM PE	50% 45% 5%
Savane arborées avec espèces d'Acacia.		
Groupement d'Acacia seyal, Acacia senegal & Acacia nilotica (Ayen) + groupement d'Acacia seyal & Balanites (BaAy)	Ayen BaAy	80% 20%
Groupement d'Acacia seyal & Balanites (BaAy) + Pterocarpus lucens & Eragrostis tremula (PE)	BaAy PE	80% 10%
Groupement de Combretum ghasalense & Acacia nilotica (GN) + groupement d'Acacia seyal, Acacia senegal & Acacia nilotica (Ayen) + groupement d'Acacia seyal & Balanites (BaAy)	GN Ayen BaAy	50% 25% 25%
Groupement d'Acacia seyal, Acacia sénégal & Acacia nilotica (Ayen) + groupement d'Acacia seyal & Balanites (BaAy) + groupement d'Acacia seyal & Adansonia (AyAd)	Ayen BaAy AyAd	30% 30% 30%
Savane arbustives et arborées avec Pterocarpus lucens et/ou Adansonia.		
Groupement d'Adansonia & Pterocarpus (AdP) + groupement de Pterocarpus & Eragrostis (PE)	AdP PE terrain nu	30% 30% 40%
Champs cultivés, itinérants et jachères.		
Culture permanente, y compris les cultures itinérantes assez intensives et jachères. Groupement d'Acacia albida (Aa) + groupement de Bauhenia & Eragrostis (Bhe) + groupement des jachères à Combretum ghasalense & Eragrostis (ge) + quelques autres comme pimi et baay	Aa Bhe ge pimi	75% 10% 10% 5%
Complexes des forêts ripicoles et forêts claires des vallées des rivières et ruissaux, y compris leurs stages de dégradation structurelle, les mares et les marais.		
Groupement d'Acacia nilotica scorpoïdes & Mitragyna inermis (AsMi) + les mares diverses	AsMi mares	80% 20%
Groupement de Piliostygma & Mitragyna (PiMi) + des marais divers	PiMi marais	80% 20%
Dominés par jachères avec les groupements de Piliostigma & Mitragyna (pimi) + groupements naturels comme AsMi et PiMi	pimi AsMi PiMi	50% 25% 25%
Dominés par legroupement de Bauhenia & Eragrostis (Bhe) et groupements naturels comme BEM, GAeD et BCM	Bhe BEM GAeD BCM	50% 15% 15% 15%
Terrain non cultivé mais avec végétation naturelle totalement dégradée. (Unités de légende I à V)		

15°10'

15°

9°50

Fragment of the VEGETATION MAP
of the KAARTA–REGION in MALI
ITC 1977

Projet de développement rural intégré de la Région Kaarta, République du Mali.
Phase de reconnaissance du volet VIII: Cartographie

© ITC-Enschede, Pays Bas, 1977

Auteurs responsables: Ir. A. de Gelder
Drs. H. A. M. J. van Gils
Dr. I. S. Zonneveld
A. Vreugdenhil

L'Ecologie de la végétation de la région du Kaarta

Caractère édaphique	Pluviosité
Sols sableux éoliens	
Sols sableux profonds sur des dunes longitudinales, GAe dans les dépressions interdunaires	moins de 600 mm
Sols sableux sur des dunes ondulées avec GAeD dans les dépressions interdunaires	plus de 600 mm
Sols sableux profonds sur des dunes longitudinales basses avec BEM dans les dépressions interdunaires et BAo et AoAy dans les mêmes dépressions et le long des ruissaux	600 mm – ± 700 mm
Des sols sableux dans les dépressions interdunaires des dunes ondulées	moins de ± 700 mm
Des sols sableux peu profonds sur sous-sols limoneux ou argileux	600 mm – ± 700 mm
Des sols limoneux très peu profonds et sols limoneux sableux sur sous-sols argileux ou rocheux	600 mm – ± 700 mm
Des sols limoneux durs à argileux et localement sur sols limoneux-sableux humides.	
Des sols argileux-limoneux sur les pédiments et les pentes de vallées	
Des sols argileux plats fortement érodés	moins de ± 700 mm
Des sols sableux pas profonds, localement sols argileux-limoneux sur les dépressions dans les pédiments	moins de ± 700 mm
Comme II.1 et II.3, spécialement sur les champs de cultivation abandonnés	moins de ± 700 mm
Sur roches doléritiques escarpées	moins de ± 700 mm
Des sols sableux divers dans les dunes, les vallées, les dépressions et sur les terraces	moins de ± 700 mm
Des sols sédimentaires limoneux-argileux	moins de ± 700 mm
Id.	moins de 700 à 800 mm
Sols limoneux-argileux dans les vallées	
Sols sableux bas et sols limoneux	
Sols divers	

* avec une nappe fréatique

** avec une nappe fréatique ou inondé

11A. — Fig. 3.

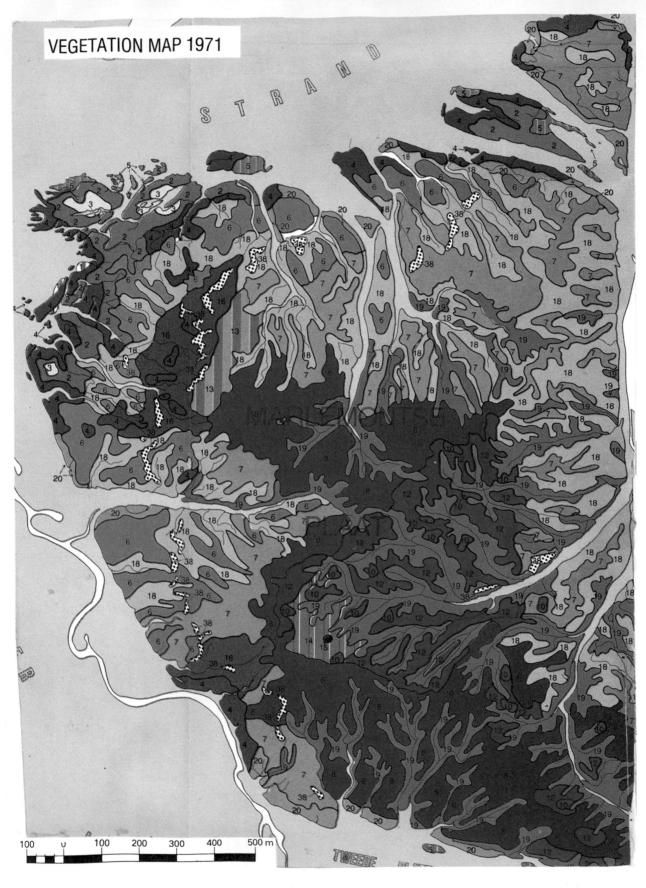

VEGETATION MAP 1971

100 0 100 200 300 400 500 m

11A. — Fig. 1.

SLIK EN LAAG SCHOR
MUD FLAT AND LOW SALT MARSH

 gemeenschap van Zeekraal
community of Salicornia europaea
(Salicornietum europaeae)

 gemeenschap van Zeeaster, lage rozetten
community of Aster tripolium, vegetative

 gemeenschap van Zeeaster met Engels slijkgras
community of Aster tripolium with Spartina anglica

 gemeenschap van Zeeaster, hoog en bloeiend
community of Aster tripolium, flowering

 gemeenschappen van Zeeaster en van Engels slijkgras
communities of Aster tripolium and of Spartina anglica
(Spartinetum anglicae)

NAT TOT DROGE KOMMEN
WET TO DRY BACK SWAMPS

 gemeenschap van Engels slijkgras
community of Spartina anglica
(Spartinetum anglicae)

 gemeenschap van Engels slijkgras met veldjes Heen*
community of Spartina anglica (Spartinetum anglicae)
with small clusters of Scirpus maritimus

gemeenschap van Engels slijkgras, dichte velden
community of Spartina anglica (Spartinetum anglicae), dense clusters

 gemeenschap van Heen, dichte velden
community of Scirpus maritimus (Scirpetum maritimi), dense clusters

gemeenschappen van Heen en van Engels slijkgras
communities of Scirpus maritimus (Scirpetum maritimi) and of Spartina anglica
(Spartinetum anglicae)

gemeenschappen van Engels slijkgras, van Heen en van Zeeaster
communities of Spartina anglica (Spartinetum anglicae), of Scirpus maritimus
(Scirpetum maritimi) and of Aster tripolium (sociation of Aster tripolium)

gemeenschappen van Engels slijkgras, van Heen en van Schorrezoutgras
communities of Spartina anglica (Spartinetum anglicae), of Scirpus
maritimus (Scirpetum maritimi) and of Triglochin maritima (sociation of Triglochin
maritima)

gemeenschap van Riet
community of Phragmites australis
(sociation of Phragmites australis)

DROGE KOMMEN EN OEVERWALLEN
DRY BACK SWAMPS AND NATURAL LEVEES

gemeenschappen van Gewoon kweldergras, van Spiesmelde en van Zeeaster
communities of Puccinellia maritima (Puccinellietum maritimae), of Atriplex hastata
(Atriplicetum hastatae), and of Aster tripolium (sociation of Aster tripolium)

gemeenschap van Spiesmelde
community of Atriplex hastata (Atriplicetum hastatae)

gemeenschap van Spiesmelde en Strandkweek
community of Atriplex hastata and Elytrigia pungens (Atriplici-Elytrigietum
pungentis)

gemeenschap van Gewoon kweldergras met Echt lepelblad e.a.
community of Puccinellia maritima (Puccinellietum maritimae) with Cochlearia
officinalis et al.

Compiled by: Drs. J. Leemans en Drs. B. Verspaandonk under supervision of Dr. Ir.
I. S. Zonneveld, I.T.C. Enschede in cooperation with Dr. Ir. W. G. Beeftink, Delta
Institute, Yerseke.
Published by the Foundation „Het Zeeuwsch Landschap", Heinkenszand,
supported by the Prins Bernhardfonds.
Cartography : S. de Silva, Intermap B.V., Enschede 1979.

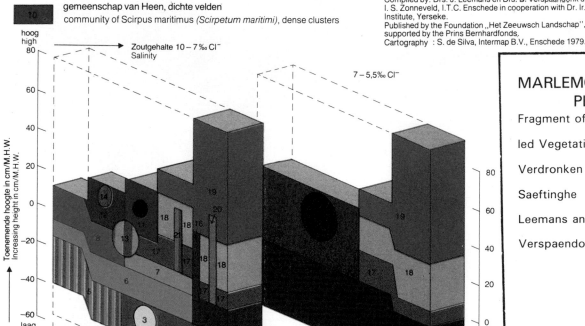

hoog
high
80
Zoutgehalte 10 – 7 ‰ Cl⁻
Salinity

7 – 5,5 ‰ Cl⁻

Toenemende hoogte in cm/M.H.W.
Increasing height in cm/M.H.W.

laag
low

Afnemende bodemvochtigheid
Decreasing soil humidity

MARLEMONTSE PLAAT
Fragment of the Detailed Vegetation Map
Verdronken Land van Saeftinghe
Leemans and Verspaendonk

See also map (Fig 11A.2)

11A. — Fig. 1.

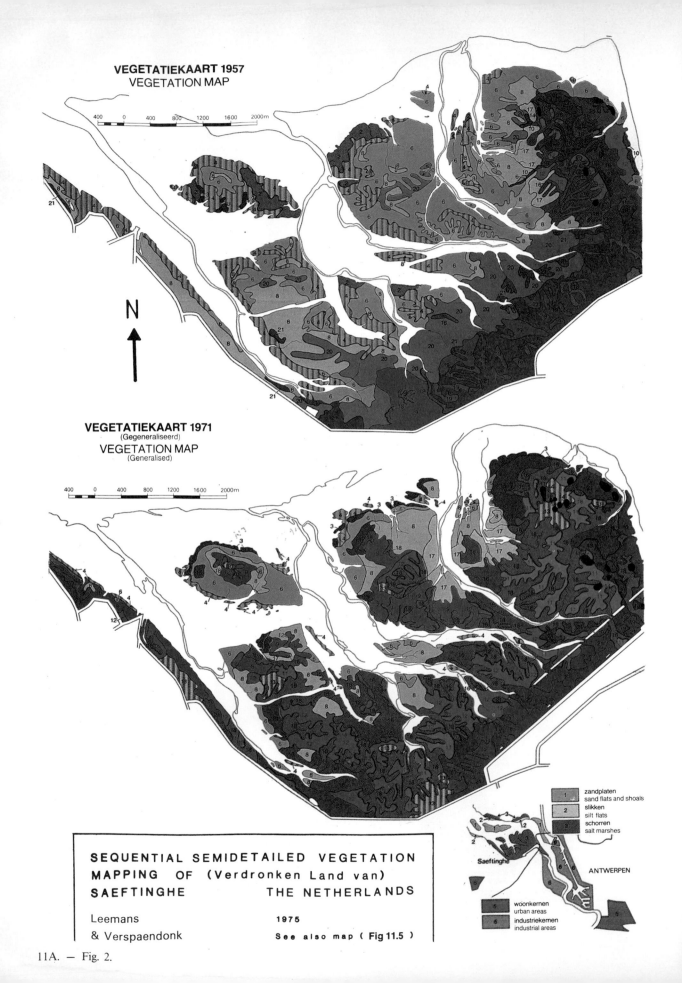

VEGETATIEKAART 1957
VEGETATION MAP

400 0 400 800 1200 1600 2000m

N

VEGETATIEKAART 1971
(Gegeneraliseerd)
VEGETATION MAP
(Generalised)

400 0 400 800 1200 1600 2000m

SEQUENTIAL SEMIDETAILED VEGETATION
MAPPING OF (Verdronken Land van)
SAEFTINGHE   THE NETHERLANDS

Leemans   1975
& Verspaendonk   See also map (Fig 11.5)

	zandplaten sand flats and shoals
1	
2	slikken silt flats
3	schorren salt marshes

Saeftinghe   ANTWERPEN

| 5 | woonkernen
urban areas |
| 6 | industriekernen
industrial areas |

11A. — Fig. 2.

SLIK EN LAAG SCHOR
MUD FLAT AND LOW SALT MARSH

 aanslibbende schorrand met ijle begroeiing van Zeeaster en Zeekraal
salt marsh shore in accretion with thin vegetation of Aster tripolium and Salicornia europaea

 opslibbing met Zeeaster
silting up with Aster tripolium

 hogere plekken in 2 met hoge bloeiende Zeeaster
higher spots in 2 with high flowering Aster tripolium

 aanslibbing met Zeeaster en Engels slijkgras
accretion with Aster tripolium and Spartina anglica

 zich vestigend Engels slijkgras
settling Spartina anglica

GEBIEDEN MET NAT TOT DROGE KOMMEN
AREAS WITH WET TO DRY BACK SWAMPS

 Zeeaster en Gewoon kweldergras met zich vestigend Engels slijkgras
Aster tripolium and Puccinellia maritima with settling Spartina anglica

 lage kommen met Engels slijkgras. Lage oeverwalfen met Gewoon kweldergras en Spiesmelde
low back swamps with Spartina anglica. Low natural levees with Puccinellia maritima and Atriplex hastata

 kommen met Gewoon kweldergras, Zeeaster, Spiesmelde en velden Heen.*
Oeverwallen met Spiesmelde
back swamps with Puccinellia maritima, Aster tripolium, Atriplex hastata and clusters of Scirpus maritimus. Natural levees with Atriplex hastata

 hogere kommen met Engels slijkgras, Spiesmelde en veldjes Heen. Oeverwallen met Spiesmelde (en Strandkweek)
higher back swamps with Spartina anglica, Atriplex hastata and clusters of Scirpus maritimus. Natural levees with Atriplex hastata (and Elytrigia pungens)

 vrij hoge natte kommen met dichte velden Engels slijkgras. Oeverwallen met Strandkweek
rather high wet back swamps with dense clusters of Spartina anglica. Natural levees with Elytrigia pungens

 kommen met dichte velden Heen. Oeverwallen met Strandkweek
back swamps with dense clusters of Scirpus maritimus. Natural levees with Elytrigia pungens

 hoge kommen met velden Heen. Hoge oeverwallen met Strandkweek
high back swamps with clusters of Scirpus maritimus. High natural levees with Elytrigia pungens

 hoge kommen met velden Heen en velden Engels slijkgras. Hoge oeverwallen met Strandkweek
high back swamps with clusters of Scirpus maritimus and of Spartina anglica. High natural levees with Elytrigia pungens

 hoge kommen met mozaïek van Heen, Engels slijkgras, Gewoon kweldergras en Schorrezoutgras. Hoge oeverwallen met Strandkweek
high back swamps with mosaic of Scirpus maritimus, Spartina anglica, Puccinelli maritima and Triglochin maritima. High natural levees with Elytrigia pungens

 rietvelden
stands of Phragmites australis

GEBIEDEN MET DROGE KOMMEN, OEVERWALLEN
AREAS WITH DRY BACK SWAMPS, NATURAL LEVEES

 hoge kommen en lage oeverwallen met Zeeaster en Spiesmelde. Invloed van vloedmerk
high back swamps and low natural levees with Aster tripolium and Atriplex hastata. Influence of tide mark

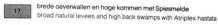 hoge kommen en lage oeverwallen met Spiesmelde, Gewoon kweldergras, Zeeaster en veldjes Heen. In laagste kommen Engels slijkgras
high back swamps and low natural levees with Atriplex hastata, Puccinellia maritima, Aster tripolium and small clusters of Scirpus maritimus. In lowest back swamps Spartina anglica

 brede oeverwallen en hoge kommen met Spiesmelde
broad natural levees and high back swamps with Atriplex hastata

 brede oeverwallen met Strandkweek
broad natural levees with Elytrigia pungens

GEBIEDEN MET BEWEIDING
AREAS WITH GRAZING

 vrij hoge natte kommen met dichte velden Engels slijkgras. Oeverwallen met Gewoon kweldergras, Spiesmelde
rather high wet back swamps with dense clusters of Spartina anglica. Natural levees with Puccinellia maritima, Atriplex hastata

 overgang tussen 3 en 19
transition between legend unit 3 and 19

grasmat van Gewoon kweldergras in de kommen en van Zilt rood zwenkgras op de oeverwallen. In het oostelijk deel van Saeftinghe met Zilt fioringras
sward of Puccinellia maritima in the back swamps and sward of Festuca rubra f. litoralis at the natural levees. In the eastern part of Saeftinghe with Agrostis stolonifera subvar. salina

11A. — Fig. 2.

596

FOREST APPEARED   FOREST DISAPPEAR   FOREST REMAINING

LANDUSE 1982 THAT REPLACED FOREST SINCE 1956

14. — Fig. 12.

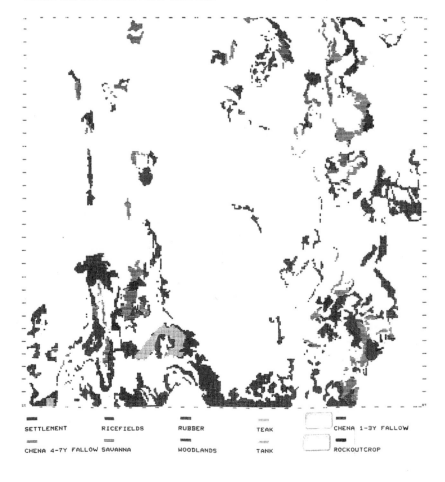

SETTLEMENT   RICEFIELDS   RUBBER   TEAK   CHENA 1-3Y FALLOW

CHENA 4-7Y FALLOW SAVANNA   WOODLANDS   TANK   ROCKOUTCROP

14. — Fig. 13.

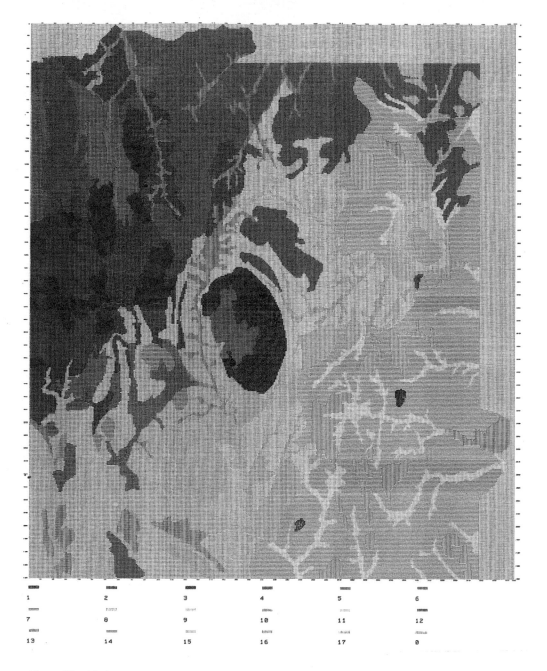

14. — Fig. 14.

598

THE LEMELE AREA: LOCATION, TOPOGRAPHY, POLYGON
MAP OF LAND UNITS AND TABLE OF LAND ATTRIBUTES

TOPOGRAPHIC MAP OF 1973

14. – Fig. 16.

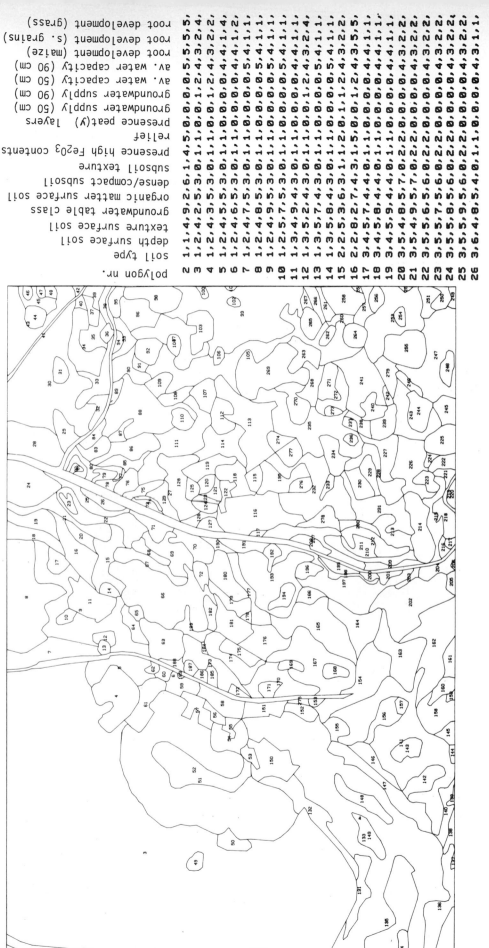

polygon nr.	soil type	depth surface soil	texture surface soil	groundwater table class	organic matter surface soil	dense/compact subsoil	subsoil texture	presence high Fe2O3 contents	relief	presence peat(y) layers	groundwater supply (50 cm)	groundwater supply (90 cm)	av. water capacity (50 cm)	av. water capacity (90 cm)	root development (maize)	root development (s. grains)	root development (grass)
2	1	1	4	9	2	6	1	4	5	0	0	5	5	5	5		
3	1	2	4	2	5	3	0	1	1	0	0	5	5	3	2	4	
4	1	2	4	3	5	3	0	1	1	0	0	4	4	4	1	4	
5	1	2	4	5	5	3	0	1	1	0	0	4	4	4	1	2	
6	1	2	4	6	5	3	0	1	1	0	0	4	4	4	1	1	
7	1	2	4	7	5	3	0	1	1	0	0	5	5	4	1	1	
8	1	2	4	8	5	3	0	1	1	0	0	5	4	1	1	1	
9	1	2	4	9	5	3	0	1	1	0	0	5	4	1	1	1	
10	1	2	5	3	4	3	0	1	1	0	0	4	4	4	1	1	
11	1	3	4	9	4	3	0	1	1	0	0	4	4	3	2	4	
12	1	3	5	2	4	3	0	1	1	0	0	4	4	3	2	2	
13	1	3	5	7	4	3	0	1	1	2	0	5	4	3	2	2	
14	1	3	8	4	3	6	3	1	1	5	0	5	4	3	5	5	
15	2	2	5	3	6	3	1	1	2	0	1	2	4	3	2	2	
16	2	2	8	2	7	4	3	0	1	1	0	2	4	4	3	5	5
17	3	4	5	8	4	4	0	1	1	0	0	4	4	4	1	1	
18	3	4	5	9	4	4	0	1	1	0	0	4	4	4	1	1	
19	3	4	5	9	4	4	0	1	1	0	0	4	4	4	1	1	
20	3	5	4	8	5	7	0	2	2	0	0	4	4	4	3	2	2
21	3	5	4	9	5	6	0	2	2	0	0	4	4	3	2	2	
22	3	5	5	7	5	6	0	2	2	0	0	4	4	3	2	2	
23	3	5	5	8	5	6	0	2	2	0	0	4	4	3	2	2	
24	3	5	5	9	5	6	0	2	2	0	0	4	4	3	2	2	4
25	3	5	5	8	5	4	0	1	1	0	0	4	4	3	2	2	
26	3	6	4	8	5	4	0	1	1	0	0	4	4	3	1	1	

DIGITISED POLYGONS OF LAND UNIT MAP

SECTION OF TABLE WITH POLYGON NUMBERS OF LAND UNIT
MAP AND CODES FOR LAND ATTRIBUTE CLASSES

14. – Fig. 16.

600

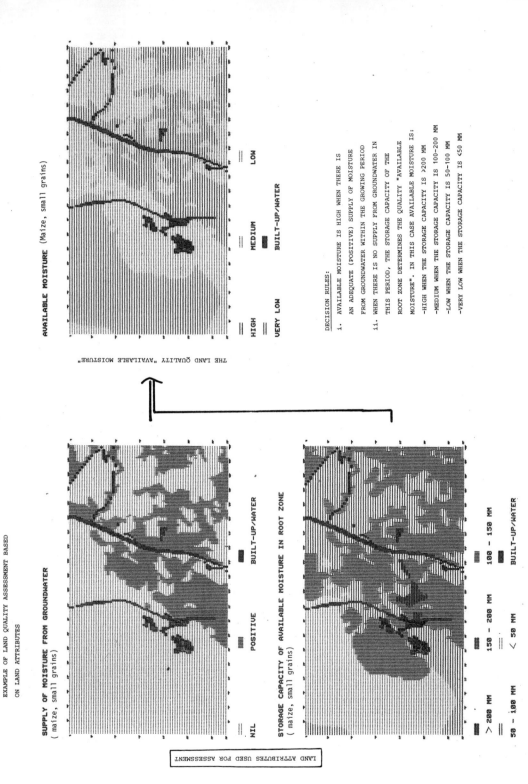

EXAMPLE OF LAND QUALITY ASSESSMENT BASED
ON LAND ATTRIBUTES

14. — Fig. 18.

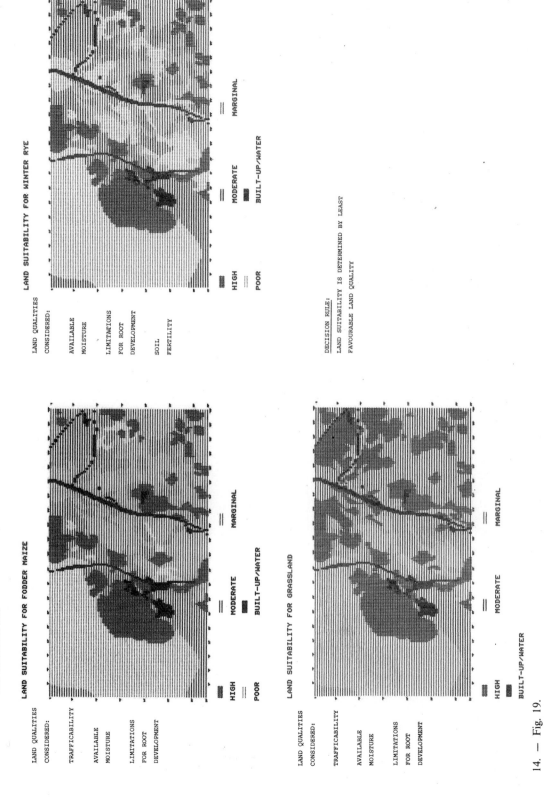

LAND SUITABILITY MAPS

LAND SUITABILITY FOR FODDER MAIZE

LAND SUITABILITY FOR WINTER RYE

LAND QUALITIES
CONSIDERED:

TRAFFICABILITY

AVAILABLE
MOISTURE

LIMITATIONS
FOR ROOT
DEVELOPMENT

LAND QUALITIES
CONSIDERED:

AVAILABLE
MOISTURE

LIMITATIONS
FOR ROOT
DEVELOPMENT

SOIL
FERTILITY

HIGH   MODERATE   MARGINAL

POOR   BUILT-UP/WATER

HIGH   MODERATE   MARGINAL

POOR   BUILT-UP/WATER

LAND SUITABILITY FOR GRASSLAND

LAND QUALITIES
CONSIDERED:

TRAFFICABILITY

AVAILABLE
MOISTURE

LIMITATIONS
FOR ROOT
DEVELOPMENT

HIGH   MODERATE   MARGINAL

BUILT-UP/WATER

DECISION RULE:
LAND SUITABILITY IS DETERMINED BY LEAST
FAVOURABLE LAND QUALITY

 14. — Fig. 19.

602

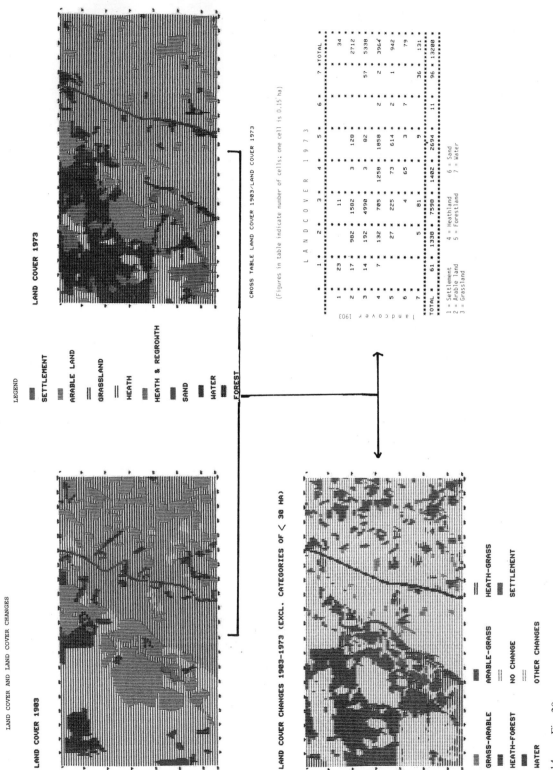

LAND COVER AND LAND COVER CHANGES

LAND COVER 1903

LAND COVER 1973

LEGEND

SETTLEMENT

ARABLE LAND

GRASSLAND

HEATH

HEATH & REGROWTH

SAND

WATER

FOREST

LAND COVER CHANGES 1903-1973 (EXCL. CATEGORIES OF < 30 HA)

GRASS-ARABLE   ARABLE-GRASS   HEATH-GRASS

HEATH-FOREST   NO CHANGE   SETTLEMENT

WATER   OTHER CHANGES

CROSS TABLE LAND COVER 1903/LAND COVER 1973

(Figures in table indicate number of cells; one cell is 0.15 ha)

L A N D C O V E R   1 9 7 3

landcover 1903	1	2	3	4	5	6	7	*TOTAL
1	23		11					34
2	17	982	1582	3	128			2712
3	14	192	4990	3	82		57	5338
4	7	132	785	1258	1858	2	2	3964
5		27	225	73	614	2	1	942
6			4	65	3	7		79
7	5	81		9			36	131
TOTAL	61	1338	7598	1402	2694	11	96	13200

1 = Settlement   4 = Heathland   6 = Sand
2 = Arable land   5 = Forestland   7 = Water
3 = Grassland

14. — Fig. 20.

603

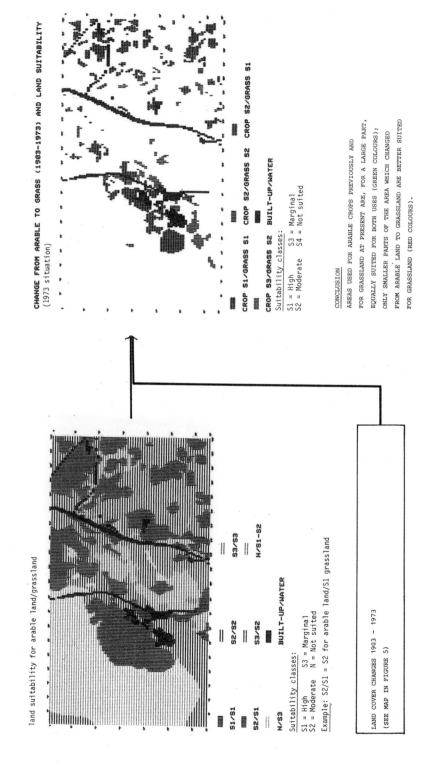

CONVERSION OF ARABLE LAND (1903) TO GRASSLAND
(1973) IN RELATION TO SUITABILITY OF LAND

land suitability for arable land/grassland

S1/S1

S2/S2

S2/S1

S3/S2

N/S3

BUILT-UP/WATER

S3/S3

N/S1-S2

Suitability classes:

S1 = High   S3 = Marginal
S2 = Moderate N = Not suited

Example: S2/S1 = S2 for arable land/S1 grassland

LAND COVER CHANGES 1903 – 1973
(SEE MAP IN FIGURE 5)

14. – Fig. 21.

CHANGE FROM ARABLE TO GRASS (1903-1973) AND LAND SUITABILITY
(1973 situation)

CROP S1/GRASS S1   CROP S2/GRASS S2   CROP S2/GRASS S1

CROP S3/GRASS S2   BUILT-UP/WATER

Suitability classes:

S1 = High   S3 = Marginal
S2 = Moderate S4 = Not suited

CONCLUSION

AREAS USED FOR ARABLE CROPS PREVIOUSLY AND
FOR GRASSLAND AT PRESENT ARE, FOR A LARGE PART,
EQUALLY SUITED FOR BOTH USES (GREEN COLOURS);
ONLY SMALLER PARTS OF THE AREA WHICH CHANGED
FROM ARABLE LAND TO GRASSLAND ARE BETTER SUITED
FOR GRASSLAND (RED COLOURS).

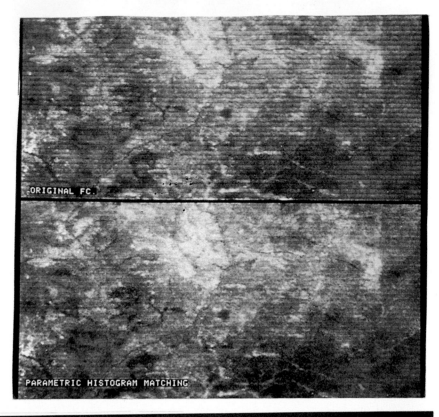

ORIGINAL FC.

PARAMETRIC HISTOGRAM MATCHING

21. — Fig. 7.

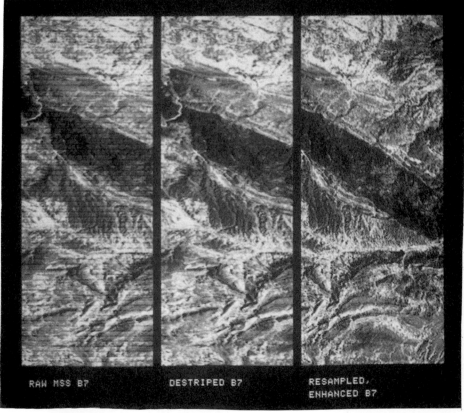

RAW MSS B7   DESTRIPED B7   RESAMPLED,
ENHANCED B7

21. — Fig. 9.

21. — Fig. 13.

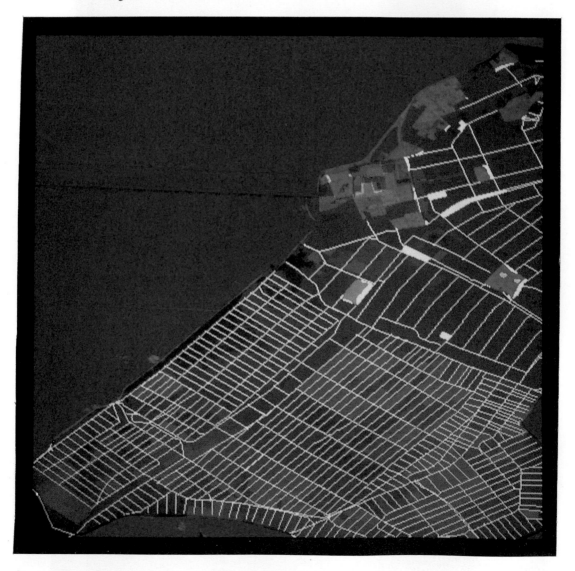

21. — Fig. 14.

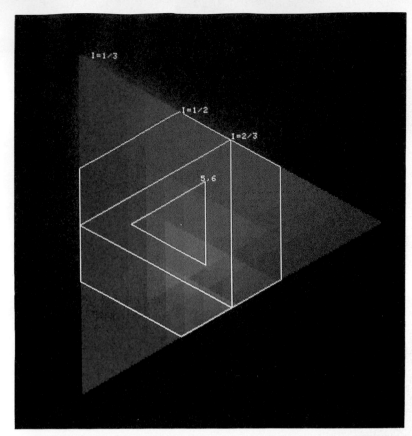

21. — Fig. 18.

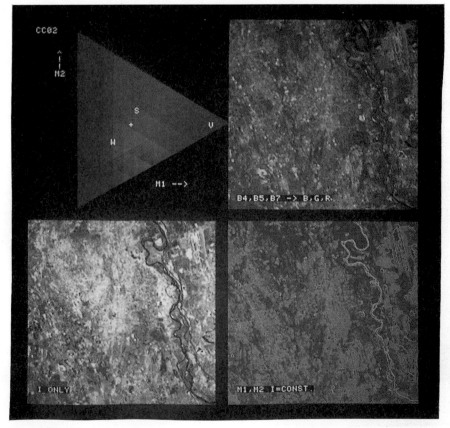

21. — Fig. 19.

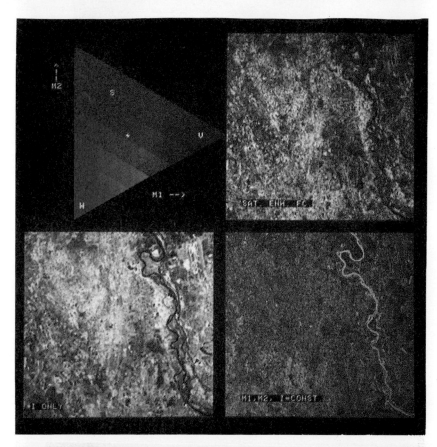

21. — Fig. 21.

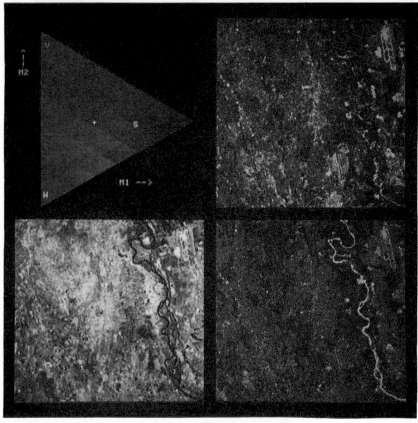

21. — Fig. 22.

608

21. — Fig. 25A.

21. — Fig. 25B.

21. — Fig. 28.

610

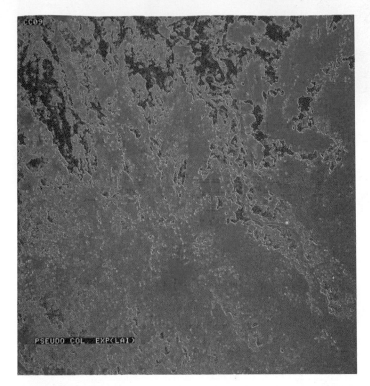

21. — Fig. 27.

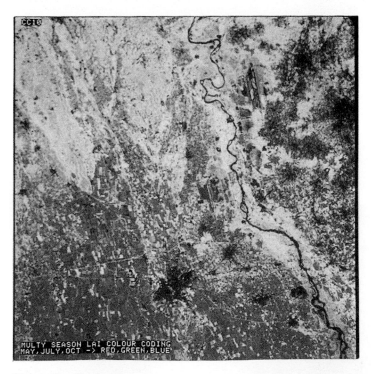

21. — Fig. 29.

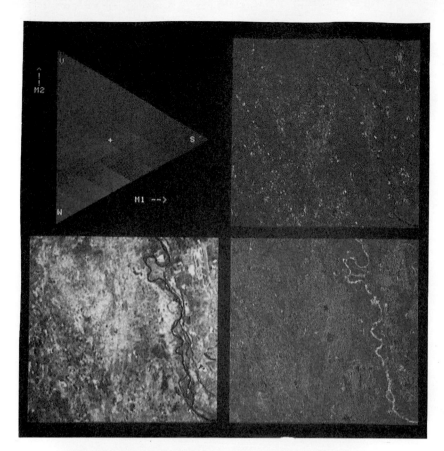

21. — Fig. 23.

21. — Fig. 31A.

21. — Fig. 31B.

612

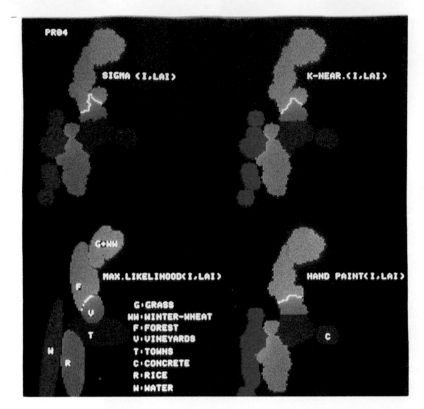

21. — Fig. 34.

21. — Fig. 35.

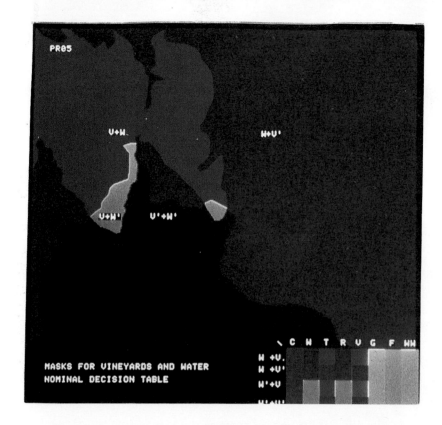

21. — Fig. 36.

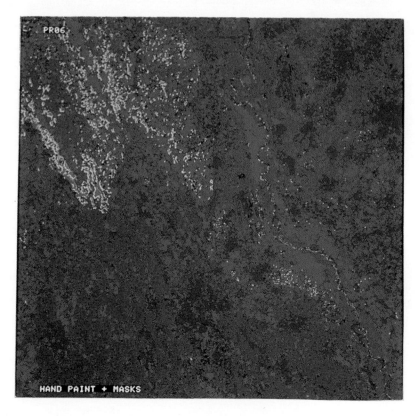

21. — Fig. 37.

Birkenbruchwald des Flachlandes, örtlich Heidemoor, Gagelgebüsch, Eichen - Birken - wald und Eichen - Buchenwald

Feuchter Eichen - Buchenwald des Flachlandes, selten Übergänge zum Eichen - Birkenwald

Feuchter Eichen - Buchenwald des Berglandes

Trockener Eichen - Buchenwald des Flachlandes, selten Übergänge zum Eichen - Birkenwald

Erlenbruchwald des Flachlandes, selten waldfreies Niedermoor (N)

Erlenbruchwald und feuchter Eichen - Buchenwald im Wechsel

Weidenwald und Mandelweidengebüsch

Traubenkirschen - Erlen - Eschenwald, stellenweise mit Erlenbruchwald und Eichen - Hainbuchenwald

Eichen - Ulmenwald (rote Punkte : auf stark entwässerten Standorten im Erfttal)

Artenreicher Sternmieren - Stieleichen - Hainbuchenwald

Stieleichen - Hainbuchen - Auenwald der Berglandtäler einschließlich der bach - und flußbegleitenden Erlenwälder

Maiglöckchen - Stieleichen - Hainbuchenwald der Niederrheinischen Bucht

Maiglöckchen - Stieleichen - Hainbuchenwald und feuchter Eichen - Buchenwald im Wechsel

Trockener Flattergras - Traubeneichen - Buchenwald mit Übergang zum Eichen - Buchenwald

Flattergras - Traubeneichen - Buchenwald

Typischer Hainsimsen - Buchenwald

Hainsimsen - Buchenwald mit Rasenschmiele

Flattergras - Hainsimsen - Buchenwald

Typischer und Flattergras - Hainsimsen - Buchenwald im Wechsel

Maiglöckchen - Perlgras - Buchenwald der Niederrheinischen Bucht, meist auf kiesigen Böden

Maiglöckchen - Perlgras - Buchenwald der Niederrheinischen Bucht, stellenweise Flattergras - Traubeneichen - Buchenwald, auf lehmigen Böden

Maiglöckchen - Buchenwälder des Villeosthanges

Maiglöckchen - Perlgras - Buchenwald und Maiglöckchen - Stieleichen - Hain - buchenwald im Wechsel

Hainsimsen - Perlgras - Buchenwald sowie Perlgras - Buchenwald und Hainsimsen - Buchenwald im Wechsel

Hainsimsen - Perlgras - Buchenwald mit Rasenschmiele

Typischer Perlgras - Buchenwald

26. — Fig. 1.

Feldaufnahme : H. Wedeck, G. Wolf, K. Meisel u. a. (1965 - 1969)

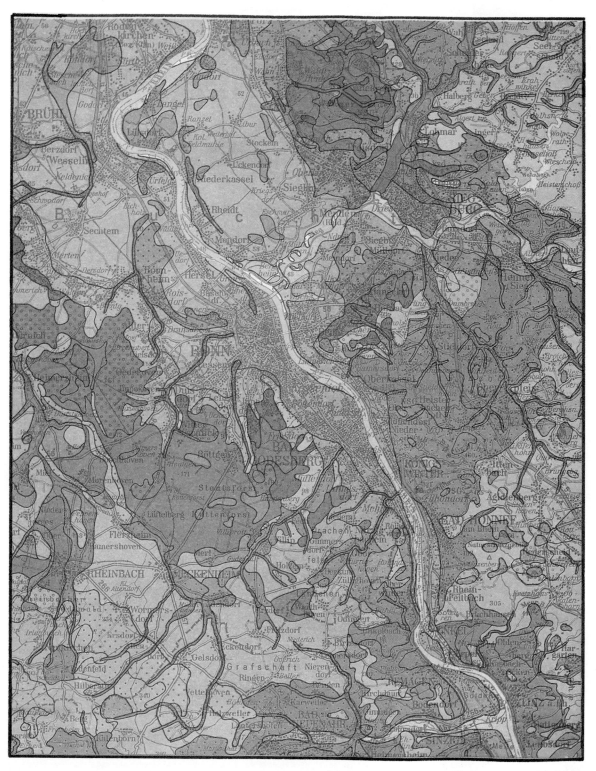

Druck: Institut für Angewandte Geodäsie, Frankfurt a. M.

0   2   4   10 km

26. – Fig. 1.

LEGEND

Vegetatieuaapt van Nederland.
Vegetation map of the Netherlands
Vegetation series, named after the natural forest types (according to Westhoff & den Held, 1969)

a	Querco roboris-Betuletum
ab	complex of Querco roboris-Betuletum and Fago-Quercetum
b	Fago-Quercetum
c	Fago-Quercetum, type poor in species
f	Fraxino-Ulmetum
g	Anthrisco-Fraxinetum
h	Circaeo-Alnion
hk	complex of Circaeo-Alnion and Alnion glutinosae
k	Alnion glutinosae
l	Alno-Padion
lm	complex of Alno-Padion and Salicetum albo-fragilis
bl	complex of Fago-Quercetum and Alno-Padion
bh	complex of Fago-Quercetum and Circaeo-Alnion
cn	complex of Fago-Quercetum and Macrophorbio-Alnetum
abhk	complex of Querdo roboris-Betuletum, Fago-Quercetum, Circaeo-Alnion and Alnion glutinosae
bhk	complex of Fago-Quercetum, Circaeo-Alnion and Alnion glutinosae

Additions

	predominantly dry vegetation types
	predominantly moist and wet vegetation types
	development to peat bog within the Alder forest series
w	open water
	urban area

Code for the actual vegetation

The code counts five positions for forest, shrub, semi-natural vegetation, cropland and marsh/water vegetation respectively. The presence of each category is expressed in % of the area of the map unit concerned: 5 = >75%, 4 = 50-75%, 3 = 25-50%, 2 = <25% or many line and point shaped elements, 1 = few line and point shaped elements..

example: code 31142

3....   25-50% forest
.1...   few hedgerows
..1..   few small patches of semi-natural vegetation
...4.   50-75% cultivated land (grassland and arable land)
....2   a mumber of patches and lines with water and marsh vegetation

26. — Fig. 2.

It depends on the colour, that is the vegetation series, what types of shrub, semi-natural and marsh/water vegetation are present. In the original study this can be checked in a table of vegetation series.

VEGETATIEKAART VAN NEDERLAND

OORSPRONKELIJK SCHAAL 1:200.000

1975

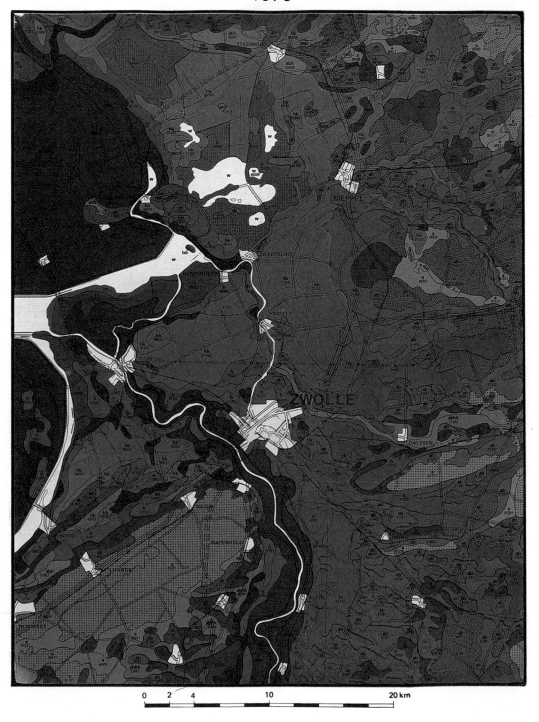

0 2 4 10 20 km

26. — Fig. 2.

618

A fragment of the map "CORSE" *(DUPIAS et al. 1965)*
A specimen of its simplified legend   scale :

D. ÉTAGE MÉDITERRANÉEN INFÉRIEUR

BOIS

3. SÉRIE DU CHÊNE-LIÈGE

Bois   Arbres épars

4. SÉRIE DU CHÊNE VERT EUMÉDITERRANÉEN

Bois   Arbres épars

5. ANNEXE PIN PIGNON

Bois   Arbres épars

6. ANNEXE PIN D'ALEP

Bois   Arbres épars

MAQUIS GARRIGUES ET PELOUSES

7. PREMIER ÉCHELON OLÉO-LENTISQUE

| 4 à 7 |
Garrigue

8. DEUXIÈME ÉCHELON A MYRTE

Maquis

m 4 à 7
Garrigue

m 2 à 3
Pelouse

9. TROISIÈME ÉCHELON A LENTISQUE

8 à 9
Maquis

l 4 à 7
Garrigue

l 2 à 3
Pelouse

E. ÉTAGE MÉDITERRANÉEN SUPÉRIEUR

BOIS

10. SÉRIE DU CHÊNE VERT PROPÉMÉDITERRANÉEN

Bois   Arbres épars

11. SÉRIE DU CHÊNE PUBESCENT

Bois   Arbres épars

12. FACIÈS A CHATAIGNIER

Bois   Arbres épars

16-17. *MAQUIS GARRIGUES ET PELOUSES*

ms 8 à 9
Maquis

ms 4 à 7
Garrigue

ms 2 à 3
Pelouse

ms 5

17. Faciès à Pteris

G. ÉTAGE MONTAGNARD

BOIS

19. SÉRIE DU PIN LARICIO

PL
Bois   Arbres épars

20. SÉRIE DU HÊTRE

H
Bois   Arbres épars

21. SÉRIE DU SAPIN

S
Bois   Arbres épars

22. FACIÈS A BOULEAU

23. *LANDES ET PELOUSES*

m 4 à 5
Lande

m 2 à 3
Pelouse

H. ÉTAGE SUBALPIN

24. SÉRIE DE L'AUNE VERT

AuV 6 à 8
Fourré

s 4 à 5
25. Lande subalpine

26. MARAIS TOURBEUX

Pozzines

I. ÉTAGE ALPIN

alp 0 à 2
27. Pelouse et éboulis

VÉGÉTATION INTRODUITE OU TRANSFORMÉE

REBOISEMENTS

28. Eucalyptus

29. Pin Pignon, Pin d'Alep, Pin mésogéen

30. Jeunes reboisements divers

PLANTATIONS SUR ROUTES

31. Eucalyptus

32. Feuillus divers

CULTURES

Labours

33. Pelouse sèche

34. Prairie de fauche

35. Bocage à Aunes et Frênes

31. — Fig. 3.

31. — Fig. 3.

A fragment of the map "South Sumatra".   Scale: 0 — 50km
The venerable concept of "series of vegetation" has desappeared
A specimen of its legend (Laumonier 1983)

II - Formations de basse altitude (300 à 1000 m) des pentes occidentales
Western slopes at low elevation, between 300 and 1000 m

5 — Forêt dense humide sempervirente de basse altitude du bloc de Bengkulu *(Shorea javanica, Shorea sp., Hopea aryobalanoides, Burseraceae, Euphorbiaceae, Myrtaceae)*
Dense moist evergreen forest, at low elevation, of the Bengkulu block

6 — Mosaïque de types dérivés, essentiellement arbustifs
Secondary and derived types mosaic, mainly shrubby

7 — Forêt dense humide sempervirente de basse altitude des Monts Pesisir
Dense moist evergreen forest, at low elevation of the Pesisir Mountains

8 — Mosaïque de types dérivés, essentiellement arbustifs
Secondary and derived types mosaic, mainly shrubby

36 — Cultures de plantes vivrières (Manihot, Brassica, riz, mais) et végétation secondaire
Food-crops (Manihot, dryland rice, corn) cultivation and secondary vegetation mosaic

37 — Plantations d'Hevea
Rubber-trees estates

III - Zonation forestière altitudinale sur la chaîne des Barisan, au-dessus de 1000 m
Forest altitudinal zonation on the Barisan Range

9 — Forêt dense humide sempervirente sub-montagnarde 1000 à 1800 m *(Fagaceae, Myrtaceae, Lauraceae, Hamamelidaceae, Podocarpus imbricatus, Shorea platyclados)*
Sub-mountain, dense, moist evergreen forest, between 1000 and 1800 m

10 — Forêt sempervirente humide de montagne 1800-2700 m *(Fagaceae, Magnoliaceae, Dacrydium elatum, Podocarpus, Vaccinium)*
Moist evergreen mountain forest (1800-2700 m)

11 — Forêt basse et fourré de la zone à Ericaceae au-dessus de 2700 m *(Ericaceae, Rosaceae, Aquifoliaceae, Symplocaceae)*
Evergreen low forest and thicket of the ericaceous zone, above 2700 m

12 — Formations herbacées dues à l'activité volcanique
Herbaceous communities caused by volcanic activities

38 — Plantations de Palmiers à huile *(Elaeis guineensis)*
Oil-Palm estates

39 — Mosaïque de formations dégradées et plantations de Canneliers *(Cinnamomum burmani)*
Degraded vegetation and Cinnamon plantations mosaic

Types de végétation des régions inondées
Types found in swampy areas

24 — Forêt haute, dense, inondée de la plaine d'Indrapura (Côte Occidentale) *Mangifera, Alstonia, Gluta, Coelostegia, Xylopia) etc...*
Tall, dense, swamp forest of the Indrapura plain (west coast)

25 — Forêt basse, dense, inondée de la plaine d'Indrapura (Côte Occidentale)
Dense low swamp forest of the Indrapura plain (West Coast)

26 — Forêt haute, dense, inondée de la plaine orientale
Tall, dense, swamp forest of the eastern plain
Tourbe/peat : *(Dyera lowii, Gonystylus, Tetramerista, Campnosperma, Shorea uliginosa, S.teijsmanniana, Durio carinatus) etc...*
Sans Tourbe/non peat : *(Shorea sumatrana, balangeran, palembanica, Xylopia, Alstonia, Mangifera) etc...*

27 — Fourré dense sempervirent *(Alstonia, Combretocarpus, Melaleuca, Cyperaceae, palms)* ou forêt primaire très dégradée
Dense evergreen thickets or very degraded primary forest

28 — Fourré dense sempervirent de Melaleuca leucadendron
Dense evergreen thickets with Melaleuca leucadendron

29 — Communautés herbacées d'hygrophytes *(Cyperaceae, Scleria, Cyperus, Gramineae, Phragmites)*
Herbaceous hygrophytic communities

Mangroves
(Rhizophoraceae, Sonneratiaceae) etc...

31 — Forêt ripicole
Riparian forest

32 — Etangs
Pools

❢ Eusideroxylon zwageri
⬆ Pinus merkusii (natural occurence)

IV - Formations de basse altitude (300 à 1000 m) des pentes orientales
Eastern slopes at low elevation, between 300 and 1000 m

13 — Forêt dense humide sempervirente du bloc Semangka-Gumai aux Monts Tembesi-Rawas *(Dipterocarpus hasseltii, Shorea, Theaceae, Bombacaceae, Podocarpus blumei)*
Dense moist evergreen forest from Semangka-Gumai block to Tembesi-Rawas Mountains

14 — Mosaïque de types dérivés essentiellement arbustifs
Secondary and derived types mosaic, mainly shrubby

15 — Savane et savane arbustive
Savanna and shrub savanna

40 — Mosaïque de formations dégradées et plantations de Caféiers
Degraded vegetation and coffee plantations mosaic

41 — Mosaïque de formations dégradées et plantations de girofliers
Degraded vegetation and clove plantations mosaic

42 — Plantations de cocotiers et de fruitiers
Coconut and fruit trees

43 — Agroforêt à Shorea javanica
Agroforest with Shorea javanica

31. — Fig. 5.

31. — Fig. 5.

SPOT simulations in the Sunderbans (Hatya island—Ganges) and their interpretation

1 - Natural dense mangrove (mainly Sonneratia apetala)
2 - Plantations (about 5 years old)
3 - Plantations (2 to 4 years old)
4 - Orchards with villages
5 - Dry harvested fields
6 - Dry fields with some dry crops including pulses,
    tomatoes, chillies
7 - Very wet fields

8 - Almost barren wet muds (intertidal zone)
9 - Recent deposits, scattered grasses, new cropland

A,A' Brackish water (sea water)
      deep (A), turbid and shallow (A')
    Embankments consolidated by trees
    Footpath and embankment under construction
    Drainage network

31. — Fig. 5.

INDEX

A.W. Küchler & I.S. Zonneveld (eds.), Vegetation mapping. ISBN 90-6193-191-6.
© 1988, Kluwer Academic Publishers, Dordrecht. Printed in the Netherlands.